Environmental Biotechnology: Principles and Applications is the essential tool for understanding and designing microbiological processes used for environmental protection and improvement. The book lays a foundation in microbiology and engineering principles and provides comprehensive coverage of all the major environmental applications, from traditional ones like activated sludge and anaerobic digestion to emerging applications like detoxification of hazardous chemical and biofiltration of drinking water. An abundance of worked examples that show in a step-by-step way how the tools are used in analysis and design enrich the discussion. *Environmental Biotechnology* is the authoritative source for learning how processes in environmental biotechnology work and how to create reliable processes to meet contemporary and emerging needs. Students, practitioners, and researchers will find this book invaluable.

Key features of this first edition include:

- Consistent backup of the fundamental principles of microbiological processes by their practical applications.
- Discussion of the traditional applications (e.g., activated sludge and anaerobic digestion) and the emerging applications (e.g., bioremediation and drinking water treatment).
- Numerous examples illustrating how the design and analysis tools are applied correctly.
- Each chapter consists of many problems, ranging in scope, that can be assigned as homework, used as supplemental examples in class, or used as study tools.
- Abundant use of figures to illustrate concepts.

Also Available from McGraw-Hill:

Water Chemistry (2001)
by Mark M. Benjamin
ISBN 0-07-238390-9

Introduction to Engineering and the Environment (2001)
by Edward S. Rubin and Cliff Davidson
ISBN 0-07-235467-4

Pollution Prevention: Fundamentals and Practice (2000)
by Paul Bishop
ISBN 0-07-36614703

ENVIRONMENTAL BIOTECHNOLOGY: PRINCIPLES AND APPLICATIONS

McGraw-Hill Series in Water Resources and Environmental Engineering

CONSULTING EDITOR

George Tchobanoglous, *University of California, Davis*

Bailey and Ollis: *Biochemical Engineering Fundamentals*
Benjamin: *Water Chemistry*
Bishop: *Pollution Prevention: Fundamentals and Practice*
Canter: *Environmental Impact Assessment*
Chanlett: *Environmental Protection*
Chapra: *Surface Water-Quality Modeling*
Chow, Maidment, and Mays: *Applied Hydrology*
Crites and Tchobanoglous: *Small and Decentralized Wastewater Management Systems*
Davis and Cornwell: *Introduction to Environmental Engineering*
deNevers: *Air Pollution Control Engineering*
Eckenfelder: *Industrial Water Pollution Control*
Eweis, Ergas, Chang, and Schroeder: *Bioremediation Principles*
LaGrega, Buckingham, and Evans: *Hazardous Waste Management*

Linsley, Franzini, Freyberg, and Tchobanoglous: *Water Resources and Engineering*
McGhee: *Water Supply and Sewage*
Mays and Tung: *Hydrosystems Engineering and Management*
Metcalf & Eddy, Inc.: *Wastewater Engineering: Collection and Pumping of Wastewater*
Metcalf & Eddy, Inc.: *Wastewater Engineering: Treatment, Disposal, Reuse*
Peavy, Rowe, and Tchobanoglous: *Environmental Engineering*
Rittmann and McCarty: *Environmental Biotechnology: Principles and Applications*
Rubin: *Introduction to Engineering and the Environment*
Sawyer, McCarty, and Parkin: *Chemistry for Environmental Engineering*
Tchobanoglous, Thiesen, and Vigil: *Integrated Solid Waste Management: Engineering Principles and Management Issues*
Wentz: *Safety, Health, and Environmental Protection*

ENVIRONMENTAL BIOTECHNOLOGY: PRINCIPLES AND APPLICATIONS

Bruce E. Rittmann
Northwestern University, Evanston, Illinois

Perry L. McCarty
Stanford University, Stanford, California

Boston   Burr Ridge, IL   Dubuque, IA   Madison, WI   New York   San Francisco   St. Louis
Bangkok   Bogotá   Caracas   Kuala Lumpur   Lisbon   London   Madrid   Mexico City
Milan   Montreal   New Delhi   Santiago   Seoul   Singapore   Sydney   Taipei   Toronto

McGraw-Hill Higher Education

A Division of The McGraw-Hill Companies

ENVIRONMENTAL BIOTECHNOLOGY: PRINCIPLES AND APPLICATIONS

3 4 5 6 7 8 9 20 ANL 20 9

Library of Congress Catalog Card Number: 00-034882

www.mhhe.com

Printed in Singapore

To Marylee and Martha for their patience and understanding

PREFACE

Environmental biotechnology utilizes microorganisms to improve environmental quality. These improvements include preventing the discharge of pollutants to the environment, cleaning up contaminated environments, and generating valuable resources for human society. Environmental biotechnology is essential to society and truly unique as a technical discipline.

Environmental biotechnology is historic and eminently modern. Microbiological treatment technologies developed at the beginning of the 20th century, such as activated sludge and anaerobic digestion, remain mainstays today. At the same time, new technologies constantly are introduced to address very contemporary problems, such as detoxification of hazardous chemicals. Important tools used to characterize and control processes in environmental technology also span decades. For example, traditional measures of biomass, such as volatile suspended solids, have not lost their relevance, even though tools from molecular biology allow us to explore the diversity of the microbial communities.

Processes in environmental biotechnology work according to well established principles of microbiology and engineering, but application of those principles normally requires some degree of empiricism. Although not a substitute for principles, empiricism must be embraced, because materials treated with environmental biotechnology are inherently complex and varying in time and space.

The principles of engineering lead to quantitative tools, while the principles of microbiology often are more observational. Quantification is essential if processes are to be reliable and cost-effective. However, the complexity of the microbial communities involved in environmental biotechnology often is beyond quantitative description; unquantifiable observations are of the utmost value.

In *Environmental Biotechnology: Principles and Applications,* we connect these different facets of environmental biotechnology. Our strategy is to develop the basic concepts and quantitative tools in the first five chapters, which comprise the principles part of the book. We consistently call upon those principles as we describe the applications in Chapters 6 through 15. Our theme is that *all microbiological processes behave in ways that are understandable, predictable, and unified.* At the same time, each application has its own special features that must be understood. The special features do not overturn or sidestep the common principles. Instead, they complement the principles and are most profitably understood in the light of principles.

Environmental Biotechnology: Principles and Applications is targeted for graduate-level courses in curricula that exploit microbiological processes for environmental-quality control. The book also should be appropriate as a text for upper-level undergraduate courses and as a comprehensive resource for those engaged in professional practice and research involving environmental biotechnology.

The material in *Environmental Biotechnology: Principles and Applications* can be used in one or several courses. For students not already having a solid background in microbiology, Chapter 1 provides a foundation in taxonomy, metabolism, genetics, and microbial ecology. Chapter 1 addresses the microbiology concepts that are most essential for understanding the principles and applications that follow. Chapter 1 can serve as the text for a first course in environmental microbiology, or it can be used as a resource for students who need to refresh their knowledge in preparation for a more process-oriented course, research, or practice.

The "core" of the principles section is contained in Chapters 2, 3, 4 and 5. Chapter 2 develops quantitative tools for describing the stoichiometry and energetics of microbial reactions: what and how much the microorganisms consume and produce. Stoichiometry is the most fundamental of the quantitative tools. Chapters 3 and 4 systematically develop quantitative tools for kinetics: how fast are the materials consumed and produced. Reliability and cost-effectiveness depend on applying kinetics properly. Chapter 5 describes how principles of mass balance are used to apply stoichiometry and kinetics to the range of reactors used in practice.

Chapters 6 through 15 comprise the applications section. Each chapter includes information on the stoichiometry and kinetics of the key microorganisms, as well as features that are not easily captured by the stoichiometric or kinetic parameters. Each chapter explains how processes are configured to achieve treatment objectives and what are the quantitative criteria for a good design. The objective is to link principles to practice as directly as possible.

In one sense, the applications chapters are arranged more or less in order from most traditional to most modern. For example, Chapters 6, 7, and 8 address the aerobic treatment of wastewaters containing biodegradable organic matter, such as the BOD in sewage, while Chapters 14 and 15 address biodegradation of hazardous chemicals. Aerobic treatment of sewage can be traced back to the early 20th century, which makes it quite traditional. Detoxification of hazardous chemicals became a major treatment goal in the 1980s. On the other hand, Chapters 6 to 8 describe newly emerging technologies for attaining the traditional goal. Thus, while a goal may be traditional, the science and technology used to attain it may be very modern.

We prepared a chapter on "Complex Systems" that does not appear in the book in an effort to keep the book to a reasonable length. The website chapter extends principles of Chapters 1 to 5 by systematically treating nonsteady-state systems (suspended and biofilm) and systems having complex multispecies interactions. McGraw-Hill agreed to put this chapter on a web site so that it would be available to those who are interested. Having an official web site for the book provides another advantage: We will now have a convenient location to post corrections to the inevitable errors that remain in the book. Perhaps there will be other book-related items that we may wish to post as times go by; we encourage the reader to occasionally check the web page. The web site URL for *Environmental Biotechnology* is http://www.mhhe.com/engcs/civil/rittmann/.

One important feature of *Environmental Biotechnology: Principles and Applications* is that it contains many example problems. These problems illustrate the step-by-step procedures for utilizing the tools in order to understand how microbial systems work or to design a treatment process. In most cases, learning by example is the most effective approach, and we give it strong emphasis.

Each chapter contains many problems that can be assigned as "homework," used as supplemental examples in class, or used as study tools. The problems range in scope. Some are simple, requiring only a single calculation or a short expository response. At the other extreme are extensive problems requiring many steps and pages. Most problems are of intermediate scope. Thus, the instructor or student can gradually advance from simple, one-concept problems to comprehensive problems that integrate many concepts. Computer spreadsheets are very helpful in some cases, particularly when complex or iterative solutions are needed.

In an effort to promote uniformity in notation, we have elected to adapt the "Recommended Notation for Use in the Description of Biological Wastewater Treatment Processes," agreed upon internationally and as published in *Water Research* **16**, 1501–1505 (1982). We hope this will encourage others to do the same, as it will facilitate much better communication among us.

This text is too brief to do justice to general principles, applications of environmental biotechnology, and the numerous specific mechanical details that one must consider in the overall design of biological systems. We have chosen to focus on the principles and applications. For the specific design details, we suggest other references, such as the two-volume *Design of Municipal Wastewater Treatment Plants,* published jointly by the Water Environment Federation (Manual of Practice No. 8) and the American Society of Civil Engineers (Manual and Report on Engineering Practice No. 76).

We take this opportunity to thank our many wonderful students and colleagues, who have taught us new ideas, inspired us to look farther and deeper, and corrected our frequent errors. The numbers are too many to list by name, but you know who you are. We especially thank all of the students in our environmental biotechnology classes over the past few years. These students were subjected to our chapter first drafts and provided us with much welcomed feedback and many corrections. Thank you for everything.

A few individuals made special contributions that led directly to the book now in print. Viraj deSilva and Matthew Pettis provided the model simulations in the website chapter on "Complex Systems." Drs. Gene F. Parkin and Jeanne M. VanBriesen provided extensive suggestions and corrections. Pablo Pastén and Chrysi Laspidou provided solutions to many of the problems in the Solutions Manual. Janet Soule and Rose Bartosch deciphered BER's handwriting to create the original electronic files for all or parts of Chapters 1, 3, 4, 6, 8, 9, 10, 11, 12, and 15. Dr. Saburo Matsui and the Research Center for Environmental Quality Control (Kyoto University) provided a sabbatical venue for BER so that he could finish all the details of the text and send it to McGraw-Hill on time.

Finally, we thank Marylee and Martha for loving us, even when we became too preoccupied with the "book project."

Bruce E. Rittmann
Evanston, Illinois

Perry L. McCarty
Stanford, California

CONTENTS

Brief Contents

chapter

1

BASICS OF MICROBIOLOGY

Environmental biotechnology applies the principles of microbiology to the solution of environmental problems. Applications in environmental microbiology include

- Treatment of industrial and municipal wastewaters.
- Enhancement of the quality of drinking water.
- Restoration of industrial, commercial, residential, and government sites contaminated with hazardous materials.
- Protection or restoration of rivers, lakes, estuaries, and coastal waters from environmental contaminants.
- Prevention of the spread through water or air of pathogens among humans and other species.
- Production of environmentally benign chemicals.
- Reduction in industrial residuals in order to reduce resource consumption and the production of pollutants requiring disposal.

Although this textbook can cover only some of the numerous topics that can be categorized under environmental biotechnology, the principles of application in one area of the environmental field often apply equally to other environmental problems. What is required in all cases is a linking of the principles of microbiology with engineering fundamentals involving reaction kinetics and the conservation of energy and mass.

The purpose of this first chapter is to review the basic principles of microbiology. Fundamentals of reaction kinetics and mass and energy conservation are addressed in four subsequent chapters, while the last chapters in the text address important applications. Readers desiring more detailed information on microbiology are referred to texts such as Madigan, Martinko, and Parker (1997) and Alcamo (1997).

This chapter summarizes

- How microorganisms are classified (*taxonomy*).
- What they look like (*morphology*).
- How they reproduce so that their functions can be maintained.
- The biochemical reactions that they mediate (*metabolism*).
- The major divisions among microorganisms based upon their function in the environment (*trophic groups*).
- How information about structure and function of organisms is transmitted and changed (*genetics*).
- An aspect of great importance in environmental biotechnology, that is *microbial ecology*, or the interactions among organisms and their environment.

The major difference between environmental biotechnology and other disciplines that feature biotechnology is that environmental applications almost always are concerned with mixed cultures and open, nonsterile systems. Success depends on how individual microorganisms with desired characteristics can survive in competition with other organisms, how desired functions can be maintained in complex ecosystems, and how the survival and proliferation of undesired microorganisms can be prevented.

Anyone interested in environmental biotechnology needs to be familiar with organism interactions and the principles of mixed culture development and maintenance in order to obtain sound solutions to environmental problems. For example, creating novel organisms that can carry out specific reactions of interest seems like a wonderful way to solve difficult environmental problems. The question of importance then is: How can such organisms survive in competition with the thousands of other organisms in the environment that are also fighting for survival in situations that can be quite hostile to them? Developing robust microbiological systems that can carry out intended functions over time is the major challenge before those seeking to apply principals of biotechnology to the solution of environmental problems.

1.1   THE CELL

The *cell* is the fundamental building block of life. A cell is an entity that is separate from other cells and its environment. As a living entity, a cell is a complex chemical system that can be distinguished from nonliving entities in four critical ways.

1. Cells are capable of growth and reproduction; that is, they can self-produce another entity essentially identical to themselves.
2. Cells are highly organized and selectively restrict what crosses their boundaries. Thus, cells are at low entropy compared to their environment.
3. Cells are composed of major elements (C, N, O, and S, in particular) that are chemically reduced.
4. Cells are self-feeding. They take up necessary elements, electrons, and energy from their external environment to create and maintain themselves as

reproducing, organized, and reduced entities. They require sources of the elemental building blocks that they use to reproduce themselves. They require a source of energy to fuel the chemical processes leading to all three properties. In addition, they require a source of electrons to reduce their major elements. How the cells obtain elements, energy, and electrons is called *metabolism,* and it is one essential way in which we characterize cells. Understanding metabolism is a theme that runs throughout this book.

Cells are physically organized so that they can carry out the processes that make them living entities. Later in this chapter, the basic components of cells are described in more detail. At this point, the essential components of cells are identified and connected to the distinguishing features of what makes a living cell.

- The *cell membrane* is a barrier between the cell and its environment. It is the vehicle for restricting what crosses its boundaries, and it is the location of reactions that the cell needs to conduct just outside itself.

- The *cell wall* is a structural member that confers rigidity to the cell and protects the membrane.

- The *cytoplasm* comprises most of the inside of the cell. It contains water and the macromolecules that the cell needs to function.

- The *chromosome* stores the genetic code for the cell's heredity and biochemical functions.

- The *ribosomes* convert the genetic code into working catalysts that carry out the cell's reactions.

- The *enzymes* are the catalysts that carry out the desired biochemical reactions.

Cells may have other components, but these are the essential ones that define them as living entities.

Figure 1.1 shows that three major *domains* comprise all organisms. The *Bacteria* and the *Archaea* domains contain the *prokaryotes,* or cells that do not contain their chromosome inside a nucleus. The organisms within these two major domains are single cellular, because they are complete living entities that consist of only one cell. The other major domain is the *Eukarya,* which comprise organisms that may be single cellular or multicellular and have their chromosomes inside a nucleus. All higher plants and animals belong to the Eukarya domain.

All prokaryotes are microorganisms, or organisms that can only be seen with the aid of a microscope. Some of the eukaryotic life forms are microorganisms, and some are not. Eukarya range from single cellular microscopic algae and protozoa (*protista*) up to large multicellular mammals, such as the whales, and plants, such as the redwood trees. Organisms from all three domains are of importance in environmental microbiology, and thus the structure and function of all are of interest.

Some cells may undergo change in form or function through the process of *differentiation.* For example, cells within the human body act differently depending upon whether they form part of an eye, a muscle, or a strand of hair. As part of differentiation, cells can often interact with one another through various chemical

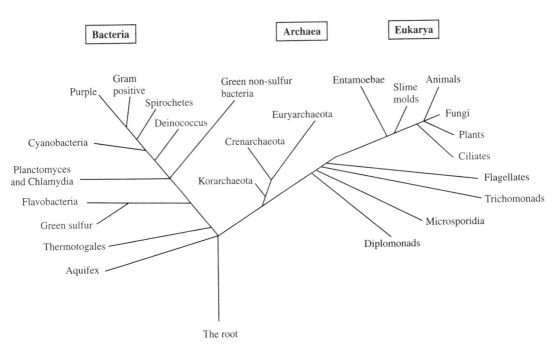

Figure 1.1   Phylogenetic tree of life as determined from comparative ribosomal RNA sequencing. SOURCE: Based on data from Carl R. Woese and Gary J. Olsen.

signals in ways that can change their form or function. It is significant that cells also can evolve into organisms that are markedly different from the parent, a process that usually is quite slow, but nevertheless of great importance to the formation of new organisms or to the development of new capabilities that may aid in organism survival.

Our interest in environmental biotechnology centers primarily on the single-celled organisms, which include the microorganisms in the Bacteria and Archaea domains. In the Eukarya domain, microorganisms of particular interest are algae and protozoa. Recently, environmental biotechnology has begun to focus on plants, too. *Phytoremediation* is a process in which plants help to bring about the destruction of toxic chemicals in soils and groundwater. Here, trees such as the poplar may uptake toxic chemicals along with water. In some cases, the plants or microorganisms associated with the plant roots transform the toxic compounds into nonharmful products. Phytoremediation is an evolving field with much yet to be learned. For the most part, however, the primary focus of this text will be with the single-celled microorganisms.

1.2   TAXONOMY AND PHYLOGENY

Taxonomy is the science of classification. Microbiologists and engineers need to classify microorganisms for many practical reasons: for example, to determine the possible presence of disease-related species in drinking water. Taxonomy relies on the observable physical properties of organisms to group microorganisms. Observable properties are called a cell's *phenotype* and may involve its appearance (*morphology*), the manner in which it interacts with dyes or staining, and its ability to use or convert a given chemical into another one (*transformation*).

Phylogeny is a different and newer method of classification that detects differences in microorganisms based upon genetic characteristics. Such characteristics are encoded in the organism's DNA (deoxyribonucleic acid), which contains the hereditary material of cells, and RNA (ribonucleic acid), which is involved in protein synthesis. Of particular importance here are the sequences of base pairs in an organism's 16S ribosomal RNA (rRNA), one of the three major categories of RNA. Through the analysis of 16S rRNA, Carl R. Woese discovered that the large group of single-celled microorganisms once collectively termed bacteria was actually comprised of two very distinct domains, the Bacteria and the Archaea.

Phylogeny relates organisms based on their evolutionary history, while taxonomy relates organisms based on observable characteristics of the cells. Both are powerful methods of classification, and they yield different kinds of information. Despite fundamental differences between taxonomic and phylogenetic classifications, some important phenotypic characteristics consistently differentiate among the three basic domains of organisms from one another. For example, Table 1.1 illustrates that only the Eukarya have a membrane-enclosed nucleus, while only the Archaea are capable of generating methane gas (*methanogenesis*).

The basic taxonomic unit is the *species*, which is the collection of strains having sufficiently similar characteristics to warrant being grouped together. Such a loose definition often makes it difficult to determine the difference between strains (members of a given species that have measurable differences between themselves) and species. Groups of species with major similarities are placed in collections called

Table 1.1 Differing features among Bacteria, Archaea, and Eukarya

Characteristic	Bacteria	Archaea	Eukarya
Membrane-enclosed nucleus	Absent	Absent	Present
Cell wall	Muramic acid present	Muramic acid absent	Muramic acid absent
Chlorophyll-based photosynthesis	Yes	No	Yes
Methanogenesis	No	Yes	No
Reduction of S to H_2S	Yes	Yes	No
Nitrification	Yes	No	No
Denitrification	Yes	Yes	No
Nitrogen fixation	Yes	Yes	No
Synthesis of poly-β-hydroxyalkanoate carbon storage granules	Yes	Yes	No
Sensitivity to chloramphenicol, streptomycin, and kanamycin	Yes	No	No
Ribosome sensitivity to diphtheria toxin	No	Yes	Yes

SOURCE: Madigan, Martinko, and Parker, 1997.

genera (or *genus* for singular), and groups of genera with sufficient similarity are collected into *families.* Microorganisms are generally given a genus and species name. For example, *Escherichia coli* is an indicator organism of fecal pollution of drinking water. The genus name, *Escherichia,* is always capitalized and placed before the species name (*coli*), which is never capitalized. Note that the entire organism name is always written in italics (or underlined if writing in italics is not possible). A common practice once the genus name is identified is to abbreviate the genus name by using only the first capitalized letter (in this case, *E. coli*), but the species name is never abbreviated. These are rules as set out in *The International Code of Nomenclature of Bacteria,* which apply to Archaea as well. Individual differences among the major groupings of microorganisms are now described in some detail.

1.3 PROKARYOTES

In discussing the microorganisms of most general importance to environmental biotechnology, the prokaryotes, little is gained by dividing their descriptions between the Bacteria and Archaea since, functionally, their similarities are greater than their differences. Indeed, until the development of genetic phylogeny, the collection of microorganisms within these two domains was simply called bacteria. Also, the Bacteria and Archaea generally are found together and often participate together to bring about the destruction or mineralization of complex organic materials, such as in the formation of methane from the decay of dead plants and animals. In this example, the Bacteria ferment and convert complex organic materials into acetic acid and hydrogen, and the Archaea convert the acetic acid and hydrogen into methane gas. The organisms must work closely together, as in an assembly line, in order to bring about the destruction of the organic matter.

Another distinction that became clearer with the development of genetic phylogeny is among the photosynthetic microorganisms, which are of great importance in natural and engineered systems. Formerly, *algae* was a term used to describe the group of single-celled organisms that behaved like plants: that is, they contain chlorophyll and derive energy for growth from sunlight. However, a group of photosynthetic prokaryotes formerly called *blue-green algae* has no nucleus, a property of bacteria, not plants. Now, they are classified within a bacterial grouping called *cyanobacteria.* Cyanobacteria and algae are commonly found together in natural waters and tend to compete for the same energy and carbon resources. Cyanobacteria are a pesky group of phototrophs that cause many water quality problems, from tastes and odors in drinking water to the production of toxins that kill cows and other ruminants that may consume them while drinking from highly infested surface waters. Despite important differences between algae and cyanobacteria, grouping together makes good practical sense when such differences are not of interest. Here, we are reminded that nature does not strive to classify things; humans do. Despite many gray areas where classification of organisms is not easy (and sometimes does not seem to make much sense), classification is essential for our organization of knowledge and for communication among scientists, practitioners, and others.

Living things cover a very broad spectrum of attributes with a great diversity of overlapping capabilities. Also, much exchange of genetic information occurs between species, not only between organisms within a given domain, but between organisms living in different domains as well. The graying of boundaries that this causes creates major problems for those seeking the systematic ordering of things. To make the picture somewhat more complicated, the interest in environmental biotechnology is often more on function than it is on classification into species. Like cyanobacteria and algae, which carry out the similar function of photosynthesis, the ability to biodegrade a given organic chemical may rest with organisms spanning many different genera. Thus, the identification of a particular species is important in some instances, such as when concerned with the spread of infectious diseases, while it is not at all of interest to us in other circumstances. Indeed, in a biological wastewater treatment system that is degrading an industrial organic waste efficiently and reliably, the dominant bacterial species may change from day to day. These treatment systems are not pure culture systems, and it would be impossible to maintain them as such. They are open, complex, mixed-culture systems that are ruled by principles of microbial ecology. Species that can find a niche for themselves and can dominate over their competitors will survive and flourish, at least until some new chemical or environmental perturbation affects the system in such a way as to push the balance in some other direction. Such competition and change in dominance with changing conditions is one characteristic that allows environmental systems to be robust and to work so well.

Bacteria comprise the first domain that we describe. A description of the Archaea then follows. However, Archaea are so similar to the Bacteria that the section on them will emphasize the differences between the two, rather than their similarities. Following that, we consider the microscopic Eukarya of interest, the single-celled algae and protozoa. Although microscopic Eukarya are sometimes termed the simplest plants and animals, they have many similarities to single-celled species in the other two domains. Next come the multicellular organisms, important members of which are the fungi. They are neither plants nor animals, but are more similar to bacteria in their important role in degrading dead organic materials in the environment. Finally, the discussion turns to another entity that is neither living nor dead: the virus, which is implicated in diseases to humans. The viruses do not have the genetic makeup to live on their own, but depend on other living forms for their reproduction. This leads into a brief discussion of microorganisms that cause the infectious diseases that can be transmitted through environmental media of air, water, and food.

1.3.1 BACTERIA

Bacteria, along with Archaea, are among the smallest of the entities that are generally agreed to be "living." Their activities are essential to the survival of all other species. Bacteria are ubiquitous, and the same or similar species can generally be found in every part of the world. A handful of soil taken from the San Joaquin Valley in California is likely to yield a similar distribution of bacterial species as one taken from a mountain in Tibet. They are present in land, water, and air and are distributed throughout the world on air currents and through water movement. Even if the same species are not present at different locations, the functions carried out by the organisms present will be similar, such as in the mineralization of dead plants and animals.

Bacteria are important to the environment, as they have the ability to transform a great variety of inorganic and organic pollutants into harmless minerals, which can then be recycled back into the environment. They have the ability to oxidize many organic chemicals synthesized industrially, as well as those produced naturally through normal biological processes, and bacteria are employed in wastewater treatment plants for this purpose. Some can convert waste organic materials into methane gas, which is a useful form of energy. Others can transform inorganic materials, such as ammonia or nitrate, which under certain circumstances may be harmful, into harmless forms, such as nitrogen gas, the main constituent of air.

The functions of bacteria are not all beneficial to humans, however. Some are the causative agents of disease and are responsible for many of the plagues of the past, and, even today, some are responsible for major sickness and misery in the world. Thus, the need is to protect humans from pathogenic bacteria in water, food, and air, while taking advantage of their important abilities to cleanse pollutants from water bodies and soil.

Morphology   The morphology of bacteria includes their shape, size, structure, and their spatial relationship to one another. Bacteria have three general shapes as illustrated in Figure 1.2. Those with a spherical shape are *cocci* (singular, *coccus*), with a cylindrical shape are rods or *bacilli* (singular, *bacillus*), and with a helical shape are *spirilla* (singular, *spirillum*). Electron microphotographs of typical bacteria are illustrated in Figure 1.3. (The photomicrographs in Figure 1.3, as well as in Figures 1.6,

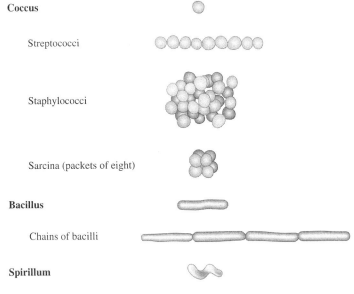

Coccus

Streptococci

Staphylococci

Sarcina (packets of eight)

Bacillus

Chains of bacilli

Spirillum

Figure 1.2   Typical morphologies of bacteria.

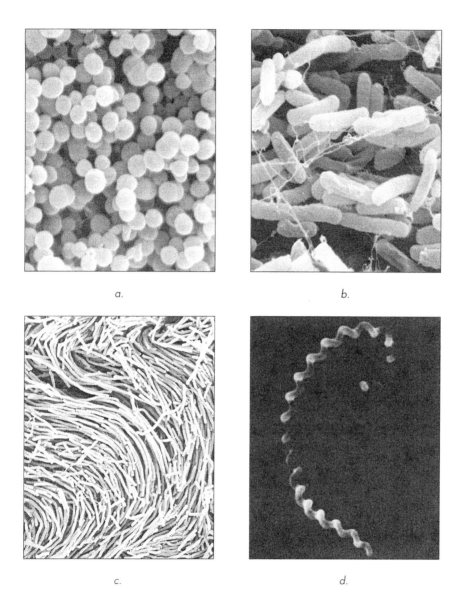

Figure 1.3   Scanning electron microphotographs of typical bacteria.
a. Staphylococcus epidermidis. b. Escherichia coli. c. Bacillus chains.
d. Leptospyra interrogans. SOURCE: With permission of the Microbe Zoo
and Bergey's Manual trust, Michigan State University. Images *a, b, c*
were created by Shirley Owens, and image *d* is from *Bergey's Manual of
Systematic Bacteriology*, Vol. 1, Williams and Wilkins, Baltimore (1984).

1.7, and 1.8, come from the Microbe Zoo, a project of the Center for Microbial Ecology at Michigan State University, which permits their use in this text. The Microbe Zoo is copyrighted by the Trustees of the Michigan State University. (Readers can find more images of microorganisms at http://commtechlab.msu.edu/sites/dlc-me.)

Bacteria generally vary in width from 0.5 to 2 μm and in length from 1 to 5 μm. Cocci generally have a diameter within the 0.5- to 5-μm range as well. There are about 10^{12} bacteria in a gram (dry solid weight). Because of the small size, the surface area represented is about 12 m^2/g. Thus, bacteria have a large exposure of surface to the outside environment, permitting rapid diffusion of food into the cell and very fast rates of growth.

Features of prokarya are illustrated in Figure 1.4, which also shows key features of Eukarya. Although not all bacteria have the same structural features, common to all are the external *cell wall;* the *cell membrane,* which is just inside the cell wall; and the internal colloidal fluid called *cytoplasm.* The cell wall is composed of a repeat-

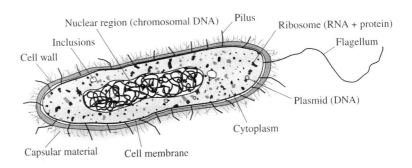

a.

b.

Figure 1.4   The structure of typical prokaryotic and eukaryotic cells. *a.* Prokaryotic cell. *b.* Eukaryotic cell.

ing building block called *peptidoglycan.* Each section of peptidoglycan consists of N-acetylglucosamine, N-acetylmuramic acid, and a small group of amino acids, such as alanine, glutamic acid, and either lysine or diaminopimelic acid. The intact cell wall often has other materials associated with it that give the cell special characteristics. For example, bacteria that cannot be stained with Gram stain (Gram-negative) have a higher content of lipopolysaccharide in the cell wall and more amino acids than Gram-positive bacteria. On the other hand, teichoic acids are characteristic of the cell walls of Gram-positive bacteria.

The cell membrane, or *cytoplasmic membrane,* is a phospholipid bilayer that lies immediately beneath the cell wall and performs several important functions for the cell. It is semipermeable, is the main barrier to the passage of large molecules, and controls the passage of nutrients into and out of the cell. It is also the location of several important enzymes, including the cytochromes, the enzymes that are involved in electron transport and energy conservation. The cell membrane of most bacteria is quite simple. It becomes more complex in autotrophic bacteria, which synthesize essentially all cellular components from inorganic materials, and even more complex in the phototrophic bacteria, which obtain energy from sunlight. In these latter organisms, the cell membrane intrudes into the internal portion of the cell at certain points, a phenomena that apparently provides more membrane surface needed for the increased complexity and intensity of functions.

Cytoplasm is the material contained within the cell membrane that is used to carry out cell growth and function. It consists of water, dissolved nutrients, enzymes (proteins that enable the cell to carry out particular chemical reactions), other proteins, and the nucleic acids (RNA and DNA). Also included are the densely packed *ribosomes,* which are RNA-protein particles that contain the enzymes for protein synthesis.

Concentrated deposits of materials called *cytoplasmic inclusions* are contained in some cells. Generally, they serve as storage for food or nutrients. Phosphates are stored as polymers in volutin granules, polysaccharide granules store carbohydrates, and other granules contain polymerized β-hydroxybutyric acid (PHB) or fatty materials. In certain sulfur-metabolizing bacteria, large amounts of sulfur accumulate in granules.

Important to all cells is the presence of the DNA molecule, a double-stranded helix-shaped molecule that contains all the genetic information required for cell reproduction. It also contains in coded form all information required to carry out the normal cell functions. This information is stored in the sequence of nucleotides (each consisting of deoxyribose connected to one of four nitrogen bases: adenine, guanine, cytosine, or thymine). The genetic information of the cell is stored, replicated, and transcribed by the DNA molecule, which may also be called a *prokaryotic chromosome.* Some bacteria also contain one or several much smaller circular DNA molecules, called *plasmids,* which also can impart additional genetic characteristics to the organism.

The information contained in DNA is "read" and carried to the ribosomes for production of proteins by RNA. RNA is single stranded, but otherwise similar to DNA. The major difference is that the sugar portion of each nucleotide is ribose, rather than deoxyribose, and uracil is substituted for the base thymine. There are three basic forms: messenger RNA (mRNA), which carries the information for protein synthesis from the DNA molecule; transfer RNA (tRNA), which carries amino acids to their proper site on the mRNA for the production of a given protein; and ribosomal RNA (rRNA), which functions as structural and catalytic components of the ribosomes, the sites where the proteins are synthesized. DNA is the blueprint for the structure and functioning of the bacteria, and the different forms of RNA can be viewed as the different workers that read these plans and carry out the construction of the bacterial cell.

Some bacteria, such as those of the genera *Bacillus* and *Clostridium,* have the capability of forming *endospores* (or spores formed inside the cell). Spore formation generally results when the environment approaches a state adverse to the growth of the bacteria, such as the depletion of a required nutrient or unfavorable temperature or pH. In the sporulated state, the organism is inactive. The spore has many layers over a core wall, within which there are the cytoplasmic membrane, cytoplasm, and nucleoid. Thus, it is similar to a vegetative cell except for the cell wall. It also contains about 10 percent dry weight of a calcium-dipicolinic acid complex and only about 10 to 30 percent of the water content of a vegetative cell. These characteristics help impart great resistance of the cell to heat and chemical attack. An endospore can remain inactive for years, perhaps centuries. When an endospore finds itself in a favorable environment for growth, then it *germinates* back to the active vegetative state. Bacteria that form endospores are quite difficult to destroy through normal sterilization techniques.

There are certain external features of bacteria that also can help in identification. Some contain a *capsule* or *slime* layer, which is believed to be excreted from the cell proper and, because of its viscous nature, does not readily diffuse away from the cell. This excreted material can increase the viscosity of the surrounding fluid and can also help to hold bacteria together in aggregates or bacterial "flocs." Such layers are also significant in bacterial attachment to surfaces, such in the formation of biofilms, which are of much significance in environmental biotechnology. Most capsule or slime layers consist of polysaccharides of several types (*glycocalyx*) and polymers of amino acids. Some slime layers are readily visible, and others are not. Removal of these external layers does not adversely affect the normal functioning of the bacterial cell.

Another external feature of some bacteria is the *flagella.* These hairlike structures are attached to the cytoplasm membrane and protrude through the cell wall into the surrounding medium. The length of the flagella may be several times that of the cell. The number and location of the flagella are different for different species. Some may contain a single terminal (polar) flagellum, some may have several at one end, some may have several at both ends, and others may have them distributed over the entire cell (peritrichous). Most species of bacilli have flagella, but they are rare in cocci. Flagella are responsible for mobility in bacteria. When they rotate at high speed, they can push or pull an organism a distance equal to several times its own length in a second. Some bacteria, however, can move on surfaces without flagella by an action called *gliding,* although this mode of motion is quite slow by comparison. Cells can be caused to move toward or away from chemical or physical conditions, which is called a *tactic* or *taxes* response. Movement in response to chemical agents is called *chemotaxis,* and that in response to light is termed *phototaxis.*

Structural features that are similar to flagella are *fimbriae* and *pili*. Fimbriae are more numerous than flagella, but are considerably shorter and do not impart mobility. They may aid organisms in attachment to surfaces. Pili are less numerous than fimbriae, but are generally longer. They too aid in organism attachment to each other and to surfaces. An example is the use of pili by *E. coli* for clumping together and for attachment to intestinal linings. Pili appear to be involved in bacterial *conjugation,* the important process by which genetic information from one cell is transferred to another. Conjugation appears to be how resistance to pesticides and antibiotics or the ability to degrade certain toxic chemicals is transferred between microorganisms.

Another characteristic property of bacteria is the way they are grouped together, a feature that is related to the manner in which they reproduce. As discussed in detail later, bacteria multiple by *binary fission,* or simply fission, which means one organism divides into two. The two may remain attached together for some time, even after each divides again. Because of varying tendencies among different species to cling together after division, and the differing ways in which they divide, characteristic groupings of bacteria emerge. For this reason bacteria may occur singly or in pairs (*diplococci*). They may form short or long chains (*streptococci* or *streptobacilli*), may occur in irregular clusters (*staphylococci*), may form tetrads (*tetracocci*), or may form cubical packets of eight cells (*sarcinae*). These particular groupings can be observed with the optical microscope and can aid in identification.

Chemical Composition   In order for bacteria to grow and maintain themselves, they must have available essential nutrients, such as carbon, nitrogen, phosphorus, sulfur, and other elements for synthesis of proteins, nucleic acids, and other structural parts of the cell. These requirements must be met whether the bacteria are in nature, treating a wastewater, or detoxifying hazardous contaminants. If these elements are not present in available forms in a wastewater to be treated, as an example, they must be provided as part of normal treatment plant operation. An idea of the quantity of different nutrients needed to grow the cells can be obtained by combining an estimate of the rate of cell production with the quantity of each element in the cell.

Table 1.2 contains a summary of the general chemical characteristics of bacteria. They contain about 75 percent water, which is useful knowledge when attempting to dispose of or incinerate waste biological solids produced during wastewater treatment. The water internal to the cells is extremely difficult to remove by normal dewatering processes. One result is that the "dewatered" biological solids or sludge still contains at least 75 percent water. A second result is that most "dewatered" sludges have too much water to be incinerated without the addition of supplemental fuel.

The dried material—that remaining after all cellular water is evaporated by heat drying at 105 °C—contains about 90 percent organic matter, about half of which is the element carbon, one quarter is oxygen, and the remainder is divided between hydrogen and nitrogen. The nitrogen is an important element in proteins and nucleic acids, which together represent about three-fourths of the total organic matter in the cell. Nitrogen must be added to many industrial wastewaters in order to satisfy the need for this element for bacteria. While air is about 80 percent nitrogen, it is present in the zero-valent form (N_2), which is not available to most bacteria, except for a few with the ability to fix atmospheric N_2. *Nitrogen fixation* is a rather slow process

Table 1.2   Chemical and macromolecular characteristics of prokaryotic cells

Chemical Composition

Constituent	Percentage		
Water	75		
Dry Matter	25		
Organic		90	
C			45–55
O			22–28
H			5–7
N			8–13
Inorganic		10	
P_2O_5			50
K_2O			6.5
Na_2O			10
MgO			8.5
CaO			10
SO_3			15

Macromolecular Composition — *E. coli* **and** *S. typhimurium*[a]

	Percentage[b]	Percentage	Molecules per cell
Total	100	100	24,610,000
Proteins	50–60	55	2,350,000
Carbohydrates	10–15	7	
Lipids	6–8	9.1	22,000,000
Nucleic Acids			
DNA	3	3.1	2.1
RNA	15–20	20.5	255,500

[a]Data from Madigan, Martinko, and Parker (1997) and G.C. Neidhardt et al. (1996). *E. coli* dry weight for actively growing cells is about 2.8×10^{-13} g.
[b]Dry weight.

and usually is not adequate to satisfy nitrogen needs in engineered treatment systems. When grown in the presence of sufficient nitrogen, the bacterial cell will normally contain about 12 percent nitrogen. The content can be reduced to about one-half of this value by nitrogen starvation, although this will reduce the possible rate of growth. Under such conditions, the carbohydrate and lipid portions of the cell tend to increase.

Another necessary element for bacterial growth is phosphorus, which is an essential element in nucleic acids and certain key enzymes. Phosphorus may be added in the form of orthophosphates to satisfy this need, the amount required being one-fifth to one-seventh of that for nitrogen on a weight basis.

As indicated in Table 1.2, certain other elements are needed, but in trace amounts. Of particular importance here are sulfur and iron. Some elements are not required by all bacteria, but may be for some as critical components of key enzymes, such as molybdenum for nitrogen fixation or nickel for anaerobic methane production. Some bacteria require specific organic growth factors that they cannot produce themselves, such as vitamins. Often, these are produced by other organisms living in mixed cultures, so that they need not be added externally except when growing as a pure culture.

Empirical formulas for bacterial cells are useful when making mass balances for biological reactors. Empirical formulas based upon the relative masses of the five major elements in cells are summarized in Chapter 2, with $C_5H_7O_2N$ being a commonly used formulation for design computations. The empirical molecular weight with this formulation is 113 g, for which nitrogen represents 12.4 percent and carbon 53 percent, on weight bases.

Reproduction and Growth   Knowledge of rates of bacterial growth and reproduction is essential for design of engineered biological treatment systems, as well as for understanding bacteria in nature. Bacteria normally reproduce through *binary fission,* in which a cell divides into two after forming a transverse cell wall or *septum* (Figure 1.3). This asexual reproduction occurs spontaneously after a growing cell reaches a certain size. After reproduction, the parent cell no longer exists, and the two daughter cells normally are exact replicates (i.e., *clones*) of each other, both containing the same genetic information as the parent. The interval of time required for the formation of two cells from one is the generation time, which varies considerably depending on the particular organism and the environmental conditions. It may be as short as 30 min, as with *E. coli.* Here, there would be 2 organisms after 30 min, 4 after 1 h, 8 after 1.5 h, and 16 after 2 h. After 28 h and 56 divisions, the total population, if unrestricted in growth, would equal more than 10^{16} and would have a dry weight of about 18 kg. In principle, the weight would increase from 10^{-13} g to that of a human child in a single day! Environmental and nutritional limitations of the culture flask generally limit the growth long before such a mass increase occurs, but the potential for such rapid growth is inherent in the bacterial cell.

There are other, but less common, methods of reproduction among bacteria. Some species of the genus *Streptomyces* produce many reproductive spores per organism, and each can give rise to a new organism. Bacteria of the genus *Nocardia* can produce extensive filamentous growth, which if fragmented can result in several filaments, each of which can give rise to a new cell. Some bacteria can reproduce by budding, and the stalk that develops from the parent can separate and form a new cell. Even sexual reproduction of a type can occur in bacteria. This process of conjugation requires contact between the mating bacteria, during which genetic material, usually in the form of a plasmid, rather than a whole chromosome, is transferred from one to another. Such conjugation between different species can result in offspring with characteristics that are a combination of the characteristics of the two parents. This can result in the transfer of resistance to antibiotics or in the ability to completely degrade a compound that could not be degraded by either parent alone. For the most part, however, asexual reproduction by binary fission is the usual mode of reproduction among bacteria.

Energy and Carbon-Source Classes of Bacteria   An important characteristic of bacteria is the very wide variety of energy sources they can use for growth and maintenance. Organisms that use energy from light are termed *phototrophs,* and those obtaining energy from chemical reactions are termed *chemotrophs.* There are two types of chemotrophs: those that use organic chemicals for energy are *chemoorganotrophs* and those that use inorganic chemicals for energy as *chemolithotrophs*. The phototrophs, which generally use carbon dioxide as a source of cell carbon, are also classified into two types based on the chemical from which electrons are obtained to reduce carbon dioxide to form cell organic matter. Those that use water and convert it photochemically into oxygen and hydrogen, the electron source, are termed *oxygenic phototrophs*. They produce oxygen as an oxidized waste product, as do plants. The others are *anoxygenic phototrophs,* and they generally live only in the absence of oxygen. Instead of water, they extract electrons from reduced sulfur compounds, such as H_2S or elemental sulfur; H_2; or organic compounds, such as succinate or butyrate. When H_2S is used, it is converted into H_2 and S, somewhat akin to the photochemical conversion of water, with H_2 being used as the electron source for synthesis and S being an oxidized product. Different chlorophylls are used in anoxygenic photosynthesis, compared with oxygenic photosynthesis, to capture the energy from the light and extract the electrons.

Other common terms used to describe microorganisms are related to the carbon source used for cell synthesis. *Autotrophs* use inorganic carbon, such as CO_2, for cell synthesis, while *heterotrophs* use organic compounds for cell synthesis. Chemolithotrophic bacteria are commonly also autotrophic, and chemoorganotrophs are commonly heterotrophic. For this reason, the different terms relating to the energy source and to the carbon source for chemotrophs are often used interchangeably.

Environmental Conditions for Growth   In addition to the nutritional needs for energy and cell synthesis, bacteria require a proper physical and chemical environment for growth. Factors of importance here are temperature, pH, oxygen partial pressure, and osmotic pressure. The rate of all chemical reactions is influenced by temperature, and since bacterial growth involves a series of chemical reactions, their rate of growth is also greatly influenced by temperature. Over a limited range for each bacterial species, growth rates increase with temperature, and, in general, they roughly double for each 10 °C rise. At temperatures above the normal range for a given species, key enzymes are destroyed, and the organism may not survive.

Bacteria can be classified into four different groups based upon their normal temperature range for growth:

Temperature Class	Normal Temperature Range for Growth (°C)
Psychrophile	−5 to 20
Mesophile	8 to 45
Thermophile	40 to 70
Hyperthermophile	65 to 110

Figure 1.5 illustrates the effect of temperature on the growth rate for these different temperature classes of bacteria. In general, organisms with a higher temperature range have a higher maximum growth rate than organisms in a lower range. Also, within

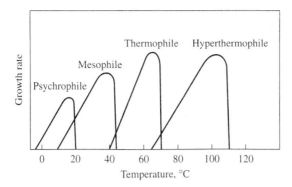

Figure 1.5   Effect of temperature on growth rate of different temperature classes of bacteria.

each range, the maximum growth rate occurs over a small temperature range. Once this temperature range is exceeded, growth rate drops off rapidly due to denaturation of key proteins. The structural characteristics of bacteria in the different temperature classes are different in order to provide protection from either cold temperatures or hot temperatures. Psychrophiles are commonly found in Arctic regions, where temperatures are commonly cold. Hyperthermophiles, on the other hand, tend to thrive in hot springs, where temperatures are near boiling.

Knowledge of temperature classes of bacteria is useful in the design and operation of wastewater treatment systems. Higher treatment rates are associated with higher growth rates; thus treatment at higher temperatures may have the advantage of a smaller tank size. However, higher temperatures may require energy inputs that counter the savings from a smaller tank. If the temperature of a treatment system varies too much from day to day, poor operation may result, as the system may not be optimal for any particular temperature class of bacteria. Thus, treatment of wastewaters by methane fermentation is usually conducted near the optimum of 35 °C for mesophilic organisms, or at 55 to 60 °C for thermophilic operation, but not at 45 °C, which is suboptimal for either group.

Most species of bacteria also have a narrow pH range for growth, and for most organisms this range lies between 6 and 8. For some species, the operating range is quite broad, while for others, it can be quite narrow. In addition, some bacteria, particularly chemolithotrophs that oxidize sulfur or iron for energy, thrive best under highly acidic conditions. This characteristic enhances their chance for survival, since an end product of their energy metabolism is generally a strong acid, such as sulfuric acid. The design and operation of a treatment system must consider the optimum pH conditions required for growth of the bacteria of interest.

Another important environmental characteristic that differentiates bacterial species is their ability to grow either in the presence or absence of molecular oxygen (O_2). *Aerobic* bacteria require the presence of oxygen to grow, and they use it as an electron acceptor in energy-yielding reactions. *Anaerobic* bacteria live in the absence of oxygen. Their energy-yielding reactions do not require O_2. Bacteria that can live either in the presence of oxygen or its absence are *facultative* bacteria. Finer classifications sometimes are used to describe organisms that lie within these three groups. *Obligate anaerobes* are killed by the presence of oxygen. *Aerotolerant anaerobes* can tolerate oxygen and grow in its presence, but cannot use it. Organisms that prefer oxygen, but can survive without it are called *facultative aerobes,* while bacteria that grow in the presence of minute quantities of molecular oxygen are called *microaerophiles.* The distinctions have practical implications. For example, a treatment system that relies on methane fermentation requires obligate anaerobes, and oxygen must be excluded from the system. On the other hand, a system that depends upon aerobic bacteria must be provided with O_2 at a sufficient rate.

Another classification of bacteria is based upon their tolerance of salt. Those that grow best under salt conditions similar to seawater (about 3.5 percent sodium chloride) are called *halophiles,* while those that live well in a saturated sodium chloride solution (15–30 percent) are termed *extreme halophiles.* Natural environment homes for the latter organisms are the Great Salt Lake in Utah and the salt ponds ringing San Francisco Bay, where seawater is evaporated for commercial use of the salt it contains.

These various classifications of bacteria illustrate the very wide range of environmental conditions in which bacteria not only survive, but also thrive. Bacteria evolved over time, resulting in a wide range of species able to take advantage of available energy under conditions that appear quite extreme to us. Of course, strains adapted to one set of conditions are not necessarily adapted for other conditions that are hospitable to other species. Thus, the organisms one finds in dominance at any particular location are generally those best adapted to survive under the particular physical and chemical environment present there. If that environment changes, then the dominant species will likely change with it.

Phylogenetic Lineages   As illustrated in Figure 1.1, bacteria can be grouped into different lineages based on rRNA sequencing. The lineages also follow to some extent *phenotypic* properties based on energy source and the organism's tolerance to different environmental conditions. Typical characteristics of the different lineages are summarized in Table 1.3. Some characteristics are unique to one lineage, whereas others are found in only a few of the domains. Bacteria within the three oldest lineages, *Aquifex/Hydrogenobacter, Thermotoga,* and Green nonsulfur bacteria, are all thermophiles, perhaps reflecting the fact that the earth at the time of their evolution was much warmer than it is today. The *Aquifex/Hydrogenobacter* group is the most ancient and is composed of hyperthermophilic chemolithotrophs that oxidize H_2 and sulfur compounds for energy, perhaps reflecting the dominant conditions present when they evolved. The next oldest group, the *Thermotoga,* is not only hyperthermophilic, but also anaerobic chemoorganotropic, representing perhaps another stage in the earth's evolution toward the presence of more organic matter, but still with an atmosphere devoid of oxygen. The third group, the Green nonsulfur bacteria, is merely thermophilic, but contain among them a group of anoxygenic phototrophs (*Chloroflexus*).

Table 1.3 Characteristics of the 12 phylogenic lineages of bacteria

Phylogenic Group	Characteristics
Aquifex/Hydrogenobacter	Hyperthermophilic, chemolithotrophic
Thermotoga	Hyperthermophilic, chemoorganotrophic, fermentative
Green nonsulfer bacteria	Thermophilic, phototrophic and nonphototrophic
Deinococci	Some thermophiles, some radiation resistant, some unique spirochetes
Spirochetes	Unique spiral morphology
Green sulfur bacteria	Strictly anaerobic, obligately anoxygenic phototrophic
Bacteroides-Flavobacteria	Mixture of types, strict aerobes to strict anaerobes, some are gliding bacteria
Planctomyces	Some reproduce by budding and lack peptidoglycan in cell walls, aerobic, aquatic, require dilute media
Chlamydiae	Obligately intracellular parasites, many cause diseases in humans and other animals
Gram-positive bacteria	Gram-positive, many different types, unique cell-wall composition
Cyanobacteria	Oxygenic phototrophic
Purple bacteria	Gram-negative; many different types including anoxygenic phototrophs and nonphototrophs; aerobic, anaerobic, and facultative; chemoorganotrophic and chemolithotrophic

The nine other lineages, which evolved later, do not contain obligate thermophiles and tend to fan out in many different directions. Phototrophs are represented in three of these groups. The purple and the green sulfur bacteria include anoxygenic phototrophs among their members, while the cyanobacteria are oxygenic. The major differences among these phototrophic groups are in the particular type of chlorophyll they contain and the photosystem used to capture the light energy.

The *Deinococcus* group is characterized by extreme tolerance to radiation. These Gram-positive cocci are able to survive up to 3 million rad of ionizing radiation, an amount that would destroy the chromosomes of most organisms. About 500 rad can kill a human. Some strains have been found living near atomic reactors. *Spirochetes* are recognized by their unique morphology: The cells are helical in shape and often quite long. They are widespread in aquatic environments and vary from being strict aerobes to strict anaerobes.

Bacteroides represent a mixture of physiological types from strict aerobes to strict anaerobes, including the bacteria that move by gliding on surfaces. The *Planctomyes* and relatives contain organisms that reproduce by budding and lack peptidoglycan in their cell walls.

Chlamydias are degenerate organisms that cannot survive on their own because of their loss of metabolic functions. They are obligate parasites to animals and humans and the causative agents in a number of respiratory, sexually transmitted, and other diseases.

Perhaps the most studied and certainly the most important group of bacteria for environmental biotechnology belong to the three major groupings of Gram-positive bacteria, cyanobacteria, and purple bacteria. The *Gram-positive bacteria* include cocci and rods that take on the purple color of the gram stain (as do *Deinococcus*). They differ from other bacteria in that their cell walls consist primarily of peptido-glycan, a feature that causes the positive Gram-staining reaction. They also lack the outer membranes of Gram-negative cells.

The *purple bacteria* (also called *Proteobacteria*) present a very diverse group of microorganisms and include phototrophs (anoxygenic) and chemotrophs, including chemolithotrophs and chemoorganotrophs. They are divided into five genetically distinct groups (based on evolutionary divergence) as listed in Table 1.4, which also indicates some of the common genera within each group. Phototrophs exist within three of these groups. *Pseudomonads,* a broad classification of microorganisms important in organic degradation, also span three of these groupings. Pseudomonads include different genera (e.g, *Pseudomonas, Commamonas,* and *Burkholderia*) and are straight or slightly curved rods with polar flagella. They are Gram-negative chemoorganotrophs that show no fermentative metabolism.

The gamma group contains enteric bacteria, a large group of Gram-negative rods that are facultative aerobes and includes *E. coli* among its members. Also included among the enteric bacteria are many strains that are pathogenic to humans, animals, and plants. Other pathogenic organisms within the gamma group are strains from the *Legionella* and *Vibrio* genera. Because of their importance to health, as well as to the biotechnology industry, this group of microorganisms has been widely studied. Other organisms within the category of purple bacteria of particular importance to environmental biotechnology are the autotrophic nitrifying bacteria that oxidize ammonia to nitrate, including *Nitrobacter* from the alpha group and *Nitrosomonas* from the beta group.

Bacteria that use oxygen or nitrate as electron acceptors in energy metabolism reside in the alpha, beta, and gamma groups, while those that use sulfate as an electron acceptor reside in the delta group. Bacteria that oxidize reduced compounds of

Table 1.4   Major groupings among the purple bacteria and the common genera for each group

Alpha	*Rhodospirillum,* Rhodopseudomonas,* Rhodobacter,* Rhodomicrobium,* Rhodovulum,* Rhodopila,* Rhizobium, Nitrobacter, Agrobacterium, Aquaspirillum, Hyphomicrobium, Acetobacter, Gluconobacter, Beijerinckia, Paracoccus, Pseudomonas* (some species)
Beta	*Rhodocyclus,* Rhodoferax,* Rubrivivax,* Spirillum, Nitrosomonas, Sphaerotilus, Thiobacillus, Alcaligenes, Pseudomonas, Bordetella, Neisseria, Zymomonas*
Gamma	*Chromatium,* Thiospirillum,** other purple sulfur bacteria,* *Beggiatoa, Leucothrix, Escherichia* and other enteric bacteria, *Legionella, Azotobacter,* fluorescent *Pseudomonas* species, *Vibrio*
Delta	*Myxococcus, Bdellovibrio, Desulfovibrio* and other sulfate-reducing bacteria, *Desulfuromonas*
Epsilon	*Thiovulum, Wolinella, Campylobacter, Helicobacter*

*Phototrophic representatives.
SOURCE: Madigan, Martinko, and Parker, 1997.

sulfur, involved in such problems as acid mine drainage and concrete corrosion, reside among the gamma group. Such organisms are also of importance to environmental biotechnology. Thus, for various reasons, the purple bacteria are of great interest to us.

1.3.2 ARCHAEA

The Archaea also include prokaryotic microorganisms of great interest in environmental biotechnology. Of particular importance are the microorganisms that convert hydrogen and acetate to methane (*methanogens*), a gas that is of significant commercial value, as well as environmental concern. Hydrogen and acetate are the normal end products from bacterial transformation of organic matter when the usual external electron acceptors—for example, oxygen, nitrate, and sulfate—are absent. While CO_2 is the electron acceptor used by methanogenic bacteria for hydrogen oxidation, it is generally produced in excess during the oxidation of organic matter; thus external electron acceptors are not needed for organic conversion to methane to occur.

Methane is an insoluble gas that readily leaves the aqueous environment and passes into the gas phase. If that gas phase is the atmosphere, the methane can be oxidized by methanotrophic bacteria or through photochemical reactions. Because photochemical destruction of methane is a slow process, large quantities of methane emitted from swamps, rice fields, ruminants, sanitary landfills, and other sources eventually reach the stratosphere. Methane serves as a potent greenhouse gas, one that is twenty times more effective as carbon dioxide in absorbing infrared radiation. On the other hand, under controlled conditions in biological treatment systems or sanitary landfills, the methane produced can be captured for use as a fuel, thus reducing the need for fossil fuels.

The Archaea are similar in many ways to Bacteria; indeed they were not recognized as distinctly different lineages until molecular methods became available to differentiate organisms based upon genetic and evolutionary characteristics. One defining difference between these major phylogenetic domains is that the cell wall of Bacteria, while varying widely, almost always contains peptidoglycan, while that of the Archaea does not. Certain of the Archaea (including some of the methanogens) have a cell-wall polysaccharide that is similar to that of peptidoglycan, a material referred to as pseudopeptidoglycan, but the two materials are not the same. As with Bacteria, the cell walls of Archaea differ in composition considerably from group to group. Another defining difference is in the membrane lipids. Bacteria and Eukarya have membrane backbones consisting of fatty acids connected in *ester* linkages to glycerol. In contrast, Archaea lipids consist of *ether*-linked molecules. The bacterial membrane fatty acids tend to be straight chained, while the archaeal membrane lipids tend to be long-chained, branched hydrocarbons. Other differences are in the RNA polymerase. That of the Bacteria is of single type with a simple quaternary structure, while RNA polymerases of Archaea are of several types and structurally more complex. As a result, some aspects of protein synthesis are different between the two lineages. Several other small differences between them also exist. These overall major differences in structure and operation again emphasize the different evolutionary histories of the major lineages.

Table 1.5   Major groups and subgroups for the Archaea

Group	Subgroups
Crenarchaeota	*Desulfurococcus, Pyrodictium, Sulfolobus, Thermococcus, Thermoproteus*
Korarchaeota	
Euryarchaeota	*Archaeroglobus, Halobacterium, Halococcus,* Halophilic methanogen, *Methanobacterium, Methanococcus, Methanosarcina, Methanospirillum, Methanothermus, Methanopyrus, Thermoplama*

Table 1.5 contains a summary of the three major groupings of the Archaea. The Korarchaeota grouping contains hyperthermophilic Archaea that have not yet been obtained in pure culture; thus, good characterization has not yet been done. As generic names in the other two groups suggest, thermophiles are present throughout the Archaea. Also halophilic organisms are common, including the acidophilic, thermophilic cell-wall-less *Thermoplasma*. The Archaea are thus characterized by a large number of *extremophiles,* or organisms that live under extreme environmental conditions, compared with the world as we have come to know it. Such conditions were perhaps not extreme, but more normal at the time these organisms evolved. The methanogens, which are of particular interest to us, belong to the *Euryarchaeota*. They occur in many shapes and forms and vary from those preferring the type of environment enjoyed mostly by humans, to those preferring extremes in temperature and/or salinity. Extremophiles could be useful for biological treatment of industrial wastewaters that may contain extremes in salt concentration or temperature, but have been exploited little for such potential benefits in the past. They are finding new uses in industrial applications and in biotechnology.

1.4   EUKARYA

The eukaryotic microorganisms of particular interest in environmental biotechnology are the fungi, algae, protozoa, and some of the other multicellular microscopic Eukarya, such as rotifers, nematodes, and other zooplankton. All are characterized by having a defined nucleus within the cell. The larger of the Eukarya can often be divided into the categories of animals and plants, but the smaller organisms often cannot be so divided, since the boundaries between plants and animals at this scale can become quite blurred.

1.4.1   FUNGI

The fungi, together with Bacteria and Archaea, are the primary decomposers in the world. The decomposers are responsible for oxidation of dead organic material, which returns the resulting inorganic elements back into the environment to be recycled again by other living forms. Fungi do not have the wide-ranging metabolic characteristics of Bacteria: Essentially all are organolithotrophic, and none are phototrophic. Most fungi tend to favor terrestrial habitats, although some prefer aquatic systems. They are important in the decomposition of leaves, dead plants and trees, and other *lignocellulosic* organic debris, which accumulates in soil.

Because of the significant role fungi play in the soil, they are important in the decomposition of dry organic matter, a process that is used for the stabilization of refuse and organic sludges from wastewater treatment. Fungi are also known to decompose a great variety of organic materials that tend to resist bacterial decay. Important here is the ability of some to decompose lignin, a natural organic aromatic polymer that binds the cellulose in trees, grasses, and other similar plants, giving them structure and strength. Bacteria do not have this ability, as they lack the key oxidative enzyme, peroxidase, which helps to break the linkages between the aromatic groups lignins. This enzyme also gives some fungi special abilities to degrade some resista but hazardous industrial organic chemicals.

It would seem that the versatile ability of fungi for organic destruction would have been exploited fully for degradation of toxic compounds, but it has not. This is due partly to the slow rate at which fungal decomposition occurs, making fungi less attractive for engineered systems. The relative unimportance of fungi in detoxification also is due to a lack of understanding of how best to capture their potential. Perhaps the increased concern for the presence of toxic organic molecules in the environment will lead to more research in this potentially important area.

Morphology   Morphological characteristics alone identify more than 50,000 different species of fungi. Included are groups with common names such as molds, yeast, mildews, rusts, smuts, puffballs, and mushrooms. The cells of fungi are eukaryotic, and this is the major distinguishing feature that separates them from heterotrophic bacteria.

Except for some single-celled forms, such as yeast, fungi are composed of masses of filaments. A single filament is called a *hypha,* and all the *hyphae* of a single fungus are together called a *mycelium.* The mycelium is generally 5 to 10 μm wide, may be branched, and may be either on the surface or hidden beneath the surface of the nutrient or soil on which the fungi grow. The cell walls are generally composed of *chitin,* which is not found either in bacteria or higher plants, but is found in the hard outer coverings of insects.

Perhaps the largest and most important fungi in environmental biotechnology are the molds. This group of fungi reproduces by means of spores, and reproduction may be either sexual or asexual. Most molds are nonmotile, although some of the reproductive cells may be motile. Spores and the manner in which they are formed are quite varied, as illustrated in Figure 1.6, and these differences are commonly used for classification of molds.

Another important group of fungi is the yeasts, which reproduce by budding. While generally known for their usefulness in making bread and wine, they are common in the soil, and in this environment, like the molds, they can utilize a large number of different organic compounds. The yeasts consist of single elongated cells varying from 3 to 5 μm in length. Thus, they are slightly bigger than the typical bacterial cell.

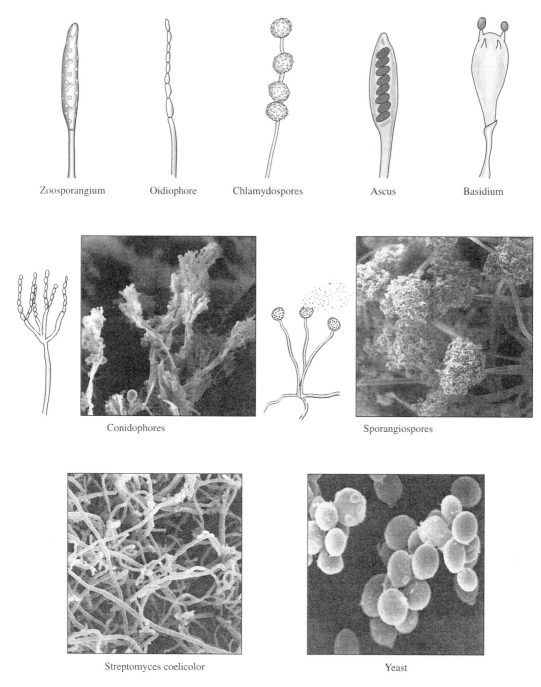

Zoosporangium   Oidiophore   Chlamydospores   Ascus   Basidium

Conidophores   Sporangiospores

Streptomyces coelicolor   Yeast

Figure 1.6   Typical morphologies of fungi and their spores. SOURCE: Photoimage with permission of the Microbe Zoo, Michigan State University. Images created by Catherine McGowan, Shirley Owens, and Steven Rozewell.

Classification Fungi are sometimes classified into three major groups. The true fungi, which are of special interest here, belong to the division *Eumycota*. The second division is the *Mycophycomycota*, or slime molds, which have an amoeba-like motile stage and a fungus-like spore-producing stage. The third division is the *Myxophycomycota*, which are the lichens that consist of a fungus and an alga growing together. The true fungi are divided into five classes having the characteristics listed in Table 1.6.

The *Ascomycetes* are the largest class of fungi with about 30,000 species. The hyphae are divided by cross walls or septa. Spores are formed asexually or sexually, and the latter always involves the formation of an ascus, or "small sac" that contains eight spores. Among the Ascomycetes are yeasts used in baking and brewing, the powdery mildews, the causative agents of Dutch elm disease, the sources of many antibiotics, the common black and blue-green molds, and many fungi that reduce dead plants to humus and digest cellulose in decomposing refuse.

The *Basidiomycetes* have a spore-bearing structure known as a basidium from which basidiospores are derived. This group also includes many species of decomposers. Some are responsible for destruction of wooden structures such as fences, railroad ties, and telephone poles; some are parasitic and destroy wheat and fruit crops; and some are macroscopic in size, such as the mushrooms.

The *Deuteromycetes* include all fungi that do not have sexual reproduction. This is a diverse group with perhaps little else in common besides asexual reproduction. As with most other groups of fungi, many are useful to humans, producing cheeses, such as Roquefort and Camembert, and antibiotics such as penicillin. Others are parasitic to plants and animals, including humans, causing skin diseases such as ringworm and "athlete's foot."

Many of the fungi of the class *Oomycetes* are aquatic and thus are known as *water molds*. The aquatic and the terrestrial forms produce spores that are flagellated and swim. This is the only group of fungi that has cellulose rather than chitin in the cell wall. This class is of great economic importance, as it includes many parasites

Table 1.6 Classification of the true fungi, **Eumycota**

| Class | Characteristics | | | |
	Common Name	Hyphae	Type of Sexual Spore	Habitats
Ascomycetes	Sac fungi	Septate	Ascopore	Decaying plant material, soil
Basidiomycetes	Club fungi, mushrooms	Septate	Basidiospore	Decaying plant material, soil
Deuteromycetes	Fungi imperfecti	Septate	None	Decaying plant material, soil, animal body
Oomycetes	Water molds	Coenocytic	Oospore	Aquatic
Zygomycetes	Bread molds	Coenocytic	Zygospore	Decaying plant material, soil

that harm fish, the organisms responsible for the great potato famine in Ireland, and the mildew that threatened the entire French wine industry in the late 1800s.

The *Zygomycetes* produce zygospores for sexual reproduction and include primarily terrestrial forms that live on dead plant and animal matter. Some are used for chemical production in the fermentation industry; others are parasitic to insects, growing fruits, and microscopic animals; and still others attack stored fruit and bread.

This brief summary indicates that fungi can be of economic benefit or a detriment to humans. The reproductive features that distinguish one group from another are perhaps not as important to environmental microbiology as is the potential for degradation of a wide variety of highly resistant organic materials of natural and anthropogenic origin. Also of interest are the different environmental requirements of the different species.

Nutritional and Environmental Requirements   All fungi decompose organic materials for energy. Most can live on simple sugars such as glucose, and, as a group, they have the ability to decompose a great variety of organic materials. Some can satisfy requirements for nitrogen from inorganic sources such as ammonium and nitrate; some require and almost all can satisfy their nitrogen needs from organic sources. Because fungi contain less nitrogen than bacteria, their requirement for this element is less. In general, all molds are aerobic and require a sufficient supply of oxygen for survival. Yeasts, on the other hand, are largely facultative. Under anoxic conditions, they obtain energy from fermentation such as in the conversion of sugar to alcohol.

Fungi generally grow slower than bacteria, but they generally can tolerate extreme environmental conditions better. They can live in relatively dry climates, as they can obtain water from the air, as well as from the medium on which they grow. Molds can survive in dry climates that are inhibitory to vegetative bacteria. Under extremely dry conditions, fungi also can produce protective spores. Molds can grow on concentrated sugars with high osmotic pressure, and they can live under relatively acidic conditions that would be detrimental to most bacteria. Optimum pH for growth is generally about 5.5, but a range from 2 to 9 can often be tolerated. The optimal temperature for growth of most fungi is in the mesophilic range, from 22 °C to 30 °C, but some cause food spoilage under refrigerated conditions near 0 °C, and others can live under thermophilic conditions with temperatures as high as 60 °C.

As a broad general characterization of the differences between fungi and heterotrophic bacteria, the fungi generally favor a drier, more acidic environment than bacteria, and their growth is less rapid. Fungi prefer terrestrial environments and high concentrations of organic matter, while bacteria prefer aquatic environments. Bacteria and the Archaea, on the other hand, can carry out more complete decomposition of organic materials under anaerobic conditions, which are only tolerated by yeasts. These differences are important when one wishes to capitalize on the degradative abilities of these organisms.

1.4.2 A LGAE

Algae are of great importance in water quality and water pollution control. Microscopic free-floating forms, or *phytoplankton,* are the major primary producers of organic matter in natural bodies of water, and they convert light energy into cellular organic matter that is used for food by protozoa, crustacea, and fish. Being oxygenic phototrophs, algae are among the main sources of oxygen in natural water bodies. Engineers take advantage of the oxygenic photosynthesis of algae in stabilization lagoons for wastewater treatment. Although their situation at the base of the food chain makes algae indispensable, discharge of wastes may stimulate algal growth to such an extent that the algae become a nuisance. Problems from too much algal growth include tastes and odors in water supplies, clogged filters at water treatment plants, decreased clarity of lakes, floating mats that interfere with boating and swimming and decrease property values along the shore line, and increased sedimentation in lakes and estuaries. Decomposition of algae can deplete oxygen resources, which are required by fish and other aerobic forms. Thus, a balanced population of algae is required to maintain a desirable ecosystem in natural waters.

During growth, algae consume inorganic minerals in the water, resulting in chemical changes such as to pH, hardness, and alkalinity. They also excrete organic materials that stimulate the growth of bacteria and associated populations. Some algae produce toxins that can kill fish and make shellfish unsafe to eat. Thus, the growth and decay of algae profoundly affect water quality.

Morphology   Algae represent a large and diverse group of eukaryotic organisms that contain chlorophyll and carry out photosynthesis. Cyanobacteria, which are prokaryotic organisms, are often classified among the algae, since they have so many of the same characteristics; cyanobacteria commonly were called *blue-green algae.* Most true algae are microscopic in size, but a number of forms, such as the marine *Macrocystis,* or giant kelp, is seaweed and may grow to several hundred feet in length. It is somewhat difficult to differentiate such large forms from nonalgal plants, except that the large algal forms produce spores within unicellular structures, but the spores of plants are all produced within multicellular walls. The microscopic forms that are mainly of interest here are distinctly different from higher plants, and many are unicellular and very small.

The algae as a group occur everywhere—in the oceans, lakes, rivers, salt ponds, and hot springs; on trees and rocks; in damp soil; and in other plants and animals. Some grow on and impart a red color to snow at high mountain elevations, others can flourish in hot springs at temperatures up to 90 °C, and still others prefer concentrated brines. Some algae grow together with fungi to produce the *lichens* that grow on rocks and trees and in very dry or cold climates. In lichens, the algae convert sun energy into organic matter, which is needed by fungi, and the fungi in turn, extract water and minerals from the environment and provide them to the algae.

In terms of form, some algae occur as single cells that may be spherical, rod-shaped, spindle-shaped, or club-shaped. They may occur as membranous colonies, filaments grouped singly or in clusters, or as individual strands that are either branched or unbranched. Some colonies may be aggregates of single identical cells, while others may be composed of different kinds of cells with particular functions. Thus, it is difficult to describe a "typical" alga.

Algae contain different kinds of *chlorophyll,* each of which is efficient in absorbing a characteristic range of the light spectrum. All algae contain chlorophyll *a.* Some contain other chlorophylls in addition, and it is on this basis that one group may be distinguished from another. Although the chlorophyll of cyanobacteria is distributed throughout the cell, that of the eukaryotic algae is contained within membranes to form *chloroplasts,* which may occur singly or in multiples within any one cell.

Table 1.7 summarizes the seven different taxonomic groups of algae (we include cyanobacteria here), based largely upon the different chlorophylls and photosynthetic pigments they contain. Some are illustrated in Figure 1.7. Their common names often are related to their characteristic color, which is a function of the photosynthetic pigments they contain. The first six groups are of particular importance in water quality control, and their characteristics are described in more detail below.

Chlorophyta   Green algae are common in many freshwaters and are among the most important algae in stabilization lagoons for wastewater treatment. Many are single celled, and some are motile by means of flagella. Some are large, such as sea lettuce, which sometimes grows excessively along coastlines because of nutrient pollution. The green algae commonly have one chloroplast per cell. Important genera that frequently predominate in the nutrient conditions of stabilization ponds are *Chlorella, Scenedesmus,* and *Chlamydomonas.* The latter is motile by means of flagella.

Chrysophyta   The *diatoms* are the most important among the Chrysophyta. They have a two-part shell that fits together like a pillbox and is impregnated with silica, giving the shell structure. They contain a yellow-brown pigment that provides their characteristic brown color, making it difficult upon casual observation to tell whether turbidity in natural waters containing them is due to algae or minerals such as silt and clay. When diatoms decompose, the siliceous shell is left behind, forming the huge deposits of uniformly sized, microscopic shells that are mined as diatomaceous earth, a material widely used as a filter aid at water treatment plants and industry.

Diatoms are present in fresh- and marine waters. They, together with dinoflagellates, are the major marine photosynthetic forms, serving as food for organisms varying in size from microscopic crustacean to whales.

Table 1.7   Characteristics of the different groups of algae

Algal Group	Common Name	Chlorophylls	Storage Products	Structural Details	Distribution
Cyanophyta	Blue-Green	*a*	Starch	Procaryotic, no flagella	Marine, freshwater, soil
Chlorophyta	Green	*a, b*	Starch	0 to several flagella	Marine, freshwater, soil
Chrysophyta	Golden, brown diatoms	*a, c, e*	Lipids	0 to 2 flagella, silica covering	Marine, freshwater, soil
Euglenophyta	Euglenoids Motile green	*a, b*	Polysaccharide	1, 2, 3 flagella, gullet	Mostly freshwater
Phaeophyta	Brown	*a, c*	Carbohydrate	2 flagella, many celled	Marine
Pyrrophyta	Dinoflagellates	*a, c*	Starch	2 flagella, angular plates with furrows	Marine, freshwater
Rhodophyta	Red	*a, d*	Starch, oils	No flagella	Marine

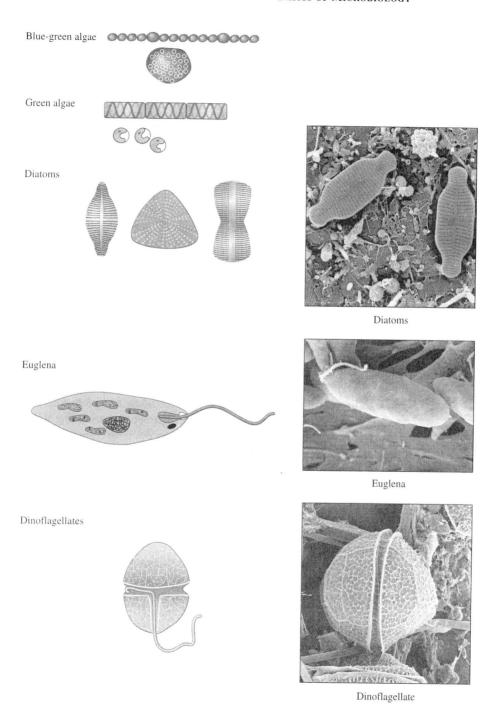

Blue-green algae

Green algae

Diatoms

Diatoms

Euglena

Euglena

Dinoflagellates

Dinoflagellate

Figure 1.7 Typical morphologies of algae. SOURCE: Photoimages with permission of the Microbe Zoo, Michigan State University. Images created by Shirley Owens.

Euglenophyta   Euglenophyta represent a rather small group of single-celled algae that are motile by means of flagella and live in freshwaters. The group name is derived from one of its most common members, *Euglena,* which is frequently a dominant organism in stabilization ponds. *Euglena* has a characteristic red spot or stigma that appears to act as a photoreceptor so that the organism can swim toward the surface of the water when in need of sunlight, or away from the surface when the sunlight is too intense. Such vertical movement by masses of *Euglena* is readily apparent in stagnant stabilization ponds by the clearing of the surface layers in midday and clouding of the surface in the morning and late afternoon.

Pyrrophyta   The *dinoflagellate* members of the Pyrrophyta have a stiff cellulose wall that looks like a helmet or shield. Two flagella beat within grooves and cause the organisms to spin as they move through the water (Figure 1.7). One species, *Gonyaulax catanella,* can suddenly appear in large numbers in waters off the southern coasts of California and Florida, causing the infamous red tide, which can kill thousands of fish. These deaths result from an extremely poisonous toxin excreted by the algae. Mussels, which ingest these algae and concentrate the poison, become dangerous to humans, and, for this reason, shellfish consumption often is banned during summer months, when the dinoflagellates tend to be most productive. Increased red tide severity may result from discharge of nutrients in wastes, but probably the greatest cause is periodic upwelling of nutrients from deep waters or other favorable natural conditions for the organisms.

Cyanophyta   This oxygenic bacterium commonly competes with eukaryotic algae and is properly considered together with algae when considering algal-associated water quality characteristics and problems. *Cyanobacteria* commonly are called *blue-green algae* because of the color of their pigments. Cyanobacteria occur in a variety of shapes, including long filaments.

Cyanobacteria justifiably receive blame for many detrimental water quality conditions. Tastes and odors in freshwater supplies are often associated with their presence. In addition, they frequently form large floating mats on waters polluted by phosphates from domestic or agricultural wastes, causing unsightly conditions. An important property of this group is the ability among many of its members to use nitrogen gas (N_2) from the atmosphere through nitrogen fixation to satisfy their needs for this element for cellular protein and nucleic acids. Thus, unlike their eukaryotic cousins, they can grow even in nitrogen deficient waters, albeit slowly.

Reproduction and Growth   Algae reproduce sexually and asexually in manners similar to bacteria and fungi. Growth is generally quite rapid under favorable conditions, and generation times for unicellular species can be as low as a few hours. Although some species of algae can live heterotrophically or phototrophically, most species are strictly phototrophs. Most are autotrophs, but some can use simple carbon sources, such as acetate, for cell synthesis.

The essential elements for *autotrophic growth,* such as carbon, hydrogen, oxygen, nitrogen, phosphorus, sulfur, and iron, generally come from the minerals dissolved in water or the water itself. Algae generally have a composition quite similar to bacteria.

Algae are about 50 percent carbon, 10 percent nitrogen, and 2 percent phosphorus. Under nitrogen or phosphorus limiting conditions, the rate of growth is reduced, and the relative contents of these two essential nutrients may drop to as low as 2 and 0.2 percent, respectively, although such low contents are frequently observed more under laboratory rather than field conditions. In general, they have a relatively high protein content, which suggests their use as food for humans and animals. They generally are not too tasty, but may be eaten by animals if mixed with other more palatable food.

The concentration of either nitrogen or phosphorus is often limiting to the growth of algae in natural waters, although there are circumstances where carbon, iron, or some other element may be in more limited supply. Diatoms also require silica, which is common in most freshwaters, but often limited in seawater. An understanding of the inorganic nutrient requirements is essential when attempting either to encourage their growth, as in stabilization lagoons, or discouraging their growth in rivers, lakes, estuaries, and reservoirs in order to prevent the nuisance conditions they may create.

Because of the great diversity of algal species, making generalizations about environmental needs is difficult. While some species can grow at 0 °C and others at 90 °C, growth for most is more optimal at intermediate temperatures. Diatoms in general prefer colder waters than green algae, which in turn prefer colder waters than cyanobacteria. These differences in temperature optima, as well as changing nutrient requirements, result in changes in dominant algal species in lakes throughout the year.

Algae generally prefer a near neutral to alkaline pH. This may be related to the availability of inorganic carbon in the form of bicarbonates and carbonates and the associated relationship among these inorganic species, carbon dioxide, and pH. As algae grow and extract carbon dioxide from water, the pH tends to increase, as indicated by the following greatly simplified autotrophic-synthesis reactions:

$$H_2CO_3 \rightarrow CH_2O + O_2 \qquad\qquad \textbf{[1.1]}$$

or

$$HCO_3^- + H_2O \rightarrow CH_2O + O_2 + OH^- \qquad\qquad \textbf{[1.2]}$$

The first reaction illustrates that algal growth (represented as CH_2O) results in oxygen production and the consumption of carbonic acid, thus causing the pH to rise. Similarly, the second equation indicates that, if bicarbonate is the carbon source, then hydroxide is produced, also with a resulting increase in pH. Generally, a pH greater than 8.5 to 9 is detrimental to algal growth. However, a few species can continue to extract inorganic carbon from water until the pH reaches values as high as 10 to 11. The extreme salt tolerance of some algal species has already been mentioned, again illustrating the extreme conditions under which some of the algal species can survive.

1.4.3 PROTOZOA

Protozoa are single-celled, heterotrophic eukaryotes that can pursue and ingest their food. They are common members of the complex ecological systems of which most aerobic and some anaerobic biological wastewater treatment systems are comprised. They lack a true cell wall and vary from the size of a large bacterium to an organism that can be seen with the unaided eye. As indicated in Table 1.8, four major groups

Table 1.8   Characteristics of the different groups of Protozoa

Group	Common Name	Method of Locomotion	Reproduction	Other Characteristics
Sarcodina	Amoeba	Pseudopodia	Asexual by mitosis, sexual sometimes	Generally free living, no spore formation
Mastigophora	Flagellates	One or more flagella	Asexual by mitosis, sexual uncommon	No spore formation, many parasites
Ciliophora	Ciliates	Cilia	Asexual by mitosis, sexual	Generally free living
Sporozoa		Generally no flagella or cilia, nonmotile	Asexual and sexual	Typically has spores, all parasitic

can be differentiated according to their method of locomotion. Some are illustrated in Figure 1.8.

Protozoa commonly reproduce asexually, often by *mitosis* (similar to binary fission in prokaryotes), but sexual reproduction is also common. Protozoa can be found in most freshwater and marine habitats. They generally feed on bacteria and other small organic particulate matter; in turn, microscopic and large animals prey upon them. Many varieties live free in nature, while others function only as parasites, living off of nutrients provided by other living forms. Some parasitic protozoa cause disease in humans. In biological treatment systems, protozoa act to "polish" effluent streams by helping to cleanse them of fine particulate materials that would otherwise leave in the effluent. Protozoa also can serve as indicators of the presence of toxic materials, to which many protozoa are quite sensitive.

As with most other large, heterogeneous groups of microorganisms, some species of protozoa can survive under fairly extreme environmental conditions. Some can tolerate a pH range as low as 3.2 or as high as 8.7. Most, however, survive best in the neutral pH range of 6 to 8. Some can live in warm springs at temperatures as high as 55 °C, but most have temperature optima between 15 °C and 25 °C, with temperature maxima between 35 °C and 40 °C. Perhaps most protozoa are aerobic, but many flourish in anaerobic environments, such as the rumen of animals. Some biologists prefer to classify the Euglenophyta group of algae as protozoa because of the many similarities between them. As usual, the distinctions among different classes of organisms are not always clear. Because of their presence in most biological treatment systems and the key indicator roles that they play, the different characteristics of protozoa within each group are of interest.

Sarcodina   The sarcodina are *amoeba-like organisms* that generally move and feed by means of a *pseudopod,* or "false foot," that is a temporary projection of the cytoplasm in the direction of motion. The cell mass tends to flow over a surface by this slow method of movement. The pseudopods also surround and capture food particles, which are then taken inside the cytoplasm and digested.

Entamoeba histolytica is a member of this group and an intestinal parasite that causes amoebic dysentery in humans. They are transmitted through fecal contam-

Amoeba

Zooflagallates

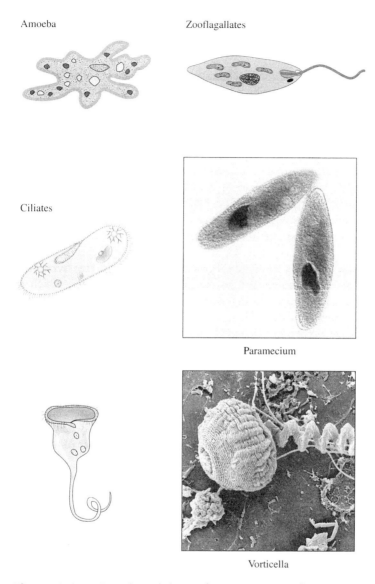

Ciliates

Paramecium

Vorticella

Figure 1.8   Typical morphologies of protozoa. SOURCE: Photoimages with permission of the Microbe Zoo, Michigan State University. Images created by Shirley Owens.

ination of water and food. This amoeba can form a cyst that can survive drying and normal disinfection. Some sarcodina can form flagella during particular stages of their life cycle, and others can cover their cells with bright shells or a sticky organic coating that picks up sand and other debris to form a protective coating. The Forminifera form a calcium carbonate shell in seawater. Accumulations of discarded shells are common in ocean deposits and are responsible for the white cliffs of Dover and other similar chalky deposits.

Mastigophera The zooflagellates are closely related to algae and are similar in characteristics to *Euglena,* which can live as a phototroph, like other algae. However, the zooflagellates are part of the Mastigophera, or heterotrophic protozoa. Most have either one or two flagella, which permits them to move rapidly in a circular pattern through water. Most forms are free living, but many are parasites. Members of the genus *Trichonymphas* live symbiotically in the digestive tracts of termites, where they digest wood and cellulose ingested by the termite. Some members of the family Trypanosomatidae are pathogenic to humans: *Trypanosom gambiense* is transmitted by the tsetse fly and causes the usually fatal African sleeping sickness. Zooflagellates are also common in biological treatment systems where they survive on the plentiful supply of bacteria produced in these systems.

Ciliophora Almost all of the Ciliophora, or *ciliates,* are free living (i.e., nonparasitic). They are characterized by the presence of many cilia, which move in a coordinated way to propel the organism or to create water currents that convey food to the oral region or mouth.

Biological wastewater treatment processes show two distinct types of ciliate: the *free-swimming ciliates,* which move through water seeking organic particulate matter, and the *stalked ciliates,* which attach themselves to large clumps or surfaces with long thin filaments called *trichocysts. Vorticella* are characteristic of the stalked ciliates. The stalked ciliates are sessile (attached) forms that bring food to themselves by creating water currents through cilia movement. Their presence in abundance in an aerobic biological wastewater treatment system is generally considered a sign of a stable, good operation, and one that is free from the influence of toxic substances.

Sporozoa All Sporozoa are parasites. They usually are not motile and do not ingest food, but absorb it in a soluble form through the cell membrane. While not of importance in biological wastewater treatment systems, the Sporozoa are of great importance to public health. Of particular interest to public health is *Plasmodium vivax,* the causative agent of malaria. It carries out part of its life cycle in humans and the other part in the *Anopheles* mosquito, which inhabits the warmer parts of the world.

1.4.4 OTHER MULTICELLULAR MICROORGANISMS

The microscopic members of the animal world, which are common in natural waters, are generally not covered in courses on microbiology, but a brief description of the morphology and function of some of the most common members is appropriate here. Some are illustrated in Figure 1.9. These include members of the phylum *Aschelminthes,* commonly referred to as sac worms, the most predominant members belonging to the classes *Rotifera* and *Nematoda.* Other common microscopic animals found in fresh- and marine waters and frequently in stabilization lagoons belong to the class *Crustacea* under the phylum *Arthropoda.* The Crustacea also include large

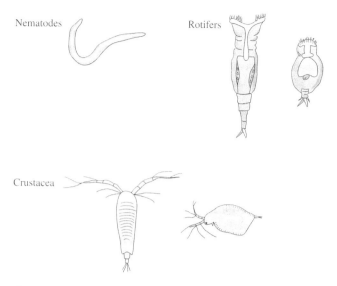

Figure 1.9   Multicellular organisms.

animals such as shrimp, lobsters, and crabs. The microscopic animals are multicellular, strictly aerobic, and ingest small particulate organic materials, such as bacteria,
algae, and other living or dead organic particles of similar size. This group of organisms borders on the line between microscopic and macroscopic. They are sufficiently
small to impart turbidity to water, but in several cases their presence and movement
may be detected without a microscope.

Nematodes, or roundworms, are perhaps the most abundant animals in numbers
of individuals and numbers of species. They are abundant in soils and in natural
waters, where they are free living, and in higher animals, where they live as parasites.
Nematodes are long and slender (Figure 1.9) and have a mouth at one tapered end
and an anus at the other. They grow in abundance in aerobic treatment systems
and can be discharged in high numbers to receiving streams if chlorination is not
practiced. While these free-living forms are not in themselves harmful, some concern
has arisen over their presence in water supplies because of the possibility that they
can ingest pathogenic bacteria, which then become protected from the lethal action
of disinfectants. While transfer of pathogens to humans in this manner is a distinct
possibility, no real health problems from this source have been demonstrated. Some
nematodes, however, are themselves pathogenic to humans.

Rotifers are soft-bodied animals having a head, a trunk, and a tapered foot (Figure
1.9). The head generally has a wheel organ or wreath of cilia, which it rotates for
swimming locomotion and to create currents that draw in food particles. The foot
is generally bifurcated into two toes that allow the organism to anchor itself to large
clumps of debris while it eats. When viewed with a microscope, it is common to
see just below the head the internal action of the chewing organ or *mastax,* which
contains jaws studded with teeth. In biological systems, rotifers can frequently be
seen tearing at pieces of biological floc.

Microscopic crustacea are similar to their larger cousins in that they have a hard exterior covering, usually sport a pair of antennae, and breathe through gills. The microscopic forms generally have a larger head, a shortened trunk, and fewer segments in the trunk appendages than in the larger forms (Figure 1.9). Representative members are the water flea (*Daphnia*) and the copepods (*Cyclops*), which look like miniature lobsters. They are present more frequently in stabilization ponds than in other types of biological treatment systems.

1.5   VIRUSES

Viruses are generally not considered to be "living" entities, as they are unable, on their own, to replace their parts or to carry out metabolism. They can be replicated only when in association with a living cell, which translates the genetic information present in the virus and causes its replication. However, their great importance to living entities is without question, as they cause disease and death. In general, viruses are submicroscopic genetic elements consisting of nucleic acid (DNA or RNA) surrounded by protein and, occasionally, other components. When the viral DNA or RNA is inserted into a proper host cell, it can cause the host cell to redirect its metabolic machinery into the production of duplicate viral cells. When the viral number reaches a sufficient size, the host cell dies and breaks open, and the new virus particles are released for infection of new host cells.

Viruses range in size from about 15 to 300 nm, the latter being just at the limit of resolution of the light microscope. Thus, an electron microscope or other instrument permitting higher resolution is required for viewing viruses. The number of different host species that a specific virus can infect is very limited. Since each species can be infected by a variety of viruses, the total number of viruses is large. Viruses that infect prokaryotic cells are called *bacteriophages* (or *phage* for short). Phages do not differ significantly from other viruses, except in their choice of a host cell. Various viruses of interest are illustrated in Figure 1.10.

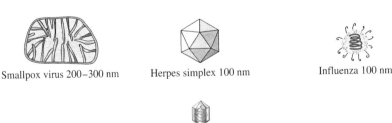

Smallpox virus 200–300 nm   Herpes simplex 100 nm   Influenza 100 nm

Adenovirus 75 nm   Bacteriophage 80 nm   Tobacco mosiac virus 15 × 280 nm

Figure 1.10   Typical structures of viruses.

Phages are prevalent in biological wastewater treatment systems and at times have been suspected of causing process upsets by killing needed bacteria, although this has not been well documented. They may be a factor causing changes in dominance of one bacterial species over another in mixed culture systems, but this aspect of wastewater treatment has not been adequately explored. A virus infection can occur quite rapidly. Within about 25 min after a bacterium is infected with a phage, about 200 new phages can be produced. The bacterial cell bursts open, releasing them for infection of other cells, and the infection can spread at an increasingly rapid rate.

1.6 INFECTIOUS DISEASE

The environmental engineering and science fields arose during the last half of the nineteenth century as a result of the finding that water was a carrier of some of the worst diseases then plaguing humans. Engineers who were responsible for delivering water for human consumption then began pursuing ways to insure that the water they delivered was safe and wholesome. Major triumphs were the use of coagulation and filtration in the late nineteenth century and then disinfection with chlorine in the early twentieth century. Such steps, together with other methods to protect food, such as pasteurization, resulted in dramatic reductions in water- and food-related disease. In spite of these triumphs, water-related disease outbreaks are still common in the United States and, for the developing world, still represent the major cause of sickness. In addition, some of the large water-related disease outbreaks in the United States are caused by organisms that were of no concern in the past. They represent the emergence of new pathogens for which there is no history on which to base the design of preventive measures. In some cases, no suitable methods for detection of the causative agents are available.

The growing world population and its increasing demand on water resources has promoted increased interest in wastewater reclamation for potable use. The major question of such growing practice is: How can water so highly contaminated with pathogenic microorganisms be adequately and reliably cleaned for domestic use? Reuse represents a major challenge for the water professional. Treatment and monitoring techniques of unparalleled reliability are needed.

Because of obvious importance of water-related pathogenic microorganisms, engineers and scientists who are involved in the treatment of water and wastewater need to know about the pathogenic microorganisms that are transmitted from fecal material or food.

Table 1.9 is a summary of common diseases related to water and/or human wastes. Many of the diseases listed are sometimes referred to as intestinal or filth diseases, because they are often caused by drinking water or eating food contaminated with human feces. Such contamination can be prevented by proper means of sanitation. These diseases are especially prevalent in developing countries, because individuals either do not adequately understand disease transmission or lack economic means for prevention. Many of the organisms listed cause *gastroenteritis,* a term often

Table 1.9   Causative agents of waterborne and/or human-waste-related disease

Microorganism Class	Group	Organism Name	Disease and Symptoms
Virus	Viscerotropic	Coxsackie virus, Norwalk virus, Rotavirus, Echovirus	Gastroenteritis
	Viscerotropic	Hepatitis A virus	Infectious hepatitis, liver inflammation
	Neurotrophic	Polio virus	Poliomyelitis
Bacteria (Purple Group)	Epsilon	*Campylobacter jejuni*	Gastroenteritis, diarrhea, fever, abdominal pain
	Epsilon	*Helicobacter pylori*	Peptic ulcers
	Gamma	*Escherichia coli* O157:H7	Diarrhea, hemorrhagic colitis
	Gamma	*Legionella pneumophilia*	Legionellosis, fever, headache, respiratory illness
	Gamma	*Salmonella typhi*	Typhoid fever, blood in stools
	Gamma	*Shigella dysenteriae*	Dysentery, abdominal cramps, bloody stools
	Gamma	*Vibrio cholerae*	Cholera, severe diarrhea, rapid dehydration
Algae	Dinoflagellate	*Gambierdiscus toxicus*	Ciguatera fish poisoning
	Dinoflagellate	*Gonyaulax catanella*	Shellfish poisoning
	Dinoflagellate	*Pfiesteria piscicida*	Fish poisoning, memory loss, dermatitis
Protozoa	Mastigophora	*Giardia lamblia*	Giardiasis, diarrhea, bloating
	Sarcodina	*Entamoeba histolytica*	Amebiasis, sharp abdominal pain, bloody stools
	Sporozoa	*Cryptosporidium parvum*	Cryptosporidiosis, diarrhea
Multicellular parasites		*Schistosoma mansoni*	Schistosomiasis, fever, diarrhea, dermatitis

used to designate a water- or foodborne disease for which the causative agent has not been determined. Gastroenteritis refers to the inflammation of the stomach and intestines, with resultant diarrhea and extreme discomfort. The responsible organisms for water and food associated disease span the breadth of microorganism classes from virus particles to multicellular organisms. Since all the microorganisms are particles, all can be removed from the water by coagulation and filtration. Some of the pathogens—most particularly the bacteria—also can be inactivated by chemical disinfectants, such as chlorine or ozone. Whenever possible, physical removal by filtration and chemical disinfection should be used together as "multiple barriers" against the pathogens.

Only a few of the viruses that infect humans are actually waterborne. The poliovirus has long been suspected to be waterborne because of its long survival in water, but its main mode of transmission is undoubtedly direct human contact. Thanks to the development by Jonas Salk of an inactivated polio vaccine in the 1950s, the incidence of the dreaded poliomyelitis or infantile paralysis dropped dramatically from over 1,000 cases per 100,000 population previously to zero cases in the United States since about 1990. The Salk vaccine was replaced by the oral Sabin attenuated vaccine in the 1960s. Of particular interest to water scientists is that use of the oral vaccine resulted in large numbers of harmless polio virus particles being discharged to municipal wastewaters, permitting the testing of the ability of treatment processes to remove a known human virus from wastewaters.

Today, the waterborne virus of most concern is the causative agent of *infectious hepatitis*, the Hepatitis A virus. The greatest problem has been in the eating of raw shellfish, such as oysters, which can concentrate virus particles through filter feeding of waters contaminated with human feces. A major outbreak of infectious hepatitis that infected close to 30,000 people occurred in Delhi, India, in 1955, when sewage discharged to the river backed up into the water intake of the water treatment plant. Apparently, the disinfection practiced was inadequate to kill the virus sufficiently, and widespread disease resulted. The Coxsackie, Norwalk, Rota, and Echo viruses are associated with gastroenteritis and the associated diarrhea. Contaminated food, shellfish, or water may transmit them. Resulting disease is so common from these sources among travelers and young children that knowledge of the general carrier responsible is often not sought.

Bacterial waterborne pathogens fall within the Purple group of bacteria, mostly from among the Gamma class of enteric organisms. This is a relatively homogeneous phylogenetic group of bacteria that are Gram-negative, nonsporulating, facultative-aerobic, oxidase-negative rods that ferment sugars to yield a variety of end products. Most survive well in the human intestine, the pathogenic ones causing problems such as dysentery and diarrhea.

The most well known member of the Purple group is *Escherichia coli,* which lives in the intestines of all humans. *E. coli* is generally harmless, but because of its large numbers in human fecal discharge, its presence is commonly used as an indicator of fecal contamination of water supplies. The common assumption is that a drinking water is safe if the number of *E. coli* and related coliform bacteria is less than a prescribed level, such as to about 2 per 100 ml.

Recently, we have found some strains of *E. coli* are pathogenic, and their presence in food and water is a growing concern. A particular problem here is *E. coli* 0157:H7, which has been implicated in a rash of food-related diseases in the United States during the 1990s, with numerous deaths from the bloody diarrhea that results. In 1998, an outbreak among small children attending a water park in California was believed to have resulted from defecation in a pool by an infected child. The common "traveler's diarrhea" is believed to be caused by a pathogenic strain of *E. coli,* which causes a diarrhea that lasts one to ten days. The bacteria adhere to the intestinal lining with pili, as do many of the other enteropathogenic bacteria, and produce enterotoxins, which induce water loss in humans.

The most common and feared waterborne infectious diseases are typhoid fever, dysentery, and cholera. Contaminated food is perhaps the most prevalent carrier of these diseases today, but water-associated transmission still often occurs. The causative bacteria can be readily eliminated from water supplies by adequate chemical coagulation, filtration, and chemical disinfection. The waterborne outbreaks that still commonly occur throughout the world are generally caused by inadequate treatment of water. Sometimes the source of contamination is not known. For example, a gastroenteritis outbreak occurred in Riverside, California, in the early 1960s and infected 18,000 people. All patients were carriers of *Salmonella typhimurium,* and the water supply was implicated, because *S. typhimurium* was found present in several water samples. This finding underscores the limitation of using only the coliform indicator of a water's sanitary safety, because coliform bacteria were not detected in "excessive" levels in the distribution system.

Another organism causing diarrhea comes from the Epsilon group, *Campylobacter jejuni.* It is more frequently carried through contaminated poultry or infected domestic animals, especially dogs. It is responsible for the majority of cases of diarrhea in children. *Helicobacter pylori,* closely related to *C. jejuni,* is implicated as a cause of ulcers, and water is one of the suspected routes for its transmission.

A relatively new water-related disease is legionellosis, or *Legionnaires' disease.* The first recognized major outbreak occurred during a convention of the American Legion in 1976. Many of the Legionnaires were housed in a hotel with a water-cooled air-conditioning system. The organisms appear to thrive in such air-conditioning systems and are spread to the victims through the cool air circulating from the towers to the rooms. Thus, the actual route of infection is through the air, rather than through water. Nevertheless, the infection represents a water-related problem.

Essentially all major algal-related waterborne diseases have dinoflagellates as the causative agents. However, sickness in animals that have consumed water from ponds containing heavy growths of cyanobacteria, the phototrophically related bacterial species, is well known. Dinoflagallates do not generally cause direct harm to humans, but act through the contamination of shellfish. *Gonyaulax* is responsible for the red tides that occur off the Pacific Coast of California and the Gulf Coast of Florida. Their sudden growth as a result of nutrient upwelling and summer temperatures often imparts a red color to the water. They produce a toxin that is sometimes lethal to fish and is concentrated by shellfish. Human consumption of toxin-contaminated fish or shellfish can lead to severe illness. For this reason, the consumption of shellfish during summer months is commonly banned. While nutrient enrichment of coastal waters through human activity has sometimes been blamed for red tide occurrence, there is little evidence that it results from other than natural causes.

A recently emerging problem is *Pfiesteria piscidida,* discovered by JoAnn Burkholder in 1991. This single-celled alga has 24 different life forms, one of which releases toxins that have killed billions of fish off the North Carolina coast. Subsequently, fish kills resulting from growth of *Pfiesteria* occurred in Chesapeake Bay and have led to broader concerns and new studies. The hypothesis gaining support is that the toxic form is stimulated by the discharge of nutrient-bearing agricultural runoff, such as from animal farms. A report from the Maryland Department of Health and Mental Hygiene concluded that exposure to *Pfiesteria* toxins also has caused human health problems such as memory loss, shortness of breath, and skin rashes. Studies to better understand the nature of this new water-related problem are underway.

Protozoan-related diseases are well known worldwide and are emerging as perhaps the most serious water-related problem in the developed world. *Amebiasis* has long been associated with drinking contaminated water and eating food in tropical areas. Estimates are that 10 to 25 percent of the native populations of tropical countries harbor the causative agents.

A more recently emerged protozoan disease, *giardiasis,* was recognized as a problem disease only in the 1960s and is associated with the contamination of numerous water supplies in colder climate, such as in the United States, Canada, and Russia. Between 1965 and 1981, 53 waterborne outbreaks of giardiasis, affecting over 20,000 people, were reported in the United States. A resting-stage cyst of the flagellated *Giardia lamblia* is the primary mode of water transmission. The cysts germinate in the gastrointestinal tract and bring about a foul-smelling diarrhea, intestinal cramps, nausea, and malaise. *Giardia* cysts are resistant to disinfection, but can be removed by good water treatment consisting of chemical coagulation, filtration, and disinfection. Animals, such as muskrats and beavers, are major carriers of the *Giardia* cells and cysts and often contaminant seemingly pristine waters in wilderness areas. For this reason, giardiasis is often called *beaver fever.* Wilderness campers used to take pleasure in drinking cool untreated drinking water from streams remote from human influence, but now the large number of incidences of giardiasis from such practice has resulted in warnings to campers to boil or filter water taken from seemingly pristine environments.

An even more recently emerging protozoan disease is caused by *Cryptosporidium parvum,* which is responsible for the largest single outbreak of waterborne disease ever recorded. In 1993, about 370,000 people in Milwaukee, Wisconsin, developed a diarrheal illness that was traced to the municipal water supply. About 4,000 people were hospitalized, and as many as 100 died from complications of the disease. *C. parvum* is a common intestinal pathogen in dairy cattle, and its sudden presence in the Milwaukee water supply is believed to have been partly caused by heavy spring rains and runoff from farmlands that greatly increased the *C. parvum* load to the water treatment system. At the same time, the coagulation-filtration processes were not operating optimally, allowing turbidity to break through. The *C. parvum* oocysts, a type of cyst formed by the organism, is highly resistant to chlorination, and thus must be removed by coagulation and filtration to prevent human infection. The Milwaukee experience underscores why water-treatment practices are being changed in the United States to achieve greater turbidity removals through filtration.

Certain other protozoan diseases are also of great public-health significance and are water related. These include malaria, which is caused by four different protozoan species belonging to the *Plasmodium* genera. The protozoa are transmitted by the *Anopheles* mosquito, which consumes infected human blood to provide chemical components for its eggs. Malaria differs from yellow fever, which is caused by an arbovirus, one of the smallest viruses known, and is transmitted by a different mosquito, either *Aedes aegypti* or *Haemogogus.* Malaria remains one of the most widespread diseases, especially in Africa. The World Health Organization (WHO) estimates that over a million children under the age of five die from malaria annually. International organizations thought in the 1960s that malaria would be controlled through the use of the drug chloroquine and the application of DDT to control mosquitoes. However, protozoa resistant to chloroquine and mosquitoes resistant to DDT began to emerge; now, over 300 million cases per year result.

Many multicellular parasites are associated with human wastes and food. Three species of flukes invade the bloodstream in humans: *Schistosoma mansoni, S. japonicum,* and *S. haematobium.* The resulting disease in all cases is called *schistosomiasis,* although the nature of the disease is different. The WHO estimates that 250 million people worldwide are infected. Male and female schistosomes mate in the human liver and produce eggs that are released in the feces. The eggs hatch to miracidia in water, which make their way to snails, where conversion to other forms called sporo-cysts and cercariae occurs. The cercariae escape from the cells into water, where they attach themselves to the bare skin of humans, for example, who are planting or tending rice in rice paddies. The cercariae then become schistosomes, which infect the bloodstream, causing fever and chills, and are carried to the liver, where the reproductive process is repeated. The eggs damage the liver and can gather in the intestinal wall, causing ulceration, diarrhea, and abdominal pain. A milder form of schistosomiasis that occurs in North America is commonly called swimmer's itch, but the intermediate host is birds, rather than humans. Human contact with water contaminated with the resulting cercariae causes dermatitis, a result of allergenic substances produced by the body's immune response to destroy the cercariae.

Other multicellular organisms of worldwide significance are the intestinal roundworm (*Ascaris lumbricoides*), the tapeworm (*Taenia solium* and *T. saganata*), and the hookworm (*Ancylostoma duodenale* and *Necator americanus*). These organisms live in human intestines, are discharged with feces, and contaminate food and soil, especially when human feces are used as night soil for fertilization of vegetables that are eaten raw. As this is common practice in many countries of the world, infection with parasitic worms is widespread, infecting 40 to 50 percent of the population in many countries. Another parasite of concern is trichinosis, caused by the roundworm *Trichinella spiralis*. This worm lives in the intestines of pigs, and, when humans eat poorly cooked pork, the cysts pass into the human intestine. Later, the worms emerge, causing intestinal pain, vomiting, nausea, and constipation. The cysts are passed back into nature with human feces to contaminate garbage and other food eaten by the pigs.

While environmental engineers and scientists have been highly successful in the control of some water and human-waste-related disease, this summary indicates that major problems from improper sanitation still occur throughout the world. Although we appear to have successfully controlled certain diseases, others tend to emerge, requiring new knowledge and efforts. Also, controls that involve the use of chemicals such as pesticides and antibiotics are often very temporary. Organisms have repeatedly demonstrated an ability to develop strains that are resistant to our chemicals. At the same time, the increasing world population and demand for water means that we will increasingly need to draw water supplies from more contaminated sources. Thus, the challenges to providing safe water for human consumption will continue to manifest themselves. We can never be complacent.

1.7 BIOCHEMISTRY

All organisms grow at the expense of potential energy stored in inorganic or organic molecules or present in radiant energy, such as sunlight. In order to obtain the energy, organisms transform the chemicals through oxidation and reduction reactions. They invest some of that energy into reproducing themselves. In Environmental Biotechnology, the oxidation/reduction reactions and the production of new biomass carried out by microorganisms remove materials that we consider pollutants.

Examples of environmental benefits that can be achieved by microbial systems are:

•   Elimination of nitrogen and phosphorus nutrients from wastewaters to avoid stimulating unwanted algal growth in surface waters

•   The conversion of organic wastes into methane, a useful fuel

•   The conversion of putrescible organic matter into harmless inorganic compounds

•   The degradation of toxic chemical contaminants in soil and groundwater.

Only through an understanding of the microorganisms' energy-yielding and energy-consuming reactions can we create the conditions that lead to environmental protection and improvement. In short, we must understand the microorganisms' *biochemistry*.

Enzymes catalyze all the key reactions—whether the oxidation/reduction reactions involved in energy capture or the many types of reactions involved in synthesizing new cells. In the next section, we describe the nature and function of enzymes that microorganisms use. This section lays the foundation for the following sections, which are on different aspects of metabolism, the reactions by which the cells capture energy and electrons and then invest them to create and maintain new cells.

1.8   ENZYMES

Enzymes are organic catalysts produced by microorganisms and used by them to speed the rate of the thousands of individual energy-yielding and cell-building reactions that occur within the cell. Enzymes comprise the largest and most specialized group of protein molecules within the cell. Proteins are polymers of a mixture of the 20-plus different amino acid subunits. Enzymes are large macromolecules, with molecular weights generally between 10,000 and a million. A computer-generated picture of an enzyme is illustrated in Figure 1.11. Enzymes have a *primary structure* that is dictated by the sequence of the chain of attached amino acids. Enzymes also have a *secondary structure* that forms when the amino acid chain twists itself into a three-dimensional configuration held together by hydrogen bonding between nearby amino acids and sulfur-to-sulfur (disulfide) linkages that are common to some of the amino acids. Many proteins also have a *tertiary* structure, in which the protein is folded back upon itself and again held together by hydrogen bonding. This tertiary structure is generally broken easily by heat or chemicals, causing enzyme *denaturation* or inactivation. Enzyme structure is quite vulnerable to changes in the physical environment, especially temperature and pH, and to many chemicals that may bind to the protein and change its secondary or tertiary structure.

The two important characteristics of an enzyme are its specificity and the rate of the reaction that it catalyzes. *Specificity* means that an enzyme channels the transformation of a chemical along a desired pathway. *Rate* refers to the speed at which the desired reaction occurs when the enzyme is present. In a given mixture of organic and inorganic compounds, such as occur in wastewaters, many reactions are thermodynamically possible, but do not occur by themselves at finite rates under the normal

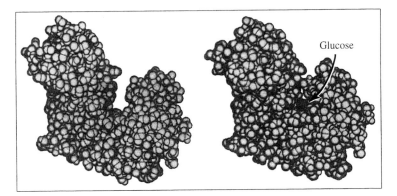

Glucose

Figure 1.11   Computer-generated structure of the enzyme hexokinase that converts glucose and ATP into glucose 6-phosphate during the first step of glycolysis. Shown is the location where the glucose molecule fits into the enzyme structure and the resulting change in the enzyme configuration. SOURCE: L. Stryer, 1995.

conditions of temperature and pressure. However, by having the right enzymes in the correct amounts, microorganisms can gain control and direct the pathways through which chemical reactions occur. For example, they can take a simple sugar molecule and oxidize it in a stepwise manner to carbon dioxide and water. For each enzymatically controlled oxidation step, they capture electrons and energy in small "packets." Those "packets" can then be used to convert other sugar molecules, as well as sources of nitrogen, phosphorus, and other elements, into the proteins, lipids, carbohydrates, and nucleic acids needed for the microorganism to reproduce and grow. Enzymes do not increase the amount of energy released from a given reaction, but they minimize the diversion of resources (electrons, energy, and elements) into nonproductive pathways. Through the control of enzyme catalysis, the cells maximize the benefits that they receive from the resources available to them.

A single enzyme molecule can effect from 1,000 to 100,000 molecular transformations per second. Enzymes are not consumed by the reaction, and this, together with the high rate at which they can cause reactions to occur, explains why the cell requires only very small quantities of each enzyme.

Some enzymes depend for their activity only upon their structure as proteins. Others require, in addition, a nonprotein component for their activity. When the nonprotein portion is a metal ion, it is called a *cofactor*. Table 1.10 lists various metal-ion cofactors and some of their associated enzymes. If the nonprotein structure is organic, it is called either a *coenzyme* or a *prosthetic* group. Prosthetic groups are bound very tightly to their enzymes, usually permanently. Coenzymes are bound rather loosely, and a given coenzyme may combine with several different enzymes at different times. Coenzymes often serve as intermediate carriers of small molecules from one enzyme to another. Table 1.11 contains a summary of some of the principal coenzymes and the particular enzymatic reactions in which they participate. Coenzymes generally serve in the transfer of electrons, elements, or functional groups from one molecule

Table 1.10 Metal cofactors, the enzymes they activate, and the enzyme function

Metal Cofactor	Enzyme or Function
Co	Transcarboxylase, Vitamin B_{12}
Cu	Cytochrome c oxidase, proteins involved in respiration, some superoxide dismutases
Fe	Activates many enzymes, catalases, oxygenases, cytochromes, nitrogenases, peroxidases
Mn	Activates many enzymes, oxygenic photosynthesis, some superoxide dismutases
Mo	Nitrate reductase, formate dehydrogenases, oxotransferases, molybdenum nitrogenase
Ni	Carbon monoxide dehydrogenase, most hydrogenases, coenzyme F_{430} of methanogens, urease
Se	Some hydrogenases, formate dehydrogenase
V	Vanadium nitrogenase, some peroxidases
W	Oxotransferases of hyperthermophiles, some formate dehydrogenases
Zn	RNA and DNA polymerase, carbonic anhydrase, alcohol dehydrogenase

Table 1.11 Coenzymes involved in group-transferring reactions

Group Transferred	Coenzyme	Acronym
Hydrogen atoms (electrons)	Nicotinamide adenine dinucleotide	NAD
	Nicotinamide adenine dinucleotide phosphate	NADP
	Flavin adenine dinucleotide	FAD
	Flavin mononucleotide	FMN
	Coenzyme Q	CoQ
	Coenzyme F_{420}	F_{420}
Acyl groups	Lipoamide	
	Coenzyme A	HSCoA
One-carbon units	Tetrahydrofolate	
	Methanofuran	
	Tetrahydromethanopterin	
	Coenzyme M	CoM
Carbon dioxide	Biotin	
Methyl	S-Adenosylmethionine	
Glucose	Uridinediphosphate glucose	
Nucleotides	Nucleotide triphosphates	
Aldehyde	Thiamine pyrophosphate	

to another or from one place in the cell to another. Many coenzymes contain an active portion that is a trace organic substance, normally called a *vitamin*. Vitamins must be supplied to the cells when required for coenzymes, but not manufactured by the cell itself.

Enzymes are commonly named by adding the suffix *-ase* to a root that is either the reaction catalyzed or the substrate (reactant) transformed. For example, a *dehydrogenase* removes two hydrogen atoms and associated two electrons from a molecule, an *hydroxylase* removes two electrons and inserts an -OH group from H_2O, and a *proteinase* acts in the hydrolysis of proteins to form amino acids. A more formal system for naming enzymes is used for more precise classification. Of particular importance are the *hydrolases,* which split complex carbohydrate, lipid, and protein polymers into their simpler building blocks through water addition, and the *oxido-reductases,* which are involved in the very important oxidation and reduction reactions that provide cells with energy and are essential in synthesis.

The major source of energy is provided through oxidation-reduction reactions, which involve the transfer of electrons from one atom to another or from one molecule to another. Electron carriers move the electrons from one compound to another. When such carriers are involved, we refer to the initial donor as the *primary electron donor* and the final acceptor as the *terminal electron acceptor.* For example, in the aerobic oxidation of acetate, electrons and hydrogen are removed from acetate through a complex set of enzymatic reactions involving various electron carriers that pass the electrons through a cytochrome system where the energy is captured and passed to energy carriers. After the energy is extracted, the spent electrons finally combine with molecular oxygen to form water. In this case, acetate is the primary electron donor and oxygen is the terminal electron acceptor.

Hydrolytic reactions normally yield very little energy and are not used as a primary energy source by the cell. Instead, they are used to disassemble complex polymers so that the parts can be reused to make new polymers or for energy.

Hydrolytic enzymes may be *extracellular,* or *exoenzymes* that act outside the cell wall in order to break large molecules into smaller ones that can pass through the cell wall. They may also be *intracellular,* or *endoenzymes* that function within the cell. Oxido-reductase enzymes are intracellular, although they often are associated with or span the cytoplasmic membrane. In some cases, they react with materials on both sides of the membrane. Some enzymes are closely associated with the cell wall and can act as exoenzymes or endoenzymes, depending on where the reactions they mediate occur.

1.8.1   ENZYME REACTIVITY

The reactivity of an enzyme toward a substrate involves specificity and kinetics. The conventional view is that a substrate molecule or a portion of it fits into the *active site* of the enzyme in a lock-and-key fashion. The enzyme, which is much larger, positions itself around the smaller substrate molecule, which just fits within the three-dimensional protein structure (Fig. 1.11). Coenzymes, cofactors, or cosubstrates are brought into contact with the substrate in such a way that electrostatic or vibrational forces between them reduce the activation energy and permit a given transformation to take place at a specific site on the substrate molecule.

The high degree of specificity of some enzymes and lack of it in others deserves some comment. Some enzymes are nearly absolute in specificity for a given molecule and will not act upon molecules that are very similar in structure. Other enzymes can act on an entire class of molecules. For enzymes that can cause reactions with a variety of different molecules, the rate often differs with each substrate molecule, presumably because of differences in the substrate's fit in the active site.

The rate (or kinetics) of an enzyme-catalyzed reaction is, in general, governed by the same principles that govern other chemical reactions. In many cases, the rates of growth and substrate utilization by microorganisms can be described quite well by kinetics developed to describe the rate of transformation of a single substrate by a single molecule. Hence, the kinetics of enzyme catalysis are of interest for describing the individual reactions and, often, complex biochemical processes.

A characteristic of enzyme reactions is substrate saturation. Figure 1.12 illustrates the observed effect of substrate concentration on the rate of an enzyme-catalyzed reaction. At very low substrate concentrations, the rate increases in direct proportion to substrate concentration; thus, the rate here is first-order with respect to substrate concentration. As substrate concentration increases, however, the rate of increase begins to decline, giving a mixed-order reaction. At higher substrate concentrations, the enzyme becomes saturated with substrate, and the rate increases no further. This rate is the maximum rate for the reaction. Here, the rate becomes zero order with respect to substrate concentration. This saturation effect is typical of all enzyme-catalyzed reactions.

L. Michaelis and M. L. Menten recognized this saturation phenomenon and, in 1913, developed a general theory of enzyme action and kinetics, which was later extended by G. E. Briggs and J. B. S. Haldane. This theory suggests that enzyme E first reacts with substrate S to form an enzyme-substrate complex ES, which then

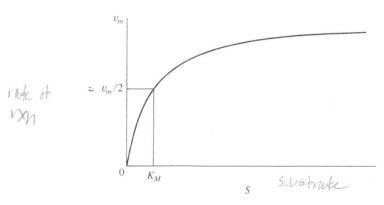

Figure 1.12   Effect of substrate concentration (S) on enzymatic transformation rate (v) based upon Michaelis-Menten kinetics. When S equals K_M, v is one-half of the maximum rate v_m.

breaks down to form free enzyme and products P:

$$E + S \underset{k_{-1}}{\overset{k_1}{\rightleftharpoons}} ES \qquad \textbf{[1.3]}$$

$$ES \underset{k_{-2}}{\overset{k_2}{\rightleftharpoons}} E + P \qquad \textbf{[1.4]}$$

Both reactions are considered reversible, and the various k values are rate coefficients for each of the four possible reactions. In the Briggs and Haldane development, E equals the total enzyme concentration, ES is the concentration of enzyme-substrate complex, and the difference between the two, $E - ES$, is the concentration of free enzyme.

The rate of formation of ES from $E + S$ is thus given by

$$\frac{dES}{dt} = k_1(E - ES)S \quad + k_2(E+P) \qquad \textbf{[1.5]}$$

The rate of formation of ES from $E + P$ is <u>very small and neglected</u>. The rate of breakdown of ES is thus given by:

$$-\frac{dES}{dt} = k_{-1}ES + k_2 ES \qquad \textbf{[1.6]}$$

When the rate of formation of ES just equals its rate of breakdown, the system is at steady state with respect to ES concentration, and

$$k_1(E - ES)S = k_{-1}ES + k_2 ES \qquad \textbf{[1.7]}$$

Rearranging gives

$$\frac{S(E - ES)}{ES} = \frac{k_{-1} + k_2}{k_1} = K_M \qquad \textbf{[1.8]}$$

The coefficient K_M, which represents a composite of the three rate coefficients, is called the *Michaelis-Menten coefficient*. Equation 1.8 may be solved for the concentration of the ES complex,

$$ES = \frac{E \cdot S}{K_M + S} \qquad \textbf{[1.9]}$$

Our interest is in the overall rate of the reaction, in other words, the rate of formation of product P. The velocity of this reaction v is given by

$$v = k_2 ES \qquad \textbf{[1.10]}$$

Combining this with Equation 1.9 yields

$$v = \frac{k_2 \cdot E \cdot S}{K_M + S} \qquad \textbf{[1.11]}$$

Equation 1.11 is most useful for describing the rate of an enzyme reaction, as the two constants, the enzyme concentration, and the substrate concentration are readily measurable quantities. If the substrate concentration is very high so that essentially all the enzyme is present as the ES complex, that is, $ES = E$, then the maximum velocity, v_m, is obtained,

$$v_m = k_2 E^\circ \qquad \text{[1.12]}$$

Dividing Equation 1.11 by Equation 1.12 then yields

$$v = v_m \frac{S}{K_M + S} \qquad \text{[1.13]}$$

Equation 1.13 is the Michaelis-Menten equation, which defines the quantitative relationship between the substrate concentration and the reaction rate in relation to the maximum possible rate.

For the important case where $v = 1/2 v_m$,

$$\frac{1}{2} = \frac{S}{K_M + S} \quad \text{or} \quad v = \frac{v_m}{2} \qquad \text{[1.14]}$$

which shows that

$$S = K_M, \text{ when } v = \frac{v_m}{2} \qquad \text{[1.15]}$$

Thus, the coefficient K_M equals the substrate concentration at which the velocity of the reaction is one-half of the maximum velocity. K_M represents the affinity between the substrate and the enzyme. A low value of K_M indicates a very strong affinity, such that the maximum rate is reached at a relatively low substrate concentration. A large value of K_M, on the other hand, shows a poor affinity.

When an enzyme transforms more than one substrate, the values for K_M and v_m are different for each substrate. However, these coefficients are independent of the enzyme concentration. The principles of affinity and maximum rate are illustrated in Figure 1.12.

The derivation from first principles leading to Equations 1.11 through 1.15 can serve as a model for obtaining many other useful relationships. Although the simple case just treated is adequate for describing many enzyme-catalyzed reactions, more complex reactions may involve several substrates and enzymes. They can be approached mathematically in a similar way, even though the final results may be complicated and may require computer solutions.

Enzyme-catalyzed reactions depend upon pH and temperature, which strongly affect the secondary and tertiary structures of enzymes. Many enzymes have optimal activity at neutral pH, and the activity decreases with either increasing or decreasing pH from this optimal point. Others have optima at higher and some at lower pH. Still others may be affected little by changes in pH. The pH sensitivity of microorganisms, mentioned earlier in this chapter, is to some degree a composite result of the multitude of enzyme reactions involved in its overall metabolism. Temperature also affects individual enzyme activity in a way similar to its overall effect on microorganisms. Up until a temperature is reached at which denaturation of the enzyme protein begins to occur, rates of reaction roughly double for each 10 °C increase in temperature. This positive effect of temperature is similar to what occurs for all reactions. However, the rate of reaction reaches a maximum within an optimal temperature range. For temperatures greater than the optimum, the enzyme begins to denature, and the enzyme's activity then deteriorates and eventually ceases.

Chemical agents may also reduce enzyme reactivity, which is one way in which toxic chemicals can adversely affect a biological treatment process. The chemical agent does not destroy the enzyme, and the reactivity can be reversed if the agent is removed. Two common types of reversible inhibition are competitive and noncompetitive.

TOXICITY

In *competitive inhibition,* a chemical that is similar in structure to the normal enzyme substrate competes with the substrate for the active site on the enzyme. For example, trichloroethene complexed at the active site of methane monooxygenase prevents the enzyme from complexing with methane, causing the rate of methane oxidation to decrease. However, if the methane concentration is increased, its reaction rate also increases, because methane then displaces trichloroethene from the enzyme. A competitive-inhibition model, developed from fundamental principles similar to the treatment above, yields the following result:

$$v = v_m \frac{S}{K_M \left(1 + \frac{I}{K_I}\right) + S} \qquad [1.16]$$

where I is the concentration of the competitive inhibitor, and K_I is a competitive inhibition coefficient having the same concentration units as I. As I increases, the rate decreases in a manner similar to that as if K_M were increased in Equation 1.14. If S is large enough, the inhibitor is removed from the active site, and the maximum rate of substrate transformation is achieved.

In *noncompetitive inhibition,* the chemical agent acts by complexing a metallic activator or by binding at a place on the enzyme other than the active site. The enzyme is then less reactive toward its substrate. For example, cyanide affects enzymes that require iron for activation, because it forms a strong complex with this metal. Metals such as Cu(II), Hg(II), and Ag(I) combine with sulfhydryl groups (-SH) of cysteine, a common amino acid in proteins, and thus affect enzyme activity. In noncompetitive inhibition, an increase in substrate concentration does not counteract the effect of the inhibitor. The model for noncompetitive inhibition, developed from fundamental principles, is

$$v = v_m \cdot \frac{S}{K_M + S} \cdot \frac{1}{1 + \frac{I}{K_I}} \qquad [1.17]$$

An increase in I here effectively reduces the value of v_m, if Equation 1.13 were used to describe the enzyme activity.

By comparing the manner in which an inhibitor affects the enzyme reaction rate, whether altering the effective K_m or the effective v_m, one can determine if the inhibitor is competitive or noncompetitive. However, other types of inhibition are combinations of the two, sometimes making the exact nature of the inhibition difficult to ascertain.

1.8.2 REGULATING THE ACTIVITY OF ENZYMES

Microorganisms can produce hundreds of different enzymes, and the production of each must be regulated in some coordinated fashion so that the organism can properly respond to changes in substrate types and concentrations, environmental conditions, and its needs of energy for movement, growth, and reproduction. A cell cannot afford to produce all possible enzymes at all times. The organism must be able to synthesize sufficient quantities of certain enzymes when needed and to turn off enzyme production when the need is no longer present in order to prevent wasteful diversions of energy and to conserve limited space. The right amount of each enzyme must also be produced, so that the correct amount of each substance needed by the cell is produced. Overproduction is wasteful, while underproduction prevents the cell from carrying out necessary tasks. Thus, the cell must have broad capability for enzyme regulation. This regulatory function is so important that a large proportion of the information contained in a cell's DNA must be assigned to enzyme regulation.

Details of how the cells regulate the production of enzymes are presented later in this chapter, in the section on Genetics. In short, they have biochemical means to control whether or not an enzyme is produced from its code on the cell's DNA. In order to conserve the cell's resources, enzymes are continually broken down to yield the amino-acid components. If an enzyme is not resynthesized from the instructions on the DNA, it is broken down and disappears within minutes to hours.

Cells often need to control the activity of existing enzymes in time frames much shorter than can be achieved by regulating enzyme production and decay. Therefore, the activity of existing enzymes can be controlled by reversible or irreversible means. Among the reversible processes are product inhibition and feedback inhibition. In product inhibition, an accumulation of the product of an enzyme slows the activity of that enzyme. Usually, the product reacts with the enzyme through noncompetitive inhibition. Feedback inhibition is similar, but the accumulation of product several steps along a chain of reaction steps slows the activity of an initial enzyme in the chain. When a cell wants to increase enzyme activity quickly, it can degrade proteins that inhibit the enzyme of interest. Irreversible processes include accelerating the degradation of the enzyme of interest.

1.9 ENERGY CAPTURE

All living organisms, including the microorganisms, capture energy released from oxidation-reduction reactions. Electrons are removed from the primary electron donor and transferred to intracellular electron carriers. The carriers transport the electrons to the terminal electron acceptor, which is reduced to regenerate the carrier. The transfer steps have a free-energy release that the cells capture in the form of energy carriers.

1.9.1 ELECTRON AND ENERGY CARRIERS

The electron carriers can be divided into two different classes, those that are freely diffusible throughout the cell's cytoplasm and those that are attached to enzymes in the cytoplasmic membrane. The freely diffusible carriers include the coenzymes nicotinamide-adenine dinucleotide (NAD^+) and nicotinamide-adenine dinucleotide phosphate ($NADP^+$). NAD^+ is involved in energy-generating (*catabolic*) reactions, while $NADP^+$ is involved in biosynthetic (*anabolic*) reactions. Electron carriers attached to the cytoplasmic membrane include NADH dehydrogenases, flavoproteins, the cytochromes, and quinones. The particular electron carriers that operate in a given cell depend upon the relative energy levels of the primary electron donor and the terminal electron acceptor. More electron carriers can participate when the energy levels are very different.

The reactions of NAD^+ and $NADP^+$ are

$$NAD^+ + 2\,H^+ + 2\,e^- = NADH + H^+ \qquad \Delta G^{0'} = 62\,kJ \qquad \textbf{[1.18]}$$
$$NADP^+ + 2\,H^+ + 2\,e^- = NADPH + H^+ \qquad \Delta G^{0'} = 62\,kJ \qquad \textbf{[1.19]}$$

NAD^+ (or $NADP^+$) extracts two protons and two electrons from a molecule being oxidized and in turn is converted to its reduced form, NADH. The reaction free energy is positive, meaning that energy must be taken from the organic molecule in order for NADH to be formed. When the NADH in turn gives up the electrons to another carrier and is reduced back to NAD^+, it also gives up the chemical energy, which may be converted to other useful forms.

If oxygen is the terminal electron acceptor, the energy released as the electrons are passed through a chain of electron carriers to oxygen can be determined from the overall free energy change of the NADH and O_2 half reactions:

$$NADH + H^+ = NAD^+ + 2\,H^+ + 2\,e^- \qquad \Delta G^{0'} = -62\,kJ \qquad \textbf{[1.20]}$$
$$\frac{1}{2}\,O_2 + 2\,H^+ + 2\,e^- = H_2O \qquad \qquad \Delta G^{0'} = -157\,kJ \qquad \textbf{[1.21]}$$

$$\text{Net: } NADH + \frac{1}{2}\,O_2 + H^+ = NAD^+ + H_2O \qquad \Delta G^{0'} = -219\,kJ \qquad \textbf{[1.22]}$$

Thus, the energy transferred along with electrons from an organic chemical to NADH is released to subsequent electron carriers and ultimately to oxygen in aerobic respiration, yielding in this case -219 kJ per mole of NADH for use by the organism.

How is this energy captured? It is accomplished by transferring the energy from intermediate electron carriers to energy carriers. The primary example of an *energy carrier* is adenosine triphosphate (ATP). When energy is released from an electron carrier (such as Equation 1.22), it is used to add a phosphate group to adenosine diphosphate (ADP):

$$ADP + H_3PO_4 = ATP + H_2O \qquad \Delta G^{0'} = 32\,kJ \qquad \textbf{[1.23]}$$

or in simplified form:

$$ADP + P_i = ATP + H_2O \qquad \Delta G^{0'} = 32\,kJ \qquad \textbf{[1.24]}$$

One mole of ADP picks up only 32 kJ in this reaction, while one mole of NADH has given up over six times that amount of energy when oxygen is the terminal electron acceptor (Equation 1.22). Thus, theoretically about six moles of ATP could be formed under aerobic conditions from each mole of NADH. However, only three moles are actually formed, because real reactions do not capture 100 percent of the standard free energy. We thus see that roughly 50 percent of the energy in NADH is actually captured in the transfer of its energy content to ATP.

One question of interest is how many ATPs can be formed from NADH under anaerobic conditions. This might be estimated by computing the overall free energy released when other known electron acceptors accept the electrons from NADH:

$$NO_3^-: \quad NADH + \frac{2}{5} NO_3^- + \frac{7}{5} H^+$$
$$= NAD^+ + \frac{1}{5} N_2 + \frac{6}{5} H_2O \qquad \Delta G^{0'} = -206 \text{ kJ} \quad \textbf{[1.25]}$$

$$SO_4^{2-}: \quad NADH + \frac{1}{4} SO_4^{2-} + \frac{11}{8} H^+$$
$$= NAD^+ + \frac{1}{8} H_2S + \frac{1}{8} HS^- + H_2O \quad \Delta G^{0'} = -20 \text{ kJ} \quad \textbf{[1.26]}$$

$$CO_2: \quad NADH + \frac{1}{4} CO_2 + H^+$$
$$= NAD^+ + \frac{1}{4} CH_4 + \frac{1}{2} H_2O \qquad \Delta G^{0'} = -15 \text{ kJ} \quad \textbf{[1.27]}$$

This energy analysis indicates that the energy available with nitrate as the electron acceptor is similar to that with oxygen, but sulfate and carbon dioxide yield much less energy per NADH. Since the energy required to generate one mole of ATP is 32 kJ, the energy available in the latter cases is too low to produce even one mole of ATP per mole of NADH. This puzzled biochemists for some time and, indeed, led many to believe that sulfate and carbon dioxide reduction could not support energy generation. However, growth of sulfate and carbon-dioxide reducers does occur. The puzzle was solved in 1961 by the English scientist Peter Mitchell, who won the Nobel Prize for his concept that ATP formation is linked to a *proton motive force* produced across the cell membrane that is established through the electron transport reactions, rather than directly to NADH oxidation.

Equation 1.20 shows that, when NADH gives up its electrons and is oxidized to NAD^+, it also releases a proton. This proton is discharged outside of the cell membrane, causing a charge imbalance and a pH gradient across the membrane. This is similar to what happens when charging a battery. The chemical energy stored in the proton gradient is used by the cells for such things as ion transport across the membrane, flagella movement, and the formation of ATP from ADP. The key is that the transport of an electron from NADH to the terminal acceptor is not directly linked to formation of an ATP molecule. Instead, ATP formation occurs independent of an individual electron-transfer process. The proton motive force builds up with as many electron transfers as needed until it is sufficient for ATP formation to occur. The function then of the movement of electrons from one carrier to the next is to produce the external protons needed to create this potential across the membrane.

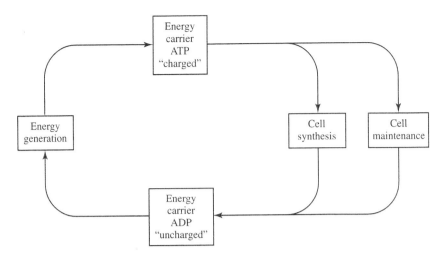

Figure 1.13   Transfer of energy from energy generation to cell synthesis or maintenance via an energy carrier, represented by ATP.

The chemical energy captured in ATP is used by the cell for cell synthesis and maintenance. ATP diffuses through the cell, and whenever the energy it contains is needed, the cell extracts the energy, releasing a phosphate molecule and converting the ATP back into ADP. This process is illustrated in Figure 1.13. The uses of ATP to fuel synthesis are discussed later, under Anabolism.

1.9.2   ENERGY AND ELECTRON INVESTMENTS

ATP and NADH are required as an "investment" to initiate some energy-generating reactions. For example, coenzyme A (CoA), which "activates" fatty acids so that they can be oxidized further by other enzymes, requires ATP to create a complex with the fatty-acid substrate (R-COOH):

$$R\text{-COOH} + HCoA + ATP = R\text{-COCoA} + ADP + H_3PO_4 \qquad \textbf{[1.28]}$$

A second example of an investment is a group of reactions of great importance in initiating the degradation of hydrocarbons: *oxygenations.* Oxygenase enzymes catalyze the direct incorporation of oxygen from O_2 into the organic molecule. This is a key first step in the oxidation of aliphatic hydrocarbons, such as those comprising most of gasoline and natural gas, and many aromatic compounds, such as benzene, polycyclic aromatic hydrocarbons (PAHs) like naphthalene and pyrene, and most halogenated aromatic compounds. Oxygenation reactions insert oxygen into the molecule, which makes it accessible for subsequent oxidation steps and more water soluble.

There are two kinds of oxygenases: *dioxygenases,* which catalyze the incorporation of both atoms of O_2 into a molecule, and *monooxygenases,* which catalyze the transfer of only one atom of O_2 into a molecule. Introducing the first oxygen into a hydrocarbon is chemically difficult to do. While the oxidation itself releases energy, the cells must invest energy to bring it about. This important first insertion of oxygen makes the compound accessible for further oxidation and generation of NADH.

Monooxygenations require investment of NADH and energy. The example of methane monooxygenation (MMO) illustrates the investments. If methane were oxidized to methanol by a conventional hydroxylation reaction, one NADH would be generated:

$$CH_4 + NAD^+ + H_2O = CH_3OH + NADH + H^+ \qquad \textbf{[1.29]}$$

Instead, in a monooxygenation, NADH and O_2 are consumed:

$$CH_4 + NADH + H^+ + O_2 = CH_3OH + NAD^+ + H_2O \qquad \textbf{[1.30]}$$

In either case, the organism then oxidizes methanol to obtain energy in the form of three NADHs through stepwise oxidations to formaldehyde, formic acid, and carbon dioxide. For the conventional hydroxylation, a total of four NADHs are generated. For monooxygenation, one of the NADHs formed must be invested in the first step; thus, a net of only two NADHs is produced.

In dioxygenations, two -OH groups are inserted, and no net NADH is produced or consumed. For example, dioxygenation of toluene (C_7H_8) produces a catechol:

$$C_7H_8 + O_2 = C_7H_8O_2 \qquad \textbf{[1.31]}$$

On the other hand, two hydroxylations of toluene to create catechol theoretically would release sufficient energy to generate two NADHs:

$$C_7H_8 + 2\,NAD^+ + 2\,H_2O = C_7H_8O_2 + 2\,NADH + 2\,H^+ \qquad \textbf{[1.32]}$$

Thus, the dioxygenation reaction from toluene to catechol was an investment of two NADHs and the ATP that theoretically could have been generated from them.

1.10   METABOLISM

Metabolism is the sum total of all the chemical processes of the cell. In can be separated into *catabolism*, which is all the processes involved in the oxidation of substrates or use of sunlight in order to obtain energy, and *anabolism*, which includes all processes for the synthesis of cellular components from carbon sources. Thus, catabolism furnishes the energy required for anabolism. Catabolism also furnishes the energy required for motion and any other energy-requiring processes.

During catabolism, the energy-yielding substrate usually is oxidized stepwise through intermediate forms (*metabolites*) before the final end products are produced. Accompanying this oxidation, the chemical energy released is conserved by transfer of electrons to electron carriers (such as NADH) and the formation of energy-rich phosphate-to-phosphate bonds (such as ATP). The electrons and the phosphate-bond energy may then be transferred to other parts of the cell where they are needed for cell synthesis, maintenance, or motility.

The transfer of energy between catabolism and anabolism is termed *energy coupling*. Figure 1.13, shown earlier, is a simplified diagram indicating the relationships among catabolism, anabolism, and energy coupling. Cells obtain energy for growth and maintenance either through oxidation of chemicals (*chemotrophs*) or through photosynthesis (*phototrophs*). In catabolism by chemotrophs, either organic materials (*chemoorganotrophs*) or inorganic materials (*chemolithotrophs*) are oxidized for energy. In either case, the chemical energy released is transferred by energy coupling, usually through a coenzyme such as ATP. The ATP energy is then given up for anabolic processes such as cell growth or maintenance.

Description of catabolic processes for chemoorganotrophs can be quite complicated, since so many different organic compounds can be oxidized to yield electrons and energy. Likewise, the anabolic processes involve the synthesis of many different organic compounds—proteins, carbohydrates, lipids, nucleic acids, and more. However, the complexity of catabolism and anabolism can be transcended when a "big picture" view of the processes is taken; most of the details of the individual steps along each of the metabolic paths are not essential for gaining the "big picture."

Figure 1.14 represents such an overall view, one of catabolism presented by Hans A. Krebs, a pioneer who 50 or so years ago made several important contributions to the biochemical knowledge of organic catabolism. Figure 1.14 illustrates the three basic stages to organic catabolism. Any organic material that can be used as a substrate to provide electrons and energy can fit into this general scheme. Catabolic processes that gain electrons and energy from inorganic compounds or from photosynthesis do not follow the catabolic steps shown.

In the first stage of organic catabolism, large or complex molecules are degraded to their basic building blocks, usually through hydrolysis, which yields little if any energy for the cells. In stage two, these smaller molecules are converted into a smaller number of simpler compounds. Fatty acids and amino acids are converted for the most part to acetyl-CoA, which is a coenzyme A complex of acetic acid. The hexose and pentose sugars, as well as glycerol, are converted into three-carbon compounds, glyceraldehyde-3-phosphate and pyruvate, which can be converted into acetyl-CoA also. Amino acids are also converted to acetyl-CoA, plus a few other products such as α-ketoglutarate, succinate, fumarate, or oxaloacetate. Some energy may be released for use by the cell during the stage-two conversions. Finally, in stage three, the end products of the second stage enter a common pathway in which they are oxidized ultimately to carbon dioxide and water. This final stage three pathway, called the citric acid cycle or the Krebs cycle, generates the largest amounts of electrons and energy for the cells.

Anabolism in chemoorganolithotrophs, such as bacteria, also occurs in three stages that are analogous to those shown in Figure 1.14, but in the reverse direction. Beginning with the small number of compounds at the end of stage two or in stage three, a cell can work backward to synthesize the building blocks. From these, the lipid, polysaccharide, protein, and nucleic-acid macromolecules that comprise the basic cellular components are constructed.

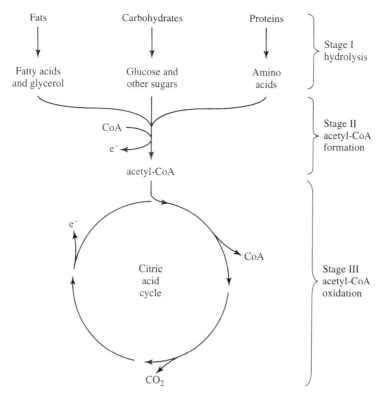

Figure 1.14 The three general stages of catabolism of fats, carbohydrates, and proteins under aerobic conditions. Reversing the processes gives anabolism.

Chemoorganolithotrophs frequently can use a single organic compound—such as one fatty acid, simple sugar, or amino acid—as its sole energy and carbon source. Here, a portion of the simple substrate is oxidized through all or some of the three stages for energy, depending upon where it enters the overall catabolic scheme. Another portion of the same simple substrate is then converted to stage-two and stage-three intermediates, which in turn serve as the starting materials for synthesis of all the lipids, polysaccharides, proteins, and other materials required by the cell.

In some cases, however, the pathway for synthesis of a given building block may not be open, perhaps because the organism lacks the ability to synthesize a key enzyme. In this case, that particular building block must be furnished to the organism from some other source if it is to grow. For example, some cells lack the ability to synthesize key vitamins or amino acids, and these must then be furnished for the cell, perhaps by another organism or in the growth medium. Many higher animals, including humans, lack the ability to synthesize many key organic compounds. All of these essential vitamins and growth factors must be present in the animal's food. Despite their small size and relative simplicity, a great number of bacteria can produce all needed materials themselves.

While the big-picture view of catabolism and anabolism, shown in Figure 1.14, seems to imply that anabolism is simply a reversal of catabolism, this is not wholly accurate. Some of the enzymes involved in the conversions in one direction along a given path are different from those involved in conversions in the reverse direction. This may have resulted because an enzyme that is efficient in taking a molecule apart may not be the most efficient for putting it back together again, thus leading to the evolution of different and more specialized enzymes for the two separate processes. Also, the catabolic and anabolic processes may take place at different locations within the cell, especially in eukaryotic cells. Also, the organism's need for the various building blocks for synthesis is independent of the substrate used for energy. Thus, the number of enzymes required for synthesis of a given cellular component may be quite different than the number required for degradation, which suggests an obvious advantage for having separate pathways for catabolism and anabolism.

Although the catabolic and anabolic pathways are somewhat separate, the similar metabolites at the end of stage two and in stage three provide a common meeting ground for the two processes. The overall design of the three-stage system also suggests how organisms can collectively adapt to the degradation of such a broad range of organic materials. The requirement essentially is to convert the molecule into a form that can enter at some point in the common pathways of the organism's metabolic system.

We now explore some of the catabolic, anabolic, and energy-coupling systems in more detail in order to obtain a more sophisticated idea of how the overall system works and can be put to use in microbiological systems for environmental control.

1.10.1　CATABOLISM

Catabolism in chemolithotrophs depends on the oxidation and reduction of chemicals in the organism's environment. Oxidation removes electrons, and reduction adds electrons. Materials that are oxidized are called *electron donors*, and those being reduced are *electron acceptors*. Thus, for a chemolithotrophic organism to obtain energy, it must be furnished with an electron donor and an electron acceptor. In some cases, as will be illustrated later, a single compound can serve both of these functions.

Generally, the electron donor is considered to be the energy substrate or "food" for the microorganisms. Common electron donors are compounds containing carbon in a reduced state (organic chemicals) or containing other elements in a reduced state (reduced inorganic compounds, such as ammonia, hydrogen, or sulfide). The Earth contains an innumerable variety of electron donors for microorganisms.

The electron acceptors, by comparison, appear to be quite few and include primarily oxygen, nitrate, nitrite, Fe(III), sulfate, and carbon dioxide. In recent years, however, the number of known electron acceptors for energy metabolism has been growing, and it now includes such species as chlorate, perchlorate, chromate, selenate, and chlorinated organics such as tetrachloroethylene and chlorobenzoate. These compounds often are of environmental concern, and their potential conversion through use by organisms as electron acceptors is of growing interest.

In principle, microorganisms should be able to exploit any oxidation-reduction reaction that yields energy. Microorganisms, especially Bacteria, are quite versatile as a group in the variety of reactions from which they can capture energy.

The quantity of energy released per electron transferred depends upon the chemical properties of the electron donor and the electron acceptor. The energy produced by coupled reactions is best illustrated through the use of half reactions, as indicated below for acetate oxidation with oxygen:

Acetate and Oxygen (Aerobic Oxidation of Acetate)

$$\Delta G^{0\prime}, \text{kJ}$$

Donor: $\frac{1}{8} CH_3COO^- + \frac{3}{8} H_2O = \frac{1}{8} CO_2 + \frac{1}{8} HCO_3^- + H^+ + e^-$ -27.40

Acceptor: $\frac{1}{4} O_2 + H^+ + e^- = \frac{1}{2} H_2O$ -78.72

Net: $\frac{1}{8} CH_3COO^- + \frac{1}{4} O_2 = \frac{1}{8} CO_2 + \frac{1}{8} HCO_3^- + \frac{1}{8} H_2O$ -106.12

The value $\Delta G^{0\prime}$ represents the free energy released under standard conditions and at pH $= 7$. The half reactions and overall reaction are written for one electron equivalent (e$^-$ eq). For example, the oxidation of 1/8 mole of acetate donates one mole of electrons (i.e., one e$^-$ eq), while 1/4 mole of oxygen accepts that electron equivalent. The energy released by the oxidation of acetate coupled to the reduction of oxygen is 106.12 kJ/e$^-$ eq, the release indicated by the negative sign on the net overall reaction.

The energy released by other reactions of interest is discussed in detail in Chapter 2. A few examples show how the overall reaction energy differs for different possible electron donors and acceptors.

Acetate and Carbon Dioxide (Methanogenesis of Acetate)

$$\Delta G^{0\prime}, \text{kJ}$$

Donor: $\frac{1}{8} CH_3COO^- + \frac{3}{8} H_2O = \frac{1}{8} CO_2 + \frac{1}{8} HCO_3^- + H^+ + e^-$ -27.40

Acceptor: $\frac{1}{8} CO_2 + H^+ + e^- = \frac{1}{8} CH_4 + \frac{1}{4} H_2O$ 23.53

Net: $\frac{1}{8} CH_3COO^- + \frac{1}{8} H_2O = \frac{1}{8} CH_4 + \frac{1}{8} HCO_3^-$ -3.87

Glucose and Carbon Dioxide (Methanogenesis from Glucose)

$$\Delta G^{0\prime}, \text{kJ}$$

Donor: $\frac{1}{24} C_6H_{12}O_6 + \frac{1}{4} H_2O = \frac{1}{4} CO_2 + H^+ + e^-$ -41.35

Acceptor: $\frac{1}{8} CO_2 + H^+ + e^- = \frac{1}{8} CH_4 + \frac{1}{4} H_2O$ 23.53

Net: $\frac{1}{24} C_6H_{12}O_6 = \frac{1}{8} CH_4 + \frac{1}{8} CO_2$ -17.82

Hydrogen and Oxygen (Aerobic Oxidation of Hydrogen)

$\Delta G^{0\prime}$, kJ

Donor: $\qquad \frac{1}{2} H_2 = H^+ + e^-$ $\qquad\qquad\qquad$ -39.87

Acceptor: $\frac{1}{4} O_2 + H^+ + e^- = \frac{1}{2} H_2O$ $\qquad\qquad$ -78.72

_____

Net: $\qquad \frac{1}{2} H_2 + \frac{1}{4} O_2 \quad = \frac{1}{2} H_2O$ $\qquad\qquad$ -118.59

These examples illustrate three trends that are key for understanding microbiology. First, methanogenesis, an anaerobic process, provides much less energy per electron equivalent than does aerobic oxidation. Second, glucose contains more energy than acetate. Third, the chemolithotrophic oxidation of hydrogen produces more energy than chemoorganotrophic oxidation of acetate, but less than that of glucose ($-41.35 - 78.72 = -120.07$ kJ/e$^-$ eq). Microorganism growth should be and is greater when the oxidation/reduction couple provides more energy rather than less energy. The impacts of the differences in energy release are quantified in Chapter 2.

Figure 1.15 illustrates graphically the relative free energy available from oxidation/reduction couples exploited by microorganisms. Two vertical scales are shown: one with energy expressed in kJ, and the other in volts. Since all reactions are written for one electron equivalent, the two different methods are related directly through Faraday's constant, 96.485 kJ/volt.

In order that energy is released from an oxidation/reduction reaction, the donor half reaction must lie above the acceptor half reaction, as seen for glucose and oxygen. The glucose and oxygen half reactions define the extremes of the typical half reactions experienced in natural systems. The difference between the two half reactions is -120 kJ/e$^-$ eq. Energy related to the electron acceptors oxygen, Fe(III), and nitrate are not too different from one another and are near the bottom of the scale. Many compounds can be electron donors with these three acceptors. Sulfate and carbon dioxide (in methane fermentation) lie far higher on the scale. Much less energy is released when they accept electrons from donors like glucose.

Some microorganisms can use organic compounds as electron acceptors and donors, a process called *fermentation*. Part of the molecule is oxidized, while another part is reduced. The energy available can be estimated from half reactions involving the products, such as shown in Figure 1.15. For example, facultative or anaerobic bacteria commonly ferment glucose. The end products vary, but may include ethanol, acetate, hydrogen, or, in many cases, a mixture of such products. The energy available from glucose fermentation to ethanol is the difference between the glucose and ethanol half-reaction energies, which equals about -10 kJ per e$^-$ eq. If glucose were fermented to acetate instead of ethanol, the energy available would be even higher, about -14 kJ per e$^-$ eq. If the end products of fermentation are a mixture of compounds, then the net free energy per electron equivalent is a weighted average of the different free energies. Fermentation requires that the starting compound have a large positive $\Delta G^{0\prime}$.

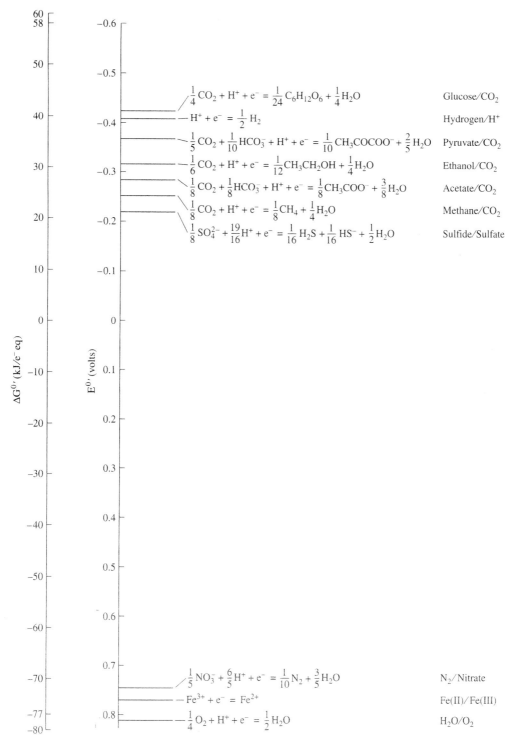

Figure 1.15 Energy scale for various oxidation/reduction couples. Energy scales can be expressed as kJ/e⁻ eq or volts, where the relationship between the two is given as one volt equals −96.485 kJ.

From the standpoint of free-energy changes for the transfer of electrons from the electron-donor to the electron-acceptor, it makes no difference which pathway is followed in carrying out the reaction. However, from the viewpoint of the organisms, the pathway does matter, as it can affect how much of the free energy to be released can be captured. One objective of catabolic reactions is to capture as much of the energy released as possible. However, sometimes "investment" objectives supersede energy capture. These include activating molecules (like via monooxygenations) and producing certain catabolic intermediates for use in synthesis.

The broad strategy that all microorganisms use for energy capture was outlined in the earlier section on Energy Capture. The enzymatic steps by which organic matter is oxidized in order to capture this energy are provided in somewhat more detail here. The oxidations of four major groups of organic compounds are illustrated: (1) saturated hydrocarbons, alcohols, aldehydes, and ketones, (2) fatty acids, (3) carbohydrates, and (4) amino acids. These collectively represent the building blocks for fats, carbohydrates, and proteins.

Saturated Hydrocarbons, Alcohols, Aldehydes, and Ketones

Hydrocarbons contain only carbon and hydrogen and may be either aliphatic (no benzene ring involved) or aromatic (compounds containing a benzene ring). Hydrocarbons are difficult to attack, chemically and biologically, because of the strength of the carbon-hydrogen and carbon-carbon bonds. The initial attack on the very stable hydrocarbon structure generally involves an *oxygenation,* such as methane monooxygenation in Equation 1.30 or toluene dioxygenation in Equation 1.31. Molecular oxygen is required for both oxygenations, while NADH is a cosubstrate for monooxygenation. Thermodynamically, oxygenation reactions are costly, since electrons that might have been used to reduce NAD^+ to NADH are used to reduce O_2 to H_2O, a process that does not capture the electrons or the energy for the cells. Because of this large energy loss, organism yield per electron equivalent of hydrocarbon oxidized is generally lower than with other organic compounds. On the other hand, the oxygenations convert the hydrocarbons to compounds that are more readily oxidized by NADH-generating reactions.

The difficulty initiating oxidation of hydrocarbons and the role played by oxygenases and molecular oxygen have been known for some time. Oil and natural gas are hydrocarbons for the most part, and their accumulation in deposits was thought to have occurred because oxygen, necessary for this first oxygenation step, was absent in marshy areas where petroleum precursors were deposited. For this reason, great skepticism emerged within scientific circles over the first reports in the mid-1980s that aromatic hydrocarbons, such as toluene, xylene, ethylbenzene, and benzene, were being biodegraded anaerobically, i.e., in the absence of molecular oxygen. However, several pure cultures that can anaerobically oxidize aromatic hydrocarbons have since emerged, and novel biochemical pathways for their oxidation are being found. (Chapter 14 describes this new phenomenon.) Reports of anaerobic biodegradation of aliphatic hydrocarbons are also emerging. Thus, the known ways that organisms can obtain energy from organic compounds is enlarging. In all cases, oxygen is added to the hydrocarbon to create an alcohol. When molecular oxygen is not available, organic compounds containing oxygen or water can be used to supply the initial oxygen. It is important that these newer pathways be recognized, as they help explain the disappearance of hydrocarbons observed in anaerobic contaminated groundwaters, sediments, and soils. Such pathways might also be exploited in engineered processes for contaminant biodegradation. However, our general discussion here focuses on the older, well-recognized mechanisms, which are applicable in most cases.

The general steps in oxidation of a hydrocarbon are illustrated in Figure 1.16 for a typical straight-chain alkane. Here, R represents an alkyl hydrocarbon chain of some undefined length. The oxidation of organic compounds generally involves the removal of two atoms of hydrogen and associated electrons and their transfer to the electron carrier NADH. However, we indicated previously that the oxidation of an alkane to an alcohol is generally carried out by the direct addition of molecular oxygen with the aid of an oxygenase enzyme and NADH as a cosubstrate. Because of the difficulty of this oxidation, the organism sacrifices the electrons and energy available in the first 2-electron oxidation step with an alkane. However, in the subsequent oxidation steps from an alcohol to an aldehyde and then to an organic acid, NADH is generated. If

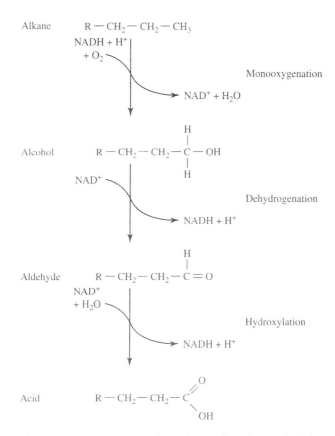

Figure 1.16 Steps in the oxidation of an alkane, alcohol, or aldehyde to an organic acid.

the original compound were an alcohol or an aldehyde, it would enter the oxidation scheme illustrated in Figure 1.16 at the appropriate point. An alkene (not shown here) enters this scheme by being converted to an alcohol by the enzymatic addition of water across the double bond; this is not an oxidation, and no NADH is formed.

Figure 1.16 illustrates a recurring pattern: alternating dehydrogenation and hydroxylation steps. In *dehydrogenation,* an alcohol is oxidized by removing two hydrogen ions and two electrons to produce an aldehyde. NAD^+ is reduced to $NADH_2^+$ (or $NADH + H^+$). The alcohol and the aldehyde have an O substituent, but the alcohol is changed from a single-bond hydroxyl to a double-bond aldehyde. In *hydroxylation,* the aldehyde is oxidized by removing two hydrogens and two electrons and adding water to form a carboxylic acid. NAD^+ is again reduced to NADH. H_2O is the source of the O, which is added to make the acid group.

Fatty acids, either present as the original substrate or formed from other alkanes as illustrated in Figure 1.16, are in turn oxidized by the process of β-oxidation, shown in Figure 1.17. The second carbon from the carboxyl carbon of the organic acid is the β-carbon atom, and it is this carbon that is oxidized through β-oxidation. The initial step in this oxidation is addition of coenzyme A, here represented as HS-CoA, to form an acyl CoA. This activates the fatty acid for the subsequent steps of oxidation, and this activation requires some energy in the form of the energy carrier ATP. ATP gives up two of its phosphate atoms in this process and is converted to AMP, or adenosine monophosphate. Electrons and protons are removed from the activated acid two at a time, and water is added to the molecule in alternate steps, leading to the oxidation of the β-carbon to form a keto group. Coenzyme A is then added to the molecule, splitting it into acetyl CoA and an acyl-CoA compound two carbon atoms shorter than the original. The shorter acyl-CoA group then undergoes β-oxidation again and again until it has been completely converted to a group of acetyl CoA molecules. A mole of a 16-carbon fatty acid, such as palmitic acid, is converted in this manner to eight moles acetyl CoA and seven moles each of $FADH_2$ and NADH. Figure 1.18 also shows that one ATP is converted to AMP to help fuel the first step.

The protons and electrons removed are thus transferred to the electron carriers FAD and NAH^+, forming $FADH_2$ and NADH. The electrons involved in the first oxidative step are transferred to FAD rather than NAD^+, since the energy yield is somewhat less, and a less energetic carrier is required for electron transfer to be effected.

Similar processes of electron removal and water addition occur for fatty acids of odd carbon length, and one end product is the 3-carbon propionic acid. Branched fatty acids also undergo similar β oxidations to form acetyl CoAs. The acetyl CoA atoms formed from any of the reactions can enter the citric acid cycle, which was overviewed in Figure 1.14 and will be described in more detail later in this section.

Carbohydrates   Carbohydrates are polysaccharides, such as cellulose, starch, and complex sugars. Enzymatic hydrolysis of carbohydrates generally produces 6-carbon hexoses or 5-carbon pentoses. These simple sugars are higher in energy content per electron equivalent than acetate or most other simple organic molecules. For this reason, microorganisms frequently can obtain energy from carbohydrates by

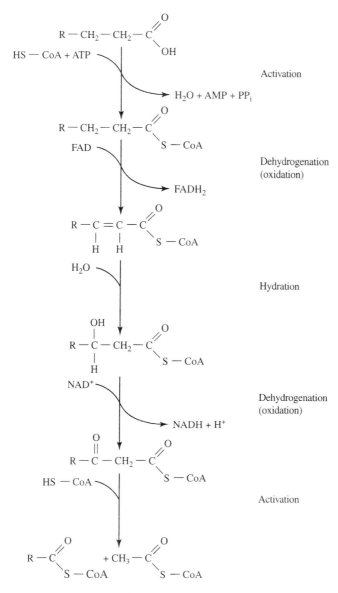

Figure 1.17 β-oxidation of fatty acids to acetyl-CoA.

anaerobic fermentative pathways that lead to less energetic end products. However, when a terminal electron acceptor is available, carbohydrates follow pathways leading to formation of acetyl CoA, which then enters the citric acid cycle, as with other organic substrates. We first show the transformations of a simple sugar, represented by glucose, to acetyl CoA. We then give a brief discussion of fermentative pathways.

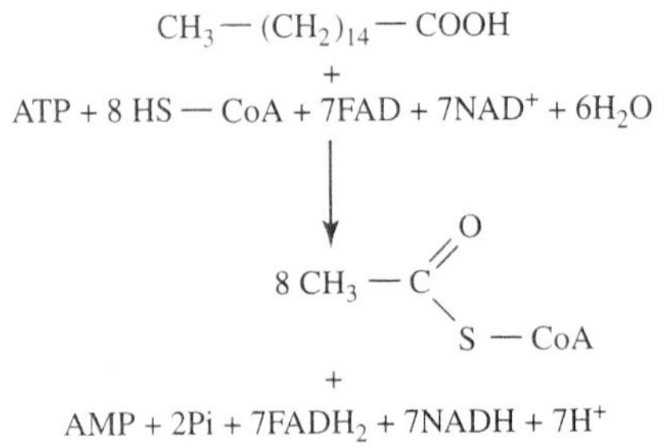

$$CH_3 - (CH_2)_{14} - COOH$$
$$+$$
$$ATP + 8\,HS - CoA + 7FAD + 7NAD^+ + 6H_2O$$

$$\downarrow$$

$$8\,CH_3 - C \overset{O}{\underset{S - CoA}{\diagup\!\!\!\diagdown}}$$

$$+$$
$$AMP + 2Pi + 7FADH_2 + 7NADH + 7H^+$$

Figure 1.18   Overall stoichiometry for conversion of palmitic acid into acetyl CoA through β-oxidation.

Figure 1.19 gives a brief summary of the several steps for conversion of glucose to acetyl CoA. Through the introduction of energy in the form of two moles ATP, a mole of the 6-carbon glucose is split in half to form two moles of a 3-carbon intermediate, glyceraldehyde 3-phosphate. This compound undergoes oxidation (electron removal to NADH) and energy transfer (ATP formation) in a step-wise conversion to two moles of pyruvate, a 3-carbon compound that is more oxidized than carbohydrates. Then acetyl CoA is formed with the release of more NADH and CO_2. The acetyl CoA formed then undergoes the normal oxidations involved in the citric acid cycle. The overall change resulting from this glucose conversion is

$$C_6H_{12}O_6 + 2\,HSCoA + 4\,NAD^+ + 2\,ADP + 2\,P_i \rightarrow$$
$$2\,CH_3COSCoA + 4\,NADH + 2\,ATP + 2\,CO_2 + 4\,H^+ \qquad \textbf{[1.33]}$$

If a terminal electron acceptor such as oxygen is not available, many facultative microorganisms can obtain energy from glucose or other simple sugars by using the organic substrate as the electron donor and the electron acceptor. This fermentation process involves the steps shown in Figure 1.19, but it ends at pyruvate production, or before the formation of acetyl CoA:

$$C_6H_{12}O_6 + 2\,NAD^+ + 2\,ADP + 2\,P_i \rightarrow$$
$$2\,CH_3COCOO^- + 2\,NADH + 2\,ATP + 4\,H^+ \qquad \textbf{[1.34]}$$

Equation 1.34 shows formation of 2 ATP, and this is the way in which the fermentation yields energy. Formation of ATP directly through the oxidation of an electron-donor substrate is called *substrate-level phosphorylation.*

However, 2 NADHs also are formed in Equation 1.34. Since the organism has no terminal electron acceptor for regenerating the NAD^+ it needs to carry out this overall reaction, the cell must somehow get rid of the electrons present in the two moles of NADH. The organism accomplishes this through the transfer of the electrons back to pyruvate, leading to the formation of a variety of possible compounds. Possible end products are ethanol, acetate, or perhaps any one or a mixture of other simple organic compounds, such as propanol, butanol, formate, propionate, succinate, butyrate, and,

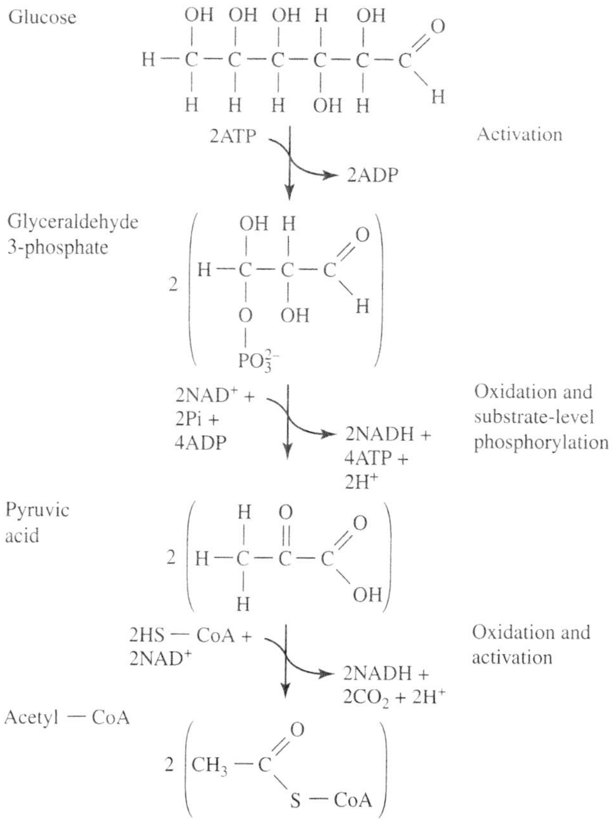

Glucose

$$H-\underset{\underset{H}{|}}{\overset{\overset{OH}{|}}{C}}-\underset{\underset{H}{|}}{\overset{\overset{OH}{|}}{C}}-\underset{\underset{H}{|}}{\overset{\overset{OH}{|}}{C}}-\underset{\underset{OH}{|}}{\overset{\overset{H}{|}}{C}}-\underset{\underset{H}{|}}{\overset{\overset{OH}{|}}{C}}-C{\overset{\diagup O}{\diagdown H}}$$

2ATP ⟶ 2ADP   Activation

Glyceraldehyde
3-phosphate

$$2\left(H-\underset{\underset{O}{|}}{\overset{\overset{OH}{|}}{C}}-\underset{\underset{OH}{|}}{\overset{\overset{H}{|}}{C}}-C{\overset{\diagup O}{\diagdown H}} \right)$$
$$PO_3^{2-}$$

2NAD$^+$ +
2Pi +   ⟶ 2NADH +   Oxidation and
4ADP   4ATP +   substrate-level
                    2H$^+$ phosphorylation

Pyruvic
acid

$$2\left(H-\underset{\underset{H}{|}}{\overset{\overset{H}{|}}{C}}-\overset{\overset{O}{\|}}{C}-C{\overset{\diagup O}{\diagdown OH}} \right)$$

2HS — CoA +
2NAD$^+$ ⟶ 2NADH +   Oxidation and
                    2CO$_2$ + 2H$^+$ activation

Acetyl — CoA

$$2\left(CH_3-C{\overset{\diagup O}{\diagdown \,S-CoA}} \right)$$

Figure 1.19   Conversion of carbohydrates, represented here by glucose, to acetyl CoA.

perhaps, H$_2$. The mixture of end products that results depends upon the organism involved and the environmental conditions present.

An example is the well-known overall fermentation of sugars to ethanol:

$$C_6H_{12}O_6 \rightarrow 2\,CH_3CH_2OH + 2\,CO_2 \qquad \textbf{[1.35]}$$

The ethanol is formed from pyruvate and NADH in two steps:

Acetaldehyde formation from pyruvate:

$$CH_3COCOO^- + H^+ \rightarrow CH_3CHO + CO_2 \qquad \textbf{[1.36]}$$

Ethanol formation from acetaldehyde and NADH:

$$CH_3CHO + NADH + H^+ \rightarrow CH_3CH_2OH + NAD^+ \qquad \textbf{[1.37]}$$

The net result of combining Equations 1.36 and 1.37 with Equation 1.34 is:

$$C_6H_{12}O_6 + 2\,ADP + 2\,P_i \rightarrow 2\,CH_3CH_2OH + 2\,CO_2 + 2\,ATP \qquad \textbf{[1.38]}$$

Through alcohol fermentation, two moles of ATP are formed per mole of glucose fermented, and all the NADH is regenerated to NAD^+ through the reduction of pyruvate to ethanol.

Figure 1.15 indicates that more energy could be obtained if the organism fermented glucose to acetate, rather than to ethanol. In fact, some microorganisms have devised schemes for obtaining the greater energy from pyruvate fermentation to fatty acids. Suffice it to say here that many fermentative pathways are possible, and each is exploited by many different organisms. In the anaerobic fermentation of mixed wastes containing carbohydrates by mixed cultures of soil organisms, the simpler fatty acids rather than ethanol are generally formed, because the dominating microorganisms are generally those that have found the best routes for extracting the energy available from carbohydrates under the conditions imposed.

Amino Acids   The concept of oxidation through electron removal and water addition is common for amino acids, just as it is for fatty acids and simple sugars. In addition, the amino group must be removed. All amino acids have an amino group bonded to the alpha carbon next to the carboxylate terminus:

$$R\text{-}CHNH_2COOH$$

Deamination involves several steps but can be summarized as

$$R\text{-}CHNH_2COOH + NAD^+ + H_2O = R\text{-}COCOOH + NH_3 + NADH + H^+$$

$$[1.39]$$

Here, ammonia removal results in oxidation of the alpha carbon to a keto-group, release of ammonia, and the formation of a mole of NADH. The organic acids that are formed from deamination of the 20 or so different amino acids then enter the citric acid cycle at different points, generally after conversion to acetyl CoA. However, some amino acids are converted into products such as α-ketoglutarate (arginine, glutamate, glutamine, histidine, proline), succinate (isoleucine, methionine, threonine, valine), fumarate (aspartate, phenylalanine,thyrosine), or oxaloacetate (asparagine, aspartate). These products are components of the citric acid cycle and are oxidized by directly entering the cycle.

Citric Acid Cycle   We have now seen how the basic food categories of fats, carbohydrates, and proteins are hydrolyzed (if in polymerized form) and then partially oxidized to produce NADH and acetyl CoA or some component of the citric acid cycle. Essentially all organic compounds fit into this metabolic scheme in some manner. The acetyl CoA enters the citric acid cycle when a terminal electron acceptor is available.

The steps of the citric acid cycle are summarized in Figure 1.20. To enter the cycle, acetyl CoA is first combined with oxaloacetic acid and water to form citric acid. The most important features of the citric acid cycle are these:

- The eight electrons in acetate are pairwise removed in four steps that generate three NADHs and one $FADH_2$.
- The two carbons in acetic acid are removed in two steps that produce CO_2.

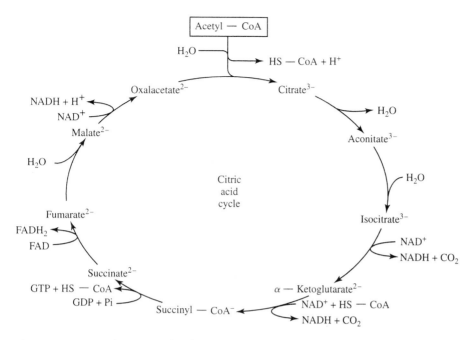

Figure 1.20 The citric acid cycle.

- One step is a substrate-level phosphorylation that gives one GTP (guanosine triphosphate, an analog of ATP).

- Four steps add H_2O, and one step removes H_2O.

- In the last step, malate is oxidized to oxalacetate, which is then available to combine with acetyl CoA and begin the cycle again.

The net result of all the reactions involved in the citric acid cycle for acetyl-CoA oxidation is

$$CH_3COSCoA + 3\ NAD^+ + FAD + GDP + P_i + 3\ H_2O \rightarrow$$
$$2\ CO_2 + 3\ NADH + FADH_2 + GTP + 3\ H^+ + HSCoA \qquad \textbf{[1.40]}$$

If we return to the catabolism of glucose and follow the pathway all the way through the citric acid cycle, the net reaction becomes

$$C_6H_{12}O_6 + 10\ NAD^+ + 2\ FAD + 2\ ADP + 2\ GDP + 4\ P_i + 6\ H_2O \rightarrow$$
$$6\ CO_2 + 10\ NADH + 2\ FADH_2 + 2\ GTP + 2\ ATP + 10\ H^+$$

$$\textbf{[1.41]}$$

In the overall oxidation of the glucose carbon to carbon dioxide, 10 NADH, 2 $FADH_2$, 2 GTP, and 2 ATP are generated. While substrate-level phosphorylation generates four high-energy phosphate bonds, the bulk of the energy from glucose is held in the reduced electron carriers, NADH and $FADH_2$. In order to obtain maximum energy for cell growth and maintenance, the energy in the electron carriers must be captured in a process called *oxidative phosphorylation*.

Oxidative Phosphorylation   An organism needs a great deal of ATP for its synthesis and maintenance processes, and this ATP can be obtained by converting the large quantities of potential energy stored in NADH and $FADH_2$ to ATP. How do the organisms make use of the electron carriers to fuel growth, reproduction, movement, and maintenance? The process through which this energy transfer takes place is termed *oxidative phosphorylation.* A closely related term is *respiration,* which is the reduction of a terminal electron acceptor. Oxidative phosphorylation is the means by which respiration generates energy for the cell. The potential energy that can be captured depends upon the terminal electron acceptor used. If it is oxygen, then the energy available is quite high, perhaps sufficient to form three moles of ATP per mole of NADH. If the electron acceptor is sulfate, then one mole of ATP at the most might be formed.

We can readily see the potential quantity of energy available through electron transfer from NADH to ATP by examining the free-energy differences between NADH and various possible electron acceptors (Figure 1.21). Comparing those energy differences with that required to produce one mole of ATP,

$$ATP = ADP + P_i \qquad \Delta G^{0'} = -32 \text{ kJ/mol ATP} \qquad \textbf{[1.42]}$$

gives an idea of how much ATP can be generated from an NADH. With actual intracellular concentrations of the reactants, the actual ΔG value is higher, about -50 kJ/mol ATP.

The vertical distance between the NADH and oxygen half reactions in Figure 1.21 shows that the free energy released by NADH oxidation with oxygen is about -110 kJ/e$^-$ eq or 220 kJ/mol. This is sufficient for generation of three to four moles of ATP, if the standard free energies are representative. With nitrate, the quantity of free energy available is somewhat less (about -105 kJ/e$^-$ eq) and barely sufficient to generate three moles of ATP. Tetrachloroethene, a chlorinated solvent, also can serve as an electron acceptor in energy metabolism by some organisms. The free energy available from this reaction is even less, about -84 kJ/e$^-$ eq. For sulfate, the energy available is very low, only about -9 kJ/e$^-$ eq. It suggests that less than one-third mole of ATP could be generated by one mole of NADH.

In reality, microorganisms are able to generate ATP and grow for all of these examples, as well as others that span the energy spectrum. The next question is: What is the mechanism by which NADH free energy is transferred to ATP so that this huge variation in energy potential is accommodated?

The first part of the answer is that the electrons in NADH travel to the terminal electron acceptor via a cascade of membrane-bound proteins and cytochromes. Figure 1.22 gives an example of the longest cascade, the one when oxygen is the acceptor. Each protein or cytochrome is lower on the energy scale than the one from which it receives the electrons. In cases with other terminal electron acceptors or with different organisms, the particular protein complexes and cytochromes involved may be different than shown in Figure 1.22 and perhaps fewer in number. When the terminal acceptor has a more positive $\Delta G^{0'}$ value than does oxygen, the cascade terminates before cytochrome aa$_3$.

The second part of the answer is that the electron transfers from NADH to the terminal acceptor results in the release of protons that are passed to the outside of the

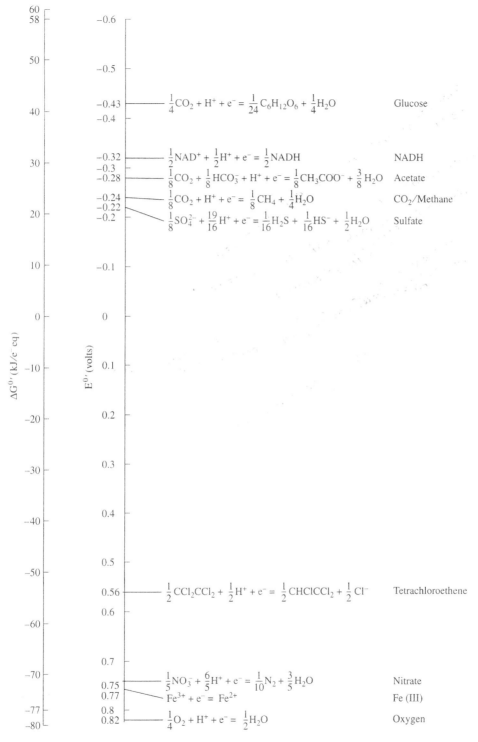

$$\frac{1}{4}CO_2 + H^+ + e^- = \frac{1}{24}C_6H_{12}O_6 + \frac{1}{4}H_2O \qquad \text{Glucose}$$

$$\frac{1}{2}NAD^+ + \frac{1}{2}H^+ + e^- = \frac{1}{2}NADH \qquad \text{NADH}$$

$$\frac{1}{8}CO_2 + \frac{1}{8}HCO_3^- + H^+ + e^- = \frac{1}{8}CH_3COO^- + \frac{3}{8}H_2O \quad \text{Acetate}$$

$$\frac{1}{8}CO_2 + H^+ + e^- = \frac{1}{8}CH_4 + \frac{1}{4}H_2O \qquad \text{CO}_2/\text{Methane}$$

$$\frac{1}{8}SO_4^{2-} + \frac{19}{16}H^+ + e^- = \frac{1}{16}H_2S + \frac{1}{16}HS^- + \frac{1}{2}H_2O \quad \text{Sulfate}$$

$$\frac{1}{2}CCl_2CCl_2 + \frac{1}{2}H^+ + e^- = \frac{1}{2}CHClCCl_2 + \frac{1}{2}Cl^- \qquad \text{Tetrachloroethene}$$

$$\frac{1}{5}NO_3^- + \frac{6}{5}H^+ + e^- = \frac{1}{10}N_2 + \frac{3}{5}H_2O \qquad \text{Nitrate}$$

$$Fe^{3+} + e^- = Fe^{2+} \qquad \text{Fe (III)}$$

$$\frac{1}{4}O_2 + H^+ + e^- = \frac{1}{2}H_2O \qquad \text{Oxygen}$$

Figure 1.21 Comparison between the energy content of the $NAD^+/NADH$ couple with that of various potential electron acceptors couples. Energy content is expressed as standard free energy (kJ/e$^-$ eq) or standard potential (V) at pH = 7.

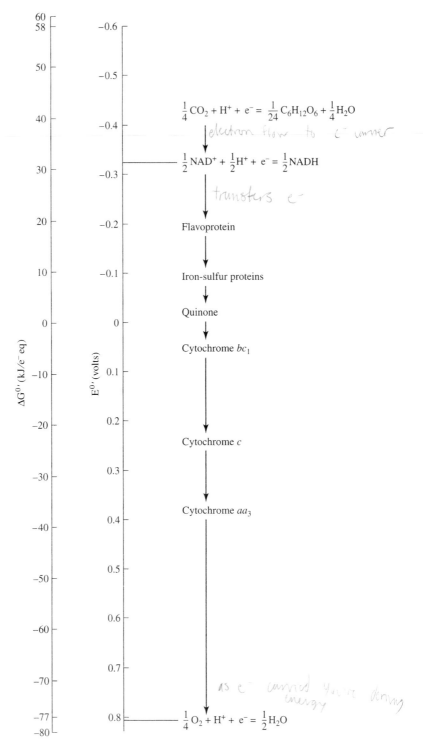

$$\frac{1}{4}CO_2 + H^+ + e^- = \frac{1}{24}C_6H_{12}O_6 + \frac{1}{4}H_2O$$

electron flow to e⁻ carrier

$$\frac{1}{2}NAD^+ + \frac{1}{2}H^+ + e^- = \frac{1}{2}NADH$$

transfers e⁻

Flavoprotein

Iron-sulfur proteins

Quinone

Cytochrome bc_1

Cytochrome c

Cytochrome aa_3

as e⁻ carried you're driving energy

$$\frac{1}{4}O_2 + H^+ + e^- = \frac{1}{2}H_2O$$

Figure 1.22 Energetics of electron transfer from glucose to NADH, and from there through several possible electron carriers to the terminal electron acceptor, oxygen. Energy scales are the same as in Figure 1.21.

cell membrane or hydroxide ions that are concentrated on the inside of the membrane. For example,

$$NADH(in) + H^+(in) + FP(ox) = NAD^+(in) + 2 H^+(out) + FP(red)^{2-} \quad \textbf{[1.43a]}$$

$$2H_2O(in) = 2H^+(out) + 2OH^-(in) \quad \textbf{[1.43b]}$$

in which (in) and (out) indicate on which side of the cell membrane the component is located. For the electron-transport train shown in Figure 1.22, most of the ion separation occurs as the electrons are passed from the flavoprotein to cytochrome bc_1. Pumping the H^+ and OH^- ions to opposite sides of the cell membrane creates a *proton motive force (PMF)* across the cell membrane. The PMF is a gradient in free energy that can be exploited to drive ATP formation from ADP and phosphate.

$$H^+(out) + ADP + P_i = H^+(in) + ATP \quad \textbf{[1.44]}$$

If the terminal electron acceptor lies above oxygen on the energy scale of Figure 1.22, then less energy is available to create the PMF. Then, more NADH must be oxidized to create a PMF large enough to drive Equation 1.44. Readers wishing more detail of oxidative phosphorylation should refer to the advanced texts on microbiology or biochemistry listed at the end of the chapter.

Returning now to our example with glucose, we see from Equation 1.41 that ten moles NADH and two moles $FADH_2$ are produced per mole of glucose oxidized. The electrons contained in these carriers are transferred through oxidative phosphorylation to the terminal electron acceptor. If this is oxygen, current estimates are that the ATP yield from the proton motive force generated is about 2.5 for NADH and 1.5 for $FADH_2$. Using these values, we can write an overall stoichiometry for glucose oxidation coupled with ATP generation:

$$C_6H_{12}O_6 + 6 O_2 + 30 ADP + 2 GDP + 32 P_i \rightarrow$$
$$6 CO_2 + 6 H_2O + 30 ATP + 2 GTP \quad \textbf{[1.45]}$$

We can gain an idea as to how efficient microorganisms are in transferring energy from substrate oxidation or fermentation to the energy carriers ATP and GTP by considering the three cases for glucose fermentation and oxidation indicated in Table 1.12. We see that about 50 percent of the standard free energy released by substrate conversion is transferred over to the energy carriers. We can expect that when the electron carrier energy is transferred back to the production of proteins, carbohydrates, fats, and nucleic acids—the basic components of the cell structure— perhaps the same efficiency of energy transfer can be obtained. Then, the fraction of total energy transferred into cell synthesis would be 50 percent times 50 percent, or about 25 percent.

From such energy calculations, one can approximate how many ATP molecules are formed from substrate conversion simply by free-energy release for transferring electrons from the donor to the acceptor and then assuming that about 50 percent is converted into ATP. In Chapter 2, we advance this approach in a systematic way and link it with the energy costs of synthesis.

Not all organisms are equally efficient in capturing the energy from substrate oxidation. An energy-transfer efficiency of about 60 percent should be considered a maximum value. When a microorganism's enzymes or electron carriers have not

Table 1.12   Efficiency of transfer of energy from glucose conversion to ATP formation

Glucose Reaction	$\Delta G^{0'}$ (kJ/mol Glucose)	Number ATP Formed	Total ATP/ADP Energy* (kJ)	Energy Transfer Efficiency (%)
$C_6H_{12}O_6 + O_2 \rightarrow 6\ CO_2 + 6\ H_2O$	$-2,882$	32**	1,600	56
$C_6H_{12}O_6 \rightarrow 2\ CH_3CH_2OH + 2\ CO_2$	-244	2	100	41
$C_6H_{12}O_6 \rightarrow 3\ CH_3COO^- + 3\ H^+$	-335	3	150	45

*Assuming physiological energy content of ATP is 50 kJ/mol.
**Summation of 30 mol ATP and 2 mol GTP formed from glucose oxidation.

evolved to take advantage of the energy released, the net transfer of energy may be lower. Likewise, lower energy-capture efficiency occurs when an environmental inhibitor uncouples energy capture from electron flow or when the microorganism is forced to divert electron flow to detoxify its environment. For example, a recent finding [Holliger et al., 1993; Scholz-Muramatsu et al., 1995] is that some bacteria can obtain energy for growth through use of either acetate or hydrogen as the electron donor and the chlorinated solvent, tetrachloroethylene (sometimes called perchloroethylene or PCE) as an electron acceptor. The reduction of PCE to cis-1,2-dichloroethlyene follows:

$$2\ H_2 + CCl_2{=}CCl_2 \rightarrow CHCl{=}CHCl + 2\ H^+ + 2\ Cl^-, \Delta G^{0'} = -374\ kJ$$

[1.46]

Biochemical studies on one of the organisms involved indicated that only one mole of ATP is formed per mole of hydrogen oxidized, yielding about 100 kJ from the above reaction. In this case, then, the energy transfer efficiency is only 27 percent. These factors need to be considered in energetic calculations for bacterial growth.

Microorganisms not only use energy for growth, but also for maintenance functions, such as movement and repairing their macromolecules. Interesting questions are: How much energy does it take for a bacterium to move and how is the energy from substrate catabolism converted into cell movement? Many bacteria move through circular motion of their flagella. An organism such as *Salmonella typhimurium,* a food- and waterborne pathogen, has six flagella. Through counterclockwise rotation of the six flagella together at about 100 revolutions per second, the organism is propelled in a forward direction at a speed of about 25 μm/s. This is about 10 body lengths per second! The proton motive force established across the membrane drives the flagella-rotating motors, contained in the cytoplasmic membrane. It takes about 1,000 protons to cause a single rotation of a flagellum. Thus, a flow of about 600,000 protons per second is required to propel a bacterium forward. While this appears to be a very large number, we must remember that a mole of a substance contains 6.02×10^{23} molecules; thus we are not talking about a lot of substrate energy. In reality, the cell spends less than 1 percent of the energy it obtains on motion. The fact that the cell can use the proton motive force directly for motion, rather than using an energy carrier such as

ATP, underscores the efficient design of microorganisms for energy usage. It is efficient because it minimizes the number of electron and energy carriers. The fewer the transfers required, the more efficient is the overall reaction of energy capture and use.

Phototrophic Energy Transfer   Many phototrophic microorganisms obtain their energy from sunlight. Here, we need to consider that the energy in sunlight is carried by quanta of electromagnetic radiation or *photons,* which have a characteristic energy content that is related to the wavelength of the light involved:

$$E_{photon} = \frac{hc}{\lambda} \qquad \text{[1.47]}$$

where h is Planck's constant and equals 6.63×10^{-34} J-sec, c is the velocity of light or 3×10^{10} cm/s, and λ is the wavelength of the radiation (cm). The energy per "mole" or einstein ($N = 6.023 \times 10^{23}$) of photons thus is

$$E = \frac{hc}{\lambda} N = \frac{12}{\lambda} \frac{\text{J-cm}}{\text{einstein}} \qquad \text{[1.48]}$$

With the wavelength of visible light lying between the violet range of 400 nm (4×10^{-5} cm) and the far red range of 700 nm, the energy content lies between 170 and 300 kJ/einstein.

Next, we consider how light is converted into chemical energy that the cell can use. Chlorophylls are the main light-trapping pigments of cells. Generally, chlorophyll is green, because chlorophyll normally traps energy in the red and purple ends of the light spectrum, but not in the middle or green part of the spectrum. The poorly trapped green light is what then reflects off the photosynthetic pigment. However, not all photosynthetic organisms are green. Phototrophs that have other classes of light-trapping pigments or have accessory pigments that absorb light in other parts of the spectrum can have colors that vary from brown through red.

Although the light-trapping reactions differ somewhat between oxygenic and anoxygenic photosynthesis, both lead to a common outcome: The phototroph uses the trapped energy from the light to generate ATP, NADPH, and an oxidized product.

In oxygenic photosynthesis, two separate light reactions interact, as indicated in Figure 1.23. In photosystem I, a light photon with wavelength less than 700 nm is absorbed, and the energy so released generates a strong reductant that leads to the formation of NADPH. In photosystem II, more energetic light of a shorter wavelength (< 680 nm) produces a strong oxidant that results in the formation of oxygen from water. In addition, photosystem II produces a weak reductant, while photosystem I generates a weak oxidant. Electrons flow from photosystem II to photosystem I, and the electron flows within each photosystem generate a transmembrane proton gradient (or PMF) that drives the formation of ATP. The details of the many intermediate reactions involved are described in general textbooks on microbiology (Madigan et al., 1997) and biochemistry (Stryer, 1995) .

The net overall reactions that occur in oxygenic photosynthesis can be summarized as

$$\text{H}_2\text{O} + \text{NADP}^+ + \text{ADP} + \text{P}_i = \text{NADPH} + \text{H}^+ + \text{ATP} + 0.5\,\text{O}_2 \qquad \text{[1.49]}$$

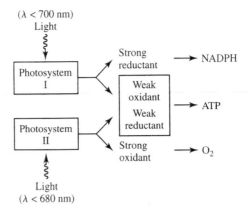

Figure 1.23   Interaction of photosystems I and II in oxygenic photosynthesis. SOURCE: After Stryer, 1995.

Note that the two electrons used to reduce $NADP^+$ to $NADPH + H^+$ came from the O in H_2O, which forms the oxidized product $0.5\ O_2$. This reaction requires two light quanta, or 400 to 500 kJ per reaction as written, depending upon the energy of the absorbed radiation. We can again evaluate the efficiency of energy transfer involved if we consider that Equation 1.49 is essentially the reversed summation of Equations 1.22 (219 kJ) and 1.23 (32 kJ), giving a combined free-energy content of the NADPH and ATP formed of 251 kJ. On this basis, the efficiency of energy transfer from light to chemical energy is in the range of 50 to 62 percent, or similar to that with other catabolic reactions of the cell.

Anoxic photosynthesis, such as by purple and green bacteria, occurs under anaerobic conditions and generally makes use only of photosystem I. Certain algae and cyanobacteria also are able to grow under anaerobic conditions using photosystem I alone. In these cases, an alternative reductant to water, such as H_2 or H_2S, is utilized. For example, two electrons can be extracted from H_2S, just as it is in the above reaction with H_2O, resulting in the formation of NADPH and leaving S rather than $0.5\ O_2$ as an oxidized end product. These interesting variations in chemistry again emphasize the versatility that microorganisms have in capturing energy from a great variety of potential chemical and photochemical sources.

1.10.2   ANABOLISM

Anabolism is the suite of metabolic processes leading to cell synthesis. In a simple sense, anabolism is the reverse of catabolism. Simple chemical precursors (like acetate) are converted into a set of more complex building blocks (like glucose), which are then assembled into the macromolecules that comprise proteins, carbohydrates, lipids, nucleic acids, and any other cell component. However, the enzymes involved and details of pathways for the two general metabolic processes differ somewhat, because a system that is especially good at disassembling and oxidizing complex molecules is not necessarily good at reducing and assembling them.

Two basic options for anabolism are heterotrophy and autotrophy. In *heterotrophy*, an organic compound, generally one with two or more carbon atoms, is used as the main source of cell carbon. In *autotrophy*, inorganic carbon is used as the sole basic carbon source, although small amounts of organic compounds such as vitamins may also be required. Some organisms can grow either autotrophically or heterotrophically, a situation called *mixotrophy*.

Chemoorganotrophs normally are heterotrophs; indeed, the two terms are often used interchangeably. Chemolithotrophs commonly are autotrophs. Phototrophs most commonly are autotrophs as well (*photoautotrophy*), although some may be heterotrophic (*photoheterotrophy*).

The energy required to synthesize cellular components from organic carbon is very much less than that for synthesis from inorganic carbon. Therefore, heterotrophs have an advantage over autotrophs, as long as organic carbon is present. On the other hand, when inorganic carbon is the only carbon available, autotrophs can dominate. Again, the wide-ranging ability of microorganisms to obtain energy and carbon for growth under considerably different environmental conditions is emphasized.

Since cellular synthesis can be either heterotrophic or autotrophic, we consider these as two separate processes. In both cases, energy is required. The difference can be illustrated by considering how much energy is required for the conversion of a simple organic compound, such as acetate, into glucose, one of the major building blocks for carbohydrates. This can be compared with the energy required for conversion of carbon dioxide into the same building block:

$$3 \, CH_3COO^- + 3 \, H^+ \rightarrow C_6H_{12}O_6 \quad \Delta G^{0'} = 335 \text{ kJ} \qquad \textbf{[1.50]}$$

$$6 \, CO_2 + 6 \, H_2O \rightarrow C_6H_{12}O_6 + 6 \, O_2 \quad \Delta G^{0'} = 2,880 \text{ kJ} \qquad \textbf{[1.51]}$$

Over eight times more energy is required to synthesize a 6-carbon sugar from carbon dioxide than from acetate. The advantage of heterotrophy is obvious.

The energy required for cellular growth directly comes from the ATP synthesized during catabolism. However, at times and especially with autotrophic growth, reducing power in the form of NADH or NADPH is used directly to reduce inorganic carbon into organic carbon. The consumption of NADH or NADPH during synthesis is an energy cost, because those reduced carriers cannot be sent to the terminal electron acceptor to generate ATP.

The pathways and processes involved in converting carbon dioxide into glucose by phototrophs was demonstrated by Melvin Calvin and his colleagues beginning in 1945 in what has become known as the *Calvin cycle* (Figure 1.24). Carbon dioxide is brought into the cycle where it is joined to ribulose 1,5-bisphosphate to form 3-phosphoglycerate. Through the addition of energy in the form of ATP and reducing power in the form of NADPH, the 6-carbon fructose 6-phosphate is formed and can be used for biosynthesis. In the cycle, glyceraldehyde 3-phosphate is formed as well, and it in turn is converted with the addition of more ATP energy to produce the ribulose

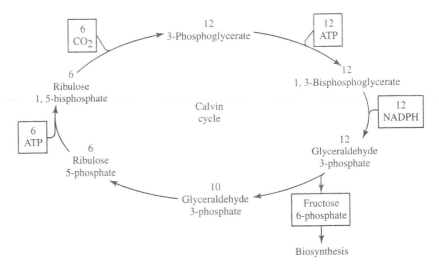

Figure 1.24   The Calvin cycle. Numbers above the compounds in the cycle indicate the number of moles of each compound involved in cycling of six moles CO_2.

1,5-bisphosphate that is needed to again react with carbon dioxide, completing the cycle. The net result of the Calvin cycle is

$$6\,CO_2 + 18\,ATP + 12\,NADPH + 12\,H^+ \rightarrow \\ C_6H_{12}O_6 + 18\,ADP^+ + 12\,NADP^+ + 18\,P_i + 6\,H_2O \quad \text{[1.52]}$$

We can readily evaluate the energy input needed to form the six-carbon sugar from carbon dioxide if we consider that Equation 1.52 is equivalent to combining Equation 1.51 with Equations 1.22 and 1.24 for NADPH and ATP changes, respectively:

$$18\,ATP + 18\,H_2O \rightarrow 18\,ADP + 18\,P_i \quad \Delta G^{0'} = -576\,kJ \quad \text{[1.53]}$$

$$12\,NADPH + 6\,O_2 + 12\,H^+ \rightarrow 12\,NADP^+ + 12\,H_2O \quad \Delta G^{0'} = -2{,}628\,kJ$$

$$\text{[1.54]}$$

Here, the total energy released from ATP and NADPH is 3,204 kJ. The energy content of the hexose formed (Equation 1.51) of 2,880 kJ is 90 percent of this value, which suggests a very high efficiency of energy transfer. Most of the energy for hexose formation comes from NADPH, rather than from ATP. If the conversion of light energy in oxygenic photosynthesis to ATP and NADPH is 50 to 60 percent efficient, while that from ATP and NADPH energy to hexose formation is about 90 percent efficient, then the overall conversion of light energy to hexose is about 50 percent efficient.

While the Calvin cycle is used by oxygenic autotrophs, carbon-dioxide fixation does not occur by this cycle for several groups of autotrophs, such as the phototrophic green sulfur bacteria, *Chloroflexus,* homoacetic bacteria, some sulfate-reducing bacteria, and methanogenic Archaea. These organisms in general produce acetyl CoA from carbon dioxide, and the acetyl CoA so formed is used as the starting material for synthesis of cellular materials by these organisms. The acetyl CoA synthesis pathway, which is found only in certain obligate anaerobes, involves the reduction of one molecule of CO_2 to form the methyl group of acetate and another molecule to form the carbonyl group. These reductions require H_2 as an electron donor. A key enzyme in this pathway is carbon monoxide (CO) dehydrogenase, a complex enzyme that contains the metals Ni, Zn, and Fe as cofactors. Thus, sufficient quantities of these metals are required for organism growth. The reaction results in the formation of CO from CO_2, and the CO becomes the carbonyl carbon of acetyl CoA. A coenzyme, tetrahydrofolate, participates in the reduction of CO_2 to form the methyl group of acetyl CoA. Acetyl CoA then becomes the building block for synthesis of all the cellular components of the organisms.

Synthesis of cellular material by heterotrophs is simpler than for autotrophs, as the carbon for synthesis is already in a reduced form. The catabolism of most organic compounds results in the formation of acetyl CoA, which can then be used as the general building block for cell synthesis, as is the case for certain autotrophs as discussed above. In an alternative case, microorganisms termed *methylotrophs* use organic compounds that consist of a single-carbon compound, such as methane, methanol, formate, methylamine, and even the reduced inorganic compound, carbon monoxide. No pathway converts these compounds directly to acetyl CoA. Instead, most methylotrophs use what is termed the *serine pathway* to combine formaldehyde, which is formed from the single-carbon organic compound, with carbon dioxide to form acetyl CoA. This is true of the Type II methanotrophs, which are among those that use methane for energy and growth. However, another group, the Type I methanotrophs, use the ribulose monophosphate cycle for assimilation of methane. These organisms lack the citric acid cycle into which acetyl CoA feeds for cell synthesis, and so an alternative strategy is required. Instead of acetyl CoA, the Type I methanotrophs generate glyceraldehyde 3-phosphate, which is used for synthesis of cellular components. The position of this compound in the overall metabolic scheme of microorganisms is noted in Figure 1.19.

Once the cells have the basic materials, such as acetyl CoA, they can use them to create all the major types of molecules needed. For example, two acetyl CoAs can be joined (via a multistep process) to form the 4-carbon oxaloacetate, which leads to glucose-6-phosphate though a reversal of the pathway of glucose fermentation, or glycolysis. The glucose-6-phosphate is a precursor to the 6-carbon glucose used for carbohydrate synthesis and the 5-carbon ribulose –5-phosphate, which is used to construct ribose and deoxyribose nucleotides needed for DNA and RNA.

This brief description of anabolic pathways points out that the synthesis strategies used by all organisms are quite similar. All must convert a suite of common intermediates (the simple building blocks) into the many different proteins, fats, carbohydrates, and nucleic acids of which organisms consist. Noteworthy differences occur when an organism must use inorganic carbon or 1-carbon organic molecules, versus organic carbon, as the carbon source to create the common intermediates. Once the starting carbon sources are converted into common intermediates or building blocks, however, the subsequent biosynthetic processes in all organisms are similar.

Table 1.13   Trophic classification of major microbial types according to their electron donor, electron acceptor, carbon source, and domain

Microbial Group	Electron Donor	Electron Acceptor	Carbon Source	Domain*
Aerobic Heterotrophs	Organic	O_2	Organic	B & E
Nitrifiers	NH_4^+	O_2	CO_2	B
	NO_2^-	O_2	CO_2	B
Denitrifiers	Organic	NO_3^-, NO_2^-	Organic	B
	H_2	NO_3^-, NO_2^-	CO_2	B
	S	NO_3^-, NO_2^-	CO_2	B
Methanogens	Acetate	Acetate	Acetate	A
	H_2	CO_2	CO_2	A
Sulfate Reducers	Acetate	SO_4^{2-}	Acetate	B
	H_2	SO_4^{2-}	CO_2	B
Sulfide Oxidizers	H_2S	O_2	CO_2	B
Fermenters	Organic	Organic	Organic	B & E
Dehalorespirers	H_2	PCE	CO_2	B
	Organic	PCE	Organic	B
Phototrophs	H_2O	CO_2	CO_2	E & B
	H_2S	CO_2	CO_2	B

*The domains are **B**acteria, **A**rchaea, and **E**ukarya.

1.10.3   METABOLISM AND TROPHIC GROUPS

The microorganisms of importance in environmental biotechnology almost always are significant because of the electron donors and acceptors that they utilize. In many cases, one of these energy-generating substrates is a pollutant from the point of view of humans. Table 1.13 lists key types of microorganisms according to their electron donor and acceptor substrates, as well as their carbon source and domain. Thus, Table 1.13 organizes the microorganisms according to their *trophic groups*. Trophic refers to how an organism "feeds" itself. Microorganisms need to consume the donor, acceptor, and C source, and they are the most fundamental determinants of a trophic group.

Table 1.13 provides a handy reference for the reader who wants to keep a perspective on where the microorganisms that are discussed in subsequent chapters fit into the large scheme of microbial metabolism. Not every microbial type is included, and new types are constantly being discovered.

1.11   GENETICS AND INFORMATION FLOW

Like all living organisms, microorganisms employ a complex network of *information flow* to control who they are and what they do. Information flow is synonymous with *genetics*, because the repository of information is in the *gene*. However, the gene is only the starting point for genetics and information flow. Several other kinds of molecules are essential.

The information that tells the cells what macromolecules to assemble, what energy-generating and synthesis reactions to use, and how to interact with their environment is coded in a large macromolecule of *deoxyribonucleic acid, or DNA.* The gene coded on the DNA does not actually do any of the cell's work, such as energy generation or synthesis. Instead, elaborate, multistep machinery is used to decode the information on the DNA and eventually lead to production of the working molecules, the enzymes needed to catalyze all the key reactions. Between the genetic code on the DNA and the working enzymes are three different types of macromolecules comprised of *ribonucleic acids, or RNA:* messenger RNA, ribosomal RNA, and transfer RNA.

Figure 1.25 summarizes how information in a gene is converted to an enzyme. The gene is a segment of DNA, or a sequence of deoxyribonucleotides. The sequence of deoxyribonucleotides is faithfully copied to a complementary (i.e., matching) sequence of ribonucleotides, which constitutes the *messenger RNA, or mRNA.* By copying the DNA code onto mRNA, the cells can use the information on the DNA without destroying or distorting it.

The mRNA molecule then interacts with the *ribosome,* a large multicomponent molecule that contains another form of RNA: *ribosomal RNA, or rRNA.* Within the ribosome, the mRNA directs the assembly of the protein by indicating the order in which the amino acids are to be joined. The code on the mRNA identifies which type of *transfer RNA, or tRNA,* should enter the ribosome and supply a particular amino acid to the growing protein chain. Different tRNA molecules bond specifically to the range of amino acids needed to make enzymes and match specifically with the code on the mRNA. The mechanism of bringing tRNA and mRNA together in a ribosome

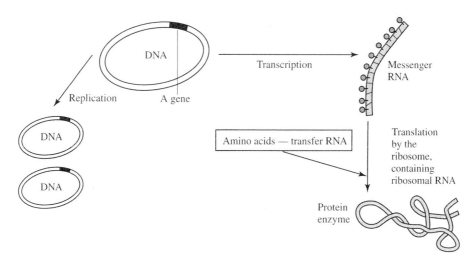

Figure 1.25   Summary of information flow from the gene (in DNA) to the working enzyme catalyst.

is a versatile and efficient way for the cells to be able to synthesize any and all types of enzymes as they need them.

The left side of Figure 1.25 shows that the basic information is replicated when the cell needs another copy, such as when the cell divides to form two cells. What is important here is that replication creates an exact copy of the DNA, not a complementary copy of RNA.

The next sections describe the nucleic acids and how they carry out the steps shown in Figure 1.25. Prokaryotes and eukaryotes have several important differences in the details of how they manage information flow, but the basic phenomena are the same and as outlined in Figure 1.25. The next section emphasizes the common features and approaches used by prokaryotes.

1.12 DEOXYRIBONUCLEIC ACID (DNA)

Genes are coded by sequences of purine and pyrimidine nucleotides that are linked into a larger polymer through phosphodiester bonds. The basic nucleotide has three parts: a 5-carbon carbohydrate, or deoxyribose; an N-containing base; and a phosphate group. Figure 1.26 shows the structure of the basic deoxyribose units and the four N-containing bases that can be used to form DNA. Of particular importance in the figure are locating the 1, 3, and 5 positions on the deoxyribose and distinguishing the two *purine bases* (adenine and guanine) from the two *pyrimidine bases* (cytosine and thymine). The purines are N-containing heterocycles with two rings and 4 N atoms in the rings, while the pyrimidines are single-ring heterocycles with 2 N atoms in the ring. Both groups are called bases, because the N atoms can attract the H^+ ion from H_2O and release strong base (OH^-).

Figure 1.27 illustrates how joining the heterocycle and a phosphate group through the ribose creates the deoxyribonucleotide. The example is for adenosine, and the pattern holds for guanosine, cytosine, and thymosine. The pattern is that the heterocycle and the ribose join at the ribose's 1 carbon and the bottom-left heterocyclic N (Figure 1.26), while the phosphate group joins at the ribose's 5 carbon. The linkage between the heterocyclic base and the ribose's 1 carbon is a glycosidic bond, in which the OH from the C and the H from the N are released to form H_2O. The phosphate and the 5 carbon form an ester bond, which releases OH from the C and H from the phosphates to create H_2O. If a phosphate group is absent, the molecule is called a *ribonucleoside*.

Two further reactions can occur to the basic ribonucleotide illustrated in Figure 1.27, and both involve the phosphate group. The first reaction is addition of one or two more phosphate groups to form adenosine disphosphate deoxynucleotide and adenosine triphosphate deoxynucleotide. The multiple phosphate groups are critical for energy conservation and transport in all cells. ATP and ADP, the nondeoxy analogs, are the key energy transfer units in cells. The second reaction is polymerization of the ribonucleotides to form a chain, or a polynucleotide. Figure 1.28 illustrates how forming phosphodiester bonds between the deoxyribose's 3 and 5 carbons creates a polynucleotide. The figure shows four bases that form the sequence ACTG. The 5 or 3 carbon notes the ends of a sequence: $5'$ end or $3'$ end.

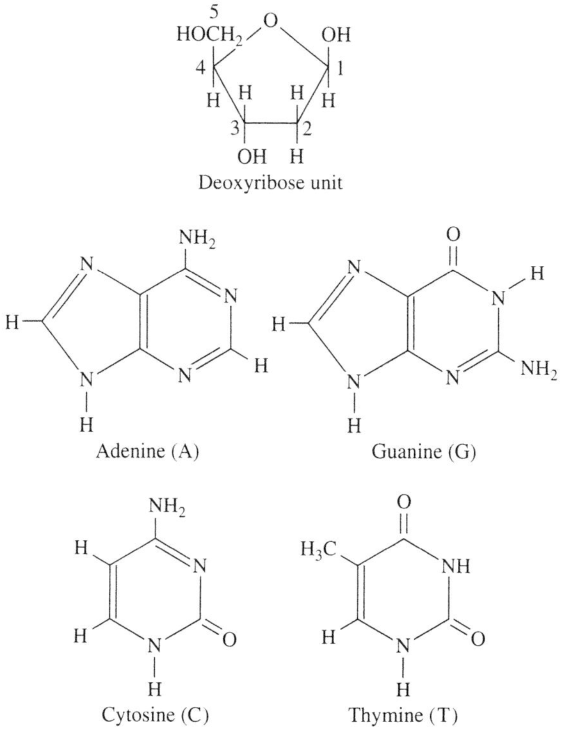

5
HOCH₂   O   OH
4 | | 1
  H   H   H   H
3 | | 2
  OH   H
Deoxyribose unit

Adenine (A)

Guanine (G)

Cytosine (C)

Thymine (T)

Figure 1.26   The structures of the basic ribose unit (top), purine bases (middle), and pyrimidine bases (bottom) that constitute DNA. The five carbons are numbered for the ribose unit.

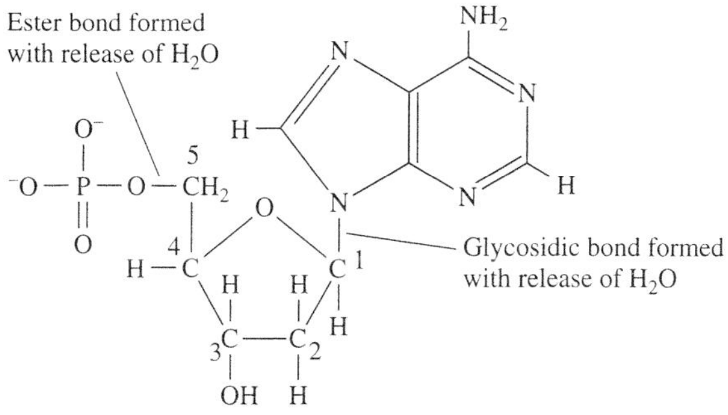

Ester bond formed with release of H_2O

Glycosidic bond formed with release of H_2O

Figure 1.27   The adenosine monophosphate deoxynucleotide is created by joining adenine and a phosphate group to the ribose unit. Note the glycosidic and ester bonds to the ribose's 1 and 5 carbons, respectively.

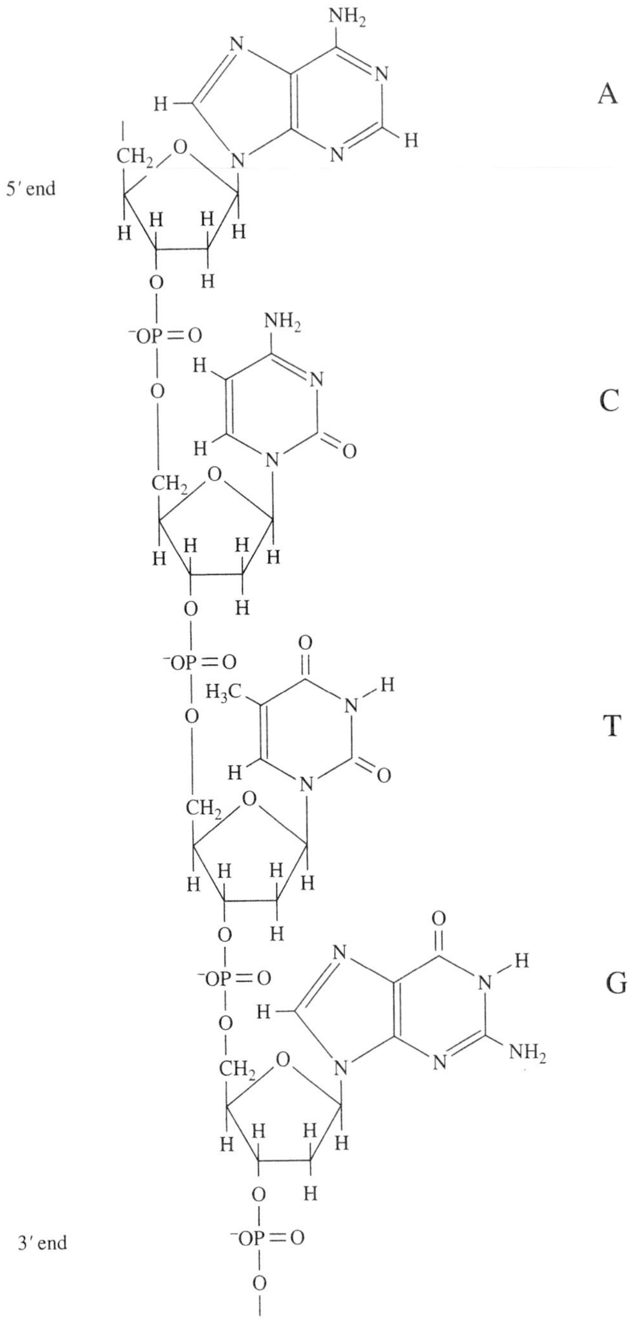

Figure 1.28   Creation of a DNA polynucleotide through a series of phosphodiester bonds linking the 3 and 5 carbons of the deoxyribose units. The bases give the sequence ACTG.

A short sequence, like the one in Figure 1.28, is called an *oligonucleotide.* On the other hand, the genes used to encode information are much longer nucleotides, hundreds or thousands of bases long. The size of a nucleotide used in information flow is measured in kilobases, or 1,000s of nucleotide bases. Since each nucleotide has a molecular weight around 400 daltons, the molecular weight of a gene sequence can be a million daltons or more.

Because it is tedious to write out the full chemical formulas for all the nucleotide monomers, we normally express the sequence in a shorthand notation that uses the letter A, T, G, and C for adenine, thymine, guanine, and cytosine bases, respectively. For example, an oligonucleotide can have the sequence $5'$-ATCGGATTCGCGTAC-$3'$. The sequence is directional, and it is essential to distinguish the $5'$ end from the $3'$ end.

One of the most important features of DNA is *complementary bonding.* Due to their chemical structures, the bases C and G form three hydrogen bonds, while A and T form two hydrogen bonds. The bonding patterns are shown in Figure 1.29. When two strands of nucleotides have complementary sequences that run in the opposite directions, they form a series of hydrogen bonds that hold the strands together very strongly. Figure 1.30 illustrates the hydrogen bonding between complementary strands. Because C and G share three hydrogen bonds, while A and T share two bonds, C-G complementarity forms a stronger or more stable double-strand structure. For example, a double-stranded DNA that has 60 percent C+G remains double stranded at a higher temperature than does DNA with 40 percent C+G.

The genetic information encoded by genes almost always is stored in a double-stranded form. This structure has important implications. First, having two complementary strands provides redundancy, which makes it possible for the cells to replace or correct damaged DNA. Second, the strong DNA-DNA bonding through complementarity means that the cells require special apparatuses to disrupt temporarily the hydrogen bonding when the DNA must be replicated or transcribed.

1.12.1 THE CHROMOSOME

The genetic information that defines a microorganism is encoded on the chromosome. In all cases, the chromosome is double-stranded DNA that contains information to tell the cell how to carry out its most essential functions: e.g., energy generation; synthesis of the cell wall, membrane, and ribosomes; and replication. When the cell reproduces itself, the chromosome must be replicated so that each daughter cell contains a chromosome. The replication of the chromosome so that it can be passed along to a daughter cell is called *vertical transfer* of DNA, because vertical refers to a new generation of cells.

The prokaryotic chromosome is circular, double-stranded DNA that is located in a nuclear area, but is not contained inside its own membrane. A typical chromosome for a Bacteria species has 5×10^6 base pairs (or 10^7 nucleotide bases). Species from the Archaea typically have 2×10^6 base pairs.

The eukaryotic chromosome is much more complicated than the bacterial chromosome. First, it is enclosed inside a phospholipid membrane that forms the nucleus.

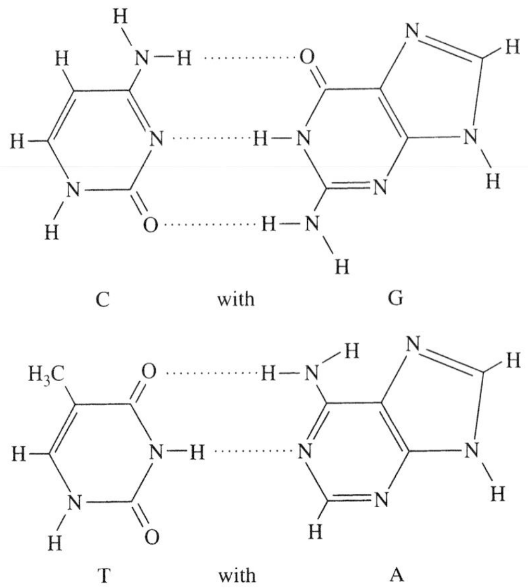

Figure 1.29   The hydrogen-bonding pattern of C with G and T with A is the basis for complementary bonding.

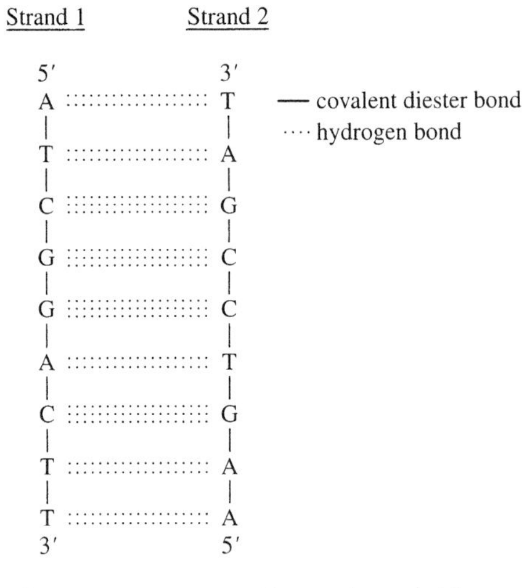

Figure 1.30   DNA strands are held together very strongly by hydrogen bonds when the bases are complementary in the opposing directions.

The nucleus communicates with the rest of the cell through transport pores in the membrane and directs the cell's reproduction. Second, the amount of chromosomal DNA is much larger, up to 10,000 times more than in a prokaryote. Third, each Eukaryote has two to hundreds of chromosomes. Fourth, the DNA in the chromosome is associated closely with proteins called histones. Finally, eukaryotic DNA is famous for having major regions that do not code for genes; the purpose of these noncoding regions, called *introns,* is an area of current study.

1.12.2 PLASMIDS

Prokaryotes can contain DNA that is not part of the chromosome. The most important example is the *plasmid,* which usually is a circular, double strand of DNA that is smaller than and separate from the chromosome. A typical plasmid contains 10^5 base pairs, although sizes vary widely. Plasmids often contain genes for functions that are not essential for the microorganism under its normal environmental conditions. Instead, genes present on a plasmid code for functions that are needed to overcome stressful events: e.g., resistance to antibiotics, degradation of unusual and often inhibitory organic molecules, and reducing the availability of heavy metals.

Prokaryotes can have one or several copies of a plasmid. They also can gain or lose a plasmid without affecting their essential genetic characteristics and functions.

In the environmental field, the replication and transfer of plasmids is the most important type of *horizontal transfer* of DNA. Horizontal transfer means that an existing cell gains the DNA; so, cell reproduction is not involved. *Conjugation* requires direct cell-to-cell contact between a plasmid-containing donor cell and a recipient cell that does not contain the plasmid. In an energy-requiring process, the plasmid DNA is replicated and transferred to the recipient cell in such a way that both cells contain the plasmid after the conjugation event is complete. Through conjugation, the plasmid is replicated and proliferated independently from cell fission. Thus, conjugation is a means to circulate genetic information among different prokaryotic cells without growing new cells.

1.12.3 DNA REPLICATION

In order to achieve faithful transmission of the genetic information to a subsequent generation of daughter cells, the DNA strands must be replicated exactly. The five critical steps of DNA replication are:

1. The DNA double strand is separated in a region called the origin of replication. A special enzyme and binding proteins are responsible for opening the strands and keeping them open. The two separated strands going from the double strand core resemble a tuning fork, and the site where replication occurs is called the *replication fork.*

2. A *DNA polymerase* enzyme binds to one strand in the fork and moves from base to base along both strands in the $3'$ to $5'$ direction.

3. A complementary strand of DNA is generated by the polymerase, which links the deoxyribonucleoside triphosphate complementary to the base at which the polymerase is stationed to the previous base on the new, growing chain. The polymerase obtains the nucleotide triphosphate from the cytoplasm and links its 3 phosphate to the 5' OH of the preceding nucleotide. A diphosphate is released in the process to fuel the formation of the phosphodiester bond. The trailing end of the new strand is the 3' end and has an OH terminus. The phosphate for the next linkage originates from the 5' triphosphate of the incoming base.

4. Both strands are replicated simultaneously, and the replication fork moves to expose more of the DNA to the polymerase. One strand, called the *leading strand,* can be replicated continuously in the 3' to 5' direction. The other, the *lagging strand,* is replicated in a piecewise manner, and the pieces are linked together by DNA ligase. In the end, replication produces two double strands, each one with the original and one new strand.

5. Since the DNA must be replicated very accurately, the cells have elaborate mechanisms for proofreading and correcting errors. The polymerase has an *exonuclease* activity that detects errors, excises the incorrect base, and replaces it with the correct one. The result is that the error rate normally is very small, from 10^{-8} to 10^{-11} errors per base pair.

1.13   RIBONUCLEIC ACID (RNA)

Ribonucleic acids are used in three ways to convert the genetic code on the DNA to working proteins. The code is transcribed to form a strand of messenger RNA, or mRNA. The mRNA is translated by the ribosome, of which the ribosomal RNA (rRNA) is a key component. The amino acids are transported to the ribosome by RNA carrier molecules, called transfer RNA (tRNA).

The basic structure of the RNA is the same in each case and is very similar to that of DNA: A ribose is linked to a purine or pyrimidine base at the ribose's 1 carbon, while phosphate is linked at the 5 carbon. However, RNA differs from DNA in two ways. First, the ribose has an -OH group at its 2 carbon, not an -H group as with the deoxyribose for DNA. Second, uracil replaces thymine as one of the two pyrimidine bases. Figure 1.31 shows the ribose unit and the uracil base. Uracil forms a double hydrogen bond with adenine.

RNA nucleotides form polymers with the same structure as for DNA (Figure 1.28). The only differences are that the ribose is not deoxy, and U occurs in place of T. This similar structure is what makes it possible for the code in DNA to be copied to RNA during transcription.

1.13.1   TRANSCRIPTION

The information code in DNA is packaged as genes, or segments along the DNA strand. During *transcription,* the code in one strand of the DNA is used as a template to form a complementary single strand of RNA. Five major steps take place during transcription.

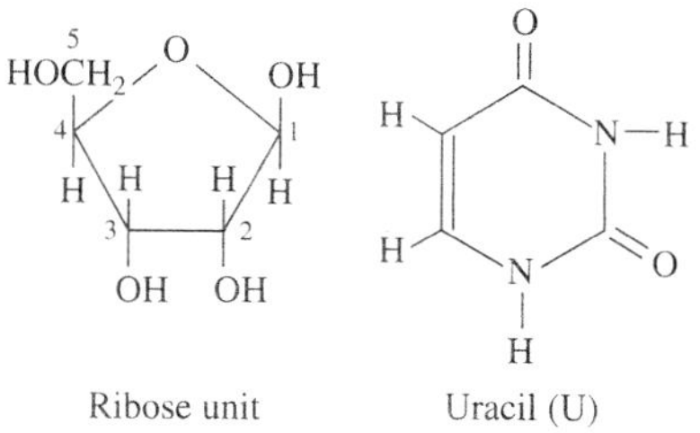

Ribose unit   Uracil (U)

Figure 1.31   The components of RNA. The ribose sugar on the left has an -OH group on its 2 carbon. The uracil base on the right replaces the thymine base found in DNA.

1. An enzyme called *RNA polymerase* binds to one strand of the DNA at the gene's *promoter region,* which extends about 35 bases ahead of where transcription begins. The sequences from about 5 to 10 and about 30 to 35 bases before the start of transcription are similar in all promoter regions and are used by the RNA polymerase to recognize where the promoter is located. A very important feature of the polymerase binding at the promoter is that the promoter region must be free. Normally, the promoter region is blocked by proteins whose job it is to prevent binding of the polymerase and transcription. This blocking occurs because cells cannot afford to *express* all genes all the time. Thus, they *repress* the transcription of genes whose proteins are not needed by blocking the promoter region. The promoter sequence between 10 and 50 bases before initiation determines what repressor protein binds. When a protein is needed, the cells have ways to remove the repression protein, which allows the RNA polymerase to bind and begin transcription.

2. The DNA double strand separates, and the RNA polymerase moves from base to base along one strand in its 3′ to 5′ direction.

3. At each base, the polymerase links the complementary ribonucleoside triphosphate to the growing RNA chain. The complementarity is

DNA	RNA
A	U
T	A
C	G
G	C

The entering triphosphate ribonucleoside releases a diphosphate to fuel the reaction. As with DNA replication, the new strand grows in the 5′ to 3′ direction. In other words, the 5′ phosphate of the new nucleotide is added to the 3′ -OH group of the old chain, thereby extending the 3′ end of the RNA.

4. The DNA double strand closes together behind the point at which transcription occurs.

5. When the RNA polymerase reaches the end of the gene being transcribed (or a series of related genes), the RNA breaks away, and the RNA polymerase releases from the DNA. The RNA contains a termination sequence that signals that transcription is to terminate.

Unlike replicated DNA, transcribed RNA is single-stranded. This is necessary because the different types of RNA must interact directly with each other.

1.13.2   MESSENGER RNA (mRNA)

When the transcribed RNA is used to generate a protein, it is *messenger RNA,* or *mRNA.* The mRNA is transported to a ribosome, where its code is translated into an amino acid polymer. The translation is directed by the rRNA of the ribosome and involves a code that is shared with the tRNA.

1.13.3   TRANSFER RNA (tRNA)

Transfer RNA is an important example of a hybrid molecule. In this case, it is a hybrid that contains RNA and amino acid parts. Figure 1.32 illustrates the two parts. The RNA part contains 73 to 93 nucleotides that form three loops and an acceptor stem formed by hydrogen bonding between complementary ribonucleotides (C-G, A-U). The *anticodon* contains three nucleotides that create a code. The acceptor binds to one amino acid, which is specific to the code of the anticodon.

The three-base anticodon complements a three-base sequence in the mRNA. In this way, the mRNA directs the synthesis of proteins by telling the order of amino acids. Table 1.14 lists the amino acids that correspond to the mRNA codons. For example, UCU on mRNA, which matches AGA on the tRNA anticodon, means serine. Likewise, GCG on the mRNA signifies the alanine amino acid. Most of the amino acids have more than one mRNA code, a tract called *degeneracy.* In these cases, the third base can vary or *wobble.*

Four three-base codes have special functions. AUG indicates exactly where reading of the three-base codes should begin. Because the bases on the mRNA are read as triplets, and almost all triplets code an amino acid, it is essential that the starting point be exact. AUG also codes for methionine within the RNA code, but N-formylmethionine is coded when AUG is used as the start signal. UAA, UAG, and UGA are the stop signals.

The start and stop codons are very useful for identifying genes within DNA sequences. The DNA complements to the start and stop codons are: TAC (start) and ATT, ATC, and ACT (stop). Significant stretches of DNA between these sequences are likely to code for expressible DNA, or genes. Such genes are called *open reading frames.*

The tRNAs are shuttles for amino acids. Much like the formation of RNA and DNA polymers, linking the amino acids to the tRNA is an energy-consuming process that releases a diphosphate from ATP to activate the amino acid.

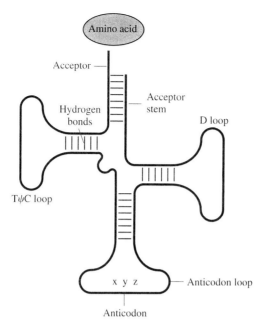

Figure 1.32   Cloverleaf representation of the tRNA molecule shows the three loops, the anticodon, the acceptor, and an amino acid attached to the acceptor.

1.13.4   TRANSLATION AND THE RIBOSOMAL RNA (rRNA)

The matching of the mRNA codon and the tRNA anticodon occurs within the ribosome, which is a complex assembly of rRNA and proteins. The three-base code of the mRNA is *translated* into an amino-acid polymer, or a protein. Starting with the initial AUC sequence, the ribosome *reads* the mRNA code three bases at a time. It finds the matching tRNA and binds the codon and anticodon through hydrogen bonding. The ribosome keeps two sites side-by-side. This allows it to covalently link the incoming amino acid to the last amino acid of the growing chain via a peptide bond that releases H_2O. Then, the ribosome shifts its position three bases further to the 3' end of the mRNA and repeats the process. The ribosome keeps elongating the protein until it reaches one of the stop codons.

A cell has thousands of ribosomes, and some microorganisms increase the number when they are growing rapidly. The ribosome is a large structure, on the order of 150 to 250 Å. They are approximately 60 percent RNA and 40 percent protein. The rRNA is made up of two major subunits that are identified by their size as measured by sedimentation rate: for prokaryotes, they are 50S and 30S. The 30S subunit is further subdivided into 5S, 16S, and 23S components that are approximately 120, 1,500, and 2,900 nucleotides long, respectively. The 16S subunit in Figure 1.33 gives

Table 1.14   Three-base codon sequences for amino acids

Amino Acid	mRNA Codons
Phenylalanine	UUU, UUC
Leucine	UUA, UUG, CUU, CUC, CUA, CUG
Isoleucine	AUU, AUC, AUA
Methionine	AUG
Valine	GUU, GUC, GUA, GUG
Serine	UCU, UCC, UCA, UCG, AGU, AGC
Proline	CCU, CCC, CCA, CCG
Threonine	ACU, ACC, ACA, ACG
Alanine	GCU, GCC, GCA, GCG
Tyrosine	UAU, UAC
Histidine	CAU, CAC
Glutamine	CAA, CAG
Asparagine	AAU, AAC
Lysine	AAA, AAG
Aspartic Acid	GAU, GAC
Glutamic Acid	GAA, GAG
Cysteine	UGU, UGC
Tryptophan	UGG
Arginine	CGU, CGC, CGA, CGG, AGA, AGG
Glycine	GGU, GGC, GGA, GGG
Stop Signal	UAA, UAG, UGA
Start Signal	AUG

the sequence and secondary structure of the 16S rRNA from *E. coli.* Eukaryotes have a similar rRNA component with size 18S. The 16S and 18S rRNAs are called the *small system subunit,* or *SSU.* All of the SSU rRNAs share a very similar secondary structure, although the base sequences (called the primary structure) differ. The differences in primary structure are very useful for finding the genetic *relatedness* of different organisms, a subject covered in a later section on Phylogeny.

Although the ribosome contains protein, as well as rRNA, the catalytic agent appears to be the RNA. This is an example of a *ribozyme.* When translation of an mRNA is to begin, the 50S and 30S subunits come together to form a functioning ribosome.

1.13.5 TRANSLATION

Translation involves all three forms of RNA and occurs through five steps that occur in three sites (or positions) within the ribosome. The three sites are adjacent to each other and are called A, P, and E. A stands for arrival site, P stands for polymerization site, and E stands for exit site.

First, the mRNA positions itself at the P-site of the 30S rRNA unit such that the starting codon (AUG) is in the P site. The corresponding tRNA (anticodon UAC) arrives at the P-site with a formylmethionine attached and hybridizes with the AUG

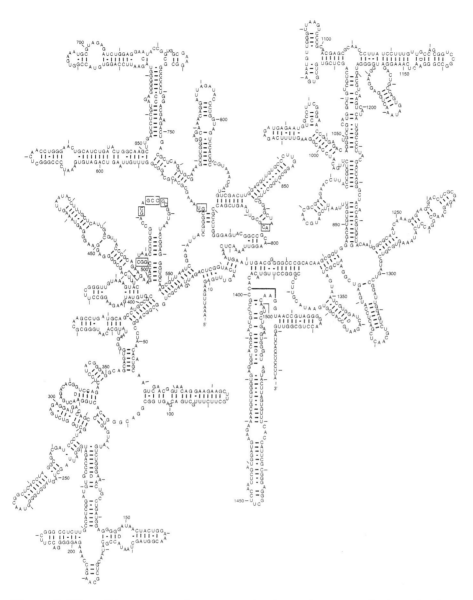

Figure 1.33 Primary structure (base sequence) and secondary structure (hydrogen bonding and folding) of the 16S rRNA from *E. coli*. The bases are numbered at 50-base intervals.

codon. Second, a tRNA complementary to the next (toward the 3′ end) three-base sequence arrives at the adjacent A-site of the rRNA and hybridizes. Third, the amino acids form a peptide bond, and the formylmethionine releases from the tRNA. Fourth, the first and second tRNA, and their RNA, move to the E- and P-sites of the rRNA.

Fifth, the first tRNA releases from the E-site, while the tRNA from the next codon on the mRNA arrives and hybridizes. Then, steps three, four, and five are repeated. This repetition continues until one of the Stop codons is detected in the mRNA. Proteins known as release factors separate the polypeptide chain from the last tRNA. The ribosome subunits then dissociate, letting the mRNA go free and enabling the subunits to recombine when another translation is to commence.

1.13.6   REGULATION

A cell contains many more genes than it expresses at any one time. The means by which a cell controls what genes are expressed is called *regulation*. The most common approach to regulation is through the binding of regulatory proteins to the promoter region of a gene. The bound protein prevents the RNA polymers from transcribing the gene into mRNA. Such a binding protein is called a *repressor protein,* and transcription is *repressed.* When the cell wants to transcribe the gene, it alters the chemical structure of the repressor protein in such a way that it no longer binds the gene.promoter. The molecule that changes the structure of the repression protein is called an *inducer;* the inducer is said to *induce* or to *derepress* enzyme transcription. Inducers often are substrates for the gene product or chemically similar (analogs) to the substrate.

In some cases, regulation of transcription involves *activator proteins* that allow the RNA polymerase to bind to the promoter. In this case, the inducer molecule alters the structure of the activator protein so that it enables the RNA polymerase to bind.

Some genes are expressed all the time. They are called *constitutive* genes and generally produce proteins very central to the cell's basic metabolism. The expression of constitutive genes in not regulated, but the activities of these gene products often are modulated through inhibition mechanisms discussed earlier in the chapter.

1.14   PHYLOGENY

Phylogeny is the name given to a systematic ordering of the species into larger group-ings (higher taxa) based on inheritable genetic traits. Phylogeny is rooted in the concept of genetic evolution, which states that incremental changes in the sequences of chromosomal DNA lead gradually to new species. Fundamental to our modern understanding of phylogeny is that all species have one evolutionary ancestor that forms the root of a phylogenetic tree. All the species branch off from the root.

Figure 1.34 is a very simple form of the universal phylogenetic tree as we now understand it. The basis for the tree is presented later. The tree shows that all known life on planet earth can be placed into three domains: Eukarya, Archaea, and Bacteria. Two of the domains—Archaea and Bacteria—comprise the prokaryotes. The lines to these three domains form the *trunks* of the phylogenetic tree. As shown for the Eukarya in Figure 1.35, branches spring off from the trunks as we seek greater detail in the tree. Figure 1.35 shows detail to the level of kingdom. Each of the kingdom branches then split into smaller branches that reflect finer detail among related species.

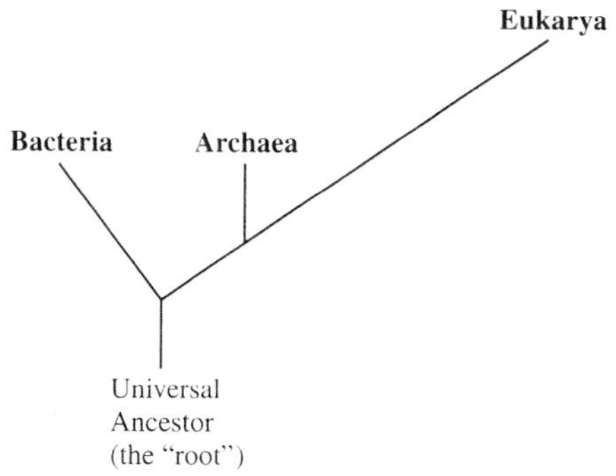

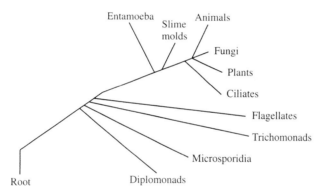

Figure 1.34   A very simple version of the universal phylogenetic tree shows the three domains.

Figure 1.35   A more detailed tree for the Eukarya illustrates branching of kingdoms from the trunk. The tree shown indicates approximate relatedness and is not quantitative in terms of evolutionary distance.

The lineal distance from one species to another along the tree quantifies their genetic distance. For instance, Figure 1.35 shows that plants and fungi are quite closely related, while plants and animals are more distant. However, the protozoan diplomonads are much more different from plants than animals are from plants.

The distance from the root indicates the evolutionary position of taxa. Presumably, taxa closer to the root evolved earlier than those farther away. Evolutionary time refers to the gradual changes in the sequences in the chromosomal DNA. Evolutionary time does not necessarily correspond exactly to calendar time. The rates of change along the different branches are not the same.

Figure 1.36 shows the kingdom breakdown for the domain Bacteria. It also shows how one kingdom, the Proteobacteria, branches off into five groups that are closely

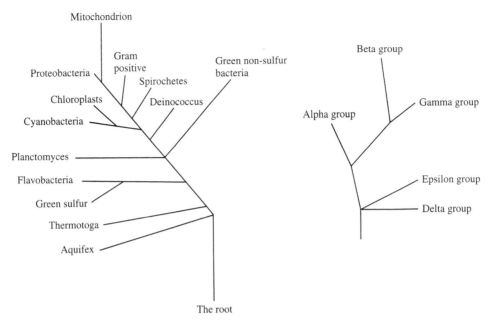

Figure 1.36   Kingdoms of the Bacteria (left) and branching of the Proteobacteria into five closely related groups (right). (The trees shown indicate approximate relatedness and are not quantitative in terms of evolutionary distance.)

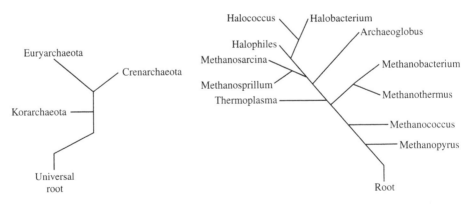

Figure 1.37   The domain Archaea is broken into Euryarchaeota and Crenarchaeota. The former is shown in detail to the genus level on the right side. (The trees shown indicate approximate relatedness and are not quantitative in terms of evolutionary distance.)

related genetically, although they are quite different in terms of their metabolism. For example, the beta group contains phototrophs and chemotrophs that oxidize ammonium, ferrous iron, reduced sulfur, and a wide range of organic compounds.

Figure 1.37 presents similar information for the domain Archaea, which is subdivided into Euryarchaeota and Crenarchaeota kingdoms. The Euryarchaeota are shown in more detail in the bottom panel. The incredible diversity of the Euryarchaeota is illustrated by the genus names, which represent a wide range of methanogens, halophiles, and thermophiles. As for the Bacteria domain, genetic similarity does not necessarily mean that the organisms are similar in terms of metabolism or other phenotype expressions.

1.14.1   THE BASICS OF PHYLOGENETIC CLASSIFICATION

A phylogenetic classification depends on identifying a molecule that is present in every type of cell and that is directly related to the cell's genetic heritage. At first glance, the chromosome seems like the most likely candidate, since it is the ultimate repository for the cell's genetic makeup. However, the chromosome does not work well because it is too large. On the one hand, this largeness creates a practical problem if we want to have its entire sequence. Although advances in molecular biology are making sequencing easier, determining the entire sequence of even a simple bacterium still requires a massive research effort. Largeness also creates a second problem that will exist no matter how fast DNA sequencing becomes in the future. The second problem is that different parts of the chromosome change at very different rates, and, probably, with very different patterns. Thus, comparing sequences for the entire chromosome likely will give a spectrum of answers concerning how organisms are phylogenetically related.

By far, the most widely accepted chronometer for phylogeny is the ribosomal RNA. All cells must have rRNA to perform translation, and the structure of the rRNA is remarkably consistent for all types of cells. Therefore, the rRNA has proven an excellent molecule for doing phylogeny.

In most cases, the sequence of the SSU (i.e., 16S or 18S) rRNA serves as the basis for phylogenetic comparisons. The SSU rRNA works well because its 1,500 base pairs are neither too few nor too many. The 1,500 bases provide enough genetic diversity to distinguish one species from another, but they are not so many that sequencing is overly burdensome.

Figure 1.38, which shows the secondary structure of the 16S rRNA, identifies which bases are the same for all organisms and which bases vary from organism to organism. The regions that do not change are called *conserved regions* and are very useful for aligning RNA sequences for comparisons. The *variable regions* are useful for differentiating among different organisms

The sequences for the 16S rRNA are compared using computer-based statistical methods that generate a tree that optimally arranges the sequences into the trunk-branch pattern shown earlier. Depending on the number of sequences evaluated and details of how the algorithm works, phylogenetic trees can differ in how they place organisms. Trees can be represented graphically in the fan format shown in Figures 1.35 to 1.37. Other presentation forms are more useful when the number of strains increases, and they are utilized later in this chapter.

The sequences of the SSU rRNA can be obtained in two ways. The more traditional way is to extract the rRNA from the cells to be evaluated. An extraction with phenol separates the RNA from the protein. The ribosomal RNA is then separated by alcohol precipitation. A DNA primer specific to the 16S rRNA and reverse

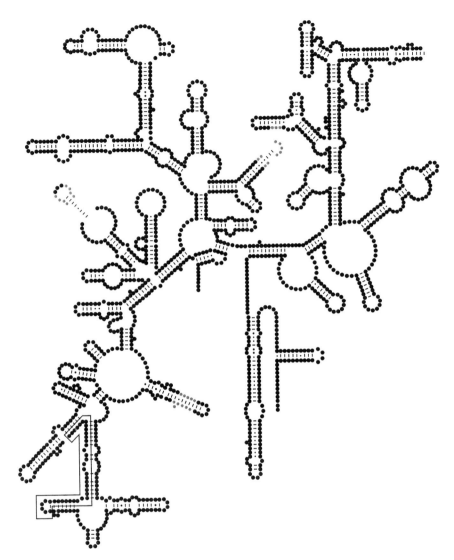

Figure 1.38   Secondary structure of the SSU(16S) rRNA shows the degree to which bases in certain regions are conserved. The very dark circles indicate completely conserved. The very faint dots indicate regions of very high variability. SOURCE: Stahl, 1986.

transcriptase are then used to make DNA fragments complementary to the 16S rRNA. The DNA fragments are sequenced, and the rRNA sequence is deduced from the DNA sequences. In the second method, a technique called the *polymerase chain reaction,* or PCR, is used to replicate the gene for DNA, which is then sequenced. The polymerase chain reaction is described later in this chapter in the section on Tools to Study Microbial Ecology.

Once the sequence of the 16S rRNA is known, it can be added to the database of sequences. The conserved regions are used to align the new sequence with existing sequences. Then, the variable regions are evaluated to compute the new organism's evolutionary distance from other organisms and its place in the tree.

Further computer analysis of the variable regions can identify stretches that are unique to a given organism or a set of similar organisms. These unique sequences are very useful for designing oligonucleotide probes that can be used to identify and quantify different microorganisms present in complex microbial communities, the subject of the remainder of this chapter.

1.15   MICROBIAL ECOLOGY

Except in highly specialized research or industrial settings, microorganisms exist in communities that include a great amount of genetic and phenotypic diversity. The interactions among the different microbial types and their environment are what we call *microbial ecology*. To understand the microbial ecology of a community, we must answer these basic questions:

1.   What microorganisms are present?

2.   What metabolic reactions could the microorganisms carry out?

3.   What reactions are they carrying out?

4.   How are the different microorganisms interacting with each other and the environment?

The first question refers to *community structure*, which has at least three facets. The first is an enumeration of the number of distinct microbial types present. Just as important is the second facet, the abundance of each type. The third facet of structure is the spatial relationships among the different populations. Because communities present in environmental settings usually are aggregated, the spatial relationships can be quite stable and of great importance.

The second and third questions describe different aspects of *community function*, or what the organisms individually or collectively do. Question 2 relates to potential activity and the range of behaviors possible. Question 3 indicates what the behaviors actually occur under the circumstances.

Question 4 requires an integration of the information represented by the first questions. The physical relationships among different microorganisms and materials that they produce, consume, and (especially) exchange define their *interactions*, the essence of microbial ecology.

Design and operation of engineered processes in environmental biotechnology are practical ways in which microbial ecology is manipulated so that a microbial community achieves a human goal. As applied microbial ecologists, engineers create systems (reactors) in which the right kinds of microorganisms are present (community structure), they are accumulated to quantities sufficient to complete a desired biochemical task (community function), and they work together to perform their tasks stably over time (integrated community ecology). To achieve these goals, engineers devise means to supply the proper nutrition to the desired microorganisms, retain those microorganisms in the system, and contact the community of microorganisms with the stream to be treated.

The next three sections introduce the key ecological concepts that underlie how engineers control a system's microbial ecology. Those concepts are selection, exchange of materials, and adaptation. Then, several interesting examples of how these concepts have been applied successfully are outlined. Finally, the newly emerging tools for studying microbial ecology are described.

1.15.1 SELECTION

Within a microbial community, all the individual microorganisms are not the same in terms of the biochemical reactions they carry out and other phenotypic traits. *Selection* is a process in which those individuals who are most fit to survive in their environment generate the greatest number of descendants. Over time, the selected microorganisms establish a stable foothold in the community, continuing to carry out their biochemical reactions and maintain their heritable genetic information.

From the point of view of the microorganism, its primary goal is to maintain its genetic heritage within a community. The way it accomplishes the primary goal is by finding or creating a *niche,* which is a multidimensional space in which energy supply, nutrients, pH, temperature, and other conditions allow it to sustain itself and its genetic heritage. Often, a number of different microorganisms compete for common resources, such as an electron acceptor or a nutrient. The selected microorganisms are the ones able to capture the resources available to them. Microorganisms that lose the competition for resources are *selected against* and may be excluded from the community.

Engineers who design and operate microbiological processes to achieve environmental goals manipulate the environment so that the most desired microorganisms are selected. In most cases in environmental biotechnology, the treatment goal is to remove a pollutant. Fortunately, the majority of pollutants are electron-donor or electron-acceptor substrates for some prokaryotes. For example, the biochemical oxygen demand (BOD) is a measure of organic electron donors. To remove BOD from a waste, heterotrophic bacteria are selected by supplying them with an appropriate electron acceptor (e.g., O_2) at a rate commensurate with the rate at which the BOD must be removed. As long as nutrients are present and the environment is hospitable in terms of pH, temperature, and salinity, heterotrophs will gain energy from the oxidation of BOD and *self-select.* Thus, selection of microorganisms able to utilize an electron-donor or -acceptor substrate is straightforward, since the utilization reaction provides the electron and energy flows that fuel the growth of the desired microorganism.

Heterotrophic bacteria offer an excellent example of the *selection hierarchy* according to electron acceptor. For many types of organic electron-donor substrates, different prokaryotes are able to couple donor oxidation to the reduction of a wide range of electron acceptors. These different microorganisms are competing for a common resource, the electron donor. One microorganism's relative competitiveness to establish its niche, which involves oxidizing the common donor, depends on the energy yield it obtains from the electron transfer from donor to acceptor. Chapter 2 provides a methodology to quantify the relative energy yields. Figure 1.15 summarizes the relative energy yields for the common electron acceptors. In a direct competition for an organic electron donor, cells carrying out aerobic respiration gain more energy and have a growth-rate advantage. For example, competitiveness declines in the order $O_2 > NO_3^- > SO_4^{2-} > CO_2$. Microorganisms using different electron acceptors can coexist as long as the donor resource is not fully consumed by electron transfer to the more favorable acceptor.

Selection is not based solely on the availability of an electron acceptor (or donor). Selection in treatment reactors also is strongly determined by an organism's ability to be retained. In environmental biotechnology, the microbial communities are almost always aggregated in flocs or biofilms. These aggregates are much more easily retained in the process than are individual, or dispersed cells. Hence, the ability to form or join into *microbial aggregates* is critical for selection in some environmental systems. In some cases, the ability of a microbial strain to create a niche by locating in its most favorable location in an aggregate is key to its selection.

Other factors that can favor selection include resistance to predation, the ability to sequester and store valuable resources, and the ability to exchange materials with other microorganisms. The value of resisting predation is obvious. Sequestering valuable resources, such as a common electron donor, is particularly useful when environmental conditions fluctuate widely and regularly. Then, organisms able to sequester and store the resource during periods of *feast* can thrive during the *famine* periods by utilizing their storage materials. Engineers have become quite adept at taking advantage of resource-storing bacteria. The last advantage—exchange of materials—is the subject of the upcoming section.

Although research is just beginning to shed light on the relationships between community structure and function, it appears that communities often contain a significant amount of *functional redundancy*. In other words, the community structure contains several different strains that can carry out the same or very similar biochemical functions. Over time, the relative abundance of the different strains may vary, but the overall community function hardly changes. For treatment applications, functional redundancy appears to be an advantage in that system performance is stable, even when community structure shifts.

A concept complementary to redundancy is *versatility*. Certain groups of prokaryotes are famous for being metabolically diverse. Many heterotrophs, particularly in the genus *Pseudomonas,* are able to oxidize a very wide range of organic molecules. This ability allows them to scavenge many and diverse sources of electrons and carbon, and it greatly reduces their vulnerability to starvation. The sulfate-reducing bacteria also are famous for their metabolic versatility. When sulfate is present, they reduce it and fully oxidize H_2 or a range of organic electron donors. When SO_4^{2-} is absent, they can shift to fermentation, in which H_2, acetate, and other fermentation products are released. Thus, sulfate reducers are ubiquitously present in microbial communities, even when SO_4^{2-} is absent.

1.15.2 EXCHANGE OF MATERIALS

One of the most important strategies for survival and selection in microbial communities is exchange of materials among different types of microorganisms. Three broad exchange strategies are available: exchange of substrates, exchange of genetic material, and exchange of communication signals.

Exchange of Substrates   Microbial communities have food chains that share some characteristics of food chains well known for macroorganism ecology. Figure 1.39 sketches the structure of a classical macroorganism ecosystem. At the top of the food chain are primary producers, such as plants and algae, that capture sunlight energy to create biomass via phototrophy. The biomass of the primary producers is then consumed by primary consumers, which directly eat the primary producers. A cascade of secondary, tertiary, etc. consumers is possible. The consumers are heterotrophs. The consumed cells are the electron-donor substrate. In general, the mass of the population decreases with each level of consumption, although the physical size of the individual organisms often increases.

Some aspects of microbial ecosystems resemble what is depicted in Figure 1.39. A perfect example is *predation* by protozoa or simple worms on bacteria, cyanobacteria, or algae. In this case, the protozoa consume the smaller cells and use them directly as food. With predation, the entire cell is the exchange material. This form of predation, termed *grazing,* is of great importance in aerobic treatment processes and is discussed in detail in Chapter 7.

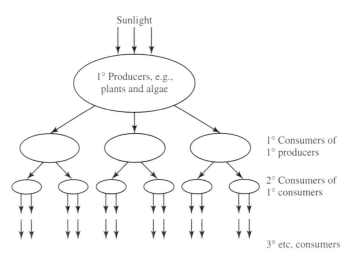

Figure 1.39   Schematic of a simple macroorganism ecosystem in which primary producers (e.g., plants) are consumed directly by 1° consumers, which are then consumed by 2° consumers, and so on. Note that the mass of the population decreases with each level of consumption.

On the other hand, exchange in most microbial ecosystems is not by consumption of an entire cell. Instead, one cell releases a molecule into the environment, and another cell takes it up in such a way that one or both cells benefit. Thus, the main connections in a microbial ecosystem involve the exchange of molecules, not the consumption of one cell by another.

Three characteristic patterns for microbial exchange of substrate molecules can be identified: reduction of CO_2 to organic (reduced) carbon by autotrophs, release of partially oxidized organic intermediates, and cycling of inorganic elements. Each pattern is described here, and chapters that contain more detailed information are noted.

Photoautotrophs (e.g., algae and cyanobacteria) and chemoautotrophs (e.g., bacteria that oxidize ammonium, sulfide, and hydrogen gas) reduce inorganic carbon (oxidation state +4) to approximately oxidation state zero. Their main purpose is to synthesize their own new biomass. At the same time and as a normal part of their metabolism, the autotrophs release part of the organic carbon as soluble molecules that are electron-donor substrates for a range of heterotrophic bacteria. In special cases, such as when photoautotrophs are exposed to strong light, but severe limitation of nutrients N or P, the autotrophs release a large flow of organic molecules to the environment. This appears to be a mechanism for dumping electrons (phototrophically extracted from H_2O or H_2S) when they cannot be invested in biomass synthesis, due to a nutrient limitation. Under more normal conditions, autotrophs release small, but steady flows of cellular macromolecules to the environment. These normal releases are called *soluble microbial products* and are discussed in detail in Chapters 3, 4, 6, and 7. In reality, all microorganisms, including heterotrophs, release soluble microbial products that can be a substrate for some heterotrophic bacteria. The impact is most dramatic when the autotrophs generate them, because this expands the pool of reduced carbon available to heterotrophs.

The formation and exchange of organic intermediates is a second widespread phenomenon in microbial ecosystems. Any system in which fermentation occurs provides a classic example. Figure 1.40 illustrates the complex flow for a simple anaerobic ecosystem in which carbohydrate ($C_6H_{12}O_6$) is fermented through a range of organic intermediates to acetate and H_2, which are then utilized by methanogens. This exchange web is exploited in methanogenic processes in anaerobic treatment, the topic of Chapter 13. In Figure 1.40, acidogenic fermenting bacteria produce propionic and acetic acids, hydrogen gas, and inorganic carbon in such a way that the 24 electron equivalents and carbon equivalents are conserved. Different fermenting bacteria convert the propionic acid to more acetic acid, hydrogen gas, and inorganic carbon. The acetic acid is fermented to methane and inorganic carbon by specialized acetoclastic methanogens, while the H_2 gas is oxidized to generate more methane by specialized hydrogen-oxidizing methanogens. The net result is that all of the 24 electron equivalents in $C_6H_{12}O_6$ not used for synthesis end up in CH_4. However, four distinct groups of microorganisms were able to extract energy and create a niche through the stepwise formation and consumption of metabolic intermediates.

The microorganisms that utilize the intermediate products depend on the metabolic action of those that generate the intermediate, and their ecological dependency is

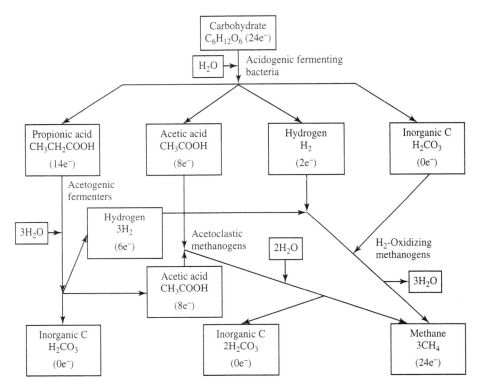

Figure 1.40   Flow of intermediate molecules in an anaerobic ecosystem that starts with carbohydrate, forms intermediate organic acids and H_2, and ultimately generates CH_4. The net reaction is $C_6H_{12}O_6 + 3\ H_2O \rightarrow 3\ CH_4 + 3\ H_2CO_3$, but four unique microbial groups are involved.

obvious. For the situation in Figure 1.40 and many similar ones in anaerobic ecosystems, the microorganisms performing the early steps also are ecologically dependent on the action of the microorganisms that consume the intermediate products. As is described in more detail in Chapters 2 and 13, the early fermentation steps have low energy yields and are thermodynamically possible only when their consumers maintain their products (particularly H_2) at a very low concentration. This situation is an excellent example of *synergistic interactions,* in which all microorganisms benefit. The type of synergy illustrated by Figure 1.40 is an example of *syntrophy,* in which the cooperation is through exchange of electron donor substrates. Many anaerobic ecosystems are *obligately syntrophic,* which means that the individual organisms cannot be selected independently of their syntrophic partners. Figure 1.41 shows the intimate physical relationship among different anaerobic microorganisms in a mixed methanogenic culture.

Another way in which reduced intermediates are generated occurs when a biochemical step part way through a mineralization pathway is slowed or blocked. Then, partially reacted intermediates build up inside the cell and are excreted to the environ-

Figure 1.41   Scanning electron photomicrograph of a mixed methanogenic culture growing on wheat straw; large rods are acetate–using methanogens. Photo is courtesy of Xinggang Tong.

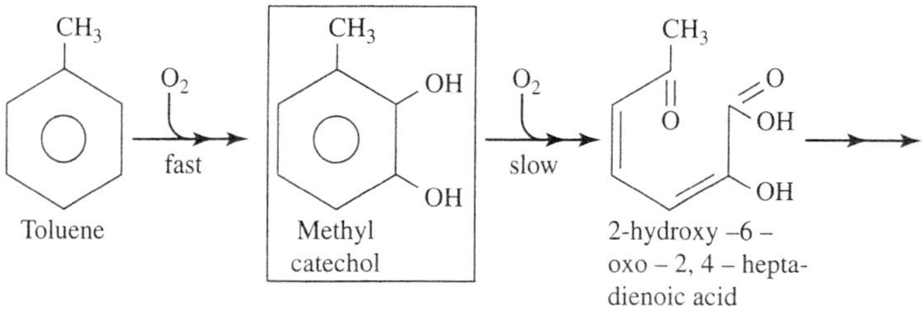

Figure 1.42   Example of a catabolic pathway in which an intermediate accumulates. Build up of methyl catechol from toluene occurs when its dioxygenation is blocked, such as by O_2 depletion.

ment. Nutrient or cosubstrate limitation is a prime cause for the accumulation of partly oxidized intermediates. Figure 1.42 shows an example in which an intermediate is known to accumulate. In this case, methyl catechol accumulates when its dioxygenation is blocked, such as by a low dissolved oxygen concentration or intrinsically slow kinetics. Release of methyl catechol to solution provides an electron-donor substrate to other heterotrophs that are not limited by a low dissolved-oxygen concentration or that catabolize the methyl catechol by another pathway.

Many microbial communities thrive through *recycling elements* that can be an electron donor for one group of microorganisms and an electron acceptor for another group. Sulfur and iron are the classic examples. For sulfur, a large number of heterotrophic and autotrophic bacteria use sulfate as a terminal electron acceptor, reducing it to sulfides or elemental sulfur. Aerobic, autotrophic bacteria are able to use these reduced sulfur forms as electron donors. For iron, heterotrophic bacteria can use the ferric ($+3$) form as an electron acceptor; typically, the ferric iron is present as an iron hydroxide solid (e.g., $Fe(OH)_3(s)$) and must be dissolved as part of the process. Iron-oxidizing bacteria can use ferrous ($+2$) iron as an inorganic electron donor to carry out aerobic autotrophy. The pattern here is that the organisms using the reduced forms as an electron-donor substrate are autotrophs. The organisms that respire the oxidized forms can be heterotrophs or autotrophs that oxidize H_2.

For cycling to work as an exchange process, the oxidizing and reducing conditions must alternate in time or space. The diurnal cycle creates a natural time variation in which strongly oxidized conditions occur when sunlight drives phototrophic O_2 production during the day, while reducing conditions occur at night. Microbial aggregates set up strong spatial gradients. When the elements can diffuse or otherwise be transported between reducing and oxidizing regions, the elements can shuttle back and forth between oxidizing and reducing zones.

Table 1.15 summarizes the ways in which microbial communities exploit substrate exchange. It also summarizes exchange of genetic information, growth factors, and signals, the next topics.

Exchange of Genetic Information   Microorganisms can exchange genetic information in three ways: conjugation, transformation, and transduction. By far the most important mechanism is conjugation, which was described in detail is the section on Genetics. In brief, *conjugation* involves the replication and transfer of plasmid DNA from a donor cell to a recipient cell. The result of conjugation is that both cells contain the plasmid: The donor cell retains the plasmid, while the recipient cell gains the plasmid and becomes a transconjugant. Transfer of genetic information by conjugation amplifies and proliferates genes throughout a community, even when the community has no net growth. Plasmids contain genes that code for resistance to antibiotics and other microbial toxicants, including the detoxification of many

Table 1.15   Summary of exchange mechanisms used by microbial communities

Exchange Type	Exchanged Material	Its Role for the Receiving Cell
Predation	The cell itself	Electron-donor substrate
Intermediates	Reduced products of fermentation or partial oxidation	Electron-donor substrate
Element Cycling	Reduced or Oxidized elements	Electron-acceptor or Electron-donor substrate
Conjugation	Plasmid DNA	Source of genes
Transformation	Free DNA	Source of genes
Transduction	Virus-borne DNA	Source of genes
Growth Factors	Vitamins, amino acids	Allows replication
Signaling	Fatty acids, pheromones, proteins	Turns on a physiological function

hazardous compounds and elements. More information on detoxification is given in Chapter 14.

Transformation occurs when free DNA is incorporated into the chromosome of a recipient cell. Cells strictly regulate their uptake of DNA, and the *competence* to take-up free DNA and integrate it into the chromosome is not the normal situation. The DNA not transformed is rapidly degraded in the environment or by nucleases inside the nontransforming cell.

Transduction moves DNA from cell to cell by incorporating it into the DNA of a bacteriophage. The virus first infects the donor bacterial cell and incorporates a piece of the bacterial chromosome into its DNA. The virus DNA is then replicated inside the infected bacterium. When the bacterium lyses, the released bacteriophages contain the donor DNA. Infection of a different bacterial cell can result in recombination of the donor's bacterial DNA into the recipient's chromosome.

Growth Factors   Some prokaryotes require amino acids, fatty acids, or vitamins for replication. Examples include vitamin B_{12}, thiamine, biotin, riboflavin, and folic acid. Normally, these *growth factors* are released to the environment by other microorganisms in the community.

Exchange of Chemical Signals   In some special cases, microorganisms receive chemical signals. These signal molecules bind to receptors on the cell membrane and trigger a physiological response. A good example is the sex pheromone excreted by *Streptococci.* The pheromone, a small polypeptide, signals a nearby microorganism that does not harbor a plasmid to engage in conjugation. A related, but different example is *Aerobacterium,* which senses products from plants that it infects. These signal molecules accelerate the conjugative proliferation of plasmids, which code for the ability to infect the plant. Other important signaling molecules are the homoserine lactones, which are implicated in *quorum sensing,* in which cells alter their phenotype when they are closely aggregated, such as in biofilms. The exchange of chemical signals is an exciting new research topic.

1.15.3   ADAPTATION

Because of their complex nature, microbial ecosystems are able to respond in a very dynamic manner to changes in their environment, particularly changes that stress the community. Examples of stresses include changes in temperature, pH, or salinity; exposure to a toxic material; exposure to xenobiotic organic molecules; and large changes in the availability of substrates.

The term used to describe the community's response to a stress is *adaptation.* Sometimes, the term *acclimation* is used as a synonym. Adaptation is any response that ultimately leads the community to eliminate the stress or find a way to maintain its function despite the stress. The *adaptation period* is the time interval between the initial exposure to the stress and when the community has adapted. A very common and important example is adaptation to a recalcitrant, xenobiotic chemical. Little or no biotransformation occurs during adaptation periods ranging from a few hours to several months. At the end of the adaptation period, the community is transforming the xenobiotic rapidly, and it normally continues rapid transformation for subsequent exposures.

Communities adapt by using one or more of five main mechanisms: selective enrichment, enzyme regulation, exchange of genetic information, inheritable genetic change, and alteration of their environment.

In *selective enrichment,* microbial types uniquely capable of benefiting from the stress selectively grow and become a larger proportion of the total biomass. In the case of exposure to a recalcitrant compound, microorganisms able to metabolize the compound and gain some benefit from it are enriched. For exposure to a toxicant, microorganisms that have resistance measures gain a selective advantage.

A very important and interesting selective-enrichment response is to different substrate-loading patterns. The two extremes are a perfectly steady loading and *feast* and *famine* loading. With a very steady loading to a process in environmental biotechnology, the rate-limiting substrate (usually the election donor) comes to a low, steady-state value. (This is developed quantitatively in Chapters 3, 4, and 5.) The microorganisms have a constant and low specific growth rate. These conditions favor microorganisms that are called *oligotrophs* or *K-strategists.* Oligotrophs normally do not have the fastest maximum specific growth rate, but they have high affinities for the rate-limiting substrate, and they usually have very small specific loss rates. The high substrate affinity means that they are able to scavenge substrates present at very low concentrations, or *oligotrophic.* The term K-strategist reflects that a high substrate affinity can be represented by a very small value of K, the half-maximum-rate concentration (details are in Chapter 3), which is analogous to the K_M of Equation 1.12.

The opposite of oligotrophy is called *copiotrophy.* Copiotrophs are well suited to the feast and famine lifestyle, in which substrate is loaded at an extremely high rate for a period, but then is loaded at a zero or very low rate for other times. Copiotrophs can use one or more of three strategies to cope advantageously with feast and famine loading. First, the classic copiotroph has a very fast maximum specific growth rate, compared to oligotrophs. Thus, they are able to outgrow the oligotrophs during the feast period and, thereby, capture most of the substrate for themselves. Copiotrophs that employ this rapid-growth strategy are called *r-strategists,* because they rely on a very large reaction rate, or a large r. Second, copiotrophs may rapidly take up and sequester the substrate during the feast period. By sequestering the substrate as an internal storage product (more details in Chapters 6 and 11), this type of copiotroph does not need to have a rapid growth rate during the feast period. Instead, they have a relatively steady growth rate that is based on the gradual utilization of internal storage products during the famine period. Third, some copiotrophs go into a dormant state, such as spores during starvation. The third category is the least important for processes in environmental biotechnology.

Selective enrichment is a very important adaptation mechanism in environmental biotechnology. The duration of the adaptation period depends on the enriched cell's doubling time (under the conditions present) and its starting inoculum size. As a general rule, adaptation by selective enrichment requires a few days to several months, and it should be marked by a significant alteration in community structure.

Enzyme regulation, the second adaptation mechanism, does not require a change in community structure, and it usually involves a short adaptation period, measured in hours. Adaptation by enzyme regulation occurs when the community already contains a substantial number of microorganisms capable of responding, and that response is coded on one or more regulated enzymes. Induction or depression of enzyme synthesis occurs rapidly in response to the environmental stress.

The third adaptation mechanism is *exchanged genetic material,* which includes conjugation, transformation, and transduction. Each of these mechanisms was described earlier in this chapter. Conjugation is the most rapid and general exchange mechanism. Proliferation of critical genetic information throughout a community could take place very rapidly with adaptation periods of hours to days. Adaptation via genetic exchange need not involve an alteration to community structure.

Inheritable genetic change to community members can come about through mutation, duplications, and recombination. The changes are permanent for the affected microorganisms and can be considered a form of community evolution. In most cases, inheritable genetic change occurs through infrequent, random events. Thus, adaptation via genetic change usually has a long adaptation period and may not be reproducible.

Microbial communities often *alter their environment* in such a way that the community is better able to cope with or benefit from the stress. Examples of environmental alterations include depletion of preferred substrates, supply of deficient substrates or nutrients, changes in the redox status, changes in pH, and elimination of toxicity.

Depletion of preferred substrates has been well studied. A classic example is *diauxie,* in which a highly favorable substrate, usually glucose, represses the enzymes needed for metabolism of other substrates. *Substrate inhibition* is a more general phenomenon in which the presence of one substrate inhibits the activity of a catabolic enzyme for another substrate. The applicability of diauxie to processes in environmental biotechnology is questionable, but general substrate inhibition is of widespread importance, particularly for the biodegradation of xenobiotic molecules (more in Chapters 3 and 14). Clearly, removing an inhibitory substrate by the community frees the capable microorganisms to catabolize the substrate of interest.

A subtler, but very important form of substrate depletion involves microorganisms that have highly versatile metabolism. For example, anaerobic communities adapt to biodegrade chlorinated aromatics only after other easily available organic electron donors are depleted. Selected members of the community then reductively dechlorinate the chlorinated aromatics, presumably so that the dechlorinated aromatic becomes available as an electron donor. Thus, the community is forced to reduce the chlorinated aromatics only after other electron donors are depleted.

Release of substrates or nutrients also can spur growth or activity in response to a stress. Lysis of some community members release nutrients and growth factors that can spur the activity of other members. Heterotrophic activity releases inorganic carbon, which is needed by autotrophs. Increased autotrophic activity releases soluble microbial products useful as electron donors by heterotrophs. Increased cycling of elements that can serve as electron donors and acceptors by different microbial types can spur more activity in general or by specific groups.

Table 1.16   Summary of adaptation mechanisms

Mechanism	Expected Period	Change in Community Structure?
Selective Enrichment	Days to months	Significant
Exchange of Genetic Information	Weeks to years	Not necessarily needed
Inheritable Genetic Change	Hours to days	Not necessarily needed
Enzyme Regulation	Hours	Not needed
Altering the Environment	Highly variable: Hours to months	Highly variable: Not needed to be significant

Sequential consumption of electron acceptors normally follows the thermodynamic ·order shown in Figure 1.15. In order to bring about methanogenesis, for example, a community needs first to alter the redox status by consuming O_2, NO_3^-, SO_4^{2-}, and Fe^{3+}.

Many microbial reactions produce strong acid or base. Chapters 2, 9, 10, and 14 describe more on these reactions. The pH change may favor one microbial group or change the thermodynamics of key reactions.

Changes in redox status, changes in pH, supply of nutrients, or direct action by microorganisms can reduce toxicity. The precipitation of heavy metals as hydroxide, carbonate, or sulfide solids is an outstanding example. Direct biotransformation also can reduce toxicity, particularly from xenobiotic organic molecules.

Table 1.16 summarizes the adaptation mechanisms. In practice, more than one mechanism may act. For example, genetic exchange, genetic change, or environment alteration could precede selective enrichment of the microorganisms benefiting from the former change.

Adaptation is not yet well understood. Fortunately, the development of new tools to study microbial ecosystems promises to expand our understanding of adaptation in the upcoming decade. Table 1.16 helps us keep in mind that adaptation can involve different mechanisms that take effect over periods from only a few hours to many months or years. In some cases, the community structure changes drastically, but community structure can remain intact, even though the community's function is greatly altered by adaptation.

1.16   TOOLS TO STUDY MICROBIAL ECOLOGY

The start of the twenty-first century marks a period of revolutionary advancement in our ability to study microbial ecology. At the forefront of the revolution are tools adapted from molecular biology. For the first time, we are able to get direct answers to the three fundamental questions about structure and function in microbial ecosystems: What microorganisms are present? What metabolic reactions could they carry out? What metabolic reactions are they carrying out?

This section describes the tools that can be used to study microbial ecology. We first outline traditional tools, which are based on enrichment culture. We then provide an introduction to the powerful new tools coming from molecular biology. Finally, we describe a second emerging area that is key for studying microbial ecology: namely, multispecies mathematical modeling.

1.16.1 TRADITIONAL ENRICHMENT TOOLS

For many decades, microbiologists have tried to gain insight into microbial communities by isolating single strains based on a key phenotypic characteristic. For instance, nitrifying bacteria can be enriched by providing a medium rich in NH_4^+-N, dissolved oxygen, and bicarbonate, but lacking other electron donors. As illustrated in Figure 1.43, successive exposure and dilutions in such a strongly selective growth medium eventually can lead to isolation of a single strain able to carry out ammonium oxidation, because other types of microorganisms have been diluted out. The time and amount of dilution needed to obtain a pure isolate depend on the complexity of the starting culture, the starting concentration and growth rate of the enriched strain, and the degree to which the enriched strain exchanges materials with other microbial types. In some cases of synergism, it is practically impossible to obtain a true pure isolate.

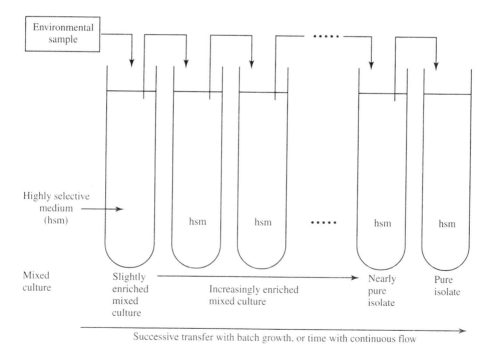

Figure 1.43   Selective enrichment involves successive dilutions of an environmental sample in highly selective medium. Each transfer occurs after growth and involves a major dilution. In practice, the dilution need not occur via successive transfers, as shown here, but can be achieved by a continuous flow of selective medium.

Once a pure isolate is available, it can be characterized for its genetics (e.g., DNA or rRNA sequences), metabolic potentials (e.g., what electron donors and acceptors it uses), temperature and pH optima, catabolic and growth kinetics, and morphology. Most of the information currently used to construct phylogenetic trees comes from isolated organisms.

Although enrichment has provided much valuable information over many years, it has three serious limitations. First, it requires that we know *a priori* the physiological characteristic that will enrich for the right microorganisms. Since the microorganism's traits are not fully known before it is isolated, we must make major assumptions about the right answers in order to do the enrichment. Second, microorganisms that live in tightly coupled synergistic relationships may never be isolated. Third and probably most important, the enrichment approach often biases the results. This bias comes about because important strains in their normal environment often are oligotrophs, but the enrichment conditions favor copiotrophs. High concentrations of an electron donor, such as NH_4^+-N, give fast growing copiotrophs a selective advantage over their oligotrophic relations, who may be more important in the natural setting. Because microbial communities contain functional redundancy, an enrichment experiment need not enrich for the microorganisms actually of the greatest importance. When copiotrophic *weeds* are selected instead of more important oligotrophs, we gain a distorted view of the particular community and of the diversity of microbial life in general.

Despite their limitations, traditional methods provide valuable information, as long as their results are interpreted properly. They are most useful when we are interested in a *functional assessment* of the community. For example, if we want to know whether or not nitrifying bacteria are significant in a mixed culture, we can expose the community to NH_4^+-N and look for its loss, as well as the production of NO_2^--N or NO_3^--N. These evidences verify that the nitrification function is present, and the rates give us an approximate measure of the amount of biomass capable of carrying out nitrification. These data do not give us any details on the strains responsible for nitrification. For engineers interested in ensuring that a desired function is present, this traditional approach is quite valuable.

1.16.2   MOLECULAR TOOLS

Numerous molecular tools are being developed to overcome the biases of traditional enrichment culturing. Instead of enriching for cells based on some phenotypic trait, molecular tools directly interrogate the community's genetic information. Specifically, the molecular techniques assay the base sequences of the cell's DNA or RNA.

Depending on what information is desired, the molecular assay targets different types of DNA or RNA. Table 1.17 summarizes the targets and what information each provides. To determine the *phylogenetic identity* of cells in a community, the rRNA (normally the SSU rRNA) or the gene that codes for it is the target. The *phenotypic potential*, such as for degradation of a particular substrate, is assayed by finding the genes on the DNA. The *expressed phenotypic potential* can be demonstrated by detecting the mRNA or the protein product, such as an enzyme.

Table 1.17   Targets for molecular methods

Target	Information Gained
rRNA	Phylogenetic identity
Genes for rRNA on DNA	Phylogenetic identity
Other genes on DNA	Phenotypic potential
mRNA	Expressed phenotypic potential
Protein product (reporter)	Expressed phenotypic potential

In the remainder of this section on molecular methods, we introduce most of the molecular methods in use today. The field is advancing at a remarkable pace, and technical improvements continually render methodological details obsolete. Therefore, we do not focus on the "how to" of the methods. Instead, we highlight the broad principles underlying the methods.

When we want to understand community structure, we need to identify and enumerate the different microorganisms according to their inheritable genetic content. The common approach is to target the SSU rRNA, which is a powerful phylogenetic marker. *Oligonucleotide probing* directed toward the SSU rRNA is the most direct approach. An oligonucleotide probe is single-stranded DNA comprised of 15 to 25 bases whose sequence is complementary to a region in the target-cell's SSU rRNA. Under strictly controlled assay conditions, the probe DNA hybridizes to the complementary region of the target-cell's RNA, but it does not hybridize the RNA from any other cells, due to mismatches in the sequence. If the RNA is fixed in place, unhybridized probe is washed away, leaving only probe hybridized to the target RNA. As long as the hybridized probe can be detected, the presence and quantity of target rRNA can be obtained.

Oligonucleotide hybridization can be carried out in two basic formats. The more traditional format is called *slot blotting,* and it requires that RNA be extracted from the sample. Extraction typically is performed with phenol and chloroform, and the RNA in the aqueous phase is precipitated with an alcohol. The RNA pellets are resuspended, denatured, and applied to nylon membranes in a grid pattern, or slot blots. The oligonucleotide probe, labeled with ^{32}P, is contacted with the membranes in a hybridization buffer, usually overnight. The membranes are then washed at a carefully controlled temperature, dried, and assayed for radioactivity.

The wash temperature is of critical importance, because it determines the specificity of the probe. Each probe is tested for hybridization efficiency against its target RNA and other RNA. The temperature of dissociation, T_d, is the temperature at which 50 percent of the probe binds with its complementary RNA. If the probe is designed well, hybridization with similar, but different RNA should be minimal at T_d.

The radioactivity can be measured in three ways. The most traditional way is to expose photographic film to the membranes. The intensity of the image is proportional to the hybridized probe for each blot. More recently, storage phosphor screens, which can be scanned automatically, are replacing film. The third method is to place each blot into a scintillation vial, add scintillation cocktail, and perform scintillation counting.

The second format for oligonucleotide probing is *fluorescence in situ hybridization, or FISH.* FISH and slot blotting differ in key ways. First, the oligonucleotide probe in FISH is labeled with a molecule that fluoresces when excited by light of a given wavelength. Therefore, detection is through fluorescence microscopy, in which the hybridized sample is illuminated with the exciting light, and the emitted light is observed through the microscope. Second, the RNA is not extracted with FISH, but remains inside the cells (*in situ*), which are fixed and made porous to the probe. Because the cells are not destroyed, FISH is able to provide information on the spatial relationships among different types of cells.

Other features of FISH resemble slot blotting. Hybridization is based on sequence complementarity, and the unhybridized probe is removed by washing at a carefully chosen temperature.

Figure 1.44 is a dramatic example of what FISH can do. The photomicrograph is part of an activated sludge floc. The black cells are hybridized to a probe specific for ammonium-oxidizing bacteria. The grey cells are hybridized to a probe specific for nitrite-oxidizing bacteria. These FISH results show that the ammonium oxidizers form a very dense cluster within the floc. The nitrite oxidizers, which use the NO_2^- generated by ammonium oxidizers, are found in smaller clusters surrounding the larger cluster of ammonium oxidizers.

Similar information on community structure can be obtained from the chromosomal DNA. The DNA is first extracted from the sample in a manner similar to that for the RNA. It is then amplified selectively using the *polymerase chain reaction (PCR)* and primers specific for the gene that codes for the SSU rRNA. The amplified DNA is fixed to a membrane and hybridized with the oligonucleotide probes of interest.

One of the most powerful features of oligonucleotide probing against the SSU rRNA is that the probes can be designed for a wide range of specificity. In other words, a probe can be designed to hybridize only to one strain, to a group of similar strains (e.g., a genus), to an entire domain, or to all life. Figure 1.45 illustrates this

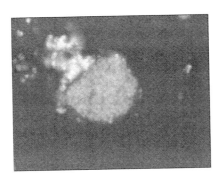
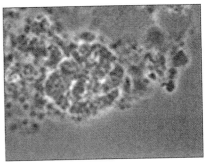

a.   *b.*

Figure 1.44   FISH photomicrograph (a) of clusters of ammonium oxidizers (grey) and nitrite oxidizers (white) inside an activated sludge floc (b). SOURCE: Photo by Bruce Mobarry.

ORDER I: *METHANOBACTERIALES*
 Family I: *Methanobacteriaceae*
  Genus I: *Methanobacterium*
  Genus II: *Methanobrevibacter* — MB310, MB1174
  Genus III: *Methanosphaera*
 Family II: *Methanothermaceae*
  Genus I: *Methanothermus*

ORDER II: *METHANOCOCCALES*
 Family I: *Methanococcaceae*
  Genus I: *Methanococcus* — MC1109

ORDER III: *METHANOMICROBIALES*
 Family I: *Methanomicrobiaceae*
  Genus I: *Methanomicrobium*
  Genus II: *Methanogenium*
  Genus III: *Methanoculleus*
  Genus IV: *Methanospirilum* — MG1200
 Family II: *Methanocorpusculaceae*
  Genus I: *Methanocorpusculum*
 Family III: *Methanoplanaceae*
  Genus I: *Methanoplanus*
 Family IV: *Methanosarcinaceae*
  Genus I: *Methanosarcina* — MS821; can use acetate and other substrates (H_2/CO_2, methanol, and methylamines)
  Genus II: *Methanococcoides*
  Genus IV: *Methanolobus*
  Genus V: *Methanohalophilus* — can use methanol and methylamines
  Genus III: *Methanosaeta* — MX825; can only use acetate

(MS1414, MSMX860 group the Methanosarcinaceae probes)

Probe	Sequence (5'–3')	Target site (*E. coli* numbering)	T_d (°C)
MC1109	GCAACATAGGGCACGGGTCT	1128–1109	55
MB314	GAACCTTGTCTCAGGTTCCATC*	335–314	
MB310	CTTGTCTCAGGTTCCATCTCCG	331–310	57
MB1174	TACCGTCGTCCACTCCTTCCTC	1195–1174	62
MG1200	CGGATAATTCGGGGCATGCTG	1220–1200	53
MSMX860	GGCTCGCTTCACGGCTTCCCT	880–860	60
MS1414	CTCACCCATACCTCACTCGGG	1434–1414	58
MS1242	GGGAGGGACCCATTGTCCCATT*	1263–1242	
MS821	CGCCATGCCTGACACCTAGCGAGC	844–821	60
MX825	TCGCACCGTGGCCGACACCTAGC	847–825	59
ARC915	GTGCTCCCCCGCCAATTCCT	934–915	56
ARC344	TCGCGCCTGCTGCTCCCCGT	363–344	54

* underlined sequences indicate regions of internal complementarity

Figure 1.45 Oligonucleotide probes designed for the methanogens and all Archaea. SOURCE: Raskin et al., 1994.

approach of probe nesting for methanogenic Archaea. The figure shows the phylogenetic organization of the methanogens, by order, family, and genus. It indicates which oligonucleotide probes, designated by letters that correspond to the genus and a number that corresponds to the 5′ end of the target site in the SSU rRNA. The figure also gives the sequence of the probe in the 5′ to 3′ direction (the target is the complementary sequence in the 3′ to 5′ direction), the location of the target site on the rRNA (3′ to 5′), and the T_d value. Also listed are two probes (ARC915 and ARC344) that hybridize to all known Archaea. Not listed on the figure is an example of a universal probe, which hybridizes the SSU rRNA of all known life. An example is UNIV1392, which has the sequence 5′-ACGGGCGGTGTGAG-3′ and a T_d of 44 °C (Lane et al., 1985).

Figure 1.46 illustrates how the different probes relevant to the family *Methanosarcinaceae* are used to track these acetate-cleaving methanogens to different levels of specificity in a mixed methanogenic culture. Of the SSU rRNA for all life, tracked by UNIV1392, 22 percent is associated with all Archaea. Half of the Archaea, or 11 percent, comes from *Methanosarcinaea,* which are acetate fermenters, while 5 percent come from *Methanosaeta* and *Methanosarcina,* respectively. If mass balance holds on the rRNA hybridizations, the 78 percent not associated with ARC915 is distributed between the domains for Bacteria and Eukarya. In a strongly methanogenic system,

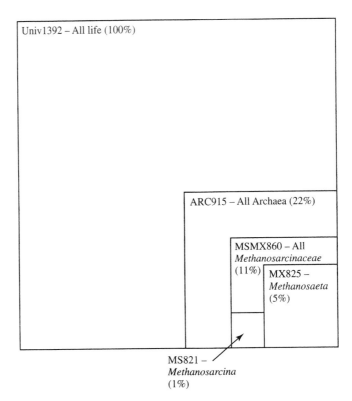

Univ1392 – All life (100%)

ARC915 – All Archaea (22%)

MSMX860 – All
Methanosarcinaceae
(11%)

MX825 –
Methanosaeta
(5%)

MS821 –
Methanosarcina
(1%)

Figure 1.46   Illustration of how nested probes lead to different
specificity relevant to the family
Methanosarcinaceae, or the acetate-cleaving
methanogens.

the 11 percent of ARC915 RNA not hybridized by MSMX960 should reflect the methanogens that are strictly H_2 oxidizers.

Figure 1.47 shows nested probes for nitrifying bacteria. The nesting is particularly deep for the ammonium oxidizers. Figure 1.47 also illustrates a second format for showing phylogenetic trees. This format is easier to read when the genetic relationships are broken down in detail for closely related strains. The total horizontal distances between strains is a measure of their genetic difference expressed as a fraction of the base pairs differing. The scale bar shows a distance of 0.1, or 10 percent difference. Not shown is a probe for the domain bacteria, of which the nitrifiers are a part. An example of a Bacterial probe is EUB338, which is 5′-GCTGCCTCCCGTAGGAGT-3′, with $T_d = 54\,°C$ (Amann, Krumholz, and Stahl, 1990).

One drawback of oligonucleotide probing is that it can be used confidently only for strains that have been isolated and sequenced. Microbial ecologists estimate that only a few percent of microbial strains have been isolated. Furthermore, they may not be good representations of the most important strains in the natural setting, due to

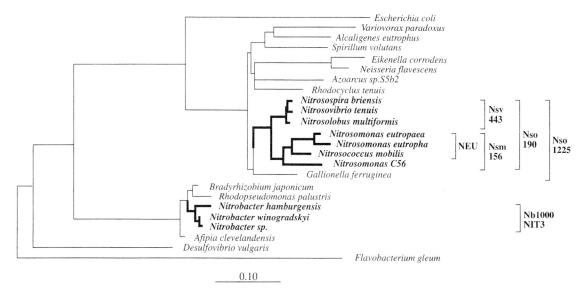

Probe	*E. coli* 16S rRNA position	Probe sequence
Nb1000	1000–1012	5′-TGCGACCGGTCATGG-3′
NIT3	1035–1048	5′-CCTGTGCTCCATGCTCCG-3′
NEU	653–670	5′-CCCCTCTGCTGCACTCTA-3′
Nso190	190–208	5′-CGATCCCCTGCTTTTCTCC-3′
Nso1225	1225–1244	5′-CGCCATTGTATTACGTGTGA-3′
Nsm156	156–174	5′-TATTAGCACATCTTTCGAT-3′
Nsv443	444–462	5′-CCGTGACCGTTTCGTTCCG-3′

Figure 1.47   Nested probes for ammonium-oxidizing (prefix *Nitroso*) and nitrite-oxidizing (prefix *Nitro*) bacteria. SOURCE: Mobarry, Wagner, Urbain, Rittmann, and Stahl, 1996.

the biases discussed above. Therefore, it is very useful to have a molecular technique that provides a *fingerprint* of the community's diversity whether or not the key strains have been isolated and sequenced.

New fingerprinting techniques are being developed. The basic principle underlying them is that the DNA coding for a specified and universal function of the microorganisms of interest is selectively amplified by PCR. This step requires that the typical promoter region for the function be well known, so that a primer specific to that region can be designed. Then, the PCR amplifies only DNA associated with that function.

The function selected can be general or highly specific. The gene that produces the 16S rRNA is a perfect example of the choice for amplification. There are primers known to be general for large groupings, such as all Bacteria. On the other hand, primers specific to smaller groups or individual strains can be designed. Another target for PCR application is a gene that codes for catabolic genes restricted to one strain or a related group of strains. The gene for dissimilatory bisulfite reductase is an example of one useful for characterizing sulfate reducers.

Before we discuss how the amplified DNA is analyzed to interpret the fingerprint, we must mention the limitations of using PCR as the first step for obtaining a fingerprint of community structure. Biases can be introduced if the extraction efficiency varies among strains, if some DNA is more sensitive to shear breakage during handling, or if the primer does not work equally well for all genes of the same type. Although less problematic than traditional enrichment methods, PCR amplification can introduce its own biases, particularly if minor components in the community have a gene that is amplified much more efficiently than genes of other related strains.

The amplified DNA must now be analyzed to create and interpret the fingerprint. A method whose use is growing is *denaturing gradient gel electrophoresis (DGGE)*. In DGGE, the DNA is applied at one end of an electrophoresis cell containing a polyacrylamide gel. Electrodes at each end create an electric field that draws the negatively charged DNA toward the positive electrode located at the end opposite where the DNA is placed. The rate at which a DNA strand moves toward the positive electrode depends on the ratio of its size and charge. A larger ratio means slower movement.

In DGGE, the acrylamide gel is prepared in such a way that it has a continuous gradient of a denaturing agent, typically urea plus formamide. As the DNA moves toward the positive electrode, it encounters increasingly denaturing conditions. Depending mainly on the DNA's G+C content, it denatures, greatly expands in size, and stops at different locations between the electrodes. This creates a pattern of DNA bands that constitute the fingerprint. Figure 1.48 is an example of bands for two different communities from which the 16S rRNA was amplified with a bacterial primer. Many of the bands are the same in these two anaerobic communities, which demonstrates that the overall community structures are similar. On the other hand, bands a and b are unique for the two communities and reflect that each has at least one important microbial type that is not important in the other.

For further analysis of the DGGE fingerprint, the bands can be cut out of the gels so that the DNA can be further amplified and sequenced. Particularly when the SSU rRNA gene is amplified for DGGE, the sequences can be compared to the database of SSU rRNA sequences to determine if it matches an isolated strain and where it fits into the phylogenetic tree. The sequence also can be used to design an oligonucleotide probe.

A second fingerprinting approach is *restriction fragment length polymorphism (RFLP)*. The amplified DNA is cut into fragments using a restriction endonuclease enzyme that does not cut the DNA within the region of the target gene, such as for the SSU rRNA. The flanking regions around the target gene have a variable number of restriction sites. Thus, the fragments after restriction are of different sizes and can be separated by electrophoresis.

The *phenotypic potential* within a community can be assayed by methods that mirror those used for community structure. The pattern is to selectively amplify the DNA for a gene or gene set known to confer the phenotypic trait, such as metabolism of a particular substrate. Then, the amplified products can be slot blotted or separated

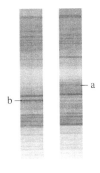

a.   b.

Figure 1.48
Example of DNA bands that form a fingerprint using DGGE. The DNA was amplified using a bacterial 16S rRNA primer. The positive electrode is at the bottom. (DGGE bands by Jennifer Becker.)

by gel electrophoresis, detected, quantified, and identified, as appropriate. The key to this kind of evaluation is having enough sequence information on the gene of interest so that a promoter can be designed.

The *expressed phenotypic potential* is assayed most directly by targeting the mRNA from the critical gene or gene set. The mRNA is extracted and slot blotted. An oligonucleotide probe is hybridized and detected in a manner similar to slot blotting for rRNA. The basic sequence of the gene must be known so that the oligonucleotide probe can be designed to complement the mRNA in an appropriate region.

One additional molecular method, called a *reporter,* has been gaining popularity and provides another way to assess the expressed phenotypic potential. To use a reporter, recombination techniques are used to insert a reporter gene within the region of DNA of interest. When the DNA sequence is being transcribed, the reporter gene also is transcribed, since it follows the promoter and is within the stretch of DNA to be transcribed. The mRNA from the reporter is translated to an enzyme product that catalyzes some easily detected reaction. Light emission, or luminescence, is the most common reporter response. The principle is that the reporter response is observed only when the entire sequence is transcribed; thus, expression of the reporter implies expression of the target gene(s).

Reporters have considerable promise, but they are not without their biases. Although the hypothesis that transcription of reporter mRNA is proportional to the transcription of target gene mRNA seems sound, translation of the protein product is not necessarily proportional. If the reporter mRNA is translated much more or less rapidly, a quantitative correlation is impossible. A potential advantage of reporters is that they give a real-time measure of expressed phenotypes. This advantage is realized only if a real-time detection device is available.

1.16.3   MULTISPECIES MODELING

The second powerful tool that can revolutionize microbial ecology is mathematical modeling that explicitly captures multispecies diversity and the exchanges of materials within the community. By systematically quantifying the interactions that occur in a microbial ecosystem, a model can help us understand how the interactions occur and why the system behaves as it does. Armed with this fundamentally based and quantitative tool, engineers can design processes that take advantage of ecological responses.

Chapters 2, 3, and 4 lay the foundation for building models of complex microbial ecosystems. Information provided on the website for this book (see the Preface) explicitly addresses multispecies modeling and illustrates several important insights gained from modeling. Today, multispecies models already exist for activated sludge, anaerobic digestion, and a range of biofilm processes. Increasing powerful computers and rapidly advancing tools from molecular biology are opening the door for vast improvement in multispecies modeling. The first decade in the twenty-first century should bring forth a new generation of models that are needed to complement the molecular tools.

1.17   BIBLIOGRAPHY

Alcamo, I. E. (1997). *Fundamentals of Microbiology,* 5th ed., Menlo Park, CA: Addison-Wesley-Longman.

Alexander, M. (1971). *Microbial Ecology.* New York: John Wiley & Sons.

Alm E. W.; D. B. Oerther; N. Larson; D. A. Stahl; and L. Raskin (1996). "The oligonucleotide probe database." *Appl. Environ. Microb.* 62, pp. 3557–3559.

Amann, R. I.; L. Krumholz; and D. A. Stahl (1990). "Fluorescent oligonucleotide probing of whole cells for determinative, phylogenetic, and environmental studies in microbiology." *J. Bacteriology* 172, pp. 762–770.

Brusseau, G. A.; B. E. Rittmann; and D. A. Stahl (1997). "Addressing the microbial ecology of marine biofilms." *Molecular Approaches to the Study of the Oceans,* ed. K. E. Cooksey, pp. 449–470.

Davies, D. G.; M. R. Parsek; J. P. Pearson; B. H. Iglewski; J. W. Costerton; and E. P. Greenberg (1998). "The involvement of cell-to-cell signaling in the development of a bacterial biofilm." *Science* 280, pp. 295–298.

Devereux, R.; M. D. Kane; J. Winfrey; and D. A. Stahl (1992). "Genus- and group-specific hybridization probes for determinative and environmental studies of sulfate-reducing bacteria." *Syst. Appl. Microbiol.* 15, pp. 601–609.

Furumai, H. and B. E. Rittmann (1992). "Advanced modeling of mixed populations of heterotrophs and nitrifiers considering the formation and exchange of soluble microbial products." *Water Sci. Technol.* 26(3–4), pp. 493–502.

Holliger, C.; G. Schraa; A. J. M. Stams; and A. J. B. Zehnder (1993). "A highly purified enrichment culture couples the reductive dechlorination of tetrachloroethene to growth." *Appl. Environ. Microb.* 59, pp. 2991–2997.

Lane, D. J.; B. Pace; G. J. Olsen; D. A. Stahl; M. L. Sogin; and N. R. Pace (1985). "Rapid determination of 16S ribosomal RNA sequences for phylogenetic analysis." *Proc. Natl. Acad. Sci. USA* 82, pp. 6955–6959.

Linkfield, T. G.; J. M. Suflita; and J. M. Tiedje (1989). "Characterization of the acclimation period before anaerobic dehalogenation of halobenzoates." *Appl. Environ. Microb.* 55, pp. 2773–2778.

Madigan, M. T.; J. M. Martinko; and J. Parker (1997). *Brock Biology of Microorganisms,* 8th ed. New York, Prentice-Hall.

Mobarry, B. K.; M. Wagner; V. Urbain; B. E. Rittmann; D. A. Stahl (1996). "Phylogenetic probes for analyzing abundance and spatial organization of nitrifying bacteria." *Appl. Environ. Microb.* 62, pp. 2156–2162.

Neidhardt, F. C. et al., Eds. (1996). *Escherichia coli and Salmonella typhimurium—Cellular and Molecular Biology,* 2nd ed. Washington DC: American Society for Microbiology.

Pace, N. R.; D. A. Stahl; D. J. Lane; and G. J. Olsen (1986). "The analysis of natural microbial populations by ribosomal RNA sequences." *Adv. Microb. Ecol.* 9, pp. 1–55.

Raskin, L.; B. E. Rittmann; and D. A. Stahl (1996), "Competition and coexistence of sulfate-reducing and methanogenic populations in anaerobic biofilms." *Appl. Environ. Microb.* 62, pp. 3847–3857.

1.29. What will be the additional fermentation products if one one mole of glucose is fermented to:

(*a*) two moles acetate and two moles CO_2
(*b*) one mole propanal (CH_3CH_2CHO) and two moles CO_2?

1.30. Compare and contrast how light is used by cyanobacteria versus green and purple bacteria.

1.31. Which of the following respiratory carriers carry electrons, but not protons:

(*a*) Cytochrome c
(*b*) Flavin adenine dinucleotide
(*c*) Ferredoxin
(*d*) Nicotinamide adenine dinucleotide
(*e*) None of the above

1.32. Gluconeogenesis is best defined as:

(*a*) A synonym for glycolysis
(*b*) The reverse of glycolysis
(*c*) The formation of glucose from non-carbohydrate precursers
(*d*) None of the above

1.33. Which of the following does *not* occur in anoxygenic photosynthesis:

(*a*) The system operates under anaerobic conditions
(*b*) Water is split into oxidized and reduced parts
(*c*) ATP is formed via cyclic photophosphorylation
(*d*) Bacteriochlorophyll is involved in the light reaction.

1.34. In the typical bacterial cell, about 12–15 percent of the cellular dry weight is _____ .

1.35. Listed below are genus names for different kinds of prokaryotes. After each genus name are spaces for the electron donor, electron acceptor, and carbon source. One of the three is filled in already. You must fill in the other two.

Genus Name	Electron Donor	Electron Acceptor	Carbon Source
Methanococcus			CO_2
Nitrosolobus			CO_2
Pseudomonas	C_6H_6O (phenol)		
Desulfomonas	$C_6H_{12}O_6$ (glucose)		
Thiobacillus	Fe^{2+}		

1.36. Present a reasonable pathway for the full oxidation of propionate (CH_3CH_2COOH). How many ATP equivalents will be generated per mole of propionate if the cells carry out aerobic respiration?

1.37. What are the major ways in which a mixed microbial community can adapt? Which of these mechanisms is available to a pure culture?

1.38. Extreme pH is well-known for inhibiting enzyme catalysis. How does an extreme pH affect enzyme reactivity?

chapter

2

STOICHIOMETRY AND BACTERIAL ENERGETICS

Probably the most important concept in the engineering design of systems for biological treatment is the *mass balance.* For a given quantity of waste, a mass balance is used to determine the amount of chemicals that must be supplied to satisfy the energy, nutrient, and environmental needs of the microorganisms. In addition, the amounts of end products generated can be estimated. Examples of chemicals are oxygen as an electron acceptor, nitrogen and phosphorus as nutrients for biomass growth, and lime or sulfuric acid to maintain the pH in the desired range. Examples of end products of importance are excess microorganisms (sludge), which is a costly disposal problem, and methane from anaerobic systems, which can be a useful source of energy.

Balanced chemical equations are based upon the concept of *stoichiometry,* which is an aspect of chemistry concerned with mole relationships among reactants and products in chemical reactions. Although anyone who has taken even the most rudimentary chemistry course is familiar with writing balanced reactions, several features inherent to microbial reactions complicate the stoichiometry. First, microbial reactions often involve oxidation and reduction of more than one species. Second, the microorganisms have two roles, as catalysts for the reaction, but also as products of the reaction. Third, microorganisms carry out most chemical reactions in order to capture some of the energy released for cell synthesis and for maintaining cellular activity. For this reason, we must consider reaction energetics, as well as balancing for elements, electrons, and charge.

We take an approach that is fundamentally based and also very useful for practice. In this chapter, we demonstrate how to write stoichiometric equations that balance elements, electrons, charge, and energy. This approach integrates all the factors that control microbial growth and its relationships to the materials that the cells consume and produce.

2.1 AN EXAMPLE STOICHIOMETRIC EQUATION

One of the first balanced equations for biological oxidation of a wastewater was that presented by Porges, Jasewicz, and Hoover (1956) for a casein-containing wastewater:

$$C_8H_{12}O_3N_2 + 3\,O_2 \rightarrow C_5H_7O_2N + NH_3 + 3\,CO_2 + H_2O$$

"casein"   "bacterial
                    cells"

Formula Weight: 184   96   113   17   132   18
                $\sum = 280$ $\sum = 280$

[2.1]

Equation 2.1 indicates that, for each 184 g of casein consumed by microorganisms, 96 g of oxygen must be supplied for the reaction to proceed properly. The reaction produces 113 g of new microbial cells, 17 g of ammonia (or 14 g of ammonia-N), 132 g of carbon dioxide, and 18 g of water. Such knowledge is essential for designing a biological treatment system for casein treatment. For example, when we treat 1,000 kg/d of casein, we must supply 520 kg/d of oxygen via aeration, and 610 kg/d of biomass solids (i.e., dry sludge) must be dewatered and disposed of.

In Equation 2.1, some of the carbon in casein is fully oxidized to CO_2. Thus, casein is the electron-donor substrate. The rest of the carbon in casein is incorporated into the newly synthesized biomass, because casein also is the carbon source. The complex protein-containing mixture in casein is represented through the empirical formula $C_8H_{12}O_3N_2$. This formula was constructed from knowledge of the relative mass proportions of organic carbon, hydrogen, oxygen, and nitrogen contained in the wastewater, values that can be obtained from normal organic chemical analyses for each element present.

The same approach is taken for the empirical formula for bacterial cells of $C_5H_7O_2N$. Bacterial cells are highly complex structures containing a variety of carbohydrates, proteins, fats, and nucleic acids, some with very high molecular weights. Indeed, microorganisms contain many more than the four elements indicated by the above equation, such as phosphorus, sulfur, iron, and many other elements that are generally present in only trace amounts. An empirical formula could contain as many elements as desired, as long as the relative proportions on a mass basis are known. However, Porges et al. (1956) chose to represent only the four major elements, and this is generally satisfactory for most practical purposes. The mass requirements for elements not shown in the formula can be determined once the mass of bacterial cells formed from a given reaction is known. For example, phosphorus normally represents about 2 percent of the bacterial organic dry weight. Thus, if 1,000 kg/d of casein were consumed in a treatment process, then 0.02(610) or 12 kg/d of phosphate-phosphorus must be present in the wastewater or added to it in order to satisfy bacterial needs for this element.

Porges et al. (1956) obtained Equation 2.1 from empirical data. Could we predict the stoichiometry of such a reaction? The answer is yes, and the rest of the chapter develops and applies an approach for doing this. To achieve this goal, we need three things:

1. Empirical formula for cells

2. Framework for describing how the electron-donor substrate is partitioned between energy generation and synthesis

3. Means to relate the proportion of the electron-donor substrate that is used to synthesize new biomass to the energy gained from catabolism and the energy needed for anabolism.

These three items are presented in order in the rest of this chapter.

2.2 EMPIRICAL FORMULA FOR MICROBIAL CELLS

The empirical formula noted earlier for cells ($C_5H_7O_2N$) was one of the first used in the balancing of biological reactions. However, the relative proportion of elements actually present in cells depends on the characteristics of the microorganisms involved, the substrates being used for energy, and the availability of other nutrients required for microbial growth. If microorganisms are grown in a nitrogen-deficient environment, they tend to produce more fatty material or carbohydrates, and the resulting empirical cell formula should reflect a smaller proportion of nitrogen. Table 2.1 is a summary of empirical cell formulas reported by others, some for anaerobic and some for aerobic growth, some for mixed cultures of microorganisms and others for pure cultures, and some for cultures grown on different organic substrates. The nitrogen content ranges from 6 to 15 percent, although it averages the "typical" 12 percent.

An extremely important way to compare empirical cell formulas is by the ratio of oxygen required for full oxidation of the cellular carbon per unit weight of cells. This oxygen requirement is termed *COD'*, which we define as the *calculated oxygen demand*. This normally is equal to the *chemical oxygen demand (COD)*, which is a standard chemical procedure for evaluating this quantity and is based on the reduction of dichromate in a boiling acid solution. The COD' can be found from the empirical formula by

$$C_nH_aO_bN_c + \left(\frac{2n + 0.5a - 1.5c - b}{2}\right)O_2 \rightarrow nCO_2 + cNH_3 + \frac{a - 3c}{2}H_2O \quad \textbf{[2.2]}$$

and

$$COD'/\text{Weight} = \frac{(2n + 0.5a - 1.5c - b)16}{12n + a + 16b + 14c} \quad \textbf{[2.3]}$$

where

$$n = \%C/12T, a = \%H/T, b = \%O/16T, \text{ and } c = \%N/14T$$

and

$$T = \%C/12 + \%H + \%O/16 + \%N/14 \quad \textbf{[2.4]}$$

If the mass distribution of the four different organic elements in a biological culture is known, then the empirical formula for the cells can readily be constructed and the COD' computed.

Table 2.1 Empirical chemical formulas for prokaryotic cells

Empirical Formula	Formula Weight	COD' Weight	% N	Reference	Growth Substrate and Environmental Conditions
Mixed Cultures					
$C_5H_7O_2N$	113	1.42	12	1	casein, aerobic
$C_7H_{12}O_4N$	174	1.33	8	2	acetate, ammonia N source, aerobic
$C_9H_{15}O_5N$	217	1.40	6	2	acetate, nitrate N source, aerobic
$C_9H_{16}O_5N$	218	1.43	6	2	acetate, nitrite N source, aerobic
$C_{4.9}H_{9.4}O_{2.9}N$	129	1.26	11	3	acetate, methanogenic
$C_{4.7}H_{7.7}O_{2.1}N$	112	1.38	13	3	octanoate, methanogenic
$C_{4.9}H_9O_3N$	130	1.21	11	3	glycine, methanogenic
$C_5H_{8.8}O_{3.2}N$	134	1.16	10	3	leucine, methanogenic
$C_{4.1}H_{6.8}O_{2.2}N$	105	1.20	13	3	nutrient broth, methanogenic
$C_{5.1}H_{8.5}O_{2.5}N$	124	1.35	11	3	glucose, methanogenic
$C_{5.3}H_{9.1}O_{2.5}N$	127	1.41	11	3	starch, methanogenic
Pure Cultures					
$C_5H_8O_2N$	114	1.47	12	4	bacteria, acetate, aerobic
$C_5H_{8.33}O_{0.81}N$	95	1.99	15	4	bacteria, undefined
$C_4H_8O_2N$	102	1.33	14	4	bacteria, undefined
$C_{4.17}H_{7.42}O_{1.38}N$	94	1.57	15	4	*Aerobacter aerogenes,* undefined
$C_{4.54}H_{7.91}O_{1.95}N$	108	1.43	13	4	*Klebsiella aerogenes,* glycerol, $\mu = 0.1\ h^{-1}$
$C_{4.17}H_{7.21}O_{1.79}N$	100	1.39	14	4	*Klebsiella aerogenes,* glycerol, $\mu = 0.85\ h^{-1}$
$C_{4.16}H_8O_{1.25}N$	92	1.67	14	5	*Escherichia coli,* undefined
$C_{3.85}H_{6.69}O_{1.78}N$	95	1.30	15	5	*Escherichia coli,* glucose
Highest	218	1.99	15		
Lowest	92	1.16	6		
Median	113	1.39	12		

References: [1]Porges et al. (1956); [2]Symons and McKinney (1958); [3]Speece and McCarty (1964); [4]Bailey and Ollis (1986); [5]Battley (1987).

EMPIRICAL BIOMASS FORMULA A sample of a biological culture is submitted to a chemical laboratory for analysis of the percentage by weight of each major element present in the organic portion. The laboratory evaporates the sample to dryness and then places it in an oven overnight at 150 °C to drive off all water present. The organic portion of the residue remaining is then analyzed, after which the sample is burned in a Muffle furnace at 550 °C to determine the weight of the ash remaining. The ash contains the phosphorus, sulfur, iron, and other inorganic elements present in the sample. The composition of the cells by weight is found to be 48.9% C, 5.2% H, 24.8% O, 9.46% N, and 9.2% ash. Prepare an empirical formula for the cells, letting $c = 1$, and determine the COD'/organic weight ratio for the cells.

Example 2.1

$$T = 48.9/12 + 5.2 + 24.8/16 + 9.46/14 = 11.50$$

and

$$n = 48.9/(12 \times 11.5) = 0.354, a = 5.2/11.5 = 0.452,$$
$$b = 24.8/(16 \times 11.5) = 0.135, \text{ and } c = 9.46/(14 \times 11.5) = 0.0588.$$

In order to normalize to $c = 1$, divide by 0.0588, in which case the following result is obtained as the empirical formula for the cells:

$$C_{6.0}H_{7.7}O_{2.3}N$$

This formula illustrates that fractional coefficients (e.g., the 2.3 on O) are normal and correct. Rounding fractional coefficients to whole numbers introduces errors and should not be done. The COD$'$ to organic weight ratio is then equal to

$$(2 \times 6.0 + 0.5 \times 7.7 - 1.5 - 2.3)16/(12 \times 6 + 7.7 + 16 \times 2.3 + 14) = 1.48 \text{ g COD}'/\text{g cells}$$

2.3   SUBSTRATE PARTITIONING AND CELLULAR YIELD

When microorganisms use an electron-donor substrate for synthesis, a portion of its electrons (f_e^0) is initially transferred to the electron acceptor to provide energy for conversion of the other portion of electrons (f_s^0) into microbial cells, as illustrated in Figure 2.1. The sum of f_s^0 and f_e^0 is 1. Cells also decay because of normal maintenance or predation; part of the electrons in f_s^0 are transferred to the acceptor to generate more energy, and another part is converted into a nonactive organic cell residue.

The portions initially converted into cells, f_s^0, and used to generate energy, f_e^0, provide the framework for partitioning the substrate between energy generation and synthesis. A very important facet of the partitioning framework is that it is in terms

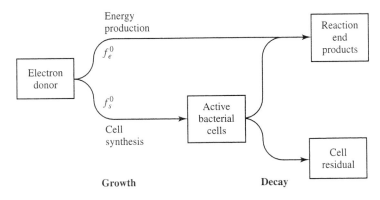

Figure 2.1   Utilization of electron donor for energy production and synthesis.

of electron equivalents (e$^-$ eq). Because electron flows generate the cell's energy, it is essential that the partitioning be expressed as electron equivalents.

The fraction f_s^0 can be converted into mass units, such as g cell produced/g COD′ consumed. When expressed in mass units, it is termed the true yield and given the symbol Y. The conversion from f_s^0 to Y is

$$Y = f_s^0 (M_c \text{ g cells/mol cells})/[(n_e e^- \text{ eq/mol cells})(8 \text{ g COD/e}^- \text{ eq donor})]$$

in which $M_c =$ is the empirical formula weight of cells, n_e is the number of electron equivalents in an empirical mole of cells, and the donor mass is expressed as COD. When cells are represented by $C_5H_7O_2N$ and ammonium is the nitrogen source, $M_c = 113$ g cells/mol cells, and $n_e = 20$ e$^-$ eq/mol cells. Then, the conversion is $Y = 0.706 f_s^0$, and Y is in g cells/g COD. The numbers used in the conversion change if the cell formula differs or if the cells use oxidized nitrogen sources, such as NO_3^-; these changes are discussed later.

Mathematically, the growth rate of microbial cells is frequently expressed as

$$\frac{dX_a}{dt} = Y\left(\frac{-dS}{dt}\right) - bX_a \qquad \textbf{[2.5]}$$

where dX_a/dt represents the net growth rate (M/L^{3}T) of active organism (X_a, M/L^3), $-dS/dt$ represents the rate of disappearance (M/L^{3}T) of substrate (S, M/L^3), b is the decay rate (T^{-1}) of the organisms, and Y is the true yield of microorganisms (M/M). (A full development of Equation 2.5 and the equations that follow is given in Chapter 3. A short development is used here to support the concept of yield.) The net growth rate is equal the difference between the growth from substrate consumption minus decay due to self (endogenous) respiration or predation. The net yield (Y_n, M/M) can be found by dividing Equation 2.5 by the rate of substrate utilization:

$$Y_n = \frac{dX_a/dt}{-dS/dt} = Y - b\frac{X_a}{-dS/dt} \qquad \textbf{[2.6]}$$

The net yield is less than Y, because some of the electrons originally present in the substrate must be consumed for energy of maintenance. When considering net yield, the portion of electrons used for synthesis is f_s rather than f_s^0, and the portion for energy generation is f_e rather than f_e^0. Still, the sum of f_s and f_e equals 1, and $f_s < f_s^0$, while $f_e > f_e^0$.

Equation 2.6 indicates that this decayed portion becomes larger as the decay rate or organism concentration increases or as the rate of substrate consumption declines. For the casein example (Equation 2.1), the cellular yield of 0.61 g cells/g casein actually represents a net yield, as it was developed from experimental data in which a mixed culture of organisms was growing from substrate utilization and, at the same time, decaying.

When the rate of substrate utilization per unit mass of cells is sufficiently low, the right side of Equation 2.6 can become zero, meaning that the net yield of cells, Y_n, also equals zero. The substrate utilization rate is then just sufficient to maintain the cells, and no net growth of active cells results. Under these conditions:

$$Y_n = 0, \quad \frac{-dS/dt}{X_a} = \frac{b}{Y} = m \qquad \text{[2.7]}$$

For this case, the substrate utilization rate per unit mass of organisms is termed the maintenance energy (m, M/MT). From Equation 2.7, m is proportional to b and inversely proportional to Y. When the substrate utilization rate is less than m, the substrate available is insufficient to satisfy the total metabolic needs of the microorganisms. This represents a form of starvation.

Example 2.2 | **GROWTH RATE AND NET YIELD**   A reactor containing 500 mg/L of active microorganisms is consuming acetate at a rate of 750 mg L^{-1} d^{-1}. For this culture, $Y = 0.6$ g cells per g acetate and $b = 0.15$ d^{-1}. Determine the specific growth rate of the microorganisms $[(dX_a/dt)/X_a]$, the specific rate of substrate utilization $[(-dS/dt)/X_a]$, and the net yield of cells.

From Equation 2.5, $(dX_a/dt)/X_a = [Y(-dS/dt)/X_a - b] = [0.6(750)/500] - 0.15 = 0.75$ d^{-1}. This means that the population of organisms under these conditions is increasing at a rate of 75 percent per day. Next, $(-dS/dt)/X_a = 750/500 = 1.5$ g acetate per g cells per day. In other words, the microorganisms are consuming 1.5 times their weight in food each day. The net yield is given by Equation 2.6, $Y_n = 0.6 - 0.15/1.5 = 0.5$ g cells per g acetate per day. This is $0.5/0.6$ or 83 percent of the true yield Y.

2.4   ENERGY REACTIONS

Microorganisms obtain their energy for growth and maintenance from oxidation-reduction reactions. Even the photosynthetic microorganisms, which obtain energy from electromagnetic radiation or sunlight, use oxidation-reduction reactions to convert the light energy into ATP and NADH.

Oxidation-reduction reactions always involve an electron donor and an electron acceptor. Generally, we think of the electron donor as being the "food" substrate for the organisms. The most common electron donor for all nonphotosynthetic organisms, except some prokaryotes, is organic matter. The chemolithotrophic prokaryotes, however, use reduced inorganic compounds, such as ammonia and sulfide, as electron donors in energy metabolism. Prokaryotes are thus exceptionally versatile. The common electron acceptor under aerobic conditions is diatomic or molecular oxygen (O_2). However, under anaerobic conditions, some prokaryotes can use other electron acceptors in energy metabolism, including nitrate, sulfate, and carbon dioxide. In some cases, organic matter is used as the electron acceptor, as well as the electron donor, and the reaction is then termed fermentation.

Examples of different energy reactions in which glucose is the electron donor, listed below, show that the energy gained from one mole of glucose varies widely, depending on the electron acceptor.

ΔG^0 comparison for glucose as donor and various electron acceptors.

	Free Energy kJ/mol glucose	

Aerobic oxidation:
$$C_6H_{12}O_6 + 6\,O_2 \rightarrow 6\,CO_2 + 6\,H_2O \qquad -2{,}880 \qquad [2.8]$$

Denitrification:
$$5\,C_6H_{12}O_6 + 24\,NO_3^- + 24\,H^+ \rightarrow 30\,CO_2 + 42\,H_2O + 12\,N_2 \qquad -2{,}720 \qquad [2.9]$$

Sulfate reduction:
$$2\,C_6H_{12}O_6 + 6\,SO_4^{2-} + 9\,H^+ \rightarrow 12\,CO_2 + 12\,H_2O + 3\,H_2S + 3\,HS^- \qquad -492 \qquad [2.10]$$

Methanogenesis:
$$C_6H_{12}O_6 \rightarrow 3\,CO_2 + 3\,CH_4 \qquad -428 \qquad [2.11]$$

Ethanol Fermentation:
$$C_6H_{12}O_6 \rightarrow 2\,CO_2 + 2\,CH_3CH_2OH \qquad -244 \qquad [2.12]$$

Obviously, microorganisms would like to obtain as much energy from a reaction as possible; therefore, they would prefer to use oxygen as an electron acceptor. Not all microorganisms can use oxygen as an electron acceptor, and these anaerobic microorganisms are unable to compete with the aerobes when oxygen is available. On the other hand, when oxygen is absent, anaerobic microorganisms may dominate. The order of preference for electron acceptors based upon energy considerations alone would be oxygen, nitrate, sulfate, carbon dioxide (methanogenesis), and, finally, fermentation. Some organisms, such as *E. coli*, have the ability to use several different electron acceptors, including oxygen, nitrate, or organic matter in fermentation. Others, such as methanogens, can only carry out the methane-forming reactions. In fact, methanogens are strict anaerobes, and oxygen is harmful to them.

The above reaction energetics suggest that aerobic organisms need to send relatively few electrons from their donor to oxygen in order to generate the energy required to synthesize a given amount of new biomass. In the terms of our partitioning framework, f_e^0 is small and f_s^0 is large. Since Y and f_s^0 are proportional, the aerobic microorganisms should have a higher true yield (Y) value than anaerobic microorganisms. Thus, knowledge of reaction energetics, as well as energy reactions, gives valuable insight into reaction stoichiometry.

Constructing energy reactions is a useful beginning in the development of an overall stoichiometric equation for microbial growth. Writing equations such as those for glucose oxidation is relatively straightforward. However, more complex reactions, such as those for nitrate and sulfate reduction, are more difficult. Various approaches can be used to do this, including methods described by Battley (1987) and Bailey and Ollis (1986). The approach that we use focuses on half-reactions. The half-reaction approach is the most straightforward to use, particularly for very complex reactions, and it is totally consistent with the energetics of reactions as used here.

The example that follows leads to construction of Equation 2.9, glucose oxidation with nitrate. The oxidation half-reaction for glucose, written on a one-electron-equivalent basis, is

electron donor

$$\frac{1}{24}\,C_6H_{12}O_6 + \frac{1}{4}\,H_2O \rightarrow \frac{1}{4}\,CO_2 + H^+ + e^- \qquad [2.13]$$

The reduction half-reaction for nitrate, also written for one electron equivalent, is

e^- acceptor

$$\frac{1}{5} NO_3^- + \frac{6}{5} H^+ + e^- \rightarrow \frac{1}{10} N_2 + \frac{3}{5} H_2O \qquad \text{[2.14]}$$

Adding Equations 2.13 and 2.14 gives the overall balanced reaction, in which no free electrons (e^-) are present:

$$\frac{1}{24} C_6H_{12}O_6 + \frac{1}{5} NO_3^- + \frac{1}{5} H^+ \rightarrow \frac{1}{4} CO_2 + \frac{7}{20} H_2O + \frac{1}{10} N_2 \qquad \text{[2.15]}$$

If Equation 2.15 is multiplied by the least common denominator of 120, Equation 2.9 results.

Tables 2.2 and 2.3 summarize half-reactions, written as one-electron reduction reactions (e^- on the left), for a series of oxidation-reductions of interest in environmental biotechnology. When the half-reaction is to be used as an oxidation, the left and right sides are switched, so that the e^- appears on the right. The sign of the free energy changes when the half-reactions are written as oxidations.

Reduction half-reactions not in the lists can be written readily if a set of simple steps is followed. Consider the reduction half-reaction for the amino acid alanine (CH_3CHNH_2COOH). Only one element is allowed to have its oxidation state changed in a half-reaction. In this case, the element is carbon, which represents 12 e^- eq equivalents in alanine (7 e^- eq in the left-hand CH_3 group, 4 in the middle $CHNH_2$ group, and 1 in the COOH group). The other elements (H, O, and N) must retain their oxidation state the same on both sides of the equation, in this case, +1, −2, and −3, respectively. In half-reactions for organic compounds, the oxidized form is always carbon dioxide, or, in some cases, bicarbonate or carbonate. In the following case, carbon dioxide is used.

Step 1   Write the oxidized form of the element of interest on the left and the reduced form on the right.

$$CO_2 \qquad \rightarrow CH_3CHNH_2COOH$$

Step 2   Add other species that are formed or consumed in the reaction. In oxidation-reduction reactions, water is almost always a reactant or a product; here it will be included as a reactant in order to balance the oxygen present in the organic compound. As a reduction half-reaction, electrons must also appear on the left side of the equation. Since the nitrogen is present in the reduced form as an amino group in alanine, N must appear on the left side of the equation in the reduced form, either as NH_3 or NH_4^+. In this case, we arbitrarily select NH_3 for illustration.

$$CO_2 + H_2O + NH_3 + e^- \qquad \rightarrow CH_3CHNH_2COOH$$

Step 3   Balance the reaction for the element that is reduced and for all elements except oxygen and hydrogen. In this case, carbon and nitrogen must be balanced.

$$3 CO_2 + H_2O + NH_3 + e^- \qquad \rightarrow CH_3CHNH_2COOH$$

Table 2.2   Inorganic half-reactions and their Gibb's standard free energy at pH = 7.0

Reaction Number	Reduced-oxidized Compounds	Half-reaction		$\Delta G^{0'}$ kJ/e$^-$ eq
I-1	Ammonium-Nitrate:	$\frac{1}{8} NO_3^- + \frac{5}{4} H^+ + e^-$	$= \frac{1}{8} NH_4^+ + \frac{3}{8} H_2O$	−35.11
I-2	Ammonium-Nitrite:	$\frac{1}{6} NO_2^- + \frac{4}{3} H^+ + e^-$	$= \frac{1}{6} NH_4^+ + \frac{1}{3} H_2O$	−32.93
I-3	Ammonium-Nitrogen:	$\frac{1}{6} N_2 + \frac{4}{3} H^+ + e^-$	$= \frac{1}{3} NH_4^+$	26.70
I-4	Ferrous-Ferric:	$Fe^{3+} + e^-$	$= Fe^{2+}$	−74.27
I-5	Hydrogen-H$^+$:	$H^+ + e^-$	$= \frac{1}{2} H_2$	39.87
I-6	Nitrite-Nitrate:	$\frac{1}{2} NO_3^- + H^+ + e^-$	$= \frac{1}{2} NO_2^- + \frac{1}{2} H_2O$	−41.65
I-7	Nitrogen-Nitrate:	$\frac{1}{5} NO_3^- + \frac{6}{5} H^+ + e^-$	$= \frac{1}{10} N_2 + \frac{3}{5} H_2O$	−72.20
I-8	Nitrogen-Nitrite:	$\frac{1}{3} NO_2^- + \frac{4}{3} H^+ + e^-$	$= \frac{1}{6} N_2 + \frac{2}{3} H_2O$	−92.56
I-9	Sulfide-Sulfate:	$\frac{1}{8} SO_4^{2-} + \frac{19}{16} H^+ + e^-$	$= \frac{1}{16} H_2S + \frac{1}{16} HS^- + \frac{1}{2} H_2O$	20.85
I-10	Sulfide-Sulfite:	$\frac{1}{6} SO_3^{2-} + \frac{5}{4} H^+ + e^-$	$= \frac{1}{12} H_2S + \frac{1}{12} HS^- + \frac{1}{2} H_2O$	11.03
I-11	Sulfite-Sulfate:	$\frac{1}{2} SO_4^{2-} + H^+ + e^-$	$= \frac{1}{2} SO_3^{2-} + \frac{1}{2} H_2O$	50.30
I-12	Sulfur-Sulfate:	$\frac{1}{6} SO_4^{2-} + \frac{4}{3} H^+ + e^-$	$= \frac{1}{6} S + \frac{2}{3} H_2O$	19.15
I-13	Thiosulfate-Sulfate:	$\frac{1}{4} SO_4^{2-} + \frac{5}{4} H^+ + e^-$	$= \frac{1}{8} S_2O_3^{2-} + \frac{5}{8} H_2O$	23.58
I-14	Water-Oxygen:	$\frac{1}{4} O_2 + H^+ + e^-$	$= \frac{1}{2} H_2O$	−78.72

Step 4   Balance the oxygen through addition or subtraction of water. Elemental oxygen is not to be used here, as oxygen must not have its oxidation state changed.

$$3 CO_2 + NH_3 + e^- \rightarrow CH_3CHNH_2COOH + 4 H_2O$$

Step 5   Balance hydrogen by introducing H$^+$.

$$3 CO_2 + NH_3 + 12 H^+ + e^- \rightarrow CH_3CHNH_2COOH + 4 H_2O$$

Table 2.3   Organic half-reactions and their Gibb's free energy

Reaction Number	Reduced Compounds	Half-reaction	$\Delta G^{0\prime}$ kJ/e⁻ eq
O-1	Acetate:	$\frac{1}{8} CO_2 + \frac{1}{8} HCO_3^- + H^+ + e^- = \frac{1}{8} CH_3COO^- + \frac{3}{8} H_2O$	27.40
O-2	Alanine:	$\frac{1}{6} CO_2 + \frac{1}{12} HCO_3^- + \frac{1}{12} NH_4^+ + \frac{11}{12} H^+ + e^- = \frac{1}{12} CH_3CHNH_2COO^- + \frac{5}{12} H_2O$	31.37
O-3	Benzoate:	$\frac{1}{5} CO_2 + \frac{1}{30} HCO_3^- + H^+ + e^- = \frac{1}{30} C_6H_5COO^- + \frac{13}{30} H_2O$	27.34
O-4	Citrate:	$\frac{1}{6} CO_2 + \frac{1}{6} HCO_3^- + H^+ + e^- = \frac{1}{18}(COO^-)CH_2COH(COO^-)CH_2COO^- + \frac{4}{9} H_2O$	33.08
O-5	Ethanol:	$\frac{1}{6} CO_2 + H^+ + e^- = \frac{1}{12} CH_3CH_2OH + \frac{1}{4} H_2O$	31.18
O-6	Formate:	$\frac{1}{2} HCO_3^- + H^+ + e^- = \frac{1}{2} HCOO^- + \frac{1}{2} H_2O$	39.19
O-7	Glucose:	$\frac{1}{4} CO_2 + H^+ + e^- = \frac{1}{24} C_6H_{12}O_6 + \frac{1}{4} H_2O$	41.35
O-8	Glutamate:	$\frac{1}{6} CO_2 + \frac{1}{9} HCO_3^- + \frac{1}{18} NH_4^+ + H^+ + e^- = \frac{1}{18} COOHCH_2CH_2CHNH_2COO^- + \frac{4}{9} H_2O$	30.93
O-9	Glycerol:	$\frac{3}{14} CO_2 + H^+ + e^- = \frac{1}{14} CH_2OHCHOHCH_2OH + \frac{3}{14} H_2O$	38.88
O-10	Glycine:	$\frac{1}{6} CO_2 + \frac{1}{6} HCO_3^- + \frac{1}{6} NH_4^+ + H^+ + e^- = \frac{1}{6} CH_2NH_2COOH + \frac{1}{2} H_2O$	39.80

O-11	Lactate:	$\frac{1}{6}CO_2 + \frac{1}{12}HCO_3^- + H^+ + e^- = \frac{1}{12}CH_3CHOHCOO^- + \frac{1}{3}H_2O$	32.29
O-12	Methane:	$\frac{1}{8}CO_2 + H^+ + e^- = \frac{1}{8}CH_4 + \frac{1}{4}H_2O$	23.53
O-13	Methanol:	$\frac{1}{6}CO_2 + H^+ + e^- = \frac{1}{6}CH_3OH + \frac{1}{6}H_2O$	36.84
O-14	Palmitate:	$\frac{15}{19}CO_2 + \frac{1}{92}HCO_3^- + H^+ + e^- = \frac{1}{92}CH_3(CH_2)_{14}COO^- + \frac{31}{92}H_2O$	27.26
O-15	Propionate:	$\frac{1}{7}CO_2 + \frac{1}{14}HCO_3^- + H^+ + e^- = \frac{1}{14}CH_3CH_2COO^- + \frac{5}{14}H_2O$	27.63
O-16	Pyruvate:	$\frac{1}{5}CO_2 + \frac{1}{10}HCO_3^- + H^+ + e^- = \frac{1}{10}CH_3COCOO^- + \frac{2}{5}H_2O$	35.09
O-17	Succinate:	$\frac{1}{7}CO_2 + \frac{1}{7}HCO_3^- + H^+ + e^- = \frac{1}{14}(CH_2)_2(COO^-)_2 + \frac{3}{7}H_2O$	29.09
O-18	Domestic Wastewater:	$\frac{9}{50}CO_2 + \frac{1}{50}NH_4^+ + \frac{1}{50}HCO_3^- + H^+ + e^- = \frac{1}{50}C_{10}H_{19}O_3N + \frac{9}{25}H_2O$	*
O-19	Custom Organic Half Reaction:	$\frac{(n-c)}{d}CO_2 + \frac{c}{d}NH_4^+ + \frac{c}{d}HCO_3^- + H^+ + e^- = \frac{1}{d}C_nH_aO_bN_c + \frac{2n-b+c}{d}H_2O$ where, $d = (4n + a - 2b - 3c)$	*
O-20	Cell Synthesis:	$\frac{1}{5}CO_2 + \frac{1}{20}NH_4^+ + \frac{1}{20}HCO_3^- + H^+ + e^- = \frac{1}{20}C_5H_7O_2N + \frac{9}{20}H_2O$	*

1 * Equations O-18 to O-20 do not have $\Delta G^{0\prime}$ values because the reduced species is not chemically defined.

general formula for organic matter: $C_nH_aO_bN_c$

Step 6   Balance the charge on the reaction by adding sufficient e^- to the left side of the equation.

$$3\,CO_2 + NH_3 + 12\,H^+ + 12\,e^- \longrightarrow CH_3CHNH_2COOH + 4\,H_2O$$

The coefficient on electrons should equal the number of electron equivalents in the reduced compound, or $12\,e^-$ eq in alanine, which it does.

Step 7   Convert the equation to the electron-equivalent form by dividing by the coefficient on e^-.

$$\frac{1}{4}\,CO_2 + \frac{1}{12}\,NH_3 + H^+ + e^- \to \frac{1}{12}\,CH_3CHNH_2COOH + \frac{1}{3}\,H_2O \qquad \textbf{[2.16]}$$

Inorganic oxidation-reduction half-reactions can be written in a similar fashion. The reduced and oxidized form of the element must be known. For example, chromium commonly exists in aqueous solution in either the oxidized Cr(VI) form or the reduced Cr(III) form. The chemical species associated with Cr(VI) in water are the oxyanion forms, either CrO_4^{2-} at neutral pH, or as the dichromate form, $Cr_2O_7^{2-}$, in acid waters. Cr(III) would exist in water as a cation, Cr^{3+}. Assuming that neutral-pH conditions apply and following the above seven steps, we derive the reduction half-reaction for chromate by following the seven steps.

Step 1

$$CrO_4^{2-} \to Cr^{3+}$$

Step 2

$$CrO_4^{2-} + H^+ + e^- \to Cr^{3+} + H_2O$$

Step 3

$$CrO_4^{2-} + H^+ + e^- \to Cr^{3+} + H_2O \quad \text{(Cr balanced in Step 2; thus no change here)}$$

Step 4

$$CrO_4^{2-} + H^+ + e^- \to Cr^{3+} + 4\,H_2O$$

Step 5

$$CrO_4^{2-} + 8\,H^+ + e^- \to Cr^{3+} + 4\,H_2O$$

Step 6

$$CrO_4^{2-} + 8\,H^+ + 3\,e^- \to Cr^{3+} + 4\,H_2O$$

Step 7

$$\frac{1}{3}\,CrO_4^{2-} + \frac{8}{3}\,H^+ + e^- \to \frac{1}{3}\,Cr^{3+} + \frac{4}{3}\,H_2O$$

One half-reaction of note is the last one in Table 2.3. It is for the formation of bacterial cells based on the empirical formula $C_5H_7O_2N$:

$$\frac{1}{5} CO_2 + \frac{1}{20} NH_4^+ + \frac{1}{20} HCO_3^- + H^+ + e^- \rightarrow \frac{1}{20} C_5H_7O_2N + \frac{9}{20} H_2O$$

[2.17]

Also listed in Table 2.3 is a generic equation for organic compounds in general (reaction O-19) based on an assumed formula for organic matter of $C_nH_aO_bN_c$. Organic compounds may contain other elements, such as P and S, and these may be added if desired. Also, organic compounds may have charge, such as negative charge for carboxylate anions or positive charge for some nitrogen-containing organic compounds. Charge could be added as well if needed. Another representation in Table 2.3 is an empirical formula for organic matter in domestic sewage (reaction O-18). Such representations can be developed if knowledge is available on the relative composition of a wastewater in terms of carbon, oxygen, hydrogen, and nitrogen in the organic matter.

If Equation 2.16 is compared with its counterpart in Table 2.3 (reaction O-2), there are differences, particularly in the distribution of CO_2, HCO_3^-, H_2O, and H^+. Yet, the number of moles of alanine given by one electron equivalent (1/12) remains the same. Both equations balance. Thus, questions may be asked, such as: Why the difference? Which equation is correct? Which one should I use for my particular problem? One answer is that both are correct from the standpoint of being balanced half-reactions. The difference results from the assumed species that are involved in the reaction.

For example, in one equation, NH_3 is assumed to be the species of nitrogen involved, while in the other, the ionized form, NH_4^+, is assumed. The best form to include depends on the problem being addressed. If we are describing the biodegradation of alanine in a biological treatment reactor operating at neutral pH, then an assumption that NH_4^+ is formed in the reaction is perhaps the better one, since this is the species that dominates in aqueous solutions at pH below about 9.3. If the interest were in a reaction occurring at pH well above this value, then NH_3 would be the species expected to dominate. When NH_4^+ is released following oxidation of an organic compound, a negatively charged species must also be formed to balance the charge. With organic oxidation at neutral pH, the species thus formed is generally HCO_3^-. It is for this reason that, for each half-reaction in Table 2.3 that involves organic nitrogen, HCO_3^- enters into the reaction. The significance of this may be illustrated by considering the overall oxidation-reduction reaction that occurs in the anaerobic methanogenic fermentation of alanine, which can be obtained by subtracting reaction O-2 from reaction O-12:

Reaction O-12 $\quad \frac{1}{8} CO_2 + H^+ + e^- \rightarrow \frac{1}{8} CH_4 + \frac{1}{4} H_2O$ [2.18]

$-$ Reaction O-2 $\quad \frac{1}{12} CH_3CHNH_2COOH + \frac{5}{12} H_2O \rightarrow$

$$\frac{1}{6} CO_2 + \frac{1}{12} NH_4^+ + \frac{1}{12} HCO_3^- + H^+ + e^-$$ [2.19]

Sum $\quad \frac{1}{12} CH_3CHNH_2COOH + \frac{1}{6} H_2O \rightarrow \frac{1}{8} CH_4 + \frac{1}{24} CO_2 + \frac{1}{12} NH_4^+ + \frac{1}{12} HCO_3^-$ [2.20]

This equation indicates that, when 1/12 mole (or 7.42 g) of alanine is converted for energy during methane fermentation, 1/8 mole methane and 1/24 mole carbon dioxide are formed as gases, and 1/12 mole each of ammonium and bicarbonate are released to the solution. This correlation between NH_4^+ and HCO_3^- is very important for pH control in anaerobic treatment and is emphasized again in Chapter 13.

Example 2.3 | **PRODUCTS FROM METHANOGENESIS OF ALANINE**   An industrial wastewater contains 100-mM alanine. If it were all converted for energy during methane fermentation at a pH near neutral: (a) what would be the composition of the gas produced, and (b) what would be the concentrations of ammonium and bicarbonate in solution?

(a) Since the gas produced has three times more methane than carbon dioxide, according to Equation 2.20, then the composition of the gases produced is 75% methane and 25% carbon dioxide.

(b) Since Equation 2.20 indicates that 1/12 mole each of NH_4^+ and HCO_3^- are produced for each 1/12 mole of alanine fermented to methane for energy, then 100-mM alanine produces 100 mM each of ammonia nitrogen and bicarbonate. With a formula weight of 14, the ammonium-nitrogen concentration would increase by 1,400 mg/l. Also, the bicarbonate alkalinity, when expressed as equivalent calcium carbonate (equivalent weight of 50), increases by 5,000 mg/l. The bicarbonate formed produces a natural pH buffer to help maintain a near-neutral pH.

For half reactions that involve organic chemicals with a negative charge, such as the carboxylate group for acetic acid, HCO_3^- can be used in the reaction to provide charge balance. Several examples are contained in Table 2.3; that for pyruvate is repeated here as an example:

$$\frac{1}{5}CO_2 + \frac{1}{10}HCO_3^- + H^+ + e^- \rightarrow \frac{1}{10}CH_3COCOO^- + \frac{2}{5}H_2O \qquad [2.21]$$

Here, the moles of bicarbonate on the left side of the equation are made to just equal the moles of carboxylate on the right side. Carbon dioxide is then added in sufficient amount to provide a carbon balance. We use HCO_3^- when NH_4^+ nitrogen or organic anions are involved because the species conform to the dominant species in solution at neutral pH. Some authors prefer other forms for different reasons. All are technically correct as long as the equations are written in a balanced form. Different half-reaction forms sometimes used are illustrated in the following half-reactions for alanine:

$$\frac{1}{4}CO_2 + \frac{1}{12}NH_3 + H^+ + e^- \rightarrow \frac{1}{12}CH_3CHNH_2COOH + \frac{1}{3}H_2O \qquad [2.22]$$

$$\frac{1}{6}CO_2 + \frac{1}{12}NH_4^+ + \frac{1}{12}HCO_3^- + H^+ + e^- \rightarrow \frac{1}{12}CH_3CHNH_2COOH + \frac{5}{12}H_2O \qquad [2.23]$$

$$\frac{1}{6}CO_2 + \frac{1}{12}NH_4^+ + \frac{1}{12}HCO_3^- + \frac{11}{12}H^+ + e^- \rightarrow \frac{1}{12}CH_3CHNH_2COO^- + \frac{5}{12}H_2O \qquad [2.23a]$$

$$\frac{1}{6}CO_2 + \frac{1}{12}NH_4^+ + \frac{1}{12}HCO_3^- + \frac{1}{2}H_2 \rightarrow \frac{1}{12}CH_3CHNH_2COOH + \frac{5}{12}H_2O \qquad [2.24]$$

$$\frac{1}{12}NH_4^+ + \frac{1}{4}HCO_3^- + \frac{13}{12}H^+ + e^- \rightarrow \frac{1}{12}CH_3CHNH_2COO^- + \frac{7}{12}H_2O \qquad [2.25]$$

Still other forms can be written through various combinations of the above. Regardless of the form selected, 1/12 mole of alanine gives one electron equivalent.

2.5 OVERALL REACTIONS FOR BIOLOGICAL GROWTH

Bacterial growth involves two basic reactions, one for energy production and the other for cellular synthesis. The electron donor provides electrons to the electron acceptor for energy production. The previous section showed how to combine half-reactions for the energy part. The next step is to combine half-reactions to represent *synthesis*. With this information and knowledge of how the electrons from the donor are partitioned, we can write an overall, balanced reaction such as Equation 2.1.

Before we can do this, we need to have a half-reaction for synthesis, R_c. Table 2.4 lists the key synthesis half-reactions. We now consider the first synthesis half-reaction (C-1), which is used when ammonium is the nitrogen source for formation of proteins and nucleic acids. Ammonium is the preferred nitrogen source. If it is not available, microorganisms may be able to use the other nitrogen sources indicated in the table. Also contained in Table 2.4 for convenience are the acceptor half-reactions (R_a) for the five most common electron acceptors: O_2, NO_3^-, Fe^{3+}, SO_4^{2-}, and CO_2.

As an example, we assume that benzoate is available as an electron donor, nitrate is available as an electron acceptor, and ammonium is available as the nitrogen source. On a net-yield basis, we also assume that 40 percent of the electron equivalents in benzoate is used for synthesis ($f_s = 0.40$), while the other 60 percent is used for energy ($f_e = 0.60$).

First, the overall energy and synthesis reactions are developed. The donor half-reaction is designated as R_d, the acceptor half-reaction as R_a, and the cell half-reaction as R_c. The energy reaction, R_e, then becomes

$$R_e = R_a - R_d \qquad \textbf{[2.26]}$$

and the synthesis reaction R_s, becomes

$$R_s = R_c - R_d \qquad \textbf{[2.27]}$$

It is important to note that R_d has a negative sign, because the donor is oxidized.

The actual half-reactions are now substituted into Equations 2.26 and 2.27. First is the energy reaction:

$$R_a: \quad \frac{1}{5} NO_3^- + \frac{6}{5} H^+ + e^- \rightarrow \frac{1}{10} N_2 + \frac{3}{5} H_2O \qquad \textbf{[2.28]}$$

$$-R_d: \quad \frac{1}{30} C_6H_5COO^- + \frac{13}{30} H_2O \rightarrow \frac{1}{5} CO_2 + \frac{1}{30} HCO_3^- + H^+ + e^- \qquad \textbf{[2.29]}$$

$$R_e: \quad \frac{1}{30} C_6H_5COO^- + \frac{1}{5} NO_3^- + \frac{1}{5} H^+ \rightarrow \frac{1}{5} CO_2 + \frac{1}{10} N_2 + \frac{1}{30} HCO_3^- + \frac{1}{6} H_2O \qquad \textbf{[2.30]}$$

Table 2.4   Cell formation (R_c) and common electron acceptor half-reactions (R_a)

Reaction Number	Half-reaction	$\Delta G^{0'}$ kJ/e^- eq
Cell Synthesis Equations (R_c)		
Ammonium as Nitrogen Source		
C-1	$\frac{1}{5} CO_2 + \frac{1}{20} HCO_3^- + \frac{1}{20} NH_4^+ + H^+ + e^- = \frac{1}{20} C_5H_7O_2N + \frac{9}{20} H_2O$	
Nitrate as Nitrogen Source		
C-2	$\frac{1}{28} NO_3^- + \frac{5}{28} CO_2 + \frac{29}{28} H^+ + e^- = \frac{1}{28} C_5H_7O_2N + \frac{11}{28} H_2O$	
Nitrite as Nitrogen Source		
C-3	$\frac{5}{26} CO_2 + \frac{1}{26} NO_2^- + \frac{27}{26} H^+ + e^- = \frac{1}{26} C_5H_7O_2N + \frac{10}{26} H_2O$	
Dinitrogen as Nitrogen Source		
C-4	$\frac{5}{23} CO_2 + \frac{1}{46} N_2 + H^+ + e^- = \frac{1}{23} C_5H_7O_2N + \frac{8}{23} H_2O$	
Common Electron-Acceptor Equations (R_a)		
I-14 Oxygen	$\frac{1}{4} O_2 + H^+ + e^- = \frac{1}{2} H_2O$	-78.72
I-7 Nitrate	$\frac{1}{5} NO_3^- + \frac{6}{5} H^+ + e^- = \frac{1}{10} N_2 + \frac{3}{5} H_2O$	-72.20
I-9 Sulfate	$\frac{1}{8} SO_4^{2-} + \frac{19}{16} H^+ + e^- = \frac{1}{16} H_2S + \frac{1}{16} HS^- + \frac{1}{2} H_2O$	20.85
O-12 CO$_2$	$\frac{1}{8} CO_2 + H^+ + e^- = \frac{1}{8} CH_4 + \frac{1}{4} H_2O$	23.53
I-4 Iron (III)	$Fe^{3+} + e^- = Fe^{2+}$	-74.27

Similarly, the synthesis reaction is

$$R_c: \quad \frac{1}{5} CO_2 + \frac{1}{20} NH_4^+ + \frac{1}{20} HCO_3^- + H^+ + e^- \rightarrow \frac{1}{20} C_5H_7O_2N + \frac{9}{20} H_2O \qquad \textbf{[2.17]}$$

$$-R_d: \quad \frac{1}{30} C_6H_5COO^- + \frac{13}{30} H_2O \rightarrow \frac{1}{5} CO_2 + \frac{1}{30} HCO_3^- + H^+ + e^- \qquad \textbf{[2.29]}$$

$$R_s: \quad \frac{1}{30} C_6H_5COO^- + \frac{1}{20} NH_4^+ + \frac{1}{60} HCO_3^- \rightarrow \frac{1}{20} C_5H_7O_2N + \frac{1}{60} H_2O \qquad \textbf{[2.31]}$$

Second, and in order to obtain an overall reaction that includes energy generation and synthesis, Equation 2.30 is multiplied by f_e, Equation 2.31 is multiplied by f_s, and they are summed:

$$f_e R_e: \quad 0.02\, C_6H_5COO^- + 0.12\, NO_3^- + 0.12\, H^+ \rightarrow \qquad \textbf{[2.32]}$$
$$0.12\, CO_2 + 0.06\, N_2 + 0.02\, HCO_3^- + 0.1\, H_2O$$

$$f_s R_s: \quad 0.0133\, C_6H_5COO^- + 0.02\, NH_4^+ + 0.0067\, HCO_3^- \rightarrow \qquad \textbf{[2.33]}$$
$$0.02\, C_5H_7O_2N + 0.0067\, H_2O$$

$$R: \quad 0.0333\, C_6H_5COO^- + 0.12\, NO_3^- + 0.02\, NH_4^+ + 0.12\, H^+ \rightarrow$$
$$0.02\, C_5H_7O_2N + 0.06\, N_2 + 0.12\, CO_2 + 0.0133\, HCO_3^- + 0.1067\, H_2O \qquad \textbf{[2.34]}$$

both energy and synthesis

Equation 2.34 provides an overall equation for net synthesis of bacteria that are using benzoate as an electron donor and nitrate as an electron acceptor. The bacteria here also use ammonium as a nitrogen source for synthesis.

We can see that Equation 2.34 is obtained by summation of Equation 2.32 and 2.33, that is,

$$R = f_e(R_a - R_d) + f_s(R_c - R_d) \qquad \textbf{[2.35]}$$

Since the fractions of electrons used for energy generation and synthesis must equal 1.0,

$$f_s + f_e = 1.0 \quad \text{and} \quad R_d(f_s + f_e) = R_d \qquad \textbf{[2.36]}$$

which converts Equation 2.35 to

$$R = f_e R_a + f_s R_c - R_d \qquad \textbf{[2.37]}$$

Equation 2.37 is a general equation that can be used for constructing a wide variety of stoichiometric equations for microbial synthesis and growth. The result is an equation written on an electron-equivalent basis. In other words, the equation represents the net consumption of reactants and production of products when the microorganisms consume one electron equivalent of electron donor.

The bacteria in the above case with benzoate used ammonium as a nitrogen source for synthesis. In some situations, ammonium is not available, but oxidized sources of nitrogen are available. The oxidized sources are NO_3^-, NO_2^-, and N_2, although NO_3^- is the most common one. Table 2.4 gives the cell-formation half-reactions for all four nitrogen sources. An important feature of the half-reactions is that the cell formula has a different stoichiometric coefficient for each nitrogen source: 1/20 for NH_4^+, 1/28 for NO_3^-, 1/26 for NO_2^-, and 1/23 for N_2. This difference reflects the number of electron equivalents that must be added to the 20 for C when the N source for $C_5H_7O_2N$ must be reduced: 0 extra for NH_4^+, 8 extra for NO_3^-, 6 extra for NO_2^-, and 3 extra for N_2. When an oxidized N source is required, the f_s value also may change.

The use of an oxidized nitrogen source is illustrated here for NO_3^-. We write an overall equation for bacterial growth with benzoate as the electron donor and carbon source, while NO_3^- is the electron acceptor and nitrogen source. For this case, we assume that f_s is 0.55. The appropriate equations are O-3 from Table 2.3 for benzoate and R_d, C-2 from Table 2.4 for R_c, and I-7 from Table 2.4 for R_a. The value of f_e becomes $1 - f_s = 0.45$. The overall reaction is then determined by the same pattern:

$f_e R_a$: $0.09\,NO_3^- + 0.54\,H^+ + 0.45\,e^- \rightarrow 0.045\,N_2 + 0.27\,H_2O$

$f_s R_c$: $0.0196\,NO_3^- + 0.0982\,CO_2 + 0.5696\,H^+ + 0.55\,e^- \rightarrow 0.0196\,C_5H_7O_2N + 0.2161\,H_2O$

$-R_d$: $0.0333\,C_6H_5COO^- + 0.4333\,H_2O \rightarrow 0.2\,CO_2 + 0.0333\,HCO_3^- + H^+ + e^-$

R: $0.0333\,C_6H_5COO^- + 0.1096\,NO_3^- + 0.1096\,H^+ \rightarrow$
$0.0196\,C_5H_7O_2N + 0.045\,N_2 + 0.0333\,HCO_3^- + 0.1018\,CO_2 + 0.0528\,H_2O$

We see that 0.09 mole nitrate is converted to nitrogen gas, and 0.0196 mole is converted into the organic nitrogen of the cells.

Most of the examples provided up to this point are for oxidations of organic matter. However, chemolithotrophs are important microorganisms that use reduced inorganic chemicals for energy and most frequently use inorganic carbon (CO_2) for organism synthesis. The inorganic half-reactions listed in Table 2.2 are examples from the many different inorganic oxidation-reduction reactions that can be mediated by lithotrophic microorganisms in order to obtain energy for growth. An important aspect to be noted from Equation 2.37 and its development is that the electron equivalents from the electron donor are divided between the energy reaction and cell synthesis. While the preceding examples illustrate this for heterotrophic reactions, it is equally true for lithotrophic reactions.

Example 2.4 | **NITRIFICATION STOICHIOMETRY**   Lithotrophic microorganisms are employed to oxidize ammonium in a wastewater to nitrate under aerobic conditions in order to reduce the oxygen consumption by nitrification in a receiving stream. If the concentration of ammonium in the wastewater, expressed as nitrogen, is 22 mg/l, how much oxygen will be consumed for nitrification in the treatment of 1,000 m^3 of wastewater, what mass of cells in kg dry weight will be produced, and what will be the resulting concentration of nitrate-nitrogen in the treated water? Assume f_s equals 0.10 and that inorganic carbon is used for cell synthesis.

For the reaction involved, ammonium serves as the electron donor and is oxidized to nitrate, oxygen is the electron acceptor since it is an aerobic reaction, and ammonium likewise serves as the source of nitrogen for cell synthesis. In addition, $f_e = 1 - f_s = 0.90$. Selecting the appropriate half-reactions from Tables 2.2 and 2.4 and employing Equation 2.37 gives the following overall biological reaction:

$$f_e R_a: \ 0.225\,O_2 + 0.9\,H^+ + 0.9\,e^- \rightarrow 0.45\,H_2O$$

$$f_s R_c: \ 0.02\,CO_2 + 0.005\,NH_4^+ + 0.005\,HCO_3^- + 0.1\,H^+ + 0.1\,e^- \rightarrow 0.005\,C_5H_7O_2N + 0.045\,H_2O$$

$$-R_d: \ 0.125\,NH_4^+ + 0.375\,H_2O \rightarrow 0.125\,NO_3^- + 1.25\,H^+ + e^-$$

$$R: \ 0.13\,NH_4^+ + 0.225\,O_2 + 0.02\,CO_2 + 0.005\,HCO_3^- \rightarrow$$
$$0.005\,C_5H_7O_2N + 0.125\,NO_3^- + 0.25\,H^+ + 0.12\,H_2O$$

For each $0.13(14) = 1.82$ g ammonium-nitrogen, $0.225(32) = 7.2$ g oxygen is consumed. Also, $0.005(113) = 0.565$ g cells and $0.125(14) = 1.75$ g NO_3^--N are produced. The amount of ammonium-nitrogen treated $= (22\ \text{mg/l})(1000\ m^3)(10^3\ \text{liters}/m^3)(\text{kg}/10^6\ \text{mg}) = 22$ kg. Thus,

$$\text{Oxygen consumption} = 22\ \text{kg}(7.2\ \text{g}/1.82\ \text{g}) \qquad = 87\ \text{kg}$$
$$\text{Cell dry weight produced} = 22\ \text{kg}(0.565\ \text{g}/1.82\ \text{g}) \quad = 6.83\ \text{kg}$$
$$\text{Effluent}\ NO_3^-\text{-N conc.} = 22\ \text{mg/l}(1.75\ \text{g}/1.82\ \text{g}) \quad = 21\ \text{mg/l}$$

Example 2.5 | **METHANOGENESIS STOICHIOMETRY**   Based upon analysis of its organic carbon, hydrogen, oxygen, and nitrogen content, the representative empirical formula of $C_8H_{17}O_3N$ was developed for organic matter in an industrial wastewater, and, from the same analysis, the organic concentration was found to be 23,000 mg/l. For a wastewater flow of 150 m^3/d, what is the daily methane production from the wastewater in m^3 at 35 °C and 1 atm and methane percentage in the gas produced if the waste were treated anaerobically by methane fermentation? Assume f_s is 0.08, the process is 95 percent efficient at removal of organic matter, and all gases formed evolve to the gas phase.

First, an electron-donor half-reaction for the wastewater needs to be developed. Applying Equation O-19 of Table 2.3 and the empirical formula ($C_8H_{17}O_3N$) give R_d:

$$R_d: \quad \frac{1}{40} NH_4^+ + \frac{1}{40} HCO_3^- + \frac{7}{40} CO_2 + H^+ + e^- \rightarrow \frac{1}{40} C_8H_{17}O_3N + \frac{7}{20} H_2O$$

The quantity of organic matter removed per day = $0.95(23 \text{ kg/m}^3)(150 \text{ m}^3/\text{d}) = 3{,}280$ kg/d.

Using $f_e = 1 - f_s = 1 - 0.08 = 0.92$ and the half-reaction for CO_2 to CH_4 as the electron acceptor reaction O-12 from Table 2.3, we develop R:

$f_e R_a$: $0.115 \, CO_2 + 0.92 \, H^+ + 0.92 \, e^- \rightarrow 0.115 \, CH_4 + 0.23 \, H_2O$

$f_s R_c$: $0.016 \, CO_2 + 0.004 \, NH_4^+ + 0.004 \, HCO_3^- + 0.08 \, H^+ + 0.08 \, e^- \rightarrow 0.004 \, C_5H_7O_2N + 0.036 \, H_2O$

$-R_d$: $0.025 \, C_8H_{17}O_3N + 0.35 \, H_2O \rightarrow 0.025 \, NH_4^+ + 0.025 \, HCO_3^- + 0.175 \, CO_2 + H^+ + e^-$

R: $0.025 \, C_8H_{17}O_3N + 0.084 \, H_2O \rightarrow$

$0.004 \, C_5H_7O_2N + 0.115 \, CH_4 + 0.044 \, CO_2 + 0.021 \, NH_4^+ + 0.021 \, HCO_3^-$

The formula weight of $C_8H_{17}O_3N$ is 175, and the equivalent weight is 0.025(175), or 4.375 g. Methane fermentation of one equivalent of organic matter produces 0.115 mol methane and 0.044 mol carbon dioxide. Thus,

Methane produced = $[(273 + 35)/273][0.0224 \text{m}^3 \text{ gas/mol}][3{,}280{,}000\text{g/d}][0.115 \text{ mol}/4.375 \text{ g}]$

$= 2{,}180 \text{ m}^3/\text{d}$

Percent methane = $100[0.115/(0.115 + 0.044)] = 72$ percent

2.5.1   FERMENTATION REACTIONS

In fermentations, an organic compound serves as electron donor and electron acceptor. A simple example is ethanol fermentation from glucose, as given by Equation 2.12. Here, one mole of glucose is converted into two moles ethanol and two moles carbon dioxide. We would like to develop a procedure for writing such an equation that is compatible with Equation 2.37, which we have been using to write overall reactions for biological growth. To do this, we again rely on half-reactions. We need to select appropriate half-reactions for electron donors and electron acceptors.

The complex pathways by which glucose is converted into ethanol were described in Chapter 1. From the viewpoint of writing an overall balanced reaction resulting from this process, one need not consider all the intermediate processes involved. No matter how complicated are the pathways by which chemical A is transformed into chemical B, the laws of energy and mass conservation must be obeyed. Knowledge of all reactants and products in a given case is all that is required to construct a balanced equation for the reaction, and we do not need to know the intermediates along the path, as long as they do not persist.

Simple Fermentations   A simple fermentation has only one reduced product, such as ethanol from glucose. All the electrons starting in glucose must end up in ethanol. Our first task is to select an electron donor half-reaction to use in Equation 2.37. The donor obviously is glucose, and the half-reaction we will use is that from Table 2.3 for conversion of CO_2 to glucose (reaction O-7). The second task is to identify the electron-acceptor reaction. It also is rather simple: the half-reaction for CO_2 to ethanol from Table 2.3 (reaction O-5). For the energy reaction alone, we simply use Equation 2.26:

$$R_a: \quad \frac{1}{6} CO_2 + H^+ + e^- \rightarrow \frac{1}{12} CH_3CH_2OH + \frac{1}{4} H_2O$$

$$-R_d: \quad \frac{1}{24} C_6H_{12}O_6 + \frac{1}{4} H_2O \rightarrow \frac{1}{4} CO_2 + H^+ + e^-$$

$$R_e: \quad \frac{1}{24} C_6H_{12}O_6 \rightarrow \frac{1}{12} CH_3CH_2OH + \frac{1}{12} CO_2$$

Multiplying R_e by 24 equivalents per mole gives Equation 2.12. Also, the selection of electron donor and acceptor half-reactions in this manner is compatible with using Equation 2.37 to build an overall reaction.

Example 2.6	**SIMPLE FERMENTATION STOICHIOMETRY** Write the overall biological reaction for ethanol fermentation from glucose, assuming that f_s equals 0.22.

Using the preceding ethanol and glucose half-reactions, coupled with the cell synthesis reaction from Table 2.3, and $f_e = 1 - f_s = 0.78$:

$$0.78 R_a: \quad 0.13 CO_2 + 0.78 H^+ + 0.78 e^- \rightarrow 0.065 CH_3CH_2OH + 0.195 H_2O$$

$$0.22 R_c: \quad 0.044 CO_2 + 0.011 NH_4^+ + 0.011 HCO_3^- + 0.22 H^+ + 0.22 e^- \rightarrow 0.011 C_5H_7O_2N + 0.099 H_2O$$

$$-R_d: \quad 0.0417 C_6H_{12}O_6 + 0.25 H_2O \rightarrow 0.25 CO_2 + H^+ + e^-$$

$$R: \quad 0.0417 C_6H_{12}O_6 + 0.011 NH_4^+ + 0.011 HCO_3^- \rightarrow$$
$$0.011 C_5H_7O_2N + 0.065 CH_3CH_2OH + 0.076 CO_2 + 0.044 H_2O$$

This reaction indicates that for every equivalent of glucose fermented, 0.065 mol ethanol is formed. Also, 0.011 mol ammonium is required to produce 0.011 empirical mol of microbial cells. Carbon dioxide is also produced in the process, which, if the fermentation is carried out in a closed bottle, will provide carbonation, if that is desired!

Mixed Fermentations   Many fermentations form more than a single product. For example, *E. coli* generally ferments glucose to a mixture of acetate, ethanol, formate, and hydrogen. In methane fermentation, a consortium of Bacteria and Archaea converts organic material into methane, and the products of incomplete fermentation, such as acetate, propionate, and butyrate, often are present. Once the relative proportions of reduced end products are known, the energy reaction can be constructed. The reduced products include all organic products plus hydrogen (H_2). The carbon dioxide formed is not as important in analyzing fermentations, since it is fully oxidized.

The critical step is determining the relative proportion of electron equivalents represented by each of the reduced end products. The electron equivalents of each product are computed, the sum is taken, and the fraction that each reduced end product represents of the total is then computed. That fraction is then used as a multiplier for the half-reaction for each reduced product, and the resulting equations are added together to produce a half-reaction for the electron acceptor, R_a. This series of operations is written in mathematical form as

$$R_a = \sum_{i=1}^{n} e_{ai} R_{ai} \qquad \textbf{[2.38]}$$

where

$$e_{ai} = \frac{\text{equiv}_{ai}}{\displaystyle\sum_{j=1}^{n} \text{equiv}_{aj}} \quad \text{and} \quad \sum_{i=1}^{n} e_{ai} = 1$$

Here, e_{ai} is the fraction of the n reduced end products that is represented by product a_i. Equiv$_{ai}$ represents the equivalents of a_i produced. The sum of the fractions of all reduced end products equals 1.

Some cases could have mixed electron donors. Certainly this is the case in the treatment of municipal and industrial wastewaters. Here, the electron-donor reaction, R_d, is written similar to that above for the electron acceptor:

$$R_d = \sum_{i=1}^{n} e_{di} R_{di} \qquad \textbf{[2.39]}$$

where

$$e_{di} = \frac{\text{equiv}_{di}}{\displaystyle\sum_{j=1}^{n} \text{equiv}_{dj}} \quad \text{and} \quad \sum_{i=1}^{n} e_{di} = 1$$

CITRATE FERMENTATION TO TWO REDUCED PRODUCTS   *Bacteroides* sp. con- | **Example 2.7**
verts 1 mol citrate into 1 mol formate, 2 mol acetate, and 1 mol bicarbonate. Write the overall balanced energy reaction (R_e) for this fermentation.

The reduced end products are formate and acetate. Bicarbonate, like carbon dioxide, is an oxidized end product and not considered in constructing the electron balance. The first step is to determine the number of equivalents (equiv$_{ai}$) formed for each reduced product. From reaction O-1, Table 2.3, there are 8 e$^-$ eq in a mole of acetate; so 2 mol acetate represents 16 e$^-$ eq. Similarly, there are 2 e$^-$ eq of formate per mole. Then,

$$e_{\text{formate}} = 2/(2 + 16) \text{ or } 0.111, \text{ and } e_{\text{acetate}} = 16/(2 + 16) \text{ or } 0.889.$$

The sum of e_{formate} plus e_{acetate} equals 1.0.

Using half-reactions from Table 2.3:

0.111 $R_{formate}$: $0.0555\,HCO_3^- + 0.111\,H^+ + 0.111\,e^- \rightarrow 0.0555\,HCOO^- + 0.0555\,H_2O$

0.889 $R_{acetate}$: $0.111\,CO_2 + 0.111\,HCO_3^- + 0.889\,H^+ + 0.889\,e^- \rightarrow 0.111\,CH_3COO^- + 0.333\,H_2O$

R_a: $0.111\,CO_2 + 0.166\,HCO_3^- + H^+ + e^- \rightarrow 0.0555\,HCOO^- + 0.111\,CH_3COO^- + 0.388\,H_2O$

The overall energy reaction (R_e) is then found using Equation 2.37, or $R_a - R_d$, where R_d represents the half-reaction for citrate from Table 2.3. Combining these produces the following for R_e:

$0.0555\,(COO^-)CH_2COH(COO^-)CH_2COO^- + 0.056\,H_2O \rightarrow$

$0.0555\,HCOO^- + 0.111\,CH_3COO^- + 0.056\,CO_2$

If this equation is normalized by dividing through by 0.0555, the moles of citrate in one equivalent, the following mole-normalized equation is obtained:

$(COO^-)CH_2COH(COO^-)CH_2COO^- + H_2O \rightarrow$

$HCOO^- + 2\,CH_3COO^- + CO_2$

It is readily apparent that the equation satisfies the requirement that 1 mol formate and 2 mol acetate are formed from 1 mol citrate. This is a fairly simple example that could have been balanced by other procedures. However, the next example provides a more complicated case, one of mixed reactants and products.

Example 2.8 | **FERMENTATION WITH MIXED DONORS AND PRODUCTS**   In the methane fermentation of a mixture containing 1.1 mol lactate and 1.1 mol glucose, the distribution of reduced end products was: 3.6 mol methane, 0.21 mol acetate, and 0.42 mol propionate. Write a balanced energy reaction for this case.

Based upon the electron equivalents noted in Table 2.3, the following table can be constructed:

Electron-Donor Substrates	Moles	e⁻ eq/mol	Equiv$_{di}$	e_{di}
lactate	1.1	12	$1.1 \times 12 = 13.2$	$13.2/39.6 = 0.33$
glucose	1.1	24	$1.1 \times 24 = 26.4$	$26.4/39.6 = 0.67$
			$\sum = 39.6$	$\sum = 1.00$

Electron Acceptor Products	Moles	e⁻ eq/mol	Equiv$_{ai}$	e_{ai}
methane	3.6	8	$3.6 \times 8 = 28.8$	$28.8/36.22 = 0.796$
acetate	0.21	8	$0.21 \times 8 = 1.68$	$1.68/36.22 = 0.046$
propionate	0.41	14	$0.41 \times 14 = 5.74$	$5.74/36.22 = 0.158$
			$\sum = 36.22$	$\sum = 1.000$

The sum of the electron equivalents for the donor is about 10 percent larger than the sum of the equivalents for the acceptor. Part of the donor in a biological reaction is associated with cell synthesis; when net synthesis occurs, the donor equivalents should be larger, as is the case here.

The acceptor reaction is then constructed first:

$0.796 R_{meth}$: $0.0995\ CO_2 + 0.796\ H^+ + 0.796\ e^- \rightarrow 0.0995\ CH_4 + 0.199\ H_2O$

$0.046 R_{acet}$: $0.0058\ CO_2 + 0.0058\ HCO_3^- + 0.046\ H^+ + 0.046\ e^- \rightarrow 0.0058\ CH_3COO^- + 0.0172\ H_2O$

$0.158 R_{prop}$: $0.0226\ CO_2 + 0.0113\ HCO_3^- + 0.158\ H^+ + 0.158\ e^- \rightarrow$
$$0.0113\ CH_3CH_2COO^- + 0.0564\ H_2O$$

R_a: $0.128\ CO_2 + 0.017\ HCO_3^- + H^+ + e^- \rightarrow$
$$0.0995\ CH_4 + 0.0058\ CH_3COO^- + 0.0113\ CH_3CH_2COO^- + 0.273\ H_2O$$

The donor reaction is constructed similarly:

$0.33 R_{lact}$: $0.055\ CO_2 + 0.0275\ HCO_3^- + 0.33\ H^+ + 0.33\ e^- \rightarrow 0.0275\ CH_3CHOHCOO^- + 0.11\ H_2O$

$0.67 R_{glu}$: $0.168\ CO_2 + 0.67\ H^+ + 0.67\ e^- \rightarrow 0.0279\ C_6H_{12}O_6 + 0.168\ H_2O$

R_d: $0.223\ CO_2 + 0.0275\ HCO_3^- + H^+ + e^- \rightarrow$
$$0.0275\ CH_3CHOHCOO^- + 0.0279\ C_6H_{12}O_6 + 0.278\ H_2O$$

Finally, we use the relationship that $R_e = R_a - R_d$ to obtain the overall balanced energy reaction:

$$0.0275 CH_3CHOHCOO^- + 0.0279 C_6H_{12}O_6 + 0.005 H_2O \rightarrow$$
$$0.0995 CH_4 + 0.0058 CH_3COO^- + 0.0113 CH_3CH_2COO^- + 0.095 CO_2 + 0.0105 HCO_3^-$$

A quick mass-balance check shows that the different species are consistent with the problem statement if it is assumed that 1 mol each of lactate and glucose is converted to the reduced end products. However, 1.1 mol of each was consumed. The difference is due to net biomass synthesis. If this is true, then $f_s = 0.1/1.1$ or 0.091, and $f_e = 1 - 0.091$ or 0.909.

The overall reaction including energy and synthesis is constructed by adding 0.909 of the energy reaction we just formed to 0.091 of the synthesis reaction from the donors to biomass. The result is the following:

$$0.0275\ CH_3CHOHCOO^- + 0.0279\ C_6H_{12}O_6 + 0.0046\ NH_4^+ \rightarrow$$
$$0.0046\ C_5H_7O_2N + 0.0904\ CH_4 + 0.00527\ CH_3COO^- +$$
$$0.0103\ CH_3CH_2COO^- + 0.088\ CO_2 + 0.0075\ HCO_3^- + 0.011\ H_2O$$

Such an equation would be difficult to construct from the information provided without the formal procedure as presented here.

The approach used to analyze fermentations is applicable for analyzing any situation in which reduced products are formed. An excellent example is sulfate reduction, in which sulfate is reduced to sulfide, just as methane is in the reaction above. If reduced organic products are present along with the sulfide, then the above procedure can be used to construct an energy reaction or overall biological reaction in a similar manner. An example of such an energy reaction is lactate conversion to acetate by *Desulfovibrio desulfuricans* through sulfate reduction:

$$0.084CH_3CHOHCOO^- + 0.042SO_4^{2-} + 0.063H^+ \rightarrow$$
$$0.084CH_3COO^- + 0.021H_2S + 0.021HS^- + 0.084CO_2 + 0.084H_2O$$

The reaction shows that $e_{acetate} = 0.67$ and $e_{sulfide} = 0.33$ for this case.

2.6   ENERGETICS AND BACTERIAL GROWTH

Microorganisms carry out oxidation-reduction reactions in order to obtain energy for growth and cell maintenance. The amount of energy released per electron equivalent of an electron donor oxidized varies considerably from reaction to reaction. It is not surprising then that the amount of growth that results from an equivalent of donor oxidized varies considerably as well. The purpose of this section is to explore relationships between reaction energetics and bacterial growth.

Figure 1.13 is a simple illustration of how cells capture and transport energy that is released from a catabolic oxidation-reduction reaction to synthesize new cells or to use in cell maintenance. Cell maintenance has energy requirements for activities such as cell movement and repair of cellular proteins that decay because of normal resource recycling or through interactions with toxic compounds. When cells grow rapidly in the presence of nonlimiting concentrations of all factors required for growth, cells make the maximum investment of energy for synthesis. However, when an essential factor, such as the electron-donor substrate, is limited in concentration, then a larger portion of the energy obtained from substrate oxidation must be used for cell maintenance. This is illustrated by Equation 2.6, which indicates that the net yield of cells decreases with a decrease in the rate of substrate utilization. Equation 2.7 further indicates that the net yield becomes zero when the energy supplied through substrate utilization is just equal to m, the maintenance energy. Under these conditions, all energy released is used just to maintain the integrity of the cells. If the substrate available for the microorganisms decreases further, then food is insufficient to maintain the microorganisms, and they go into net decay. Conversely, when substrate and all other required factors are unlimited in amount, the rate of substrate utilization will be at its maximum, and the net yield, Y_n, will approach (but not quite reach) the true yield, Y.

Over the years, many approaches have been taken to describe how the true yield relates to the energy released from electron-donor oxidation. An extensive summary, historical review, and discussion of relationships presented between energy production and cell yield are provided by Battley (1987). Although no unified approach for relating growth to reaction energetics is yet widely accepted, a method that is based on electron equivalents and that differentiates between the energy portion of an overall biological reaction and the synthesis portion, as exemplified by Equation 2.37, has proven highly useful. This is the approach that is used here, and it is based on broader discussions presented elsewhere (McCarty, 1971, 1975; Christensen and McCarty, 1975).

In addition to being fundamentally sound, a practical advantage of the approach presented here is that electron equivalents are easily related to measurements of widespread utility in environmental engineering practice, that is, the common expression of waste strength in terms of oxygen demand, or OD. Examples are the biochemical oxygen demand (BOD), calculated oxygen demand (COD'), and chemical oxygen demand (COD). Since one equivalent of oxygen is 8 g of O_2, one equivalent of any electron donor is equivalent to an OD of 8 g as O_2. Thus, numbers of equivalents per liter can be directly converted into an OD concentration for that substance. This makes it very easy to compute Y values that contain the widely used units of COD, COD', or BOD.

COMPUTING COD'   A wastewater contains 12.6 g/l of ethanol. Estimate the e^- eq/l and the COD' (g/l) for this wastewater.

Example 2.9

From reaction O-5 in Table 2.3, there are 12 e^- eq/mol ethanol. Since 1 mol of ethanol weighs 46 g, the equivalent weight is 46/12 or 3.83 g/e^- eq. The ethanol concentration in the wastewater is thus 12.6/3.83 or 3.29 e^- eq/l. The COD' thus becomes (8 g OD/e^- eq)(3.29 e^- eq/l) = 26.3 g/l.

2.6.1   FREE ENERGY OF THE ENERGY REACTION

Tables 2.2 and 2.3 summarize the standard free energies corrected to pH 7 ($\Delta G^{0'}$) for various inorganic and organic half-reactions, respectively. A student can readily determine standard free energies of other half-reactions by using values of free energy of formation for individual constituents, as listed in Appendix A. For illustration, the half-reaction for oxidation of 2-chlorobenzoate is obtained in the following manner.

Balanced half-reaction for 2-chlorobenzoate formation:

$$\frac{1}{28}\ HCO_{3(aq)}^- + \frac{3}{14}\ CO_{2(g)} + \frac{1}{28}\ Cl_{(aq)}^- + \frac{29}{28}\ H_{(aq,10^{-7})}^+ + e^-$$

$$\rightarrow \frac{1}{28}\ C_6H_4ClCOO^- + \frac{13}{28}\ H_2O_{(l)}$$

The free energies of formation for each species from Appendix A are (in kJ/e^- eq)

$$\frac{1}{28}(-586.85),\ \frac{3}{14}(-394.36),\ \frac{1}{28}(-31.35),\ \frac{29}{28}(-39.87),\ 0,$$

$$\frac{1}{28}(-237.9),\ \frac{13}{28}(-237.18)$$

The half-reaction free energy is calculated as the sum of the product free energies minus the sum of the reactant free energies, or $\Delta G^{0'} = 29.26$ kJ/e$^-$ eq.

In order to obtain the free energy for a full energy reaction, ΔG_r, the free energy for the donor half-reaction is added to the free energy for the acceptor half-reaction. This simple procedure is parallel to the way the overall energy reaction was constructed from half-reactions in the previous examples. The free energy for the donor half-reaction is the negative of the value listed in the tables, since a donor half-reaction must be written as the reverse of that listed in the table. For example, the aerobic oxidation of ethanol is constructed from half-reactions having standard free energy values in the tables.

		$\Delta G^{0'}$ kJ/e$^-$ eq
Reaction I-14:	$\dfrac{1}{4} O_2 + H^+ + e^- = \dfrac{1}{2} H_2O$	-78.72
$-$ Reaction O-5:	$\dfrac{1}{12} CH_3CH_2OH + \dfrac{1}{4} H_2O = \dfrac{1}{6} CO_2 + H^+ + e^-$	-31.18
Result:	$\dfrac{1}{12} CH_3CH_2OH + \dfrac{1}{4} O_2 = \dfrac{1}{6} CO_2 + \dfrac{1}{4} H_2O$	-109.90

The resulting $\Delta G_r^{0'}$ of -109.90 kJ/e$^-$ eq is specifically for the standard conditions indicated in the tables (i.e., 1 M ethanol aqueous concentration, 1 atm partial pressures of oxygen and carbon dioxide, and liquid water). The pH also is fixed at 7.0. Since H^+ does not appear on either side of the overall ethanol oxidation equation, the fixed pH is of no consequence in this case. However, pH can have a significant effect in other cases.

Figure 2.2 illustrates the relationships among various electron donors and acceptors and the resulting reaction free energy, assuming all constituents are at unit activity except pH $= 7.0$. This figure illustrates that, for the range of organic materials from methane to glucose, $\Delta G_r^{0'}$ for aerobic oxidation (oxygen as electron acceptor) or denitrification (nitrate as electron acceptor) varies relatively little, covering a range from about -96 to -120 kJ/e$^-$ eq. With inorganic electron acceptors, on the other hand, the range is very large, varying from about -5 kJ/e$^-$ eq with iron oxidation to -119 kJ/e$^-$ eq with hydrogen oxidation under aerobic conditions. However, if one considers organic electron donors under anaerobic conditions, either with carbon dioxide or sulfate as electron acceptors, the relative range of $\Delta G_r^{0'}$ is also quite large. For example, with carbon dioxide as the electron acceptor (methanogenesis), $\Delta G_r^{0'}$ is -3.87 kJ/e$^-$ eq for acetate oxidation and -17.82 kJ/e$^-$ eq for glucose oxidation, a factor of 4.6 difference. The great differences in reaction free energies for aerobic versus anaerobic and organic versus inorganic reactions have great effects on resulting bacterial yields, as will be demonstrated later.

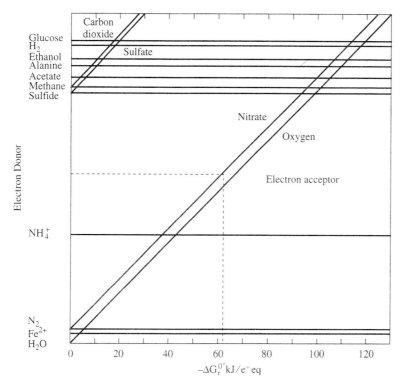

Figure 2.2   Relationship between various electron donors and acceptors and resulting reaction free energy.

One can adjust the reaction free energy for nonstandard concentrations of reactants and products. We first consider a generic reaction r that involves n different constituents:

$$v_1 A_1 + v_2 A_2 + \cdots \rightarrow v_m A_m + v_{m+1} A_{m+1} + \cdots + v_n A_n \qquad \textbf{[2.40]}$$

which can be written in an even more general form:

$$0 = \sum_{i=1}^{n} v_{ir} A_i \qquad \textbf{[2.41]}$$

The value of v_{ir} is negative if constituent A_i appears on the left side of Equation 2.40 and positive if it appears on the right side.

The nonstandard free energy change for this reaction can be determined from

$$\Delta G_r = \Delta G_r^0 + RT \sum_{i=1}^{n} v_{ir} \ln a_i \qquad \textbf{[2.42]}$$

Here, v_{ir} represents the stoichiometric coefficient for constituent A_i in reaction r, and a_i represents the activity of constituent A_i. T is absolute temperature (K).

The terms are easily identified from the final reaction of the ethanol example, which is repeated here,

$$\frac{1}{12} CH_3CH_2OH + \frac{1}{4} O_2 = \frac{1}{6} CO_2 + \frac{1}{4} H_2O$$

For this energy reaction, ΔG_r^0 equals $\Delta G_r^{0\prime}$ because H^+ is not a component of the equation; $n = 4$; and v_{ik} equals $-1/12$, $-1/4$, $1/6$, and $1/4$ for ethanol, oxygen, carbon dioxide, and water, respectively. Substituting the information into Equation 2.42 gives

$$\Delta G_r = \Delta G_r^0 + RT \ln \frac{[CO_2]^{1/6}[H_2O]^{1/4}}{[CH_3CH_2OH]^{1/12}[O_2]^{1/4}} \qquad \textbf{[2.43]}$$

We assume that the concentration of ethanol in an aqueous solution is 0.002 M, the oxygen partial pressure is that in the normal atmosphere at sea level (0.21 atm), the carbon dioxide concentration is that in normal air (0.0003 atm), and the temperature is 20 °C. We also assume that the activities of the three constituents are equal to their molar concentration or partial pressure. (These values could be corrected if one knew the respective activity coefficients for each constituent. In this case the activity coefficient is likely to be close to 1.0.) With aqueous solutions in which water is the major solvent, the activity of H_2O is sufficiently close to 1. Then,

$$\Delta G_r = -109,900 + 8.314(273 + 20) \ln \frac{[0.0003]^{1/6}[1]^{1/4}}{[0.002]^{1/12}[0.21]^{1/4}}$$

$$= -111,000 \text{ J/e}^- \text{ eq or } -111 \text{ kJ/e}^- \text{ eq}$$

A conclusion from this exercise is that the reaction free energy, corrected for concentrations that are within a typical range for biological systems of interest, is within 1 percent of the standard free energy of -109.9 kJ/e$^-$ eq. This is frequently the case, providing a correction has been made for pH.

While the exercise with aerobic oxidation of ethanol indicates that corrections for constituent concentrations may not be necessary, this is not always the case. Concentration corrections become important when the concentration of the electron donor or acceptor is very low or when the magnitude of ΔG_r^0 for a reaction is on the order of -10 kJ/e$^-$ eq or less. The latter situation is common for some anaerobic electron acceptors and inorganic electron donors.

The $\Delta G^{0\prime}$ values in Tables 2.2 and 2.3 already have the pH correction for pH $= 7.0$. If pH is significantly different from 7, then corrections for the effect of H^+ concentration may be necessary. When pH is significantly different than 7, or in cases where corrections to $\Delta G^{0\prime}$ are deemed necessary, one can convert to ΔG^0, which considers $\{H^+\} = 1$, with

$$\Delta G_r^0 = \Delta G_r^{0\prime} - RT v_{H^+} \ln[10^{-7}] \qquad \textbf{[2.44]}$$

2.7 YIELD COEFFICIENT AND REACTION ENERGETICS

The microbial yield from substrate utilization takes place in two steps. First, the energy reaction creates high-energy carriers, such as ATP. Second, the energy carriers are "spent" to drive cell synthesis or cell maintenance (recall Figure 1.13). As with all reactions, a certain amount of thermodynamic free energy is lost with each transfer. In this section, we describe how to compute the energy costs of cell synthesis and how to account for the energy lost in transfers. By combining all three aspects, we are able to estimate f_s^0 and the true yield (Y) based on thermodynamic principles. For determining the true yield and in accordance with Equation 2.5, the energy for maintenance is set to zero, so that all energy is used for cell synthesis.

First, we specify that the energy required to synthesize one equivalent of cells from a given carbon source is ΔG_s when the nitrogen source is ammonium. We first must determine the energy change resulting from the conversion of the carbon source to the common organic intermediates that the cell uses for synthesizing its macromolecules. We use pyruvate as a representative intermediate. The half-reaction for pyruvate is given as reaction O-16 in Table 2.3, and it has a free energy of 35.09 kJ/e$^-$ eq. The energy required to convert the carbon source to pyruvate is ΔG_p and is computed as the difference between the free energy of the pyruvate half-reaction and that of the carbon source:

$$\Delta G_p = 35.09 - \Delta G_c^{0'} \qquad \text{[2.45]}$$

For heterotrophic bacteria, the carbon source almost always is the electron donor. Thus, $\Delta G_c^{0'}$ is a value taken from Table 2.3 for the given electron donor. For example, if the electron donor were acetate, $\Delta G_c^{0'}$ would equal 27.4 kJ/e$^-$ eq. In autotrophic reactions, inorganic carbon is used as the carbon source. Considerable energy is required to reduce inorganic carbon to pyruvate. In photosynthesis, the hydrogen or electrons for reducing carbon dioxide to form cellular organic matter comes from water. By analogy to this case, we can determine the energy involved if we take $\Delta G_c^{0'}$ to equal that for the water-oxygen reaction from Table 2.2, or -78.72 kJ/e$^-$ eq. Thus, ΔG_p for the autotrophic case is always taken to equal $35.09 - (-78.72) = 113.8$ kJ/e$^-$ eq.

Next, pyruvate carbon is converted to cellular carbon. The energy required here (ΔG_{pc}) is based on an estimated value of 3.33 kJ per gram cells (McCarty, 1971). From Table 2.4, an electron equivalent of cells is $113/20 = 5.65$ grams when ammonia is used as the nitrogen source. Thus, ΔG_{pc} is $3.33 \cdot 5.65 = 18.8$ kJ/e$^-$ eq for the case with ammonium nitrogen.

Finally, energy is always lost in the electron transfers. The loss is considered by including a term for energy-transfer efficiency ε. In sum, the energy requirement for cell synthesis becomes

$$\Delta G_s = \frac{\Delta G_p}{\varepsilon^n} + \frac{\Delta G_{pc}}{\varepsilon} \qquad \text{[2.46]}$$

We note that an exponent n is used for the energy transfer efficiency for conversion of carbon to pyruvate. The n accounts for the fact that ΔG_p for some electron donors, such as glucose, is negative, meaning energy is obtained by its conversion to pyruvate. Some of this energy is lost, and for such cases, n is taken to equal -1. In other cases, such as with acetate, ΔG_p is positive, meaning that energy is required in its conversion to pyruvate. Here, more energy is required than the thermodynamic amount, and for this case, $n = +1$.

Now that we have an estimate of how much energy is needed to synthesize an equivalent of cells, we can estimate how much electron donor must be oxidized to supply the energy. We define that, in order to supply this amount of energy, A equivalents of electron donor must be oxidized. The energy released by this oxidation is $A\Delta G_r$, where ΔG_r is the free energy released per equivalent of donor oxidized for energy generation. When this energy is transferred to the energy carrier, a portion again is lost through transfer inefficiencies. If the transfer efficiency here is the same as that for transferring energy from the carrier to synthesis (ε), then the energy transferred to the carrier will is $\varepsilon A\Delta G_r$.

An energy balance must be maintained around the energy carrier at steady state:

$$A\varepsilon\Delta G_r + \Delta G_s = 0 \qquad \textbf{[2.47]}$$

Solving for A gives:

$$A = -\frac{\dfrac{\Delta G_p}{\varepsilon^n} + \dfrac{\Delta G_{pc}}{\varepsilon}}{\varepsilon\Delta G_r} \qquad \textbf{[2.48]}$$

This equation indicates that the equivalents of donor used for energy production per equivalent of cells formed (A) increases as the energy required for synthesis from the given carbon source increases and as the energy released by donor oxidation decreases.

Our ultimate goal is developing balanced stoichiometric equations. As Equation 2.48 does not include energy of maintenance, the resulting value of A is for the situation of the true yield or Y. Thus, f_s is its maximum value or f_s^0, and the portion of donor used for energy is at its minimum value, f_e^0. Since part of the donor consumed is used for energy (A equivalents in this case) and the other part for synthesis (1.0 equivalents in this case), the total donor used is $1 + A$. Thus, we can compute f_s^0 and f_e^0 from A by

$$f_s^0 = \frac{1}{1 + A} \quad \text{and} \quad f_e^0 = 1 - f_s^0 = \frac{A}{1 + A} \qquad \textbf{[2.49]}$$

The energy-transfer efficiency is the key factor that needs to be assumed in solving Equation 2.48. Under optimum conditions, transfer efficiencies of 55 to 70 percent are typical, and a ε value of 0.6 frequently is employed. However, some reported cell yields, especially with pure cultures that may be grown on substrates for which they do not have highly efficient enzyme systems, suggest transfer efficiencies lower than 0.6. In general, we would not expect microorganisms to survive well in mixed cultures if they had a low efficiency. On the other hand, some enzymatic reactions, such as the initial oxidation of hydrocarbons (information in Chapters 1 and 14), require an energy input, such as in the form of NADH, even though the overall reaction being carried out may result in net energy release. Those reactions should have a lower energy yield due to the initial energy investment, but these energetic calculations do not directly take into account such activation reactions.

EFFECTS OF ε ON HETEROTROPHIC YIELD   Compare estimates for f_s^0 and Y for aerobic oxidation of acetate, assuming $\varepsilon = 0.4$, 0.6, and 0.7, that pH = 7, and that all other reactants and products are at unit activity. Ammonium is available for synthesis. | **Example 2.10**

Since the reaction is heterotrophic, Equation 2.45 applies. Using reaction O-1 from Table 2.3, $\Delta G_d^{0'} = 27.40$ kJ/e$^-$ eq. Therefore,

$$\Delta G_p = 35.09 - 27.40 = 7.69 \text{ kJ/e}^- \text{ eq}$$

Since this is an aerobic reaction, $\Delta G_a^{0'} = -78.72$ kJ/e$^-$ eq. Thus,

$$\Delta G_r = \Delta G_a^{0'} - \Delta G_d^{0'} = -78.72 - 27.40 = -106.12 \text{ kJ/e}^- \text{ eq}$$

Since ΔG_p is positive, $n = +1$. Also, since ammonium is available for cell synthesis, ΔG_{pc} equals 18.8 kJ/e$^-$ eq. Hence,

$$A = -\frac{\dfrac{7.69}{\varepsilon+1} + \dfrac{18.8}{\varepsilon}}{-106.12\varepsilon}$$

Letting $\varepsilon = 0.4$, 0.6, and 0.7, $A = 1.56$, 0.69, and 0.51, respectively. Using Equation 2.49, values for f_s^0 that result are:

ε	f_s^0	Y g cells/mol donor	Y g cells/g donor	Y g cells/g COD$'$
0.4	0.39	18	0.30	0.28
0.6	0.59	27	0.45	0.42
0.7	0.66	30	0.51	0.47

In order to determine bacterial yield, it is well to write a balanced stoichiometric equation. This will be done for illustration only for the case where $\varepsilon = 0.6$, for which $f_s^0 = 0.59$ and $f_e^0 = 1 - 0.59 = 0.41$:

$-$ Reaction O-1: $0.125\ CH_3COO^- + 0.375\ H_2O \rightarrow 0.125\ CO_2 + 0.125\ HCO_3^- + H^+ + e^-$

0.41(Reaction I-14): $0.1025\ O_2 + 0.41\ H^+ + 0.41\ e^- \rightarrow 0.205\ H_2O$

0.59(Reaction C-1): $0.118\ CO_2 + 0.0295\ HCO_3^- + 0.0295\ NH_4^+ + 0.59\ H^+ + 0.59\ e^- \rightarrow$

$$0.0295\ C_5H_7O_2N + 0.2655\ H_2O$$

Overall: $0.125\ CH_3COO^- + 0.1025\ O_2 + 0.0295\ NH_4^+ \rightarrow$

$$0.0295\ C_5H_7O_2N + 0.007\ CO_2 + 0.0955\ HCO_3^- + 0.0955\ H_2O$$

From this balanced reaction:

$$Y = 0.0295(113)/0.125 = 27 \text{ g cells/mol acetate}$$
$$= 0.0295(113)/0.125(59) = 0.45 \text{ g cells/g acetate}$$
$$= 0.0295(113)/8 = 0.42 \text{ g cells/g COD}'$$

Example 2.11 | **EFFECTS OF ELECTRON ACCEPTORS AND DONORS FOR HETEROTROPHS**
Compare estimated values of f_s^0 for glucose and acetate when oxygen, nitrate, sulfate, and carbon dioxide are the electron acceptors. The calculated values will equal the total net synthesis, which may include more than one microbial type. Assume $\varepsilon = 0.6$ and that the nitrogen source is ammonium.

The $\Delta G_d^{0'}$ and $\Delta G_a^{0'}$ values are in Tables 2.3 and 2.4, respectively. Since ammonium is the nitrogen source, ΔG_{pc} is 18.8 kJ/e⁻ eq.

Acetate ($\Delta G_d^{0'}$ = 27.40 kJ/e⁻ eq; ΔG_p = 7.69 kJ/e⁻ eq; n = +1):

Electron Acceptor	$\Delta G_a^{0'}$	ΔG_r	A	f_s^0
Oxygen	−78.72	−106.12	0.69	0.59
Nitrate	−72.20	−99.60	0.74	0.57
Sulfate	+20.85	−6.55	11.2	0.08
CO₂	+23.53	−3.87	19.0	0.05

Glucose ($\Delta G_d^{0'}$ = 41.35 kJ/e⁻ eq; ΔG_p = −6.26 kJ/e⁻ eq; n = -1):

Electron Acceptor	$\Delta G_a^{0'}$	ΔG_r	A	f_s^0
Oxygen	−78.72	−120.07	0.38	0.72
Nitrate	−72.20	−113.55	0.40	0.71
Sulfate	+20.85	−20.50	2.24	0.31
CO₂	+23.53	−17.82	2.58	0.28

The yield, as indicated by f_s^0, is lower for acetate than for glucose. Carbohydrates have more positive $\Delta G_d^{0'}$ values due to their more ordered (lower entropy) structure. This makes ΔG_p negative. Oxygen and nitrate give much higher yields than do sulfate and carbon dioxide, which is due to the different values of $\Delta G_a^{0'}$. These trends are consistent with and help explain empirical findings for the different types of microorganisms.

Example 2.12 | **YIELDS FOR VARIOUS AEROBIC CHEMOLITHOTROPHS**   Compare estimated f_s^0 values for bacterial chemolithotrophs that oxidize ammonium to nitrate, sulfide to sulfate, Fe(II) to Fe(III), and H_2 to H_2O under aerobic conditions. Carbon dioxide and ammonium are the C and N sources, respectively. Assume $\varepsilon = 0.6$.

Since all the microorganisms are autotrophs (carbon dioxide as C source), ΔG_s is the same: $113.8/0.6 + 18.8/0.6 = 221$ kJ/e$^-$ eq. The values of $\Delta G_d^{0'}$ come from Table 2.2, while $\Delta G_a^{0'} = -78.72$ kJ/e$^-$ eq for O_2. The results are:

Electron Donor	$\Delta G_d^{0'}$	ΔG_r	A	f_s^0
Ammonium	-35.11	-43.61	8.45	0.11
Sulfide	$+20.85$	-99.57	3.70	0.21
Fe(II)	-74.27	-4.45	82.8	0.012
Hydrogen	$+39.87$	-118.59	3.11	0.24

Compared to the f_s^0 values in Example 2.11, all values for the chemolithotrophs are low. This comes about mainly from the high cost of autotrophic biomass synthesis, which is shown quantitatively by the large ΔG_s. Despite the low yields, all of these electron donors are exploited by chemolithotrophs under aerobic conditions. Interestingly, the low yield with iron oxidation is increased under low-pH conditions when a correction is made to ΔG_r for pH (Equation 2.44). Iron oxidizers that are tolerant to very low pH (e.g., 2) are well known.

2.8 OXIDIZED NITROGEN SOURCES

Microorganisms prefer to use ammonium nitrogen as an inorganic nitrogen source for cell synthesis, because it already is in the $(-III)$ oxidation state, the status of organic nitrogen within the cell. However, when ammonium is not available for synthesis, many prokaryotic cells can use oxidized forms of nitrogen as alternatives. Included here are nitrate (NO_3^-), nitrite (NO_2^-), and dinitrogen (N_2). When an oxidized form of nitrogen is used, the microorganisms must reduce it to the $(-III)$ oxidation state of ammonium, a process that requires electrons and energy, thus reducing their availability for synthesis.

Using an oxidized nitrogen source affects the energy cost of synthesis, or ΔG_s. We extend the energetics method presented in the previous section to include the added synthesis costs. The accounting approach assumes that all electrons used for synthesis are first routed to the common organic intermediate, pyruvate. Thus, the ΔG_p part of ΔG_s remains the same (Equation 2.45):

$$\Delta G_p = 35.09 - \Delta G_c^{0'}$$

The electrons needed to reduce the oxidized form of nitrogen to ammonium are assumed to come from pyruvate and transferred to the oxidized nitrogen source. Reducing nitrogen for synthesis is not part of respiration, and it does not yield any energy. Therefore, reducing the nitrogen source is an energy cost in that the electrons that otherwise could have been transferred to the acceptor to give energy are diverted to nitrogen reduction for synthesis. The ΔG_{pc} part of ΔG_s depends on the nitrogen source, since a mole of $C_5H_7O_2N$ contains different numbers of electron equivalents, depending on how many electrons had to be invested in nitrogen reduction: 20 e^- eq/mole for NH_4^+, 28 for NO_3^-, 26 for NO_3^-, and 23 for N_2. The number of electron equivalents invested in a mole of cells can be seen from half-reactions C-1 to C-4 in Table 2.4. The energy cost to synthesize cells from the common intermediates (i.e., ΔG_{pc}) is the same in units of kJ/gram, that is, 3.33 kJ/g cells (McCarty, 1971), but it varies with the number of electron equivalents per mole of cells: $\Delta G_{pc} = 18.8$ kJ/e^- eq for NH_4^+, 13.5 kJ/e^- eq for NO_3^-, 14.5 kJ/e^- eq for NO_2^-, and 16.4 kJ/e^- eq for N_2. Equation 2.48 is used to compute A as before, but the value of ΔG_{pc} in the numerator depends on the nitrogen source, as shown in the previous sentence.

Once A is computed from Equation 2.48, f_s^0 and f_e^0 are computed in the usual manner from Equation 2.49. Then, the overall stoichiometric reaction is generated in the normal manner from the appropriate donor (R_d), acceptor (R_a), and synthesis (R_c) half-reactions. The key is that the synthesis half-reaction be selected from C-2 to C-4, instead of C-1 (Table 2.4), when the nitrogen source is oxidized. The entire process is illustrated for nitrate as the nitrogen source in the following example, which shows that using an oxidized nitrogen source decreases the true yield (Y) due to the electron and energy investments to reduce it.

Example 2.13 | **EFFECT OF AN OXIDIZED NITROGEN SOURCE**   Estimate f_s^0 and Y (in g cells/g COD') for acetate utilization under aerobic conditions when NO_3^- is used as the source of nitrogen for cell synthesis. Write the stoichiometric equation for the overall reaction. Determine the oxygen requirement in g O_2/g COD', and compare with the result from Example 2.10, when NH_4^+ was used for cell synthesis. Assume $\varepsilon = 0.6$.

From Example 2.10, $\Delta G_p = 7.69$ kJ/e^- eq, $n = +1$, and $\Delta G_r = -106.12$ kJ/e^- eq. When nitrate is the nitrogen source, $\Delta G_{pc} = 13.5$ kJ/e^- eq. Then,

$$A = -\frac{\dfrac{7.69}{0.6^{+1}} + \dfrac{13.5}{0.6}}{0.6(-106.12)} = 0.55$$

and

$$f_s^0 = \frac{1}{1+A} = \frac{1}{1+0.55} = 0.65$$

With $f_e^0 = 1 - f_s^0 = 0.35$, the overall reaction for biological growth is then developed as follows:

$0.35 R_a$: $0.0875\,O_2 + 0.35\,H^+ + 0.35\,e^- \rightarrow 0.175\,H_2O$

$0.65 R_c$: $0.0232\,NO_3^- + 0.1161\,CO_2 + 0.6732\,H^+ + 0.65\,e^- \rightarrow$

$\qquad\qquad\qquad\qquad\qquad 0.0232\,C_5H_7O_2N + 0.2554\,H_2O$

$-R_d$: $0.125\,CH_3COO^- + 0.375\,H_2O \rightarrow 0.125\,CO_2 + 0.125\,HCO_3^- + H^+ + e^-$

Net: $0.125\,CH_3COO^- + 0.0875\,O_2 + 0.0232\,NO_3^- + 0.0232\,H^+ \rightarrow$

$\qquad\qquad 0.0232\,C_5H_7O_2N + 0.0089\,CO_2 + 0.125\,HCO_3^- + 0.0554\,H_2O$

From this equation,

$$Y = 0.0232(113)/0.125 = 21 \text{ g cells/mol acetate}$$
$$= 0.0232(113)/0.125(59) = 0.36 \text{ g cells/g acetate}$$
$$= 0.0232(113)/8 = 0.33 \text{ g cells/g COD}'$$

The oxygen requirement is $0.0875 (32)/8 = 0.35$ g O_2/g COD$'$.

For comparison, when ammonium is the nitrogen source:

$$Y = 0.42 \text{ g cells/g COD}'$$

$$\text{Oxygen required} = 0.1025(32)/8 = 0.41 \text{ g } O_2/\text{g COD}'$$

The comparison indicates that, when nitrate is used as the nitrogen source, the yield of bacterial cells and the requirement for oxygen are less than when ammonia is the nitrogen source. The diversion of electrons from acetate for the reduction of nitrate to ammonium for cell synthesis reduces the acetate available for energy generation and synthesis.

2.9 BIBLIOGRAPHY

Bailey, J. E. and D. F. Ollis (1986). *Biochemical Engineering Fundamentals.* 2nd ed. New York: McGraw-Hill.

Battley, E. H. (1987). *Energetics of Microbial Growth.* New York: Wiley.

Christensen, D. R. and P. L. McCarty (1975). "Multi-process biological treatment model." *J. Water Pollution Control Fedn.* 47, pp. 2652–2664.

McCarty, P. L. (1971). "Energetics and bacterial growth." In *Organic Compounds in Aquatic Environments,* eds. S. D. Faust and J. V. Hunter. New York: Marcel Dekker.

McCarty, P. L. (1975). "Stoichiometry of biological reactions." *Prog. Water Technol.* 7, pp. 157–172.

Porges, N.; L. Jasewicz; and S. R. Hoover (1956). "Principles of biological oxidation." In *Biological Treatment of Sewage and Industrial Wastes.* eds. J. McCabe and W. W. Eckenfelder. New York: Reinhold Publ.

Sawyer, C. N.; P. L. McCarty; and G. F. Parkin (1994). *Chemistry for Environmental Engineering.* New York: McGraw-Hill.

Speece, R. E. and P. L. McCarty (1964). "Nutrient requirements and biological solids accumulation in anaerobic digestion." In *Advances in Water Pollution Research.* London: Pergamon Press, pp. 305–322.

Symons, J. M. and R. E. McKinney (1958), "The biochemistry of nitrogen in the synthesis of activated sludge." *Sewage and Industrial Wastes* 30, no. 7, pp. 874–890.

2.10 PROBLEMS

2.1. Which of the following electron donor/electron acceptor pairs represent potential energy reactions for bacterial growth? Assume that all reactants and products are at unit activity, except that pH $= 7$.

Case	Electron Donor	Electron Acceptor
a.	Acetate	carbon dioxide (methanogenesis)
b.	Acetate	Fe^{3+} (Reduction to Fe^{2+})
c.	Acetate	H^+ (reduction to H_2)
d.	Glucose	H^+ (reduction to H_2)
e.	H_2	carbon dioxide (methanogenesis)
f.	H_2	nitrate (denitrification to N_2)
g.	S (oxidized to sulfate)	NO_3^- (denitrification to N_2)
h.	CH_4	NO_3^- (denitrification to N_2)
i.	NH_4^+ (oxidation to NO_2^-)	SO_4^{2-} (reduction to $H_2S + HS^-$)

2.2. Estimate f_s^o for each of the following, assuming all constituents are at unit activity, except that pH = 7.0, and that the energy transfer efficiency (ε) is 0.6:

Case	Electron Donor	Electron Acceptor	Nitrogen Source
a.	Ethanol	oxygen	ammonia
b.	Ethanol	oxygen	nitrate
c.	Ethanol	sulfate	ammonia
d.	Ethanol	carbon dioxide (methanogenesis)	ammonia
e.	Propionate	carbon dioxide (methanogenesis)	ammonia
f.	Sulfur (oxidized to sulfate)	nitrate (denitrification to nitrogen gas)	ammonia
g.	Ammonia (oxidized to nitrite)	oxygen	ammonia

2.3. Organic matter is converted in sequential steps by different bacterial species to methane in anaerobic methanogenesis of organic wastes. One important step is the conversion of the intermediate butyrate to acetate, for which the following electron donor and acceptor half-reactions apply:

Electron donor:

$$\frac{1}{2} \underset{\text{acetate}}{CH_3COO^-} + \frac{1}{4} CO_2 + H^+ + e^- = \frac{1}{4} \underset{\text{butyrate}}{CH_3CH_2CH_2COO^-} + \frac{1}{4} HCO_3^- + \frac{1}{4} H_2O$$

Electron acceptor half-reaction:

$$H^+ + e^- = \frac{1}{2} H_{2(g)}$$

Determine ΔG_r for the resulting energy reaction under the following conditions:

1. (a) All constituents are at unit activity
   (b) All constituents are at unit activity, except pH = 7.0
   (c) The following typical activities under anaerobic conditions apply:

   $$[CH_3COO^-] = 10^{-3}M, [CO_2] = 0.3 \text{ atm}, pH = 7.0,$$
   $$[CH_3CH_2CH_2COO^-] = 10^{-2}M, \text{ and } [HCO_3^-] = 10^{-1}M.$$
   $$P_{H2(g)} = 10^{-6} \text{ atm}.$$

2. Under which of the above three conditions is it possible for bacteria to obtain energy for growth?

2.4. Per equivalent of electron donor oxidized, would you expect to obtain more cell production from anaerobic conversion of ethanol to methane or from oxidation of acetate through reduction of sulfate to sulfide? Why?

2.5. Estimate through use of reaction energetics the yield coefficient Y in g cells per gram substrate for benzoate when used as an electron donor and when sulfate is the electron acceptor (sulfate reduced to sulfide). Assume ammonia is present for cell synthesis and that the energy transfer efficiency (ε) equals 0.6.

2.6. How many grams of oxygen are required per electron equivalent of lactate in a wastewater when subjected to aerobic biological treatment if f_s is 0.4 (assume sufficient ammonium is present for cell synthesis)?

2.7. For each of the following, write an oxidation half-reaction and normalize the reaction on an electron-equivalent basis. Add H_2O as appropriate to either side of the equations in balancing the reactions.
   (a) $CH_3CH_2CH_2CHNH_2COO^-$ oxidation to CO_2, NH_4^+, and HCO_3^-
   (b) Cl^- to ClO_3^-

2.8. You wish to consider the addition of methanol to groundwater for anaerobic biological removal of nitrate by denitrification. If the nitrate nitrogen concentration is 84 mg/l, what minimum concentration of methanol should be added to achieve complete nitrate reduction to nitrogen gas? Assume no ammonium is present and that f_s for the reaction is 0.30.

2.9. Some facultative bacteria can grow on acetate while using tetrachloroethylene (PCE) as an electron acceptor, and that here, $f_s = 0.16$. Determine the energy-transfer efficiency for cell synthesis considering the following to be the electron acceptor reaction by which PCE is reduced to cis-1,2-dichloroethylene:

$$\frac{1}{4} CCl_2 = CCl_2 + \frac{1}{2} H^+ + e^- = \frac{1}{4} CHCl = CHCl + \frac{1}{2} Cl^-$$

$$\Delta G^{\circ\prime} = -53.55 \text{ kJ/e}^- \text{ eq}$$

2.10. Use energetics to compute f_s° and Y (in g cells/g COD′) for the four situations in which the electron donor is acetate, the electron acceptor is oxygen, the carbon source is acetate, and the nitrogen source is either ammonium, nitrate, nitrite, or dinitrogen. Assume $\varepsilon = 0.6$.

2.11. Repeat problem 2.10, except that glucose replaces acetate as the electron donor and carbon source.

2.12. Nitrification is carried out by autotrophs that utilize ammonium as the electron donor, oxygen as the electron acceptor, ammonium as the nitrogen source, and inorganic carbon as the carbon source. First, you are to use energetics to compute f_s° and Y (g cells per g NH_4^+-N) for this normal case when ammonium is oxidized to nitrate. Second, compute the changes to f_s° and Y if the bacteria were able to use acetate as the carbon source, instead of inorganic carbon. Assume $\varepsilon = 0.6$.

2.13. Use energetics to compute f_s° and Y (g cells per g H_2) for hydrogen-oxidizing bacteria that use either oxygen or sulfate as the electron acceptor (two sets of answers). In both cases, the cells are autotrophs. Assume $\varepsilon = 0.6$.

2.14. Write the balanced, overall reaction for the situation in which acetate is the donor and carbon source, nitrate is the acceptor and nitrogen source, and $f_s = 0.333$.

2.15. Write the balanced, overall reaction for the situation in which $Cu(s)$ is the donor (it is oxidized to Cu^{2+}), oxygen is the acceptor, inorganic carbon is the carbon source, ammonium is the nitrogen source, and $f_s^\circ = 0.036$.

2.16. Write the balanced overall reaction for the situation in which ammonium is the donor and is oxidized to nitrite, oxygen is the acceptor, the cells are autotrophs, ammonium in the nitrogen source, and $f_e^\circ = 0.939$.

2.17. Write the balanced overall reaction for the situation in which dichloromethane (CH_2Cl_2) is the donor and carbon source, oxygen is the acceptor, ammonium is the nitrogen source, and $f_s^\circ = 0.31$.

chapter

3

MICROBIAL KINETICS

The previous chapters emphasized that microorganisms fuel their lives by performing oxidation/reduction reactions that generate the energy and reducing power needed to construct and maintain themselves. Because redox reactions are nearly always very slow unless catalyzed, microorganisms produce enzyme catalysts that increase the kinetics of their essential reactions to rates fast enough for them to exploit the chemical resources available in their environment. Engineers want to take advantage of these microbially catalyzed reactions, because the chemical resources of the microorganisms usually are the pollutants that the engineers must control. For example, the biochemical oxygen demand (BOD) is an organic electron donor for heterotrophic bacteria, NH_4^+-N is an inorganic electron donor for nitrifying bacteria, NO_3^--N is an electron acceptor for denitrifying bacteria, and PO_4^{3-} is a nutrient for all microorganisms.

In trying to employ microorganisms for pollution control, engineers must recognize two interrelated principles: First, metabolically active microorganisms catalyze the pollutant-removing reactions. The rate of pollutant removal depends on the concentration of the catalyst, or the active biomass. Second, the active biomass is grown and sustained through the utilization of its energy- and electron-generating primary substrates, which are its electron donor and electron acceptor. The rate of production of active biomass is proportional to the utilization rate of the primary substrates.

The connection between the active biomass (the catalyst) and the primary substrates is the most fundamental factor needed for understanding and exploiting microbial systems for pollution control. Because those connections must be made systematically and quantitatively for engineering design and operation, mass-balance modeling is an essential tool. That modeling is the foundation for the subject of this chapter.

3.1 BASIC RATE EXPRESSIONS

At a minimum, a model of a microbial process must have mass balances on the active biomass and the primary substrate that limits the growth rate of the biomass. In the vast majority of cases, the rate-limiting substrate is the electron donor. That convention is used here, and the term *substrate* now refers to the *primary electron-*

donor substrate. To complete the mass-balance equations, rate expressions for the growth of the biomass and utilization of the substrate must be supplied. Those two rate expressions are presented first.

The relationship most frequently used to represent bacterial growth kinetics is the so-called *Monod equation,* which was developed in the 1940s by the famous French microbiologist Jacques Monod. His original work related the specific growth rate of fast-growing bacteria to the concentration of a rate-limiting, electron-donor substrate,

$$\mu_{syn} = \left(\frac{1}{X_a} \frac{dX_a}{dt} \right)_{syn} = \hat{\mu} \frac{S}{K + S} \qquad \textbf{[3.1]}$$

in which

μ_{syn} = specific growth rate due to synthesis (T^{-1})

X_a = concentration of active biomass ($M_x L^{-3}$)

t = time (T)

S = concentration of the rate-limiting substrate ($M_s L^{-3}$)

$\hat{\mu}$ = maximum specific growth rate (T^{-1})

K = concentration giving one-half the maximum rate ($M_s L^{-3}$)

This equation is a convenient mathematical representation for a smooth transition from a first-order relation (in S) at low concentration to a zero-order relation (in S) at high concentration. The Monod equation is sometimes called a saturation function, because the growth rate saturates at $\hat{\mu}$ for large S. Figure 3.1 shows how μ varies with S and that $\mu = \hat{\mu}/2$ when $K = S$. Although Equation 3.1 is largely empirical, it has widespread applicability for microbial systems.

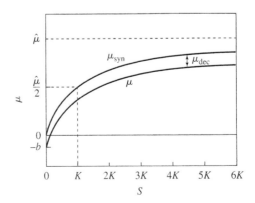

Figure 3.1   Schematic of how the synthesis and net specific growth rates depend on the substrate concentration. At $S = 20K$, $\mu_{syn} = 0.95\hat{\mu}$.

People who studied more slowly growing bacteria (environmental engineers are among the key examples of these people) discovered that active biomass has an energy demand for maintenance, which includes cell functions such as motility, repair and resynthesis, osmotic regulation, transport, and heat loss. Environmental engineers usually represent that flow of energy and electrons required to meet maintenance needs as *endogenous decay*. In other words, the cells oxidize themselves to meet maintenance-energy needs.

The rate of endogenous decay is

$$\mu_{dec} = \left(\frac{1}{X_a}\frac{dX_a}{dt}\right)_{decay} = -b \qquad [3.2]$$

in which

b = endogenous-decay coefficient (T^{-1})

μ_{dec} = specific growth rate due to decay (T^{-1}).

Equation 3.2 says that the loss of active biomass is a first-order function. However, not all of the active biomass lost by decay is actually oxidized to generate energy for maintenance needs. Although most of the decayed biomass is oxidized, a small fraction accumulates as inert biomass. The rate of oxidation (or true respiration for energy generation) is

$$\left(\frac{1}{X_a}\frac{dX_a}{dt}\right)_{resp} = -f_d b \qquad [3.3]$$

in which f_d = fraction of the active biomass that is biodegradable. The rate at which active biomass is converted to inert biomass is the difference between the overall decay rate and the oxidation decay rate,

$$-\frac{1}{X_a}\frac{dX_i}{dt} = \left(\frac{1}{X_a}\frac{dX_a}{dt}\right)_{inert} = -(1-f_d)b \qquad [3.4]$$

in which X_i = inert biomass concentration $(M_x L^{-3})$.

Overall, the net specific growth rate of active biomass (μ) is the sum of new growth (Equation 3.1) and decay (Equation 3.2):

$$\mu = \frac{1}{X_a}\frac{dX_a}{dt} = \mu_{syn} + \mu_{dec} = \hat{\mu}\frac{S}{K+S} - b \qquad [3.5]$$

Figure 3.1 also shows how μ varies with S and that μ can be negative for low enough S.

Because of our ultimate interest in substrate removal (e.g., BOD) and because biomass growth is "fueled" by substrate utilization, environmental engineers often prefer to regard the rate of substrate utilization as the basic rate, while cell growth is derived from substrate utilization. Then, the Monod equation takes the form

$$r_{ut} = -\frac{\hat{q}S}{K+S}X_a \qquad [3.6]$$

in which

r_{ut} = rate of substrate utilization ($M_s L^{-3} T^{-1}$)

$\hat{q}$ = maximum specific rate of substrate utilization ($M_s M_x^{-1} T^{-1}$)

Substrate utilization and biomass growth are connected by

$$\hat{\mu} = \hat{q} Y \qquad \textbf{[3.7]}$$

in which Y = true yield for cell synthesis ($M_x M_s^{-1}$). Yes, the true yield is exactly the same Y as you studied in Chapter 2 on *Stoichiometry*. It represents the fraction of electron-donor electrons converted to biomass electrons during synthesis of new biomass. The net rate of cell growth becomes

$$r_{net} = Y \frac{\hat{q} S}{K + S} X_a - b X_a \qquad \textbf{[3.8]}$$

in which r_{net} = the net rate of active-biomass growth ($M_x L^{-3} T^{-1}$). Of course,

$$\mu = r_{net}/X_a = Y \frac{\hat{q} S}{K + S} - b \qquad \textbf{[3.9]}$$

Some prefer to think of cell maintenance as being a shunting of substrate-derived electrons and energy directly for maintenance. This is expressed by

$$\mu = Y \left(\frac{\hat{q} S}{K + S} - m \right) \qquad \textbf{[3.10]}$$

in which m = maintenance-utilization rate of substrate ($M_s M_x^{-1} T^{-1}$), which also was introduced in Chapter 2. When systems go to steady state, there is no difference in the two approaches to maintenance, and $b = Ym$. We use the endogenous-decay approach exclusively.

3.2 PARAMETER VALUES

The parameters describing biomass growth and substrate utilization cannot be taken as "random variables." They have specific units and ranges of values. In several cases, the values are constrained by the cell's stoichiometry and energetics, described in Chapter 2.

The true yield, Y, is proportional to f_s^o, presented under *Stoichiometry*. The thermodynamic method presented at the end of Chapter 2 is an excellent means to obtain a good first estimate of Y if experimental values are not available. Table 3.1 lists f_s^o and Y values for the most commonly encountered bacteria in environmental biotechnology. Table 3.1 illustrates that f_s^o ranges from its highest values (0.6 to 0.7 e$^-$ eq cells/e$^-$ eq donor) for aerobic heterotrophs to its lowest values (0.05 to 0.10 e$^-$ eq cells/e$^-$ eq donor) for autotrophs and acetate-oxidizing anaerobes. These values reflect the balancing of energy costs for synthesis with the energy gains from

Table 3.1 Typical f_s^0, Y, $\hat{q}$, and $\hat{\mu}$ values for key bacterial types in environmental biotechnology

Organism Type	Electron Donor	Electron Acceptors	C-Source	f_s^0	Y	$\hat{q}$	$\hat{\mu}$
Aerobic, Heterotrophs	Carbohydrate BOD	O_2	BOD	0.7	0.49 gVSS/gBOD$_L$	27 gBOD$_L$/gVSS-d	13.2
	Other BOD	O_2	BOD	0.6	0.42 gVSS/gBOD$_L$	20 gBOD$_L$/gVSS-d	8.4
Denitrifiers	BOD	NO_3^-	BOD	0.5	0.25 gVSS/gBOD$_L$	16 gBOD$_L$/gVSS-d	4
	H_2	NO_3^-	CO_2	0.2	0.81 gVSS/gH$_2$	1.25 gH$_2$/gVSS-d	1
	$S(s)$	NO_3^-	CO_2	0.2	0.15 gVSS/gS	6.7 gS/gVSS-d	1
Nitrifying Autotrophs	NH_4^+	O_2	CO_2	0.14	0.34 gVSS/gNH$_4^+$-N	2.7 gNH$_4^+$-N/gVSS-d	0.92
	NO_2^-	O_2	CO_2	0.10	0.08 gVSS/gNO$_2^-$-N	7.8 gNO$_2^-$-N/gVSS-d	0.62
Methanogens	acetate BOD	acetate	acetate	0.05	0.035 gVSS/gBOD$_L$	8.4 gBOD$_L$/gVSS-d	0.3
	H_2	CO_2	CO_2	0.08	0.45 gVSS/gH$_2$	1.1 gH$_2$/g VSS-d	0.5
Sulfide Oxidizing Autotrophs	H_2S	O_2	CO_2	0.2	0.28 gVSS/gH$_2$S-S	5 gS/gVSS-d	1.4
Sulfate Reducers	H_2	SO_4^{2-}	CO_2	0.05	0.28 gVSS/gH$_2$	1.05 gH$_2$/gVSS-d	0.29
	acetate BOD	SO_4^{2-}	acetate	0.08	0.057 gVSS/gBOD$_L$	8.7 gBOD$_L$/gVSS-d	0.5
Fermenters	sugar BOD	sugars	sugars	0.18	0.13 gVSS/gBOD$_L$	9.8 gBOD$_L$/gVSS-d	1.2

Y is computed assuming a cellular VSS$_a$ composition of $C_5H_7O_2N$, and NH_4^+ is the N source, except when NO_3^- is the electron acceptor; then NO_3^- is the N source. The typical units on Y are presented.

$\hat{q}$ is computed using $\hat{q} = 1e^-$ eq/gVSS$_a$-d.

$\hat{\mu}$ has units of d^{-1}.

the donor-to-acceptor energy reaction. Chapter 2 showed that Y values are computed directly from f_s^o as a unit conversion. For example,

Aerobic heterotrophs:

$$Y = 0.6 \frac{e^- \text{ eq cells}}{e^- \text{ eq donor}} \cdot \frac{113 \text{ gVSS}}{20 \text{ e}^- \text{ eq cells}} \cdot \frac{1 \text{ e}^- \text{ eq donor}}{8 \text{ gBOD}_L}$$

$$= 0.42 \text{ gVSS/gBOD}_L$$

Denitrifying heterotrophs:

$$Y = 0.5 \frac{e^- \text{ eq cells}}{e^- \text{ eq donor}} \cdot \frac{113 \text{ gVSS}}{28 \text{ e}^- \text{ eq cells}} \cdot \frac{1 \text{ e}^- \text{ eq donor}}{8 \text{ gBOD}_L}$$

$$= 0.25 \text{ gVSS/gBOD}_L$$

H_2-Oxidizing Sulfate Reducers:

$$Y = 0.05 \frac{e^- \text{ eq cells}}{e^- \text{ eq donor}} \cdot \frac{113 \text{ gVSS}}{20 \text{ e}^- \text{ eq cells}} \cdot \frac{2 \text{ e}^- \text{ eq donor}}{2 \text{ gH}_2}$$

$$= 0.28 \text{ gVSS/gH}_2$$

The wide range of Y values reflects a combination of changes in f_s^o values and different units for the donor.

The true yield also can be estimated experimentally from batch growth. A small inoculum is grown to exponential phase and harvested. The true yield is estimated from $Y = -\Delta X / \Delta S$, where ΔX and ΔS are the measured changes in biomass and substrate concentration from inoculation until the time of harvesting. The batch technique is adequate for rapidly growing cells, but can create errors when the cells grow slowly so that biomass decay cannot be neglected.

For the cells' usual primary substrates, the maximum specific rate of substrate utilization, $\hat{q}$, is controlled largely by electron flow to the electron acceptor. For 20 °C, the maximum flow to the energy reaction is about 1 e$^-$ eq/gVSS-d. If this flow is termed $\hat{q}_e$, $\hat{q}$ can be computed from

$$\hat{q} = \hat{q}_e / f_e^0 \qquad\qquad \textbf{[3.11]}$$

Table 3.1 also lists typical $\hat{q}$ values when $\hat{q}_e = 1$ e$^-$ eq/gVSS-day. The range of $\hat{q}$ values again reflects variations in f_e^o and different units for the donor. Table 3.1 also lists $\hat{\mu} = Y\hat{q}$ values. The table makes it clear that the fast growing cells are those that have a large f_s^o, which directly gives a large Y and indirectly gives a large $\hat{q}$. Thus, $\hat{\mu}$ is controlled mainly by the microorganism's stoichiometry and energetics.

Temperature affects $\hat{q}$. For temperatures up to the microorganism's optimal temperature, the substrate-utilization rate roughly doubles for each 10 °C increase in temperature. This phenomenon can be approximated by

$$\hat{q}_T = \hat{q}_{20}(1.07)^{T-20} \qquad\qquad \textbf{[3.12]}$$

where T is in °C and $\hat{q}_{20}$ is the $\hat{q}$ value for 20 °C. If $\hat{q}_{20}$ is not known, the relationship can be generalized to

$$\hat{q}_T = \hat{q}_{T^R}(1.07)^{(T-T^R)} \qquad\qquad \textbf{[3.13]}$$

in which T^R is any reference temperature (°C) for which $\hat{q}_{T^R}$ is known.

The endogenous decay rate (b) depends on species type and temperature. The b values tend to correlate positively with $\hat{\mu}$ values. For example, aerobic heterotrophs have b values of 0.1 to 0.3/d at 20 °C, while the slower-growing species have $b <$ 0.05/d. The temperature effect on b can be expressed by a $(1.07)^{T-T^R}$ relationship parallel to Equation 3.13. Endogenous decay coefficients normally encompass several loss phenomena, including lysis, predation, excretion of soluble materials, and death.

McCarty (1975) found that the biodegradable fraction (f_d) is quite reproducible and has a value near 0.8 for a wide range of microorganisms.

The Monod half-maximum-rate concentration (K) is the most variable and least predictable parameter. Its value can be affected by the substrate's affinity for transport or metabolic enzymes. In addition, mass-transport resistances, usually ignored for suspended growth, often are "lumped" into the Monod kinetics by an increase in K. When mass transport is not included in K and simple electron-donor substrates are considered, K values tend to be low, less than 1 mg/l and often as low as the μg/l range. For more difficult to degrade compounds and when mass-transport resistance is de facto included, K values range from a few mg/l to 100s of mg/l for electron donors. Terminal electron acceptors for respiration often have very low K values, well below 1 mg/l. On the other hand, direct use of O_2 as a cosubstrate in oxygenation reactions may have a higher K value, around 1 mg/l. The information provided here for K should be taken as general guidance only. More exact values require analyzing experimental results.

3.3   BASIC MASS BALANCES

Writing mass balances requires specifying a control volume. All the mass balances, their solutions, and important trends can be obtained by considering one of the simplest systems: a steady-state chemostat. Figure 3.2 illustrates the key features of a chemostat: a completely mixed reactor having uniform and steady concentrations of active cells (X_a), substrate (S), inert biomass (X_i), and any other constituents we wish to consider. Substrate (S) includes soluble compounds or suspended materials that are hydrolyzed to soluble compounds within the biological reactor. The chemostat's volume is V, and it receives a constant feed flow rate Q having substrate concentration S^0. For now, we preclude inputs of active or inert biomass. The effluent has flow rate Q and concentrations X_a, X_i, and S.

We must provide mass balances on the active biomass and the rate-limiting substrate (assumed to be the electron donor), because they are the active catalysts and the material responsible for the accumulation of the catalysts, respectively. The steady-

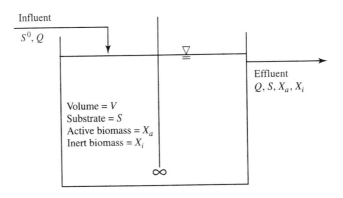

Figure 3.2   Schematic of a chemostat.

state mass balances are:

$$\text{Active Biomass:} \quad 0 = \mu X_a V - Q X_a \qquad \textbf{[3.14]}$$

$$\text{Substrate:} \quad 0 = r_{ut} V + Q(S^0 - S) \qquad \textbf{[3.15]}$$

Equations 3.6 and 3.9 provide the functions for μ and r_{ut}, respectively. Substituting them gives

$$0 = Y\frac{\hat{q}S}{K+S}X_a V - bX_a V - QX_a \qquad \textbf{[3.16]}$$

for biomass and

$$0 = -\frac{\hat{q}S}{K+S}X_a V + Q(S^0 - S) \qquad \textbf{[3.17]}$$

for substrate.

Equations 3.16 and 3.17 can be solved to yield the steady-state values of S and X_a. The strategy is to solve Equation 3.16, the active-biomass balance, first for S.

$$S = K\frac{1 + b\left(\dfrac{V}{Q}\right)}{Y\hat{q}\left(\dfrac{V}{Q}\right) - \left(1 + b\left(\dfrac{V}{Q}\right)\right)} \qquad \textbf{[3.18]}$$

Once we know S, we solve the substrate balance, Equation 3.17. To do this, we first rearrange Equation 3.16 to obtain $[\hat{q}SX_a V/(K+S)]$, which is substituted into Equation 3.17 to yield

$$X_a = Y(S^0 - S)\frac{1}{1 + b\left(\dfrac{V}{Q}\right)} \qquad \textbf{[3.19]}$$

This two-step strategy is a generally useful one and can be used for more complicated systems or when the growth and substrate-utilization relationships differ from those in Equations 3.6 and 3.9.

The ratio V/Q is not the most convenient form. It can be substituted immediately by either of two alternate forms:

$$\text{hydraulic detention time } (T) = \theta = V/Q \qquad \textbf{[3.20]}$$

or

$$\text{dilution rate } (T^{-1}) = D = Q/V \qquad \textbf{[3.21]}$$

Engineers generally are more comfortable with using θ to simplify Equations 3.18 and 3.19, while microbiologists usually use D.

A much more general and powerful substitution is the *solids retention time (SRT)*, which also is commonly called the *mean cell residence time (MCRT)* or the *sludge age*. These three terms normally refer to the same quantity, which is denoted θ_x, has units of time, and always is defined as

$$\theta_x = \frac{\text{active biomass in the system}}{\text{production rate of active biomass}} = \mu^{-1} \qquad \textbf{[3.22]}$$

Equation 3.22 illustrates the two critical features of θ_x. First, it is the reciprocal of the net specific growth rate. Thus, θ_x is a fundamental descriptor of the physiological status of the system, because it gives us direct information about the specific growth rate of the microorganisms. Second, the word definition in Equation 3.22 must be converted to quantitative parameters to apply θ_x to any system.

For our chemostat, any biomass produced in the system must exit in the effluent: thus, the denominator is QX_a. The total active biomass (the numerator) is VX_a. Thus,

$$\theta_x = \frac{VX_a}{QX_a} = \theta = \frac{1}{D} \qquad \textbf{[3.23]}$$

which emphasizes that $\theta_x = \theta$ in our chemostat, as long as the chemostat is at steady state.

Rewriting Equations 3.18 and 3.19 in the (most useful) θ_x form gives

$$S = K \frac{1 + b\theta_x}{Y\hat{q}\theta_x - (1 + b\theta_x)} \qquad \textbf{[3.24]}$$

$$X_a = Y \left(\frac{S^0 - S}{1 + b\theta_x} \right) \qquad \textbf{[3.25]}$$

Figure 3.3 illustrates how S and X_a are controlled by θ_x. Although the chemostat is a simple system, several key trends shown in Figure 3.3 occur for all suspended-growth processes.

1. When θ_x is very small, $S = S^0$, and $X_a = 0$. This situation, termed *washout*, has no substrate removal and, therefore, no accumulation of active biomass. The θ_x value at which washout begins is called θ_x^{min}, which forms the boundary between having steady-state biomass and washout. θ_x^{min} is computed by letting $S = S^0$ in Equation 3.24 and solving for θ_x:

$$\theta_x^{min} = \frac{K + S^0}{S^0(Y\hat{q} - b) - bK} \qquad \textbf{[3.26]}$$

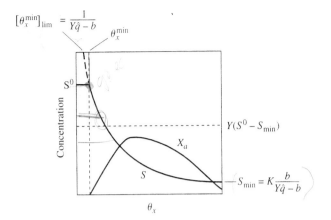

Figure 3.3   Sketch of how S and X_a vary with θ_x and what are the limiting values for θ_x, S, and X_a.

$\theta_x^{\min}$ increases with increasing S^0, but asymptotically reaches a limiting value

$$\left[\theta_x^{\min}\right]_{\lim} = \frac{1}{Y\hat{q} - b} \qquad \textbf{[3.27]}$$

which defines an absolute minimum θ_x (or maximum μ) boundary for having steady-state biomass. $[\theta_x^{\min}]_{\lim}$ is a fundamental delimiter of a biological process: washout at low θ_x.

2. For all $\theta_x > \theta_x^{\min}$, S declines monotonically with increasing θ_x. Equation 3.24 is used to compute S.

3. For very large θ_x, S approaches another key limiting value: $S_{\min}$, the minimum substrate concentration capable of supporting steady-state biomass. $S_{\min}$ can be computed by letting θ_x approach infinity in Equation 3.24:

$$S_{\min} = K\frac{b}{Y\hat{q} - b} \qquad \textbf{[3.28]}$$

If $S < S_{\min}$, the cells net specific growth rate is negative (recall Equation 3.9), and biomass will not accumulate or will gradually disappear. Therefore, steady-state biomass can be sustained only when $S > S_{\min}$. $S_{\min}$ is a fundamental delimiter of biological process performance for large θ_x.

4. When $\theta_x > \theta_x^{\min}$, X_a rises initially, because $S^0 - S$ increases as θ_x becomes larger. However, X_a reaches a maximum value in a chemostat and then declines as decay becomes dominant for large θ_x. If θ_x were to extend to infinity, X_a would approach zero.

In summary, the simple graph in Figure 3.3 tells us that we can reduce S from S^0 to $S_{\min}$ as we increase θ_x from $\theta_x^{\min}$ to infinity. The exact value of θ_x we pick depends on a balancing of substrate removal, biomass production (equals QX_a), and other factors we will discuss later. In practice, engineers often specify a microbiological safety factor, which is defined as θ_x/θ_x^{min}. Safety factors, which typically range from around five to hundreds, are discussed in Chapter 5.

3.4 MASS BALANCES ON INERT BIOMASS AND VOLATILE SOLIDS

Because some fraction of newly synthesized biomass is refractory to self-oxidation, endogenous respiration leads to the accumulation of inactive biomass. In addition, real influents often contain refractory volatile suspended solids that we cannot differentiate easily from inactive biomass. So, we must expand our chemostat analysis to account for inactive biomass (and other nonbiodegradable volatile suspended solids).

A steady-state mass balance on inert biomass is

$$0 = (1 - f_d)bX_aV + Q(X_i^0 - X_i) \qquad \textbf{[3.29]}$$

in which

X_i = concentration of inert biomass in the chemostat (M_xL^{-3})

X_i^0 = input concentration of inert biomass (or indistinguishable refractory volatile suspended solids) (M_xL^{-3}).

Thus, we are relaxing the original requirement that only substrate enters in the influent. Also, the first term in Equation 3.29, the formation rate of inert biomass from active-biomass decay, comes from Equation 3.4.

Solution of Equation 3.29 is

$$X_i = X_i^0 + X_a(1 - f_d)b\theta \qquad \textbf{[3.30]}$$

Equation 3.30 emphasizes that X_i is comprised of influent inerts (X_i^0) and inerts formed from decay of X_a (i.e., $X_a(1 - f_d)b\theta$). Figure 3.4 shows that X_i increases monotonically from X_i^0 to a maximum of $X_i^0 + Y(S^0 - S_{min})(1 - f_d)$. Thus, operation at a large θ results in greater accumulation of inert biomass.

The sum of X_i and X_a is called X_v, the volatile suspended solids concentration, or VSS. Letting $\theta_x = \theta$ for the chemostat, X_v can be computed directly as

$$X_v = X_i^o + X_a(1 + (1 - f_d)b\theta_x) = X_i^0 + Y(S^0 - S)\frac{1 + (1 - f_d)b\theta_x}{1 + b\theta_x} \qquad \textbf{[3.31]}$$

Figure 3.4 demonstrates that X_v generally follows the trend of X_a, but it does not equal zero; when X_a goes to zero, X_v equals X_i.

The second term on the right-hand side of Equation 3.31 represents the net accumulation of biomass from synthesis and decay. It is constituted by the change in substrate concentration $(S^0 - S)$ multiplied by the net yield (Y_n):

$$Y_n = Y\frac{1 + (1 - f_d)b\theta_x}{1 + b\theta_x} \qquad \textbf{[3.32]}$$

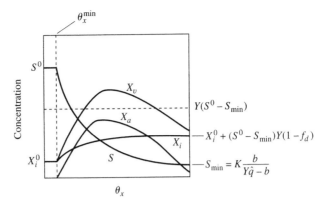

Figure 3.4   General variations when X_i and X_v are added to the chemostat analysis.

Equation 3.32 is parallel to the relationship between f_s and f_s^0:

$$f_s = f_s^0 \frac{1 + (1 - f_d)b\theta_x}{1 + b\theta_x}$$ **[3.33]**

Sometimes the net yield also is given the name observed yield and symbolized as Y_{obs}.

3.5   SOLUBLE MICROBIAL PRODUCTS

Besides consuming substrates and making new biomass, bacteria also generate *soluble microbial products (SMP)*. SMP (M_pL^{-3}) appear to be cellular components that are released during cell lysis, diffuse through the cell membrane, are lost during synthesis, or are excreted for some purpose. They have moderate formula weights (100s to 1,000s) and are biodegradable. SMP generally do not include intermediates of catabolic pathways, which behave differently than SMP and are system specific when they occur. SMP are important because they are present in all cases and form the majority of the effluent COD and BOD in many cases. SMP also can complex metals, foul membranes, and cause color or foaming.

SMP can be subdivided into two subcategories. *UAP* (short for *substrate-utilization-associated products*) are produced directly during substrate metabolism. Their formation kinetics can be described by

$$r_{UAP} = -k_1 r_{ut}$$ **[3.34]**

in which

$$r_{UAP} = \text{rate of UAP-formation } M_pL^{-3}T^{-1}$$
$$k_1 = \text{UAP-formation coefficient, } M_pM_s^{-1}$$

The second category is *BAP* (for *biomass-associated products*). They are formed directly from biomass, presumably as part of maintenance and decay, but perhaps for other reasons, as well. The BAP formation rate expression is

$$r_{\text{BAP}} = k_2 X_a$$ **[3.35]**

in which

r_{BAP} = rate of BAP formation, $M_p L^{-3} T^{-1}$

k_2 = BAP-formation rate coefficient, $M_p M_x^{-1} T^{-1}$

From several lines of research (summarized in Rittmann, Bae, Namkung, and Lu (1987) and Noguera, Araki, and Rittmann (1994)), we know that SMP originally formed is relatively biodegradable. However, being a highly heterogeneous mixture, the SMP has a range of biodegradation kinetics. The details of SMP-biodegradation kinetics are not fully resolved. However, the most recent work suggests that UAP and BAP have sufficiently distinct degradation kinetics that they can be described with separate Monod-degradation expressions:

$$r_{\text{deg-UAP}} = \frac{-\hat{q}_{\text{UAP}}\text{UAP}}{K_{\text{UAP}} + \text{UAP}} X_a$$ **[3.36a]**

$$r_{\text{deg-BAP}} = \frac{-\hat{q}_{\text{BAP}}\text{BAP}}{K_{\text{BAP}} + \text{BAP}} X_a$$ **[3.36b]**

in which $\hat{q}_{\text{UAP}}$ and $\hat{q}_{\text{BAP}}$ are maximum specific rates of UAP and BAP degradation ($M_p M_x^{-1} T^{-1}$), respectively; K_{UAP} and K_{BAP} are half-maximum rate concentrations for UAP and BAP ($M_p L^{-3}$), respectively; and UAP and BAP stand for their concentrations ($M_p L^{-3}$). Note that only active biomass (X_a) degrades SMP.

The steady-state mass balances on UAP and BAP are

$$0 = -k_1 r_{\text{ut}} V - \frac{\hat{q}_{\text{UAP}}\text{UAP}}{K_{\text{UAP}} + \text{UAP}} X_a V - Q\text{UAP}$$ **[3.37a]**

$$0 = k_2 X_a V - \frac{\hat{q}_{\text{BAP}}\text{BAP}}{K_{\text{BAP}} + \text{BAP}} X_a V - Q\text{BAP}$$ **[3.37b]**

These are solved for UAP and BAP as

$$\text{UAP} = -\frac{(\hat{q}_{\text{UAP}} X_a \theta + K_{\text{UAP}} + k_1 r_{\text{ut}}\theta)}{2}$$
$$+ \frac{\sqrt{(\hat{q}_{\text{UAP}} X_a \theta + K_{\text{UAP}} + k_1 r_{\text{ut}}\theta)^2 - 4K_{\text{UAP}}k_1 r_{\text{ut}}\theta}}{2}$$ **[3.38]**

$$\text{BAP} = \frac{-(K_{\text{BAP}} + (\hat{q}_{\text{BAP}} - k_2)X_a\theta)}{2}$$
$$+ \frac{\sqrt{(K_{\text{BAP}} + (\hat{q}_{\text{BAP}} - k_2)X_a\theta)^2 + 4K_{\text{BAP}}k_2 X_a\theta}}{2}$$ **[3.39]**

In Equations 3.38 and 3.39, the hydraulic detention time (θ) appears, and r_{ut} takes a negative value and can be computed from

$$r_{\text{ut}} = -(S^0 - S)/\theta = -\frac{\hat{q} S}{K + S} X_a$$ **[3.40]**

Information to define the SMP kinetic parameters is sparse. Noguera (1991) analyzed aerobic data and obtained the following best-fit values:

$$k_1 = 0.12 \text{ g COD}_p/\text{g COD}_s$$
$$k_2 = 0.09 \text{ g COD}_p/\text{g VSS}_a\text{-d}$$
$$\hat{q}_{\text{UAP}} = 1.8 \text{ g COD}_p/\text{g VSS}_a\text{-d}$$
$$K_{\text{UAP}} = 100 \text{ mg COD}_p/\text{l}$$
$$\hat{q}_{\text{UAP}}/K_{\text{UAP}} = 18 \text{ l/g VSS}_a\text{-d}$$
$$\hat{q}_{\text{BAP}} = 0.1 \text{ g COD}_p/\text{g VSS}_a\text{-d}$$
$$K_{\text{BAP}} = 85 \text{ mg COD}_p/\text{l}$$
$$\hat{q}_{\text{BAP}}/K_{\text{BAP}} = 1.2 \text{ l/g VSS}_a\text{-d}$$

While these values can be considered only provisional, they do point out several key features of SMP. First, a significant fraction of substrate COD is shunted to UAP formation; $k_1 = 0.12$ g COD$_p$/mg COD$_s$ means that 12 percent of the substrate "utilized" is neither sent to the electron acceptor nor converted to biomass, but is released as UAP. Second, the formation of BAP constitutes a significant fraction of biomass loss; $k_2 = 0.09$ g COD$_p$/g VSS$_a$-d converts to a first-order "decay" rate of approximately 0.06/d. Third, UAP is biodegraded considerably faster than is BAP; thus, we should expect to see preferential buildup of BAP in most situations.

Noguera et al. (1994) estimated SMP parameters for a methanogenic system:

$$k_1 = 0.21 \text{ g COD}_p/\text{g COD}_s$$
$$k_2 = 0.035 \text{ g COD}_p/\text{g VSS}_a\text{-d}$$
$$\hat{q}_{\text{UAP}}/K_{\text{UAP}} = 2.4 \text{ l/g VSS}_a\text{-d}$$
$$\hat{q}_{\text{BAP}}/K_{\text{BAP}} = 0.31 \text{ l/g VSS}_a\text{-d}$$

Most notable is the relatively slower degradation kinetics, compared to the aerobic system.

Since k_2 can be a significant fraction of b, we shall assume that b implicitly includes BAP generation, as well as biomass oxidation and conversion to inert biomass. By differences, the portion of endogenous decay that results in direct biomass oxidation is $b - f_d b - k_2 = (1 - f_d)b - k_2$.

Example 3.1 | **ANALYSIS OF A CHEMOSTAT'S PERFORMANCE**   A chemostat having volume $V = 2,000$ m³ receives a constant flow of $Q = 1,000$ m³/d of wastewater containing only biodegradable organic material having BOD$_L$ of $S^0 = 500$ mg BOD$_L$/l. Also included is inert VSS with $X_i^0 = 50$ mg VSS/l. From past research, we know that the electron donor is rate limiting, and we have the following kinetic and stoichiometric parameters for aerobic biodegradation:

$\hat{q} = 20$ g BOD$_L$/g VSS$_a$-d	$k_2 = 0.09$ g COD$_p$/g VSS$_a$-d
$Y = 0.42$ g VSS$_a$/g BOD$_L$	$\hat{q}_{\text{UAP}} = 1.8$ g COD$_p$/g VSS$_a$-d
$K = 20$ mg BOD$_L$/l	$K_{\text{UAP}} = 100$ mg COD$_p$/l
$b = 0.15$/d	$\hat{q}_{\text{BAP}} = 0.1$ g COD$_p$/g VSS$_a$-d
$f_d = 0.8$	$K_{\text{BAP}} = 85$ mg COD$_p$/l
$k_1 = 0.12$ g COD$_p$/g BOD$_L$	

The goal is to analyze the performance of the chemostat with regard to its effluent quality. First, we compute the delimiting values of S_{min}, $[\theta_x^{min}]_{lim}$, and θ_x^{min}:

$$S_{min} = K \frac{b}{Y\hat{q} - b} = \frac{20 \text{ mg BOD}_L}{1} \frac{0.15/d}{0.42 \frac{\text{g VSS}_a}{\text{g BOD}_L} \cdot 20 \frac{\text{g BOD}_L}{\text{g VSS}_a\text{-d}} - 0.15/d}$$

$$= 0.36 \text{ mg BOD}_L/l$$

$$[\theta_x^{min}]_{lim} = \frac{1}{Y\hat{q} - b} = \frac{1}{0.42 \frac{\text{g VSS}_a}{\text{g BOD}_L} \cdot 20 \frac{\text{g BOD}_L}{\text{g VSS}_a\text{-d}} - 0.15/d}$$

$$= 0.12 \text{ d}$$

$$\theta_x^{min} = \frac{K + S^0}{S^0(Y\hat{q} - b) - bK}$$

$$= \frac{(20 + 500) \text{ mg BOD}_L/l}{500 \frac{\text{mg BOD}_L}{1}(0.42 \cdot 20/d - 0.15/d) - 20 \text{ mg BOD}_L/l \cdot 0.15/d}$$

$$= \frac{520 \text{ mg BOD}_L/l}{(4,125 - 3) \text{ mg BOD}_L/l\text{-d}}$$

$$= 0.126 \text{ d}$$

We see that the substrate concentration could be driven well below S^0, as long as the SRT is significantly greater than 0.126 d. Note also that the product $Y\hat{q} = \hat{\mu} = 8.4 \text{ d}^{-1}$ recurs frequently.

Second, we compute θ_x and the safety factor (SF)

$$\theta_x = \theta = V/Q = \frac{2,000 \text{ m}^3}{1,000 \text{ m}^3/d} = 2 \text{ d}$$

$$SF = \theta_x/\theta_x^{min} = \frac{2 \text{ d}}{0.126 \text{ d}} = 16$$

We should have θ_x and SF large enough to take S much below S^0 and to give stable biomass accumulation.

Third, we compute the effluent substrate concentration.

$$S = K \frac{1 + b\theta_x}{Y\hat{q}\theta_x - (1 + b\theta_x)}$$

$$= 20 \frac{\text{mg BOD}_L}{1} \frac{1 + 0.15/d \cdot 2 \text{ d}}{8.4/d \cdot 2 \text{ d} - (1 + 0.15/d \cdot 2 \text{ d})}$$

$$= 20 \text{ mg/l} \frac{1.3}{16.8 - 1.3}$$

$$= 1.7 \text{ mg/l}$$

Thus, almost all of S^0 is removed. Also, $1 + b\theta_x = 1.3$ is another recurring parameter grouping.

Fourth, we determine the effluent (and reactor) concentrations of active, inert, and total volatile solids,

$$X_a = Y(S^0 - S)\frac{1}{1 + b\theta_x}$$

$$= 0.42 \frac{\text{g VSS}_a}{\text{g BOD}_L} \left((500 - 1.7)\frac{\text{mg BOD}_L}{1} \right) \frac{1}{1.3}$$

$$= 161 \text{ mg VSS}_a/1$$

$$X_i = X_i^0 + X_a(1 - f_d)b\theta_x$$

$$= 50 \text{ mg VSS}_i/1 + 161\frac{\text{mg VSS}_a}{1} \left((1 - 0.8)\frac{\text{g VSS}_i}{\text{g VSS}_a} \right) \cdot 0.15/\text{d} \cdot 2 \text{ d}$$

$$= 50 + 9.7 = 59.7 \text{ mg VSS}_i/1 \simeq 60 \text{ mg VSS}_i/1$$

$$X_v = X_a + X_i = 161 + 60 = 221 \text{ mg VSS/l}$$

These calculations illustrate that the chemostat, while removing nearly all 500 mg BOD_L/l of substrate, produces 171 mg VSS/l of biomass and passes through the 50 mg VSS_i/l of inert volatile solids.

Fifth, we estimate the effluent SMP. The individual terms required for Equations 3.38 and 3.39 are computed as

$$r_{ut} = -(S^0 - S)/\theta = -((500 - 1.7) \text{ mg BOD}_L/1)/2 \text{ d} = -249 \text{ mg BOD}_L/1 \cdot \text{d}$$

$$(\hat{q}_{UAP} X_a \theta + K_{UAP} + k_1 r_{ut}\theta) = (1.8 \cdot 161 \cdot 2 + 100 + 0.12 \cdot (-249) \cdot 2)\text{mg COD}_p/1$$

$$= 620 \text{ mg COD}_p/1$$

$$-4K_{UAP}k_1 r_{ut}\theta = -4 \cdot 100 \cdot 0.12 \cdot (-249) \cdot 2 = 23,900 \ (\text{mg COD}_p/1)^2$$

$$K_{BAP} + (\hat{q}_{BAP} - k_2)X_a\theta = 85 + (0.1 - 0.09) \cdot 161 \cdot 2 = 88.2 \text{ mg COD}_p/1$$

$$4K_{BAP}k_2 X_a\theta = 4 \cdot 85 \cdot 0.09 \cdot 161 \cdot 2 = 9,850 \ (\text{mg COD}_p/1)^2$$

Then, UAP and BAP are

$$UAP = \frac{-620 + \sqrt{(620)^2 + 23,900}}{2}$$

$$= 9.5 \text{ mg COD}_p/1$$

$$BAP = \frac{-88.2 + \sqrt{(88.2)^2 + 9,850}}{2}$$

$$= 22.3 \text{ mg COD}_p/1$$

$$SMP = UAP + BAP = 9.5 + 22.3 = 31.8 \text{ mg COD}_p/1$$

Sixth, we compute the total effluent quality. The soluble COD is equal to the sum of substrate and SMP

$$\text{Soluble COD} = S + \text{SMP}$$

$$= 1.7 + 31.8$$

$$= 33.5 \text{ mg COD/l}$$

As is often the case, the soluble COD is dominated by SMP, not original substrate. The soluble

BOD_L will equal the soluble COD if all of the SMP is biodegradable, which we assume is true:

$$\text{Soluble } BOD_L = 33.5 \text{ mg } BOD_L/l$$

The total effluent COD is the sum of the soluble COD and the COD of all the volatile solids.

$$\text{Total COD} = \text{Soluble COD} + (1.42 \text{ g COD/gVSS}) \cdot X_v$$

$$= 33.5 + 1.42 \cdot 221$$

$$= 347 \text{ mg COD/l.}$$

The total BOD_L is the sum of the soluble BOD_L (we assume that the SMP is degradable) and the oxygen demand of the biodegradable fraction of the *active* biomass.

$$\text{Total } BOD_L = \text{Soluble } BOD_L + (1.42 \text{ g BOD/g VSS}_a) \cdot X_a \cdot f_d$$

$$= 33.5 + 1.42 \cdot 161 \cdot 0.8$$

$$= 216 \text{ mg } BOD_L/l$$

We see that most of the effluent COD and BOD_L in a chemostat come from the volatile suspended solids. Also, the total COD and BOD_L computations utilize the conversion factor 1.42 g BOD_L (or COD)/gVSS, which comes from the oxygen demand of C in $C_5H_7O_2N$.

DESIGN OF A CHEMOSTAT Design requires that certain effluent criteria be met. This example proposes several types of design criteria and shows how the choice of the "best" design θ_x depends on the criterion used. The influent flow and quality remain the same as in Example 3.1, but V can be altered to control θ_x. The design problem is solved by first creating for a set of θ_x values (from 0.2 to 100 d) the effluent concentrations for S, X_a, X_v, SMP, soluble

Example 3.2

Table 3.2 Summary of effluent quality for the parameters of Example 3.1 and the range of θ_x values for Example 3.2

θ_x, d	S, mg BOD_L/l	X_a, mg VSS$_a$/l	X_v, mg VSS/l	SMP, mg COD$_p$/l	BOD_L, mg BOD_L/l Soluble	Total
0.20	31.7	191	242	40.8	72.4	289
0.25	19.5	195	246	39.1	58.7	280
0.5	6.9	193	246	32.6	39.5	258
1	3.2	181	236	29.4	32.6	239
2	1.7	161	221	31.8	33.5	216
3	1.2	144	207	35.4	36.7	201
5	0.87	120	188	41.4	42.3	178
10	0.61	84	159	49.9	50.5	146
15	0.53	65	144	54.2	54.8	126
20	0.49	52	134	56.9	57.4	117
30	0.45	38	122	59.9	60.3	104
50	0.41	25	112	62.7	63.1	91
100	0.39	13	102	65.0	65.4	80

Recall that $S_{min} = 0.36$ mg BOD_L/l and $\theta_x^{min} = 0.125$ d.

Table 3.3   Selection of the most appropriate θ_x value (from Table 3.2) for different design criteria

Criteria	Design θ_x
Reduce original substrate to 1 mg/l or less	$\theta_x >\sim 4$ d
95% removal of original substrate	$\theta_x >\sim 0.22$ d
99% removal of original substrate	$\theta_x >\sim 0.7$ d
Minimize soluble BOD_L	$\theta_x \sim 1$ d
95% removal of soluble BOD_L	not feasible
90% removal of soluble BOD_L	$0.4 < \theta_x < 9$ d
Minimize total BOD_L	θ_x large
Maximize the generation of VSS	θ_x from 0.25–0.5 d
Minimize the generation of SMP	$\theta_x \sim 1$ d
Maximize the fraction of VSS that is active	$\theta_x \sim 0.25$ d

BOD_L, and total BOD_L. Each is computed in the same manner as was done in Example 3.1. Table 3.2 summarizes the computations.

Table 3.3 summarizes how different design criteria use the results to choose the most appropriate θ_x. These results show that θ_x chosen from effluent BOD_L criteria are not controlled by original substrate alone. Instead, SMP and VSS_a are more important than S when real-world effluent criteria are used.

Example 3.3 | **EFFLUENT BOD₅**   The results in Tables 3.2 and 3.3 use BOD_L as the BOD criterion, not BOD_5, which is most often used in standards and regulations. It is quite likely that the ratios of BOD_5:BOD_L are very different for original substrate, SMP, and active biomass. Although much needs to be learned about the exertion of BOD by the different sources, this example shows how conversions can be made.

The BOD exertion equation is

$$BOD_t = BOD_L(1 - \exp\{-kt\})\qquad\textbf{[3.41]}$$

in which BOD_t is the BOD exerted (i.e., measured by dissolved-oxygen uptake) at time t, k is the first-order rate constant for BOD exertion (T^{-1}), and exp is the exponential function. From Equation 3.41, the ratio BOD_5:BOD_L is

$$BOD_5:BOD_L = 1 - \exp\{-k \cdot 5\text{ d}\}\qquad\textbf{[3.42]}$$

We shall assume that $k = 0.23$/d for original substrate, 0.1/d for active biomass, and 0.03/d for SMP. These three values are based, respectively, on the typical BOD-exertion kinetics for readily biodegraded substrates, decay rate of active biomass, and BAP-utilization kinetics. While not universal, they give a good first estimate of the relative biodegradability of each of the three components of the effluent BOD_L. The BOD_5:BOD_L ratios are then 0.68 for original substrate, 0.40 for active biomass, and 0.14 for SMP.

These ratios are combined with the data in Table 3.2 to compute the weight-averaged soluble and total BOD_5 values shown in Table 3.4. When soluble BOD_5 is used, the minimum value occurs for an SRT of 2 to 3 d, instead of 1 d (Table 3.3). Furthermore, the soluble BOD_5

Table 3.4   Estimated weight-averaged BOD_5 concentrations for the results in Table 3.2

θ_x, d	Soluble BOD_5, mg/l	Total BOD_5, mg/l
0.20	27.3	114
0.25	18.8	107
0.50	9.2	97
1	6.3	89
2	5.6	79
3	5.8	71
5	6.4	61
10	7.4	46
15	8.0	37
20	8.3	32
30	8.7	26
50	9.1	20
100	9.4	15

concentrations are close to those observed in practical aerobic systems, in which soluble BOD_5 normally is less than 10 mg/l and is relatively insensitive to changes in θ_x within its usual range (2 to 50 d).

3.6   NUTRIENTS AND ELECTRON ACCEPTORS

A process in environmental biotechnology must provide sufficient nutrients and electron acceptors to support biomass growth and energy generation. Nutrients, being elements comprising the physical structure of the cells, are needed in proportion to the net production of biomass. Active and inert biomass contain nutrients, as long as they are produced microbiologically. The electron acceptor is consumed in proportion to electron-donor utilization multiplied by the sum of exogenous and endogenous flows of electron to the terminal acceptor.

Nutrient and acceptor requirements can be determined from the stoichiometric equations developed in Chapter 2. The net yield fraction, f_s, is used to construct the overall reaction including synthesis and endogenous decay. Equation 3.33 relates f_s to f_s^0 and θ_x.

Another approach obtains nutrient requirements directly from the reactor mass balances. This approach is advantageous when materials other than biomass and substrate are important. SMP are key examples of other important materials. In terms of our chemostat model, the rate of nutrient consumption is

$$r_n = \gamma_n Y r_{\text{ut}} \frac{1 + (1 - f_d)b\theta_x}{1 + b\theta_x} \qquad \textbf{[3.43]}$$

in which

r_n = rate of nutrient consumption $(M_n L^{-3} T^{-1})$

γ_n = the stoichiometric ratio of nutrient mass to VSS for the biomass $[M_n M_x^{-1}]$.

Recall that r_{ut} is the rate of substrate utilization and has a negative sense (Equation 3.40), which also makes r_n negative. The most important nutrients are N and P. The empirical formula for bacterial VSS, $C_5H_7O_2N$, has $\gamma_N = 14$ g N/113 g VSS = 0.124 g N/g VSS. Generally, the P requirement is 20 percent of the N requirement, which makes $\gamma_P = 0.025$ g P/g VSS. An overall mass balance on a nutrient is then

$$0 = QC_n^0 - QC_n + r_n V \qquad \text{[3.44]}$$

in which C_n and C_n^0 are the effluent and influent nutrient concentrations $(M_n L^{-3})$, respectively. Solution for C_n gives

$$C_n = C_n^0 + r_n \theta \qquad \text{[3.45]}$$

The supply of nutrients must be supplemented if C_n takes a negative value in Equation 3.45.

A parallel means to compute the use rate of electron acceptor (denoted $\Delta S_a / \Delta t$ and having units $M_a T^{-1}$) is by a mass balance on electron equivalents expressed as oxygen demand. For the chemostat, the input of oxygen equivalents or oxygen demand (OD) includes the electron-donor substrate and the input VSS:

$$\text{OD inputs} = QS^0 + 1.42 \frac{\text{g COD}}{\text{g VSS}} X_v^0 Q \qquad \text{[3.46]}$$

The outputs are residual substrate, SMP, and all VSS:

$$\text{OD outputs} = QS + Q(\text{SMP}) + 1.42 \frac{\text{g COD}}{\text{g VSS}} X_v Q \qquad \text{[3.47]}$$

The acceptor consumption as O_2 equivalents is the difference between the inputs and the outputs.

$$\frac{\Delta S_a}{\Delta t} = \gamma_a Q \left[S^0 - S - \text{SMP} + 1.42(X_v^0 - X_v) \right] \qquad \text{[3.48]}$$

in which γ_a is the stoichiometric ratio of acceptor mass to oxygen demand. For oxygen, $\gamma_a = 1$ g O_2/g COD; for NO_3^--N, γ_a is 0.35 g NO_3^--N/g COD. Other γ_a ratios can be computed from stoichiometry of the electron-acceptor half-reactions presented in Chapter 2.

The acceptor can be supplied in the influent flow or by other means, such as aeration to provide oxygen. This can be expressed as

$$\frac{\Delta S_a}{\Delta t} = Q \left[S_a^0 - S_a \right] + R_a \qquad \text{[3.49]}$$

in which S_a and S_a^0 are the effluent and influent concentrations of the acceptor, and R_a is the required mass rate of acceptor supply $(M_a T^{-1})$.

NUTRIENT AND ACCEPTOR CONSUMPTION   For Example 3.1, we computed the | **Example 3.4**
following:

$$S^0 = 500 \text{ mg BOD}_L/\text{l}$$
$$S = 1.7 \text{ mg BOD}_L/\text{l}$$
$$X_i^0 = 50 \text{ mg VSS}_i/\text{l}$$
$$X_v = 221 \text{ mg VSS/l}$$
$$\text{SMP} = 31.8 \text{ mg BOD}_L/\text{l}$$
$$r_{ut} = -249 \text{ mg BOD}_L/\text{l}.$$

In this example, the input concentrations of N, P, and O_2 are 50 mg NH_4^+-N/l, 10 mg PO_4^{3-}-P/l, and 6 mg O_2/l, respectively.

The N and P consumption rates are

$$r_N = 0.124 \frac{\text{g N}}{\text{g VSS}} \cdot 0.42 \frac{\text{g VSS}}{\text{g BOD}_L} \cdot \left(-249 \frac{\text{mg BOD}_L}{\text{l-d}}\right) \cdot \frac{1 + (1-0.8) \cdot 0.15 \cdot 2}{1 + 0.15 \cdot 2}$$

$$= -10.6 \text{ mg N/l-d}$$

$$r_P = r_N \cdot 0.2 \frac{\text{g P}}{\text{g N}} = -10.6 \cdot 0.2 = -2.1 \text{ mg P/l-d}$$

The mass balances give the effluent concentrations

$$C_N = 50 \text{ mg N/l} - 10.6 \frac{\text{mg N}}{\text{l-d}} \cdot 2 \text{ d}$$

$$= 28.8 \text{ mg NH}_4^+\text{-N/l}$$

$$C_P = 10 \text{ mg P/l} - 2.1 \frac{\text{mg P}}{\text{l-d}} \cdot 2 \text{ d}$$

$$= 5.8 \text{ mg PO}_4^{3-}\text{-P/l}$$

In both cases, sufficient nutrients are contained in the influent.

For the electron acceptor, O_2 in this case, the required overall supply rate is

$$\frac{\Delta S_a}{\Delta t} = 1 \frac{\text{g O}_2}{\text{g COD}} \cdot \left(1,000 \frac{\text{m}^3}{\text{d}}\right)$$

$$\cdot [500 - 1.7 - 31.8 + 1.42(50 - 221)] \text{ mg COD/l} \cdot \frac{10^3 \text{l}}{\text{m}^3} \cdot 10^{-3} \frac{\text{g}}{\text{mg}}$$

$$= 2.24 \cdot 10^5 \text{ g O}_2/\text{d}.$$

Clearly, the influent concentration of 6 mg/l is not going to supply the O_2 required. Therefore, oxygen must be supplied at a rate (R_{O_2}) of

$$R_{O_2} = \frac{\Delta S_{O_2}}{\Delta t} - Q[S_a^0 - S_a]$$

For an aerobic system, S_a is typically 2 mg/l. Therefore,

$$R_{O_2} = 2.24 \cdot 10^5 \frac{\text{g O}_2}{\text{d}} - [1,000 \text{m}^3/\text{d}][6 \text{mg/l} - 2 \text{ mg/l}] \cdot \frac{10^3}{\text{m}^3} \frac{1}{\text{mg}} \cdot \frac{10^{-3} \text{ g}}{\text{mg}}$$

$$= 2.20 \cdot 10^5 \text{ g O}_2/\text{d}$$

If R_{O_2} is divided by Q, the rate of acceptor supply is expressed as a concentration in the influent flow: Here it is 220 g O_2/m^3 = 220 mg O_2/l. Clearly, normal oxygen in the influent cannot supply the needed oxygen, and aeration is essential.

3.7   INPUT ACTIVE BIOMASS

Whether by design or by happenstance, some biological processes receive significant inputs of biomass active in degradation of the substrate. Three circumstances provide practical examples. First, when microbial processes are operated in series, the downstream process often receives significant biomass from the upstream process. Second, microorganisms may be discharged in a waste stream or grown in the sewers. Third, bioaugmentation is the deliberate addition of microorganisms to improve some aspect of process performance.

When active biomass is input, the steady-state mass balance for active biomass, Equation 3.16, must be modified.

$$0 = QX_a^0 - QX_a + Y\frac{\hat{q}S}{K+S}X_aV - bX_aV \qquad \text{[3.50]}$$

The other mass balances remain the same. Equation 3.50 can be solved for S when the SRT is redefined mathematically to maintain its definition as the reciprocal of the net specific growth rate:

$$S = K\frac{1+b\theta_x}{Y\hat{q}\theta_x - (1+b\theta_x)} \qquad \text{[3.51]}$$

where

$$\theta_x = \mu^{-1} = \frac{X_aV}{QX_a - QX_a^0} \qquad \text{[3.52]}$$

The solution for S is exactly the same as it was for the chemostat before (Equation 3.24), but θ_x is now defined with the denominator being the gross active biomass output rate (QX_a) minus the input rate (QX_a^0). Thus, the denominator remains the net production rate of new active biomass.

Figure 3.5 illustrates how X_a^0 affects S and washout. In the example, S^0 is fixed at 100 mg/l, and the parameter values used are Y = 0.44 mg VSS_a/mg, $\hat{q}$ = 5 mg/mg VSS_a-d, K = 20 mg/l, and b = 0.2/d. Without input active biomass, washout occurs for θ_x of about 0.6 d. As X_a^0 increases, complete washout is eliminated, because the reactor always contains some biomass. Increasing X_a^0 also makes the substrate concentration lower, and the effect is most dramatic near washout.

Although the mass balances for substrate, inert, and volatile biomass remain the same as before, their solutions differ, due to the change in definition of θ_x. In particular, mass balance Equations 3.17 and 3.50, together with Equation 3.52, give

$$X_a = \frac{\theta_x}{\theta}\left[Y(S^0 - S)\frac{1}{1+b\theta_x}\right] \qquad \text{[3.53]}$$

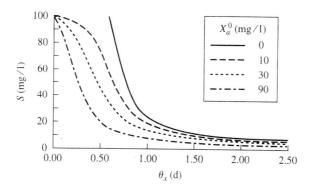

Figure 3.5 Effect of influent active biomass and θ_x on effluent substrate concentration for a chemostat at steady state. The parameters used are $Y = 0.44$ mg VSS$_a$/mg, $\hat{q} = 5$ mg/mg VSS$_a$-d, $K = 20$ mg/l, and $b = 0.2$/d.

Equation 3.29 then yields

$$X_i = X_i^0 + (1-f_d)bX_a\theta \qquad\qquad \textbf{[3.54]}$$

$$X_v = X_i + X_a = X_i^0 + (1 + (1 - f_d)b\theta)X_a \qquad\qquad \textbf{[3.55]}$$

The concentration of active biomass is "built up" by input of active biomass, and this "build up" is quantified by the ratio θ_x/θ. Rearrangement of Equation 3.52 shows that the θ_x/θ ratio depends on the relative concentrations of X_a and X_a^0.

$$\frac{\theta_x}{\theta} = \frac{1}{1 - X_a^0/X_a} \qquad\qquad \textbf{[3.56]}$$

When X_a^0 approaches, but is less than X_a, θ_x is substantially greater than θ. It is even possible to have $X_a^0 > X_a$, if X_a^0 is very large. In such a case, θ_x is negative, and the process is in net biomass decay. Supplying a very large X_a^0 is a means to make the steady-state μ negative and sustain $S < S_{\min}$.

Once S and X_a are determined from the proper definition of θ_x, UAP and BAP concentrations are computed from Equations 3.38 and 3.39, as usual.

INPUT ACTIVE BIOMASS   We recompute the results of Examples 3.1 and 3.4 if the system also receives input active biomass at a concentration of $X_a^0 = 40$ mg VSS$_a$/l. To do the computations, Equations 3.51, 3.53, and 3.56 must be solved iteratively. The strategy is to pick a $\theta_x \geq \theta$, compute S from Equation 3.51, compute X_a from Equation 3.53, compute θ_x from Equation 3.56, and compare the computed θ_x to the original θ_x. If they do not agree, pick another θ_x and iterate. Applying that strategy yields:

$\qquad \theta_x = 2.53$ d $\qquad$ (up from 2.0 with no X_a^0)

$\qquad\quad S = 1.4$ mg BOD$_L$/l $\qquad$ (down from 1.7)

$\qquad\, X_a = 191$ mg VSS$_a$/l $\qquad$ (up from 161)

Example 3.5

Further solution with Equations 3.54 and 3.55 gives

$$X_i = 61 \text{ mg VSS}_i/l \qquad \text{(up from 60)}$$
$$X_v = 252 \text{ mg VSS/l} \qquad \text{(up from 221)}$$

Thus, addition of 40 mg VSS_a/l increases X_a by 30 mg VSS_a/l and X_v by 31 mg VSS/l, while θ_x goes up by 26.5 percent. The increase in θ_x means that biomass decay increases, and X_a increases by less than X_a^0. These are the intuitively expected effects of adding active biomass.

Computation of UAP and BAP is altered only by the changes in r_{ut} and X_a.

$$
\begin{aligned}
r_{ut} &= -(500 - 1.4 \text{ mg BOD}_L/l)/2 \text{ d} \\
&= -249 \text{ mg BOD}_L/l \qquad \text{(not changed perceptibly)} \\
\text{UAP} &= 8.1 \text{ mg COD}_p/l \qquad \text{(down from 9.5)} \\
\text{BAP} &= 25.6 \text{ mg COD}_p/l \qquad \text{(up from 22.3)} \\
\text{SMP} &= 33.7 \text{ mg COD}_p/l \qquad \text{(up from 31.8)}
\end{aligned}
$$

The increased active biomass causes greater BAP formation and accumulation of SMP.

For the total effluent quality,

$$
\begin{aligned}
\text{Soluble BOD}_L \text{ or COD} &= 1.4 + 33.7 \\
&= 35.1 \text{ mg BOD}_L/l \qquad \text{(up from 33.5)} \\
\text{Total COD} \qquad &= 35.1 + 1.42 \cdot 252 \\
&= 393 \text{ mg COD/l} \qquad \text{(up from 348)} \\
\text{Total BOD}_L \qquad &= 35.1 + 1.42 \cdot 191 \cdot 0.8 \\
&= 252 \text{ mg BOD}_L/l \qquad \text{(up from 216)}
\end{aligned}
$$

So, although adding active biomass increases X_a and θ_x and decreases S, the effluent quality has higher soluble and total BOD_L and COD. This is the result of adding active biomass.

The oxygen requirement is

$$\frac{\Delta S_{O_2}}{\Delta t} = 1 \cdot 1{,}000 \cdot [500 - 1.4 - 33.7 + 1.42(90 - 252)]$$

$$= 2.35 \cdot 10^5 \text{ g O}_2/\text{d} \qquad \text{(up from } 2.24 \cdot 10^5 \text{)}$$

Thus, the endogenous decay of the added biomass slightly increases the oxygen required.

Nutrient requirements are

$$r_N = 0.124 \cdot 0.42 \cdot (-249) \frac{1 + (1 - 0.8) \cdot 0.15 \cdot 2.53}{1 + 0.15 \cdot 2.53}$$

$$= -10.1 \text{ mg N/l-d} \qquad \text{(down from } -10.6)$$

$$r_P = -10.1 \cdot 0.2 = -2.0 \text{ mg P/l-d (down from } -2.1)$$

Thus, the net synthesis of biomass is less and requires less N and P as nutrients.

3.8 HYDROLYSIS OF PARTICULATE AND POLYMERIC SUBSTRATES

Several important applications in environmental biotechnology involve organic substrates that originate as particles or large polymers. For example, more than one-half of the BOD in typical sewage is comprised of suspended solids, and sludges undergoing digestion are nearly 100 percent particulate BOD. Before the bacteria can begin the oxidation reactions characteristic of the catabolism of organic substrates, the particles or large polymers must be hydrolyzed to smaller molecules that can be transported across the cell membrane. Extracellular enzymes catalyze these hydrolysis reactions.

Despite its great importance in many situations, hydrolysis has not been thoroughly researched. The best way to represent hydrolysis kinetics is not settled. Part of the problem comes about because the hydrolytic enzymes are not necessarily associated with or proportional to the active biomass, although the active biomass produces them. Exactly what controls the level of hydrolytic enzymes is not established, and their measurement is neither simple, nor has it been carried out frequently in systems relevant to environmental biotechnology.

A simple, but reasonably reliable approach for describing hydrolysis kinetics is a first-order relationship with respect to the particulate (or large-polymer) substrate:

$$r_{hyd} = -k_{hyd}S_p \qquad\qquad \textbf{[3.57]}$$

in which

r_{hyd} = rate of accumulation of particulate substrate due to hydrolysis $(M_sL^{-3}T^{-1})$

S_p = concentration of the particulate (or large-polymer) substrate (M_sL^{-3})

k_{hyd} = first-order hydrolysis rate coefficient (T^{-1})

In principle, k_{hyd} is proportional to the concentration of hydrolytic enzymes, as well as to the intrinsic hydrolysis kinetics of the enzymes. Some workers include the active biomass concentration as part of k_{hyd} (i.e., $k_{hyd} = k'_{hyd}X_a$, in which k'_{hyd} is a specific hydrolysis rate coefficient $(L^3M_x^{-1}T^{-1})$. An advantage of using $k'_{hyd}X_a$ is that the hydrolysis rate automatically drops to zero when biomass is not present. On the other hand, it also implies that the extracellular enzymes are linearly proportional to the biomass, an assertion that is not proven.

When Equation 3.57 is used for the hydrolysis rate, the steady-state mass balance on particulate substrate in a chemostat is

$$0 = Q(S_p^0 - S_p) - k_{hyd}S_pV \qquad\qquad \textbf{[3.58]}$$

in which S_p^0 = the effluent concentration of particulate substrate (M_sL^{-3}). Solving Equation 3.58 gives

$$S_p = \frac{S_p^0}{1 + k_{hyd}\theta} \qquad\qquad \textbf{[3.59]}$$

Here, θ represents the liquid detention time and should not be substituted by θ_x.

The destruction of particulate substrate results in the formation of soluble substrate with conservation of the electron equivalents, or BOD_L. When both substrate types have the same mass measure (generally oxygen demand as a surrogate for electron equivalents), the formation rate of soluble substrate is simply $k_{hyd}S_p$. Then, the steady-state mass balance on soluble substrate is

$$0 = (S^0 - S) - \frac{\hat{q}S}{K + S} X_a V/Q + k_{\text{hyd}} S_p V/Q \qquad \textbf{[3.60]}$$

Because Equation 3.60 shows an additional source of substrate from hydrolysis, S^0 effectively is increased by $k_{\text{hyd}} S_p V/Q$; the amount of biomass accumulated should be augmented.

Other constituents of particulate substrates also are conserved during hydrolysis. Good examples are the nutrients nitrogen, phosphorus, and sulfur. The formation rate for soluble forms of these nutrients is

$$r_{\text{hyd}n} = \gamma_n k_{\text{hyd}} S_p \qquad \textbf{[3.61]}$$

in which

$r_{\text{hyd}n}$ = rate of accumulation of a soluble form of nutrient n by hydrolysis $(M_n L^{-3} T^{-1})$

γ_n = stoichiometric ratio of nutrient n in the particulate substrate $(M_n M_s^{-1})$.

Example 3.6 | **EFFECT OF HYDROLYSIS**   Example 3.1 showed that a chemostat fed 500 mg BOD_L/l with a liquid detention time of 2 d produced an effluent quality of:

$$S = 1.7 \text{ mg } BOD_L/l$$
$$X_a = 161 \text{ mg } VSS_a/l$$
$$X_v = 221 \text{ mg } VSS/l$$
$$SMP = 32 \text{ mg } COD_p/l$$
$$\text{Soluble } BOD_L = 33.5 \text{ mg } BOD_L/l$$
$$\text{Total COD} = 348 \text{ mg COD}/l$$

We consider now that the influent also contains biodegradable particulate organic matter with a concentration of 100 mg COD/l, and the hydrolysis rate coefficient is $k_{\text{hyd}} = 3$/d.

The computations to predict the new effluent quality proceed with the following steps:

1. S_p is computed from Equation 3.59:

$$S_p = \frac{100 \text{ mg COD}/l}{1 + (3/d)(2 \text{ d})}$$

$$= 14 \text{ mg COD}/l$$

2. Since $\theta_x = \theta$ remains 2 d, and no active biomass is input, $S = 1.7$ mg BOD_L/l.

3. The effluent and reactor biomass concentrations are determined in parallel to example 3.1, except that the effective S^0 is now:

$$S^0 = 500 \text{ mg}/l + (100 - 14)\text{mg}/l$$

$$= 586 \text{ mg } BOD_L/l$$

$$X_a = 0.42 \frac{\text{g VSS}_a}{\text{g BOD}_L} \left((586 - 1.7) \frac{\text{mg BOD}_L}{1} \right) \frac{1}{1.3}$$

$$= 189 \text{ mg VSS}_a/1$$

$$X_i = 50 \text{ mg VSS}_i/1 + 189 \frac{\text{mg VSS}_a}{1} (1 - 0.8)(0.15/d)(2 \text{ d})$$

$$= 61 \text{ mg VSS}_i/1;$$

$$X_v = X_a + X_i + S_p = 189 + 61 + \frac{14}{1.42} = 264 \text{ mg VSS/1}$$

Thus, the amount of active biomass is augmented by the hydrolysis of particulate COD, while the VSS also is augmented by the remaining biodegradable particulate COD.

4. The detailed computations for SMP are omitted, as they are exactly analogous to Example 3.1. The result is

    UAP $= 9.8$ mg $COD_p/1$
    BAP $= 25.3$ mg $COD_p/1$
    SMP $= 35.1$ mg $COD_p/1$

5. The effluent concentrations of COD and BOD_L are affected by the changes in X_a, X_i and SMP, as well as by the residual particulate organic substrate S_p. All increase.

    Soluble COD and $BOD_L = S + $ SMP
    $\qquad\qquad\qquad = 1.7 + 35.1$
    $\qquad\qquad\qquad = 36.8$ mg COD/1
    Total COD $\qquad = S + $ SMP $+ 1.42 \cdot X_v$
    $\qquad\qquad\qquad = 1.7 + 35.4 + 1.42 \cdot 264$
    $\qquad\qquad\qquad = 412$ mg COD/1
    Total $BOD_L \qquad = S + $ SMP $+ 1.42 \cdot f_d \cdot X_a + S_p$
    $\qquad\qquad\qquad = 1.7 + 35.4 + 1.42 \cdot 0.8 \cdot 189 + 14$
    $\qquad\qquad\qquad = 266$ mg $BOD_L/1$

3.9 INHIBITION

The rates of substrate utilization and microbial growth can be slowed by the presence of inhibitory compounds. Examples of inhibitors are heavy metals, pesticides, antibiotics, aromatic hydrocarbons, and chlorinated solvents. Sometimes these materials are called toxicants, and their effects termed toxicity. Here, we use the terms inhibitor and inhibition, because they imply one of several different general phenomena affecting metabolic rates.

The range of possible inhibitors and their different effects on the microorganisms can make inhibition a confusing topic. In some cases, the inhibitor affects a single enzyme active in substrate utilization; in such a case, utilization of that substrate is slowed. In other cases, the inhibitor affects some more general cell function, such as respiration; then, indirect effects, such as reduced biomass levels may slow utilization of a particular substrate. Finally, some reactions actually are increased by inhibition, as the cell tries to compensate for the negative impacts of the inhibitor.

Figure 3.6 shows the key places that inhibitors affect the primary flows of electrons and energy. Inhibition of a particular degradative enzyme may occur during the initial oxidation reactions of an electron-donor substrate. The immediate effect is a slowing of the degradation rate. In addition, the reduced electron flow can lead to a loss of biomass or slowing of other reactions requiring electrons (ICH$_2$) or energy (ATP). At the other end of the electron-transport chain, inhibition of the acceptor reaction prevents electron flow and energy generation, thereby leading to a loss of biomass. Interestingly, inhibition of the acceptor reaction can lead to a buildup of reduced electron carriers (ICH$_2$) and increased rates for reactions requiring reduced electron carriers as a cosubstrate (Wrenn and Rittmann, 1995). Decouplers reduce or eliminate energy generation, even though electrons flow from the donors to the acceptor. Decoupling inhibits cell growth and other energy-requiring reactions. In some cases, decoupling inhibition leads to an increase in acceptor utilization per unit biomass, as the cells attempt to compensate for a low energy yield by sending more electrons to the acceptor.

How an inhibitor affects growth and substrate-utilization kinetics can be expressed succinctly by using *effective kinetic parameters*. The kinetic expressions for substrate utilization and growth remain the same as before (i.e., Equations 3.6 and 3.9), but the effective kinetic parameters depend on the concentration of the inhibitor. Written out with effective parameters, Equations 3.6 and 3.9 are

$$r_{ut,\ eff} = -\frac{\hat{q}_{eff}\,S}{K_{eff} + S}X_a \qquad\qquad \textbf{[3.62]}$$

$$\mu_{eff} = Y_{eff}(-r_{ut,\ eff}) - b_{eff} \qquad\qquad \textbf{[3.63]}$$

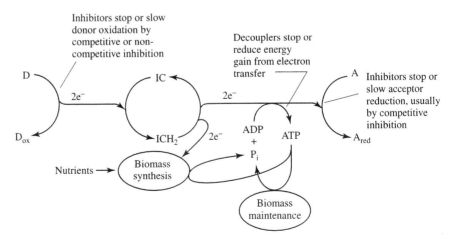

Figure 3.6   Illustration of how inhibitions can affect the primary flow of electrons and energy. D = electron donor, A = electron acceptor, and IC = intracellular electron carrier, such as NAD$^+$.
(After Rittmann and Sáez, 1993).

How $\hat{q}_{\text{eff}}$, K_{eff}, Y_{eff}, and b_{eff} are controlled by the inhibitor concentrations depends on the location and mode of the inhibition phenomenon. The most common types of inhibition and their effective parameters are reviewed here. Some aspects of inhibition were introduced in Chapter 1, and more details can be found in Rittmann and Sáez (1993).

A common type of inhibition for aromatic hydrocarbons and chlorinated solvents is *self-inhibition,* which also is called *Haldane* or *Andrews kinetics.* In this case, the enzyme-catalyzed degradation of the substrate is slowed by high concentrations of the substrate itself. It is not clear whether the self-inhibition occurs directly through action on the degradative enzyme or indirectly through hindering electrons or energy flow after the original donor reaction. In either situation, the effective parameters for self-inhibition are

$$\hat{q}_{\text{eff}} = \frac{\hat{q}}{1 + \dfrac{S}{K_{\text{IS}}}} \qquad \textbf{[3.64]}$$

$$K_{\text{eff}} = \frac{K}{1 + S/K_{\text{IS}}} \qquad \textbf{[3.65]}$$

where K_{IS} = an inhibition concentration of the self-inhibitory substrate $(M_s L^{-3})$. Y_{eff} and b_{eff} are not affected and remain Y and b, respectively.

Figure 3.7 contains two graphical presentations of the effect of substrate self-inhibition on reaction kinetics. The left figure illustrates how the reaction rate $(-r_{\text{ut}})$ varies with substrate concentration. At low substrate concentrations, the rate increases with an increase in S. However, a maximum rate is reached, and substrate concentrations beyond this become inhibitory, causing a decrease in reaction rate. When the Haldane reaction rate model is substituted for the Monod relation in the CSTR mass balances, the effect of θ_x on effluent substrate concentration is illustrated by the right graph in Figure 3.7. This graph indicates that, with θ_x greater than about 2 d, two values of S are possible for each value of θ_x. Which is the correct one? The answer depends on how the reactor arrives at the steady state.

First, we assume that the influent concentration is 45 mg/l, and that the reactor is being operated at a θ_x of 10 d, conditions indicated by the dotted lines in the figure. Under such conditions, the influent concentration is quite inhibitory to the microorganisms. If the reactor were initially filled with the untreated wastewater, seeded with microorganisms, and then operated at the θ_x of 10 d, the reactor would fail, because the specific growth rate of the organisms under these conditions is less than 0.1/d $(= 1/\theta_x)$. However, if the wastewater were initially diluted so that the concentration in the reactor was about 33 mg/l, then the specific growth rate would be greater than 0.1/d, and the bacteria would reproduce faster than they are removed from the reactor. The reactor population would continue to increase, and the value of S in the reactor would continue to decrease until the microbial population size and substrate concentration reached their steady-state concentrations, about 1 mg/l for the substrate. The value of Figure 3.7, then, is not only to indicate what this steady-state concentration would be for a well-operating reactor, but also to indicate upper limits on substrate concentration beyond which reactor startup cannot proceed.

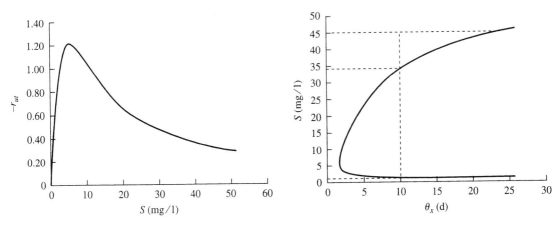

Figure 3.7   Haldane kinetics and the relationship between S and reaction rate (top) and between S and θ_x. The example is for orthochlorophenol with $K = 20$ mg/l, $K_{IS} = 1.5$ mg/l, $\hat{q} = 10$ mg/mg VSS_a-d, $Y = 0.6$ g VSS_a/g, $b = 0.15$/d, and $X_a = 1$ mg VSS/l.

The second type of inhibition is *competitive,* and a separate inhibitor is present at concentration I $(M_T L^{-3})$. The competitive inhibitor binds the catalytic site of the degradative enzyme, thereby excluding substrate binding in proportion to the degree to which the inhibitor is bound. The only parameter affected by I in competitive inhibition is K_{eff}:

$$K_{\text{eff}} = K \left(1 + \frac{I}{K_I} \right) \qquad \textbf{[3.66]}$$

where $K_I =$ an inhibition concentration of the competitive inhibitor $(M_I L^{-3})$. A small value of K_I indicates a strong inhibitor. The rate reduction caused by a competitive inhibitor can be completely offset if S is large enough, because $\hat{q}_{\text{eff}}$ remains equal to $\hat{q}$. Competitive inhibitors usually are substrate analogues.

One example of competitive inhibition that is of interest in environmental engineering practice is that of cometabolism of trichloroethylene (TCE) by microorganisms that grow on substrates such as methane. Here, the first step in methane utilization is its oxidation to methanol by an enzyme called a methane monooxygenase (MMO), which replaces one hydrogen atom attached to the methane carbon with an -OH group. With TCE, MMO interacts to place an oxygen atom between the two carbon atoms to form an epoxide. The key factor is that methane and TCE compete for the same enzyme. The presence of TCE affects the rate at which methane is consumed and, in turn, the presence of methane affects the reaction rate of MMO with TCE. This is illustrated in Figure 3.8. The presence of 20 mg/l TCE reduces the rate of methane utilization considerably over that given by the Monod model, which does not involve competitive inhibition. Similarly, the rate of TCE utilization is greatly reduced as the methane concentration increases. The rate expression for TCE is similar to that for methane, except that the roles of the substrate and the inhibitor are reversed.

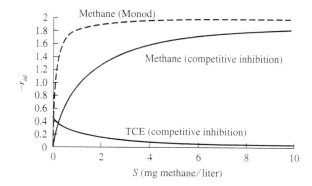

Figure 3.8   Reaction rates for methane and TCE as governed by competitive inhibition kinetics. Shown for comparison is the rate of methane oxidation in the absence of inhibition. $X_a = 1$ mg VSS_a/l, $\hat{q}$ (methane) = 2 mg/mg VSS_a-d, K (methane) = 0.1 mg/l, $\hat{q}$ (TCE) = 0.5 mg/mg VSS_a-d, K_I (TCE) = 2 mg/l, and I (TCE) = 20 mg/l.

A third type of inhibition is *noncompetitive inhibition* by a separate inhibitor. A noncompetitive inhibitor binds with the degradative enzyme (or perhaps with a coenzyme) at a site different from the reaction site, altering the enzyme conformation in such a manner that substrate utilization is slowed. The only parameter affected is $\hat{q}_{\text{eff}}$:

$$\hat{q}_{\text{eff}} = \frac{\hat{q}}{1 + I/K_I} \qquad \textbf{[3.67]}$$

In the presence of a noncompetitive inhibitor, high S cannot overcome the inhibitory effects, since the maximum utilization rate is lowered for all S. This phenomenon is sometimes called allosteric inhibition, and allosteric inhibitors need not have any structural similarity to the substrate.

Figure 3.9 illustrates the different effects that competitive and noncompetitive inhibitors have on reaction rates, in comparison with the Monod model. The values of I, K, and K_I are assumed to be the same, 1 mg/l, making the value $(1 + I/K_I) = 2$. With a competitive inhibitor, the impact is primarily on K, and the inhibitor causes the effective K to increase (from 1 to 2 as indicated by the horizontal line in the middle of the graph). If the substrate concentration (S) is high enough, the reaction rate eventually approaches $\hat{q}$. With the noncompetitive inhibitor, the effect is on $\hat{q}$, causing the apparent value to decrease as the inhibitor concentration increases (from 2 to 1 as indicated by the upper and middle horizontal lines in the illustration). The value of K, the substrate concentration at which the rate is one-half of the maximum value, in effect, remains unchanged (shown as 1 in the figure by the vertical line). By noting which coefficient in the Monod reaction (K or $\hat{q}$) appears to change when an

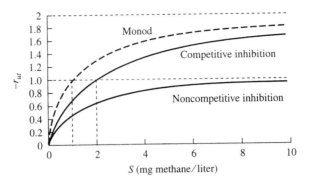

Figure 3.9 Effect of noncompetitive versus competitive inhibition on reaction kinetics. $K = 1$ mg/l, $K_I = 1$ mg/l, $X_a = 1$ mg VSS$_a$/l, $\hat{q} = 2$ mg/mg VSS$_a$-d, and $I = 1$ mg/l.

inhibitor is added, one can determine whether the inhibitor is acting in a competitive or a noncompetitive manner.

In some cases, competitive and noncompetitive impacts occur together. This situation is termed *uncompetitive inhibition*. Both effective parameters, $\hat{q}_{\text{eff}}$ and K_{eff}, vary as they do for the individual cases:

$$\hat{q}_{\text{eff}} = \frac{\hat{q}}{1 + I/K_I} \qquad \textbf{[3.68]}$$

$$K_{\text{eff}} = K(1 + I/K_I) \qquad \textbf{[3.69]}$$

Mixed inhibition is a more general form of uncompetitive inhibition in which the K_I values in Equations 3.68 and 3.69 can have different values.

The last inhibition type considered is *decoupling*. Decoupling inhibitors, such as aromatic hydrocarbons, often act by making the cytoplasmic membrane permeable for protons. Then, the proton-motive force across the membrane is reduced, and ATP is not synthesized in parallel with respiratory electron transport. Sometimes, decouplers are called protonophores. A decrease in Y_{eff} and/or an increase in b_{eff} can model the effects of decoupling inhibition:

$$Y_{\text{eff}} = \frac{Y}{1 + I/K_I} \qquad \textbf{[3.70]}$$

$$b_{\text{eff}} = b(1 + I/K_I) \qquad \textbf{[3.71]}$$

The other parameters are not necessarily changed. An interesting aspect of decoupling is that the rate of electron flow to the primary acceptor and per unit active biomass can increase. This is shown mathematically for steady state:

$$r_{A,\,\text{eff}} = \frac{\hat{q}_{\text{eff}} S}{K_{\text{eff}} + S}\left[1 - Y_{\text{eff}}\left(1 - \frac{f_d b_{\text{eff}} \theta_x}{1 + b_{\text{eff}} \theta_x}\right)\right] \qquad \textbf{[3.72]}$$

in which $r_{A,\,\text{eff}}$ = specific rate of electron flow to the acceptor ($M_s M_x^{-1} T^{-1}$), and all units of mass are proportional to electron equivalents (e.g., COD). Equation 3.72 is valid for all types of inhibition and shows how self-, competitive, and noncompetitive inhibitions normally slow $r_{A,\,eff}$. However, $r_{A,\,eff}$ increases when decoupling increases b_{eff} and/or decreases Y_{eff}.

Sometimes products of a reaction act as inhibitors. The classic example of product inhibition is in alcohol fermentation, where ethanol is produced as an end product from sugar fermentation. In wine manufacture, fermentation of sugar can occur until the ethanol concentration reaches 10 to 13 percent, at which point the alcohol becomes toxic to the yeast, and fermentation stops. This is why the alcohol content of wines normally is in the range of 10 to 13 percent.

3.10 OTHER ALTERNATE RATE EXPRESSIONS

The Monod model for microbial growth and substrate utilization, as represented by Equations 3.1 and 3.6, respectively, is the most widely used model for kinetic analysis and reactor design. However, other models sometimes are used for special circumstances. The previous section reviewed models for inhibition. This section reviews alternate models when inhibition is not the issue.

One popular alternative is the Contois model, which is represented by

$$r_{ut} = -\frac{\hat{q}\,S}{B X_a + S} X_a \qquad\qquad \textbf{[3.73]}$$

in which B = a constant $[\mathrm{M}_s \mathrm{M}_x^{-1}]$. The Contois equation shows a dependence of the specific reaction rate on the concentration of active organisms present. A high organism concentration slows the reaction rate, which approaches a first-order reaction that depends on S, but not X_a:

$$-r_{ut} = \frac{\hat{q}}{B} S \qquad X_a \to \infty \qquad\qquad \textbf{[3.74]}$$

The Contois equation is useful for describing the rate of hydrolysis of suspended particulate organic matter, such as are present in primary and waste activated sludges (Henze et al., 1995). It has been noted that hydrolysis rates of biodegradable sludge particles tend to follow first-order kinetics, as is shown by comparing Equations 3.57 and 3.74. The hydrolysis rate is relatively independent of microorganism concentration, even at fairly small organism concentrations. This independence may occur because extracellular enzymes, not the bacteria, carry out the hydrolysis reactions. For hydrolysis, typical values for the ratio $\hat{q}/B$ are on the order of 1 to 3 d^{-1}.

Two other alternate equations are the Moser and Tessier equations, shown by Equations 3.75 and 3.76, respectively.

$$r_{ut} = -\frac{\hat{q}\,S}{K + S^{-\gamma}} X_a \qquad\qquad \textbf{[3.75]}$$

$$r_{ut} = -\hat{q}\left(1 - e^{S/K}\right) X_a \qquad\qquad \textbf{[3.76]}$$

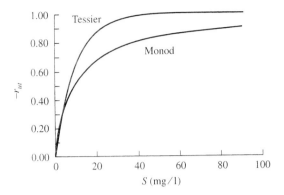

Figure 3.10   Relationship between the reaction rate and S for the Tessier and Monod equations.

in which γ is a constant [unitless]. Like the Contois model, the Moser model reverts to the Monod relationship when $-\gamma = 1$ (or $BX_a = K$ for Contois). The Tessier equation, however, provides quite a different response, as illustrated in Figure 3.10. When the same K and $\hat{q}$ are used in both equations, they show similar responses for S near zero, but the Tessier equation approaches the maximum rate much more quickly than does the Monod equation.

One last alternative rate form is used when more than one substrate is rate limiting, a situation called *dual limitation*. The most common way to represent dual limitation is by the *multiplicative Monod* approach, shown in Equation 3.77:

$$r_{ut} = -\hat{q}\,\frac{S}{K + S}\,\frac{A}{K_A + A}\,X_a \qquad\qquad \textbf{[3.77]}$$

in which A = the concentration of the second substrate $[\mathrm{M}_A \mathrm{L}^{-3}]$, and K_A = the half-maximum-rate concentration for the second substrate $[\mathrm{M}_A \mathrm{L}^{-3}]$. In most cases, the second substrate is the electron acceptor, which is the reason for the symbol A. Bae and Rittmann (1996) demonstrated through fundamental biochemical studies that the multiplicative Monod model accurately represents dual limitation by the donor and acceptor. A key feature of Equation 3.77 is that the rate slows in a Monod fashion if either S or A is below saturation, or zero-order. When both substrates are below saturation, the rate decreases by the product of the Monod fractions and can become quite small. For example, if $S = 0.1K$ and $A = 0.1K_A$, r_{ut} is only 1 percent of $\hat{q}X_a$.

3.11   BIBLIOGRAPHY

Andrews, J. F. (1968). "A mathematical model for the continuous culture of microorganisms utilizing inhibitory substrates." *Biotechnol. Bioengr.* 10, pp. 707–723.

Bae, W. and B. E. Rittmann (1996). "Responses of intracellular cofactors to single and dual substrate limitations." *Biotechnol. Bioengr.* 49, pp. 690–699.

Gottschalk, G. (1986). *Bacterial Metabolism.* 2nd ed. New York: Springer-Verlag.

Henze, M.; C. P. L. Grady, Jr.; W. Gujer; G. R. Marais; and T. Matsuo (1987). *Activated Sludge Model No. 1, Scientific and Technical Report No. 1.* London: International Association on Water Quality. ISSN 10: (0-7071).

Henze, M. et al. (1995). *Activated Sludge Model No. 2. Scientific and Technical Report No. 3.* London: International Association on Water Quality.

Herbert, D. (1960). "A theoretical analysis of continuous culture systems." *Soc. Chem. Ind. Monograph* 12, pp. 21–53.

Lawrence, A. W. and P. L. McCarty (1970). "A unified basis for biological treatment design and operation." *J. Sanitary Engr., ASCE* 96 (SAE), p. 757.

McCarty, P. L. (1975). "Stoichiometry of biological reactions." *Progress in Water Technology* 7: pp. 157–172.

Monod, J. (1949). "The growth of bacterial cultures." *Ann. Rev. Microbiol.* 3, pp. 371–394.

Namkung, E. and B. E. Rittmann (1986). "Soluble microbial products formation kinetics by biofilms." *Water Res.* 20, pp. 795–806.

Noguera, D. R.; N. Araki; and B. E. Rittmann (1994). "Soluble microbial products in anaerobic chemostats." *Biotechnol. Bioengr.* 44, pp. 1040–1047.

Noguera, D. R. (1991). "Soluble microbial products (SMP) modeling in biological processes." M. S. Thesis, Dept. Civil Engr., University of Illinois at Urbana-Champaign, Urbana, Illinois.

Rittmann, B. E. (1987). "Aerobic biological treatment." *Environ. Sci. Technol.* 21, pp. 128–136.

Rittmann, B. E. (1996). "How input active biomass affects the sludge age and process stability." *J. Environ. Engr.* 122, pp. 4–8.

Rittmann, B. E.; W. Bae; E. Namkung; and C.J. Lu (1987). "A critical evaluation of microbial product formation in biological processes." *Water Sci. Technol.* 19, pp. 517–528.

Rittmann, B. E. and P. B. Sáez (1993). "Modeling biological processes involved in degradation of hazardous organic substrates." In *Biotreatment of Industrial and Hazardous Wastes,* eds. J. M. Levin and M. Gealt. New York: McGraw-Hill, pp. 113–136.

Wrenn, B. A. and B. E. Rittman (1995). "Evaluation of a model for the effects of substrate interactions on the kinetics of reductive dechlorination." *Biodegradation* 7: pp. 49–64.

3.12 PROBLEMS

3.1. A wastewater is treated in an aerobic CSTR with no recycle. Determine θ_x min, $[\theta_x^{min}]_{lim}$, and S_{min} if the following parameters hold: $K = 50$ mg/l, $\hat{q} = 5$ mg/mg VSS$_a$-d, $b = 0.06$/d, $Y = 0.60$ g VSS$_a$/g, $S^0 = 220$ mg/l, and $f_d = 1.0$ or 0.8 (two answers).

3.2. Consider a chemostat operating at a detention time of 2 h. The following growth constants apply: $\hat{q} = 48$ g/g VSS$_a$-d; $Y = 0.5$ gVSS$_a$/g; $K = 100$ mg/l; $b = 0.1$/d; and $f_d = 1.0$

(a) What are the steady-state concentrations of substrate (S) for influent substrate concentrations (S^0) of 10,000; 1,000; and 100 mg/l?

(b) What are the steady-state concentrations of cell (X_a) for each S^0?

(c) What are the steady-state soluble-microbial product concentrations for each S^0?

(d) What is the minimum detention time below which washout would occur for each S^0? What is the limiting value of this washout detention time?

3.3. You have been treating a wastewater with bacteria that have the following kinetic coefficients:

$\hat{q} = 16$ mg BOD_L/mg VSS_a-d

$K = 10$ mg/l

$b = 0.2$/d

$Y = 0.35$ mg VSS_a/mg

$f_d = 0.8$

A genetic engineer comes to you and claims he can "fix" your bacteria so they are more efficient. The fixed bacteria will have $b = 0.05$/d and $f_d = 0.95$, while all other coefficients will be the same.

Using a chemostat reactor, you were successfully using a safety factor of 30, and you plan to continue doing so. How would the genetic engineer's plan affect your sludge production (QX_v) and oxygen requirements ($\Delta\, O_2/\Delta_t$)? (Assume $X^0 = 0$ and $S^0 = 300$ mg BOD_L/l). In what ways are the new bacteria more efficient?

3.4. Tabulate the value of S, X_a, X_i, SMP, and S + SMP for a chemostat when $\theta_x = 1, 5, 10$, and 30 d. Use the following data:

$\hat{q} = 16$ mg BOD_L/mg VSS_a-d

$K = 20$ mg/l

$b = 0.15$/d

$Y = 0.5$ mg VSS_a/mg

$f_d = 0.8$

$X_i^0 = 50$ mg VSS_i/l

$S^0 = 400$ mg BOD_L/l

and the SMP parameters of Example 3.1.

3.5. An industry has a two-stage treatment for its wastewater. The first stage is a chemostat with $\theta = 1$ d. The second stage is a lagoon, which can be approximated as a chemostat with a detention time of 10 d. The kinetic parameters are

$Y = 0.7$ g VSS_a/g

$\hat{q} = 18$ g/g VSS-d

$K = 10$ mg BOD_L/l

$b = 0.25$/d

$f_d = 0.9$

Assume that the SMP formation can be ignored completely. The total flow rate is 3,785 m³/d, and $S^0 = 1{,}000$ mg BOD_L/l.

If $X_v^0 = 40$ mg/l, $X_a^0 = 0$, and all input COD and VSS are biodegradable, estimate the effluent quality (COD, BOD_L, VSS) from the whole system. Compute the required N and P supplies; express the answer in mg/l of wastewater flow.

3.6. The kinetics of substrate utilization sometimes are approximated by the Eckenfelder relation, $-r_{ut} = k' X_a S$. Solve for the values of S and X_a in a chemostat reactor using the Eckenfelder relation.

3.7. An industry has an aerated lagoon for treatment of its wastewater. The wastewater has the characteristics:

$$Q = 10^4 \text{ m}^3/\text{d}$$
$$S^0 = 200 \text{ mg } BOD_L/\text{l}$$
$$X_a^0 = 0$$
$$X_i^0 = 30 \text{ mg VSS}_i/\text{l}$$

The lagoon is well aerated and has a total volume of 4×10^4 m³. Tracer studies have shown that the lagoon can be described as two CSTRs in series. Therefore, assume that the lagoon is divided into two CSTRs in series. Compute S, X_a, and X_i in the effluent from both CSTRs.
Use the following parameters:

$$\hat{q} = 5 \text{ mg } BOD_L/\text{mg VSS}_a\text{-d}$$
$$K = 350 \text{ mg } BOD_L/\text{l}$$
$$b = 0.2/\text{d}$$
$$Y = 0.40 \text{ mg VSS}_a/\text{mg } BOD_L$$
$$f_d = 0.8$$

3.8. Compute the effluent COD and BOD_L for a chemostat having

$$\theta = 1 \text{ d}$$
$$S^0 = 1{,}000 \text{ mg/l } BOD_L.$$
$$X_a^0 = X_i^0 = 0$$
$$\hat{q} = 10 \text{ mg } BOD_L/\text{mg VSS}_a\text{-d}$$
$$K = 10 \text{ mg } BOD_L/\text{l}$$
$$b = 0.1/\text{d}$$
$$Y = 0.5 \text{ mg VSS}_a/\text{g } BOD_L$$
$$f_d = 0.8$$

What is the percent removal of COD? BOD_L? Do not forget all types of effluent COD and BOD_L.

3.9. You have developed a novel biological treatment process that removes BOD from wastewater by using heterotrophic bacteria that utilize SO_4^{2-} as the electron acceptor. The process is novel and exciting because the SO_4^{2-} is reduced to elemental sulfur ($S(s)$), which ought to be capturable as a valuable end product. Note from the acceptor half-reaction

$$SO_4^{2-} + 6e^- + 8H^+ \rightarrow S(s) + 4H_2O$$

that formation of each mole of $S(s)$ consumes 6 e$^-$ eq, or each gram of $S(s)$ produced consumes 1.5 g of oxygen demand.

You know the following kinetic and stoichiometric parameters for the bacteria carrying out the reaction:

$$f_s^0 = 0.08e^- \text{ eq to cells/e}^- \text{ eq from donor}$$
$$Y = 0.057 \text{ g VSS}_a/\text{g BOD}_L$$
$$\hat{q} = 8.8 \text{ g BOD}_L/\text{g VSS}_a\text{-d}$$
$$b = 0.05/\text{d}$$
$$K = 10 \text{ mg BOD}_L/\text{l}$$
$$f_d = 0.8$$
$$k_1 = 0.18 \text{ g BOD}_p/\text{g BOD}_L$$
$$k_2 = 0.2 \text{ g BOD}_p/\text{g VSS}_a\text{-d}$$
$$\hat{q}_{UAP} = 1.8 \text{ g BOD}_p/\text{g VSS}_a\text{-d}$$
$$\hat{q}_{BAP} = 0.1 \text{ g BOD}_p/\text{g VSS}_a\text{-d}$$
$$K_{UAP} = 100 \text{ mg BOD}_p/\text{l}$$
$$K_{BAP} = 85 \text{ mg BOD}_p/\text{l}$$

Note that BOD$_p$ represents the BOD$_L$ of the soluble microbial products. Your test wastewater has

$$Q = 1,000 \text{ m}^3/\text{d}$$
$$S^0 = 500 \text{ mg BOD}_L/\text{l}$$
$$X_i^0 = X_a^0 = 0$$

To assess the feasibility of your process, you must compute:

(a) If the safety factor is 10, what is the chemostat volume (V) in m^3?

(b) What is the rate of total VSS production in kg/d?

(c) What is the effluent BOD$_L$ of organic materials? [Note that effluent $S(s)$ could exert a BOD, but it is not organic.]

(d) What is the rate of $S(s)$ production in kg S/d?

3.10. A wastewater is found to have the following characteristics:

COD:	soluble	100 mg/l
	particulate	35 mg/l
BOD$_L$:	soluble	55 mg/l
	particulate	20 mg/l
Suspended Solids:	volatile	20 mg/l
	nonvolatile	10 mg/l

Estimate S^0 in mg BOD$_L$/l and X_i^0 in mg SS/l.

3.11. The rate of consumption of a rate-limiting substrate was found to be 4 mg/d per mg of cells when $S = 5$ mg/l and 9 mg/d per mg cells when $S = 20$ mg/l. What are $\hat{q}$ and K?

3.12. (a) Determine the effluent concentrations of S, X_a, X_i, X_v, and X (the total suspended solids) for a 10 m³ reactor operated as a chemostat that is fed 3 m³/d of a wastewater containing BOD_L of 2,000 mg/l, 20 mg/l of nonbiodegradable suspended organic solids, and 15 mg/l of inorganic suspended solids. Assume aerobic conditions exist and that $Y = 0.65$ mg cells/mg BOD_L, $\hat{q} = 16$ mg BOD_L/d per mg cells, $K = 25$ mg/l, $b = 0.2$/d, and $f_d = 0.8$.

(b) Calculate the percent substrate (S) removal efficiency by the reactor.

3.13. You are interested in estimating the maximum growth rate for bacteria ($\hat{\mu}$) for aerobic oxidation of acetate. You have found that the yield of bacteria is 0.45 g per gram of acetate. Also, you have determined that when the acetate concentration is 5 mg/l the rate of utilization is 3 g acetate/g bacteria-d and when the concentration is 15 mg/l, the rate of utilization is 5 g acetate/g bacteria-d. Estimate $\hat{\mu}$.

3.14. Assume you are treating a wastewater and that $\hat{q} = 10$ g BOD_L/g VSS_a-d, $K = 10$ mg BOD_L/l, and $b = 0.08$ d^{-1}. Say that when operating at θ_x of 4 d the efficiency of substrate removal is 99 percent and oxygen consumption by the microorganisms is found to be 5,000 kg/d, which is determined to be equivalent to an f_e of 0.55. Estimate the oxygen consumption if θ_x is doubled to 8 d.

3.15. A wastewater being considered for biological treatment has the following average values obtained for your evaluation:

BOD_5(total) = 765 mg/l
BOD_5(soluble) = 470 mg/l
k_1 = 0.32 d^{-1}
COD:
  Total = 1,500 mg/l
  Suspended Solids = 620 mg/l
Suspended Solids Concentration:
  Total = 640 mg/l
  Volatile = 385 mg/l

From this information, estimate S^0 (mg BOD_L/l), X^0, X_v^0, X_{in}^0, and X_i^0 (mg/l) for the wastewater. Note that X_{in}^0 is the inorganic (or nonvolatile) suspended solids.

3.16. The characteristics of a wastewater were found to be the following:

Q = 150,000 m³/d
X = 350 mg/l
X_v = 260 mg/l
COD (total) = 880 mg/l
COD (soluble) = 400 mg/l
BOD_L (total) = 620 mg/l
BOD_L (soluble) = 360 mg/l

Estimate the following: S^0, Q^0, X^0, and X_i^0.

3.17. A Haldane modification of the Monod reaction is often used to describe the kinetics of substrate utilization when the substrate itself is toxic at high concentrations (phenol is an example of this). Calculate the effluent concentration for a substrate exhibiting Haldane kinetics from a CSTR and compare this with the value if the substrate were not inhibitory ($S/K_I = 0$) for the following conditions:

$$V = 25 \text{ m}^3$$
$$Q = 10 \text{ m}^3/\text{d}$$
$$S^0 = 500 \text{ mg/l}$$
$$Y = 0.5 \text{ mg/mg}$$
$$\hat{q} = 8 \text{ g/g-d}$$
$$K = 7 \text{ mg/l}$$
$$b = 0.2/\text{d}$$
$$K_I = 18 \text{ mg/l}$$

3.18. You have evaluated the rate of aerobic degradation of organic matter in an industrial wastewater and found that it does not follow normal Monod kinetics. After some testing you have found the substrate degradation rate seems to follow the relationship:

$$-r_{ut} = \hat{q}^{0.5} X_a S^{0.5}$$

where

$$-r_{ut} = \text{degradation rate in mg/l-d}$$
$$\hat{q} = 4 \text{ l/mg-d}^2$$
$$S = \text{substrate concentration in mg/l}$$
$$X_a = \text{active microorganism concentration in mg/l}$$

You have also determined coefficients of bacterial growth on this substrate under aerobic conditions to be as follows:

$$Y = 0.25 \text{ g cells/g substrate}$$
$$b = 0.08/\text{d}$$

Estimate the effluent substrate concentration when treating the above wastewater in a CSTR while operating at a θ_x of 4 d.

3.19. Fill out the following table to indicate what change an increase in each variable listed would have on the given operating characteristics of an aerobic CSTR. Assume all other characteristics of the wastewater and the other listed variables remain the same. Use: $(+)$ = increase, $(-)$ = decrease, (0) = no change, and $(\pm)$ = not determinant.

Variable	θ_x (days)	Operating Characteristic	
		Sludge Production (kg/d)	Oxygen Consumption (kg/d)
Y			
Q			
S^0			
b			
X_i^0			

3.20. A biological treatment plant (CSTR) is being operated with a θ_x of 8 d and is treating a waste consisting primarily of substrate A. The effluent concentration of A is found to be 0.3 mg/l, and its affinity constant (K) is known to be 1 mg/l. A relatively small concentration of compound B is then added to the treatment plant influent and θ_x is maintained constant. Compound B is partially degraded by the microorganisms present through cometabolism, but acts as a competitive inhibitor to substrate A degradation. If the effluent concentration of compound B is then found to be 1.5 mg/l and its affinity constant (K_I) is 2 mg/l, what will now be the effluent concentration of Substrate A?

3.21. An aerobic CSTR operating at θ_x of 8 d is treating a wastewater consisting primarily of soluble organic compound A. Under steady-state operation the effluent concentration is 1.4 mg/l and its K value is 250 mg/l. Compound B is then added to the wastewater and the reactor volume is increased appropriately to maintain the same θ_x. Different microorganisms use each of the two different substrates. Under the new steady-state conditions, the effluent concentration of B is found to equal 0.8 mg/l. What would you then expect the effluent concentration of compound A to be for the following conditions:

(a) Normal Monod kinetics apply for each substrate.
(b) Competitive inhibition between substrates occurs, and K_I (related to compound B concentration) is 0.8 mg/l.
(c) Noncompetitive inhibition between substrates occurs, and K_I (related to compound B concentration) is 0.8 mg/l.
(d) Compound B exhibits substrate toxicity (Haldane kinetics), and $K_{IS} = 0.8$ mg/l.

3.22. The initial rate of oxidation of an organic chemical A, when added to an adapted microbial mixed culture (1,000 mg volatile suspended solids/l), was as follows as a function of chemical A concentration:

Concentration (S), mg/l	1	2	4	8	16
Reaction rate $(-r_{ut})$, mg/l-d	1.5	2.0	2.0	1.5	1.0

What rate equation might be used to describe the relationship between reaction rate $(-r_{ut})$ and substrate concentration (S)? Why?

3.23. You wish to design a biological treatment system for iron contained in drainage water from mine tailings. The water contains no particulate material. You wish to consider aerobic oxidation of the soluble Fe(II) contained in the wastewater to Fe(III), which you will then precipitate in a subsequent reactor by adjustment of pH. For Fe(II) oxidation, $[\theta_x^{min}]_{lim} = 2.2$ d, $f_s^0 = 0.072$, and $b = 0.1/$d.

   (a) Estimate the detention time for a CSTR reactor with an SF of 10, and X_v of 1,000 mg/l.
   (b) Estimate the efficiency of Fe(II) oxidation if $S^0 = 10$ mg/l and $K = 0.8$ mg/l.
   (c) Write a stoichiometric equation for the reaction occurring.
   (d) Estimate the grams oxygen required per gram Fe(II) oxidized. The molecular weight of Fe is 55.8.
   (e) Estimate the grams of ammonia-N per gram Fe(II) oxidized that should be added to satisfy bacterial needs.

3.24. Methanotrophic bacteria are being used to degrade a mixture of trichloroethylene (TCE) and chloroform (CF) by cometabolism. Here, the enzyme MMO initiates the oxidation of these two compounds. Since they compete with one another for MMO, the rate of their degradation in the mixture is governed by competitive inhibition kinetics. Under the following conditions and with $X_a = 120$ mg/l, estimate the individual rates of transformation of TCE and CF in mg/l-d

$$S_{TCE} = 6.4 \text{ mg TCE/l}$$
$$\hat{q}_{TCE} = 0.84 \text{ gTCE/gVSS}_a\text{-d}$$
$$K_{TCE} = 1.5 \text{ mg TCE/l}$$
$$S_{CF} = 5 \text{ mg CF/l}$$
$$\hat{q}_{CF} = 0.34 \text{ gCF/gVSS}_a\text{-d}$$
$$K_{CF} = 1.3 \text{ mg CF/l}$$

Assume K_I for each substrate equals its respective K value.

4

BIOFILM KINETICS

4.1 MICROBIAL AGGREGATION

Microorganisms well adapted to processes in environmental biotechnology are almost always found in naturally occurring aggregates. Engineers exploit natural aggregation as means to separate the microorganisms from the effluent water, thereby providing a good effluent quality while concurrently retaining a large biomass concentration in the treatment system. The two types of aggregates are suspended flocs and attached biofilms. Flocs and biofilms differ in that biofilms adhere to a solid substratum, while flocs are formed without a solid substratum. Biomass retention is achieved by settling the large flocs from the liquid flow under quiescent conditions or by passing the liquid flow past the biofilms, which are immobilized to the solid surfaces.

Although flocs and biofilms differ in key ways, they have one common characteristic that often needs to be considered explicitly in microbial kinetics: Aggregation can create significant gradients in substrate concentrations. Because the mass transport of substrates from outside the aggregate to its inside is driven by concentration differences, bacteria on the inside of aggregates often are exposed to substrate concentrations substantially lower than at the outer surface. Therefore, the rates of substrate utilization and cell growth are not uniform, but depend on the cell's location within the aggregate.

Considering concentration gradients within aggregates increases the complexity of kinetic models. Fortunately, the modeling of biofilm kinetics has become an advanced art, and many useful tools are available for engineers to use in design and analysis, as well as research. The following sections present the basics of biofilm kinetics and then develop several practical tools that can be used with only slightly more difficulty than encountered for the chemostat kinetics of Chapter 3. The principles of biofilm kinetics also are relevant for suspended flocs.

4.2   WHY BIOFILMS?

Biofilms are layerlike aggregations of microorganisms and their extracellular polymers attached to a solid surface. Biofilms, which are naturally immobilized cells, occur ubiquitously in nature and are increasingly important in engineered processes used in pollution control, such as trickling filters, rotating biological contactors, and anaerobic filters. Biofilm processes are simple, reliable, and stable because natural immobilization allows excellent biomass retention and accumulation without the need for separate solids-separation devices.

Before we develop mathematical tools for predicting substrate removal by biofilms, we should consider a fundamental issue about biofilms (indeed, all aggregated systems): Why should microorganisms want to be part of a biofilm when being in the biofilm usually confers to them the disadvantage of being exposed to lower substrate concentrations? Many answers are possible, and the correctness of any one of them depends upon the situation. Here are some possibilities.

1. The biofilms, being fixed in space, are continually exposed to a fresh substrate due to advection of substrate past the biofilm. In other words, the substrate concentration is higher when the biofilm is fixed near the source of substrate supply.

2. Different strains of bacteria must live together in obligate consortia for substrate transport or some other synergistic relationships; the close juxtaposition of cells in a biofilm is necessary for the exchanges.

3. The biofilms create an internal environment (e.g., pH, O_2, or products) that is more hospitable than the bulk liquid. In other words, the biofilm generates unique, self-created microenvironments that benefit the cells.

4. The surface itself creates a unique microenvironment, such as by adsorption of toxins or corrosive release of Fe^{2+}, which is an electron donor.

5. The surface triggers a physiological change in the bacteria.

6. The tight packing of cells in the aggregate alters the cells' physiology.

Possibility 1 appears to be generally true for flowing systems, especially when substrate concentrations are low, but liquid flow velocities are fast. Possibilities 2, 3, and 4 involve microenvironment effects and seem to occur in specific instances. They are forms of ecological selection, and biofilm formation is one tool for ecological control of a process. Evidence to support possibility 5 is sparse for systems of relevance to environmental biotechnology, although it may be important for specific interactions between bacteria and living surfaces (such as plants and animal organs). Possibility 6 often is called "quorum sensing," and evidence of its role in biofilms and other aggregates is just emerging.

4.3   THE IDEALIZED BIOFILM

One major characteristic of a biofilm that must be captured by a model is the establishment of concentration gradients caused by the need to transport materials within the film. That characteristic is represented in a tractable, yet realistic way through an *idealized biofilm* having these properties (Figure 4.1a):

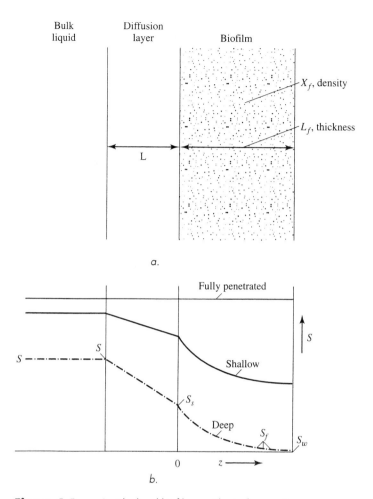

Figure 4.1 An idealized biofilm. *a.* Physical properties.
b. Characteristic concentration profiles.

- The biofilm has a uniform biomass density X_f $(M_x L^{-3})$.

- It has a locally uniform thickness of L_f.

- Mass-transport resistance can be important inside the biofilm and to the biofilm. An effective diffusion layer of thickness L represents external mass-transport resistance. Internal mass-transport resistance is represented by molecular diffusion. The most important result of mass-transport resistance is that the substrate concentrations that the bacteria "see" in the biofilm (denoted S_f) usually are lower than in the bulk liquid (denoted S).

As illustrated in Figure 4.1, the substrate-concentration profiles are nonlinear. One characteristic type of concentration profile is for a *deep biofilm,* for which the substrate concentration approaches zero at some point in the film. Deep biofilms are an important special case, because biomass "seeing" a zero substrate concentration is not active in substrate utilization; once a biofilm is deep, further increasing the biofilm thickness does not increase the overall rate of substrate utilization. When S_f remains above zero at all points in the film, the biofilm is called *shallow*. A special case of shallow is called *fully penetrated* and occurs when the substrate concentration has negligible gradient; in other words, the substrate concentration at the outer surface (S_s) and at the attachment surface (S_w) are virtually identical.

Similar to suspended growth in a chemostat, a complete model for a biofilm must include mass balances and rate expressions for the rate-limiting substrate and the active biomass. In the case of biofilms, these balances and expressions must be written for different positions in the biofilm, since substrate gradients give different concentrations and different rates throughout the film. Thus, models must begin by looking inside the biofilm.

Although the models must begin within the biofilm, their outputs need to be appropriate for use in reactor mass balances. This means that the model outputs ought to give rates of reaction integrated over the biofilm. Therefore, when the integrated rates are multiplied by the amount of biofilm in the reactor, a single mass-per-time value, analogous to $r_{ut}V$ in Equation 3.15, is obtained.

The following sections first look inside the biofilm, developing rate expressions and mass balances for the substrate and the active biomass. Then, those mass balances (inside the biofilm) are solved simultaneously so that the model outputs are the integrated rates of substrate utilization and biomass growth. Finally, the model outputs are incorporated into simple reactor mass balances (analogous to the chemostat) to illustrate the key trends in biofilm kinetics and reactor response. The focus is on steady-state biofilms, but other situations are considered.

4.3.1   SUBSTRATE PHENOMENA

At any position inside the biofilm, substrate is utilized in the same manner as in suspended growth:

$$r_{ut} = -\frac{\hat{q}\,X_f\,S_f}{K + S_f} \qquad \textbf{[4.1]}$$

in which

X_f = active-biomass density within the biofilm ($M_x L^{-3}$)

S_f = substrate concentration at that point in the film ($M_s L^{-3}$)

To get substrate into the biofilm, it must be transported by molecular diffusion, given by Fick's second law,

$$r_{\text{diff}} = D_f \frac{d^2 S_f}{dz^2} \qquad \textbf{[4.2]}$$

in which

r_{diff} = rate of substrate accumulation due to diffusion ($M_s L^{-3} T^{-1}$)

D_f = molecular diffusion coefficient of the substrate in the biofilm ($L^2 T^{-1}$)

z = depth dimension normal to the biofilm surface (L)

Since diffusion and utilization occur simultaneously, we combine Equations 4.1 and 4.2 to give the overall mass balance on substrate. For a steady-state concentration profile in the biofilm, the substrate mass balance is

$$0 = D_f \frac{d^2 S_f}{dz^2} - \frac{\hat{q} X_f S_f}{K + S_f}$$ [4.3]

Equation 4.3 requires two boundary conditions. The first is no flux into the attachment surface, or

$$\left. \frac{d S_f}{dz} \right|_{z=L_f} = 0$$ [4.4]

The other boundary condition is at the biofilm/water interface, where substrate must be transported from the bulk liquid to the outer surface. This external mass transport is described according to Fick's first law,

$$J = \frac{D}{L}(S - S_s) = D_f \left. \frac{d S_f}{dz} \right|_{z=0} = D \left. \frac{d S}{dz} \right|_{z=0}$$ [4.5]

in which

J = substrate flux into the biofilm ($M_s L^{-2} T^{-1}$)

D = molecular diffusion coefficient in water ($L^2 T^{-1}$)

L = thickness of the effective diffusion layer (L)

S, S_s = substrate concentrations in the bulk liquid and at the biofilm/liquid interface, respectively ($M_s L^{-3}$)

Finally, continuity requires that S_s on the liquid side of the interface equals S_s on the biofilm side of the interface.

Integration of Equation 4.3 one time yields $d S_f / dz$, which can be multiplied by D_f to yield the substrate flux J [$ML^{-2}T^{-1}$], the rate of substrate utilization per unit surface area of biofilm. A second integration yields the S_f profile as a function of distance z in the biofilm. In most cases, the desired information is J, which is used in a reactor mass balance.

Solution of Equation 4.3 requires knowledge of all the kinetic and mass-transport parameters ($\hat{q}$, K, D_f, D, and L), as well as the biofilm properties X_f, L_f, and the product $X_f L_f$, which is the amount of biomass per unit surface area ($M_x L^{-2}$). Unless the biofilm properties are known a priori, they must be obtained with a mass balance on active biomass in the biofilm, a subject we discuss later in this chapter.

Illustration for First-Order Kinetics When S_f is sufficiently smaller than K in all parts of the biofilm, the substrate flux and substrate-concentration profile can be expressed by truly closed-form analytical solutions. The differential mass balance for first-order kinetics (in S_f) in the biofilm is:

$$0 = D_f \frac{d^2 S_f}{dz^2} - k_1 X_f S_f$$ [4.6]

in which

$$k_1 = \text{rate coefficient } (L^3 M_x^{-1} T^{-1})$$
$$= \hat{q}/K$$

Integration of Equation 4.6 yields closed-form analytical solutions for flux and S_f:

$$J_1 = \frac{D_f S_s \tanh(L_f/\tau_1)}{\tau_1} \qquad \textbf{[4.7]}$$

$$S_f = S_s \frac{\cosh((L_f - z)/\tau_1)}{\cosh(L_f/\tau_1)} \qquad \textbf{[4.8]}$$

in which

$$J_1 = \text{substrate flux } [M_s L^{-2} T^{-1}] \text{ into a first-order biofilm}$$
$$\tau_1 = \text{first-order, standard biofilm depth dimension (L)}$$
$$= \sqrt{D_f/k_1 X_f}$$
$$\tanh(x) = \text{hyperbolic tangent of } x = (e^x - e^{-x})/(e^x + e^{-x})$$
$$\cosh(x) = \text{hyperbolic cosine of } x = 0.5(e^x + e^{-x}).$$

The standard depth dimension (τ_1) results from dividing D_f by $k_1 X_f$ and taking the square root to create a grouped parameter that compares the diffusion rate to the biodegradation rate. The ratio L_f/τ_1 is a dimensionless biofilm thickness that indicates *deepness* of the biofilm. A deep biofilm has $L_f/\tau_1 > 1$, while a fully penetrated biofilm has $L_f/\tau_1 \ll 1$.

Figure 4.2 shows concentration profiles for two biofilms having the same physical thickness ($L_f = 100 \ \mu\text{m}$), but distinctly different τ_1 values. In case a, the substrate

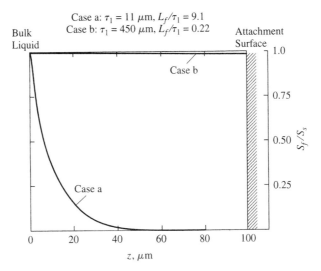

Figure 4.2   Substrate concentration profiles for characteristic deep (case a) and nearly fully penetrated (case b) biofilms. The ratio L_f/τ determines if the biofilm is deep. Many values of k_1, D_f, and X_f can give the same τ_1 value, and the main text illustrates how this affects J_1.

concentration declines to zero approximately one-half way into the biofilm, and the inside half of the biofilm is not active in substrate utilization. Case b is quite the opposite. Bacteria at all positions in the biofilm are exposed to the same concentration and have the same utilization rate. Figure 4.2 illustrates that the importance of mass-transport resistance depends on how diffusion (D_f) compares to the substrate utilization potential $(k_1 X_f$ for first order). Any combination that gives a small τ_1 tends to make the biofilm deep and more affected by diffusion resistance.

Many combinations of k_1, X_f, and D_f can give the same τ_1, and the values of the parameters affect the absolute value of J_1. If case b has parameters $D_f = 0.1$ cm^2/d, $X_f = 50$ mg/cm^3, and $k_1 = 1.0$ cm^3/mg-d (to yield $\tau_1 = 0.045$ cm), the flux for $S_s = 0.1$ mg/cm^3 is 0.048 mg/cm^2-d. The smaller τ_1 for case a can be obtained from a smaller D_f or a larger $k_1 X_f$ product. In the former case, $\tau_1 = 0.0011$ cm requires $D_f = 6 \cdot 10^{-5}$ cm^2/d. In this case, J_1 declines greatly to 5.5×10^{-3} mg/cm^2-d. On the other hand, an increase in $k_1 X_f$ to 8.3×10^5-d gives $J_1 = 91$ mg/cm^2-d. Therefore, a deep biofilm does not imply a large or a small flux, because a small τ_1 can be obtained from fast kinetics or from slow diffusion.

General Solution When S_w Is Known   An analytical solution to Equation 4.3 is available when concentrations at both boundaries of the biofilm can be specified a priori: i.e., S_s and S_w are known. In that case,

$$J = \left[2\hat{q} X_f D_f \left(S_s - S_w + K \ln \left(\frac{K + S_w}{K + S_s} \right) \right) \right]^{1/2} \qquad \textbf{[4.9]}$$

For deep biofilms, S_w approaches zero, leading to a very useful result:

$$J_{\text{deep}} = \left[2\hat{q} X_f D_f \left(S_s + K \ln \left(\frac{K}{K + S_s} \right) \right) \right]^{1/2} \qquad \textbf{[4.10]}$$

Near the end of this chapter, we present criteria for knowing whether or not a biofilm is deep, and Equation 4.10 can be used. Except for deep biofilms, Equation 4.9 seldom can be used, because we do not know S_w a priori. Ad hoc assumptions of S_w have been put forth as a means to use Equation 4.9, but lead to erroneous fluxes.

To illustrate the effects of substrate gradients in a biofilm, we compute J values from Equations 4.9 and 4.10. The fixed parameters are: $D_f = 1$ cm^2/d, $X_f = 40$ mg/cm^3, $\hat{q} = 10$ mg/mg-d, and $K = 0.001$ mg/cm^3. If the biofilm is deep ($S_w = 0$), then Equation 4.10 tells us that $J_{\text{deep}} = 2.5$ mg/cm^2-d for $S_s = 0.01$ mg/cm^3. If the biofilm is thinner so that $S_w = 0.005$ mg/cm^3, then Equation 4.9 gives $J = 1.9$ mg/cm^2-d. If S_w were to increase to 0.009 mg/cm^3, then $J = 0.85$ mg/cm^2-d.

4.3.2   THE BIOFILM ITSELF

At a position inside the biofilm, a mass balance on active biomass is

$$\frac{d(X_f \, dz)}{dt} = Y \frac{\hat{q} S_f}{K + S_f} (X_f \, dz) - b' X_f \, dz \qquad \textbf{[4.11]}$$

in which

t = time (T)

b' = an overall biofilm specific loss rate (T^{-1})

dz = the thickness of a differential section of biofilm (L)

The left-hand side is the change in biofilm mass per unit surface area ($M_x L^{-2} T^{-1}$). The terms on the right-hand side are the rates of new synthesis and the losses, both expressed as biomass per unit surface area per time ($M_x L^{-2} T^{-1}$). Since S_f changes with position in the film, the first term in the right side changes value with location. Thus, the left side seldom is constant or zero. This means that, for any position in the biofilm, the biomass is not at steady state. Near the outer surface, the substrate concentration is high (recall Fig. 4.1), the first term on the right side of Equation 4.11 is positive, and the biomass has a net positive growth rate (i.e., $(d(X_f \, dz))/dt > 0$). Conversely, the net growth rate is negative deep in the film, where the substrate concentration is low.

4.4 THE STEADY-STATE BIOFILM

Even though the biomass at any given point in the biofilm has net growth or loss, the concept of steady state still is applicable and of great value for biofilms. The key to applying the steady-state concept to biofilms is that it be applied to the biofilm as a whole. Thus, the fundamental understanding of a *steady-state biofilm* is that the biomass per unit surface area ($X_f L_f$) is constant in time, even though the biomass at any one location is not at steady state. In other words, a biofilm is steady state when Equation 4.11, integrated over the entire biofilm thickness, is equal to zero.

$$0 = \int_0^{L_f} \frac{d(X_f \, dz)}{dt} = \int_0^{L_f} Y \frac{\hat{q} S_f}{K + S} X_f \, dz - \int_0^{L_f} b' X_f \, dz \qquad \textbf{[4.12]}$$

Each integral can be evaluated, as follows, if we assume that X_f, $\hat{q}$, K, and b' are constant. According to the definition of a steady-state biofilm,

$$\int_0^{L_f} \frac{d(X_f \, dz)}{dt} = \frac{d(X_f L_f)}{dt} = 0 \qquad \textbf{[4.13]}$$

The integral of $r_{ut} dz$ is the sum of all reaction rates per unit area, which equals the flux, or the total utilization rate per unit surface area ($M_s L^{-2} T^{-1}$). Multiplying the flux by Y gives the growth rate per unit surface area:

$$\int_0^{L_f} Y \frac{\hat{q} S_f}{K + S_f} X_f \, dz = Y \int_0^{L_f} (-r_{ut}) \, dz = YJ \qquad \textbf{[4.14]}$$

Biomass loss is averaged across the biofilm:

$$\int_0^{L_f} b' X_f \, dz = b' X_f L_f \qquad \textbf{[4.15]}$$

When Equations 4.14 and 4.15 are substituted into Equation 4.12, we obtain

$$0 = YJ - b' X_f L_f \qquad \textbf{[4.16]}$$

which is the fundamental "law" of the steady-state biofilm. It says that the growth of new biomass per unit area (YJ) is just balanced by the losses per unit area ($b'X_fL_f$). Equation 4.16 can be rewritten in other useful forms. Biomass per unit area is given by

$$X_fL_f = \frac{JY}{b'} \qquad \textbf{[4.16a]}$$

while biofilm thickness is obtained by dividing by X_f.

$$L_f = \frac{JY}{X_fb'} \qquad \textbf{[4.16b]}$$

The concept of a steady-state biofilm is of a dynamic steady state. Near the outer surface, substrate concentrations are high, and $d(X_fL_f)/dt$ is positive. Near the attachment surface, S_f is low, and $d(X_fL_f)/dt$ is negative. Hence, the locations with positive growth rates are exporting biomass to the locations with negative biomass, such that the net effect overall is steady state.

4.5 THE STEADY-STATE-BIOFILM SOLUTION

The model of a steady-state biofilm requires simultaneous solution of mass-balance Equations 4.3 for substrate in the biofilm, 4.5 for substrate transport to the biofilm, and 4.16 for active biomass in the biofilm, as well as boundary and continuity conditions 4.4 and 4.5. Rittmann and McCarty (1980a) originally solved these three equations to give a steady-state-biofilm solution that computes values of J and X_fL_f when inputs are $\hat{q}$, K, D_f, D, L, Y, b', X_f, and S. Subsequent modifications by Sáez and Rittmann (1988, 1992) improved the accuracy of the steady-state solution, but maintained the key features of the original version. The mathematical form of the solution is provided later.

Figure 4.3 shows graphically how steady-state biofilms respond to changes in substrate concentration S. The horizontal axis is the logarithm of the bulk-liquid substrate concentration, and the vertical axis is the logarithm of the substrate flux. The vertical axis also is proportional to the biomass accumulation per unit surface area, as $X_fL_f = YJ/b'$ (from Equation 4.16a). Both axes are plotted on logarithmic scales in order to show the full range of response.

The key trends shown in Figure 4.3 are:

1. For $S < S_{min}$, the value of $\log_{10} J$ goes to negative infinity. This means that $J = 0$ and $X_fL_f = 0$. As with any steady-state biological process, having $S < S_{min}$ gives a negative growth rate and no possibility for sustaining steady-state biomass. So, S_{min} remains a very key factor for steady-state biofilms. S_{min} is defined with the total biomass loss rate coefficient, b':

$$S_{min} = K\frac{b'}{Y\hat{q} - b'} \qquad \textbf{[4.17]}$$

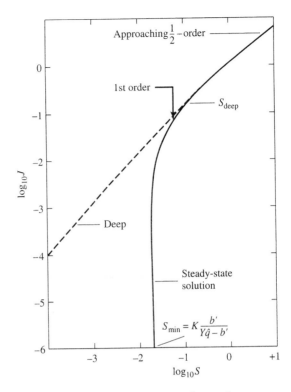

Figure 4.3   Typical response of a steady-state biofilm to changes in S. For this example, $S_{min} = 0.0204$ mg/cm³. S has units mg/cm³, and J has units mg/cm²-d.

2. J (and $X_f L_f$) increases very sharply as S increases slightly above S_{min}. This rapid escalation in J occurs because J and $X_f L_f$ increase together. A small increase in S allows greater J, which allows greater $X_f L_f$; however, greater $X_f L_f$ also increases J. Thus, the positive feedback among S, J, and $X_f L_f$ allows a very rapid increase in J for S slightly above S_{min}.

3. At some value of $S > S_{min}$, the slope of J versus S declines from near infinity and approaches 1.0. This occurs at approximately $S = 0.07$ mg/cm³ for the example in Figure 4.3.

4. For S large enough, the flux becomes equal to that of a deep biofilm. This occurs when S is approximately equal to 0.11 mg/cm³ in Figure 4.3, which is noted S_{deep}. For all $S \geq S_{deep}$, the steady-state biofilm is deep, or S_f approaches zero before the attachment surface. The practical significance of having a deep biofilm is that J no longer depends on $X_f L_f$, because additional biomass does not increase the reaction rate (for the same S) when all of the added thickness has $S_f = 0$. On the other hand, J (and $X_f L_f$) increases for

larger S, because the substrate concentrations inside the biofilm rise in response to a higher S.

5. For very large S, the slope of J versus S declines gradually and eventually approaches the limiting case of one-half-order kinetics, or $J = k_{1/2}S^{1/2}$. Half-order kinetics is a well-known special case for deep biofilms. The practical impact of a declining reaction order is that increases in S give less than proportional increases in J, and marginal increases in substrate removal go down when the system is operated well into the deep region.

The simultaneous solution of Equations 4.3 to 4.5 and 4.16 can be represented with relatively simple algebraic equations that give J and $X_f L_f$ as functions of S and the various kinetic and mass-transport parameters. While a strictly analytical solution is impossible, due to the nonlinearity of Equation 4.3, *pseudo-analytical solutions* were obtained by fitting appropriately chosen algebraic equations to thousands of numerical solutions of the equations comprising the steady-state-biofilm model (Rittmann and McCarty, 1980a; Sáez and Rittmann, 1988, 1992).

The pseudo-analytical solution of Sáez and Rittmann (1992) is the latest and most accurate version. The solution is expressed using three dimensionless master variables: S_{min}^*, K^*, and S^*. Besides consolidating the information of eight variables ($\hat{q}$, K, Y, b', D_f, D, L, and S) into only three variables, the dimensionless variables also provide insight into kinetic properties of the biofilm system.

The three dimensionless variables are:

$$S_{min}^* = \frac{b'}{Y\hat{q} - b'} = \frac{S_{min}}{K} \qquad \textbf{[4.18]}$$

S_{min}^* represents the *growth potential*; S_{min}^* much less than 1.0 indicates a very high growth potential, as the maximum net positive growth rate $(Y\hat{q} - b')$ is much greater than the loss rate (b'). $S_{min}^* > 1$ indicates very low growth potential and difficulty to maintain stable, steady-state biomass.

$$K^* = \frac{D}{L}\left[\frac{K}{\hat{q}X_f D_f}\right]^{1/2} \qquad \textbf{[4.19]}$$

K^* compares external mass transport (D/L) to the maximum internal utilization and transport potential. A small value of K^* (e.g., less than 1) means that external mass transport is slow and exerts significant control on the flux.

$$S^* = S/K \qquad \textbf{[4.20]}$$

S^* is the dimensionless substrate concentration, normalized to K. A large value of S^* (i.e., $\gg 1$) indicates that the utilization reaction is saturated, at least in the outer portions of the biofilm.

The pseudo-analytical solution is expressed in the form:

$$J = f J_{deep} \qquad \textbf{[4.21]}$$

in which

J = the actual steady-state flux, with typical units being mg_s/cm^2-d

J_{deep} = the flux into a deep biofilm having the same S_s concentration, mg_s/cm^2-d

f = the ratio expressing how much the actual flux is reduced because the steady-state biofilm is not deep. The range is $0 \leq f \leq 1$.

From fitting numerical solutions, Sáez and Rittmann (1992) found that

$$f = \tanh\left[\alpha\left(\frac{S_s^*}{S_{min}^*} - 1\right)^\beta\right]$$

[4.22]

in which

$\tanh(x)$ = hyperbolic tangent operator = $(e^x - e^{-x})/(e^x + e^{-x})$

α, β = coefficients that depend on S_{min}^* and are listed below as Equations 4.23 and 4.24.

The procedure for using the pseudo-analytical procedure is as follows.

1. Compute S^*, K^*, and S_{min}^* from Equations 4.18 to 4.20.

2. From S_{min}^*, compute α and β.

$$\alpha = 1.5557 - 0.4117 \tanh[\log_{10} S_{min}^*]$$

[4.23]

$$\beta = 0.5035 - 0.0257 \tanh[\log_{10} S_{min}^*]$$

[4.24]

3. From α, β, K^*, and S^*, iteratively compute S_s^*, the dimensionless substrate concentration at the biofilm/liquid boundary.

$$S_s^* = S^* - \frac{\left(\tanh\left[\alpha\left(\frac{S_s^*}{S_{min}^*} - 1\right)^\beta\right]\right)\left(2\left[S_s^* - \ln(1 + S_s^*)\right]\right)^{1/2}}{K^*}$$

[4.25]

Since S_s^* appears on both sides, you must iterate on S_s^* until both sides are equal and have the same S_s^*. If desired, f can be computed from Equation 4.22 to determine if the biofilm is deep (f approaches 1.0) or shallow ($f < 1$). The term $(2[S_s^* - \ln(1 + S_s^*)])^{1/2}$ in Equation 4.25 is the dimensionless form of Equation 4.10 for J_{deep}, or J_{deep}^*.

4. Compute J^*, the dimensionless flux, from

$$J^* = K^*(S^* - S_s^*)$$

[4.26]

5. Convert from J^* to J with

$$J = J^*(K\hat{q}X_f D_f)^{1/2}$$

[4.27]

6. Compute $X_f L_f$ from Equation 4.16a or L_f from Equation 4.16b.

Although the solution involves several steps, it involves only arithmetic operations that can be carried out easily by hand or programmed into a spreadsheet or equation-solving software. Some key things to remember when using the steady-state-biofilm model are:

• $J = 0$ for all $S < S_{min}$, but J has a unique finite value for all $S > S_{min}$.

- $X_f L_f$ (or L_f) is an output of the model; there is a unique $X_f L_f$ for any $S > S_{min}$.

- S_{min}^* and K^* are fundamental descriptors of what controls process performance. For example, a small K^* value indicates that external mass transport is a major determinant of substrate flux, while a large S_{min}^* indicates that biofilm accumulation strongly limits substrate utilization.

USING THE PSEUDO-ANALYTICAL SOLUTION FOR A STEADY-STATE BIOFILM | **Example 4.1**

The steady-state substrate flux (J), biofilm accumulation ($X_f L_f$), and biofilm thickness (L_f) are to be computed when the substrate concentration (S) is 0.5 mg/l and the following kinetic and mass-transport parameters hold:

$$L = 0.01 \text{ cm}$$
$$K = 0.01 \text{ mg}_s/\text{cm}^3$$
$$X_f = 40 \text{ mg}_a/\text{cm}^3$$
$$\hat{q} = 8 \text{ mg}_s/\text{mg}_a\text{-d}$$
$$b' = 0.1/\text{d}$$
$$D = 0.8 \text{ cm}^2/\text{d}$$
$$D_f = 0.64 \text{ cm}^2/\text{d}$$
$$Y = 0.5 \text{ mg}_a/\text{mg}_s$$

Note that mg_a and mg_s refer to milligrams of active biomass and substrate, respectively. The listing of parameter values emphasizes one of the most important cautions about using the model solution: All units must be consistent! Using mass in mg, length in cm, and time in days is the convention used here and recommended for general use. In these units, the substrate concentration then is expressed as $S = 0.0005 \text{ mg/cm}^3$.

The following stepwise solution is the most efficient approach for computing the dimensionless parameters, using them to obtain a dimensionless flux (J^*), and converting J^* into dimensional values for J, $X_f L_f$, and L_f. Again, note carefully that computations are carried out with all units in mg, cm, and d.

1. Compute S^*, K^*, and S_{min}^* from Equations 4.18 through 4.20

$$S^* = 0.0005/0.01 = 0.05$$

$$K^* = \frac{0.8}{0.01} \left[\frac{0.01}{8 \cdot 40 \cdot 0.64} \right]^{1/2} = 0.559$$

$$S_{min}^* = \frac{0.1}{(0.5 \cdot 8) - 0.1} = 0.02564$$

The low S_{min}^* value indicates that this process has a high growth potential and will not be limited by biofilm accumulation unless S is close to S_{min}. K^* has a moderate value that indicates some, but not dominant control by external mass transport.

2. Compute α and β from Equations 4.23 and 4.24

$$\alpha = 1.5557 - 0.4117 \tanh[\log 0.02564]$$
$$= 1.9346$$
$$\beta = 0.5035 - 0.0257 \tanh[\log 0.02564]$$
$$= 0.5272$$

3. Compute S_s^* from Equation 4.25. This requires iteration in S_s^*, yielding $S_s^* = 0.02754$. Although not necessary, Equation 4.22 can be used to compute $f = 0.46$, which means the biofilm is significantly shallow and, therefore, at least partly limited by biofilm accumulation.

4. Compute J^* from Equation 4.26

$$J^* = 0.559(0.05 - 0.02754)$$
$$= 0.01256$$

5. Convert J^* to J with Equation 4.27

$$J = 0.01256\{(0.01)(8)(40)(0.64)\}^{1/2}$$
$$= 0.0179 \text{ mg}_s/\text{cm}^2\text{-d}$$

6. Compute $X_f L_f$ from Equation 4.16a and L_f by dividing it by $X_f = 40 \text{ mg}_a/\text{cm}^3$.

$$X_f L_f = \frac{0.0179 \cdot 0.5}{0.1} = 0.0895 \text{ mg}_a/\text{cm}^2.$$

$$L_f = 0.00224 \text{ cm} = 22.4 \ \mu\text{m}$$

4.6 ESTIMATING PARAMETER VALUES

Values for the strictly microbial parameters ($\hat{q}$, K, Y, and b) are obtainable in the same way and generally have the same magnitude for biofilm bacteria and suspended bacteria. On the other hand, biofilm-specific parameters X_f, D, D_f, L, and nondecay portions of b' have their own bases for estimation. The measure of biomass solids for biofilm systems is volatile solids, or VS. Active biofilm mass is expressed in VS_a, while the total volatile solids are VS. VS is used for biofilms instead of VSS used for suspended-growth systems, because the biofilm mass is not suspended, but is attached.

The *biomass density, X_f*, can vary widely, depending on physical conditions and characteristics of the organisms. High mechanical stresses on the biofilm tend to increase biofilm density, while anaerobic biofilms often are denser than aerobic ones. A "typical" density value, based on dry weight or volatile solids (VS), is 40 mg VS/cm^3. However, volatile solids densities can range from about 5 mg VS/cm^3 in low-stress aerobic films to as much as 200 mg VS/cm^3 in high-stress anaerobic films. In addition to variation in the gross volatile solids density, the fraction of the volatile solids that is active biomass also varies widely. Some very thick biofilms are dominated by inert biomass that accumulates near the attachment surface, while biofilms kept very thin by detachment can approach 100 percent active biomass. Thus, the ratio of active to total volatile solids can vary. The distribution of active biomass in biofilms is a current area of research.

The *diffusion coefficient in water, D*, often can be found listed in handbooks of chemical properties. For compounds not listed in handbooks, the Wilke-Chang equation is useful for making a good first approximation based on the compound's molar volume. For aqueous solutions at 20 °C, the Wilke-Chang equation is

$$D = 1.279(V_b)^{-0.6} \qquad \textbf{[4.28]}$$

in which D has units cm^2/d, and V_b is the molar volume of the solute at its boiling point (ml/mol). Methods to estimate V_b are contained in the *Chemical Engineers' Handbook* (Perry, Green, and Maloney, 1984).

Although considerable debate has centered on the value of the *diffusion coefficient in the biofilm*, D_f, a consensus is emerging that the ratio D_f/D ranges from 0.5 to 0.8 for small solutes that do not adsorb to the biofilm matrix. When no definitive information exists, a value of $D_f = 0.8D$ often is appropriate (Williamson and McCarty, 1976).

The *thickness of the effective diffusion layer, L*, often can be estimated from correlations developed by chemical engineers for mass-transport coefficients in porous media or to other regular surfaces. Most of the correlations, which are empirically determined, are expressed in terms of a mass-transfer coefficient, k_m, having units MT^{-1} in general and cm/d for our units convention. L is easily computed from k_m through the relationship

$$L = D/k_m \qquad \textbf{[4.29]}$$

Many k_m correlations exist in handbooks and the chemical engineering literature. Only correlations appropriate to the media being considered and the experimental condition of interest should be used. Usually, the conditions are described with Reynolds and Schmidt numbers, which express, respectively, the ratios of advective to viscous rates and viscous to diffusive rates.

A particularly useful correlation for spherical porous media also illustrates the typical form for L as a function of Reynolds number, Schmidt number, D, and liquid velocity (Jennings, 1975).

$$L = \frac{D(\mathrm{Re}_m)^{0.75}\mathrm{Sc}^{0.67}}{5.7u} \qquad \textbf{[4.30]}$$

in which

$\mathrm{Sc} = \dfrac{\mu}{\rho D} = $ Schmidt number

$\mathrm{Re}_m = 2\rho d_p u/(1-\varepsilon)\mu = $ a modified Reynolds number

$\mu = $ absolute viscosity, g/cm-d

$\rho = $ water density, g/cm^3

$u = Q/A_c = $ superficial flow velocity, cm/d

$d_p = $ diameter of solid medium, cm

$\varepsilon = $ porosity of medium bed

$Q = $ volumetric flow rate, cm^3/d

$A_c = $ cross-sectional area of the flow stream, cm^2

This relationship was found appropriate for $1 \leq \mathrm{Re}_m \leq 30$ and for Sc typical of water.

The overall *biofilm-loss coefficient, b′*, is comprised of three parts: decay, predation, and detachment. Decay behaves similarly to suspended growth and can be represented by the usual first-order decay coefficient, b. Although it probably is important in some settings, predation of biofilms by protozoa or other higher life forms has not been researched in a manner leading to any quantitative representation. Therefore, we usually are forced to ignore predation as a directly included loss term. Biofilm detachment, on the other hand, is a critical biofilm-loss mechanism that has been quantified for certain circumstances. Thus, we usually quantify $b′$ by

$$b' = b + b_{\text{det}} \qquad \textbf{[4.31]}$$

in which

b_{det} = specific biofilm-detachment loss coefficient (T^{-1}).

In many cases, b_{det} is larger than b, making $S_{\min}$, $S_{\min}^*$, biofilm accumulation, and substrate flux largely dependent on factors controlling detachment rates.

Although detachment remains a field ripe for new insights and quantification, certain useful principles and tools have emerged. One primary principle is that mechanical forces acting at the biofilm's outer surface affect biofilm detachment. Although the force effects are complicated and not necessarily the only phenomena affecting detachment, the general pattern is that b_{det} increases when the tangential forces, represented by the shear stress, or the axial forces, represented by pressure fluctuations or physical abrasion, increase.

Rittmann (1982b) developed simple equations for the detachment rate from smooth surfaces in which shear stress is the main force causing detachment. When the biofilms are thin (i.e., $L_f < 0.003$ cm), b_{det} is related to the shear stress acting tangentially to the biofilm surface.

$$b_{\text{det}} = 8.42 \cdot 10^{-2} \sigma^{0.58} \qquad \textbf{[4.32]}$$

in which b_{det} has units d^{-1} and σ = liquid shear stress in units of dyne/cm^2. The shear stress is computed from the energy dissipation by friction per unit biofilm surface. It can be computed for fixed beds of porous media from

$$\sigma = \frac{200 \, \mu u (1 - \varepsilon)^2}{d_p^2 \varepsilon^3 a \left(7.46 \cdot 10^9 \frac{s^2}{d^2} \right)} \qquad \textbf{[4.33]}$$

and for fluidized beds of porous media from

$$\sigma = \frac{[(\rho_p - \rho_w)(1 - \varepsilon)g]}{a} \qquad \textbf{[4.34]}$$

in which

ρ_p = particle density, g/cm^3

ρ_w = water density, g/cm^3

$g = 980$ cm/s^2

ε = bed porosity

u = superficial liquid velocity, cm/d

μ = absolute viscosity, g/cm-d

σ = shear stress in dyne/cm^2, which is the same as g/cm-s^2

a = specific surface area of the biofilm carrier, cm^{-1}

When biofilms become thick enough, the bacteria deep inside the biofilm are protected from detachment, and b_{det}, which is an average for the entire biofilm, declines. Rittmann (1982b) showed that

$$b_{det} = 8.42 \cdot 10^{-2} \left(\frac{\sigma}{1 + 433.2(L_f - 0.003)} \right)^{0.58} \quad \textbf{[4.35]}$$

for $L_f > 0.003$ cm. Again, Equation 4.35 applies to smooth surfaces.

When surfaces are not smooth, biofilms tend to accumulate first in crevices that are protected from shear stress. In some cases, b_{det} approaches zero, as long as the biofilms remain only in the protected crevices. However, once the biofilms emerge from the crevices and "smooth out" the surface, detachment rates approach those of smooth surfaces, Equations 4.32 and 4.35.

Detachment rates caused by tangential shear stress can become significant. For example, a low shear stress of 0.02 dyne/cm^2 gives b_{det} of 0.009 d^{-1} for a thin biofilm, but a moderate shear stress (for a porous-medium reactor) of 1 dyne/cm^2 gives $b_{det} = 0.084$ d^{-1}. The latter value, being roughly equal to a typical b value for aerobic heterotrophs, clearly increases b', S_{min}, and S^*_{min} significantly.

When turbulence is high, pressure fluctuation can exert axial forces, or forces perpendicular to the biofilm surface. Furthermore, axial forces from abrasion occur when solid media can move about and have particle-to-particle collisions. These axial forces, as well as the tangential shear stress, are illustrated in Figure 4.4. Although the connections between axial forces and detachment rates have not been systematically investigated experimentally or theoretically, we have evidence that they can strongly affect b_{det}. Work with two-phase fluidized beds (Chang et al., 1991) gave b_{det} values up to 7.4 d^{-1}, even though σ, 8 dyne/cm^2, gives a predicted b_{det} of 0.28 d^{-1} from

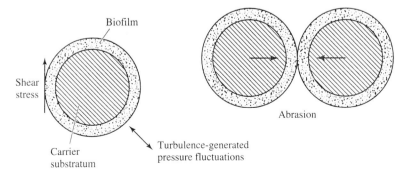

Figure 4.4 Graphical representation of tangential and axial forces at the biofilm surface.

Equation 4.32. Although their effects could not be completely separated, turbulence (represented by a Reynold's number) and abrasion (represented by the solid-particle concentration) were the dominant factors correlated positively with b_{det}. For the particular system studied, multiple linear regression gave

$$b_{det} = -3.14 + 0.0335 C_p + 19.3 \, Re - 3.46 \, \sigma \qquad \textbf{[4.36]}$$

in which

$Re = \rho_w d_p u / \mu$

C_p = solids concentration in the bed, g/l

and b_{det} and σ are in d^{-1} and dyne/cm^2, respectively.

When bed aeration was introduced, creating a three-phase fluidized bed, b_{det} values increased further, up to 15 d^{-1}. The increase in b_{det} could be attributed to added turbulence and abrasion, both effects of the added energy dissipation from introduction of the airflow. b_{det} was strongly correlated to the superficial gas velocity (u_g), with b_{det} increasing roughly by $\Delta b_{det} = (0.004/m) u_g$, where u_g is in m/d and Δb_{det} is in d^{-1} (Trinet et al., 1991).

The detachment mechanisms described above fall into the category of *erosion,* a continuous process by which relatively small pieces of biofilm are removed from the biofilm's surface. Erosion results in slight, but continuous effects on biofilm accumulation. A different detachment category is *sloughing,* an abrupt, intermittent loss of a large section of biofilm. Sloughing can result in drastic changes in the local biofilm accumulation. Figure 4.5 illustrates that erosion occurs only at the outer surface and in small pieces, while sloughing involves large segments of biofilm and can span the entire biofilm depth.

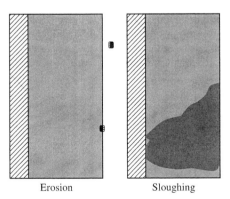

Erosion   Sloughing

Figure 4.5   Schematic of the difference between the erosion and sloughing mechanisms of detachment. Dark areas represent "holes" from which biomass was lost by detachment.

Sloughing is not well understood and seems to involve a breakdown in the structural integrity of the biofilm matrix. Anaerobic conditions, particularly deep inside the biofilm, have been implicated as one cause of sloughing. Because sloughing is neither well understood nor a universal phenomena, it is not considered quantitatively here.

4.7 AVERAGE BIOFILM SRT

Although biofilms have a range of specific growth rates, due to substrate-concentration gradients, the concept of solids retention time (SRT) can be applied to a steady-state biofilm, as long as it is recognized that the SRT is an average over the entire biofilm. As long as active biomass is not being added to the biofilm by deposition, the fundamental definition of θ_x can be applied to steady-state biofilms by realizing that the production rate of active biomass must equal the loss rate by detachment. Thus,

$$\theta_x = \frac{\text{active biomass in the biofilm}}{\text{production rate of active biomass}}$$

$$d = \frac{X_f L_f A_b}{b_{\text{det}} X_f L_f A_b} = \frac{1}{b_{\text{det}}}$$

[4.37]

in which A_b = biofilm surface area (L^2). The key conclusion is that the average θ_x for a steady-state biofilm is equal to the reciprocal of the specific detachment rate.

4.8 COMPLETELY MIXED BIOFILM REACTOR

In analogy with the chemostat, the *completely mixed biofilm reactor* (CMBR) is the simplest system for applying biofilm kinetics with mass balances on rate-limiting substrate and active biomass. Figure 4.6 illustrates the compartments of the most basic CMBR:

- influent flow rate Q carries substrate with concentration S^0.

- The total reactor volume is V, and the liquid holdup (i.e., fraction of total volume that is water) is h; this gives a total water volume of hV.

- All microbial reactions are assumed to occur in the biofilm.

- The biofilm specific surface area is denoted $a[L^{-1}]$, giving a total biofilm surface area of aV.

- The biofilm accumulation per unit surface area is $X_f L_f$, giving a total biofilm accumulation of $X_f L_f aV$.

- The effluent flow (also at Q) and the liquid volume have the same substrate concentration S.

- Active biomass in the reactor and the effluent is at concentration X_a and comes solely from detachment from the biofilm.

- Deposition of active biomass is not significant.

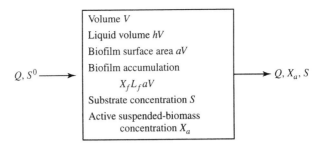

Figure 4.6   Schematic representation of a completely mixed biofilm reactor (CMBR).

The steady-state mass balance on substrate in the CMBR is

$$hV\frac{dS}{dt} = 0 = Q(S^0 - S) - J_{ss}aV \qquad \text{[4.38]}$$

in which J_{ss} is the substrate flux $[\mathrm{M}_s\mathrm{L}^{-2}\mathrm{T}^{-1}]$ into a steady-state biofilm and is calculated with the solution to the steady-state-biofilm model presented earlier. The steady-state mass balance on the biofilm's active biomass is

$$aV\frac{d(X_f L_f)}{dt} = 0 = Y J_{ss}aV - (b + b_{\text{det}})X_f L_f aV \qquad \text{[4.39]}$$

Equation 4.39 is essentially the same as Equation 4.16, which was used to derive the steady-state-biofilm model. This re-emphasizes that the steady-state-biofilm model explicitly includes the biofilm mass balance. Thus, Equation 4.39 does not need to be solved again when the steady-state solution is used.

Simultaneous solution of Equation 4.38 and the steady-state-biofilm model gives the steady-state values of S, J_{ss}, and $X_f L_f$. Then, the effluent biomass concentration (X_a) is computed from a simple mass balance on biomass in the bulk liquid,

$$hV\frac{dX_a}{dt} = 0 = b_{\text{det}}X_f L_f aV - X_a Q \qquad \text{[4.40]}$$

which yields

$$X_a = \frac{b_{\text{det}}X_f L_f aV}{Q} \qquad \text{[4.41]}$$

Example 4.2 | **PERFORMANCE OF A BASIC CMBR**   The kinetic and stoichiometric parameters used in Example 4.1 gave the following values needed to solve the steady-state-biofilm model for any substrate concentration S:

$$S_{\min} = 0.000256 \text{ mg}_s/\text{cm}^3 = 0.256 \text{ mg}_s/l$$
$$S^*_{\min} = 0.0256$$
$$K^* = 0.559$$

$$\alpha = 1.9346$$
$$\beta = 0.5272$$
$$S^* = S/K = S/0.01 \text{ mg}_s/\text{cm}^3 = S/10 \text{ mg}_s/\text{l}$$

For this example, the CMBR has a total volume of $V = 1,000 \text{ m}^3$, a specific surface area of $a = 100 \text{ m}^{-1}$, and a flow rate of $10,000 \text{ m}^3/\text{d}$. We wish to determine S, $X_f L_f$, and $X_f L_f a V$ when $S^0 = 100 \text{ mg/l}$.

An efficient strategy for solving this kind of problem is to rearrange Equation 4.38 so that S is a direct function of J_{ss}:

$$S = S^0 - \frac{J_{ss} a V}{Q} \qquad\qquad \textbf{[4.42]}$$

Substituting the numerical values of this example converts Equation 4.42 to

$$S = 100 \text{ mg/l} - \frac{(J_{ss} \text{ mg/cm}^2\text{-d})(100 \text{ m}^{-1})(1,000 \text{ m}^3)(10,000 \text{ cm}^2/\text{m}^2)}{10,000 \text{ m}^3/\text{d} \times 10^3 \text{ l/m}^3}$$

$$= 100 \text{ mg/l} - 100 \, J_{ss}$$

when J_{ss} is in its typical units of mg/cm^2-d and S is in mg/l. A straightforward trial-and-error solution involves choosing an S value, using the steady-state-biofilm model to compute J_{ss} for that S, and using $S = 100 \text{ mg/l} - 100 J_{ss}$ to check if the S is correct. The procedure is repeated until the chosen S equals the computed S. In a practical sense, two levels of iteration are required. For a given value of S, the steady-state-biofilm model is used iteratively to give J_{ss} for that S. Then, the (S, J_{ss}) pairs are used iteratively with the reactor mass balance until the correct (S, J_{ss}) pair is obtained. [This procedure also can be carried out graphically by finding the intersection of $S = 100 - 10 J_{ss}$ with a plot of J_{ss} versus S from the steady-state-biofilm model. The graphic approach is shown later.] Once J_{ss} is known, $X_f L_f$, $X_f L_f a V$, and X_a are computed directly. These steps are illustrated here.

First, guess an effluent concentration of 25 mg/l (or 0.025 mg/cm^3). Then, $S^* = 0.025/0.01 = 2.5$. We start iterating with $S_s^* = 1.0$ using Equation 4.25.

$$S_s^* = 2.5 - \frac{\tanh\left[1.9181\left(\frac{1.0}{0.0256} - 1\right)^{0.5231}\right](2[1.0 - \ln(1 + 1.0)])^{1/2}}{0.559}$$

$$= 2.5 - \frac{(0.99999)[0.783]}{0.559} = 1.1$$

The resulting S_s^* is a bit larger than the starting value of 1.0. Therefore, we iterate on S_s^* until we converge to $S_s^* = 1.05$, for which Equation 4.26 gives

$$J_{ss}^* = 0.559(2.5 - 1.05) = 0.8106$$

and Equation 4.27 gives

$$J_{ss} = 0.8106 \cdot (0.01 \cdot 8 \cdot 40 \cdot 0.64)^{1/2}$$
$$= 1.16 \text{ mg}_s/\text{cm}^2\text{d}$$

From the mass balance,

$$S = 100 - (100 \cdot 1.16) = -16 \text{ mg/l}$$

Clearly, we need to iterate using a new S guess. Let's try $S = 20$ mg/l. Iterating as before yields

$$S_s^* = 0.81$$
$$J_{ss}^* = 0.665$$
$$J_{ss} = 0.95 \text{ mg}_s/\text{cm}^2 \text{ d}$$
$$S = 5 \text{ mg/l}$$

Since the starting S does not equal the final S, we iterate again. Try

$$S = 17 \text{ mg/l}$$
$$S_s^* = 0.68$$
$$J_{ss}^* = 0.58$$
$$J_{ss} = 0.83 \text{ mg}_s/\text{cm}^2 \text{ d}$$
$$S = 17 \text{ mg/l}$$

Now, both S values are the same, and the solution has converged. Therefore, the effluent concentration is 17 mg/l, and $J_{ss} = 0.83$ mg/cm^2-d. We now calculate the biofilm accumulation from Equation 4.16a.

$$X_f L_f = (0.83 \text{ mg/cm}^2\text{-d}) \left(0.5 \frac{\text{mg VS}_a}{\text{mg}}\right)/(0.1/\text{d})$$

$$= 4.15 \text{ mg VS}_a/\text{cm}^2$$

and

$$L_f = (4.15 \text{ mg VS}_a/\text{cm}^2)/(40 \text{ mg VS}_a/\text{cm}^3) = 0.10 \text{ cm}$$
$$X_f L_f a = 4.15 \text{ mg VS}_a/\text{cm}^2 \cdot 100 \text{ m}^{-1} \cdot 0.01 \text{ m/cm} \cdot 1{,}000 \text{ cm}^3/\text{l} = 4{,}150 \text{ mg VS}_a/\text{l}$$

4.9   SOLUBLE MICROBIAL PRODUCTS AND INERT BIOMASS

As a first approximation, the concentrations of SMP and inert biomass can be computed directly from the steady-state results for S and $X_f L_f$. These direct solutions can be obtained by recognizing the analogies between a chemostat and a CMBR: The volumetric substrate-utilization rate is $J_{ss}a$, while the volumetric biomass concentration is $X_f L_f a$.

For the mass balances on UAP and BAP, the biofilm rate terms become

$$r_{\text{UAP}} = k_1 J_{ss} a \qquad \textbf{[4.43]}$$

$$r_{\text{BAP}} = k_2 X_f L_f a \qquad \textbf{[4.44]}$$

$$r_{\text{deg UAP}} = -\frac{\hat{q}_{\text{UAP}}\text{UAP}}{K_{\text{UAP}} + \text{UAP}} X_f L_f a \qquad \textbf{[4.45]}$$

$$r_{\text{deg BAP}} = -\frac{\hat{q}_{\text{BAP}}\text{BAP}}{K_{\text{BAP}} + \text{BAP}} X_f L_f a \qquad \textbf{[4.46]}$$

Then, the UAP and BAP solutions for the chemostat (Equations 3.38 and 3.39) are employed with the substitution of $J_{ss}a$ for $-r_{ut}$ (or $(S^0 - S)/\theta$) and $X_f L_f a$ for X_a. This substitution makes the biofilm the sole source and sink of UAP and BAP; suspended bacteria are neglected.

For the inert biomass, the biofilm's average θ_x values is $1/b_{\text{det}}$ as long as active biomass is not deposited to the biofilm. Therefore, the chemostat equations for X_i and X_v can be used with the same substitutions:

$$X_{fi} L_f a = (1 - f_d)(b/b_{\text{det}}) X_f L_f a \qquad \textbf{[4.47]}$$

and

$$X_{fv} L_f a = (X_f + X_{fi}) L_f a = (1 + (1 - f_d)(b/b_{\text{det}})) X_f L_f a \qquad \textbf{[4.48]}$$

in which

X_{fi} = the average density of inert biomass in the biofilm ($M_x L^{-3}$)

X_{fv} = the average total density of biomass in the biofilm ($M_x L^{-3}$).

These SMP and inert-biomass computations assume that the biofilm can be averaged with respect to content of active versus inert biomass and to production and degradation of SMP. However, we know that biofilms are not uniform. For example, the inert biomass tends to accumulate nearer the attachment surface. This means that the active biomass is concentrated near the outer surface, where it is exposed to the highest substrate concentrations (affecting r_{UAP}), experiences the highest detachment (affecting b_{det}), and has the highest density (affecting degradation of SMP). The effects of nonuniform biomass distribution have not been investigated systematically for its impact on net SMP formation. On the other hand, we know that inert biomass, being protected from the maximum detachment rates, has a smaller b_{det} (and larger θ_x) than does the active biomass. Thus, these simple computations tend to underestimate $X_{fi} L_f$, at least for biofilms achieving a thickness great enough to afford protection of inner layers.

EXTENSION TO INCLUDE SMP AND INERT BIOMASS   The results from Example 4.2 can be extended to estimate the total effluent quality and the total biomass accumulation. The SMP is computed from

Example 4.3

$$\text{UAP} = -\frac{(\hat{q}_{\text{UAP}} X_f L_f a\theta + K_{\text{UAP}} - k_1 J_{ss} a\theta)}{2}$$

$$+ \frac{\sqrt{(\hat{q}_{\text{UAP}} X_f L_f a\theta + K_{\text{UAP}} - k_1 J_{ss} a\theta)^2 + 4 K_{\text{UAP}} k_1 J_{ss} a\theta}}{2}$$

$$\text{BAP} = -\frac{(K_{\text{BAP}} + (\hat{q}_{\text{BAP}} - k_2) X_f L_f a\theta)}{2}$$

$$+ \frac{\sqrt{(K_{\text{BAP}} + (\hat{q}_{\text{BAP}} - k_2) X_f L_f a\theta_{\text{BAP}})^2 + 4 K_{\text{BAP}} k_2 X_f L_f a\theta}}{2}$$

Substituting the $X_f L_f$, a, and J_{ss} values from Example 4.2 and the $\hat{q}_{\text{UAP}}$, K_{UAP}, k_1, $\hat{q}_{\text{BAP}}$, K_{BAP}, and k_2 values from Example 3.1 (assuming substrate concentration is expressed as mg BOD$_L$/l) gives

$$J_{ss}a = 0.83 \frac{\text{mg BOD}_L}{\text{cm}^2 \text{ d}} \cdot 100 \text{ m}^{-1} \cdot 0.01 \frac{\text{m}}{\text{cm}} \cdot \frac{1,000 \text{ cm}^3}{1}$$

$$= 830 \frac{\text{mg BOD}_L}{1 \text{ d}}$$

$$X_f L_f a = 4.15 \frac{\text{mg VS}_a}{\text{cm}^2} \cdot 100 \text{ m}^{-1} \cdot 0.01 \frac{\text{m}}{\text{cm}} \cdot 1,000 \frac{\text{cm}^3}{1}$$

$$= 4,150 \text{ mg VS}_a/1$$

$$(\hat{q}_{\text{UAP}} X_f L_f a\theta + K_{\text{UAP}} - k_1 J_{ss} a\theta) = (1.8 \cdot 4,150 \cdot 0.1 + 100 - 0.12 \cdot 830 \cdot 0.1)$$

$$= 837 \text{ mg COD}_p/1$$

$$4K_{\text{UAP}} k_1 J_{ss} a\theta = 4 \cdot 100 \cdot 0.12 \cdot (830) \cdot 0.1$$

$$= 3,984 \text{ (mg COD}_p/1)^2$$

$$K_{\text{BAP}} + (\hat{q}_{\text{BAP}} - k_2) X_f L_f a\theta = 85 + (0.1 - 0.09) \cdot 4,150 \cdot 0.1$$

$$= 89.15 \text{ mg COD}_p/1$$

$$4K_{\text{BAP}} k_2 X_f L_f a\theta = 4 \cdot 85 \cdot 0.09 \cdot 4,150 \cdot 0.1$$

$$= 12,699 \text{ mg COD}_p/1$$

$$\text{UAP} = \frac{-837 + \sqrt{837^2 + 3,984}}{2} = 1.19 \text{ mg COD}_p/1$$

$$\text{BAP} = \frac{-89.15 + \sqrt{89.15^2 + 12,699}}{2} = 27.3 \text{ mg COD}_p/1$$

$$\text{SMP} = 1.2 + 27.3 = 28.5 \text{ mg COD}_p/1$$

$$\text{Effluent Soluble COD} = S + \text{SMP} = 17 + 28.5 = 45.5 \text{ mg COD}/1$$

To compute the biomass distribution, b and b_{det} must be separated. For this example, we let $b = b_{\text{det}} = 0.05/\text{d}$.

$$X_{fi} L_f a = (1 - 0.8) \left(\frac{0.05/\text{d}}{0.05/\text{d}} \right) 4,150 \text{ mg VS}_a/1$$

$$= 830 \text{ mg VS}_i/1$$

$$X_{fv} L_f a = 4,150 + 830 = 4,980 \text{ mg VSS}/1$$

The effluent suspended VSS_a and total VSS are then determined from Equation 4.41:

$$X_a^e = X_f L_f a\theta b_{\text{det}} = 4,150 \frac{\text{mg VS}_a}{1} \cdot 0.1 \text{ d} \cdot 0.05/\text{d} = 20.8 \frac{\text{mg VSS}_a}{1}$$

$$X_v^e = X_{fv} L_f a\theta b_{\text{det}} = 4,980 \frac{\text{mg VS}}{l} \cdot 0.1 \text{ d} \cdot 0.05/\text{d} = 24.9 \frac{\text{mg VSS}}{1}$$

$$\text{Effluent Total COD} = S + \text{SMP} + 1.42 X_v^e$$

$$= 17 + 28.5 + 1.42 \cdot 24.9$$

$$= 81 \text{ mg COD}/1$$

$$\text{Effluent Total BOD}_L = S + \text{SMP} + 1.42 f d X_a^e$$

$$= 17 + 28.5 + 1.42 \cdot 0.8 \cdot 20.8$$

$$= 69 \text{ mg BOD}_L/1$$

Similar to a chemostat, the large majority of effluent COD and BOD_L is not the original substrate, but SMP and biomass. Of course, altering the substrate loading or the detachment rate could change that situation.

4.10 TRENDS IN CMBR PERFORMANCE

Figure 4.7 gives a convenient and insightful graphical representation of the trends in CMBR performance. It is the analogy of Figure 3.3, which illustrated key chemostat trends. The thick line in Figure 4.7 is the relationship between S and J_{ss} for a steady-state biofilm. The dashed line is $X_f L_f a$. Superimposed is the substrate mass balance (Equation 4.42), which is the thin line. One clear feature of Figure 4.7 is that the

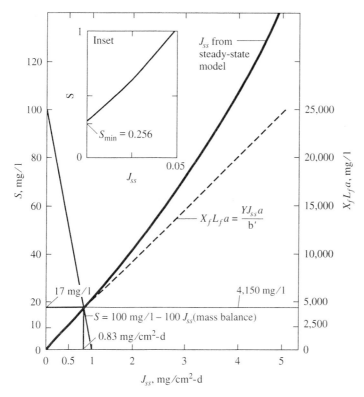

Figure 4.7   Graphical representation of CMBR performances. The thick line is the solution of the steady-state-biofilm model for the parameters of Example 4.2. The thin line is the substrate mass balance, Equation 4.42 with parameter values of Example 4.2. The dashed line is the steady-state-biofilm accumulation, Equation 4.16a with parameter values for Example 4.2. The inset shows J_{ss} for S near S_{min}.

steady-state values of S and J_{ss} are obtained at the intersection of the thick and thin lines; $X_f L_f a$ also is obtained for the steady-state J_{ss}. Thus, the figure presents a convenient graphical means for determining S, J_{ss}, and $X_f L_f a$. Note that $S = 17$ mg/l, $J_{ss} = 0.83$ mg/cm^2-d, and $X_f L_f a = 4,150$ mg/l for the solution.

An important factor for Figure 4.7 is that S and J_{ss} are related in a nonlinear manner that resembles certain aspects of the relationship between S and θ_x in a chemostat (recall Figure 3.3). The key trends of the relationship are:

- The master variable is the substrate flux, J_{ss}. Having J_{ss} be the master variable is logical, since biofilm processes are inherently surface reactions.

- For very low J_{ss}, S approaches S_{min}, the minimum concentration to sustain steady-state biomass. Here, the detachment loss rate, as well as the decay rate, determines S_{min}. For low values of J_{ss}, S is not sensitive to small changes in J_{ss}, but remains near S_{min}. This region can be considered *low loading,* and substrate removal is controlled mainly by biofilm accumulation.

- For high J_{ss}, S is very sensitive to small increases in J_{ss}, and the slope of the S versus J_{ss} curve continually increases as S becomes larger. This strongly nonlinear response is caused by the gradual dominance of mass-transport as the controlling mechanism. For the thicker biofilms obtained at high J_{ss}, the *high-loading region,* internal and/or external mass-transport resistance reduces S_s well below S and forces S to rise sharply to allow increasing J_{ss} and $X_f L_f a$.

Example 4.4

RESPONSES TO DIFFERENT LOADING   Figure 4.8 shows how different substrate loading, achieved by varying S^0 or Q, alters the steady-state values of S. The thick line remains the same as in Figure 4.7, while the different substrate mass-balance lines illustrate the effects of changing S^0 or Q. Table 4.1 summarizes the effects. The examples demonstrate that a higher effluent substrate concentration is caused by either increasing the flow rate or the influent substrate concentration. Likewise, a lower concentration results from either a reduced Q or S^0. Although $X_f L_f a$ changes proportionally with J_{ss}, S responds in a nonlinear manner.

The sensitivity of S to small changes in loading is much greater for the highest loads than for the very low loads. For example, a 10 percent increase in S^0 changes S from 58 to 65 mg/l for the highest load (situation 3), but only from 0.34 to 0.35 mg/l for the lowest load (situation 6).

Table 4.1   Summary of the effects of the changes in substrate loading

Situation Number	S^0 mg/l	Q m^3/d	J_{ss} mg/cm^2-d	S mg/l	$X_f L_f a$ mg/l	Case Type
1	100	10,000	0.83	17	4,150	base
2	100	30,000	1.79	41	8,950	high flow
3	300	10,000	2.42	58	12,100	high S^0
4	100	5,000	0.46	9	2,300	low flow
5	50	10,000	0.42	8	2,100	low S^0
6	1	10,000	0.007	0.34	33	very low S^0

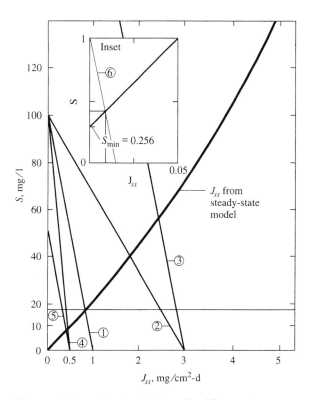

Figure 4.8   Graphical solution for different substrate loading achieved by altering Q or S^0. Situation numbers for each mass-balance line are identified in Table 4.1. Note that the y- intercept of the mass-balance line is S^0, and the slope is $-aV/Q$. As for Figure 4.7, $X_f L_f a = Y J_{ss} a/b'$.

4.11   NORMALIZED SURFACE LOADING

The pattern of how S and J_{ss} relate to each other, shown in Figure 4.7, is general for all CMBRs at steady state. That generality can be represented and exploited by properly normalizing S and J_{ss} parameters. Figure 4.9 illustrates the normalizations and the critical features of the resulting *normalized loading curve*.

First, the substrate axis is normalized by dividing S by S_{min}. Second, the flux axis is normalized by dividing J_{ss} by J_R, a reference flux that is the minimum flux giving a steady-state biofilm that is deep; the means to compute J_R is given below. Third, the curve is presented on a log-log scale in order to show a large range of loading versus concentration with a single curve.

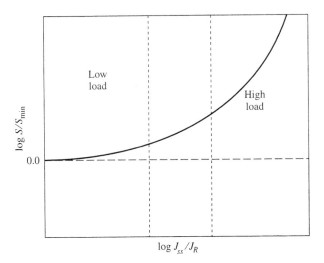

Figure 4.9   The concept of a normalized loading curve for a CMBR at steady state.

The normalized loading curve generalizes the critical trends shown before in Figure 4.7:

1. For low loads (i.e., J_{ss}/J_R is low), $\log S/S_{\min}$ goes to 0.0. In other words, $S/S_{\min}$ approaches 1.0. Thus, when J_{ss}/J_R is low, the delimiting value of $S = S_{\min}$ occurs. So, in this sense, low J_{ss}/J_R in a CMBR is similar to a large θ_x in a chemostat.

2. For low loads, the slope of the curve is very small. Thus, $S/S_{\min}$ is not sensitive to changes in J_{ss}. As long as the loading remains in the low region, changes in J_{ss} have little impact on the steady-state effluent concentration.

3. For high loads (i.e., high J_{ss}/J_R), $S/S_{\min}$ is much greater than 1. Thus, a highly loaded CMBR cannot achieve an S value near $S_{\min}$. In this way, high J_{ss}/J_R is similar to a small θ_x in a chemostat.

4. In the high-load region, $S/S_{\min}$ is very sensitive to changes in J_{ss}/J_R. Thus, relatively small changes to the loading can have profound effects on the steady-state S value. The very high sensitivity is caused by the dominance of mass-transport resistance, a factor not active for the chemostat.

For chemostats, increasing the loading by decreasing the hydraulic detention time eventually results in biomass washout. For biofilm reactors, washout is not an appropriate concept, since the biomass is attached to the solid medium. However, the upper limit on loading occurs when $S/S_{\min}$ approaches $S^0/S_{\min}$; although biofilm is present, the rate of substrate removal ($J_{ss}aV$) becomes insignificant compared to the input rate (QS^0).

Figure 4.10 presents an example of a family of normalized curves for one S^*_{min}. The entire set of curves identified by S^*_{min} ranging from 0.01 to 100 is given in an Appendix. Within each family are curves for a range of K^* values.

Using curves requires computation of S_{min}, S^*_{min}, K^*, and J_R. The first three computations were presented earlier (Equations 4.17–4.19), but are repeated here for convenience.

$$S_{min} = K \frac{b + b_{det}}{Y\hat{q} - (b + b_{det})}$$

$$S^*_{min} = S_{min}/K$$

$$K^* = \frac{D}{L} \left[\frac{K}{\hat{q} X_f D_f} \right]^{1/2}$$

J_R can be computed from the steady-state-biofilm solution by setting $f = 0.99 = \tanh[\alpha(S^*_R/S^*_{min} - 1)^\beta]$, solving for S^*_R, computing the dimensionless reference flux from $J^*_R = \sqrt{2(S^*_R - \ln(1 + S^*_R))}$, and converting J^*_R to J_R by $J_R = J^*_R \sqrt{K\hat{q} X_f D_f}$. However, Cannon (1991) discovered that J^*_R is simply related to S^*_{min} by the curve in Figure 4.11. The use of Figure 4.11 is much simpler and satisfactorily accurate.

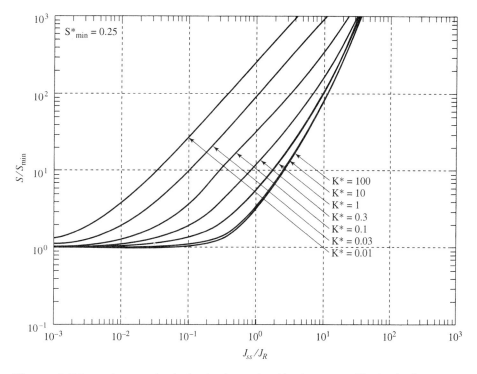

Figure 4.10 An example of a family of normalized loading curves. This family of curves is for $S^*_{min} = 0.25$. Other families of curves are found in an appendix.

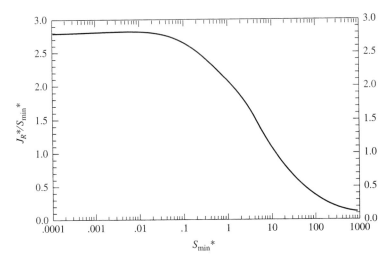

Figure 4.11   Ratio of dimensionless minimum flux (J_R^*) over minimum substrate (S_{min}^*) versus S_{min}^*. SOURCE: After Cannon, 1991.

Example 4.5 | **DESIGNING A CMBR WITH THE NORMALIZED SURFACE LOADING**   A biofilm process must be designed to treat a moderate-strength waste stream having a volumetric flow rate of 4,170 m^3/d (1.1 MGD) and an influent ultimate BOD_L of 430 mg/l. The design goal for this treatment process is to produce an effluent with a BOD_L of 5 mg/l of substrate. Table 4.2 presents reasonable estimates of the kinetic parameters for oxidation of soluble BOD_L. The key kinetic parameters are S_{min}^* and K^*, which are 0.1 and 1.2, respectively, and Figure 4.12 presents the normalized loading for these parameters values. The biofilm reactor system for this

Table 4.2   Parameter estimates of aerobic BOD_L oxidation in the biofilm-process design Example 4.5

Input Parameters

K	10 mg BOD_L/l (0.01 mg BOD_L/cm^3)	
$\hat{q}$	10 mg BOD_L/mg VS_a-d	
D	1.25 cm^2/d	
D_f	0.75 cm^2/d	
X_f	25 mg VS_a/cm^3	
Y	0.45 mg VS_a/mg BOD_L	
b	0.1 d^{-1}	
b_{det}	0.31 d^{-1}	
L	0.0078 cm (78 μm)	

Computed Parameters

S_{min}	1.0 mg BOD_L/l	(Equation 4.17)
S_{min}^*	0.10	(Equation 4.18)
K^*	1.2	(Equation 4.19)
J_R^*/S_{min}^*	2.6	(Figure 4.11)
J_R^*	0.26	
J_R	0.36 mg BOD_L/cm^2-d	$(J_R = J_R^*(K\hat{q}X_fD_f)^{0.5}$

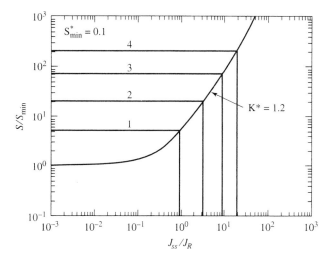

Figure 4.12 Normalized loading curve for $S^*_{min} = 0.1$ and $K^*=1.2$, which are for Examples 4.5 and 4.6.

design has an assumed specific surface area of 900 m^{-1} (2,950 ft^{-1}). The design procedure is initiated by assuming that the entire reactor is comprised of one CMBR.

To compute the volume (V) needed to treat all the flow in one CMBR, the numerical values are substituted into Equation 4.42:

$$S = S^0 - \frac{J_{ss}aV}{Q}$$

In this case, $Q = 4,170$ m^3/d, $a = 900$ m^{-1}, $S^0 = 430$ mg BOD$_L$/l, and $S = 5$ mg BOD$_L$. Thus,

$$5 \text{ mg/l} = 430 \text{ mg/l} - \frac{(J_{ss} \text{ mg/cm}^2\text{-d})(900 \text{ m}^{-1})(V \text{ m}^3) \cdot 10^4 \text{ cm}^2/\text{m}^2 \cdot 10^{-3} \text{ m}^3/\text{l}}{4,170 \text{ m}^3/\text{d}}$$

$$5 \text{ mg/l} = 430 \text{ mg/l} - 2.16 J_{ss} V$$

where J_{ss} is in mg/cm^2-d and V is in m^3. Solving for V yields

$$V = 197/J_{ss}$$

J_{ss} is determined from the normalized loading curve for $K^* = 1.2$ and $S^*_{min} = 0.1$ with the following steps. (1) S/S_{min} is computed as (5 mg BOD$_L$/l)/(1 mg BOD$_L$/l) = 5. (2) J_{ss}/J_R is determined from the loading criteria curve as 0.91 (see construction line 1 on Figure 4.12). (3) J_{ss} is computed as $0.91 \cdot 0.36$ mg BOD$_L$/cm^2-d = 0.33 mg BOD$_L$/cm^2-d. Substitution of J_{ss} into the volume equation gives $V = 197/0.33 = 596$ m^3, or approximately 600 m^3.

Thus, a single CMBR with 600 m^3 volume should be capable of reducing the influent BOD$_L$ from 430 mg/l to 5 mg/l in a flow of 4,170 m^3/d. The process is in the medium-load region. Now that J_{ss} is known, the biomass accumulation and effluent quality can be computed in the same fashion as was done in Example 4.2. In particular, $X_f L_f$ is

$$X_f L_f = Y J_{ss}/b' = 0.45 \cdot 0.33/0.41 = 0.36 \text{ mg VS/cm}^2$$

Example 4.6 | **CMBRs IN SERIES**   Many biofilm processes have a significant plug-flow nature. These processes can be represented by CMBRs in series, where each CMBR has a volume only a small fraction of that required for a single CMBR. This approach is exemplified by determining how many CMBRs of volume 20 m^3 are needed to reduce the substrate concentration from 430 mg/l to 5 mg/l when the parameters of Example 4.5 still hold. The 20 m^3 per CMBR is approximately 3 percent of the volume required for a single CMBR.

The computations can begin with the first CMBR (specifying $S^0 = 430$ mg/l) or the final CMBR (specifying $S = 5$ mg/l). In either case, the mass-balance Equation 4.42 is

$$S = S^0 - \frac{(900 \text{ m}^{-1})(20 \text{ m}^3)(J_{ss} \text{ mg}_s/\text{cm}^2\text{-d})(10^4 \text{ cm}^2/\text{m}^2)(10^{-3} \text{ m}^3/\text{l})}{4{,}170 \text{ m}^3/\text{d}}$$

or

$$S = S^0 - 43.2 J_{ss}$$

This analysis begins with the final (or effluent) CMBR. For the effluent segment (denoted segment 1), $S_1 = 5$ mg/l, and the computations are as follows:

$$S_1/S_{\min} = 5$$
$$J_1/J_R = 0.91 \text{ (line 1 in Figure 4.12)}$$
$$J_1 = 0.33 \text{ mg BOD}_L/\text{cm}^2\text{-d}$$
$$S_1^0 = 5 + 43.2\,(0.33)$$
$$= 19.3 \text{ mg BOD}_L/\text{l}$$

Thus, the concentration into the last segment and leaving the previous segment is 19.3 mg BOD$_L$/l.

For the second-to-last segment, denoted segment 2, $S_2 = 19.3$ mg BOD$_L$/l, and the computations are:

$$S_2/S_{\min} = 19.3$$
$$J_2/J_R = 3.1 \text{ (line 2, Figure 4.12)}$$
$$J_2 = 1.12 \text{ mg BOD}_L/\text{cm}^2\text{-d}$$
$$S_2^0 = 19.3 + 1.12 \cdot 43.2 = 67.7 \text{ mg BOD}_L/\text{l}$$

Since the concentration into segment 2 is still less than the influent concentration, the computation proceeds until the computed S^0 exceeds 430 mg BOD$_L$/l.

Segment 3:

$$S_3 = 67.7 \text{ mg BOD}_L/\text{l}$$
$$S_3/S_{\min} = 68$$
$$J_3/J_R = 8.8 \text{ (line 3, Figure 4.12)}$$
$$J_3 = 8.8 \cdot 0.36 = 3.2 \text{ mg BOD}_L/\text{cm}^2\text{-d}$$
$$S_3^0 = 67.7 + 3.2 \cdot 43.2 = 206 \text{ mg BOD}_L/\text{l}$$

Segment 4:

$$S_4 = 206 \text{ mg BOD}_L/l$$

$$S_4/S_{\min} = 206$$

$$J_4/J_R = 19 \text{ (line 4, Figure 4.12)}$$

$$J_4 = 6.8 \text{ mg BOD}_L/\text{cm}^2\text{-d}$$

$$S_4^0 = 206 + 6.8 \cdot 43.2 = 500 \text{ mg BOD}_L/l$$

The calculation shows that 4 segments of 20 m³ each are slightly greater than enough to reduce the BOD_L concentration from 430 to 5 mg/l. With no safety factor, the required volume interpolates to about 75 m³.

These examples demonstrate the technique for computing volumes for a single completely mixed reactor and for a reactor with a plug-flow nature. The final volume required is somewhat sensitive to the size of the segments. Use of many smaller segments gives the reactor more of a plug-flow nature and reduces the overall volume, but the effect usually is small once the number of segments exceeds 6. Choice of the correct number of reactor segments depends on the mixing characteristics of the reactor, a subject that is beyond the scope of this section. However, since the computations are simple and rapid (about 30 seconds per segment computation), the engineer easily can evaluate the impacts of different segment sizes.

4.12 NONSTEADY-STATE BIOFILMS

Biofilms are not always at steady state. Transients in loading, temperature, detachment, or other environmental conditions can cause the biofilm to experience net growth or loss. In modeling terms, Equation 4.14, the biofilm biomass balance for the entire biofilm, is not valid. However, substrate concentrations within the biofilm quickly establish steady states, which means that Equations 4.3 and 4.6 remain valid.

Rittmann and McCarty (1981), who based their work on a previous pseudo-analytical solution to Equation 4.3 alone (Atkinson and Davies, 1974), provided a pseudo-analytical solution for Equations 4.3 and 4.6. The form of the pseudo-analytical solution is

$$J = \eta \hat{q} X_f L_f \frac{S_s}{K + S_s} \tag{4.49}$$

in which J = the substrate flux into a biofilm of any thickness $(M_s L^{-2} T^{-1})$ and η = the effectiveness factor, which is the ratio of the actual flux to the flux that would occur if the biofilm were fully penetrated at concentration S_s. The η value explicitly expresses the effects of internal mass-transport resistance.

The pseudo-analytical solution is carried out in a dimensionless domain and requires intermediate estimation of η and S_s^*, the dimensionless substrate concentration at the biofilm's outer surface. The steps for nondimensionalization and solution are:

1. Compute four dimensionless parameters:

$$S^* = S/K \qquad\qquad \textbf{[4.50]}$$
$$L^* = L/\tau \qquad\qquad \textbf{[4.51]}$$
$$L_f^* = L_f/\tau \qquad\qquad \textbf{[4.52]}$$
$$D_f^* = D_f/D \qquad\qquad \textbf{[4.53]}$$

where

$$\tau = \sqrt{K D_f / \hat{q} X_f} \qquad\qquad \textbf{[4.54]}$$

(Note that the original pseudo-analytical solution (Rittmann and McCarty, 1981) defined τ slightly differently. The τ definition here (L) is consistent with more recent work (Suidan, Rittmann, and Traegner, 1987).)

2. The pseudo-analytical solution is iterative and requires an initial estimate of an effectiveness factor, η. In principle, η can take any value from 0 to 1, but the solution proceeds more rapidly if the initial estimate is close to the actual value. If the biofilm is very shallow, η approaches 1, but for deep biofilms η approaches

$$\frac{\sqrt{\dfrac{D_f(K + 2S_s)}{\hat{q} X_f}}}{L_f} = \frac{\sqrt{1 + 2S_s^*}}{L_f^*}$$

Since S_s^* is not yet known, S^* can be used as a starter estimate for $2S_s^*$. Rittmann and McCarty (1981) also suggested a starter value of

$$\eta = \frac{\tanh L_f^*}{L_f^*} \qquad\qquad \textbf{[4.55]}$$

This starter value is exact for a biofilm with strictly first-order kinetics (in S_f) within the biofilm.

3. Use the η estimate to compute a trial S_s^* from S^*

$$S_s^* = \frac{1}{2}\left[(S^* - 1 - L^* L_f^* D_f^* \eta) + \sqrt{(S^* - 1 - L^* L_f^* D_f^* \eta)^2 + 4S^*}\right] \quad \textbf{[4.56]}$$

4. Compute a trial dimensionless flux from

$$J^* = D_f^* L_f^* \eta \frac{S_s^*}{1 + S_s^*} \qquad\qquad \textbf{[4.57]}$$

5. Compute a checking S_s^*, called $S_s^{*'}$, by

$$S_s^{*'} = S^* - J^* L^* \qquad\qquad \textbf{[4.58]}$$

6. Compute ϕ from

$$\phi = \frac{L_f^*}{(1 + 2S_s^{*'})^{1/2}}$$ **[4.59]**

The ϕ parameter is $1/\eta$ for a deep biofilm having a dimensionless surface concentration of $S_s^{*'}$.

7. Compute a checking η, called η', from ϕ and Atkinson's solution (Atkinson and Davis, 1974)

$$\eta' = 1 - \frac{\tanh(L_f^*)}{L_f^*}\left(\frac{\phi}{\tanh \phi} - 1\right); \quad \phi \leq 1 \quad \text{or}$$

$$= \frac{1}{\phi} - \frac{\tanh(L_f^*)}{L_f^*}\left(\frac{1}{\tanh \phi} - 1\right); \quad \phi \geq 1$$ **[4.60]**

8. If η' is sufficiently close to η, you are finished and should go to step nine. On the other hand, if η and η' do not agree, use η' as the estimator of η, and return to the third step above. Repeat steps 3 through 8 until η' is sufficiently close to η. The sufficiency of "closeness" depends on the desired tolerance of the J prediction. For example, achieving 1 percent accuracy in J requires that $\eta' - \eta < 0.01\eta'$.

9. Compute the dimensionless flux J^* from

$$J^* = \eta D_f^* L_f^* \frac{S_s^{*'}}{1 + S_s^{*'}}$$ **[4.61]**

10. Convert to the dimensional domain:

$$J = J^*\left(\frac{KD}{\tau}\right)$$ **[4.62]**

It also is possible to compute J directly from η (after step 8) by using Equation 4.49 once $S_s^{*'}$ is converted to S_s using Equation 4.50.

Figure 4.13 shows how J^* varies with S^*. Three features are important in the figure. First, the flux continues to decline linearly as S^* decreases. In other words, J^* approaches zero when S^* approaches zero, not S_{min}^*. The lack of an S_{min}^* value comes directly from the fact that the biofilm mass balance (Equation 4.16) is not part of the model. Second, any J^* versus S^* curve starts first order (or linear) for low S^*, gradually reduces reaction order, and finally reaches zero order at high enough S^*. This behavior reflects the Monod utilization kinetics, which give saturation of a fully penetrated biofilm for very large S^*. Third, the break towards zero-order kinetics occurs at higher substrate concentrations when the biofilm thickness is larger. This third characteristic shows that it is more difficult to create a fully penetrated condition when the biofilm thickness is large.

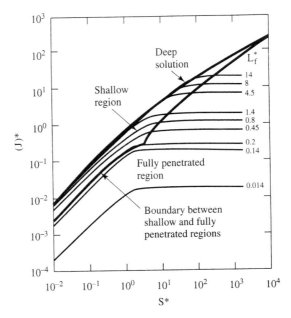

Figure 4.13   The variation of J^* with S^* for biofilms of any thickness. SOURCE: After Rittmann and McCarty, 1981.

Example 4.7 | **NONSTEADY STATE FLUXES AND CONCENTRATIONS**   Example 4.2 showed that a steady-state CMBR with a volume of 1,000 m³ reduces the substrate from 100 mg/l to 17 mg/l when the specific surface area is 100 m⁻¹, the flow rate is 10,000 m³/d, and the kinetic parameters in Example 4.1 hold. For this example, we assess the short-term performance of the same reactor when it is exposed to a different loading condition. Short term means that the substrate concentration in the bulk water and the biofilm reach steady state, but the biofilm does not have enough time to change its accumulation from the steady-state value of $X_f L_f = 4.15$ mg VS$_a$/cm² and $L_f = 0.10$ cm. The length of time that the biomass accumulation remains essentially unchanged depends on the nonsteady-state condition. In general, the duration should be much shorter than the average biofilm SRT.

The steady-state substrate mass balance for the biofilm not necessarily at steady state is

$$S = S^0 - \frac{JaV}{Q} \qquad [4.63]$$

in which J is determined by the solution for biofilms of any thickness (i.e., Equations 4.50–4.62) and has units mg/cm²-d. Key to the solution is that $L_f = 0.10$ cm must be specified as an input to the model solution, and J is not J_{ss}, since the biofilm is not necessarily at steady state. If the biofilm were at steady state, then J would equal J_{ss} from the steady-state-biofilm model.

This example considers the short-term impact of a flow-rate doubling to 20,000 m³/d, while S^0 remains at 100 mg BOD$_L$/l. Since the loading doubles, we let the effluent substrate concentration double from 17 mg/l to 34 mg/l as a first guess. Solving with Equations 4.50 to 4.62 gives:

1. Compute the dimensionless parameters:

$$S^* = \frac{34 \text{ mg/l}}{10 \text{ mg/l}} = 3.4 \qquad D_f^* = \frac{0.64 \text{ cm}^2/\text{d}}{0.8 \text{ cm}^2/\text{d}} = 0.8$$

$$\tau = \sqrt{\frac{(0.01 \text{ mg/cm}^3)(0.64 \text{ cm}^2/\text{d})}{(8 \text{ mg/mg-d})(40 \text{ mg/cm}^3)}} = 0.00447 \text{ cm} \quad (44.7 \mu\text{m})$$

$$L^* = \frac{0.01 \text{ cm}}{0.00447 \text{ cm}} = 2.24 \qquad L_f^* = \frac{0.10 \text{ cm}}{0.00447 \text{ cm}} = 22$$

2. Since L_f^* is large, use the deep-biofilm starter value for η:

$$\eta = \frac{\sqrt{1 + S^*}}{L_f^*} = \frac{\sqrt{1 + 3.4}}{22} = 0.095$$

3. Use η to compute a trial S_s^* from S^*:

$$S_s^* = 0.5 \cdot \left[(3.4 - 1 - 2.24 \cdot 22 \cdot 0.8 \cdot 0.095) \right.$$
$$\left. + \sqrt{(3.4 - 1 - 2.24 \cdot 22 \cdot 0.8 \cdot 0.095)^2 + 4 \cdot 3.4} \right]$$
$$= 1.29$$

4. Compute a trial dimensionless flux:

$$J^* = 0.095 \cdot 0.8 \cdot 22 \cdot \frac{1.29}{1 + 1.29} = 0.942$$

5. Compute a checking S_s^*, called $S_s^{*'}$.

$$S_s^{*'} = 3.4 - 0.942 \cdot 2.24 = 1.29$$

6. Compute ϕ.

$$\phi = \frac{22}{\sqrt{1 + 2 \cdot 1.29}} = 11.63$$

7. Compute a checking η, called η', from Equation 4.60:

$$\eta' = \frac{1}{11.63} - \frac{\tanh(22)}{22} \left[\frac{1}{\tanh(11.63)} - 1 \right] = 0.086 \neq 0.095$$

Since η and η' are not exactly the same (10 percent error), we let $\eta = 0.09$ and repeat the calculations:

$$S_s^* = 1.357$$
$$J^* = 0.912$$
$$S_s^{*'} = 1.357$$
$$\phi = 11.42$$
$$\eta' = 0.088$$

With one more iteration, $\eta = 0.0882$, and $S_S^* = 1.38$, making $J^* = 0.0882 \cdot 17.6 \cdot 1.38/(1 + 1.38) = 0.90$ and $J = 0.90 \cdot (0.01\ \text{mg/cm}^3) \cdot (0.8\ \text{cm}^2/\text{d})/0.00447\ \text{cm} = 1.61\ \text{mg/cm}^2\text{-d}$. The flux does not quite double when S is doubled. The mass-balance equation (Equation 4.63) gives:

$$S = 100\ \text{mg/l} - 1.61\ \frac{\text{mg}}{\text{cm}^2\text{-d}} \cdot \frac{100}{\text{m}} \cdot \frac{\text{d}}{20{,}000\ \text{m}^3} \cdot 1{,}000\ \text{m}^3 \cdot \frac{10\ \text{cm}^2 - \text{m}}{1}$$

$$= 19.5\ \text{mg/l}$$

Clearly, the S values do not agree. Therefore, we choose a new S value and iterate again. We try 29.4 mg/l, which gives $\eta = 0.083$, $S_S^* = 1.17$, and $J = 1.41\ \text{mg/cm}^2\text{-d}$. The mass balance yields $S = 29.5\ \text{mg/l}$. Thus, the actual effluent concentration is 29.5 mg/l. Thus, a doubling of the flow rate increases S by a factor of 1.74, and the flux increases by a factor of 1.7.

Example 4.8 | **DEGRADATION OF A SECONDARY-SUBSTRATE MICROPOLLUTANT**   In addition to describing the flux during short-term transients, the model solution for biofilms of any thickness is appropriate for secondary substrates, or compounds whose biodegradation contributes nothing or insignificantly towards supporting the cells' energy and electron needs for growth and maintenance. The disconnecting of secondary-substrate utilization from biomass sustenance means that Equation 4.16 is not relevant. Thus, because only Equations 4.3–4.5 apply, the model for biofilms of any thickness is the correct one.

For this example, the steady-state biofilm of Example 4.2 is exposed to a micropollutant that behaves as a secondary substrate utilized by the microorganisms grown on the input BOD_L as the primary electron donor. The secondary substrate enters in the normal influent flow of 10,000 m^3/d at a concentration of 0.1 mg/l (i.e., 100 μg/l) and has these kinetic parameters:

$$\hat{q} = 1\ \text{mg/mg VS}_a\text{-d}$$
$$K = 20\ \text{mg/l}$$
$$D = 1.2\ \text{cm}^2/\text{d}$$
$$D_f = 0.6\ \text{cm}^2/\text{d}$$
$$L = 0.008\ \text{cm}$$

X_f and L_f remain 40 mg VS$_a$/cm^3 and 0.10 cm, respectively, because the presence of the secondary substrate does not affect utilization of the primary substrate. Since the secondary substrate is not responsible for sustaining the biofilm, Y, b, and b_{det} values are not needed. (Of course, b and b_{det} remain the same, since they refer to the biomass, not the substrate.)

The computations proceed in the same manner as for Example 4.7, but with the parameters for the micropollutant. Our initial guess is 83 percent removal, or $S = 0.017\ \text{mg/l} = 1.7 \cdot 10^{-5}\ \text{mg/cm}^3$. This guess is the same percent removal as for the primary substrate.

1.

$$S^* = \frac{0.017\ \text{mg/l}}{20\ \text{mg/l}} = 8.5 \cdot 10^{-4} \qquad D_f^* = \frac{0.60\ \text{cm}^2/\text{d}}{1.2\ \text{cm}^2/\text{d}} = 0.5$$

$$\tau = \sqrt{\frac{(0.02\ \text{mg/cm}^3)(0.60\ \text{cm}^2/\text{d})}{(1\ \text{mg/mg-d})(40\ \text{mg/cm}^3)}} = 0.0173\ \text{cm} \qquad (173\mu\text{m})$$

$$L^* = \frac{0.008\ \text{cm}}{0.0173\ \text{cm}} = 0.46 \qquad L_f^* = \frac{0.10\ \text{cm}}{0.0173\ \text{cm}} = 5.78$$

2.

$$\eta = \frac{\tanh(5.78)}{5.78} = 0.173$$

3.

$$S_S^* = 0.5 \cdot \left[(8.5 \cdot 10^{-4} - 1 - 0.46 \cdot 5.78 \cdot 0.5 \cdot 0.173) \right.$$

$$\left. + \sqrt{(8.5 \cdot 10^{-4} - 1 - 0.46 \cdot 5.78 \cdot 0.5 \cdot 0.173)^2 + 4 \cdot 8.5 \cdot 10^{-4}} \right]$$

$$= 6.9 \cdot 10^{-4}$$

4.

$$J^* = 0.5 \cdot 5.78 \cdot 0.173 \cdot \frac{6.9 \cdot 10^{-4}}{1 + 6.9 \cdot 10^{-4}} = 3.45 \cdot 10^{-4}$$

5.

$$S_S^{*'} = 8.5 \cdot 10^{-4} - 3.45 \cdot 10^{-4} \cdot 0.46 = 6.9 \cdot 10^{-4}$$

6.

$$\phi = \frac{5.78}{\sqrt{1 + 2 \cdot 6.9 \cdot 10^{-4}}} = 5.78$$

7.

$$\eta' = \frac{1}{5.78} - \frac{\tanh(5.78)}{5.78} \left[\frac{1}{\tanh(5.78)} - 1 \right] = 0.173$$

So, we can accept $\eta = 0.173$, and $S_S^* = 6.9 \cdot 10^{-4}$, which gives $J^* = 3.45 \cdot 10^{-4}$ and $J = 3.45 \cdot 10^{-4}(0.02)(1.2)/0.0173 = 4.78 \cdot 10^{-4}$ mg/cm^2-d.

A mass balance on the micropollutant is:

$$S = S^0 - 100J = 0.1 \text{ mg/l} - 100 \cdot 4.78 \cdot 10^{-4} \text{ mg/l}$$

$$= 0.052 \text{ mg/l}$$

This is not yet convergence. Another iteration yields the final results of $S = 26$ μg/l and $J = 7.3 \cdot 10^{-4}$ mg/cm^2-d. Thus, the fractional removal of the micropollutants is 74 percent, somewhat less than for the primary substrate. This reflects its slower degradation kinetics.

4.13 SPECIAL-CASE BIOFILM SOLUTIONS

In some cases, the "type" of biofilm can be specified a priori. "Type" can refer to the following special cases:

- The biofilm is deep. (Equation 4.10.)
- The utilization kinetics are zero order throughout the biofilm.
- The utilization kinetics are first order throughout the biofilm (Equations 4.6 and 4.7).

- The substrate concentration is known at the outer surface and the attachment surface (Equation 4.9).

Pseudo-analytical or truly analytical solutions are available for each of the first two special cases.

4.13.1   DEEP BIOFILM

Rittmann and McCarty (1978) developed a pseudo-analytical solution for deep biofilms. It is termed the variable-order model, because it expresses J^* as

$$J^* = C^*(S^*)^q \qquad \textbf{[4.64]}$$

in which q varies from 0.5 to 1.0. The variable-order model can be used for steady-state or nonsteady-state biofilms, as long as the biofilm is deep. The equations for the variable-order model are:

$$\Lambda = (\ln S^*) - \ln(2 + (\ln D_f*)/2.303) - 1.8\ln(1 + \sqrt{2}L^* D_f^*) + 0.353 \quad \textbf{[4.65]}$$

$$q = 0.75 - 0.25\tanh(0.477\Lambda) \qquad \textbf{[4.66]}$$

$$C^* = \frac{\sqrt{2}D_f^*\left(\sqrt{2} + \sqrt{2}L^* D_f^*\right)^{(1-2q)}}{1.0 + 0.54[1 + 0.0121\ln(1 + \sqrt{2}L^*)]\left[1 - 8.325\left(\ln\left(\frac{q}{0.707}\right)\right)^2\right]} \qquad \textbf{[4.67]}$$

$$J = \frac{KD}{\tau}C^*(S^*)^q \qquad \textbf{[4.68]}$$

in which S^*, L^*, D_f^*, and τ are defined the same as for the nonsteady-state model for biofilms of any thickness (i.e., Equations 4.50–4.54). The original version of the variable-order model (Rittmann and McCarty, 1978) defined L^* slightly differently, and some constants in Equations 4.65 are changed to reflect this.

Proper use of the variable-order model requires that the biofilm is deep. For steady-state biofilms, the biofilm is deep whenever $S > S_R$ or $J_{ss} > J_R$. For biofilms not at steady state, Suidan et al. (1987) derived the following criteria for a deep biofilm:

$$L_f^* \geq \cosh^{-1}\left[\frac{L^* D_f^*\sqrt{1 + 0.01[(L^* D_f^*)^2 - 1]} - 1}{0.1[(L^* D_f^*)^2 - 1]}\right] \qquad \textbf{[4.69]}$$

if $D_f^* L^* \neq 1$ or

$$L_f^* \geq 1.15 \qquad \textbf{[4.70]}$$

if $L^* D_f^* = 1$.

4.13.2   ZERO-ORDER KINETICS

When $S_f \gg K$ throughout a biofilm, the biofilm is zero order. Analytical solutions for J_0 and S_f for a shallow biofilm are:

$$J_0 = \hat{q} X_f L_f \qquad \text{[4.71]}$$

$$S_f = S_s - \frac{\hat{q} X_f}{D_f} \left(L_f - \frac{z}{2} \right) z \qquad \text{[4.72]}$$

However, detailed analyses by Suidan et al. (1987) indicate that shallow biofilms with $S_f \gg K$ never occur.

A deep biofilm having zero-order kinetics gives the well-known "half-order" solution,

$$J = \sqrt{2 D_f \hat{q} X_f S_s} \qquad \text{[4.73]}$$

$$S_f = S_s - 2 \sqrt{\frac{\hat{q} X_f S_s}{2 D_f}} z + \frac{\hat{q} X_f}{2 D_f} z^2 \qquad \text{[4.74]}$$

Although having a deep biofilm that also is zero order may seem contradictory, the half-order solution is theoretically valid for thick enough biofilms (Suidan et al., 1987) and is observed experimentally.

4.14 BIBLIOGRAPHY

Atkinson, B. and I. J. Davies (1974). "The overall rate of substrate uptake (reaction) by microbial films. Part I. Biological rate equation." *Trans., Inst. Chem. Engr.* 52, pp. 248–259.

Atkinson, B. and S. Y. How (1974). "The overall rate of substrate uptake (reaction) by microbial films. Part II. Effect of concentration and thickness with mixed microbial films." *Trans., Inst. Chem. Engr.* 52, p. 260.

Cannon, F. S. (1991). "Discussion of simplified design of biofilm processes using normalized loading curves." *Res. J. Water Pollution Control Fedr.* 63, p. 90.

Chang, H. T.; B. E. Rittmann; D. Amar; O. Ehlinger; and Y. Lesty (1991). "Biofilm detachment mechanisms in a liquid-fluidized bed." *Biotechnol. Bioengr.* 38, pp. 499–506.

Characklis, W. G. and K. C. Marshall, eds. (1990). *Biofilms.* New York: Wiley-Interscience.

Characklis, W. G. and P. A. Wilderer, eds. (1989). *Structure and Function of Biofilms.* Winchester, England: John Wiley.

Christian, P.; L. Hollesan; and P. Harremoës (1998). "Liquid film diffusion on reaction rate in submerged biofilters." *Water Res.* 29, pp. 947–952.

Harremoës, P. (1976). "The significance of pore diffusion to filter denitrification." *J. Water Pollution Control Federation* 48, pp. 377–388.

Heath, M. S.; S. A. Wirtel; B. E. Rittmann; and D. R. Noguera (1991). "Closure to: Simplified design of biofilm processes using normalized loading curves." *Res. J. Water Poll. Control Fedr.* 63, pp. 91–92.

Heath, M. S.; S. A. Wirtel; and B. E. Rittmann (1990). "Simplified design of biofilm processes using normalized loading curves." *Res. J. Water Poll. Control Fedr.* 62, pp. 185–192.

Jennings, P. A. (1975). "A mathematical model for biological activity in expanded bed adsorption columns." Ph.D. Thesis, Department of Civil Engineering, University of Illinois, Urbana, IL.

Perry, R. H.; D. W. Green; and J. O. Maloney (1984). *Perry's Chemical Engineers Handbook,* 6th ed., New York: McGraw-Hill Book Co.

Rittmann, B. E. (1987). "Aerobic biological treatment." *Environ. Sci. Technol.* 21, pp. 128–136.

Rittmann, B. E. (1982a). "Comparative performance of biofilm reactor types." *Biotechnol. Bioengr.* 24, pp. 1341–1370.

Rittmann, B. E. (1982b). "The effect of shear stress on loss rate." *Biotechnol. Bioengr.* 24, pp. 501–506.

Rittmann, B. E. and P. L. McCarty (1978) "Variable-order model of bacterial-film kinetics." *J. Environ. Engr.* 104, pp. 889–900.

Rittmann, B. E. and P. L. McCarty (1980a). "Model of steady-state-biofilm kinetics." *Biotechnol. Bioengr.* 22, pp. 2343–2357.

Rittmann, B. E. and P. L. McCarty (1980b). "Evaluation of steady-state-biofilm kinetics." *Biotechnol. Bioengr.* 22, pp. 2359–2373.

Rittmann, B. E. and P. L. McCarty (1981). "Substrate flux into biofilms of any thickness." *J. Environ. Engr.* 107, pp. 831–849.

Sáez, P. B. and B. E. Rittmann (1988). "An improved pseudo-analytical solution for steady state-biofilm kinetics." *Biotechnol. Bioengr.* 32, pp. 379–385.

Sáez, F. B. and B. E. Rittmann (1992). "Accurate pseudo-analytical solution for steady-state biofilms." *Biotechnol. Bioengr.* 39, pp. 790–793.

Siegrist, A. and W. Gujer (1987). "Demonstration of mass-transfer and pH effects in a nitrifying biofilm." *Water Research* 21, pp. 1481–1488.

Suidan, M. T.; B. E. Rittmann; and U. K. Traegner (1987). "Criteria establishing biofilm-kinetic types." *Water Research* 21, pp. 491–498.

Trinet, F.; R. Heim; D. Amar; H. T. Chang; and B. E. Rittmann (1991). "Study of biofilm and fluidization of bioparticles in a three-phase liquid-fluidized bed reactor." *Water Sci. Technol.* 23, pp. 1347–1354.

Williamson, K. J. and P. L. McCarty (1976). "Verification studies of the biofilm model for bacterial substrate utilization." *J. Water Poll: Control Fedr.* 48, pp. 281–289.

Wirtel, S. A.; D. R. Noguera; D. T. Kampmeier; M. S. Heath; and B. E. Rittmann (1992). "Explaining widely varying biofilm-process performance with normalized loading curves." *Water Environ. Res.* 64, pp. 706–711.

4.15 PROBLEMS

4.1. You are to design a completely mixed biofilm reactor. The kinetic parameters are:

$$K = 10\frac{mg}{l} \qquad D_f = 0.75\frac{cm^2}{d} \quad Y = 0.5\frac{mg\ VS_a}{mg} \qquad L = 100\mu m$$

$$\hat{q} = 12\frac{mg}{mg\ VS_a \cdot d} \quad D = 1.0\frac{cm^2}{d} \quad b' = 0.15\ d^{-1} \qquad X_f = 10\frac{mg\ VS_a}{cm^3}$$

The input concentration is 200 mg/l, and you desire 95 percent removal for a steady-state biofilm. The flow rate is 1,000 l/d. Estimate the surface area required (in m²). Use the normalized loading curves.

4.2. The parameters for this problem are the same as for problem 4.1. For this problem, calculate the necessary surface area for a completely mixed biofilm reactor if you desire 99 percent removal. You have a steady-state biofilm. Use the steady-state biofilm model to solve this problem.

4.3. *Aerooxidans unitii* is an aerobic bacterium that grows as a biofilm. Its kinetic parameters are:

$$K = 1\frac{mg}{cm^3} \qquad D_f = 1.0\frac{cm^2}{d} \qquad Y = 1.0\frac{mg\ VS_a}{mg} \qquad L = 10^{-2}\ cm$$

$$\hat{q} = 1\frac{mg}{mg\ VS_a \cdot d} \qquad D = 1.0\frac{cm^2}{d} \qquad b' = 0.1\ d^{-1} \qquad X_f = 10\frac{mg\ VS_a}{cm^3}$$

You need to design a biofilm process that has as its prime criterion that the steady-state effluent concentration be very insensitive to changes in loading, because current knowledge of the substrate loading is very poor. So, give an estimate of design range on the substrate flux to achieve the prime criterion.

4.4. You are using a steady-state biofilm system to treat a high-strength wastewater with bacteria that have the following characteristics:

$$K = 100\frac{mg\ BOD_L}{l} \qquad Y = 0.1\frac{mg\ VS_a}{mg\ BOD_L} \qquad b = 0.1\ d^{-1}$$

$$D = 1.0\frac{cm^2}{d} \qquad \hat{q} = 10\frac{mg\ BOD_L}{mg\ VS_a \cdot d}$$

When growth in a turbulent biofilm system, you get a highly dense biofilm having:

$$D_f = 1\frac{cm^2}{d} \qquad X_f = 100\frac{mg\ VS_a}{cm^3} \qquad b_{det} = 0.1\ d^{-1} \qquad L = 10^{-2}\ cm$$

However, you can immobilize the same bacteria into alginate beads. This immobilization eliminates detachment losses, but introduces greater internal diffusion resistance and a large spacing between cells. These effects can be quantified by:

$$D_f = 0.1\frac{cm^2}{d} \qquad X_f = 10\frac{mg\ VS_a}{cm^3} \qquad b_{det} = 0\ d^{-1} \qquad L = 10^{-2}\ cm$$

Determine whether immobilization is an advantage or a disadvantage if you are to achieve effluent concentrations of 10 mg BOD$_L$/l and 1,000 mg BOD$_L$/l. (There are two comparisons required. You should quantify your conclusions by using the normalized loading curves.)

4.5. A nitrification process can easily be operated in a fixed-film mode. The typical kinetic parameters are:

$$K = 0.5 \frac{\text{mg N}}{\text{l}} \qquad D_f = 1.2 \frac{\text{cm}^2}{\text{d}} \qquad Y = 0.26 \frac{\text{mg VS}_a}{\text{mg}} \qquad L = 65 \ \mu m$$

$$\hat{q} = 2.0 \frac{\text{mg N}}{\text{mg VS}_a \cdot \text{d}} \qquad D = 1.5 \frac{\text{cm}^2}{\text{d}} \qquad b' = 0.1 \ \text{d}^{-1} \qquad X_f = 5 \frac{\text{mg VS}_a}{\text{cm}^3}$$

You are to determine the response of a complete-mix biofilm reactor to changes in detention time. The reactor has specific surface area, a, of $1 \ \text{cm}^{-1}$. The influent feed strength (S^0) is 30 mg N/l of NH_4^+-N. You can assume that NH_4^+-N completely limits the kinetics.

The steps for the solution are:

(a) Calculate $S_{\min}$, $S_{\min}^*$, and J_R.

(b) Generate a curve of J/J_R vs. $S/S_{\min}$. You may need to interpolate between given normalized curves.

(c) Set up the mass balance for substrate in the reactor.

(d) For each detention time (1, 2, 4, 8, and 24 h), solve for S. You may wish to use an iterative technique. Guess S, solve for J (from curve from part b), calculate S from the mass balance, check for agreement of S, repeat until convergence. Ignore suspended reactions. Alternatively, you can superimpose the mass-balance curve on the loading curve.

(e) Plot S vs. θ. Interpret the curve.

(f) Plot S^0/S vs. θ and see if an apparent first-order reaction seems like a good description of the process. If so, what is the apparent first-order coefficient? If not, why is the reaction not first order?

4.6. An RBC can be considered to be completely mixed biofilm reactors in series. You are to analyze the steady-state performance of the first stage of an RBC that has:

- Wastewater: $Q = 1,000 \frac{\text{m}^3}{\text{d}} \quad S^0 = 400 \frac{\text{mg BOD}_L}{\text{l}} \quad X_a^0 = 0 \frac{\text{mg VSS}_a}{\text{l}}$

- First stage: Volume of liquid $= 13.3 \ \text{m}^3$; Medium surface area $= 4,000 \ \text{m}^2$

You can assume that O_2 limitation is not important. Also, assume that only biofilm reactions occur.

Estimate S from the first stage. To make your estimate, use the following parameters:

$$K = 10 \frac{\text{mg BOD}_L}{\text{l}} \qquad D_f^* = 0.8 \qquad\qquad Y = 0.5 \frac{\text{mg VS}_a}{\text{mg BOD}_L}$$

$$L = 6 \cdot 10^{-3} \ \text{cm} \qquad \hat{q} = 11 \frac{\text{mg BOD}_L}{\text{mg VS}_a \cdot \text{d}} \qquad D = 1.0 \frac{\text{cm}^2}{\text{d}}$$

$$b' = 0.5 \text{d}^{-1} \qquad\qquad X_f = 10 \frac{\text{mg VS}_a}{\text{cm}^3}$$

Use the normalized loading curves.

4.7. You are to design a fixed-film biological process to oxidize ammonia nitrogen. The wastewater contains 50 mg NH_4^+-N/l in a flow of 1,000 m³/d. Available are modules of completely mixed biofilm reactors; each module has 5,000 m² of surface area for biofilm colonization. How many modules, operated in series, are needed to achieve an effluent concentration of 1 mg NH_4^+-N/l or less? You may use the following parameters and may assume that NH_4^+-N is the rate limiting substrate.

$$K = 1.0\frac{mg\ N}{l} \qquad D_f = 1.3\frac{cm^2}{d} \quad Y = 0.33\frac{mg\ VS_a}{mg\ N} \quad b_{det} = 0.1\ d^{-1}$$

$$\hat{q} = 2.3\frac{mg\ N}{mg\ VS_a \cdot d} \quad D = 1.5\frac{cm^2}{d} \quad b = 0.11\ d^{-1} \qquad X_f = 40\frac{mg\ VS_a}{cm^3}$$

L = a value so small that S_s approaches S.

What do you expect the actual effluent concentration to be? Use the normalized loading-criterion approach.

4.8. A rotating biological contactor can be considered to be a series of completely mixed biofilm reactors. In this case, assume that you have a system of three reactors in series. Each reactor has a total surface area of 10^4 m². The total flow is 1,500 m³/d, and it contains 400 mg/l of soluble BOD_L. The necessary kinetic parameters are:

$$K = 10\frac{mg\ BOD_L}{l} \quad D_f = 0.8\frac{cm^2}{d} \quad Y = 0.4\frac{mg\ VS_a}{mg\ BOD_L}$$

$$L = 100\ \mu m \qquad b_{det} = 0.10\ d^{-1} \quad \hat{q} = 16\frac{mg\ BOD_L}{mg\ VS_a \cdot d}$$

$$D = 1.0\frac{cm^2}{d} \qquad b = 0.10\ d^{-1} \quad X_f = 25\frac{mg\ VS_a}{cm^3}$$

Estimate the steady-state effluent substrate concentration from each complete mix segment in the series. Assume no suspended reactions. This is an example requiring the steady-state-biofilm model.

4.9. You wish to design a fixed-film treatment process that will achieve an effluent concentration of 100 mg/l of BOD_L when the influent concentration is 5,000 mg BOD_L/l. You know that your process will be a fixed-bed process that will have mixing characteristics intermediate between plug flow and completely mixed. You use four reactor segments in series to describe the mixing. Reasonable kinetic parameters for the BOD utilization are:

$$K = 200\frac{mg\ BOD_L}{l} \quad D_f = 0.4\frac{cm^2}{d} \qquad Y = 0.2\frac{mg\ VS_a}{mg\ BOD_L}$$

$$L = 150\ \mu m \qquad b = 0.04\ d^{-1} \qquad \hat{q} = 10\frac{mg\ BOD_L}{mg\ VS_a \cdot d}$$

$$D = 0.5\frac{cm^2}{d} \qquad X_f = 20\frac{mg\ VS_a}{cm^3} \quad b_{det} = 0.04\ d^{-1}$$

$$f_d = 0.9$$

If you have a total flow of 1,000 m^3/d, approximately how much fixed-film surface area (m^2) is needed to achieve your goal? Use the normalized loading curves to solve this problem.

4.10. Assume that the influent concentrations in the previous problem drops suddenly to 20 mg/l. Calculate S if all other parameters remain the same (recall that this is a nonsteady-state-biofilm).

4.11. Repeat problem 4.10, but substitute $S^0 = 40$ mg/l.

4.12. Given the following kinetic parameters and operating conditions:

$$K = 10\frac{mg}{l} \qquad D_f = 0.72\frac{cm^2}{d} \qquad D = 0.9\frac{cm^2}{d} \qquad a = 2\ cm^{-1}$$

$$\hat{q} = 5\frac{mg}{mg\ VS_a \cdot d} \qquad X_f = 40\frac{mg\ VS_a}{cm^3} \qquad S^0 = 100\frac{mg}{l} \qquad L = 0.01\ cm$$

$$V = 1\ m^3 \qquad Q = 11.9\frac{m^3}{d}$$

Use the nonsteady-state biofilm model to calculate effluent concentration and percent removal for $L_f = 10^{-4}, 10^{-3}, 10^{-2}$, and 10^{-1} cm.

4.13. RBCs have been used for denitrification, as well as for aerobic processes. Estimate the surface area needed to remove essentially all of 30 mg NO_3^--N/l from a wastewater by a complete-mix RBC having a temperature of 20 °C, methanol is the carbon source and limiting substrate, the effluent methanol concentration is 1 mg/l, and $Q = 1,000$ l/d. You may assume that the average SRT is 33 d to calculate the stoichiometry, which you may assume is constant. Also, you may assume steady-state operation, and the following parameters:

$$K = 9.1\frac{mg\ CH_3OH}{l} \qquad D_f = 1.04\frac{cm^2}{d} \qquad b_{det} = 0.03\ d^{-1}$$

$$b = 0.05\ d^{-1} \qquad L = 60\ \mu m \qquad \hat{q} = 6.9\frac{mg\ CH_3OH}{mg\ VS_a \cdot d}$$

$$D = 1.3\frac{cm^2}{d} \qquad X_f = 20\frac{mg\ VS_a}{cm^3} \qquad Y = 0.27\frac{mg\ VS_a}{mg\ CH_3OH}$$

Use the steady-state-biofilm model directly.

4.14. Hold everything the same as in problem 4.13, except that Q is increased for a short-time to 2,000 l/d. What is the effluent concentration of methanol? (You need to use the nonsteady-state-biofilm model).

4.15. You wish to use a completely mixed biofilm reactor to treat 1,000 m^3/d of wastewater with 100 mg S/l of sulfate You wish to achieve a sulfate concentration in the effluent of 1 mg S/l. You also wish to achieve an acetate concentration of 4 mg BOD_L/l. You have been very clever with your design, because you have achieved a very low b_{det} value of 0.005 d^{-1}. You also know the following information about the biofilm system and the rate-limiting substrate, which is acetate.

$$a = 100 \text{ m}^{-1} \qquad X_f = 100 \frac{\text{mg VS}_a}{\text{cm}^3} \qquad L = 0.0044 \text{ cm}$$

$$D = 1.3 \frac{\text{cm}^2}{\text{d}} \qquad D_f = 1.0 \frac{\text{cm}^2}{\text{d}} \qquad b' = 0.045 \text{ d}^{-1}$$

$$b = 0.04 \text{ d}^{-1} \qquad Y = 0.0565 \frac{\text{mg VSS}_a}{\text{mg BOD}_L}$$

$$K = 10 \text{ mg BOD}_L/\text{l} \qquad \hat{q} = 8.6 \text{ mg BOD}_L/\text{mg VSS}_a\text{-d}$$

You must complete and analyze the design.

(a) What substrate flux will you need to achieve an effluent concentration of 4 mg BOD$_L$/l? Express the flux in kg BOD$_L \cdot$ m$^{-2} \cdot$ d^{-1}.

(b) What influent BOD$_L$ concentration (in mg/l) will you need to achieve the desired effluent concentrations?

(c) What volume (in m^3) is needed for the biofilm reactor? What is the liquid detention time (in days)?

(d) What is the concentration of active biomass per unit volume of reactor? Express the result in mg VSS/l.

(e) What is the effluent concentration of active VSS? Express the answer in mg VSS/l.

(f) Do you think that the performance of the biofilm process was significantly affected by mass-transport resistance? How do you know this quantitatively?

4.16. You are operating a fluidized-bed biofilm reactor that has a high enough recycle rate to make the liquid regime completely mixed. The particle diameter is 1 mm, and it has a specific gravity of 2.5. The porosity of the medium when it is not expanded (ε_0) is 0.3. The expanded porosity is 0.46. When expanded, the specific surface area in the reactor is 3,240 m^{-1}. When not expanded, the bed has a specific surface area of 4,200 m^{-1}. Your kinetic parameters are:

$$K = 15 \frac{\text{mg}}{\text{l}} \qquad D_f = 1.28 \frac{\text{cm}^2}{\text{d}} \quad Y = 0.3 \frac{\text{mg VS}_a}{\text{mg}} \qquad L = 60 \ \mu\text{m}$$

$$\hat{q} = 8 \frac{\text{mg}}{\text{mg VS}_a \cdot \text{d}} \quad D = 1.6 \frac{\text{cm2}}{\text{d}} \quad b = 0.08 \text{ d}^{-1} \qquad X_f = 20 \frac{\text{mg VS}_a}{\text{cm}^3}$$

If the feed rate is 10 m^3/d, the total reactor volume is 10 m^3, and the effluent substrate concentration is 1.0 mg/l, what is the influent concentration? Assume that suspended reactions are unimportant, but that detachment is occurring. So, you must compute b' from b_{det}. You may assume that the specific gravity is unchanged by the biofilm. You may use normalized loading curves or the steady-state-biofilm model.

4.17. Detachment losses can be a major factor in determining the amount of attached biofilm. It is especially important when S approaches S_{min}. Use the following kinetic parameters:

$$K = 10\frac{mg}{l} \qquad D_f = 0.8\frac{cm^2}{d} \quad Y = 0.5\frac{mg\ VS_a}{mg} \qquad L = 120\ \mu m$$

$$\hat{q} = 10\frac{mg}{mg\ VS_a \cdot d} \quad D = 1.0\frac{cm^2}{d} \quad b = 0.10\ d^{-1} \qquad X_f = 10\frac{mg\ VS_a}{cm^3}$$

Calculate S_{min} and steady-state values of L_f for shear stresses of 0 and 2 dyne/cm^2 if $S = 0.5$ mg/l.

4.18. You are called as the high-priced expert because an engineering consultant cannot figure out if it is possible to design and operate a fluidized-bed biofilm reactor (with effluent recycle) to achieve his client's objectives. Here are the facts:

1. The industry must treat 1,000 m^3/d of a wastewater that contains 5,000 mg/l of BOD$_L$.

2. The goal is to reduce the original substrate BOD$_L$ to 100 mg/l.

3. The consultant has used chemostats to estimate kinetic parameters for the wastewater BOD$_L$ at 20 °C:

$$\hat{q} = 3.5\frac{mg\ BOD_L}{mg\ VS_a \cdot d} \quad K = 30\frac{mg\ BOD_L}{l}$$

$$b = 0.04\ d^{-1} \qquad Y = 0.1\frac{mg\ VS_a}{mg\ BOD_L}$$

4. The actual wastewater BOD$_L$ is a polymer and has a diffusion coefficient of 0.3 cm^2/d.

5. The consultant found that the monomer of the wastewater polymer has exactly the same values of $\hat{q}$, K, b, and Y, but its diffusion coefficient is 1.2 cm^2/d.

6. The consultant ran a pilot-scale fluidized-bed reactor. The reactor used 0.2 mm sand, $\rho_p = 2.65$ g/cm^3, with an expansion of 25 percent (i.e., the height of the bed was 1.25 times its unexpanded height). The expanded porosity was 0.5. To effect fluidization to a 25 percent expansion, effluent recycle was practiced, such that the total superficial flow velocity for the column reactor was 95,000 cm/d. The pilot reactor had an unexpanded bed volume of 250 cm^3 and an unexpanded bed height of 50 cm. It treated a steady flow of 100 l/d of monomer-containing synthetic wastewater (5,000 mg/l of BOD$_L$). The measured effluent quality, 200 mg/l of BOD$_L$, did not meet the given goal.

7. Your goal is to determine whether or not the same reactor configuration can meet the 100 mg/l goal for a full-scale design. You must decide if a design is feasible (even though the pilot unit did not meet the goal for the synthetic wastewater). If a design is feasible, you are to provide the unexpanded bed volume. If it is not feasible, you must explain to the client why the design will not work well enough. Certain things must

remain the same in the full-scale design as they were in the pilot study:

- Same medium, type, and size (0.2 mm sand)
- Same bed expansion (25 percent)
- Same total superficial flow velocity (95,000 cm/d)
- Same temperature (20 °C)
- Same influent substrate concentration (5,000 mg/l)

However, you will be treating the real wastewater BOD_L, not the synthesized monomer wastewater.

The higher priced consultant that you hired suggested that you determine the following:

(a) For both reactor systems, estimate S_{min}. Note that you need to estimate b'. Assume $L_f < 30\ \mu m$.

(b) Estimate $L(\rho_w = 1\ g/cm^3,\ \mu = 864\ g \cdot cm^{-1} \cdot d^{-1}$, and let $D_f^* = 0.8$ and $X_f = 50\ mg/cm^3$).

(c) Use the loading criteria concepts to estimate the expected performance for the pilot-scale reactor. Would you have expected the pilot unit to have the 100 mg/l effluent criterion? Did it perform the way you would have expected?

(d) Based upon your analysis so far and the loading criteria, is it feasible for a full-scale design to achieve the 100 mg/l goal?

(e) If the answer to (d) is yes, estimate the unexpanded volume needed to achieve the treatment goal. If the answer to (d) is no, explain why the analysis of the pilot study demonstrates that the design is not feasible.

4.19. You have at your disposal a completely mixed biofilm reactor of volume 100 m^3 and specific surface area 100 m^{-1}. You are going to test it out to treat a wastewater having $Q = 1,000\ m^3/d$ and containing 500 mg/l of BOD_L. Using a sulfate-reducing process that converts SO_4^{-2} to $S_{(s)}$, the bacteria have these characteristics in the biofilm system:

$$K = 10\frac{mg\ BOD_L}{1} \qquad D_f = 0.8\frac{cm^2}{d} \qquad Y = 0.057\frac{mg\ VS_a}{mg\ BOD_L}$$

$$L = 50\ \mu m \qquad \hat{q} = 8.8\frac{mg\ BOD_L}{mg\ VS_a \cdot d} \qquad D = 1.0\frac{cm^2}{d}$$

$$b = 0.05\ d^{-1} \qquad X_f = 50\frac{mg\ VS_a}{cm^3} \qquad b_{det} = 0.05\ d^{-1}$$

$$f_d = 0.8 \qquad k_1 = 0.12\frac{mg\ COD_p}{mg\ BOD_L} \qquad k_2 = 0.09\frac{mg\ COD_p}{mg\ VS_a \cdot d}$$

$$\hat{q}_{UAP} = 1.8\frac{mg\ COD_p}{mg\ VS_a \cdot d} \quad \hat{q}_{BAP} = 0.1\frac{mg\ COD_p}{mg\ VS_a \cdot d} \quad K_{UAP} = 100\frac{mg\ COD_p}{1}$$

$$K_{BAP} = 85\frac{mg\ COD_p}{1}$$

You need to estimate how well this biofilm process will perform before you commit to using it. Therefore, use the concepts from the normalized surface loading to determine the following:

(a) The effluent concentration of original substrate in mg BOD_L/l.

(b) The effluent concentration of VSS_a in mg/l (you can ignore any growth of suspended bacteria).

(c) The effluent concentration of SMP in mg/l.

(d) The effluent BOD_L concentration from organic materials, in mg BOD_L/l.

4.20. This question addresses a novel system to treat a sulfide-bearing wastewater by autotrophic bacteria that aerobically oxidize sulfides (i.e., H_2S and HS^-) to SO_4^{2-}, which is harmless. The electron acceptor is O_2, and ample NH_4^+ is present as the N source. For these sulfide oxidizers, you may utilize the following stoichiometric and kinetic parameters:

$$X_f = 40 \frac{\text{mg VS}_a}{\text{cm}^3} \qquad L = 40 \ \mu\text{m} \qquad b_{\text{det}} = 0.05 \ \text{d}^{-1}$$

$$a = 200 \ \text{m}^{-1} \qquad \hat{q} = 5.0 \frac{\text{mg S}}{\text{mg VS}_a \cdot \text{d}} \qquad f_d = 0.8$$

$$V = 162 \ \text{m}^3 \qquad f_s^0 = 0.2 \frac{e^- \text{ eq cells}}{e^- \text{ eq donor}} \qquad Y = 0.28 \frac{\text{mg VS}_a}{\text{mg S}}$$

$$D = 1.2 \frac{\text{cm}^2}{\text{d}} \qquad K = 2 \frac{\text{mg S}}{1} \qquad D_f = 1.0 \frac{\text{cm}^2}{\text{d}}$$

$$b = 0.05 \ \text{d}^{-1}$$

For the questions that follow, you should assume that the sulfides are not lost by any mechanisms other than microbially catalyzed metabolism, that no intermediate sulfur species form, and that you can use the total sulfide concentration as rate-limiting.

The influent has: $Q = 1,000 \frac{\text{m}^3}{\text{d}} \quad S^0 = 100 \frac{\text{mg S}}{1} \quad X_v^0 = 0 \frac{\text{mg VSS}}{1}$

(a) What is the effluent concentration of sulfides at steady state?

(b) What is the total biofilm accumulation expressed as mg VS/cm^2 of biofilm area and as mg VSS/l in the bulk liquid? What is the actual effluent concentration of VSS in mg VSS/l?

(c) What is the effluent organic COD concentration? Use the SMP parameters given for problem 4.19.

(d) What is the required O_2 supply rate in kg O_2/d?

4.21. Determine the flux into a steady-state biofilm (J in mg/cm^2-d) and the concentration of substrate at the biofilm surface (S_s in mg/l) for the following conditions:

$$\hat{q} = 12 \text{ mg/mg VS}_a\text{-d} \qquad X_f = 10{,}000 \text{ mg VS}_a/\text{l}$$
$$K = 20 \text{ mg/l} \qquad L = 0.005 \text{ cm}$$
$$Y = 0.6 \text{ mg VS}_a/\text{mg} \qquad D = 0.9 \text{ cm}^2/\text{d}$$
$$b' = 0.15/\text{d} \qquad D_f = 0.8 \text{ D}$$
$$S = 30 \text{ mg/l}$$

4.22. Submerged filters have been used for methane fermentation. In this problem, the volume of a reactor required to achieve 90 percent removal of acetate, the major degradable organic in a wastewater, using a submerged filter is to be determined.

The submerged filter is to be considered as a CMBR, which is a reasonable assumption for methane fermentation because of gas mixing by gas evolution. Thus, the concentration of the rate-limiting substrate is the same throughout the reactor and equal to the effluent concentration. The effluent COD concentration will be assumed to equal 10 percent of the influent concentration.

Assume the following conditions apply:

$$\hat{q} = 8 \text{ mg/mg VS}_a\text{-d} \qquad X_f = 20{,}000 \text{ mg VS}_a/\text{l}$$
$$K = 50 \text{ mg/l} \qquad L = 0.015 \text{ cm}$$
$$Y = 0.06 \text{ mg VS}_a/\text{mg} \qquad D = 0.9 \text{ cm}^2/\text{d}$$
$$b' = 0.02/\text{d} \qquad D_f = 0.8 \text{ D}$$
$$b_{\text{det}} = 0.01/\text{d}$$
$$Q = 15 \text{ m}^3/\text{d} \qquad h = 0.8$$
$$S^0 = 2{,}000 \text{ mg/l} \qquad a = 1.2 \text{ cm}^2/\text{cm}^3$$
$$X_i^0 = 0$$

Using the steady-state-biofilm model:
(a) Determine the required reactor volume, V, in m^3.
(b) Determine the hydraulic detention time (V/Q) in hours.
(c) Estimate the effluent suspended solids concentration, X_a, in mg VSS$_a$/l.
(d) Estimated the effluent inert organic suspended solids concentration, X_i, in mg VSS$_i$/l.
(e) Estimate the total effluent volatile suspended solids concentration, X_v, in mg VSS/l.

4.23. Fluidized bed reactors have been found to be excellent for removing low concentrations of petroleum hydrocarbons from groundwater. They consist of a bed of sand or granular activated carbon with wastewater flowing in an upward direction with sufficient velocity to suspend the particles. Bacteria attach to the suspended particles as a biofilm, and in this way the wastewater is treated.

The object of this problem is to determine the detention time required for reducing the concentration of BTEX compounds (benzene, toluene, ethylbenzene, and xylene) from groundwater contaminated by a gasoline spill.

Removal of the BTEX compounds down to the very low concentration of 0.1 mg/l is required.

Assume the following conditions apply:

$\hat{q} = 12$ mg/mg VS$_a$-d $X_f = 15{,}000$ mg VS$_a$/l

$K = 0.2$ mg/l $L = 0.005$ cm

$Y = 0.5$ mg VS$_a$/mg $D = 1.0$ cm^2/d

$b' = 0.14$/d $D_f = 0.8\,D$

$b_{det} = 0.05$/d

$Q = 100$ m^3/d $h = 0.75$

$S^0 = 4$ mg/l $a = 6.0$ cm^2/cm^3

$X_i^0 = 0$

(a) Estimate the empty-bed detention time (V/Q) in minutes required to achieve the required treatment for a steady-state biofilm in a CMBR.

(b) Estimate the concentration of active bacterial suspended solids (X_a), inert organic suspended solids (X_i), and volatile suspended solids (X_v) in the effluent in mg VSS/l.

4.24. From biofilm kinetics, it was found that the flux into a biofilm at steady state is 0.15 mg/cm^2-d when the bulk liquid concentration of benzoate is 2 mg/l. What reactor volume is needed to treat 100 m^3/d of wastewater containing 50 mg/l benzoate if 96 percent removal is required, and if we assume completely-mixed conditions and the specific surface area of the reactor is 3 cm^2/cm^3?

4.25. The flux of substrate into a biofilm is found to be 0.8 mg/cm^2-d when the substrate concentration at the biofilm surface is 15 mg/l. What is the bulk solution concentration for the substrate? Characteristics of the biofilm are given below.

$X_a = 20$ mg VS$_a$/cm^3 $b = 0.17$/d

$Y = 0.6$ mg VS$_a$/mg $L = 0.005$ cm

$\hat{q} = 12$ mg/mg VS$_a$-d $D = 0.8$ cm^2/d

$K = 5$ mg/l

4.26. You have been asked to design a reactor to treat a waste stream of 10^4 m^3/d containing 150 mg/l of phenol. The following coefficients apply:

$Y = 0.6$ g VS$_a$/g phenol

$\hat{q} = 9$ g phenol/g VS$_a$-d

$b = 0.15$ d^{-1}

$K = 0.8$ mg/l

You have decided to evaluate use of a fixed-film reactor in which you apply sufficient recycle so that you can assume the system will act like a completely mixed reactor. Assume that deep biofilm kinetics apply ($S_w = 0$), that the specific surface area of the reactor is 10^7 cm^2/m^3, the boundary layer thickness for the deep biofilm is 0.005 cm, X_f is 20 mg VS$_a$/cm^3, $D = 0.8$ cm^2/d, and $D_f = 0.8\,D$. Determine the volume V for this reactor in m^3.

4.27. You wish to design a plug-flow biofilm reactor without recycle for denitrification to N_2 of a wastewater containing 65 mg NO_3^-/l and no ammonia, and you have selected acetate to add as the electron donor for the reaction.

(*a*) What concentration of acetate should be added to the wastewater to achieve complete removal of nitrate by denitrification?

(*b*) Given the above concentration of acetate added to the wastewater prior to treatment, is the reaction at the entrance to the reactor rate-limited by acetate, nitrate, or neither? Show appropriate calculations to support your conclusion.

Assume that nitrate is used for cell synthesis and that the following characteristics apply for reaction kinetics and biofilm characteristics:

	Acetate	Nitrate	Biofilm
K	10 mg/l	1 mg/l	
$\hat{q}$	15 mg/mg VS_a-d		
D	0.9 cm²/d	0.7 cm²/d	
D_f	0.8 D_w	0.8 D_w	
L			0.150 cm
f_s			0.55
X_f			12 mg VS_a/cm³
b			0.15/d
Y			0.4 g VS_a/g acetate

4.28. Fill out the following table to indicate what impact a small increase in each of the variables in the left-hand column would have on the operating characteristics indicated of a completely mixed aerobic biofilm reactor. Assume all other variables in the left-hand column remain the same. Also, for this problem assume that you have a soluble waste, a fixed reactor volume and surface area for biofilm attachment, the electron donor is rate limiting, X_f is constant, and the biofilm is considered as deep. Use: $(+)$ = increase, $(-)$ = decrease, and (0) = no change.

	Operating Characteristic		
Variable	Effluent Substrate Concentration (S^e, mg/l)	Treatment Efficiency (%)	Oxygen Requirement (kg/d)
Y			
$\hat{q}$			
Q^0			
S^0			
b			
K			

4.29. A wastewater with flow rate of 5,000 liters/d and containing soluble organic material only with a BOD_L of 6,000 mg/l is being treated by methane fermentation at 35 °C in a fixed-film reactor with a volume of 2,500 liters. The reactor effluent BOD_L is found to be 150 mg/l. Estimate the total biofilm surface area in the reactor. The wastewater organic matter is primarily acetic acid (a fatty acid). Assume the reactor acts like a deep-biofilm ($S_w = 0$) completely mixed system. Also, for acetate assume D equals 0.9 cm²/d, K is 50 mg/l, $D_f = 0.8\ D$, L (boundary layer thickness) equals 0.01 cm, $\hat{q}$ is 8.4 g BOD_L/g VS_a-d, $b = 0.1$/d, and X_f equals 20 mg VS_a/cm³.

4.30. In the biological oxidation of benzene by a fixed-film reactor, the flux into the deep biofilm ($S_w = 0$) was found to be 5 mg/cm² of biofilm surface area per day when the benzene concentration at the biofilm surface was 15 mg/l. What do you estimate the flux will be if the benzene concentration at the biofilm surface is increased to 50 mg/l?

Rate coefficients for benzene are as follows: $Y = 0.6$ mg VS_a/mg benzene, $\hat{q} = 6$ mg benzene/mg VS_a/d, $K = 2$ mg/l, and $b = 0.1$/d.

4.31. (a) Draw figures showing substrate profiles from the bulk liquid through each of the following biofilm types: (1) deep biofilm, (2) shallow biofilm, and (3) fully penetrated biofilm.

(b) For a given location of biofilm used for wastewater treatment and under steady-state conditions of operation, the bulk solution concentration of the electron donor is 5 mg/l and the concentration of electron donor at the biofilm surface is 2 mg/l. Other properties of the biofilm are a boundary layer thickness $L = 0.01$ cm, diffusion coefficients $D = 0.75$ cm²/d, and D_f of 0.5 cm²/d, organism concentration $X_f = 30$ mg VS_a/cm³. Assuming the biofilm is flat as a plate and deep, estimate the electron donor flux into the biofilm.

(c) Based upon the stoichiometric equation for the reaction occurring in the above biofilm, 0.6 g of electron acceptor is required for each g of electron donor consumed. If the electron acceptor concentration in the bulk solution is 3 mg/l, and the diffusion coefficient for the acceptor is 1.5 times that of the donor, what is the concentration of the acceptor at the biofilm surface?

5

REACTORS

Many different types of reactors are used in environmental engineering practice. Individual reactors are generally designed to emphasize *suspended growth* or *biofilms*. Reactors that make use of suspended growth are also called *suspended-floc*, *dispersed-growth*, or *slurry* reactors. Reactors that make use of biofilms also are called *fixed-film* reactors, *attached-growth* reactors, or *immobilized-cell* reactors. The reactor flow regimes may be similar for the suspended growth and biofilm reactors. At times, a series of reactors may be used, some of which may be suspended growth and others may be biofilm reactors. The engineer must understand the kinetics of substrate removal by the different microbial types and the fundamental properties of different reactor types in order to select the optimum reactor or series of reactors for a given waste-treatment problem and location.

Factors influencing the choice among the different reactor types can include: the physical and chemical characteristics of the waste being considered, the concentration of contaminants being treated, the presence or absence of oxygen, the efficiency of treatment and system reliability required, the climatic conditions under which the reactor will operate, the number of different biological processes involved in the overall treatment system, the skills and experience of those who will operate the system, and the relative costs at a given location and time for construction and operation of different possible reactor configurations.

In this chapter, the different reactor types and when they are most often used are discussed. This is followed by a presentation on how to construct mass balances for reactors and how to make use of mass balances to derive basic equations that describe the relationship between reactor size and treatment performance.

5.1   REACTOR TYPES

Typical reactors used in environmental applications are illustrated in Figure 5.1. Table 5.1 contains a summary of typical applications for each type. The three basic reactors may find application as either suspended growth or biofilm reactors.

Basic reactors

Biofilm reactors

Reactor arrangements

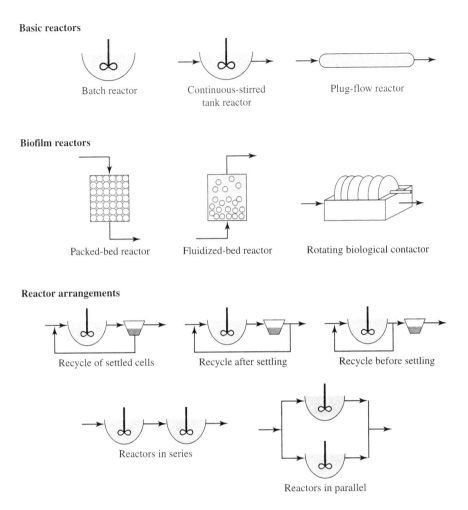

Figure 5.1   Various reactor types and arrangements

5.1.1 SUSPENDED-GROWTH REACTORS

The simplest suspended-growth reactor is the *batch reactor*. The reactor is filled with appropriate proportions of the liquid or slurry stream to be treated, the bacterial culture to be used, and required nutrients, such as nitrogen and phosphorus. Then, the reactor contents are stirred if needed to keep the reactor contents in suspension, and air or oxygen is introduced if the process is an aerobic one. The biochemical reactions then take place without new additions until the reaction is complete. Some or all of the contents are then removed, and new liquid or slurry stream and culture, etc., are added if the cycle is to be repeated. Batch reactors commonly are used in laboratory-scale basic investigations and treatability studies because of their ease of operation and the absence of mechanical pumps, which can be costly and difficult to maintain. Batch reactors are finding increasing use for treatment of slurries of soils in which

Table 5.1   Reactor types and their typical uses

Reactor Type	Typical Uses
Basic Reactors	
Batch	BOD test, high removal efficiency of individual wastewater constituents
Continuous-Flow Stirred-Tank (CSTR)	Anaerobic digestion of sludges and concentrated wastes, aerated lagoon treatment of industrial wastes, stabilization ponds for municipal and industrial wastes, part of activated sludge treatment of municipal and industrial wastewaters
Plug-Flow (PFR)	Activated sludge treatment of municipal and industrial wastes, aerated lagoon treatment of industrial wastes, stabilization ponds for municipal and industrial wastes, nitrification, high-efficiency removal of individual wastewater constituents
Biofilm Reactors	
Packed Bed	Aerobic and anaerobic treatment of municipal and industrial wastewaters, organic removal, nitrification, denitrification
Fluidized Bed	Aerobic treatment of low BOD concentration wastewaters, toxic organic biodegradation, anaerobic treatment, denitrification
Rotating Biological Contactor (RBC)	Aerobic treatment of municipal and industrial wastewaters, organic removal, nitrification
Reactor Arrangements	
Recycle	General aerobic and anaerobic treatment of municipal and industrial wastewaters, especially medium to low concentration BOD, organic removal, nitrification, denitrification
Series	BOD removal combined with nitrification or with nitrification and denitrification or combined with biological phosphorus removal, anaerobic staged treatment, stabilization pond treatment, sequential anaerobic and aerobic treatment of wastewaters such as for removal of specific toxic organic chemicals
Parallel	Generally used for redundancy and reliability in plant operation, especially with high overall wastewater flow rates
Hybrid	Used for combined forms of treatment such as organic removal and nitrification, or organic removal, nitrification, and denitrification, or organic, nitrogen, and phosphorus removal; anaerobic treatment of industrial wastewaters
Sequencing Batch	Useful for high-efficiency removal of individual constituents such as biodegradable but hazardous organics, combined removals of organics, nitrogen, and phosphorus; combination of aerobic and anaerobic processes with same microorganisms

difficult to degrade contaminants are present. The kinetics of contaminant removal in a batch reactor is similar to that of an ideal plug-flow reactor, a system that can lead to highly efficient removal of an individual contaminant. This recognition has led to the concept of *sequencing batch reactors*, a treatment system that employs several batch reactors operated in parallel. One can be filling, one can be emptying, and one or more can be treating. Thus, flow to the system of reactors can be continuous, even though treatment is batch. Indeed, with such batch operation, a single reactor may be operated aerobically for a part of the time, such as to obtain nitrification of ammonia, and subsequently operated under anoxic conditions, such as to obtain denitrification.

The second basic reactor type is the *continuous-flow stirred-tank reactor (CSTR)*, which is also commonly called a *completely mixed reactor*. When a CSTR is used to culture organisms or to study basic biochemical phenomena in the laboratory, such reactors are also called *chemostats*. Here, the liquid or slurry stream is continuously introduced, and liquid contents are continuously removed from the reactor. Microbial culture may or may not be introduced to the reactor under normal operation. If operated properly, microorganisms that grow within the reactor continuously replace the microorganisms removed from the reactor in the effluent stream. The basic characteristic of the ideal CSTR is that the concentrations of substrates and microorganisms are the same everywhere throughout the reactor. In addition, the concentrations leaving in the effluent are the same as those in the reactor. This uniformity of concentration makes analysis of CSTRs comparatively simple. CSTRs are commonly used for aerobic and anaerobic treatment of highly concentrated organic mixtures, such as waste primary and biological sludges, as well as high strength industrial wastes.

The third basic reactor type is the *plug-flow reactor (PFR)*. This is sometimes referred to as a *tubular reactor* or a *piston-flow reactor*. As in the CSTR, the liquid or slurry stream continuously enters one end of the reactor and leaves at the other. However, in the ideal PFR, we envision that the flow moves through the reactor with no mixing with earlier or later entering flows. Hence, an element of the stream entering at one time moves down the reactor as a "plug." The element thus moves downstream in the reactor in a discrete manner so that if one knows the flow rate to the reactor and its size, the location of the element at any time can be calculated. Unlike the CSTR, the concentrations of substrates and microorganisms vary throughout the reactor. An ideal PFR is difficult to realize in practice, because mixing in the direction of flow is impossible to prevent.

An ideal PFR has good and bad characteristics. Concentrations of substrates are highest at the entrance to the reactor, which tends to make rates there quite high. The high rate is bad when it exceeds the ability to supply sufficient oxygen in an aerobic system, results in excess organic acid production and pH problems in an anaerobic system, or causes substrate toxicity with some waste streams. However, if these problems are overcome, a PFR offers the advantage of highly efficient removal of individual contaminants, such as ammonium or trace organic contaminants. Because of the difficulty of obtaining ideal PFR conditions in a flow-through system, sequencing batch reactors, mentioned above, can be used as an alternative when the benefits of PFR treatment are desired. Even when PFRs do not achieve the ideal case, they still can provide many of the benefits of plug-flow. In addition, processes for in situ biodegradation of contaminants in groundwaters often operate similar to plug-flow processes. Here, mixing in the direction of flow (longitudinal direction) is generally small, making plug flow the natural outcome.

5.1.2 BIOFILM REACTORS

Biofilm reactors exhibit some of the characteristics of the three basic reactor types noted above, but most of the microorganisms are attached to a surface and, in this manner, kept within the reactor. While microorganisms detach from the biofilm and may grow in the surrounding liquid, these suspended bacteria normally play a minor role in substrate removal. Three common biofilm reactor types are illustrated in Figure 5.1.

The most common biofilm reactor is a *packed bed*, in which the medium to which the microorganisms are attached is stationary. Historically, large rocks have been used as support media, but today it is more common to use plastic media or pea-sized stones. Both are lighter and offer greater surface area and pore volume per unit of reactor volume than do large rocks.

Commonly, packed-bed reactors are used for aerobic treatment of wastewaters and are known as *trickling filters* or *biological towers*. Here, the wastewater is distributed uniformly over the surface of the bed and allowed to trickle over the surface of the rock or plastic media, giving the packed-bed reactors some plug-flow character. The void space remains open to the passage of air so that oxygen can be transferred to the microorganisms throughout the reactor.

In other applications, the reactor media are submerged in the water. If the bed is not aerated, the packed-bed reactors can be used for denitrification to remove nitrate from water supplies or wastewaters or for anaerobic treatment of more concentrated industrial wastewaters through methane fermentation in an *anaerobic filter*. To do aerobic treatment, the submerged bed normally must be aerated. Backwashing of the filter is used here to prevent clogging by excessive bacterial growth.

The *fluidized-bed reactor* depends upon the attachment of microorganisms to particles that are maintained in suspension by a high upward flow rate of the fluid to be treated. In some cases, the fluidized bed is called an *expanded-bed reactor* or a *circulating-bed reactor*. The particles often are called *biofilm carriers*. The fluidized carriers may be sand grains, granular activated carbon (GAC), diatomaceous earth, or other small solids that are resistant to abrasion. The upward velocity of the fluid must be sufficient to maintain the carriers in suspension, and this depends upon the density of the carriers relative to that of water, the carrier diameter and shape, and the amount of biomass that is attached. Normally, biomass growth increases effective carrier size, but decreases its density, with the net result that carriers with higher amounts of biomass attached tend to be lighter and move higher in the reactor. This offers an advantage for cleaning carriers with excessive biological growth, as they move into the upper regions of the reactor, where they can be separated from the bed and cleaned. Once reintroduced, the cleaned carriers drop to the lower regions of the reactor until the biofilm regrows.

Fluidized-bed reactors can lie almost anywhere between a plug-flow and a completely mixed system. When the system is operated in the once-through mode, the fluid regime has strong plug-flow character. On the other hand, effluent recycle often is required to achieve high enough upward velocities for bed fluidization. With effluent recycle, the liquid regime of the fluidized bed is more like a CSTR. Fluidization of carriers and the mixing that results provide a uniform distribution of the fluid across the cross section of the reactor and also good mass transfer from the bulk fluid to the biofilm surface.

One major disadvantage of a fluidized bed is the need to carefully control bed fluidization. The upward fluid velocity must be sufficient for fluidization, but not so high that carriers are washed from the reactor. Depending on the type of fluidized carriers used, biofilm detachment can be large due to abrasion and turbulence. This precludes using those types of carriers for microorganisms having low growth rates. Oxygen transfer also can be a problem with aerobic application for more concentrated wastewaters. Often, effluent recycle is used to oxygenate and to dilute the wastewater, as well as to maintain a constant upflow rate. Fluidized-bed reactors are used for denitrification and anaerobic wastewater treatment, processes that do not require oxygen transfer. This reactor appears to be especially good for rapid aerobic treatment of waters containing very low concentrations of organic contaminants, such as for the removal of aromatic hydrocarbons in contaminated groundwater.

A somewhat hybrid application of the fluidized-bed reactor with a dispersed-growth reactor is the *upflow anaerobic sludge bed reactor (UASBR)*, which is commonly used for anaerobic treatment of industrial wastewater. When properly operated with appropriate wastewaters, the microorganisms form granules that settle readily and serve as a biologically produced support media for additional biological growth. The rising gas bubbles, generated by rapid methanogenesis, fluidize the granules, thus effecting good mass transfer without mechanical mixing. In effect, the UASBR contains a fluidized bed of self-forming biofilm carriers.

The *rotating biological contactor (RBC)* is another approach for a biofilm reactor and, like the fluidized bed reactor, has good mixing and mass-transfer characteristics. Plastic media in a disk or spiral form are attached to a rotating shaft. Commonly used for aerobic treatment, the portion of the contactor in contact with air absorbs oxygen, and the portion within the liquid absorbs contaminants to be oxidized. Wastewater can enter from one end of the RBC and travel perpendicular to the contactors as illustrated in Figure 5.1, thus creating plug-flow character. Or, it can enter uniformly along the length of the reactor, and, in this manner, can create a completely mixed system. The RBC can be used for anoxic or anaerobic treatment as well, either by completely submerging the reactor or by placing a cover over it to exclude air.

5.1.3 REACTOR ARRANGEMENTS

Any of the three basic reactor types noted above can also be used with recycle. Recycle may involve the simple return of the effluent stream to the influent of the reactor, or it may return a concentrated stream of microorganisms after they are removed from the effluent by settling (most common), centrifugation (seldom used), membrane separation (a newer approach), or other means. Recycle has various purposes. With a CSTR, recycle of the effluent stream has little impact on reactor characteristics or performance, since the effluent stream is identical in concentration to that in the reactor itself. However, with a batch reactor or PFR, recycle dilutes chemicals in the influent stream. Dilution can be desirable to prevent excess oxygen demand, organic acid production, or substrate toxicity at the head-end of a PFR. In a biofilm reactor, the increased influent flow rate resulting from recycle can provide higher fluid velocities throughout the reactor, perhaps resulting in detachment of excess biofilm and better mass transfer kinetics. Recycle is frequently used with fluidized-bed reactors in order to maintain the required upward fluid velocity for fluidization.

Separation and recycle of microorganisms from the effluent stream, as illustrated by the left-hand drawing in Figure 5.1, are commonly practiced in the aerobic *acti-*

vated sludge process in order to maintain a large population of microorganisms within the reactor. The microorganisms are the catalysts that bring about contaminant removal, and reaction rates are proportional to the concentration of microorganisms present. Thus, organism capture and recycle provide significant benefits by greatly reducing the size of the treatment reactor, while maintaining a given efficiency of treatment. Settling tanks are most widely used for separation of microorganisms from the effluent stream for recycle, but because microorganisms do not always settle well, other separation approaches, mentioned above, are being explored.

Reactors are frequently combined in series or in parallel, as illustrated at the bottom in Figure 5.1. Reactors in series are used when different types of treatment are needed, such as organic oxidation followed by nitrification. In this example, the first reactor removes most of the organic material, and the second reactor is optimized for ammonium oxidation, or nitrification. When total-nitrogen removal is desired, a third reactor in series can be added to do denitrification. Reactors in series also are used to create plug-flow characteristics.

When connected in series, the reactors may be of the same or different types. For example, one might use a suspended-growth process for oxidizing organic matter and a biofilm process for nitrification. Another example is using a CSTR first to reduce the biodegradable toxicants below a toxic threshold, and then using a plug-flow reactor to achieve high efficiency of removal.

Reactors in parallel are used at most treatment plants to provide redundancy in the system so that some reactors can be out of service for maintenance, while others on a parallel track remain in operation. At larger treatment plants, parallel reactors also must be used, because the total flow to be treated far exceeds the capacity of the largest practical units available. Reactors in parallel also maintain more of a completely mixed nature, compared to the more plug-flow nature of reactors in series.

This section indicates that a few basic reactor systems can be operated in many different manners and connected together in many different ways. Each industrial or municipal wastewater treatment system has individual requirements. Frequently, many different overall system configurations could provide the required treatment. The design of choice for a given location depends not only upon the characteristics of the wastewater and the degree of treatment necessary, but also on local considerations, such as land availability, operator experience, designer experience, construction costs, and labor and energy costs. Municipalities frequently desire plants with higher construction costs and lower operating costs because of the difficulty in maintaining a constant supply of tax dollars for operation. On the other hand, industry generally prefers the opposite, because of the cost to borrow money and the likelihood that they will need to make periodic process changes when wastewater characteristics change. The engineer needs to work closely with the owners to determine the best reactor system to meet the particular needs, desires, and circumstances for each case.

5.2 MASS BALANCES

The *mass balance* is the key to design and analysis of microbiological processes. One type of mass balance is that provided by a balanced chemical equation such as

developed in Chapter 2. An example reproduced below is Equation 2.34 for benzoate
($C_6H_5COO^-$):

$$0.0333\ C_6H_5COO^- + 0.12\ NO_3^- + 0.02\ NH_4^+ + 0.12\ H^+ \rightarrow$$
$$0.02\ C_5H_7O_2N + 0.06\ N_2 + 0.12\ CO_2 + 0.0133\ HCO_3^- + 0.1067\ H_2O$$

This equation indicates that for each 0.0333 mol of benzoate consumed by microorganisms in denitrification, 0.12 mol nitrate is required to serve as an electron acceptor,
and 0.02 mol ammonium serves as a nitrogen source for bacterial growth. From this,
0.02 mol bacteria is produced, along with a defined amount of nitrogen and carbon
dioxide gases, bicarbonate, and water. If the required amounts of nitrate and ammonium are not present in the wastewater, then they must be added to the treatment
process. The 0.02 mol bacteria produced represents a waste product, which we call
sludge, waste biomass, waste biosolids, or *excess biosolids.* The sludge must be
removed from the system and disposed of in some acceptable manner. Knowing
the quantity of waste sludge is essential for designing the sludge-disposal facilities.
Thus, the mass balance given by such stoichiometric equations provides the critical
information on what must be added to and removed from the process.

In reactor design, we also are interested in reaction rates, as they affect the size
of the treatment system. Equations that relate reactor size to reaction rates, reaction
stoichiometry, and the required treatment efficiency also depend on mass balances.
An example of using mass balances for a chemostat system, one of the simplest
treatment systems, was provided in Chapter 3. By following a similar approach, we
can develop relevant equations for all of the biological treatment systems of interest.
The rest of this chapter formalizes the mass-balance approach and provides several
examples.

One of the first elements of a mass balance for a treatment system is a definition
of the treatment system being addressed. For example, adding a settling tank and
a recycle line to return the settled microorganisms expands the chemostat system to
a solids-recycle system, such as activated sludge. Figure 5.2 shows the expanded
system, which also has a separate sludge wasting line.

Next, a *control volume* must be defined. The dashed lines in Figure 5.2 illustrate
three possible control volumes. The upper illustration shows a control volume around
the entire treatment system, the lower left represents one around the reactor only, and
the lower right shows one around the settling tank. In some cases, an appropriate
control volume may be even smaller, such as a small volume within a plug-flow
reactor. The choice of the control volume must be consistent with the goals of the
analysis. The examples that follow illustrate why different choices are appropriate.

Once a control volume is selected, mass balances on components of interest are
made. A component may enter and/or leave the control volume. For example, in
Figure 5.2(a) the component may enter the reactor system only by way of the system
influent stream, but may leave the system either by way of the system effluent stream
or the sludge waste stream. The component may be destroyed or formed within
the reactor system. If one considers a control volume around the reactor alone, as
illustrated in Figure 5.2(b), then a given component can enter via the reactor influent

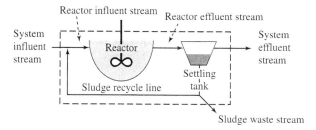

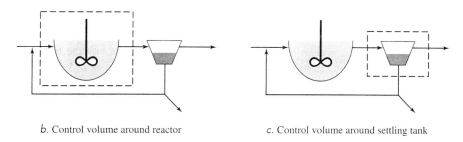

a. Control volume around entire reactor system

b. Control volume around reactor   *c.* Control volume around settling tank

Figure 5.2   Possible control volumes around a CSTR with sludge recycle.

stream and can exit via the reactor effluent stream. In this case, the destruction or production of the component in the reactor alone is considered in the mass balance. Similarly, when the control volume is taken around the settling tank (Figure 5.2(c)), the component enters through the reactor effluent stream and leaves it either by way of the settling tank effluent stream or the sludge recycle line. Any destruction or production reactions in the settling tank are considered in the mass balance.

Mass balances made around different control volumes often lead to different kinds of information about reactor performance. The choice of a control volume depends upon the kind of information being sought. In the development of equations useful for a reactor system, mass balances on several different components of interest and around several different control volumes sometimes are required. In order to develop useful equations that are no more complicated than necessary, simplifying assumptions often are made. The limitations that result from the use of such assumptions must be understood, so that the derived equations are not used for situations where they do not apply.

A very important aspect of mass balances is that each component must have its own mass balance. The components may include the chemical oxygen demand (COD) of the waste stream, which is frequently used as a measure of waste strength; total organic carbon (TOC), another measure of waste strength; biomass; oxygen, a key electron acceptor; nitrate- and ammonium-nitrogen and phosphorus, macronutrients; or others. If one has a balanced stoichiometric equation for the reaction, such as represented by Equation 2.34, the consumption of one component, such as the electron donor, can be used directly for other components, such as electron acceptor consumption or biomass production. A common mistake made by beginners is attempting to make a single mass balance equation that includes changes in mass of more than one component, such as COD and biomass. The rule that must be followed is: One mass balance for each component!

Once a reactor system, a control volume, and the components on which to do the mass balances are selected, then the mass-balance equations can be written. The mass balance always is defined in terms of rates of mass change in the control volume. In word form, that is,

Rate of mass accumulation in control volume =
rate(s) of mass in − rate(s) of mass out + rate(s) of mass generation   **[5.1]**

The rate of mass accumulation always appears on the left side. Accumulation is the total mass of the component in the system, or the product of the volume times the concentration. The rate term takes the general mathematical form of $d(VC)/dt$, in which V is the volume of the control volume, C is the component's concentration, and d/dt is the differential with respect to time.

Mass in and mass out refer to mass that crosses the control-volume boundaries. Generation refers to the formation of the component of interest within the control volume. If generation is negative, then the component is destroyed rather than being formed within the control volume. Some components may be formed by some reactions and destroyed by others. For example, bacterial cells may be produced through consumption of an electron donor or "food" source. Here, the generation rate is positive. Endogenous respiration or predation may also destroy microorganisms, making generation negative. For the terms on the right side, more than one reaction or mechanism may work. Thus, each of the three terms can be plural. However, the accumulation is not plural; only one total mass accumulation exists for a component in a control volume.

Equation 5.1 may take many mathematical forms, depending upon the nature of the control volume, the manner in which mass flows into and out of the control volume, and what kind of reactions generate or destroy the component. The best way to learn how to apply Equation 5.1 is through a series of examples, which follows.

5.3   A BATCH REACTOR

For a batch reactor operated with mixing, as illustrated in Figure 5.1, the control volume consists of the entire reactor. Components in the reactor are distributed uniformly throughout the reactor so that the concentration of any component is the same at any location within the reactor at any time. We consider the case in which the reactor liquid volume, V, does not change with time. Then, only component concentrations change with time, or the rate of mass accumulation equals $V dC/dt$.

We select the components as the bacteria and their rate-limiting substrate, which is most frequently the electron donor, as our choice here. We assume that all other bacterial requirements, such as electron acceptor and nutrients, are sufficiently high in concentration that they impose no limitations on organism growth rate. At time $= 0$, the reactor contains component concentrations of microorganisms (X^0, mg/l) and rate-limiting substrate (S^0, mg/l).

Although changes in microorganism concentration and substrate concentration are interdependent, we must construct a separate mass balance on each of these components using the form of Equation 5.1. We begin with a mass balance for substrate,

Mass rate of substrate accumulation in control volume =

rate of mass in − rate of mass out + rate of mass generation   [5.2]
$= 0$ $= 0$

While the microorganisms are consuming substrate, no substrate is added or removed from the batch reactor. Thus, over this time period the mass of substrate accumulating in the reactor equals the mass of substrate generated within the reactor. On the other hand, the substrate is consumed or destroyed by the microorganisms, and generation has a negative sign. In mathematical form, Equation 5.2 becomes:

$$V \frac{dS}{dt} = V r_{ut}$$ [5.3]

Commonly, the rate of substrate utilization is assumed to following Monod kinetics, as given by Equation 3.6. With this substitution, we obtain

$$V \frac{dS}{dt} = V \left(-\frac{\hat{q} S}{K + S} X_a \right)$$

or,

$$\frac{dS}{dt} = -\frac{\hat{q} S}{K + S} X_a$$ [5.4]

We would like to integrate Equation 5.4 to determine how S changes with time in the reactor, but X_a also changes with time. In order to determine how X_a changes, we construct our second mass balance, this time on microorganisms in the reactor. The format of Equation 5.1 converts to

Mass rate of organism accumulation in control volume =

rate of mass in − rate of mass out + rate of mass generation   [5.5]
$= 0$ $= 0$

With μ being the net specific growth rate of organisms (Equation 3.5), the mathematical form is similar to Equation 5.3:

$$V \frac{dX_a}{dt} = V(\mu X_a)$$ [5.6]

If we assume that the organism growth rate follows Monod kinetics and if decay, as well as growth, is considered, then combining Equation 5.6 with Equation 3.5 gives

$$V \frac{dX_a}{dt} = V \left(\hat{\mu} \frac{S}{K + S} - b \right) X_a$$

or

$$\frac{dX_a}{dt} = \left(\hat{\mu} \frac{S}{K + S} - b \right) X_a \qquad \text{[5.7]}$$

Here, we again see the interdependence between X_a and S, both of which vary with time. Thus, in order to solve for X_a and S as functions of time, we have to consider mass balance equations 5.4 and 5.7 together. We also need initial conditions, which have already been specified by:

$$X_a(0) = X_a^0 \qquad S(0) = S^0 \qquad \text{[5.8]}$$

Due to the nonlinear Monod forms, the system of Equations 5.4, 5.7, and 5.8 cannot be solved analytically. If it is to be solved, it must be done with a numerical solution, which can readily be accomplished with a computer program or spreadsheet. (This type of solution is demonstrated in the website chapter "Complex Systems.") However, an analytical solution can be obtained if organism decay is considered to be negligible (or not very important to the result), in which case b in Equation 5.7 can be taken to equal zero. This is a satisfactory assumption for cases of batch growth in which organism decay is a small factor while the microorganisms are growing rapidly. Ignoring decay will introduce errors whenever the cells do not grow very rapidly, such as after the exponential phase of batch growth and in most continuous processes.

In the absence of decay, the organism concentration at any time equals the initial concentration, X_a^0, plus that which results from substrate consumption during that time, $Y \Delta S$, or:

$$X_a = X_a^0 + Y \Delta S \qquad \text{or} \qquad X_a = X_a^0 + Y(S^0 - S) \qquad \text{[5.9]}$$

By substitution of Equation 5.9 into Equation 5.4, one ordinary differential equation is obtained:

$$\frac{dS}{dt} = -\frac{\hat{q} S}{K + S} \left[X_a^0 + Y(S^0 - S) \right] \qquad \text{[5.10]}$$

This equation can be integrated, subject to the boundary conditions given by Equation 5.8, to yield:

$$t = \frac{1}{\hat{q}} \left\{ \left(\frac{K}{X_a^0 + Y S^0} + \frac{1}{Y} \right) \ln(X_a^0 + Y S^0 - Y S) \right.$$
$$\left. - \left(\frac{K}{X_a^0 + Y S^0} \right) \ln \frac{S X_a^0}{S^0} - \frac{1}{Y} \ln X_a^0 \right\} \qquad \text{[5.11]}$$

It would be desirable to have an equation that explicitly gives S as a function of t, but because of the complexity of the equation, this is not possible. Here, a computer spreadsheet can be very useful for solving for S when t is known.

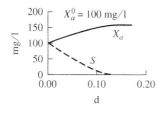 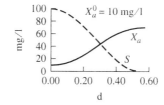 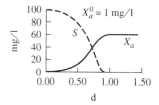

Figure 5.3   Change in S and X_a in a batch reactor with time for different concentrations of X_a^0. Here, $\hat{q} = 10$ mg/mg VSS$_a$-d, $K = 20$ mg/l, $Y = 0.6$ mg VSS$_a$/mg, $b = 0.0$/d, and $S^0 = 100$ mg/l. Note the different time scales.

Figure 5.3 illustrates how the choice of X_a^0 affects bacterial growth and the substrate concentration over the course of the batch reaction. If one adds the highest initial organism concentration (100 mg VSS$_a$/l), substrate is consumed in a much shorter period of time than if the lowest concentration (1 mg/l) is added. For example, the time to reduce the substrate concentration to near zero is reduced by a factor of 8 (from 0.8 d to 0.1 d) with a 100-fold increase in initial organism concentration. For the lowest initial organism concentration, a *lag period* occurs before the onset of significant substrate utilization. This lag reflects the time needed for the small inoculum to grow to a concentration able to consume substrate at a rate that is noticeable. Significant removal of substrate occurs when X_a is about 10 mg VSS$_a$/l. No lag time is in evidence when the starting concentration is 10 mg/l or higher. The increase in microorganism concentration between time zero and just after the substrate is depleted is the same in all cases and equals $Y S^0$, which in this case equals 60 mg/l.

5.4   A CONTINUOUS-FLOW STIRRED-TANK REACTOR WITH EFFLUENT RECYCLE

The CSTR, which is the same as a chemostat, was thoroughly discussed in Chapter 3. It differs from a batch reactor in two profound ways. First, it has a continuous flow in and out, whereas the batch reactor has no flows in or out. Second, the CSTR can reach a steady state, where the change of mass accumulation becomes zero, or $V dC/dt = 0$. Steady state is not a relevant concept for a batch reactor, as long as components are reacting. Since the steady-state mass balances for the chemostat were thoroughly developed in Chapter 3, they need not be repeated here. The reader is advised to review the set up and solution of the mass balances in Chapter 3 in light of the information on mass balances presented in this chapter. One new case is presented here to illustrate a feature not discussed in Chapter 3 and as preparation for treating more complex reactor systems.

A CSTR with effluent recycle is illustrated in Figure 5.4. The difference between this case and the normal CSTR is that some of the effluent stream, containing effluent active microorganisms and substrate, is recycled back at a flow rate Q_r and introduced

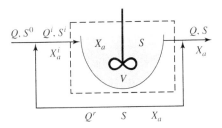

Figure 5.4   CSTR with effluent recycle

back into the reactor. The question is, how does this affect reactor performance? We will see that it does not affect reactor performance at all. In order to demonstrate this, we need to select the control volume, which might be taken around the entire system or just around the reactor itself as is the case in Figure 5.4. We can readily show that, in this case, it makes no difference to the results which is used as the control volume. But first, we need to make mass balances around the point where the influent and recycle flows come together in order to determine Q^i, S^i, and X^i. A mixing point has no reaction, and so the rate of mass flow into the mixing point equals the rate of flow out, or:

$$QS^0 + Q^r S = Q^i S^i \quad \text{and} \quad QX_a^0 + Q^r X_a^r = Q^i X_a^i$$

From which,

$$S^i = \frac{QS^0 + Q^r S}{Q^i} \quad \text{and} \quad X_a^i = \frac{QX_a^0 + Q^r X_a^r}{Q_i} \qquad \textbf{[5.12]}$$

Also,

$$Q^i = Q + Q^r \qquad \textbf{[5.13]}$$

Now, if we do the mass balance for substrate around the reactor control volume depicted in Figure 5.4, we obtain for the steady-state case

$$0 = Q^i S^i - Q^i S + r_{ut} V \qquad \textbf{[5.14]}$$

Then, by making appropriate substitutions from Equations 5.12 and 5.13 and simplifying we obtain

$$0 = Q(S^0 - S) + r_{ut} V \qquad \textbf{[5.15]}$$

Equation 5.15 is identical to Equation 3.15, the case for a chemostat without recycle. Thus, simple recycle for a CSTR does not change substrate removal compared with that obtained without recycle. A mass balance on microorganisms can be performed similarly, and the result is the same: Organism concentrations within the reactor and in the reactor effluent are not affected by effluent recycle, since the same mass flow that leaves the reactor returns to the reactor. We will see, however, that this is not the case with a plug-flow reactor, where concentrations are not the same everywhere.

5.5   A PLUG-FLOW REACTOR

With a PFR, the substrate and active-organism concentrations vary over the length of the reactor. Thus, the appropriate control volume is an incremental segment along the flow path in the reactor, as illustrated in Figure 5.5. Mass balances on substrate and on active microorganisms, following Equation 5.1, lead to:

Substrate

$$\Delta V \frac{\Delta S}{\Delta t} = QS - Q(S + \Delta S) + r_{ut} \Delta V \qquad \text{[5.16]}$$

Active microorganisms

$$\Delta V \frac{\Delta X_a}{\Delta t} = QX_a - Q(X_a + \Delta X_a) + r_{\text{net}} \Delta V \qquad \text{[5.17]}$$

where r_{ut} and r_{net} are the reaction rates for substrate (Equation 3.6) and active organisms (Equation 3.8). We consider the steady-state case, for which influent flow rate, substrate concentration, and active-organism concentration do not change with time. Thus, the left sides of Equations 5.16 and 5.17 are zero. If the reactor cross-sectional area (A) is constant throughout the reactor, then the area of the control volume is $A = \Delta V / \Delta z$, and the velocity of flow within the reactor is $u = Q/A$. With these substitutions and for steady-state conditions, Equations 5.16 and 5.17 become:

Substrate at steady state

$$u \frac{\Delta S}{\Delta z} = r_{ut} \qquad \text{[5.18]}$$

Active microorganisms at steady state

$$u \frac{\Delta X_a}{\Delta z} = r_{\text{net}} \qquad \text{[5.19]}$$

If we now let Δz approach zero and assume that the Monod reaction applies for substrate utilization (Equation 3.6) and that organism net growth represents growth and decay (Equation 3.8), then Equations 5.18 and 5.19 become:

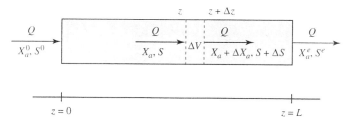

Figure 5.5   Control volume (ΔV) within a plug-flow reactor

Substrate at steady state with Monod kinetics

$$u\frac{dS}{dz} = -\hat{q}\frac{S}{K+S}X_a \qquad [5.20]$$

Active microorganisms at steady state with growth and decay

$$u\frac{dX_a}{dz} = Y\hat{q}\frac{S}{K+S}X_a - bX_a \qquad [5.21]$$

This series of equations cannot be solved analytically, and so numerical approaches must be used. However, if we again ignore organism decay ($b = 0$), then an analytical solution is possible. First, we combine Equations 5.20 and 5.21 to eliminate the Monod terms:

$$u\frac{dX_a}{dz} = -uY\frac{dS}{dz} \qquad [5.22]$$

We cancel u and take integrals to obtain

$$\int_{X_a^0}^{X_a} dX_a = -Y\int_{S^0}^{S} dS \qquad [5.23]$$

Integrating gives

$$X_a = X_a^0 + Y(S^0 - S) \qquad [5.24]$$

Substituting Equation 5.24 into Equation 5.20 gives a differential equation with only two variables, S and z:

$$u\frac{dS}{dz} = -\hat{q}\frac{S}{K+S}\left[X_a^0 + Y(S^0 - S)\right] \qquad [5.25]$$

The ratio dz/u has dimensions of time and equals the differential time, dt, for an element of water to move along the reactor a distance dz. Substituting dt for dz/u in Equation 5.25 yields a differential equation that is exactly the same as Equation 5.10 for the batch reactor. Indeed, integration of Equation 5.25 results in an equation almost identical to Equation 5.11. The only difference is that t for the batch reactor is replaced by z/u in the integrated form for the plug-flow reactor:

$$\frac{z}{u} = \frac{1}{\hat{q}}\left\{\left(\frac{K}{X_a^0 + YS^0} + \frac{1}{Y}\right)\ln\left\{X_a^0 + YS^0 - YS\right\}\right.$$
$$\left. - \left(\frac{K}{X_a^0 + YS^0}\right)\ln\frac{SX_a^0}{S^0} - \frac{1}{Y}\ln X_a^0\right\} \qquad [5.26]$$

We obtain an expression for the effluent concentration from the batch reactor by letting $z = L$. We also note that L/u is equal to V/Q, the hydraulic detention time, θ, for the reactor. With these substitutions, the following solution is identical to Equation 5.11 with θ replacing t:

$$\theta = \frac{1}{\hat{q}}\left\{\left(\frac{K}{X_a^0 + YS^0} + \frac{1}{Y}\right)\ln(X_a^0 + YS^0 - YS^e)\right.$$
$$\left. - \left(\frac{K}{X_a^0 + YS^0}\right)\ln\frac{S^e X_a^0}{S^0} - \frac{1}{Y}\ln X_a^0\right\} \qquad \textbf{[5.27]}$$

We thus see that a PFR works exactly like a batch reactor. In practice, however, it is difficult to operate a PFR according to the assumptions involved. A PFR has no mixing or short-circuiting of the fluid along the flow direction. This is impossible to achieve in a real continuous-flow reactor. At a minimum, wall effects slow the fluid near the wall boundaries relative to the velocity near the middle. Aeration or mixing to keep the biomass in suspension introduces a large amount of mixing in all directions. Methods to achieve as much of a plug-flow character as possible include using a very long, narrow reactor and using many reactors in series. These measures help somewhat, but some mixing and short-circuiting are inevitable. If achieving the reaction kinetics represented by Equation 5.27 is of paramount importance, a batch reactor is the prudent choice, although it presents its own problems. For example, time is required to fill and empty a batch reactor, time that might otherwise be used for treatment. In order to minimize downtime, a batch reactor can be operated while it is filling.

5.6   A PLUG-FLOW REACTOR WITH EFFLUENT RECYCLE

One major difference between a PFR and a CSTR is that microorganisms must be introduced at the influent end of the PFR. If no organisms are introduced, then no microorganisms are present to carry out substrate removal, and the system fails to do any treatment. One way to introduce microorganisms into a PFR is to use effluent recycle. In this manner, a portion of the microorganisms in the effluent is brought back to the influent stream. The impact of this can be seen by constructing a mass balance similar to that constructed for the case of a PFR with recycle. A PFR with effluent recycle is depicted in Figure 5.6. As with the CSTR, mass balances to determine the flow rate and concentrations of substrate and microorganisms entering the reactor yield equations identical to Equations 5.12 and 5.13, that is,

$$S^i = \frac{QS^0 + Q^r S}{Q^i} \quad \text{and} \quad X_a^i = \frac{QX_a^0 + Q^r X_a^r}{Q^i}$$

and

$$Q^i = Q + Q^r$$

It should be noted that $X^r = X^e$ and $S = S^e$.

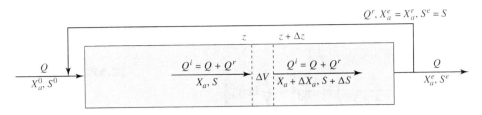

Figure 5.6   A plug-flow reactor (PFR) with effluent recycle

The reactions occurring throughout the reactor are obtained in exactly the same manner as with the simple PFR: by conducting a mass balance around a control volume taken within the reactor itself, as illustrated in Figure 5.6. Here, Equations 5.16 to 5.27 apply, but with Q^0, X^0, and S^0 now taken to equal Q^i, X^i, and S^i. For the case in which $b = 0$, an integrated form of an equation relating the effluent concentrations as a function of detention time can be obtained. In this manner, the effluent concentration of X_a can be obtained:

$$X_a^e = X_a^0 + Y(S^0 - S^e) \qquad \textbf{[5.28]}$$

With this, the series of equations can be combined and integrated to give:

$$\frac{V}{Q^i} = \frac{1}{\hat{q}} \left\{ \left(\frac{K}{X_a^i + Y S^i} + \frac{1}{Y} \right) \ln(X_a^i + Y S^i - Y S^e) \right.$$
$$\left. - \left(\frac{K}{X_a^i + Y S^i} \right) \ln \frac{S^e X_a^i}{S^i} - \frac{1}{Y} \ln X_a^i \right\} \qquad \textbf{[5.29]}$$

Of interest is the impact of recycle on the performance of a PFR. We define the recycle ratio R, as

$$R = \frac{Q^r}{Q} \qquad \textbf{[5.30]}$$

and the detention time, θ, as

$$\theta = \frac{V}{Q} = \frac{V(1+R)}{Q^i} \qquad \textbf{[5.31]}$$

The above series of equations can be solved using a spreadsheet to determine S^e as a function of θ and R. The results, illustrated in Figure 5.7, were obtained in this manner and using rate coefficients typical of aerobic treatment of organic wastewaters. The effluent concentration is illustrated using an arithmetic scale in the upper graph and a logarithmic scale in the lower graph. The upper graph in Figure 5.7 illustrates that, for each individual recycle rate, there is a detention time below which the effluent concentration equals the influent concentration; in other words, no treatment takes place. This detention time is equivalent to the washout detention time discussed in Chapter 3, or $\theta_x^{\min}$ as given by Equation 3.26 for the CSTR. Using the influent substrate concentration and kinetic coefficients illustrated in Figure 5.7, the equivalent $\theta_x^{\min}$ for the CSTR is 0.2 days, or close to the value indicated for

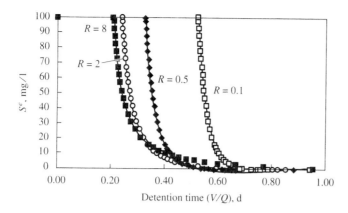

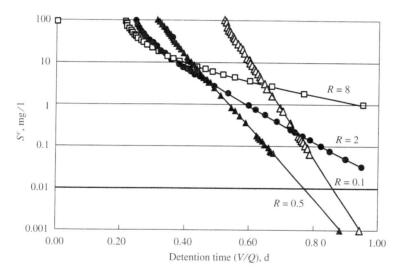

Figure 5.7   Effect of recycle ratio (R) and hydraulic detention time (θ) on effluent substrate concentration for a plug-flow reactor with effluent recycle. Case considered has $S^0 = 100$ mg/l, $Y = 0.6$ mg/mg, $K = 20$ mg/l, $\hat{q} = 10$ mg/mg-d, and $b = 0$.

$R = 8$ for plug-flow with recycle. In effect, as R increases in a plug-flow reactor with high recycle, the influent concentration to the reactor (S^i) decreases. When R approaches infinity, the PFR with recycle becomes identical to a CSTR in theoretical performance. Indeed, the performance with $R = 8$ is very close to that of a CSTR. The washout detention time is larger when R is less than 8.

Figure 5.7 helps illustrate the relative benefits of a CSTR and a plug-flow reactor with recycle. Considering that the PFR with recycle case with $R = 8$ is very similar to a CSTR's performance, one can see that the CSTR improves reliability when the system is operated near washout. This would be a good choice if contaminant removal of 80 to 90 percent were satisfactory. However, if contaminant removal of 99.9 percent were required, the lower graph shows that operation as a PFR with a low recycle ratio is much more desirable. With the recycle system, there is an optimal recycle ratio that provides reasonably reliable performance at low detention times and highly efficient contaminant removal. An important lesson is that effluent recycle with a CSTR does not change system performance, but with a PFR, recycle is essential and has a great impact on performance.

As already noted, true plug-flow operation in a continuous flow system is not possible, as mixing always occurs in the direction of flow. The actual contaminant removal in an operational plug-flow reactor with recycle lies somewhere between that of the theoretical removal and that of a CSTR.

5.7   REACTORS WITH RECYCLE OF SETTLED CELLS

One of the most widely used suspended-growth reactors employs microorganism recycle from a settling tank, as depicted in Figure 5.2. The activated sludge treatment system used at a large majority of the municipal treatment plants in the United States is of this type, as are many used for the treatment of industrial wastewaters. Although many modifications to this process are described in Chapter 6, we develop the basic mass balances for a CSTR and a PFR with settling and microorganism recycle. The primary advantage of settling and recycle of microorganisms is that a much smaller reactor volume is required, because the biomass is captured and built up to a much higher concentration than is possible in a normal CSTR or PFR. As noted in Equation 3.6, the substrate-utilization rate is directly proportional to the active microorganism concentration. Any method that increases the concentration of microorganisms in the reactor also increases the volumetric reaction rate and, in this manner, decreases the required reactor volume. Countering the advantage of a smaller reactor volume is the cost of the settler and the recycling system. The comparison is between the cost of the smaller reactor plus a settler versus a larger reactor without a settler.

5.7.1   CSTR WITH SETTLING AND CELL RECYCLING

We begin with the simplest case, a CSTR with settling and microorganism recycle, as illustrated in Figure 5.2. The results we want can be obtained by constructing a mass balance around the whole reactor, as shown in the upper illustration. Figure 5.8 identifies in detail the items of interest in our mass balance. Since the settling tank is not likely to be 100-percent efficient in capturing microorganisms, some may be present in the effluent at concentration X_a^e. The microorganisms that settle in the settling tank form a thickened biological sludge that is removed for recycle back to the reactor. The concentration of microorganisms in this recycle line is X_a^r. Because we obtain a net growth of microorganisms, the net growth—called excess sludge or waste sludge—is removed from the system for subsequent sludge treatment and disposal.

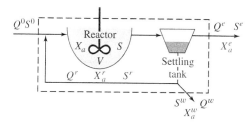

Figure 5.8   CSTR with sludge recycle

This sludge wasting of microorganisms is from the recycle line at a flow rate Q^w and a concentration X_a^w. Microorganism wasting may also occur directly from the reactor itself, but the approach we take is far more common.

In order to develop mass balances for the reactor, we again need some simplifying assumptions. Assumptions that we make here as a first exercise are: (1) biodegradation of the substrates takes place in the reactor only, no biological reactions take place in the settling tank, and the biomass in the settler is insignificant; (2) no active microorganisms are in the influent to the reactor ($X_a^0 = 0$); and (3) the substrate is soluble so that it cannot settle out in the settling tank. With these assumptions, we proceed with a mass balance for microorganisms around the control volume in Figure 5.8:

$$V\frac{dX_a}{dt} = 0 - (Q^e X_a^e + Q^w X_a^w) + [Y(-r_{ut})V - bX_aV] \qquad \textbf{[5.32]}$$

Likewise, a mass balance for substrate gives:

$$V\frac{dS}{dt} = Q^0 S^0 - (Q^e S^e + Q^w S^w) + r_{ut}V \qquad \textbf{[5.33]}$$

Equations 5.32 and 5.33 are general and can be used to describe the operation of the reactor under nonsteady- or steady-state conditions. In order to illustrate the general principles of operation of the treatment system, we consider here only the steady-state case. That is, we assume that the reactor has been operating continuously for some time with all flows and influent concentrations constant over time. At steady state, the changes in accumulation are zero, or:

$$\text{Steady-state conditions} \quad V\frac{dX_a}{dt} = 0 \quad \text{and} \quad V\frac{dS}{dt} = 0 \qquad \textbf{[5.34]}$$

We now consider the definition of solids retention time (θ_x) given by Equation 3.22, which is repeated here:

$$\theta_x = \frac{\text{active biomass in the system}}{\text{production rate of active biomass}}$$

Based upon our assumption for this exercise that no active biomass is in the settling tank, the active biomass in the system is only that in the reactor, or X_aV. Under steady-state conditions, the production rate of active biomass must just equal its rate of removal from the control volume, either in the effluent ($Q^e X_a^e$) or in the waste sludge stream ($Q^w X_a^w$). Thus,

$$\theta_x = \frac{X_a V}{Q^e X_a^e + Q^w X_a^w}$$ [5.35]

We can now rearrange Equation 5.32 for the steady-state case to give:

$$\frac{Q^e X_a^e + Q^w X_a^w}{X_a V} = \frac{Y(-r_{ut})}{X_a} - b$$ [5.36]

Seeing the similarity between the left side of Equation 5.36 and the right side of Equation 5.35, we make the appropriate substitution to give an important result:

$$\frac{1}{\theta_x} = \frac{Y(-r_{ut})}{X_a} - b$$ [5.37]

Equation 5.37 is general for a CSTR with settling and recycle and can be applied whatever the form of the biological reaction, r_{ut}, may be. If we assume that it takes the usual form of the Monod reaction (Equation 3.6), then we obtain the following expression:

$$\frac{1}{\theta_x} = Y \frac{\hat{q} S}{K + S} - b$$ [5.38]

We solve this equation explicitly for S,

$$S = K \frac{1 + b\theta_x}{\theta_x (Y\hat{q} - b) - 1}$$ [5.39]

Equation 5.39 is identical to Equation 3.24, which was developed for the chemostat without settling and recycle. So then, what is unique about the CSTR with settling and microorganism recycle? The answer is that the retention time of the microorganisms in the system (θ_x) is separated from the hydraulic detention time (θ). Thus, one can have a large θ_x, in order to obtain high efficiency of substrate removal, and at the same time have a small θ, which translates into a small reactor volume. For example, in a normal activated sludge biological treatment system used to treat municipal wastewaters, a typical solids retention time is 4 to 10 d, while the hydraulic detention time is only 4 to 10 h. Thus, while operating at the same treatment efficiency for substrate, the reactor can be one-twenty-fourth the volume that it would otherwise be if it were designed as a CSTR system without settling and recycle.

Now, we consider the concentration of active microorganisms in the reactor. We first solve Equation 5.37 for X_a,

$$X_a = \theta_x \frac{Y(-r_{ut})}{1 + b\theta_x}$$ [5.40]

Next, we return to Equation 5.33, the mass-balance equation for substrate, and apply steady-state (Equation 5.34) to obtain r_{ut} as a function of S,

$$-r_{ut} = \frac{Q^0 S^0 - Q^e S^e - Q^w S^w}{V}$$ [5.41]

For the CSTR, we see that the substrate concentration in the reactor, S, is equal to the concentration in the effluent S^e and in the waste sludge line, S^w, since no reaction occurs in the settling tank. Also, through mass balance, $Q^e + Q^w = Q^0$. With these substitutions into Equation 5.41, we obtain,

$$-r_{ut} = \frac{Q^0(S^0 - S)}{V} = \frac{(S^0 - S)}{\theta} \qquad \textbf{[5.42]}$$

Like Equation 5.37, Equation 5.42 is another general representation, this time of the utilization rate in terms of reactor characteristics and performance. Substituting Equation 5.42 into Equation 5.40 gives:

$$X_a = \frac{\theta_x}{\theta} \frac{Y(S^0 - S)}{1 + b\theta_x} \qquad \textbf{[5.43]}$$

Equation 5.43 indicates that the active biomass concentration in the reactor depends on the ratio of solids retention time to the hydraulic detention time, θ_x/θ. We call this important ratio the *solids-concentration ratio*. We might note that for a CSTR without settling and recycle, $\theta_x/\theta = 1$, and Equation 5.43 becomes identical to Equation 3.25. For an activated sludge system treating municipal wastewaters with a solids-concentration ratio of about 24 and the same θ_x, the concentration of active organisms in the reactor is about 24 times higher than it would be without biomass recycling.

Also crucially important is the quantity of waste sludge (or biosolids) produced in the system, as it must be removed continually in order to maintain the steady-state conditions. In addition, the waste sludge must be properly disposed of. Therefore, the rate of sludge wasting is essential for operating the treatment system and for determining the total cost of construction and operation for the system.

We can see from Figure 5.8 that, at steady state, the mass rate of *active biomass production* (r_{abp}, M/T) must just equal the rate at which biomass leaves the system from the effluent stream and the waste stream:

$$r_{abp} = Q^e X_a^e + Q^w X_a^w \qquad \textbf{[5.44]}$$

Substituting this equation into Equation 5.35 and rearranging yields:

$$r_{abp} = \frac{X_a V}{\theta_x} \qquad \textbf{[5.45]}$$

With the addition of Equation 5.45, we now have a series of equations that allow us to design a treatment system consisting of a CSTR with settling and recycle. The important equations are summarized in Table 5.2.

The equations in Table 5.2 can be used for a CSTR without a settler by letting $\theta_x = \theta$. Because inert biomass and total volatile solids are particles, like active biomass, the concentrations X_i and X_v take the same form as for a chemostat, but are multiplied by the solids-concentration factor, as shown in Equations 5.46 and 5.47. The minimum value of the mean cell residence time (θ_x^{min}) and its limiting value $[\theta_x^{min}]_{lim}$ are identical to the case without settling (Equations 3.26 and 3.27) and are listed in Table 5.2. The total biological-solids production rate is analogous to Equation 5.45 and is shown as Equation 5.48 in Table 5.2. The student should be

Table 5.2   Summary of applicable equations for a CSTR with settling and recycle of microorganisms (operating at steady state, treating a soluble substrate, and with no input of active biomass)

Hydraulic Detention Time (θ):

$$\theta = \frac{V}{Q^0} \qquad \textbf{[3.20]}$$

Solids Retention Time, SRT (θ_x):

$$\theta_x = \frac{X_a V}{X_a^e Q^e + X_a^w Q^w} \qquad \textbf{[5.35]}$$

SRT at which microorganism washout results (θ_x^{min}), and the limit thereto:

$$\theta_x^{min} = \frac{K + S^0}{S^0(Y\hat{q} - b) - Kb} \qquad S \to S^0 \qquad \textbf{[3.26]}$$

$$\left[\theta_x^{min}\right]_{lim} = \frac{1}{Y\hat{q} - b} \qquad S \to \infty \qquad \textbf{[3.27]}$$

Reactor or Effluent Substrate Concentration ($S = S^e$):

$$S = K \frac{1 + b\theta_x}{\theta_x(Y\hat{q} - b) - 1} \qquad \textbf{[5.39]}$$

Reactor Minimum Substrate Concentration (S_{min}):

$$S_{min} = K \frac{b}{Y\hat{q} - b} \qquad \theta_x \to \infty \qquad \textbf{[3.28]}$$

Reactor Active Microorganism Concentration (X_a):

$$X_a = \theta_x \frac{X(-r_{ut})}{1 + b\theta_x} \qquad \textbf{[5.40]}$$

$$X_a = \frac{\theta_x}{\theta} \frac{Y(S^0 - S)}{1 + b\theta_x} \qquad \textbf{[5.43]}$$

Reactor Inert Microorganism Concentration (X_i):

$$X_i = \frac{\theta_x}{\theta} \left[X_i^0 + X_a(1 - f_d)b\theta \right] \qquad \textbf{[5.46]}$$

Reactor volatile suspended solids concentration (X_v):

$$X_v = X_i + X_a$$

$$X_v = \frac{\theta_x}{\theta} \left[X_i^0 + \frac{Y(S^0 - S)(1 + (1 - f_d)b\theta_x)}{1 + b\theta_x} \right] \qquad \textbf{[5.47]}$$

Active Biological Sludge Production Rate (r_{abp}):

$$r_{abp} = \frac{X_a V}{\theta_x} \qquad \textbf{[5.45]}$$

Total Biological Solids Production Rate (r_{tbp}):

$$r_{tbp} = \frac{X_v V}{\theta_x} \qquad \textbf{[5.48]}$$

---

able to derive these equations using mass balance as performed above and using the definitions in Chapter 3.

One aspect of using the θ_x for design of a CSTR that many find baffling at first is that the effluent concentration, S, as given by Equations 3.24 and 5.39, is independent of the influent concentration S^0. Only one operational variable affects S, and it is θ_x. All other parameters in the equations are coefficients. How can this be? Why is it that when one operates a CSTR at a constant θ_x, the effluent concentration remains the same, regardless of the influent concentration? The answer is related to the fact that as the influent concentration increases, so does the concentration of active microorganisms in the reactor (see Equations 3.25 and 5.43). The increased biomass that results is sufficient to consume the additional substrate that is added to the reactor. Another way to view this is that the organisms' growth rate (μ) and θ_x are equal to the inverse of each other. By maintaining θ_x constant, μ is also held constant. If μ is constant, then S, the concentration to which the active microorganisms are exposed, must be constant, since growth rate is a direct function of S (Equation 3.9).

For those who are still troubled by this concept, we can develop another equation without using θ_x as the master variable. We proceed this time by considering the control volume around the reactor, as illustrated in Figure 5.4. At steady state, a mass balance for substrate leads to

$$V\frac{dS}{dt} = Q^i S^i - Q^i S + r_{ut} V = 0 \qquad \text{[5.49]}$$

If Monod kinetics apply, then substituting Equation 3.6 and rearranging give

$$\frac{\hat{q} S}{K + S} X_a = \frac{Q^i (S^i - S)}{V} \qquad \text{[5.50]}$$

We can define V/Q^i to equal the hydraulic detention time in the reactor itself (θ_r) and solve Equation 5.50 for S. Furthermore, we can do a similar mass balance around the entire reactor in Figure 5.4, including the recycle line. We would obtain

$$S = \frac{S^i}{1 + \dfrac{\hat{q} X_a \theta_r}{K + S}} = \frac{S^0}{1 + \dfrac{\hat{q} X_a \theta}{K + S}} \qquad \text{[5.51]}$$

Here, we see that S is directly proportional to S^i or S^0, providing X_a remains constant. Actually though, we have not solved the above equations explicitly for S, as S is also present in the denominator on the right side of the equations. Frequently, with highly efficient wastewater treatment, S is much less than K, so that on the right side S can be eliminated:

$$S = \frac{S^i}{1 + \dfrac{\hat{q} X_a \theta_r}{K}} = \frac{S^0}{1 + \dfrac{\hat{q} X_a \theta}{K}} \qquad (S << K) \qquad \text{[5.52]}$$

The result indeed shows S to be exactly proportional to S^i and to S^0, a result that should be satisfying to those who are not comfortable with the use of the θ_x concept. The important point is that either approach is based on the same principles. Which approach is more useful in a given situation? Equations similar to Equation 5.51 and 5.52 are frequently used to design biological treatment systems. One major problem with using Equation 5.51 in practice, however, is that X_a is very difficult, and often impossible, to measure with the usual wastewaters that contain many other forms of suspended solids. This is not a problem when Equations 3.24 or 5.39 are used, since prior knowledge of X_a is not needed. The key here is to control θ_x, which can be achieved through controlling the hydraulic detention time in a system without cell settling and recycle, or by controlling the rate at which suspended solids leave the system (Equation 5.35).

5.7.2 EVALUATION OF ASSUMPTIONS

Now we return to the assumptions we made in the development of Equations 5.32 to 5.52. The first assumption was that no biological reactions take place in the settling tank and that no significant biomass is present in that tank. Is either part of this assumption acceptable for the design of real treatment systems? If not, what other assumption should be made? The answer is that it depends on the particular circumstances; no general, a priori answer can be given.

The settling tank often contains a considerable mass of microorganisms; indeed, it might be as much as in the reactor. Clearly, having a large amount of biomass in the settling tank affects the total biomass accumulation, which is represented by $X_a V$ in the exercise. The assumption that no biological reactions occur in the settling tank must be considered at the same time. If the bacteria in the settling tank do not consume substrate, grow, or decay, then they do not affect the mass balances on substrate and biomass. Quantitatively, Equations 5.32 to 5.52 are still valid when no reactions occur in the settling tank.

If the microorganisms in the settling tank grow or decay significantly, the mass balances given in the exercise are in error, as are many of the subsequent equations through 5.52. For illustration, we make the other extreme assumption: that is, we assume that the rates of substrate utilization, biomass synthesis, and biomass decay are the same in the settling tank as in the reactor. Further, we assume that the average biomass concentration in the settling tank is equal to the biomass concentration in the reactor. The effect on the mass balances is that the volume term V in Equations 5.32 to 5.52 includes the combined volumes of the reactor and the settler. Hence, the solids retention time is still given by Equation 5.35, as long as V in that equation equals $V_{\text{reactor}} + V_{\text{settler}}$. The Table 5.2 equations all apply by making the simple substitution for V in all instances in which it occurs:

Is one assumption about system volume superior? The answer is not clear. The truth probably lies somewhere between the two extremes. For example, organism growth due to substrate utilization may occur to some extent in the settler, as there is some substrate carryover from the reactor. However, the carryover likely is small compared to the substrate-input rate to the reactor. How about biomass decay? It probably continues in the settler at a rate similar to that in the reactor, providing that the electron acceptor required for decay is present. In an aerobic system, the acceptor is oxygen, which can become depleted in the settler.

We consider a third case in which microorganism growth from substrate utilization is zero, but the decay rate remains the same as in the reactor. Applying these "between the extremes" assumptions to the mass balances results in equations similar to those in Table 5.2, but with b increased to account for the decay occurring in the settling tank. If we assume that the average biomass concentration in the settling tank is equal to the biomass concentration in the reactor, the effect is that all the equations in Table 5.2 remain valid with $V = V_{\text{reactor}}$, but an effective decay coefficient (b_{eff}) is required. The effective decay coefficient equals b multiplied by the ratio of the total volume divided by the reactor volume: $b_{\text{eff}} = b(V_{\text{reactor}} + V_{\text{settler}})/V_{\text{reactor}}$.

The second assumption made in the exercise is that the incoming wastewater contains no active biomass. Because most wastewaters contain bacteria, we might be tempted to conclude that active biomass is input to the treatment system. Normally, this conclusion is not warranted, and the original assumption is accurate. In many cases, the bacteria carried with the incoming water are not the same microorganisms that are selected and accumulated in the biological-treatment system. For example, enteric bacteria contained in sewage are not going to be important in an aerobic process operated at temperatures far below the temperature of the human digestive systems, under aerobic conditions, and with a long solids retention time. In such a case, the input bacteria become part of the organic substrate, or the BOD_L.

The third assumption of the exercise is that the substrate is soluble and does not settle out with the solids in the settling tank. In many situations, the input BOD_L is comprised of a significant fraction of BOD_L that is suspended. In order for the suspended BOD_L to be utilized, it must be hydrolyzed. Nearly complete hydrolysis of suspended BOD_L makes the assumption valid. Complete hydrolysis is possible when the solids retention time is long enough. For short SRTs (e.g., 2 d, Example 3.6), hydrolysis is incomplete, and the assumption of a soluble substrate is inaccurate. When the input BOD_L contains a significant fraction of suspended BOD_L and hydrolysis is incomplete, suspended BOD_L and its hydrolysis kinetics must be included in the mass-balance equations.

This discussion reflects the normal fact that models of complex processes are simplifications of reality. They contain simplifying assumptions that involve uncertainties. However, other uncertainties are inherent to the design process. These include predictions of future waste flow rates and composition, as well as changes in temperature and other characteristics of the wastewater. The prudent engineer generally accounts for these uncertainties by making appropriately conservative choices in design. However, overly conservative design comes at a financial cost. Financial risk must be weighed against the risk of system failure in performance. In order to balance these risks well, the engineer must understand the assumptions made and the limitations in the development of the models used for design.

5.7.3 PLUG-FLOW REACTOR WITH SETTLING AND CELL RECYCLE

Suspended-growth biological treatment systems often have significant plug-flow character, especially if they are large and use long, narrow tanks. For the plug-flow reactor, Equation 5.29 applies, but requires a value of X_a^i, which is difficult to obtain. However, the concentration of microorganisms maintained in the reactor is generally quite high with settling and recycle, and the change in microorganism concentration from the inlet to the outlet of the reactor tends to be small. In this case, we can assume X_a to be constant ($\bar{X}_a$) throughout the reactor, making integration of Equation 5.20 for the steady-state case much easier. The result is

$$\theta_r = \frac{1}{\hat{q}\bar{X}_a}\left[K\ln\left(\frac{S^i}{S}\right) + (S^i - S)\right] \qquad \textbf{[5.53]}$$

We would like to relate treatment efficiency to θ_x, since it is much easier to control θ_x than it is to measure X_a. We construct a mass balance for active microorganisms around the entire reactor, which provides the same results as for the CSTR with settling and recycle, repeated here for convenience:

$$\frac{1}{\theta_x} = \frac{Y(-\bar{r}_{ut})}{\bar{X}_a} - b$$

Similarly, Equations 5.42 and 5.43 also apply to the plug-flow case,

$$-\bar{r}_{ut} = \frac{Q^0(S^0 - S)}{V} = \frac{(S^0 - S)}{\theta}$$

$$\bar{X}_a = \frac{\theta_x}{\theta}\frac{Y(S^0 - S)}{1 + b\theta_x}$$

where we recognize that $\bar{r}_{ut}$ and $\bar{X}_a$ are average values for the reactor.

Substituting Equation 5.42 into 5.37, we obtain

$$\frac{1}{\theta_x} = \frac{Y Q^0(S^0 - S)}{\bar{X}_a V} - b \qquad \textbf{[5.54]}$$

Recognizing that $\theta_r = V/(Q^0 + Q^r)$, substituting this value into Equation 5.53, solving for $\bar{X}_a V$, and then substituting this into Equation 5.54 give

$$\frac{1}{\theta_x} = \frac{\hat{q}Y(S^0 - S)}{(S^0 - S) + eK} - b \qquad \textbf{[5.55]}$$

in which

$$e = (1 + R)\ln[(S^0 + RS)/(1 + R)S] \qquad \textbf{[5.56]}$$

When $R < 1$, e approximately equals $\ln(S^0/S)$, so that Equation 5.55 becomes

$$\frac{1}{\theta_x} = \frac{\hat{q}Y(S^0 - S)}{(S^0 - S) + K\ln\dfrac{S^0}{S}} - b, \qquad R < 1 \qquad \textbf{[5.57]}$$

Similarities between the plug-flow Equations 5.55 and 5.57 and the CSTR Equation 5.38 are apparent. However, in the plug-flow case, θ_x depends on S^0 and S, while S^0 is not involved with a CSTR.

One of the outcomes of using Equations 5.55 or 5.57 is that the computed value of S becomes vanishingly small when θ_x is not close to the washout value [θ_x^{min}]. This means that the biomass is in positive growth near the inlet of the plug-flow reactor, but it is in negative growth (decay) near the outlet end. Thus, the specific growth rate calculated as the reciprocal of θ_x is an average value for the whole reactor. The very low S value, less than $S_{\min}$, is possible because the cells are growing faster than the average μ near the inlet, and this balances the negative μ near the outlet.

5.8 USING ALTERNATE RATE MODELS

The last parts of Chapter 3 describe a range of alternative rate expressions that can be used for r_{ut} instead of the Monod expression, Equation 3.6. For example, when competitive inhibition occurs, Equation 3.62, along with 3.66, should be used. Or, when hydrolysis controls the removal of a particulate substrate, Equation 3.57 may be appropriate. When the situation requires a different rate expression, the correct form of r_{ut} replaces the Monod form in the mass-balance equations, such as Equation 5.3 for a batch reactor, Equation 3.15 for a CSTR, and Equation 5.18 for a plug-flow reactor. In addition, the same rate form should be used for new biomass synthesis, since $\mu_{\mathrm{syn}} = -Yr_{ut}$.

5.9 LINKING STOICHIOMETRIC EQUATIONS TO MASS-BALANCE EQUATIONS

Chapter 2 presents procedures for writing balanced stoichiometric equations for biological reactions. The mass-balance equations developed so far in this chapter relate only to suspended solids and substrate. We want to tie these two together so that mass balances can be made on electron acceptors and nutrients as well. In this section, we link the full stoichiometry to the mass balances.

The first step is to develop a relationship between f_s, the critical component of the stoichiometric equation, and biomass synthesis, as expressed in the reactor mass balances. In order to do this, we make use of volatile suspended solids Equation 3.31 for the case without settling and cell recycle or Equation 5.47 (Table 5.2) with settling and cell recycle. Our concern is only with biological solids production, which equals the net production of active cells (X_a) and inactive biomass (X_i). We exclude any inactive solids that were present in the influent stream (X_i^0), as they do not represent suspended solids produced during substrate metabolism. Thus, we can define the biomass solids concentration within the reactor (X_b) with

Reactor without settling and cell recycle:

$$X_b = Y(S^0 - S)\left[\frac{1 + (1 - f_d)b\theta_x}{1 + b\theta_x}\right] \qquad \textbf{[5.58]}$$

Reactor with settling and cell recycle:

$$X_b = \frac{\theta_x}{\theta}\left\{Y(S^0 - S)\left[\frac{1 + (1 - f_d)b\theta_x}{1 + b\theta_x}\right]\right\} \qquad \textbf{[5.59]}$$

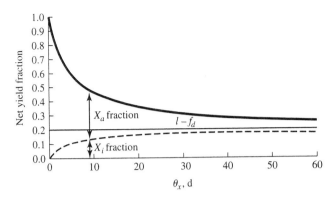

Figure 5.9   Relationship between the net yield fraction and θ_x for the case where f_d equals 0.8 and $b = 0.2\ \text{d}^{-1}$. The fractions of the net yield that are represented by active microorganisms and by inactive microorganisms are also indicated.

We next recognize that the last bracketed term in these equations is dimensionless and represents a fractional adjustment to the true yield (Y). We define the term in brackets as the *net yield fraction (NYF)*. Indeed, if the NYF is plotted as a function of θ_x, a curve as shown in Figure 5.9 results. For Figure 5.9, f_d equals the typical value of 0.8. We see that the NYF varies from 1, when θ_x equals zero, to $(1 - f_d)$, or in this case 0.2, as θ_x approaches infinity. Also shown in Figure 5.9 are the fractions of net produced biomass that equal X_a and X_i. In Equations 5.58 and 5.59, the fraction of active biomass is represented by $[1/(1 + b\theta_x)]$, and the inert biomass is $(1 - f_d)b\theta_x/(1 + b\theta_x)$.

Finally, we convert Y, which is expressed as mass of bacterial cells per unit mass of substrate consumed, into electron equivalent units. The equivalent expression for Y is then f_s^0 as discussed in Chapter 2. The result is that the net yield of active plus inactive biomass is related to the bracketed modifier in Equations 5.58 and 5.59 as indicated previously in Equation 3.32

$$f_s = f_s^0 \left[\frac{1 + (1 - f_d)b\theta_x}{1 + b\theta_x} \right]$$

Example 5.1 | **ALL STOICHIOMETRIC RATES FOR A CSTR**   A CSTR with settling and recycle is being used for aerobic treatment of a wastewater containing 600 mg/l of acetate at a flow rate of 15 m³/s, θ_x is 6 d, and X_v is 2,000 mg/l. Determine the reactor volume, biological sludge production rate, oxygen demand rate, and requirements for the biological nutrients nitrogen and phosphorus. Assume there are no suspended solids or nutrients in the influent stream. The appropriate coefficients are: $Y = 0.55$ g cells/g acetate, $b = 0.15\ \text{d}^{-1}$, $\hat{q} = 12$ g acetate/g cells-d, $K = 10$ mg acetate/l, and $f_d = 0.8$.

First, Equation 5.39 and θ_x give the effluent acetate concentration:

$$S = 10\frac{1 + 0.15(6)}{6[0.55(12) - 0.15] - 1} = 0.5\ \text{mg/l}$$

A rearrangement of Equation 5.47 makes it possible to determine the hydraulic detention time, θ, from S, θ_x, and X_v:

$$\theta = \frac{6}{2,000}\left[\frac{0.55(600 - 0.5)(1 + (1 - 0.8)(0.15)(6))}{1 + 0.15(6)}\right] = 0.61 \text{ d}$$

The reactor volume can then be determined from Q^0 and θ:

$$V = \theta Q^0 = 0.61 \text{ d}\left(15\frac{\text{m}^3}{\text{s}}\right)\left(\frac{(3,600)(24)\text{s}}{\text{d}}\right) = 791,000 \text{ m}^3$$

Now, we develop a stoichiometric equation for the reaction. First, we need to determine f_s^0, which is related to Y by units conversion:

$$f_s^0 = \frac{0.55 \text{ g cells}}{\text{g acetate}} \times \frac{59 \text{ g acetate}}{\text{mol acetate}} \times \frac{\frac{1}{8} \text{ mol acetate}}{\text{e}^- \text{ eq acetate}} \times \frac{\text{mol cells}}{113 \text{ g cells}} \times \frac{\text{e}^- \text{ eq cells}}{\frac{1}{20} \text{ mol cells}}$$

$$f_s^0 = 0.72 \frac{\text{e}^- \text{ eq cells}}{\text{e}^- \text{ eq acetate}}$$

We can then determine f_s from Equation 5.60 and f_e by difference:

$$f_s = 0.72\left[\frac{1 + (1 - 0.8)(0.15)(6)}{1 + (0.15)(6)}\right] = 0.447$$

$$f_e = 1 - f_s = 1 - 0.447 = 0.553$$

We construct the overall stoichiometric equation for this reaction by using, from Table 2.3, Reaction 0-1 (acetate) for R_d, and Reaction 0-20 (cell synthesis) for R_c; and from Table 2.2, Reaction I-14 (oxygen) the electron acceptor for R_a:

$$R = f_e R_a + f_s R_c - R_d$$

$f_e R_a$: $\quad 0.1382 \text{ O}_2 + 0.553 \text{ H}^+ + 0.553 \text{ e}^- = 0.2765 \text{ H}_2\text{O}$

$f_s R_c$: $\quad 0.0894 \text{ CO}_2 + 0.0224 \text{ HCO}_3^- + 0.0224 \text{ NH}_4^+ + 0.447 \text{ H}^+ + 0.447 \text{ e}^- =$
$\quad\quad\quad 0.0224 \text{ C}_5\text{H}_7\text{O}_2\text{N} + 0.2012 \text{ H}_2\text{O}$

$-R_d$: $\quad 0.125 \text{ CH}_3\text{COO}^- + 0.375 \text{ H}_2\text{O} = 0.125 \text{ CO}_2 + 0.125 \text{ HCO}_3^- + \text{H}^+ + \text{e}^-$

R: $\quad 0.125 \text{ CH}_3\text{COO}^- + 0.1382 \text{ O}_2 + 0.0224 \text{ NH}_4^+ =$
$\quad\quad\quad 0.0224 \text{ C}_5\text{H}_7\text{O}_2\text{N} + 0.0356 \text{ CO}_2 + 0.1027 \text{ HCO}_3^- + 0.1027 \text{ H}_2\text{O}$

This equation indicates that, for each $0.125(59) = 7.38$ g acetate consumed, $0.1382(32) = 4.42$ g O_2 is consumed, and $0.0224(113) = 2.53$ g biomass is produced. The nitrogen required is $0.0224(14) = 0.314$ g ammonium-nitrogen. Since the phosphorus requirement is about 1 g per 6 g nitrogen, the phosphorus requirement is $0.314/6 = 0.052$ g.

The acetate consumed per day can be obtained from the flow rate and the concentration of acetate consumed:

Acetate consumption rate $= Q^0(S^0 - S)$

$$= 15\frac{\text{m}^3}{\text{s}}(600 - 0.5)\frac{\text{mg}}{\text{l}} \times \frac{10^3 \text{l}}{\text{m}^3} \times \frac{\text{g}}{10^3 \text{mg}} \times \frac{3,600(24)\text{s}}{\text{d}} = 777(10^6) \text{ g acetate/d}$$

From the rate of substrate consumption, we compute all the other consumptions and productions:

$$\text{Oxygen consumption rate} \quad = \frac{4.42}{7.38} \times 777(10)^6 = 465(10)^6 \text{ g O}_2/\text{d}$$

$$\text{Biomass production rate} \quad = \frac{2.53}{7.38} \times 777(10)^6 = 266(10)^6 \text{ g biomass/d}$$

$$\text{Nitrogen required} \quad = \frac{0.314}{7.38} \times 777(10)^6 = 33(10)^6 \text{ g NH}_3\text{-N/d}$$

$$\text{Phosphorus required} \quad = \frac{0.052}{7.38} \times 777(10)^6 = 5.5(10)^6 \text{ g P/d}$$

Finally, we can compare the results of the above equation for biomass production rate, as determined from stoichiometry, with that given by Equation 5.48 (Table 5.2):

$$r_{tbp} = \frac{X_v V}{\theta_x}$$

$$= \frac{2000 \frac{\text{mg}}{\text{l}} (791{,}000 \text{ m}^3) \left(\frac{10^3 \text{l}}{\text{m}^3}\right) \left(\frac{\text{g}}{10^3 \text{ mg}}\right)}{6 \text{ d}} = 264(10)^6 \text{ g biomass/d}$$

Within round-off error, the biomass production rates obtained by the two different methods are the same, which they should be. This provides an overall check on the two approaches. In addition, it is possible to compute the same nitrogen and phosphorus requirements by multiplying r_{tbp} by the N and P contents of the biomass, as described by Equation 3.43. The same oxygen consumption rate can be obtained by mass balance on oxygen demand, as described by Equation 3.48.

5.10   ENGINEERING DESIGN OF REACTORS

The discussions to this point have been concerned with the development of mathematical equations that relate substrate removal, biological sludge production, reactor size, and reaction stoichiometry to important operational variables, such as hydraulic detention time and solids retention time. The question remains as to what are the criteria that must be considered when an engineer designs a treatment system to perform a given task. First, the engineer must specify the required efficiency of contaminant removal. This is usually based upon regulatory requirements, which may be specified as general minimum levels of performance that are applicable to all treatment systems, or they may be levels of performance that are specifically selected for the treatment system to be designed. The level of performance may be specified in different ways. For example, a minimum BOD_5 removal of 85 percent by the biological treatment process may be required. In another case, the specifications may be that the effluent BOD_5 and total suspended solids (TSS) concentrations shall not exceed 30 mg/l. Often, instead of setting absolute levels of performance such as these, some variation in performance may be allowed. For example, instead of an absolute requirement that the effluent BOD_5 or TSS concentrations not exceed 30 mg/l, the requirement may recognize that these values may be exceeded at times. One example is a statement that the 30-d average not exceed 30 mg/l. This would mean that if the concentration did exceed 30 mg/l on some days, then the concentration would have to be below 30 mg/l on other days so that the average requirement could be met. The 85-percent removal and 30-d effluent average of 30 mg/l for BOD_5 and TSS typically are minimum requirements for municipal wastewater treatment in the United States, as specified by the Federal Water Pollution Control Act Amendments of 1972 (Public Law 92-500). States may and do adopt more stringent requirements when justified in order to protect receiving waters.

Should the engineer design the treatment system to just barely meet the effluent concentration requirement? No, because the engineer must guarantee reliability in performance. Just as the structural engineer must provide a factor of safety when designing a building or bridge to take a certain loading, environmental engineers must guarantee the reliability of the treatment plants they design by applying a factor of safety. Thus, treatment plants should be designed for treatment efficiency with operational reliability. How might safety factors be assigned?

The suggested approach for applying a safety factor to treatment reactor design (Christensen and McCarty, 1975) is to select a design θ_x (i.e., θ_x^d) as a multiple of the limiting value of θ_x. The multiple is the safety factor (SF):

$$\theta_x^d = \text{SF}\left[\theta_x^{\min}\right]_{\lim} \qquad \textbf{[5.60]}$$

Typical values of SF can be inferred from the designs developed from long-term empirical practice. Since the inception of suspended-growth reactors, such as the aerobic activated sludge system, reliable designs evolved through years of trial-and-error experience, rather than from fundamentals of reactor design, since such fundamentals were not initially known. The long history of trial and error eventually resulted in designs for municipal wastewater treatment plants, for which wastewater characteristics do not vary greatly, that gave an acceptable degree of reliability in operation. Applying the more fundamental understanding, we can calculate implied safety factors using Equation 5.60, and they are listed in Table 5.3 for the typical activated sludge systems. Parallel safety factors apply to anaerobic systems, although "conventional" safety factors are somewhat lower (i.e., 10 to 30).

Historically, activated sludge treatment plants are described as "conventional," "high-rate," or "low-rate." In general, "conventional" means medium-sized treatment systems that are expected to operate reliably with fairly constant supervision by

Table 5.3 Implied safety factors for typical biological treatment design loadings

Loading	Implied SF
Conventional	10–80
High Rate	3–10
Low Rate	>80

Table 5.4 Factors to consider when selecting a safety factor for a given design loading

---

Expected temperature variations
Expected wastewater variations
  Flow rate
   Wastewater concentration
   Wastewater composition
Possible presence of inhibitory materials
Level of operator skill
Efficiency required
Reliability required
Potential penalties for noncompliance
Confidence in design coefficients

---

reasonably skilled operators. High-rate systems are plants for which the skill of the operator is unusually high and where system oversight is exceptional. An alternate view of high-rate operation is that the removal efficiency and high reliability are not as critical. The low-rate systems are generally used when operator attention is quite limited. Examples are small "extended-aeration" activated sludge plants that may be used in shopping centers or apartment complexes, where operators are present for only short periods of time and mainly to insure that the equipment such as pumps and aerators are working properly.

Table 5.3 illustrates two key points about safety factors for activated sludge processes. First, the safety factor for design is larger for systems whose operators are less skilled and attend to the process for less time. Second, the SF multiplier is much larger than 1. Even for high-rate design, the safety factor is at least 3, and most of the time it is 5 or larger. For low-rate systems, the safety factor often exceeds 100.

Within each range, the values for θ_x^d vary significantly. Here, the designing engineer must use judgment in the selection of an appropriate safety factor. Table 5.4 lists several of the factors that should be taken into consideration in making such judgments. The engineer should work with the community and regulators to determine the most appropriate safety factors to use in a given situation. Higher safety factors increase the degree of reliability of operation, but also give higher construction costs. Designs with lower safety factors require more continuous supervision and operators with increased skill, which add to the operational costs. The appropriate balance for a given situation depends upon many local factors that must be evaluated carefully by the design engineer.

An additional design decision by the engineer for a suspended-growth system with recycle is the concentration of suspended solids that is to be maintained in the treatment reactor. This depends to a large degree upon the settling characteristics of the floc and the degree of recycle to be provided. In general, using a higher suspended solids concentration makes the reactor smaller (and less expensive) for a given θ_x^d.

Table 5.5 Typical values for total suspended
solids concentration (X) in aerobic
suspended-growth reactors with
settling and recycle

Floc Type	X (mgSS/L)
Normal	1,500–3,000
Poor compaction or low recycle rate	500–1,500
Good compaction or high recycle rate	3,000–5,000

However, higher suspended-solids concentrations may require larger settling tanks, because of the increased load of suspended solids to the settling tank. The proper balance requires an approach to optimize the size for the combined reactor-settling-tank system, which is described in detail in Chapter 6. Some rules of thumb that can be useful in preliminary design calculations are provided in Table 5.5.

Settling characteristics of suspended flocs can vary significantly from location to location, from one type of design to another (e.g., CSTR vs. PFR), and with time at a given plant. A changeover from good settling floc to poor settling floc can cause failure of the treatment system. This is the most common cause of failure of activated sludge treatment plants. Designing for a smaller suspended solids concentration in the reactor generally increases the reliability of the plant, but results in larger reactor size and higher capital costs. This is another area where good judgment by the design engineer is critical. Joint decisions among the client, the regulators, and the engineer are recommended.

SELECTION OF DESIGN CRITERIA Select an appropriate volume for a CSTR reactor with settling and recycle to achieve an average 90-percent nitrification of ammonium in a municipal wastewater following biological treatment for BOD removal. The following wastewater characteristics and organism parameters apply:

Example 5.2

$$Q^0 = 10^4 \text{ m}^3/\text{d}$$
$$\text{Influent BOD}_5 = 0 \text{ mg/l}$$
$$S^0 \text{ (Influent NH}_4^+\text{-N)} = 25 \text{ mg/l}$$
$$X_i^0 = 18 \text{ mg VSS/l}$$
$$Y = 0.34 \text{ g VSS/g NH}_4^+\text{-N}$$
$$\hat{q} = 2.7 \text{ g NH}_4^+\text{-N/g VSS-d}$$
$$K = 1.0 \text{ mg NH}_4^+\text{-N/l}$$
$$b = 0.15/\text{d}$$

First, we compute the baseline value of $[\theta_x^{min}]_{lim}$ from Equation 3.27:

$$\left[\theta_x^{min}\right]_{lim} = \frac{1}{Y\hat{q} - b} = \frac{1}{0.34(2.7) - 0.15} = 1.3 \text{ d}$$

We arbitrarily select an intermediate safety factor from Table 5.3 of 15. (Factors as listed in Table 5.4 need to be weighed in selecting an SF for an actual design.) So, the design SRT becomes

$$\theta_x^d = \text{SF}\left[\theta_x^{\min}\right]_{\lim} = 15(1.3) = 20 \text{ d}$$

Second, we must select a concentration of total suspended solids. We choose a typical value for conventional treatment, 2,000 mg/l, from Table 5.5. We rearrange Equation 5.47 to select the hydraulic detention time for the reactor:

$$\theta = \frac{\theta_x^d}{X_v}\left[X_i^0 + \frac{Y(S^0 - S)(1 + (1 - f_d)b\theta_x^d)}{1 + b\theta_x^d}\right]$$

$$\theta = \frac{20}{2,000}\left[18 + \frac{0.34(25 - 0)(1 + (1 - 0.8)(0.15)(20))}{1 + 0.15(20)}\right] = 0.21 \text{ d}$$

The reactor volume is then simply the product of the flow rate and detention time:

$$V = Q^0\theta = 10^4(0.21) = 2,100 \text{ m}^3$$

Third, we must check whether or not the ammonium removal is sufficient. This requires that the effluent NH_4^+-N concentration be computed using Equation 5.39:

$$S = K\frac{1 + b\theta_x^d}{\theta_x^d(Y\hat{q} - b) - 1} = 1.0\frac{1 + 0.15(20)}{20(0.34(2.7) - 0.15) - 1} = 0.28 \text{ mg } NH_4^+\text{-N/l}$$

The efficiency of ammonia-nitrogen removal is much greater than 90 percent:

$$\text{Removal efficiency} = 100\frac{S^0 - S}{S^0} = 100\frac{20 - 0.28}{20} = 98.6 \text{ percent} \gg 90 \text{ percent}$$

The average NH_4^+-N removal efficiency is thus much greater than required, providing another degree of safety in the design. For this case, as with most designs in practice, reactor volumes tend to be governed more by reliability considerations than by considerations of removal efficiency per se.

5.11   REACTORS IN SERIES

Reactors connected in series (Figure 5.1) are commonly used in wastewater treatment. There are different reasons for doing so. Two of the most common treatment requirements for municipal wastewaters are removal of organic material (BOD) and transformation of nitrogen, such as the oxidation of ammonium to nitrate. Organic removal and nitrification are carried out by two different groups of microorganisms, but both reactions can be carried out together in a single aerobic reactor. However, the treatment can be easier to control if two reactors are connected in series: the first for organic removal and the second for nitrification. Here, a different type of reactor may be used for each process. For example, a CSTR with settling and recycle may be used for organic removal and then followed by a biofilm reactor for nitrification. The reverse has also been used, a biofilm reactor for organic removal and a suspended-growth system for nitrification. A CSTR is often adequate for organic removal, but a plug-flow reactor tends to be more efficient for nitrification. This combination can be used through reactors connected in series. In other cases, the same type of reactor may be used both for the first stage of organic removal and the second one of nitrification.

Another use of series reactors is when different electron acceptors are desired in the overall treatment scheme. As an example, nitrogen removal, rather than just transformation of ammonium to nitrate, may be required. Two reactors in series might accomplish this: the first an aerobic reactor to accomplish organic removal and ammonium oxidation and the second reactor operated under anoxic (the absence of oxygen) conditions to obtain conversion of the nitrate produced in the first reactor to N_2 through denitrification. It is also increasingly common to have a first anoxic stage followed by an aerobic stage. Here, nitrate may be produced in the aerobic stage, and through recycle, the nitrate is brought back to the first stage through recycle of second stage effluent to achieve a mixture of untreated wastewater BOD and nitrate to achieve BOD oxidation by denitrification in the first anoxic stage. These various schemes for nitrogen transformation and removal are discussed in more detail in Chapters 9 and 10.

Staged reactors also can be used to achieve highly efficient removal of toxic organic chemicals. The first stage here might be a CSTR with settling and recycle, and the second stage could be some form of plug-flow reactor, such as a suspended-growth reactor with settling, but no cell recycle. This system is illustrated in Figure 5.10. Here, excess biomass is removed for disposal only from the second settling tank. We assume that the concentration of the contaminant in the wastewater is S^0, and that this concentration is quite toxic to microorganisms. Thus, we do not want to expose the organisms to this high concentration. For example, phenol is known to be quite toxic to microorganisms at concentrations above 200 to 300 mg/l, but concentrations in some industrial wastewaters may be in the range of several thousands of milligrams per liter. In addition, phenol can be toxic to some aquatic species at concentrations in the low mg/l range, it can impart taste and odor to fish at even smaller concentrations, and it causes taste and odor in drinking water at a concentration of about 40 μg/l. Indeed, chlorination for disinfection can form chlorinated phenols that cause even more severe taste and odors problems in drinking water. The question is how can we

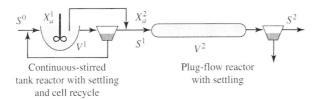

Continuous-stirred
tank reactor with settling
and cell recycle

Plug-flow reactor
with settling

Figure 5.10   Possible two-stage reactor system for treatment of high concentrations of toxic chemicals such as phenol to meet low concentration standards

design a treatment system to treat a wastewater with a phenol concentration of, say, 4,000 mg/l and reduce the concentration to 20 μg/l? We consider the reactor system illustrated in Figure 5.10 as one possibility. We also assume that no other organic substrates are present, and inorganic nutrients are available.

First, we use aerobic treatment and consider Haldane kinetics, since we are dealing with a substrate that is self-inhibitory. We assume the following wastewater treatment conditions apply for the first reactor:

$$Q^0 = 10^4 \text{ m}^3/\text{d}$$

$$S^0 \text{ (phenol)} = 4,000 \text{ mg/l}$$

$$\theta_x = 8 \text{ d}$$

$$\hat{q} = 6 \text{ d}^{-1}$$

$$K = 2 \text{ mg/l}$$

$$K_{IS} = 120 \text{ mg phenol/l}$$

$$Y = 0.35 \text{ g cells/g phenol}$$

$$b = 0.2 \text{ d}^{-1}$$

$$X_a^1 = 1,500 \text{ mgVSS}_a/\text{l}$$

For the first stage reactor, we can use Equation 5.37 for a CSTR with settling and recycle of microorganisms to determine the effluent concentration of phenol (S^1), the active organism concentration (X_a^1), and the reactor volume (V^1). For Haldane kinetics, we need to combine Equations 3.62, 3.64, and 3.65 to obtain r_{ut} in Equation 5.37:

$$\frac{1}{\theta_x} = \frac{Y(-r_{ut}^1)}{X_a^1} - b = \frac{Y\hat{q}S^1}{K + S^1 + (S^1)^2/K_{IS}} - b$$

$$\frac{1}{8} = \frac{0.35(6)S^1}{2 + S^1 + (S^1)^2/120} - 0.2$$

from which $S = 0.37$ mg phenol/l. Then,

$$\theta^1 = \frac{\theta_x^1}{X_a^1}\frac{Y(S^0 - S^1)}{1 + b\theta_x^1} = \frac{8}{1,500}\frac{0.35(4,000 - 0.37)}{1 + 0.2(8)} = 2.87 \text{ d}$$

$$V^1 = \frac{Q^0}{\theta^1} = \frac{10^4}{2.87} = 3,480 \text{ m}^3$$

We see that CSTR treatment alone, on the average, reduces the phenol concentration below 1 mg/l. In order to reduce the concentration to near our goal of 20 μg/l, θ_x would need to be increased to near infinity. However, if we use a plug-flow reactor for the second stage, then our goal can be achieved. Since the phenol concentration is now well below the K_{IS} value of 120 mg/l, we can assume the simpler Monod kinetics for further analyses.

One problem with a second-stage plug-flow reactor is that the phenol concentration in the effluent from the CSTR settling is much too low to support adequate biomass for phenol destruction. We can remedy this if we transfer the excess biological solids produced in the first stage reactor directly into the second-stage reactor, as illustrated in Figure 5.10. We might take these organisms from the recycle line of the first-stage reactor, but we will obtain better control if draw it directly from the first-stage reactor as shown. The quantity to be withdrawn daily can be determined from the equations in Table 5.2. Since our design θ_x for the first stage is 8 d, we can accomplish the required transfer of excess biological solids by transferring one-eighth of the stage-one reactor mixed liquor contents each day to the second-stage reactor: 3,480 m^3/8 d or 435 m^3/d. When we mix the remaining wastewater from stage one (10^4 m^3/d − 435 m^3/d or 9,565 m^3/day) with the excess microorganisms produced in stage one, the active microorganism concentration in the combined wastewater entering the stage two reactor is 435 m^3/d (1,500 mg/l)/10^4 m^3/d = 65 mg/l. If we assume at this point that the second-stage reactor does not have organism recycle, then we can use the desired S^2 phenol concentration of 0.02 mg/l directly to determine θ_x^2, which here also equals θ^2 for the reactor. For this simple plug-flow reactor, Equation 5.27 applies:

$$\theta^2 = \frac{1}{\hat{q}}\left\{\left(\frac{K}{X_a^2 + YS^1} + \frac{1}{Y}\right)\ln(X_a^2 + YS^1 - YS^2)\right.$$
$$\left. - \left(\frac{K}{X_a^2 + YS^1}\right)\ln\frac{S^2 X_a^2}{S^1} - \frac{1}{Y}\ln X_a^2\right\}$$

$$\theta^2 = \frac{1}{6}\left\{\left(\frac{2}{65 + 0.35(0.37)} + \frac{1}{0.35}\right)\ln(65 + 0.35(0.37) - 0.35(0.02))\right.$$
$$\left. - \left(\frac{2}{65 + 0.35(0.37)}\right)\ln\frac{0.02(65)}{0.37} - \frac{1}{0.35}\ln(65)\right\}$$

$$\theta^2 = 0.016 \text{ d}$$

from which,

$$V = \theta^2 Q^0 = 0.016(10^4 \text{m}^3) = 160 \text{ m}^3$$

We see here that the volume of the second-stage reactor is quite small compared with the first-stage reactor and is able to do a job that the first-stage CSTR could not do, even if infinite in size. We have not applied any safety factor to the second-stage reactor design, which should be done. Perhaps we can optimize the system by reducing θ_x for the first-stage and shifting more of the load to the second stage. We could also have cell recycle in the second stage to increase organism concentration within the reactor. No matter how we optimize the system, this preliminary analysis illustrates important benefits of the two-stage system. In the first reactor, the microorganisms never experience a high phenol concentration and thus are not subject to its toxicity. This would not be true if a plug-flow reactor were used here. In addition, the inefficiency of the CSTR is circumvented by use of the second-stage plug-flow reactor. A similar argument can be made for operating anaerobic systems in two stages. A first stage with a CSTR helps avoid problems associated with organic acid production and low pH with high organic concentrations, but a second-stage plug-flow reactor permits high removal efficiencies with a much smaller overall reactor size.

5.12 BIBLIOGRAPHY

Christensen, D. R. and P. L. McCarty (1975). "Multi-process biological treatment model." *J. Water Pollution Control Federation* 47, pp. 2652–2664.

5.13 PROBLEMS

5.1. You have the following parameters:

$$f_s^0 = 0.7 \qquad\qquad Y = 0.5 \text{ g VSS}_a/\text{g}$$
$$\hat{q} = 10 \text{ mg/mg VSS}_a\text{-d} \quad X_a = 1{,}500 \text{ mg VSS}_a/\text{l}$$
$$K = 10 \text{ mg/l} \qquad\qquad Q = 2{,}500 \text{ m}^3/\text{d}$$
$$f_d = 0.8 \qquad\qquad b = 0.1/\text{d}$$

Design safety factor (SF) = 30. What is the design value of f_s?

5.2. Calculate the hydraulic detention time (θ) of a CSTR with settled-solids recycle if:

$$Y = 0.6 \text{ g VSS}_a/\text{g}$$
$$\hat{q} = 20 \text{ g/g VSS}_a\text{-d}$$
$$K = 20 \text{ mg BOD}_L/\text{l}$$
$$b = 0.25/\text{d}$$
$$f_d = 0.8$$
$$S^0 = 10{,}000 \text{ mg BOD}_L/\text{l}$$
$$X_v = 4{,}000 \text{ mg/l}$$
$$X_v^0 = 0$$
$$\theta_x = 6 \text{ d}$$
$$Q = 1{,}000 \text{ m}^3/\text{d}$$

Ignore SMP formation and biomass in the settler. Discuss the practicality of treating this rather high-strength, soluble wastewater by this method.

5.3. In the laboratory you are operating two CSTRs in parallel. Both have solids recyle, the same total volume (100 liters), and the same influent conditions:

$$Q = 400 \text{ l/d}$$
$$S^0 = 1{,}000 \text{ mg BOD}_L/\text{l}$$
$$X_v^0 = 0$$

Both are being operated with the same SRT, 10 d. Both have the same effluent VSS concentration, 15 mg/l. However, system A is aerobic, while system B is anaerobic. Pertinent parameters are:

	A	B
$\hat{q}$, mg BOD_L/mg VSS_a-d	16	10
K, mg BOD_L/l	10	10
b, d^{-1}	0.2	0.05
Y, mg VSS_a/mg BOD_L	0.35	0.14

Estimate and compare the total effluent-BOD_L concentrations for systems A and B. Assume the effluent SMP contributes to the BOD_L.

5.4. You are to treat a wastewater that is contaminated with 100 mg/l of SO_4^{2-} as S. You will do the treatment biologically by sulfate reduction, in which SO_4^{2-}-S is reduced to H_2S-S, which is stripped to a gas phase. You do not need to be concerned with the stripping. The electron donor will be acetate (CH_3COO^-), which you will need to add, say, in the form of vinegar. You know the following about the wastewater: flow rate $= 1,000$ m^3/d, sulfate $= 100$ mg/l as S, volatile suspended solids (X_v^0) $= 0$.

From previous tests and theory, you have estimated the following parameters for the rate-limiting substrate, which will be acetate (expressed as BOD_L):

$$f_s^0 = 0.08$$
$$\hat{q} = 8.6 \text{ g } BOD_L/\text{g } VSS_a\text{-d}$$
$$b = 0.04/\text{d}$$
$$K = 10 \text{ mg } BOD_L/\text{l}$$
$$f_d = 0.8$$
$$\text{N-source} = NH_4^+\text{-N}$$

You also have selected these design criteria:

- Use a solids-recycle system with a CSTR reactor
- Have $\theta_x = 10$ d and $\theta = 1$ d
- Have an effluent sulfate concentration of 1 mg S/l

You must provide the following crucial design information.

(*a*) Compute the effluent concentration of acetate (in mg/l of BOD_L).

(*b*) Compute (using stoichiometric principles) the input concentration of acetate (in mg/l as BOD_L) needed to take the effluent sulfate concentration from 100 to 1 mg S/l. [Note that the formula weight of S is 32 g.]

(*c*) Compute the concentrations (in mg VSS/l) of active, inert, and total volatile suspended solids.

(*d*) Compute the waste-sludge mass flow (in kg VSS/d) and volumetric flow rate (in m^3/d) if the effluent VSS is 5 mg/l and the recycle VSS is 5,000 mg/l.

(e) Compute the recycle ratio R.

(f) Based on the (normal) multisubstrate approach, determine the effluent concentration of SMP.

(g) Do you see any problems with the design? If yes, explain the cause of any problems and ways to overcome them.

5.5. This question addresses a novel system to treat a sulfide-bearing wastewater by autotrophic bacteria that aerobically oxidize sulfides (i.e., H_2S and HS^-) to SO_4^{2-}, which is harmless. The electron acceptor is O_2, and ample NH_4^+ is present as the N source. For these sulfide oxidizers, you may utilize the following stoichiometric and kinetic parameters:

$$f_s^0 = 0.2 \text{ e}^-$$
$$Y = 0.28 \text{ mg VSS}_a/\text{mg S}$$
$$\hat{q} = 5 \text{ mg S/mg VSS}_a\text{-d}$$
$$b = 0.05/\text{d}$$
$$K = 2 \text{ mg S/l}$$
$$f_d = 0.8$$

For the questions that follow, you should assume that the sulfides are not lost by any mechanism other than microbially catalyzed metabolism, that no intermediate sulfur species form, and that you can use the total sulfide concentration as rate-limiting. The influent has:

$$Q = 1{,}000 \text{ m}^3/\text{d}$$
$$S^0 = 100 \text{ mg S/l}$$
$$X_v^0 = 0.$$

You will use a CSTR reactor with a solids-recycle system. Take the following steps.

(a) Determine the design θ_x if a safety factor of 10 is employed. Please round your θ_x value to two significant digits.

(b) You have determined from judgment that the mixed liquor volatile suspended solids concentration (X_v) must be no less than 1,000 mg/l. If you accept this minimum X_v value, what are the hydraulic residence time (hours) and volume of the system (m³)?

(c) Compute the sludge wasting flow rate (in m³/d) and the sludge recycle flow rate (also in m³/d) if the effluent VSS (X_v^e) is 15 mg VSS/l and the underflow has 8,000 mg VSS/l.

(d) What is the total effluent COD from organic materials? For this part, we need to make some reasonable assumptions about SMP. For the formation of SMP, you may assume: $k_1 = 0.04$ mg COD_p/mg COD_s and $k_2 = 0.09$ mg COD_p/mg VSS-d. Note carefully the units! As a first approximation and because the sulfide oxidizers are autotrophs, you should assume that neither the UAP nor the BAP is biodegraded. In other words, there are no heterotrophs in the system. (This is not going to be true in reality, but it keeps the problem from getting too complicated.)

(e) What is the rate of O_2 supply needed (in kg O_2/d)?

5.6. Compare the cell concentration that will be present after 24 h through batch growth on a nongrowth limiting concentration of acetate for the following four cases, assuming that Y is the only variable between the organisms. The initial cell concentration in all cases is 10 mg/l. Assume other constants of interest are $\hat{q} = 12$ mg acetate/mg VSS_a-d, $K = 2$ mg/l, and $b = 0.1$/d.

(a) $Y = 0.6$ mg VSS_a/mg acetate (aerobic growth)
(b) $Y = 0.45$ mg VSS_a/mg acetate (denitrification)
(c) $Y = 0.06$ mg VSS_a/mg acetate (sulfate reduction)
(d) $Y = 0.04$ VSS_a/mg acetate (methane fermentation)

5.7. A CSTR reactor is used for treatment of 50 m³/d of wastewater with BOD_L of 10,000 mg/l by methane fermentation. Assume $Y = 0.04$ mg VSS_a/mg BOD_L, $\hat{q} = 8$ mg BOD_L/per VSS_a-d, $K = 200$ mgBOD_L/l, and $b = 0.05$/d. What reactor detention time will result in the maximum BOD_L removal per day per unit volume of reactor? (You may find it useful to use a spreadsheet or programmable calculator, as a trial and error solution is likely to be needed). Show a graph of BOD_L removal per day per unit reactor volume versus reactor detention time.

5.8. The characteristics of a wastewater were found to be the following:

$$Q = 150{,}000 \text{ m}^3/\text{d}$$
$$X = 350 \text{ mg/l}$$
$$X_v = 260 \text{ mg/l}$$
$$\text{COD (total)} = 880 \text{ mg/l}$$
$$\text{COD (soluble)} = 400 \text{ mg/l}$$
$$BOD_L \text{ (total)} = 620 \text{ mg/l}$$
$$BOD_L \text{ (soluble)} = 360 \text{ mg/l}$$

Estimate the following: S^0, Q^0, X^0, and X_i^0.

5.9. Compare the relative volumes of the following reactors required to achieve (1) 85 percent removal and (2) 98 percent removal of ammonium by oxidation to nitrate for the following wastewater characteristics:

$Q^0 = 50$ m³/d $\hat{q} = 2.5$ g NH_4^+-N/g VSS_a-d
$S^0 = 60$ mg NH_4^+-N/l $K = 1$ mg/l
$X_i^0 = 20$ mg/l $Y = 0.34$ g VSS_a/g NH_4^+-N
                          $b = 0$

(a) CSTR
(b) CSTR with settling and solids recycle, with $\theta_x/\theta = 10$.

5.10. For the wastewater and biological characteristics in problem 5.9, determine the reactor volumes and ammonium removal efficiencies for the case where the safety factor used in design for both reactors is the same or 10:

(a) CSTR

(b) CSTR with settling and biomass recycle, and with $X_v = 1,000$ mgVSS/l

5.11. You have a given wastewater to treat and are evaluating different reactor types for treatment and the sensitivity of effluent substrate concentration to various changes in operation. Let's assume for each reactor type you have already selected a given reactor volume that results in satisfactory treatment. What effect on effluent substrate concentration (S^e) will an increase in each of the factors listed below have? Use $+$ for an increase in S^e, $-$ for a decrease, 0 for no change, and ? for undetermined.

	Reactor Type			
Factor Increased	CSTR with Recycle	CSTR with Cell Settling and Recycle (θ_x Kept Constant)	Plug Flow with Recycle	Fixed-Film Completely Mixed with Recycle
Q^0				
S^0				
X_i^0				
X_a^0				
R (recycle rate)				
Y				

5.12. You are designing a CSTR with cell settling and recycle for treatment of an organic industrial wastewater. You have selected a reactor for design that results in a given substrate concentration in the effluent. You now wish to do a sensitivity analysis to determine the effect of certain changes on the volume of the reactor, while keeping the effluent concentration unchanged. Fill out the following table to indicate how an increase in each of the variables in the left-hand column will affect reactor volume. Assume all other factors listed in the left-hand column remain constant. Use: $+$ = increase, $-$ = decrease, 0 = no change, and i = need more information to tell.

Variable Increased	Effect on Reactor Size
K	
$\hat{q}$	
Q^0	
S^0	
X_i^0	
X_a	

5.13. You have designed a CSTR with settling and solids recycle to treat a waste with flow rate (Q^0) of 100 m^3/d, and have assumed that X_i^0 is zero. Your design θ_x is 6 d, resulting in a reactor volume V of 20 m^3. However, you have now found that X_i^0 for the waste will actually be 200 mgVSS$_i$/l. If in your design you wish to maintain the previously selected θ_x and mixed liquor suspended solids concentration ($X = 2{,}000$ mg SS/l), what change, if any, in reactor volume (in m^3) is required?

5.14. Determine the volume of an aerobic CSTR with settling and solids recycle for BOD removal under the following conditions (assume typical values for aerobic noncarbohydrate BOD from Chapter 3):

(a) SF $= 40$
(b) $Q^0 = 10^3$ m^3/d
(c) $X_i^0 = 300$ mgVSS$_i$/l
(d) Influent BOD$_L = 200$ mg/l
(e) Effluent BOD$_L = 5$ mg/l
(f) $X_v = 2{,}000$ mgVSS/l

5.15. A 200 m^3/d stream flows from an abandoned mine and contains 10 mg/l of dissolved Fe(II) (formula weight $= 56$ g) and no suspended solids. When oxidized in the stream to Fe(III), the brown precipitate that forms degrades the stream. You wish to consider biological oxidation of the Fe(II) to remove at least 95 percent of the iron before discharge to the stream. It must be considered that the Fe(III) produced in the reactor precipitates there forming Fe(OH)$_3$, which becomes part of the reactor suspended solids and is removed from the system for disposal along with the waste biological solids. What reactor volume do you suggest be used for a CSTR with settling and solids recycle? Assume the reactor total suspended solids concentration (X) equals 3,000 mgSS/l and that $[\theta_x^{min}]_{lim}$ for Fe(II) oxidation is 2.2 d.

5.16. Consider a suspended-growth reactor with settling and solids recycle and with $X_a = 1{,}200$ mgVSS$_a$/l and detention time $\theta = 4$ h. A contaminant has an influent concentration S^0 to the reactor of 0.5 mg/l, and is decomposed with kinetic coefficients based upon the total X_a concentration of $\hat{q} = 0.05$ mg/mg VSS_a-d and $K = 3$ mg/l. Estimate the effluent concentration S^e from the reactor if the reactor were (1) completely mixed and (2) plug-flow.

5.17. A wastewater has a concentration of benzene equal to 30 mg/l, and no significant concentration of other biodegradable organic materials are present. The regulatory agencies require that the wastewater be treated to reduce the benzene concentration to 0.01 mg/l. Assume for benzene, that the following rate coefficients apply for aerobic treatment: $Y = 0.9$ g VSS_a/g benzene, $b = 0.2$/d, $\hat{q} = 8$ g benzene/g VSS_a-d, and $K = 5$ mg benzene/l.

(a) What is the minimum concentration of benzene that you could expect to achieve from biological treatment in a CSTR with settling and solids recycle?

(b) Assuming the above does not meet regulatory compliance for benzene, describe another biological approach that is likely to have better potential for meeting the requirements. Describe the approach and indicate why it may be better than a CSTR. Calculations are not required to support your answer; a description of the concept is all that is needed.

6

THE ACTIVATED SLUDGE PROCESS

The activated sludge process surely is the most widely used biological process for the treatment of municipal and industrial wastewaters. Normally, the activated sludge process is strictly aerobic, although anoxic variations are coming into use for denitrification, the subject of Chapter 10. In simple terms, the activated sludge process consists of a reactor called the *aeration tank, a settling tank, solids recycle* from the settler to the aeration tank, and a *sludge wasting line*. The aeration tank is a suspended-growth reactor containing microbial aggregates, or *flocs,* of microorganisms termed the *activated sludge*. The microorganisms consume and oxidize input organic electron donors collectively called the *BOD*. The activated sludge is maintained in suspension in the reactor through mixing by aeration or other mechanical means. When the slurry of treated wastewater and microbial flocs pass to the settling tank, the flocs are removed from the treated wastewater by settling and returned to the aeration tank or wasted to control the solids retention time (SRT). The clear effluent is discharged to the environment or sent for further treatment. Capturing the flocs in the settler and recycling them back to the reactor are the keys to the activated sludge process, because they lead to a high concentration of microorganisms in the reactor. Thus, the sludge is "activated" in the sense that it builds up to a much higher concentration than could be achieved without the settler and recycle. The high biomass concentration allows the liquid detention time to be small, generally measured in hours, which makes the process much more cost effective. Wasting the sludge through the separate sludge-wasting line makes the solids retention time (SRT or θ_x) separate from and much larger than the hydraulic detention time (θ).

In 1914, E. Ardern and W. T. Lockett (1914) discovered the activated sludge process in England. They noted that aeration of sewage led to formation of flocculent suspended particles. They discovered that the time to remove organic contaminants (and to achieve nitrification, the subject of Chapter 9) was reduced from days to hours when these flocculent particles were held in the system. They referred to the suspended particles, more specifically the resulting sludge from settling to collect the particles, as being "activated," and so was born the activated sludge process. By 1917, the Manchester Corporation had brought a 946-m^3/d continuous-flow plant into operation, and, in the same year, Houston, Texas, completed the construction of a 38,000-m^3/d plant. Many activated sludge plants were soon constructed at larger cities throughout England and the United States (Sawyer, 1965).

It is interesting to realize that successful application of the process occurred even though an understanding of how the process actually worked was lacking. The early literature contains many articles featuring debates over whether the removal obtained was physical or biological. By 1930, the evidence in favor of a biological process was sufficiently convincing. However, an adequate theory about factors affecting removal rates was not then available. Aeration tank sizes were selected based upon experience at other locations and using rule-of-thumb parameters, such as liquid detention time or population-equivalent loading per unit volume of reactor. These empirical strategies generally worked for municipal wastewater treatment, which is sufficiently similar everywhere in flow characteristics and organic concentration. However, problems often arose when treating wastewaters with unusual characteristics, such as municipal wastewaters containing high proportions of industrial wastewaters or industrial wastewaters themselves. Several process modifications evolved, generally through trial and error. Finally, by the 1950s and 1960s, a theory of operation had developed and was sufficient so that rational designs could be achieved based on characteristics of the wastewater to be treated. As a sound theory upon which to base design continued to develop, applications of the activated sludge process grew rapidly. The activated sludge process soon overtook trickling filters (Chapter 8) as the dominant biological treatment process for organic wastewaters.

The large-scale success of the activated sludge process does not mean that it has no problems. Achieving reliable treatment for wastewaters over which the operator has little control, either in flow rate or composition, presents a great challenge to engineers and operators today. In addition, successful operation depends upon a biological process in which the microorganisms themselves can be monitored and controlled with only coarse tools. The ecology of the system can change from day to day, leading to significant problems, such as sludge bulking, a condition under which the microbial floc does not compact well. Sludge bulking makes it difficult to capture and recycle the microorganisms fast enough to maintain the desired large biomass concentration in the reactor. It is somewhat amazing that, despite all the uncertainties and uncontrollable factors involved, the activated sludge process actually has the degree of reliability that it does have. Achieving good reliability requires design and operation by individuals who are highly knowledgeable about the theory and the practicalities of operation.

This chapter begins by discussing the fundamental characteristics of the activated sludge process. Then, it describes the various modifications that have evolved, including their uses and the loading criteria commonly used (drawing heavily upon the theoretical developments provided in Chapters 2, 3, and 5 and the website chapter "Complex Systems"). Next, significant emphasis is given to limitations of the process, such as the suspended-solids concentration, oxygen transfer, and sludge-settling problems. A comprehensive design and analysis strategy is presented in order to link the process fundamentals from previous chapters to the fruits of practical experience. In order to obtain an optimum design for the activated sludge system, the reactor and settling tank designs need to be considered together as an overall unit. Therefore, the analysis and design of the settling tank is treated systematically. Finally, some alternatives to the settling tank are described.

6.1　CHARACTERISTICS OF ACTIVATED SLUDGE

6.1.1　MICROBIAL ECOLOGY

Two crucial characteristics define the kinds of microorganisms in activated sludge. First, the activated sludge contains a wide variety of microorganisms. Prokaryotes (bacteria) and eukaryotes (protozoa, crustacea, nematodes, and rotifers) generally are present, and bacteriophage, which are bacterial viruses, probably reside in the sludge, too. Fungi seldom are important members of the community. Second, most of them are held together within flocs by naturally produced organic polymers and electrostatic forces.

The primary consumers of organic wastes are the heterotrophic bacteria, although with certain organic particles, protozoa may be involved as well. By mass, the dominant members of this community are heterotrophic bacteria (Pike and Curds, 1971). Some bacterial species can consume a variety of different organic compounds, and others are more specialized, consuming only a small fraction of the organic species present.

Although the heterotrophic bacteria are quite diverse in activated sludge, they have certain common characteristics that are key for design and analysis. The fundamental stoichiometric and kinetic parameters, presented in Chapter 3, give us a useful means of characterizing the heterotrophs. The following is a list of "generic" parameters that give us a benchmark for understanding the heterotrophic bacteria. Being generic and typical for 20 °C, they give us guidance only about l nature of the heterotrophs.

Limiting substrate	**BOD_L**
Y	0.45 mg VSS_a/mg BOD_L
$\hat{q}$	20 mg BOD_L/mg VSS_a-d
$\hat{\mu} = Y\hat{q}$	9 d^{-1}
K	1 mg BOD_L/l (simple substrates); > 10 mg/l (complex substrates)
b	0.15/d
f_d	0.8
θ_x^{min}, limiting value	0.11 d
S_{min}	0.017 mg BOD_L/l (simple substrates); > 0.17 mg BOD_L/l (complex substrates)

The generic parameters tell us that the heterotrophic bacteria in activated sludge are reasonably fast growers ($\hat{\mu}$ is relatively large, while the limiting value of θ_x^{min} is relatively small). A very conservative safety factor of 100 gives a design θ_x of 11 d, while $\theta_x = 4$ d still gives a safety factor of 36. The S_{min} value tells us that we can drive the substrate concentrations quite low, well below typical effluent BOD_5 standards of 10 to 30 mg/l.

Most of the other organisms are secondary consumers that feed off of materials released by the primary consumers. They include prokaryotes, which degrade by-products of BOD degradation and from the death and lysis of other organisms, and predators, most of which are eukaryotes that feed on bacteria and bacteriophage.

Sometimes chemolithotrophic bacteria are present and obtain their energy from oxidation of inorganic compounds such as ammonium, nitrite, sulfide, and Fe(II). The chemolithotrophs are not a focus of this chapter, which addresses the oxidation of BOD in activated sludge. Ammonium and nitrite oxidizers are discussed in Chapter 9, and sulfide oxidizers are mentioned later for their role in sludge bulking.

There is great competition between microorganisms for the various energy resources available in waste mixtures such as domestic sewage. Because of the death of some species (perhaps caused suddenly by bacteriophage or by predation) and changes to the inputs and environmental conditions (like temperature or SRT), the species composition of activated sludge can change significantly over time. Accompanying changes in the microbial composition of activated sludge can be changes in the floc's physical characteristics, such as its strength of aggregation, settling velocity, and ability to compact and form a dense sludge.

The majority of the bacterial genera in activated sludge are Gram-negative (Pike and Curds, 1971). Principal identified genera include *Pseudomonas, Arthrobacter, Comamonas, Lophomonas, Zoogloea, Sphaerotilus, Azotobacter, Chromobacterium, Achromobacter, Flavobacterium, Bacillus,* and *Nocardia. Zoogloea* was once thought to be the dominant organism holding the activated sludge flocs together, but this idea is no longer accepted, since we know that many types of bacteria generate the polymers used to hold the flocs together. Certain genera, such as *Sphaerotilus* or *Nocardia,* often are blamed for problems of poor settling, but the reality is that a large number of bacterial types can create settling problems. Recent studies using oligonucleotide probes show that Gram-positive bacteria are significant in activated sludge, too.

Many species of protozoa have been identified in activated sludge, and total numbers can be on the order of 50,000 cells/ml (Pike and Curds, 1971). Protozoa, while not the primary consumers of organic wastes in the activated sludge process, have long been known to be useful indicators of process performance. In a "good" sludge, the ciliated protozoa are prominent, and those ciliates that attach to floc particles with a stalk, rather than the free-swimming ciliates, are the best indicators of a stable sludge. The ecological reasons for this and more on how to use protozoa as indicator are discussed later under Impacts of SRT. A healthy protozoan population is also indicative of a wastewater that is relatively free of toxic chemicals. Protozoa tend to be highly sensitive to toxic chemicals, and since their presence and activity are readily observed with a low-powered microscope, their use as indicator organisms for the presence of toxic chemicals or other problems in the process is extremely useful.

Rotifers, nematodes, and other multicellular forms often are found in activated sludge systems, but their roles in the process are not obvious. They can ingest clumps containing groups of bacteria and are commonly seen within the microbial flocs chewing off bits and pieces of the floc particles. Because they generally are present when the system has a long SRT, the multicellular forms often are used as indicators.

The role of bacterial viruses, or phage, in the overall process is not well documented, but it is possible that their presence can cause rapid and large shifts in dominant bacterial species. Since multiple bacterial species always are present, if one species is decimated by a phage, another can replace it rapidly so that significant perturbations in treatment efficiency are not detected. This is an example of redundancy in microbial ecosystems and probably is a main reason why activated sludge works reliably.

Because of redundancy and the great competition for energy resources that occurs within activated sludge, subtle changes in the treatment process can result in major changes in the microbial population composition and the floc physical characteristics. For example, completely mixed systems tend to maintain consistently low substrate concentrations, while plug-flow systems tend to create more of a "feast and starve" cycle. These significantly different feeding regimes foster growth of quite different microorganisms, even when the input substrate and the SRT are identical. Other factors that affect the ecology include the dissolved-oxygen level, nutrient availability, temperature, pH, and the presence of inhibitory materials. Today, we know far too little about the processes involved to predict quantitatively the effects of different environments on the microbial ecology of activated sludge. Fortunately, aside from the problems associated with a poorly settling sludge, treatment efficiency seems to remain high through population shifts. Apparently, redundancy allows the microbial strains most capable of surviving to dominate quickly as conditions change.

6.1.2 OXYGEN AND NUTRIENT REQUIREMENTS

The activated sludge process requires oxygen and nutrients to satisfy the needs of the microorganisms. In most situations, the electron donor, or the BOD, is rate-limiting for microorganism reproduction and growth. This means that nutrients and the electron acceptor (oxygen in this case) have concentrations well above their half-saturation concentration, or K. For dissolved oxygen, K is generally much less than 1 mg/l, and a concentration of 2 mg/l or higher often keeps oxygen from being rate-limiting. The literature is not definitive about just what K is for nutrients such as nitrogen, phosphorus, iron, sulfur, and other trace constituents, but the values appear to be quite low, much less than 1 mg/l.

The oxygen consumption rate is proportional to the rate of donor-substrate utilization and biomass endogenous decay. Chapters 3 and 5 describe how to compute the oxygen consumption rate, and the design example in this chapter reiterates the most direct approach. The supply rate by aeration must be large enough that that consumption rate is met and the dissolved oxygen concentration is maintained at or near nonlimiting levels.

The consumption rate of the nutrients is proportional to the net synthesis rate of biomass. Chapters 3 and 5 also show how to compute these rates. At steady state, the net synthesis rate is equal to the biomass-wasting rate. Nutrients normally enter the system in the influent wastewater. The influent concentration of a nutrient needs to be sufficiently greater than the stoichiometric requirements for net synthesis so that its concentration does not become rate-limiting. If a nutrient is not present in the influent in a high enough concentration, it must be supplemented at a rate great enough to make up the difference.

The stoichiometric requirement for trace nutrients is more difficult to predict, as research here is limited. Frequently, these nutrients are present in most wastewaters, but this is not always the case. One might analyze a wastewater to assess the concentrations of key trace nutrients such as iron, sulfur, zinc, copper, and molybdate to be certain some is present. Another approach is to conduct side-by-side laboratory activated sludge treatment studies with the wastewater as received, perhaps with nitrogen and phosphorous additions if this is an indicated limitation, and with the same wastewater to which a suitable amount of a solution containing all normally required trace nutrients. If the unamended wastewater is treated similarly to that amended with trace nutrients, then this is evidence that the trace nutrients are present in sufficient amounts.

6.1.3   IMPACT OF SOLIDS RETENTION TIME

The solids retention time, or SRT (θ_x), is commonly used in activated sludge systems not only to control efficiency of wastewater treatment, but also in the control of sludge physical and biological characteristics. Based upon kinetic theory, a longer θ_x should provide a greater degree of substrate removal. However, other factors usually are more important. First, the formation and consumption of soluble microbial products (SMP) often dominates the soluble effluent quality as measured by BOD or COD. The SRT affects SMP concentrations in a nonlinear manner that is described in Chapter 3. Second, altering the SRT can lead to changes in sludge settling characteristics. If an increase in θ_x leads to poorer suspended-solids capture, overall removals of BOD deteriorate. Based on many decades of experience, θ_x is generally limited to a range between 4 and 10 d when BOD removal and economics are to be balanced. However, operators have flexibility to adjust the SRT in order to seek an optimum point of operation for the conditions at their treatment plant.

Activated sludge is a complex ecological system in which primary consumers of influent substrate exist together with secondary organisms and predators that live off of the primary organisms. Operation at long θ_x allows for the accumulation of slower growing organisms that are washed from the system if the SRT is short. The predator eukaryotes are among the slow growers, and they can be lost during activated sludge start-up or upset, when the bacteria often exist dispersed throughout the mixed liquor. Here, the settled effluent may be turbid, not only because of the poorly flocculated bacteria present, but also because of the protozoan population, which also does not settle well because of the small size. Protozoan populations that can move rapidly to catch such dispersed prey, such as the free-swimming ciliates, usually are the best able to survive. When operation remains stable with compact and good-settling flocs and a SRT in the range of 4 to 10 d, ciliated protozoa with stalks can attach to the floc and become important predators. They help to polish the effluent by drawing in with their cilia the few free bacteria in the system and consuming them. Since they are attached and settle with the floc, these protozoa are recycled back into the aeration tank. The presence of abundant stalked protozoa is generally considered a sign of a system that is operating well now and has been stable for some time. Page 33 shows pictures and drawings of the key eukaryotes.

With SRTs longer than about 10 d, even slower growing predators come into the picture. They include the multicellular forms, such as rotifers and nematodes, which tend to eat at the floc particles themselves. The predatory feeding on the floc may lead to the formation of small nonflocculating particles that represent the decayed remains of bacteria. For this reason, the effluent turbidity in activated sludge systems operating at long θ_x may be poorer in terms of suspended solids and BOD than is the effluent from systems with a more conventional SRT of 4 to 10 d. As discussed later under sludge-settling problems, many of the microorganisms that cause operational problems—such as sludge bulking and foaming—are relatively slow growers, compared to the bacteria that form the desirable compact floc. Finally, the chemolithotrophs, particularly the nitrifying bacteria, are slow growers that can exist in activated sludge only when the SRT is relatively long. Unless nitrification is required, having nitrifying bacteria in the activated sludge is undesirable.

For various reasons, a long SRT often is not beneficial, even if the substrate concentration can be driven lower. The lesson to be gained is that the best BOD and suspended solids removals normally are achieved with operation at intermediate θ_x values of 4 to 10 d. The design must give the operator the ability to tune the SRT to an optimum value for the conditions at hand, and the operator must be constantly vigilant to maintain the desired SRT.

An excellent auxiliary to the computation and control of the SRT is microscopic examination. The slower-growing strains that survive only when the SRT is adequate and stable are easily observed with 100-times magnification using a light microscope. For example, daily examination leads to a history of the numbers of free-swimming ciliates, stalked ciliates, rotifers, and worms. Often, sudden changes to the distribution of these higher life-forms is an early warning signal of a process upset, such as from toxicity or excess sludge loss. The floc size and structure also can be tracked, as can be the presence of filamentous forms, which are signs of settling problems and, perhaps, an excessively long SRT. While microscopic examination never is a substitute for the careful control of the SRT, it offers an outstanding complement that can forewarn or confirm trends in the SRT.

6.2 PROCESS CONFIGURATIONS

Many modifications of the basic activated sludge process have evolved since Ardern and Lockett first discovered it in 1914. These modifications stemmed largely from a long history of trial-and-error efforts to overcome problems as they developed in activated sludge operation. The modifications that are in use today represent the surviving members of many alternatives that were tried over the years. Each tends to offer some particular advantage for given circumstances.

Table 6.1 indicates important configurations that are commonly used today. They are grouped into modifications to the physical configuration of the system, modifications reflecting the way in which oxygen is added to the system, or modifications based on the organic loading. Modifications from the three categories can be combined. For example, a particular loading or one particular method of oxygen addition could be used with any of the modifications based on physical configuration. The designer can select combinations from the three different categories that are most appropriate to the conditions under consideration.

The following discussion begins with the different physical configurations. Then, modifications for oxygen addition and loading are reviewed.

6.2.1 PHYSICAL CONFIGURATIONS

The first four of the five basic process configurations listed in Table 6.1 are those most widely used in practice today (i.e., plug flow, step aeration, completely mixed, and contact stabilization). The fifth configuration, the activated sludge system with a selector, is a more recent adaptation. It is not a replacement for the four basic reactors, but an addition to the front-end of any one of them. Figure 6.1 summarizes the five different reactor configurations, which are described in more detail on page 314.

Table 6.1   Summary of activated sludge configurations

A. Modifications Based on Physical Configuration
  1. Plug Flow (Conventional)
  2. Step Aeration
  3. Complete Mix
  4. Contact Stabilization
  5. Activated Sludge with Selector

B. Modifications Based on Oxygen Addition or Distribution
  1. Conventional Aeration
  2. Tapered Aeration
  3. Pure Oxygen

C. Modification Based on Organic (BOD) Loading
  1. Conventional
  2. Modified Aeration
  3. High Rate
  4. Extended Aeration

Plug-Flow   The original or conventional activated sludge systems had long narrow aeration tanks in which the wastewater enters at one end and exits through the other. They have significant plug-flow character (Chapter 5). Intuitively, plug flow would appear to be the best approach, as kinetic theory says it should have the greatest contaminant removal within a defined treatment time. Also, contaminants in the influent stream do not short-circuit through the system without being subjected to biological action. However, the plug-flow system has its own limitations, some of which led to the other configurations.

Photo 6.1   A typical "plug flow," diffused-aeration tank for conventional activated sludge.

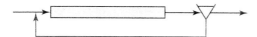

a. Plug-flow (conventional) activated sludge

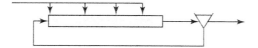

b. Step-aeration activated sludge

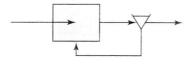

c. Completely mixed activated sludge

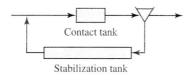

Contact tank

Stabilization tank

d. Contact-stabilization activated sludge

Selector   Aeration tank

e. Activated sludge system with anoxic selector

Figure 6.1   Activated sludge configuration
modifications.

The solid lines in Figure 6.2 illustrate how contaminant concentration and oxygen demand change with distance along the aeration tank in the plug-flow system. Contaminant concentration is highest at the entrance to the aeration tank and decreases rapidly within the first length of the aeration tank. Oxygen demand is here the highest and results from two factors. One is the oxygen demand from oxidation of the contaminant itself by the primary organisms. The other is an oxidation demand due to microbial cell respiration, either through endogenous respiration or through predation by secondary organisms within the system. The oxygen consumption from substrate oxidation occurs in the first 20 percent of the tank and is the main reason why the oxygen-consumption rate is so much higher near the inlet end. When the substrate is consumed, the oxygen demand decreases to a level close to that from cell respiration alone.

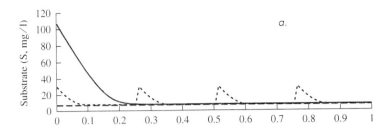

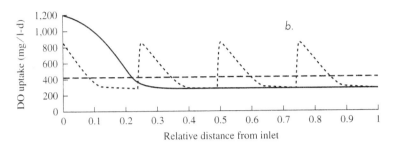

Relative distance from inlet

Figure 6.2   Changes in contaminant (substrate) concentration and oxygen (DO) uptake rate along the reactor length for plug flow (PF, solid lines), step aeration (SA, small-dash lines), and continuous-stirred tank (CSTR, large-dash lines) reactors for a typical loading with a dilute wastewater.

Possible problems with plug-flow systems result from the high concentration of contaminants at the head end of the aeration tank. First, the high rates of contaminant oxidation may lead to complete depletion of dissolved oxygen, leading to anoxic conditions. Oxygen depletion may be detrimental to some of the microorganisms and may also result in fermentation or partial oxidation of some contaminants, giving rise to organic acid production and a drop in pH. Low pH and loss of dissolved oxygen concentration can have adverse impacts on the treatment rate and health of the biological community. Second, many industrial wastes contain substances that are inhibitory to the bacteria, thus slowing down or stopping the process. The microorganisms are exposed to elevated concentrations of the inhibitor at the inlet of the plug-flow reactor. While the plug-flow system can have real and significant advantages, an influent wastewater that contains inhibitory material or gives a high BOD loading can bring the negatives of plug flow to the forefront.

Although the classical plug-flow system uses the long narrow tanks shown in Figure 6.1, two other techniques can give the plug-flow character. The first is to have several completely mixed *tanks in series*. Although each tank is completely mixed, the contents do not mix among them. Having three or more tanks in series gives about the same degree of "plug flow" as does the conventional long narrow tank. The second is to use batch reactions in what is now called a *sequencing batch reactor*

(SBR). The website chapter "Complex Systems" provides an excellent example of how an SBR reactor operates. Several batch reactors are run in parallel so that one can receive influent while the others are in various stages of the react mode. Ironically, much of the early work with activated sludge involved SBRs, which were then called *fill-and-draw* activated sludge. Fill-and-draw lost favor to continuous-flow systems due to the need for constant operator attention to make changes from fill to react to draw stages. The advent of inexpensive and reliable microprocessors in the 1970s led to a resurgence of interest in the SBR approach.

Step Aeration   *Step-aeration* (Figure 6.1), also called *step feeding* (a more accurate descriptor), was developed as a means to circumvent some of the problems of the plug-flow approach. Since the two problems result from high concentrations of contaminants at the head end of the aeration basin, high concentrations at any one location are avoided by distributing the influent along the length of the reactor in steps. The benefit is shown by the changes in contaminant concentration and oxygen demand along the reactor, as shown in Figure 6.2. While the parameters vary along the aeration tank, the changes are quite muted compared with the plug-flow system. In particular, the concentration of an influent contaminant is diluted much more, and the oxygen-uptake rate is spread out. These differences can overcome the two problems associated with plug flow. Because inhibitory materials and BOD overloading occurred frequently, step aeration became a widespread configuration.

An interesting effect of step aeration is that the *mixed liquor suspended solids (MLSS)* is highest at the inlet end, since the full sludge recycle mixes with only part of the influent flow. This feature can be exploited to increase the average MLSS concentration, which increases the SRT for the same reactor volume and sludge-wasting rate. Alternately, the volumetric loading can be increased with the same SRT, but an increased sludge-wasting rate.

Completely Mixed   The *completely mixed* activated sludge treatment process, or a CSTR with settling and recycle, evolved in the 1950s. Two factors spurred the use of completely-mixed designs. First, industrial wastewaters were being treated more, but with notable lack of success using the conventional plug-flow system. This difficulty was due to the high contaminant concentrations at the head end of the plug-flow system. A completely-mixed system is the ultimate approach for spreading the wastewater uniformly throughout the treatment system. Second, the 1950s marked the beginning of reactor modeling for biological processes. Being the simplest system to analyze, the completely-mixed system became a favorite of design engineers who wanted to apply their newly gained tools on quantitative analysis.

Figure 6.2 illustrates how contaminant concentration and oxygen demand do not vary over the reactor length. The effluent soluble contaminant concentration is uniform and low throughout the reactor. With this system, the microorganisms never are exposed to the influent concentration of a contaminant, as long as the contaminant is biodegradable by the microorganisms. A completely-mixed system is most favorable with wastewaters containing biodegradable materials that also are toxic to microorganisms at modest concentrations. This includes phenols, petroleum

Photo 6.2   A small "package plant" for a housing development is a good example of a complete-mix reactor.

aromatic hydrocarbons, and many chlorinated aromatics, such as chlorinated phenols and benzoates. The disadvantage with this system is that removal efficiency for an individual organic compound is not as high as in a well operating plug-flow system in which high concentrations are not a problem.

Contact Stabilization   The *contact stabilization process* permits high-efficiency treatment to occur in a significantly reduced total reactor volume. Waste-water is mixed with return activated sludge in a contact reactor having a relatively short detention time, on the order of 15 to 60 min. This represents the time during which the most readily biodegradable organic contaminants are oxidized or stored inside the cells, and the particulate matter is adsorbed to the activated sludge flocs. The treated wastewater then flows into the final settling tank, where the activated sludge and the treated wastewater are separated from one another. The wastewater is discharged, and the settled and concentrated activated sludge is sent to a second reactor termed the *stabilization tank*, where aeration is continued. Here, adsorbed organic particles, stored substrates, and biomass are oxidized. In fact, the majority of the oxidation occurs in the stabilization tank. The concentrated MLSS are then returned to the contact tank to treat another volume of wastewater.

What is the advantage of contact stabilization, and what are its limitations? The biggest advantage of the contact-stabilization system is the reduction in overall reactor volume. A certain θ_x for the activated sludge in the overall treatment system (not just the contact tank) is needed in order to obtain BOD removal and a good settling activated sludge. The numerator in the computation of θ_x is the sum of the total mass of MLSS in the contact and stabilization tanks combined. For example, if 1,000 kg of MLSS were produced as waste sludge per day, and the mass of MLSS in the contact tank were 2,000 kg and that in the stabilization were 6,000 kg, then θ_x equals $(2,000 + 6,000)$ kg/1,000 kg-d or 8 d. In this example, we see that 75 percent of the MLSS resides in the stabilization tank. The key to the volume savings is that the concentration of MLSS is much higher in the stabilization tank than in the contact tank, since the stabilization tank receives thickened sludge from the underflow of the settler. If 75 percent of the biomass is contained in a volume that has four-fold concentrated sludge, then the total volume of the contact-stabilization reactors is only 44 percent of that for a conventional activated sludge system. Information presented under Loading Criteria expands the discussion about the substantially smaller overall reactor volume obtainable with contact stabilization.

One disadvantage of contact stabilization is that it requires substantially more operational skill and attention. First, two mixed liquors need to be monitored, and both results are necessary to compute the SRT. Second, the small volume of the contact tank makes the effluent quality susceptible to sudden increases in loading. Contact stabilization should be utilized only when operator skills and attention are high and when the system is not subject to sudden and large swings in loading. Unfortunately, many manufacturers of prefabricated "package plants" chose to use the contact-stabilization configuration. These package plants were installed in small towns and industrial settings in which the operations and loading conditions were exactly the opposite of what is necessary. Predictably, these plants are constantly plagued by upsets and poor effluent quality. In most cases, the best strategy is to convert them to the complete-mix configuration.

Activated Sludge with a Selector   The most frequent cause of failure of activated sludge systems is *sludge bulking*, or a sludge that does not compact well in the settling tank. Many procedures used to overcome this problem are discussed later in the chapter. One of the recent innovations is to add a *selector tank* prior to the aerobic reactor, as illustrated in Figure 6.1. Return activated sludge is contacted with the waste stream for only 10 to 30 min under conditions in which complete BOD oxidation is impossible. Fermentation reactions then convert carbohydrates and perhaps some proteinaceous materials to fatty acids, which cannot be oxidized, but are stored by some microorganisms in the form of glycogen or polybetahydroxybutyric acid (PHB). The storage materials provide an ecological advantage to the bacteria when they enter the oligotrophic environment of the normal aeration tank. Fortunately for us, the bacteria able to store these materials also are good at forming compact sludge floc. Thus, the function of the selector is to change or *select* the ecology of the activated sludge system towards organisms with good settling characteristics. We provide more information on how the selector is designed and used in the section on Sludge Settling Problems.

6.2.2   OXYGEN SUPPLY MODIFICATIONS

Oxygen must be supplied to satisfy the needs of the microorganisms. Generally, this is accomplished by the introduction of air, although pure oxygen can be used. Several modifications to the activated sludge process are based on the manner by which oxygen is supplied or by which oxygen is distributed over the aeration basin. Table 6.1 identifies three generally recognized modifications of the activated sludge process that are related to the oxygen supply: conventional aeration, tapered aeration, and pure oxygen.

A corollary need in the activated sludge process is to keep the flocs in suspension and in contact with the contaminants. The aerators used to supply oxygen generally accomplish the mixing. Thus, aeration in the activated sludge process has dual purposes: oxygen supply and mixing. Both requirements must be given due attention in design.

Oxygen is supplied by one of two technologies: diffused aeration or mechanical aeration. In diffused aeration, air is compressed and pumped through diffusers located near the bottom of the aeration basin. The rising air bubbles cause mixing in the basin, and at the same time, oxygen in the bubbles is transferred to the water through normal mass transfer processes. In mechanical mixing, a mechanical mixer agitates the surface of the aeration basin. The mechanical churning at the surface effectively "pumps water droplets through the air," which brings about rapid transfer of oxygen from the air to the water. The mechanical mixing also keeps the flocs in suspension. Diffused air and mechanical aeration are common. The following discussions of the three major modifications related to oxygen distribution indicate when one or the other technology is more appropriate. More information on the design of aeration systems is contained in the section on Aeration Systems.

Conventional Aeration   Conventional aeration refers to the original way in which oxygen was distributed over the activated sludge basin. Figure 6.3 shows conventional aeration for a plug-flow system. The key here is that oxygen is distributed uniformly over the aeration basin. For plug-flow systems, the aeration technology used is almost always diffused aeration, since the diffusers can be located along the length of the plug-flow tank. For a complete-mix configuration, the geometry generally is more square or circular. Mechanical aeration is very common in complete-mix systems, although diffused aeration also is employed.

Tapered Aeration   Tapered aeration is a simple and obvious method to overcome the basic problem with conventional aeration: The oxygen uptake is high at the head of a plug-flow tank and low at the exit end. As illustrated in Figure 6.3, the air supply is increased at the head end of a plug-flow system so that the oxygen supply and demand are equalized. The air supply then tapers off with distance along the aeration tank so that supply and demand are kept in balance throughout the basin. Besides overcoming oxygen depletion, tapered aeration can reduce power costs and equipment sizes. Diffused aeration is the most common method for tapered aeration, but it is possible to taper the power of mechanical aerators to create the same effect. The plug-flow nature can be accentuated, particularly with mechanical aeration, by using reactors in series.

Pure Oxygen   In pure-oxygen activated sludge, 100% O_2, rather than air (21 percent O_2), serves as the source of oxygen to the system. The advantages of pure oxygen occur because the mass-transfer driving force is increased by as much as five times, since the partial pressure of oxygen in air is only about one-fifth of that in pure oxygen. Power costs for aeration and mixing can be reduced substantially. Just as valuable is that higher dissolved oxygen (D.O.) concentrations can be maintained in the mixed liquor. Higher D.O. is associated with higher MLSS concentrations, better settling sludge, and higher volumetric loading.

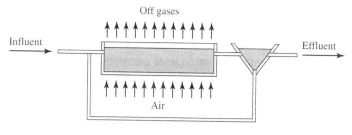

a. Conventional aeration

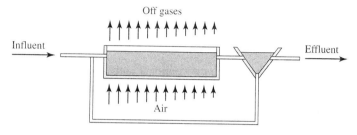

b. Tapered aeration

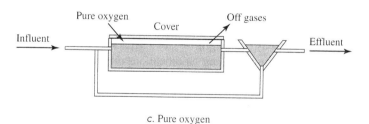

c. Pure oxygen

Figure 6.3   Comparison between conventional aeration, tapered aeration, and pure oxygen activated sludge systems.

The negative side is that pure oxygen is expensive to produce, while air is essentially free. Thus, the reduction in power cost is offset by an increase in chemical costs. The pure oxygen can be transferred either with diffusers or by mechanical mixers, but the latter are generally used, because the reduced gas-pumping rate with pure oxygen may be inadequate to keep the flocs in suspension. Mechanical mixing requires that the atmosphere above the mixed liquor be enriched in O_2, which is accomplished by covering the tank, as illustrated in Figure 6.3. Sealing the tanks with covers is an added capital expense.

Most designs of pure-oxygen systems create a plug-flow character by dividing the covered reactor into three or four chambers in series. This is the best way to use mechanical aerators and still have plug-flow character.

Do the savings in power cost and smaller size compensate for the added costs to supply O_2 and cover the tank? Apparently not, as the popularity of pure-oxygen activated sludge peaked in the 1970s. On the other hand, some new regulations require that a cover be placed over aeration basins to reduce odor or volatile organic chemical (VOC) emissions during treatment. Also, concerns are growing over worker safety from volatile organic chemicals or pathogenic microorganisms coming from aeration basins. When covers are justified by other considerations, pure oxygen systems may become economically competitive.

6.2.3   LOADING MODIFICATIONS

Modifications of the activated sludge process also are based on design loading. Two design modifications employ a higher organic loading than are used for conventional systems: modified aeration and high rate. A modification that uses a lower loading is called extended aeration.

Conventional Loading   The conventional activated sludge system for treatment of municipal wastewaters evolved through trial and error into a reliable system that removes BOD_5 and suspended solids (SS) by 85 percent or more for the usual case: municipal sewage. Municipal wastewaters are similar enough in contaminant concentration and composition that empirical experience in one city can be transferred with a good degree of confidence to another city. Having little or no theory to use as a basis for design, early engineers sized aeration tanks from hydraulic detention time, per capita BOD_5 loading, or MLSS concentration. The early literature indicated aeration basins should have a 6-h hydraulic detention time, a BOD loading of 35 pounds per 1,000 cubic feet per day (0.56 kg/m^3-d), or a MLSS concentration between 1,200 and 3,000 mg/l. The various loading modifications originally were related to these conventional criteria.

In the intervening years, a sound theoretical basis for the design and operation of activated sludge (and microbiological processes in general) developed. These principles are presented in Chapters 2 to 5. Now, we can interpret the empirical loading criteria with more fundamental parameters, such as θ_x. In the subsections that immediately follow, we define the loading modifications in terms of the traditional, empirical criteria. Later in the chapter, we revisit all the loading modifications and interpret them in terms of the SRT and other loading criteria.

Modified Aeration   Modified aeration was designed for situations in which a high degree of purification (> 85 percent BOD removal) may not be required, but something better than could be obtained with settling alone was needed. The empirical loading criteria are: the MLSS in the range of 300 to 600 mg/l, and the aeration period reduced to about 1.5 to 2 h. BOD_5 and suspended solids removal efficiencies are on the order of 65 to 75 percent. Interestingly, the relatively compact sludge dewaters well. Treatment requirements today seldom can be achieved with the modified aeration system, but it provides a low-cost alternative if the situation permits it.

High Rate   In order to reduce aeration-tank volume, efforts were made to increase the organic loading, while still maintaining the high removal efficiencies commonly required. The "trick" to doing this is to make the MLSS concentration much higher, thereby permitting shorter hydraulic detention times (with the same SRT). Raising the MLSS concentration to the range of 4,000 to 10,000 mg/l increases the BOD_5 loading up to 100 to 200 pounds per 1,000 cubic feet per day (1.6 to 3.2 kg/m^3-d). Of course, with such high loading, the oxygen uptake rate per unit aeration-tank volume increases proportionately. This requires better oxygen transfer, generally achieved by supplying more air or greater mechanical mixing, which in turn increases the turbulence in the aeration tank. Today, high-rate systems often are combined with pure oxygen to enhance the oxygen supply without the excessive turbulence that might otherwise affect sludge flocculation and settling ability.

If the oxygen supply is sufficient and problems of sludge settling can be avoided, high-rate systems can achieve 90 percent BOD_5 and SS removal efficiencies and reduce reactor volumes considerably. However, increased MLSS can force an increase in settling tank size, which partially offsets the capital savings from the smaller reactor volume. And, failure to supply enough oxygen can lead to severe settling problems, particularly bulking.

Extended Aeration   Extended aeration stemmed originally from the need to reliably treat low-volume wastewaters when full-time, knowledgeable operators could not be afforded. Typical applications are a shopping center, a condominium complex, a large hotel, a rest stop along an isolated freeway, or a small industry. In such cases, transport of the wastewater to a larger municipal treatment plant may be impractical. In the past, simple treatment systems such as septic tanks may have been used, but they can result in poor treatment and groundwater contamination. Activated sludge systems operated at low loading rates to achieve good reliability have been one answer commonly used. Detention times in extended aeration are on the order of 24 h. Relatively high MLSS concentrations of 3,000 to 6,000 mg/l and low BOD_5 loading of less than 15 pounds per day per 1,000 cubic feet of aeration capacity (≤ 0.24 kg/m^3-d) are the empirical guidelines.

A problem that often results with such systems is that they do require some knowledgeable care. The systems do not run themselves just because the pumps, compressors, and other mechanical devices are kept greased and in good running order. Such systems need routine attention by operators who understand the fundamentals of activated sludge and can judge what factors (such as a sufficient oxygen supply or a sufficiently high, but not too high MLSS concentration) to monitor and control. Typical problems are bulking sludge leading to loss of the entire microbiological population, low pH due to nitrification, and high suspended solids concentrations in the effluent. The designing engineer must emphasize the system's operating needs to the client and evaluate whether or not adequate skilled operation is likely to be available when the plant goes on line.

6.3   DESIGN AND OPERATING CRITERIA

Criteria used for the design and operation of activated sludge range from those totally empirical to those soundly based in fundamentals. In this section, we review the major design criteria. We point out fundamental bases, when they exist. We also compile these criteria for the characteristic loading modifications described earlier.

6.3.1 HISTORICAL BACKGROUND

When the activated sludge process was first invented in 1914, there was no understanding of kinetics of biological growth and substrate removal. Designs were based on empirical observations of the relationships among detention time, suspended solids concentration in the aeration basin, and BOD_5 removal. The conventional 4- to 8-h hydraulic detention time for aeration basins treating domestic wastewater came about during this period of strict empiricism. Such designs often failed when industrial wastewaters were involved, especially if of high strength. A better understanding of factors affecting the process was needed. During the late 1940s and early 1950s, other parameters were investigated. One of growing importance was the organic loading expressed in mass of BOD_5 applied per unit aeration tank volume per day, as discussed in the preceding section. Conventional loading is on the order of 35 pounds BOD_5 per day per 1,000 cubic feet of aeration capacity (0.56 kg/m^3-d). Attempts were made to increase this loading rate, as its impact on reactor size and capital cost is obvious. However, the volumetric load is not much of a step forward in terms of understanding the factors controlling the process. Later came relationships among treatment efficiency, organic loading, MLSS, and oxygen supply. Although still quite empirical, these relationships led to the observation that oxygen supply was a major limitation to the system; if higher loading was to be obtained, methods for transferring oxygen at a faster rate had to be developed.

A critical observation was that higher loading required a higher MLSS concentration (Haseltine, 1956). Finally, it was recognized that MLSS concentration represents in some manner the concentration of bacteria that are active in the system. Since it was (and still is) difficult to measure the concentration of microorganisms that are active in BOD oxidation, the MLSS became a surrogate measure for the active population. However, MLSS has many limitations. For example, the MLSS also contains suspended solids that are present in the influent waste stream, including refractory suspended solids and the remnants of the degradable suspended solids. It also contains predatory organisms that are not the primary consumers of wastewater BOD. In addition, dead cells and inactive particulate cellular components are present. Although MLSS is an imperfect measure of the active population of interest, its use as a surrogate was a major step in the right direction.

Later yet, the volatile fraction of the MLSS, the MLVSS (mixed liquor volatile suspended solids), became a better surrogate for active microorganisms. Microorganisms consist mainly of organic material, and only about 10 percent of bacterial SS is inorganic material. However, the suspended solids in MLSS often contain a higher concentration of inorganic materials, which come from silt, clay, and sand that escaped removal in primary settling tanks, and some inorganic precipitates, such as $CaCO_3$, that may form within the aeration tank.

6.3.2 FOOD-TO-MICROORGANISM RATIO

The food-to-microorganism ratio was developed in the 1950s and 1960s (Haseltine, 1956; Joint Task Force, 1967) and is still widely used because of its simplicity. It is intuitive, conceptually is easy to explain, and relies on measurements that are routinely obtained with relative ease at essentially all treatment plants. However, the food-to-microorganism ratio has limitations due to its simple format.

In equation form, the food-to-microorganism ratio (F/M) is

$$F/M = \frac{Q^0 S^0}{V X} \qquad \textbf{[6.1]}$$

in which

F/M = food-to-microorganism ratio, kg BOD or COD applied per day
      per kg of total suspended solids in the aeration tank
Q^0 = influent wastewater stream flow rate (m^3/d)
S^0 = influent wastewater concentration (BOD or COD in mg/l)
V = aeration-tank volume (m^3)
X = total suspended solids concentration in aeration tank (mg/l)

If volatile suspended solids, rather than total suspended solids are used, then Equation 6.1 is slightly modified to

$$F/M_v = \frac{Q^0 S^0}{V X_v} \qquad \textbf{[6.2]}$$

in which

F/M_v = food-to-microorganism ratio on volatile solids basis, kg BOD
      or COD per day per kg of volatile suspended solids in aeration tank
X_v = volatile suspended solids concentration in aeration tank (mg/l)

The F/M concept had some basis in theory (described in the next paragraph), but values used in practice were simply derived from empirical observations. For a conventional design for the activated sludge treatment of domestic sewage, the F/M ratio suggested is 0.25 to 0.5 kg BOD_5 per day per kg MLSS, a range that generally results in reliable operation with BOD_5 removal efficiencies of about 90 percent. This range is consistent with values obtained from Equation 6.1 when using the typical 6-h detention time (V/Q^0), an influent BOD_5 of 200 mg/l, and an MLSS concentration (X) of 1,600 mg/l. High-rate treatment is represented by an F/M ratio of 1 to 4 kg BOD_5 per day per kg MLSS, and extended aeration by an F/M of 0.12 to 0.25 kg BOD_5 per day per kg MLSS.

In order to see how the F/M can be related to basic kinetics, we rearrange the substrate mass balance using the Monod relationship (e.g., Equation 3.17) to give the definition of the volumetric loading for a CSTR:

$$\frac{\hat{q} S}{K + S} X_a = \frac{Q^0(S^0 - S)}{V} \qquad \textbf{[6.3]}$$

Equation 6.3 can be rearranged to give the following form:

$$\frac{Q^0 S^0}{V X_a} = S \left(\frac{\hat{q}}{K + S} + \frac{Q^0}{V X_a} \right) \qquad \textbf{[6.4]}$$

The left side of the equation is a type of food-to-microorganism ratio, one based upon the active organism population. We call this the F/M_a ratio, where for this purpose M_a is the same as X_a. Of the two ratios within the brackets on the right side of the equation, the left term is far larger than the right term when S is small. Indeed, if S is small with respect to K, then the left term about equals $\hat{q}/K$. Thus, for the usual case in which we have high treatment efficiency and a low effluent BOD concentration (typical of activated sludge treatment), Equation 6.4 can be simplified to

$$F/M_a = \frac{\hat{q}}{K} S \qquad\qquad \textbf{[6.5]}$$

Solving for S we obtain

$$S \approx \frac{K}{\hat{q}} \cdot F/M_a \qquad\qquad \textbf{[6.6]}$$

The reactor substrate concentration (which equals the activated sludge effluent concentration S^e in a complete-mix system) is directly related to F/M_a. Thus, if one could actually measure the M_a part of F/M_a, the effluent substrate concentration could be directly related to F/M_a. However, M_a is almost impossible to measure, which breaks the connection between a measurable F/M_a and S.

We illustrate the problem for a typical activated sludge system. Reasonable values for the coefficients in Equation 6.6 are obtained from the summary of model parameters for municipal wastewaters (Henze et al., 1995). Here, K equals 4 mg BOD_5/l and $\hat{q}$ is 10 kg BOD_5/kg VSS_a-d (based upon the ratio of $\hat{\mu}/Y$), giving a value for the ratio of $K/\hat{q}$ of 0.4 mg-d/l. If we used the F/M ratio for conventional activated sludge treatment of 0.5 kg BOD_5 per kg SS per day, the predicted effluent BOD_5 from Equation 6.6 would be 0.2 mg/l. However, if one uses the active population only (which typically may be no more than 30 percent of the total MLSS concentration ($M_a = 0.3M$)), then F/M_a would equal 1.7 instead of 0.5, and the predicted S from Equation 6.6 would be $0.4 \cdot 1.7 = 0.67$ mg/l. Although this second S value is still quite low, the sensitivity of the F/M concept to the actual value of M_a is obvious. The M_a/M ratio can vary quite widely, but it is impossible to determine by measurements that are practical at most treatment facilities.

The units of F/M are almost the same as those of the maximum specific utilization rate, $\hat{q}$: kg BOD_5 per kg MLSS per day for F/M and kg BOD_L per kg VSS_a per day for $\hat{q}$. By comparing the ratio $(F/M)/\hat{q}$, we gain some insight into how close the actual substrate utilization rate is to the maximum. To have a good safety factor, we must have $(F/M)/\hat{q}$ be far less than one. The actual F/M values are far below $\hat{q}$, which confirms that we probably are operating with a good factor of safety in terms of the substrate utilization capacity of the microorganisms. However, the measure is a very crude one, because we normally do not know exactly how the measured MLSS relates to X_a.

6.3.3 SOLIDS RETENTION TIME

Using the solids retention time, or θ_x, for the design and operation of suspended growth systems evolved from basics of bacterial growth developed by Monod (1950) and related kinetic descriptions of substrate utilization in continuously fed reactors

(Novick and Szilard, 1950; Herbert, Elsworth, and Telling, 1956). These important studies established that the hydraulic detention time of a CSTR (or dilution rate as defined by these investigators, which is the inverse of the hydraulic detention time) controls the residual substrate concentration. Of course, we now know that the hydraulic detention time in a CSTR is equal to its θ_x. They established the concept that washout of bacteria occurs at low detention times, that is, when the washout rate exceeds the growth rate.

Parallel developments were occurring in the wastewater treatment field, where attempts were being made to better understand the relationship between treatment efficiency and operating parameters for the activated sludge system. One important observation was made by Gould (1953), who noted that the operational characteristics, including that of the activated sludge itself, were related to what he termed the *sludge age*. He thought of sludge age as the time of aeration of the influent suspended solids, and thus he defined sludge age as the weight of suspended solids in the aeration tanks divided by the daily dry weight of the *incoming* suspended solids of the sewage. This definition is quite different from our definition of θ_x as the weight of suspended solids in the aeration tanks divided by the daily dry weight of the *outgoing* suspended solids in the plant effluent and activated sludge wasting lines. Since for domestic sewage, the incoming and outgoing suspended solids bear some resemblance in mass flow rate, Gould's empirical observations often had some consistency with current theory, although by coincidence. However, sludge age as defined by Gould should not be applied, because the incoming and outgoing solids are not related in any way in general. Today, in order to avoid confusion, θ_x is called the solids retention time instead of sludge age.

Garrett and Sawyer (1951) were the first to directly apply the idea of SRT as a measure of activated sludge performance. They measured the average retention time of cells in the reactor and related it to the total suspended solids concentration in a manner similar to that used in the current definition of θ_x. They found that effluent quality became poorer as θ_x decreased. Furthermore, the washout value, θ_x^{min}, was about 0.5 d at 10 °C, 0.2 d at 20 °C, and 0.14 d at 30 °C, values completely consistent with what one might expect from normal heterotrophic growth. While this recognition of θ_x came quite early after the basic theoretical developments in microbiology, general acceptance by the environmental engineering field was slower. The systematic use of this basic concept for the design of all suspended-growth biological treatment systems was not described until almost 20 years later (Lawrence and McCarty, 1970), and widespread application took more time. However, it did come.

The Water Environment Federation and the American Society of Civil Engineers, the two major United States professional organizations in the wastewater treatment field, worked together for many years to develop design manuals for municipal wastewater treatment plants. In the 1998 update of their design manual (Joint Task Force, 1998), they reported that major consulting firms in the United States used θ_x more than any other design criterion. SRT also provides the basis for the IWA model developed by an international committee (Henze et al., 1995). This IWA formulation is available in computerized form to automate design calculations. The θ_x approach has been modified and expanded over that originally presented in 1970 (Lawrence and McCarty, 1970), and the systematic developments presented in Chapters 3 and 5 and the website chapter "Complex Systems" document the numerous refinements and expansions.

In its essence, θ_x is the master variable for the design and operation of the activated sludge process. θ_x is the right master variable because it is fundamentally related to the growth rate of the active microorganisms, which in turn controls the concentration of the growth-rate-limiting substrate in the reactor. θ_x also is an excellent choice because all the parameters that comprise it can be measured accurately and consistently. The θ_x definition for a normal activated sludge system is repeated here from Chapter 5:

$$\theta_x = \frac{XV}{Q^e X^e + Q^w X^w} \qquad [6.7]$$

in which V = system volume [L^3], Q^e = effluent flow rate [L^{3}T^{-1}], Q^w = waste-sludge flow rate [L^{3}T^{-1}], and X, X^e, and X^w are the concentrations of mixed-liquor, effluent, and waste sludges in consistent mass units, which can be active volatile solids, volatile solids, or suspended solids. As long as active biomass is not an input, any of the three solids measurements can be used for the X values in Equation 6.7 and give the same correct value of θ_x. Being able to use SS and VSS, which are simply and routinely measured, to estimate θ_x is a major practical advantage.

Typical values of θ_x used for design of conventionally loaded treatment systems are in the range of 4 to 10 d. Extended aeration units generally have much longer θ_x, in the range of 15 to 30 d, and sometimes much longer. The modified aeration process, on the other hand, has a short θ_x, in the range of 0.2 to 0.5 d.

The θ_x values typically used have evolved from empirical practice over the years, rather than directly from treatment process theory. The values represent conservative designs where high reliability in performance is required. They take into account that effluent BOD measurements include the oxygen demand of effluent suspended solid and soluble microbial products, as well as any residual substrate that entered in the influent. Indeed, the oxygen demand from decay of active cells can overwhelm the soluble components if good solids separation is not achieved. Where high BOD removal efficiencies are desired, effluent suspended solids must be maintained at very low concentrations. Thus, the settling ability of the activated sludge and the efficiency of the final clarifier take on paramount importance.

This important direct relationship between effluent suspended solids and effluent BOD forms one basis upon which the typical designs leading to a θ_x of 4 to 10 d originated. At lower values of θ_x, bacterial flocs tend to disperse, and effluent suspended solids concentrations are fairly high. At long θ_x values, bacterial flocs also tend to break up and disperse (Bisogni and Lawrence, 1971). As is described more completely later in the section on Dispersed Growth and Pinpoint Floc, this may be related to reduced percentage of active bacterial population, which through polymer production tend to hold the individual particles together in the floc, or to the destruction of the floc through the action of predatory populations of protozoa, rotifers, and nematodes. Floc breakup is often noted to begin with θ_x greater than 8 d at temperatures of 20 °C and higher, or at somewhat longer times with lower temperatures.

Thus, the θ_x range of 4 to 10 d represents a zone where biological flocculation and clearer effluents appears to be optimal. It is, thus, the preferred range for design of well-operating and efficient activated sludge treatment systems.

A second basis for the conventional range of 4 to 10 d is that higher SRTs allow the growth and accumulation of slow growing microorganisms that are not desired. The nitrifying bacteria, which oxidize ammonium to nitrate, are the first slow growers in this category. Nitrification often is desired, and it is thoroughly discussed in Chapter 9. When the oxidation of ammonium is not a treatment goal, having nitrifiers is undesirable for three reasons. First, ammonium oxidation creates a very large oxygen demand that is expensive to meet and may overwhelm an aeration system not designed for it. Second, the nitrifiers release a significant amount of soluble microbial products, which increase the effluent COD and BOD. Third, the nitrifiers generate a significant amount of acid, which can be a problem in low-alkalinity waters. Filamentous bacteria, which cause sludge bulking, comprise a second group of undesired slow growers. They are discussed later in this chapter under Bulking and Other Sludge Settling Problems.

An interesting question about the computation of the SRT concerns what volume to use for V in Equation 6.7. Options were discussed in Chapter 5 in the section on Evaluation of Assumptions. Some advocate using only the aeration-tank volume, because good mixing and aeration occur there, but not in the settler. Others argue that the settling tank contains a significant amount of biomass that is within the boundaries of the activated sludge system. The actual SRT must take into account the total amount of biomass in the system, as well as all paths by which the biomass exits the system. On this basis the product XV in the numerator ought to include biomass in the settler, as well as in the aeration tank.

The practical challenge, then, is knowing what XV to use for the settler. A settler normally has a large clarified zone at the top and a smaller sludge blanket at the bottom. The clarified zone has very little biomass, but the sludge blanket has much biomass. Ideally, the operator or designer has measurements of the depth and the solids concentration in the sludge blanket. Then, the mass of sludge for the settler is easily computed. In reality, this information often is not known at all, and the characteristics of the sludge blanket change significantly with time. When the biomass in the settler cannot be ignored, a simple solution for this dilemma is to assume that the average sludge concentration in the settler is equal to that in the aeration basin (recall discussion in Chapter 5). Then, the numerator of Equation 6.7 simply is $X(V_{aer} + V_{set})$, where V_{aer} and V_{set} are the volumes of the aeration basin and the settler, respectively.

6.3.4 COMPARISON OF LOADING FACTORS

Table 6.2 summarizes typical loading factors for various activated sludge modifications. The θ_x^d indicated there represents a design value for θ_x. Only three distinct SRT ranges are practiced: 4 to 14 d for the several variations on conventional loading, greater than 14 d for extended aeration, and less than 4 d for modified aeration. Most conventional processes keep the SRT in the range of 4 to 10 d for the reasons

Table 6.2   Typical process loading factors and θ_x^d values for various activated sludge process modifications

Process Modification	Volumetric kg BOD$_5$/m^3-d	MLSS mg/l	F/M$_v$ kg BOD$_5$/ kg X$_v$-d	Typical BOD$_5$ Removal Efficiency	Typical θ_x^d d	Safety Factor*
Extended Aeration	0.3	3,000–5,000	0.05–0.2	85–95^B	>14	>70
Conventional						
Conventional	0.6	1,000–3,000	0.2–0.5	95	4–14	20–70
Tapered Aeration	0.6	1,000–3,000	0.2–0.5	95	4–14	20–70
Step Aeration	0.8	1,000–3,000	0.2–0.5	95	4–14	20–70
Contact Stabilization	1.0	A	0.2–0.5	90	4–15	20–75
Modified Aeration	1.5–6	300–600	0.5–3.5	60–85^B	0.8–4	4–20
High-Rate Aeration	1.5–3	5,000–8,000	0.2–0.5	95	4–14	20–70

*Assumed value of growth coefficients: $Y = 0.65$ g cells/g BOD$_5$, $b = 0.15$ d^{-1}.
A: Contact tank typically has 1,000–3,000 mg/l; stabilization tank typically has 5,000–10,000 mg/l.
B: Higher efficiency is based upon soluble effluent BOD$_5$.
SOURCE: Lawrence and McCarty (1970).

stated in the previous section. Many extended-aeration systems are designed very conservatively and have SRTs on the order of 25 to 50 d, and sometimes it is even larger.

Table 6.2 shows that the F/M$_v$ ratio is inversely proportional to the SRT, and this is completely logical since the F/M$_v$ ratio is a crude approximation of the cells' specific substrate-utilization rate, U. U is related to θ_x by

$$\theta_x = (YU - b)^{-1} \qquad \textbf{[6.8]}$$

where

$$U = -r_{vt}/X_a = Q(S^0 - S)/X_a V \qquad \textbf{[6.9]}$$

To the degree that F/M$_v$ is a surrogate for U, then F/M$_v$ and θ_x must be reciprocally related via Equation 6.8.

A high volumetric loading to an aeration tank can be achieved in two ways. The first is to have a small SRT, as is the case with modified aeration. The second is to maintain a high MLSS, as is the case with high-rate activated sludge and contact stabilization. No matter how the high volumetric loading is achieved, it is successful only when the aeration capacity is sufficient to meet the high volumetric oxygen-uptake rate.

6.3.5   MIXED-LIQUOR SUSPENDED SOLIDS, THE SVI, AND THE RECYLE RATIO

One of the major design decisions for the activated sludge process is what mixed-liquor suspended solids concentration, or X, to select for the aeration tank. The selection is not simple, as it depends upon many factors, including the settling characteristics of the activated sludge, the rate of recycle of sludge from the settling tank back to the aeration tank, and the design of the settling tank. These factors are discussed in detail in a following section on the Analysis and Design of Settlers. Here, a brief discussion of general practice is provided.

In general, one would like to use a high value for X in the aeration tank, as this leads to a smaller aeration basin, which translates into lower construction costs. On the other hand, the settling tank may need to be increased in size because of the greater solids flux to the settler. If a higher X is used to gain a larger volumetric loading, the cost of the aeration system may be increased to meet the more intense oxygen-uptake requirement. Increasing X also requires that the recycle sludge be returned at a faster rate. Finally, a higher X in the reactor may lead to higher effluent suspended solids and BOD. Clearly, an arbitrary choice for mixed-liquor suspended solids is very risky.

First, we develop the relationship between X and the return sludge flow rate Q^r. The flows and concentrations are illustrated in Figure 6.4. A mass balance on suspended solids around the settling tank (control volume **a**), is

$$Q^i X = Q^e X^e + Q^s X^s \qquad \textbf{[6.10]}$$

For a good-functioning settling tank and activated sludge system, the effluent suspended solids concentration X^e is very low, and over 99 percent of the suspended solids entering the settling tank settle and are removed from the bottom of the settling tank. Thus, $Q^e X^e$ should be very small relative to the other terms in Equation 6.10 and can be taken as zero for this analysis. Also, the sludge wasting flow rate, Q^w, normally is only a few percent at most of the recycle flow rate Q^r. Thus, reasonable approximations are:

$$Q^i X = Q^s X^s \qquad (X^e \to 0)$$
$$Q^r = Q^s \qquad (Q^w \to 0)$$

and

$$X^r = X^s \qquad \textbf{[6.11]}$$

Q^i equals Q^0 plus Q^r from the mass balance around control volume **b**, where the reactor influent mixes with the recycled sludge. Also, the concentration of return activated sludge in the pipeline leaving the settling tank is the same as in the recycle line and the waste activated sludge line ($X^r = X^w = X^s$). Combining these observations

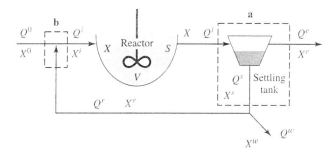

Figure 6.4 Total suspended solids flow within a complete-mix activated sludge system.

with Equations 6.11 and rearranging, we obtain for the activated sludge concentration X in the reactor:

$$X = X^r \frac{R}{1+R} \quad \text{or} \quad R = \frac{X}{(X^r - X)} \qquad \textbf{[6.12]}$$

where

$$R = \frac{Q^r}{Q^0} = \text{the recycle ratio} \qquad \textbf{[6.13]}$$

(The simplifying assumptions are eliminated in the comprehensive design section that follows, but the major trends remain the same as presented here.)

Because of sludge-settling characteristics and the practicalities of settler operation, the recycled sludge has some upper limit on concentration. We define this upper limit as X^r_m. Equation 6.12 then indicates the maximum value of X, called X_m, for a given value of R:

$$X_m = X^r_m \frac{R}{1+R} \qquad \textbf{[6.14]}$$

Operating experience has demonstrated that X^r_m is 10,000 to 14,000 mg/l for typical good-settling activated sludge. With sludge that compacts exceptionally well, this value may be as high as 20,000 mg/l. However, with bulking sludges, the value may be as low as 3,000 to 6,000 mg/l.

X^r_m can be approximated for given sludges through relatively simple tests, such as the Settled Sludge Volume Test, the Sludge Volume Index (SVI), or the Zone Settling Rate Test (Greenberg, Clesceri, Eaton, and Franson, 1992). Historically, the SVI has been used, although it underestimates the true value of X^r_m and might be considered a conservative measure. To do the SVI test, one allows activated sludge taken from the aeration basin to settle in a one-liter or two-liter graduated cylinder for 30 min. The original concentration of mixed liquor suspended solids (MLSS, in mgSS/l), the total volume of the cylinder (V_t, in liters), and the volume of the settled sludge after 30 min (V_{30}, in ml) are measured. SVI is defined as the volume in milliliters occupied by 1 g of the suspended solids after settling. It is computed by

$$\text{SVI (ml/gSS)} = V_{30} \cdot (1,000 \text{ mg/g})/[(\text{MLSS}) \cdot (V_t)] \qquad \textbf{[6.15]}$$

A first approximation to the maximum concentration of settled sludge (X^r_m) is $(10^6 \text{ mg} \cdot \text{ml/g} \cdot \text{l})/\text{SVI}$. Then, a typical good SVI, 100 ml/g, corresponds to an X^r_m of 10,000 mg/l. A bulking sludge would have an SVI of 200 or more, corresponding to an X^r_m of 5,000 mg/l or less, and a highly compact and good-settling sludge would have an SVI of 50 or less, corresponding to a X^r_m of 20,000 mg/l or more.

Settling tests carried out with slow mixing of the settling sludge, such as Settled Sludge Volume Test or the Zone Settling Rate Test, result in more compact sludges that are more representative of what is usually found in practice; they are preferred over the simple SVI. However, measurements of X^r_m made at one time by any of these procedures provides no assurance to the engineer or the plant operator that the compacting ability of an activated sludge will not be greatly different from this value at some other time. If there is no operational experience at a given plant to develop confidence in what value or range of values is likely to be experienced at a given plant, the designing engineer has little to rely on but past experience at other activated sludge treatment plants. Because of the large repercussions that may result from poor operation at a treatment plant, engineers are thus inclined to be somewhat conservative in their choices and use values for X_m^r that are somewhat low (conservative). The typical values based on SVI form a reasonable starting point for making judgments about what is conservative.

Figure 6.5 shows the effect of the recycle ratio on X_m for various values of X_m^r. For an X_m^r of 10,000 mg/l and an operational X of 2,000 mg/l, a recycle ratio of 0.25 is sufficient. If a bulking sludge develops such that X_m^r drops to only 5,000 mg/l, then increasing R to about 0.7 would still allow the operator to maintain X at 2,000 mg/l. However, if bulking were more severe such that X_m^r decreases to 2,500 mg/l, then R would need to be increased to 4, which is an extraordinarily high recycle rate.

If, in order to reduce aeration tank size, an X of 4,000 mg/l were selected for design, then a recycle ratio of about 0.7 would be required for an X_m^r of 10,000 mg/l. R would need to increase to 4 if a bulking sludge with X_m^r of 5,000 mg/l resulted. The concept of returning concentrated sludge from the settler to the aeration basin breaks down when we try to imagine maintaining an X of 4,000 mg/l when X_m^r is 2,500 mg/l. In a situation like that, the settler is irrelevant and serves no purpose.

These exercises illustrate one way in which the performance of the settler and the aeration tank are linked. They also demonstrate why the operator must be able to vary R as the sludge settleability changes. Unfortunately, such flexibility is not always provided because of budget constraints. Without being able to control R, the operator can lose control of X and, ultimately, the SRT. The section on Analysis and Design of Settlers shows the other ways that the settler and aeration basin are linked. It underscores that the operator must be able to control R in order to prevent failure of the settler.

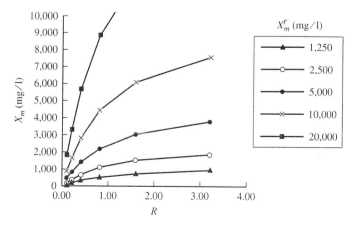

Figure 6.5   Effect of recycle ratio (R) on the maximum mixed-liquor suspended solids concentration in the aeration tank (X_m) for various maximum concentrations of sludge compaction in the settling tank (X_m^r).

6.3.6 ECKENFELDER AND McKINNEY EQUATIONS

During the late 1950s and early 1960s, McKinney and Eckenfelder developed mathematical models that relate the rate of BOD removal to basic operational parameters for the activated sludge process. These approaches are still in use. Thus, a comparison with the SRT method emphasized here is appropriate. While the approaches by these two individuals might appear different upon first glance, they are actually quite similar (Goodman and Englande, 1974). McKinney and Eckenfelder applied their equations to the completely mixed system with sludge recycle and assumed that BOD removal is first order with respect to BOD concentration in the aeration tank. These approaches are consistent with Monod when the BOD concentration is low ($S \ll K$), often an appropriate assumption for wastewater treatment.

The predicted effluent BOD concentrations (S) are

Eckenfelder:

$$S = \frac{S^0}{1 + k_E X_v \theta} \quad (S \ll K) \qquad \textbf{[6.16]}$$

McKinney:

$$S = \frac{S^0}{1 + k_M \theta} \quad (S \ll K) \qquad \textbf{[6.17]}$$

The rate equations behind the solutions relate to the Monod terms by

$$k_E = \frac{\hat{q}}{K} \cdot \frac{X_a}{X_v} \qquad \textbf{[6.18]}$$

$$k_M = \frac{\hat{q}}{K} \cdot X_a \qquad \textbf{[6.19]}$$

The Eckenfelder approach (Equations 6.16 and 6.18) assumes that the mixed-liquor volatile suspended solids concentration (X_v) can represent the active microbial population X_a, because the ratio X_a/X_v is effectively constant. Since k_E is generally determined empirically from laboratory activated-sludge treatability studies, these assumptions may not be a serious problem if the actual design is within the detention time, influent BOD, and MLVSS concentrations used to estimate k_E.

With the McKinney approach (Equations 6.17 and 6.19), the implicit assumption is that X_a remains constant, because Equation 6.17 does not explicitly include any microorganism concentration. This simplification is perhaps satisfactory if the MLVSS (X_v) is kept within a narrow range, say between 1,500 and 3,000 mg/l, and the influent loading does not vary.

The McKinney and Eckenfelder approaches are empirical in that they do not include the active microorganisms explicitly in the kinetic formulations. Instead, they rely on pseudo-first-order rate measurements from laboratory or field studies to select appropriate rate coefficients. This simple approach has a practical appeal, but will work adequately only when X_v, which can be readily measured, is related linearly to X_a, the active organism population. We know that X_a/X_v is not constant due to changes in θ_x and X_i^0. Such unreliable approximation is not required when (θ_x) is used directly for design and control of the activated sludge process.

6.4 AERATION SYSTEMS

Aeration serves a dual purpose: supply of the oxygen needed for bacterial metabolism and contaminant oxidation, and mixing of reactor contents in order to keep the mixed-liquor suspended solids in suspension and well distributed within the aeration tank. Oxygen is delivered to the aeration liquid through either diffused aeration, in which compressed gas is passed through submerged diffusers and rises through the liquid as bubbles, or by mechanical aeration, in which a mechanical mixer is used to vigorously agitate the water surface to cause oxygen transfer from the atmosphere above. There are many different designs for each, and each has its own advantages and disadvantages, which are outlined in the next section.

6.4.1 OXYGEN-TRANSFER AND MIXING RATES

An important practical limitation on the BOD loading to an activated sludge system is related to the amount of oxygen that can be transferred to the reactor economically and without destroying the sludge floc. The BOD loading cannot exceed the ability of the aeration system to deliver sufficient oxygen to satisfy the biological demand for oxygen. Insufficient oxygen slows biodegradation rates, fosters poor sludge settling, and leads to odors. Because oxygen is a sparingly soluble gas, transferring it to water is an intense and expensive proposition. The power required for its transfer from the gas phase to the liquid phase is a major cost for operation of the treatment system.

Whatever the aeration technology used, the rate of oxygen transfer between phases is governed by mass transfer from the bulk gas to the gas-liquid interface, and then from the gas-liquid interface into the liquid. For sparingly soluble gases such as oxygen, the transfer to the gas-liquid interface is fast compared with that from the interface into the liquid; thus, the liquid-side transfer is rate limiting (Bailey and Ollis, 1986). In that case, the rate of flux of oxygen from the gas to the liquid phase becomes:

$$r_{O_2} = K_L a (c_l^* - c_l) \qquad \textbf{[6.20]}$$

where

R_{O_2} = oxygen transfer rate per unit volume of reactor (mg l^{-1} d^{-1})

$K_L a$ = volumetric mass transfer rate coefficient (d^{-1}),

c_l^* = liquid phase oxygen concentration in equilibrium with bulk gas phase (mg/l),

c_l = liquid phase bulk oxygen concentration (mg/l).

c_l^* is proportional to the partial pressure of oxygen in the gas phase (c_g, atm) and

Henry's constant (H_{O_2}, atm-l/mg) for solubility of oxygen in water,

$$c_l^* = \frac{c_g}{H_{O_2}}$$ [6.21]

Henry's constant depends strongly on temperature. For clean water, the relationship is given approximately by

$$\log H_{O_2} = 0.914 - \frac{750}{T_K}$$ [6.22]

Data for the derivation of Equation 6.22 were taken from *Standard Methods* (Greenberg et al., 1992). Use in Equations 6.21 and 6.22 provides c_l^* within 0.1 mg/l of measured values for the temperature range between 5° and 45 °C ($T_K = 278$ to 318 K).

Example 6.1 | **SATURATION OXYGEN CONCENTRATION**   Determine c_l^* for aeration of clean water with air in Denver if the atmospheric pressure is 0.8 atm and the wastewater temperature is 12 °C.

$$\log H_{O_2} = 0.914 - \frac{750}{273 + 12} = -1.718$$

$$H_{O_2} = 10^{-1.718} = 0.0192 \text{ atm l mg}^{-1}$$

$$c_g = 0.2095(0.8) = 0.168 \text{ atm}$$

$$c_l^* = \frac{0.168}{0.0192} = 8.7 \text{ mg/l}$$

Oxygen solubility in wastewater often deviates from that given by the combination of Equations 6.21 and 6.22 due to the effect of salts and organic material. In order to correct for this effect, the clean water solubility for oxygen is multiplied by a value β:

$$\beta = \frac{c_l^* \text{ (wastewater)}}{c_l^* \text{ (clean water)}}$$ [6.23]

Typically, β varies from 0.7 to 0.98, and a value of 0.95 is often assumed for municipal wastewaters (Metcalf & Eddy, 1991).

Another consideration is the concentration of dissolved oxygen to maintain in the aeration tank (c_l). In the design of a treatment system, the BOD should be the kinetically limiting factor in the biodegradation rate. Thus, c_l should be maintained at a concentration that is not rate-limiting, or sufficiently in excess of the K for oxygen. Generally, K for oxygen is on the order of a few tenths of a mg/l. Keeping c_l at about 2 mg/l is generally sufficient to satisfy this criterion.

The value for $K_L a$ depends very much on the aeration system used, the power input to the system, the shape and size of the aeration basin, temperature, and the characteristics of the wastewater. As with oxygen solubility, $K_L a$ is affected by wastewater characteristics and often is less than what is determined with clean water. To express this difference, a factor α is used:

$$\alpha = \frac{K_L a \text{ (wastewater)}}{K_L a \text{ (clean water)}} \qquad [\textbf{6.24}]$$

The value for α is reported to vary from 0.35 to 0.8 for diffused aeration (Hwang and Stenstrom, 1985) and from 0.3 to 1.1 for mechanical aeration (Joint Task Force, 1998). A major factor here is the impact of synthetic detergents, which change surface tension and bubble size and characteristics. Other sources should be reviewed to gain a deeper understanding of what affects $K_L a$ (Bailey and Ollis, 1986; Hwang and Stenstrom, 1985; Joint Task Force, 1998; Metcalf & Eddy, 1991).

Regardless of the aeration type, the power input per unit volume is a key factor that affects $K_L a$ (Bailey and Ollis, 1986). The amount of power dissipated determines the scale of the eddies and the intensity of turbulent velocity fluctuations at length scales comparable to bubble sizes. In line with this observation, oxygen transfer is generally expressed as mass of oxygen transferred under standard conditions per unit of power input to the aerator. *Standard oxygen transfer efficiencies (SOTE)* are for standard conditions of 20 °C ($c_l^* = 9.2$ mg/l), zero dissolved oxygen in the liquid phase ($c_l = 0$ mg/l), and with clean water (α and $\beta = 1$). SOTEs generally are 1.2 to 2.7 kg O_2 per kWh (Joint Task Force, 1998).

For a design, the energy efficiency must be converted to field conditions. The prevailing temperature, α, β, c_l, and c_l^* must be specified. Then, the SOTE (e.g., 2 kg O_2 per kWh) can be converted to the *field oxygen transfer efficiency (FOTE)* by

$$\text{FOTE} = \text{SOTE} \cdot 1.035^{T-20} \cdot \alpha \cdot (\beta c_l^* - c_l)/9.2 \qquad [\textbf{6.25}]$$

9.2 is the mg/l O_2 saturation concentration for 20 °C and $c_g = 0.21$ atm O_2. 1.035^{T-20} is a typical correction factor for the effects of temperature on mass-transfer kinetics. For example, an SOTE of 2 kg O_2 per kWh is reduced to an FOTE of 0.74 kg O_2 per kWh when $T = 12$ °C, $c_g = 0.8(0.21)$ atm, $c_l^* = 8.7$ mg/l, $c_l = 2$ mg/l, $\alpha = 0.7$, and $\beta = 0.95$. This example illustrates the normal trend: FOTE is considerably smaller than SOTE. Manufacturers of aeration equipment normally report SOTE values, but the design engineer has the responsibility for converting that information to FOTE for use in the design. When no system-specific information is available to estimate FOTE, a reasonable value to assume for preliminary calculations is 1 kg O_2 per kWh.

For diffused-aeration systems, another measure of the oxygen-transfer efficiency is the *percent oxygen transfer (POT)*. It is the percentage ratio of the mass of O_2 transferred into the water divided by the mass of O_2 applied to the water via the compressed-air bubbles. POT can vary from very low values (< 5 percent) to quite high values (> 30 percent). Deep aeration basins, very fine bubbles, and high turbulence tend to make POT larger. For design, the POT should not be emphasized too much. It always is possible to attain a high POT if enough energy is provided to increase $K_L a$, but this often comes at the cost of a low FOTE. The engineer should focus on the FOTE, as it relates to the installed power and energy costs.

Aeration is also used to maintain the microbial flocs in suspension. The Joint Manual of Practice by ASCE and the WEF (Joint Task Force, 1998) suggests that mixed-liquor velocities in an aeration tank equal at least 0.3 m/s. This requires an air input for diffused air systems of 20 to 30 m^3 air per min per 1,000 m^3 of tank volume. For mechanical mixing with vertical mixing regime, a power input of 15 to 30 kW per 1,000 m^3 is required. Generally, the limiting factor in aerator design is oxygen demand, rather than mixing, when air is used as the source of oxygen. However, with extended aeration systems or when pure oxygen is used to supply the oxygen demand, the mixing requirement may govern.

6.4.2   DIFFUSED AERATION SYSTEMS

Compressed air is generally introduced into aeration tanks through porous diffusers, nonporous diffusers, jets, or static mixers. Porous diffusers can be made of ceramic or flexible plastic membranes cast in the form of domes, discs, tubes, or plates. Compressed air is forced out of the porous diffusers and forms small bubbles. Nonporous diffusers consist of little more than pipe perforated with holes that emit coarse bubbles. An advantage of nonporous diffusers is that they do not clog as easily as porous diffusers, but their oxygen transfer efficiency is much lower. Jets in which compressed air and water are mixed and discharged through nozzles to produce fine bubbles are also sometimes used. Jets are reported to have oxygen transfer efficiencies similar to porous diffusers. Static mixers involve tubes placed above nonporous diffusers in such a manner that an airlift pump is created. The air and water intensely mix in the tubes, causing bubble breakup and high rate oxygen transfer. Static mixers have transfer efficiencies similar to porous diffusers.

Porous and nonporous diffusers are particularly well suited for plug-flow tanks that are characterized by long tanks of relatively small cross-sectional area. Typically, the diffusers are placed near the bottom on one side of the aeration tank so that the

a.   b.

Photo 6.3   Example of porous diffusers placed throughout the tank's bottom to give good mixing. SOURCE: With permission from Sanitaire.

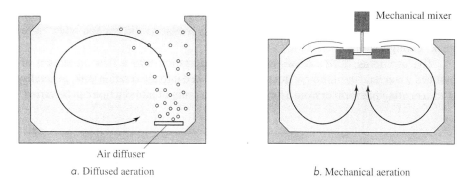

a. Diffused aeration

b. Mechanical aeration

Figure 6.6 Diffused aeration and mechanical aeration.

rising air bubbles cause the tank contents to rotate in a circular fashion, thereby keeping the mixed liquor solids in suspension (Figure 6.6). Tank depth is generally between 4.5 and 7.5 m. The width to depth ratio for the aeration tank is an important consideration to allow proper mixing without the formation of dead spaces where floc will tend to settle. The width to depth ratio is generally between 1.0 to 1 and 2.2 to 1, with 1.5 to 1 being that most generally used. Static mixers and jets are more appropriate for tanks in which complete mixing is the goal.

6.4.3 MECHANICAL AERATION SYSTEMS

Mechanical aerators are generally grouped into four different types: radial-flow low-speed, axial-flow high-speed, horizontal rotors, and aspirating devices. Radial-flow low-speed aerators are the ones generally used on aeration tanks in activated sludge. They have large turbine blades that operate typically at 20 to 100 rotations per minute. Slow-speed aerators are known for good oxygen transfer efficiency, mixing capabilities, and reliability. Low-speed mechanical aerators generally work best in tanks that are square or circular in plan view, with one aerator per tank. In larger tanks, multiple mechanical aerators are also used. Here, the length-to-width ratio should be in multiples of that which would be used with a single aerator in a tank, each aerator serving a square area of reactor surface.

Axial-flow high-speed aerators tend to be used more with aerated lagoons (Chapter 7), where the potential for disrupting floc size is not so great, but they are used sometimes in activated sludge. They have much smaller turbine blades than do the low-speed aerators, but they operate with very much higher rotating speeds. High-speed aerators have good oxygen-transfer efficiency, but their mixing range is more limited, and they are not as reliable. Multiple aerators in a large basin are normal, and the mixing zones of each aerator should overlap in order to ensure good mixing and oxygen transfer in all parts of the basin.

The horizontal rotor has a horizontal axle with radial blades that agitate the surface of the basin and move the water horizontally. The turbulence created by the blades and the horizontal motion of the water bring about oxygen transfer. Horizontal rotors are simple and reliable devices that have good oxygen transfer efficiency. However, they are effective only in shallow basins, such as oxidation ditches and other "race track" configurations.

The aspirating device draws air through aspiration into a flowing stream of pumped water, and the turbulent mixture is introduced into the aeration tank, generally from a floating platform or from a boom so that the point of introduction can be varied.

6.5   BULKING AND OTHER SLUDGE SETTLING PROBLEMS

The successful application of the activated sludge process requires that the sludge floc settles and compacts well in the settling tank. It must settle well so that the effluent has little suspended solids. High suspended solids in the effluent are the main cause of unsatisfactory effluent quality, and they also make proper SRT control difficult. The sludge must compact well so that the sludge can be returned successfully to the aeration basin. A highly compacted sludge also reduces the costs for dewatering and disposal of the wasted solids.

The major problem in activated sludge operation is the development of poor settling sludge. When this occurs, suspended solids pass to the effluent in concentrations that often exceed regulatory standards for SS, the desired SRT necessary cannot be maintained, and effluent BOD limitations are often exceeded. Many sludge settling problems can occur, and their causes are not at all the same. The cause of a particular sludge-settling problem must be recognized before a satisfactory solution can be found. The cause may be related to the particular configuration of the treatment plant, to the loading applied, to the environmental conditions (such as temperature, pH, and dissolved oxygen concentration), or to the presence of particular wastewater constituents or by the absence of others. Much has been written over the years about sludge settling problems, and significant advances have been made in understanding some of the causes of these problems. At the same time, much misleading "folklore" also exists. We provide guidance that helps the engineer apply proven principles and avoid the dangerous "folklore."

Jenkins (1992) presented a summary of the different solids-separations problems that can develop in activated sludge operation. Table 6.3 summarizes key information. Wanner and Grau (1989) provided information on the identification of the main causative organisms for sludge settling problems. Each of the sludge-separation problems is discussed here, beginning with sludge bulking.

6.5.1   BULKING SLUDGE

Bulking sludge is the most pervasive and difficult problem. *Bulking* is the term used to describe the formation of activated sludge *floc that settles slowly and compacts poorly.* Bulking makes it difficult to remove the activated sludge from the settling tank for return to the aeration basin. Higher and higher recycle rates are required

Table 6.3   Biosolids separation problems encountered in activated sludge operation

Biosolids Separation Problem	Cause of Problem	Effect of Problem
Bulking	Filamentous organisms extend from flocs into the bulk solution and interfere with compaction and settling	High sludge volume index with clear supernatant. Overflow of sludge blanket can occur. Solids handling processes become hydraulically overloaded
Viscous bulking or nonfilamentous bulking	Microorganisms present in large amounts of exocellular slime. In severe cases, the slime imparts a jelly-like consistency	Reduced settling and compaction rates. Can result in overflow of sludge blanket from secondary clarifier or formation of a viscous foam
Dispersed growth	Microorganisms do not form flocs, but are dispersed, forming only small clumps or single cells	Turbid effluent. No zone settling of sludge
Pin floc or pinpoint floc	Small, compact, weak, roughly spherical flocs. Larger aggregates settle rapidly, smaller ones slowly	Low sludge volume index and cloudy turbid effluent
Foaming/Scum formation	Caused by (i) nondegradable surfactants, or (ii) the presence of *Norcardia* sp. and/or *Microthrix parvicella*	Foams float large amounts of biosolids to surface of treatment units. Microorganism-caused foams are persistent and difficult to break. Causes solids overflow into secondary effluent and onto walkways. Anaerobic digestion foaming can also result
Blanket rising	Denitrification in settler releases poorly soluble N_2 gas, which attaches to activated sludge flocs and floats them to the clarifier surface	"Chunks" of activated sludge collect on the surface of the settler and may result in turbid effluent

SOURCE: After Jenkins (1992).

because the concentration of the settled sludge is so low. If the sludge cannot be removed fast enough from the underflow, then the sludge blanket rises until it fills the settler. Then, the activated sludge solids discharge into the effluent. This causes a massive loss of biomass, which decreases the SRT in an uncontrolled manner and destroys effluent SS and BOD quality. A prolonged siege of sludge bulking can lead to the total failure of the activated sludge process.

A compact and good-settling activated sludge results from a floc microstructure consisting of a backbone of filamentous bacteria to which zoogleal microorganisms attach to form a strong and compact macrostructure (Sezgin, Jenkins, and Parker, 1978). If there are not sufficient filamentous bacteria, the floc is weak and subject to break up into smaller particles.

On the other hand, too many filamentous bacteria are the cause of sludge bulking. The problem becomes serious when the filaments extend outside the compact floc. These *extended filaments* create bridges between the flocs. The bridging causes two serious and negative effects. First, the bridges prevent the flocs from coming close together, or compacting. Second, the bridges trap water within and between the flocs. As the sludge flocs try to move downward and compact in the settler, the water they displace must move upward. The bridging prevents the movement of the water through and away from the flocs. These two effects of extended filaments cause the bulking sludge to settle slowly and compact very poorly. These effects are quantified in the section on Analysis and Design of Settlers.

The onset of sludge bulking can be observed in three ways. The first and most direct way is microscopic examination. Regular microscopic examination by a trained technician can identify when extended filaments are increasing. A steady trend of more extended filaments is a sign that bulking is at hand. Second, the sludge volume index is well correlated to bulking. An SVI greater than about 200 ml/g usually indicates serious bulking. Very bad bulking can have SVIs much greater, perhaps as much as 500 ml/g. The third sign is the combination of a rising sludge blanket and a low concentration of suspended solids in the settler underflow. The section on Analysis and Design of Settlers describes how this combination comes about. Microscopic examination is the best detection tool, because it can point out a bulking trend before it becomes a serious problem. The other methods often give clear signs only after bulking already is a problem.

Fundamental and applied research has identified three distinct causes of sludge bulking: low dissolved oxygen, long SRT, and input reduced sulfur. We discuss each of the causes and how they can be overcome. The means to overcome the different types of bulking are totally distinct. Therefore, it is essential to understand which type of bulking is occurring before designing a solution strategy.

Low-D.O. bulking is brought about by filamentous bacteria such as *Sphaerotilus natans,* Type 021N, and Type 1701. (The famous Dutch engineer, Eikelboom, identified the species denoted by a type number, which comes from the page numbers in his notebooks.) An especially good affinity for dissolved oxygen, that is, a low K for O_2, characterizes the bacteria in this group. They begin to predominate when the dissolved oxygen concentration is not high enough to allow good oxygen penetration into the floc. Then, the low-D.O. filaments gain an advantage and extend from the floc. Palm, Jenkins, and Parker (1980) found from laboratory and field studies that the D.O. concentration needed to prevent bulking depends on the BOD specific utilization rate, U. To prevent bulking, the D.O. concentration must be greater than

$$\text{D.O. (mg/l)} > (U - 0.1)/0.22 \qquad \textbf{[6.26]}$$

in which U is in kg COD/kg MLVSS-d. The solution to overcoming low-D.O. bulking is to increase the D.O. concentration, which may require expanding the capacity of the aeration system or reducing the BOD loading.

The second form of bulking is termed *low-F/M bulking,* and it occurs when the SRT is long, such as with extended aeration. *Microthrix parvicella,* Type 0041, Type 0092, Type 0581, and *Haliscomenbacter hydrosis* are examples of filamentous species that cause low-F/M bulking. They are classic K-strategist oligotrophs in that they have a high affinity for organic substrates (a low K) and a low endogenous decay rate (low b). Thus, they have an advantage when the specific growth rate and organic substrates are low. This is just the situation with extended aeration, where low-F/M bulking is prevalent.

When an extended aeration loading is required—(such as to have nitrification (Chapter 9) or to have an especially large safety factor (Chapter 5))—the SRT cannot be reduced in an attempt to eliminate low-F/M bulking. Fortunately, the *selector* technique has proven very effective for counteracting low-F/M bulking in these situations.

The selector tank is placed in front of the aeration basin. It receives the influent and the recycled sludge. It is mixed, and it may or may not be aerated. The design goals for the selector are: (1) it must be small enough that oxidation of the entering BOD is small; and (2) it must be large enough that soluble components, like organic volatile acids, are taken up rapidly by floc-forming bacteria that store internal polymers such as polyhydroxybutyrate (PHB). The concept is that the floc formers rapidly take up and store a significant fraction of the BOD in the selector. When the cells move to the aeration basin, the floc formers gradually oxidize their sequestered storage polymers, while the filaments, which do not form storage products, are starved out of the system. Thus, the key is an ecological selection that is based on the fact that the low-F/M filaments do not form storage material, while some floc formers do.

In practice, the selector approach works remarkably well. Recovery from bulking occurs over a few days once the selector is installed correctly. The critical issue concerns the design criteria for the selector. Although no criterion is yet widely accepted, the authors have had excellent success by using a selector F/M ratio:

$$\text{F/M}_{sel} = Q \cdot \text{BOD}_5^0 / (V_{sel} \cdot \text{MLVSS}) \qquad \textbf{[6.27]}$$

in which F/M$_{sel}$ has units g BOD$_5$/g MLVSS-d, Q is the influent flow (m^3/d), BOD$_5^0$ is the influent BOD$_5$ concentration (mg/l), V_{sel} is the volume of the selector tank (m^3), and MLVSS is the VSS concentration in the mixed liquor in the selector (mg/l). From the literature and the authors' experience, good bulking abatement occurs for F/M$_{sel}$ between 30 and 40 g BOD$_5$/g MLVSS-d.

Reduced-sulfur bulking occurs whenever reduced sulfur compounds, mostly sulfides, enter the activated sludge unit. Sulfur-oxidizing species, such as *Thiothrix* and 021N, are filament formers that gain a competitive advantage from the chemolithotrophic electron donor. So far, the only reliable means to eliminate reduced-sulfur bulking is to eliminate all inputs of reduced sulfur. If the source cannot be eliminated, the reduced sulfur can be chemically oxidized before the wastewater enters the activated sludge process. Although a number of chemical oxidants can be used, the easiest one is hydrogen peroxide, which oxidizes sulfides according to

$$4\,H_2O_2 + HS^- \rightarrow SO_4^{2-} + 4\,H_2O + H^+ \qquad \textbf{[6.28]}$$

The reaction indicates that 4.25 g H_2O_2 is needed to oxidize one g S. In reality, the peroxide reacts with other reduced compounds, which means that the actual dose must be larger than given by Equation 6.28.

An interesting variation of reduced-sulfur bulking is the formation of sulfides from sulfates within the sludge floc, probably due to D.O. depletion. In this case, sulfate is the input form of sulfur, but it is reduced to sulfides within the treatment process. The formation of sulfides may be one of the mechanisms behind low-D.O. bulking. Increased concentration of D.O. (or NO_3^-) is needed to prevent sulfate reduction.

In the long term, the cause of the bulking needs to be identified so the proper solution prevents a recurrence. However, when sludge bulking strikes, an "emergency fix" often is needed to prevent disastrous discharge of suspended solids. The best emergency fix for sludge bulking is a form of *chemotherapy*—application of chlorine to the mixed liquor. The concept is to add chlorine at a proper dose rate such that it selectively kills the extended filaments, leaving the compact floc intact. Although chlorination works, it is very important to poise the chlorine dose at the right level. Too little chlorine has no effect. Too much chlorine kills the floc formers, deflocculates the sludge, and wipes out sensitive, slow-growing populations. A rule of thumb for chlorine dosing is 5 kg as Cl_2 per 1,000 kg SS per day (Jenkins, Neethling, Bode, and Paschard, 1982). One should add chlorine at this rate until the bulking problem abates. Once bulking has subsided, the chlorination should be stopped. This strategy of "titrating" the sludge with chlorine is the best way to achieve a short-term fix without harming the sludge.

In most cases, the chlorine is applied to the returned sludge, because it is simple to create a single chlorine-injection point. However, a better dosing method is to add the chlorine directly to the mixed liquor in the aeration basin. This approach is not applied often, because it requires special plumbing and distributors.

6.5.2 FOAMING AND SCUM CONTROL

Another very common problem in activated sludge systems is the formation of foam or scum on the surface of aeration tanks. Foam and scum cause many problems in plant operation, including excessive suspended solids in the effluent, unsightly and dangerous conditions, such as slippery walkways around them, and great difficulties in making a sludge inventory. In addition, the causative organisms can create foaming problems in anaerobic digesters that receive waste activated sludge.

The causative microorganisms usually belong to the genuses *Nocardia* and *Microthrix* (Pitt and Jenkins, 1990). While the many factors causing foaming and scum formation are not well known, the problem often is associated with long SRT and high wastewater temperatures. This suggests that the causative organisms are slow growers. The most common method used in attempts to control this problem is to reduce θ_x to 6 d or less (Pitt and Jenkins, 1990). In some cases, chlorination of return activated sludge helps control this problem. Since the organisms cause problems by accumulating as foam and scum on the surfaces of the tanks, another control strategy—and probably the most promising—is to implement a vigorous program to remove scum and foam. The concept is to severely reduce the SRT of the nuisance microorganisms by wasting them very aggressively from where they preferentially accumulate.

6.5.3 RISING SLUDGE

Rising sludge is a problem sometimes occurring in the settling tank of activated sludge plants in which ammonium is nitrified to nitrate. If denitrification of the nitrate to N_2 takes place in the sludge blanket of the settler, gas bubbles can attach to the settled sludge particles. "Chunks" of sludge then become buoyant and rise to the surface of the settler, where they collect. These pieces of the sludge blanket are unsightly, and they can cause a very large increase in effluent suspended solids if they escape to the effluent.

The most effective cure for rising sludge is to stop nitrification in the activated sludge. This is accomplished by reducing the SRT and washing out the slow-growing nitrifiers (Chapter 9). If no nitrate is formed by nitrification, then none can be denitrified to N_2 gas. Another way to prevent rising sludge is to promote denitrification as part of the activated sludge process. This strategy is discussed in Chapter 10. If the nitrate is removed before the mixed liquor enters the settler, then denitrification cannot occur in the settler.

Another way to reduce denitrification in the settler is by improving settler design. The idea is to prevent the sludge from "sitting" in the sludge blanket for too long. Circular settlers with vacuum-type sludge removal devices work particularly well here. Rectangular clarifiers can be effective if sludge-removal scrapers remove the sludge with sufficient speed and if accumulation of sludge in quiescent corners is prevented.

6.5.4 DISPERSED GROWTH AND PINPOINT FLOC

Dispersed growth and pinpoint floc represent sludge settling problems in which some of the microorganisms do not flocculate into particles large enough for rapid settling. *Dispersed growth* often occurs during start-up of activated sludge systems or where the SRT is too short for the formation of the normal structure of good-settling activated sludge. Viewed through the microscope, many of the microorganisms occur dispersed throughout the liquid phase either singly or in small groups. The amount of external polymer in such systems is too small to permit good flocculation. Also, during start-up and with low θ_x, organisms may be quite active and tend to resist the electrostatic forces required for their clumping together into a settleable floc. During activated sludge start-up, the diversity of organisms may include those that flocculate readily and those that do not. With time of operation, those that do not flocculate well pass out of the settling tank with the effluent water and are not returned to the aeration basin. This results in a selection for those microorganisms that have good settling ability.

Pinpoint floc is another phenomenon that results in the formation of small particles that do not settle well. It is associated with systems with long θ_x, such as extended aeration systems. The "old" floc is subject to heavy predation by eukaryotic cells, resulting in the destruction of floc characteristics and the production of biological debris, most of which is not active. Many small pinpoint floc particles are formed in this process and pass into the effluent.

6.5.5 VISCOUS BULKING

A form of nonfilamentous bulking has been identified by Jenkins (1992). Called *viscous bulking,* it is mainly associated with floc-forming bacteria and results when they produce an excess of extracellular polymer. Such polymer in moderate amounts serves to cause bacterial flocculation, which is a necessary part of good floc formation. However, if produced in excess amounts, it can be detrimental to the settling of the bacterial flocs. The sludge flocs take on a voluminous character and, in severe cases, may take on a jelly-like consistency and cause foaming and scum formation. Poor settling is caused by the high water content of the polymeric material.

Causes of nonfilamentous bulking have been described by Novack, Larrea, Wanner, and Garcia-Heras, (1993): (1) wastewaters containing high fatty and oleic compounds, (2) high sludge loading or a deficiency in nitrogen or phosphorus, (3) use of selectors to control filamentous bulking, and (4) biological phosphorus removal systems causing an excessive growth of *Acinetobacter* that produce excessive polymer. The extent of viscous bulking is not clear, and few studies have been conducted on its control.

6.5.6 ADDITION OF POLYMERS

The many sludge-separation problems that can occur with activated sludge and their devastating impacts often lead operators to find quick-fix solutions. A common example is the addition of organic polymers, usually cationic polyelectrolytes, to the mixed liquor in order to enhance flocculation, settling, and compaction. Most often, the polymers are added to the mixed liquor between the aeration basin and the settling tank. The dose normally is determined from laboratory testing and experience, often in close cooperation with the polymer manufacturer. Without question, a properly chosen polymer, delivered at the right dose, can give immediate results. It is particularly effective for relief from a rising sludge blanket (from bulking normally), dispersed growth, or pinpoint floc. As an emergency measure to prevent disastrous loss of suspended solids, polymer addition can be very effective.

Polymer addition has disadvantages, and it should not be relied upon for routine use. One disadvantage is that the required dosage frequently rises over time. Thus, the cost of the polymer continually increases. The loss of effectiveness probably is caused by the polymer being biodegraded by the microbial community, which adapts to it over time. The second disadvantage is that the normal selection process for natural floc formers is short-circuited. Thus, the community becomes less and less enriched in the good floc formers that are needed if polymer addition is to be stopped. This selection against floc formers may be a second reason why doses tend to increase over time.

6.6 ACTIVATED SLUDGE DESIGN AND ANALYSIS

The design and analysis of an activated sludge process is a creative process that integrates a series of engineering judgments with calculations that are based on the fundamentals developed in Chapters 2, 3, and 5. The best way to understand how judgment and calculation interact is through a comprehensive example. In the following example, we go through all the steps normally needed to design or analyze the reactor portion of activated sludge. We assume that the settler is performing adequately, and we use engineering judgment when information about the settler is needed. In the next section, Analysis and Design of Settlers, we show how that information can be determined from design and operating conditions of the settler. A full design must link the settler analysis to the reactor analysis, but we address that linking later in the chapter.

The example is presented as a series of steps. Some are data-gathering steps, which can include making engineering judgments when measurable data are not available. Other steps use the quantitative fundamentals of the previous chapters to compute the key information we need for a design, for example, volume of the reactor, waste-sludge flow rate, oxygen-transfer rate, and effluent quality. After the first few steps, the design/analysis need not follow the order given here. Therefore, the numbering of the steps helps order and identify the key components of design/analysis; it does not imply that a design/analysis must always go in the order shown.

Most of the equations used in the example were derived in Chapters 3 or 5. In a few cases, a new relationship is used, and its genesis is described in the example.

ACTIVATED SLUDGE DESIGN | **Example 6.2**

Step 1. Define the Influent. The influent is a wastewater with these characteristics:

$$Q = 10^3 \text{ m}^3/\text{d} = 10^6 \text{ l/d}$$
$$S^0 = 500 \text{ mg BOD}_L/\text{l}$$
$$X_a^0 = 0 \text{ mg VSS}_a/\text{l}$$
$$X_i^0 = 50 \text{ mg VSS}_i/\text{l}$$

The influent substrate should be expressed as BOD_L in order to do complete mass balances on the electron equivalents. Having no input active biomass, but significant input inert VSS is the normal situation. In this example, all the input BOD_L is soluble or is particulate material that is hydrolyzed to soluble BOD_L within the activated sludge system.

Step 2. Define the Kinetic and Stoichiometric Characteristics. The BOD is composed of complex organic molecules that have the following kinetic ($\hat{q}$, K, and b) and stoichiometric (Y and f_d) parameters:

$$\hat{q} = 10 \text{ mg BOD}_L/\text{mg VSS}_a\text{-d}$$
$$K = 10 \text{ mg BOD}_L/\text{d}$$
$$Y = 0.4 \text{ mg VSS}_a/\text{mg BOD}_L$$
$$b = 0.1/\text{d}$$
$$f_d = 0.8$$

An important thing to notice is the consistency of the units on all parameters: mg for mass, liters for volume, days for time, BOD_L for electron-donor substrate, and VSS for biomass. Since all the units are consistent, we drop the units in most cases in order to reduce the clutter in the equations. WARNING: Always use the proper and consistent units! We set the right example here, and it should be followed religiously.

Step 3. Identify Design Criteria. This system is to achieve regulated effluent standards of $BOD_5 < 20$ mg/l and SS < 20 mg/l. A key to the design is that that effluent criteria are in BOD_5 and SS. We will need to convert from our more fundamental units of BOD_L and VSS to BOD_5 and SS when we estimate the effluent quality. We should not make the conversion now, as we want to use the proper fundamental units for the design analysis.

Step 4. Compute Delimiting Values. A valuable first step is to compute the delimiting values: $[\theta_x^{min}]_{lim}$ and S_{min}. They give us a rapid "reality check" on the feasibility of the process.

$$[\theta_x^{min}]_{lim} = [Y\hat{q} - b]^{-1} = [0.4 \cdot 10 - 0.1]^{-1} = 0.26 \text{ d}$$
$$S_{min} = Kb[\theta_x^{min}]_{lim} = 10 \cdot 0.1 \cdot 0.26 = 0.26 \text{ mg BOD}_L/l$$

The limiting value of $[\theta_x^{min}]$ tells us that we can design an economical process with a reasonable safety factor. The S_{min} value tells us that the input substrate can be driven to a concentration well below the effluent standard. Thus, the design should be feasible.

Step 5. Choose a Design SRT. Because we want an economical design, but one that provides a high-quality effluent, we choose a safety factor at the low end of the conventional range (Table 6.2): SF = 20. Then, we compute the design SRT:

$$\theta_x = \text{SF} \cdot [\theta_x^{min}]_{lim} = 20 \cdot 0.26 = 5.2 \text{ d}$$

This is in the normal range for conventional loading (4 to 10 d, Table 6.2) and should make it possible to have a high-quality effluent. Since the SRT cannot be "fine-tuned" very precisely in practice, we also round the value to $\theta_x = $ **5 d**.

Step 6. Compute the Effluent Substrate Concentration S. Once we have the design SRT, we compute S from the well-known relationship:

$$S = K\frac{1 + b\theta_x}{Y\hat{q}\theta_x - (1 + b\theta_x)} = 10\frac{1 + 0.1 \cdot 5}{0.4 \cdot 10 \cdot 5 - (1 + 0.1 \cdot 5)} = 0.81 \text{ mg BOD}_L/l$$

Clearly, S is well below the effluent BOD_5 standard of 20 mg/l.

Step 7. Choose the System Hydraulic Detention Time (θ) *or the MLVSS* (X_v) *and Compute the Other.* At this stage, we must make an engineering judgment about either the system's hydraulic detention time or the MLVSS. θ and θ_x are independent in activated sludge, and they are related through the X_v by

$$X_v = \frac{\theta_x}{\theta}\left(X_i^0 + \frac{1 + (1 - f_d)b\theta_x}{1 + b\theta_x} \cdot Y \cdot [S^0 - S]\right)$$

Here, we choose an X_v that will be near the upper end of the range that normally works well for conventional loading (Table 6.2). Table 6.2 indicates that MLSS values are up to 3,000 mg SS/l. Since the ratio of MLVSS to MLSS usually is 0.8 to 0.9, the typical MLVSS is up to 2,400 to 2,700 mg VSS/l. We select $X_v = 2,500$ mg VSS/l.

$$2,500 = \frac{5}{\theta}\left(50 + \frac{1 + (1 - 0.8) \cdot 0.1 \cdot 5}{1 + 0.1 \cdot 5} \cdot 0.4 \cdot [500 - 0.81]\right)$$

Then, we solve for $\theta = $ **0.39 d, or 9.4 h**. So far, all is acceptable.

Step 8. Compute the System Volume (V), X_a, *and* X_a/X_v. The system volume is computed simply by

$$V = Q \cdot \theta = 10^6 \cdot 0.39 = \textbf{3.9} \cdot \textbf{10}^5 \text{ liters (or 390 m}^3\text{)}$$

Once we have θ, θ_x, and X_v, we can compute X_a from either of two expressions:

$$X_a = \frac{\theta_x}{\theta} Y[S^0 - S]/(1 + b\theta_x) = \frac{X_v - \frac{\theta_x}{\theta} X_i^0}{1 + (1 - f_d)b\theta_x}$$

Using either form, we compute $X_a = \textbf{1,710 mg VSS}_a\textbf{/l}$. In this case, the ratio $X_a/X_v = 1,710/2,500 = \textbf{0.68 mg VSS}_a\textbf{/mg VSS}$. We will use this information later.

It should be noted here that the system volume V may represent the volume of the reactor only, the combined volume of the reactor and the settler, or something between, depending on the assumptions made concerning biomass in the settler. These assumptions were discussed in Chapter 5 in the section on Evaluating Assumptions. If the biomass in the settler is not active, V is simply the volume of the reactor, V_{rea}. If they are considered to be equally active as in the reactor and at the same average concentration, then V is the combined volumes of the reactor and the settler, $V_{rea} + V_{set}$.

Step 9. Estimate the MLSS, X. The MLVSS has associated with it some inorganic solids. Biomass is typically about 90 percent organic and 10 percent inorganic. Thus, the MLVSS has $2,500 \cdot (10/90) = 280$ mgSS/l of inorganic solids. Also, the influent may contain inorganic solids not included in X_i^0. We assume that the input inorganic SS is 20 mg/l. It is increased by the solids-concentration factor to become in the mixed liquor $20 \cdot (5/0.39) = 260$ mg/l. In total, the MLSS is the sum of the MLVSS and the two types of inorganic SS:

$$\textbf{MLSS} = X = 2,500 + 280 + 260 = \textbf{3,040 mg SS/l}$$

This is a typical MLSS for conventional loading and gives a ratio X_a/MLSS of 56 percent.

Step 10. Determine Solids Loss Rates. At this point, many different paths are possible. We proceed by calculating the solids loss rates, but the designer could jump ahead to subsequent steps and then return to this step. However, doing solids loss is a very logical next step and one that must be accomplished.

The simplest way to compute the solids loss rates (i.e., the mass per time rate at which the different types of solids must leave the system) is to rearrange the definition of SRT to

$$\frac{\Delta X_j}{\Delta t} = \frac{X_j \cdot V}{\theta_x}$$

in which X_j is any of the suspended solids forms and $\Delta X_j/\Delta t$ is the mass per time (mg/d in our units system) loss rate. Using this form with $V = 3.9 \cdot 10^5$ liters and $\theta_x = 5$ d gives

$$\Delta X_a/\Delta t = 1.33 \cdot 10^8 \text{ mg VSS}_a\text{/d}$$
$$\Delta X_v/\Delta t = 1.95 \cdot 10^8 \text{ mg VSS/d}$$
$$\Delta X_{ss}/\Delta t = 2.37 \cdot 10^8 \text{ mg SS/d}$$

Since inert volatile suspended solids are input, not all of the VSS that leaves the system was created through net synthesis. The actual biological VSS-production rate is needed to assess nutrient needs. It is most easily computed by subtracting the mass rate of input inert and inorganic solids:

$$(\Delta X_v/\Delta t)_{\textbf{biol}} = 1.95 \cdot 10^8 - 10^6 \cdot (50) = \textbf{1.45} \cdot \textbf{10}^8 \text{ mg VSS/d}$$

Step 11. Estimate Sludge Recycle and Effluent Solids Concentrations. Since we are not doing a detailed analysis of the settler for this example, we use judgment that the settler will achieve normally "good" performance in each category. In particular, we assume that the effluent VSS is $X_v^e = 15$ mg VSS/l, and the recycled VSS is $X_v^r = 10,000$ mg VSS/l. By proportioning with the mixed-liquor concentrations, we also can estimate the active and suspended solids concentrations at $X_a^e = 10$ mg VSS$_a$/l, $X_a^r = 6,800$ mg VSS$_a$/l, $X_{ss}^e = 18$ mg SS/l, and $X_{ss}^r = 12,200$ mg SS/l. We see that the effluent SS that we assumed meets the effluent standard of 20 mg SS/l. Since we use a "conventional" design, the 20-mg SS/l level should be achievable with a good settler design, which is discussed in the next section of this chapter.

Step 12. Estimate the Sludge Wasting Rate. The solids lost from the system exit either through the waste-sludge line or in the effluent. The estimation of both requires information on how the settler performs in clarification and thickening, and those estimates were made in step 11. A mass balance on solids leaving the system can be performed on any of the solids types. We use VSS:

$$\Delta X_v / \Delta t = Q^w \cdot X_v^r + (Q - Q^w) \cdot X_v^e$$
$$1.95 \cdot 10^8 = Q^w \cdot 10,000 + (10^6 - Q^w) \cdot 15$$

Solving for Q^w gives $1.8 \cdot 10^4$ l/d, or 1.8 percent of the plant flow Q. This is an acceptably low percentage of the plant flow. This flow goes to sludge dewatering and disposal.

$Q^w \cdot X^r$ determines the mass rate of wasted sludge. For the different types of solids, it is $1.8 \cdot 10^8$ mg VSS/d, $1.2 \cdot 10^8$ mg VSS$_a$/d, and $2.2 \cdot 10^8$ mg SS/d.

The loss rates from the effluent can be obtained by difference or from $(Q - Q^w) \cdot X$. They are $1.5 \cdot 10^7$ mg VSS/d, $1.0 \cdot 10^7$ mg VSS$_a$/d, and $1.8 \cdot 10^7$ mg SS/d.

Step 13. Estimate the Nutrient Requirements. Nutrients are needed in proportion to the net biomass synthesis, which we computed in step 10 to be $1.45 \cdot 10^8$ mg VSS/d. If N is 12.4 percent of the VSS (i.e., 14 gN/113 g $C_5H_7O_2N$) and P is 2.5 percent of VSS (i.e., 20 percent of N), then the supply rates of N and P must be at least $1.8 \cdot 10^7$ mg N/d and $3.6 \cdot 10^6$ mg P/d. Dividing these mass rates by the plant flow ($Q = 10^6$ l/d) gives the minimum required concentrations in the influent: 18 mg N/l and 3.6 mg P/l.

Step 14. Estimate the Effluent SMP Concentrations. Since the UAP and BAP that comprise SMP are soluble molecules, they are not concentrated in the sludge. Instead, they behave as soluble components, like S. The chemostat solutions for UAP (Equation 3.38) and BAP (Equation 3.39) hold:

UAP
$$= \frac{-(\hat{q}_{UAP} X_a \theta + K_{UAP} + k_1 r_{ut} \theta) + \sqrt{(\hat{q}_{UAP} X_a \theta + K_{UAP} + k_1 r_{ut} \theta)^2 - 4 K_{UAP} k_1 r_{ut} \theta}}{2}$$

BAP
$$= \frac{-(K_{BAP} + (\hat{q}_{BAP} - k_2) X_a \theta) + \sqrt{(K_{BAP} + (\hat{q}_{BAP} - k_2) X_a \theta)^2 + 4 K_{BAP} k_2 X_a \theta}}{2}$$

in which $r_{ut} = -(S^0 - S)/\theta$ and the other parameters are as defined in Chapter 3. Reasonable parameter values for SMP production and biodegradation are (Chapter 3):

$\hat{q}_{UAP} = 1.8$ mg COD$_p$/mg VSS$_a$-d

$\hat{q}_{BAP} = 0.1$ mg COD$_p$/mg VSS$_a$-d

$K_{UAP} = 100$ mg COD$_p$/l

$K_{BAP} = 85$ mg COD$_p$/l

$k_1 = 0.12$ mg COD$_p$/mg BOD$_L$

$k_2 = 0.09$ mg COD$_p$/mg VSS$_a$-d.

For this example, $-r_{ut} = (500 - 0.81)/0.39 = 1{,}280$ mg BOD_L/l-d. Plugging in all the numbers gives:

UAP

$$= 0.5 \cdot \left[\frac{-(1.8 \cdot 1{,}710 \cdot 0.39 + 100 - 0.12 \cdot 1{,}280 \cdot 0.39) +}{\sqrt{(1.8 \cdot 1{,}710 \cdot 0.39 + 100 - 0.12 \cdot 1{,}280 \cdot 0.39)^2 + 4 \cdot 100 \cdot 0.12 \cdot 1{,}280 \cdot 0.39}} \right]$$

UAP = 4.8 mg COD/l

$$BAP = 0.5 \cdot \left[\frac{-(85 + (0.1 - 0.09) \cdot 1{,}710 \cdot 0.39) +}{\sqrt{(85 + (0.1 - 0.09) \cdot 1{,}710 \cdot 0.39)^2 + 4 \cdot 85 \cdot 0.09 \cdot 1{,}710 \cdot 0.39}} \right]$$

BAP = 39 mg COD/l

And,

$$\text{SMP} = \text{UAP} + \text{BAP} = 4.8 + 39 = \textbf{43.8 mg COD/l}$$

Step 15. Estimate the Effluent Quality in Terms of COD, BOD_L, and BOD_5. Since all forms of effluent organic matter are measured by the COD test, the effluent COD is computed by adding the CODs of all components containing organic matter:

Original substrate, S	0.8 mg/l
VSS, $X_v^e \cdot 1.42$ mg COD/mg VSS	21.3
SMP	43.8
Sum	65.9 mg COD/l

If the influent contained any nonbiodegradable soluble COD, it would be included in the COD sum, too. In this example, we assume that nonbiodegradable COD is zero.

The BOD_L includes the substrate, the biodegradable portion of VSS_a, and the SMP, which we assume is biodegradable.

Original substrate, S	0.8 mg/l
VSS_a, $X_a^e \cdot f_d \cdot 1.42$	11.4
SMP	43.8
Sum	56.0 mg BOD_L/l

The effluent standard is in terms of BOD_5, and we must apply the BOD_L to BOD_5 conversions derived in Chapter 3 (Example 3.3).

Original substrate, $S \cdot 0.68$	0.5 mg/l
VSS_a, $X_a^e \cdot f_d \cdot 1.42 \cdot 0.4$	4.55
SMP, SMP $\cdot 0.14$	6.1
Sum	11.1 mg BOD_5/l

Even though the COD and BOD_L are well above 20 mg/l, the standard is met by the estimated BOD_5 concentration. These low ratios of effluent BOD_5 to COD or BOD_L are normal.

Step 16. Compute the Sludge Recycle Rate. Earlier in this chapter, a simplified mass balance around the settling tank yielded an approximate relationship (Equation 6.12) for the sludge recycle ratio, R^*.

$$R = \frac{X}{X^r - X}$$

Using the VSS values of X and X^r gives

$$R = \frac{2,500}{10,000 - 2,500} = 0.33$$

Thus, the recycle ratio is about 33 percent, which is a reasonable value.

Not making all the simplifying assumptions leads to a slightly more complicated (and accurate) equation for R.

$$R = \frac{X\left(1 - \dfrac{\theta}{\theta_x}\right)}{X^r - X} \qquad \textbf{[6.14a]}$$

Substituting the numbers gives R = 0.31. The more complicated relationship allows for the net growth of new biomass in the aeration basin, which makes R slightly smaller when θ is much smaller than θ_x, as it normally is.

Step 17. Compute the Oxygen Supply Rate Needed. The largest operating expense for activated sludge normally is the electricity to power aeration. The oxygen-uptake rate needed and the efficiency of the aeration system determine the energy needed. In this step, we compute the oxygen uptake rate. Given all the information already computed, the simplest way is by a mass balance on electron equivalents, expressed as oxygen demand. We compute all rates in mg as O_2 per day.

The input oxygen demand is in the substrate and the inert VSS:

Substrate, $Q \cdot S^0$	$10^6 \cdot 500$	$5 \cdot 10^8$ mg/d
VSS, $1.42 \cdot Q \cdot X_i^0$	$1.42 \cdot 10^6 \cdot 50$	$7.1 \cdot 10^7$ mg/d
Sum		$5.7 \cdot 10^8$ mg/d

The oxygen demand leaving the system is contained in S, SMP, and VSS:

Substrate, $Q \cdot S$	$10^6 \cdot 0.81$	$8.1 \cdot 10^5$ mg/d
SMP, $Q \cdot$ SMP	$10^6 \cdot 43.8$	$4.38 \cdot 10^7$ mg/d
VSS, $1.42 \cdot \Delta X_v / \Delta t$	$1.42 \cdot 1.95 \cdot 10^8$	$2.77 \cdot 10^8$ mg/d
Sum		$3.2 \cdot 10^8$ mg/d

The oxygen-uptake rate is the difference between the equivalents entering and exiting, or

O_2 uptake rate, $\Delta O_2 / \Delta t = 5.7 \cdot 10^8 - 3.2 \cdot 10^8 = 2.5 \cdot 10^8$ mg/d.

Step 18. Estimate the Power Required for Aeration. We assume a typical FOTE of 1 kg O_2 per kWh to compute the aeration capacity to supply the computed oxygen uptake rate:

Power $= (2.5 \cdot 10^8 \text{ mg } O_2/\text{d})(10^{-6} \text{ kg/mg})/(1 \text{ kg } O_2/\text{kWh})(24 \text{ h/d})$

$= \textbf{10.4 kW}$

Thus, the installed aeration power is around 10 kW to meet the demand. Of course, a significant safety factor needs to be included to account for loading surges and redundancy.

In example 6.2, we computed all of the major factors in a process design for activated sludge, for example, SRT, system volume, MLVSS, sludge wasting and recycling rates, effluent quality, and aeration capacity. In addition, we gained insight about hard-to-measure aspects of the process, like the fraction of the MLVSS that is active and the breakdown of the effluent BOD into S, SMP, and X_a. The choices made here led to a feasible design, and we were not forced to go back and change the design decisions due to technical or economic unfeasibility. However, not all designs go so smoothly. It is common (and to be expected) that the design process is iterative, not so linear as might be implied here. Key assumptions need to be identified and justified. On the other hand, the steps shown here are the ones that need to be carried out whether the design is accomplished in one pass or requires many iterations.

6.7 ANALYSIS AND DESIGN OF SETTLERS

The dominant means for achieving solids separation in activated sludge is gravity settling of naturally flocculated biomass. A properly designed settling tank (also called a settler, clarifier, or sedimentation tank) promotes flocculation and creates quiescent conditions. When this occurs, the effluent is clear and has acceptably low SS concentration; nearly 100 percent of the suspended solids are captured, which allows independent sludge wasting to control θ_x, which should be substantially larger than θ. The recycle and waste sludges are thickened, which greatly reduces the costs for sludge pumping, dewatering, and disposal. While the first two objectives—low effluent SS and 100 percent solids capture—are closely connected, the third objective—sludge thickening—is not automatically achieved by having excellent effluent SS. In some instances, the last objective can be detrimental to the first two.

In order to ensure that the design and operation of settlers properly reconciles all three objectives, we need to apply a systematic analysis technique that is based on the physics of solids sedimentation and thickening. The general tool is called *flux theory,* and its adaptation for activated-sludge settlers is called the *state-point* approach.

In this section, we first describe key physical/chemical properties of activated-sludge solids. Second, the typical components of settlers are described, and typical *loading criteria* are reviewed. Then, the central focus of the section is on flux theory and the state-point approach to design and analysis. Through state-point analysis, we link the performance of the settler to design/operating conditions in the aeration basin and the flows between the settler and the aeration basin. State-point analysis allows us to connect the largely empirical loading criteria to physical processes occurring in the settler. It also allows the engineer to devise rational design and operating strategies for unusual, as well as typical, conditions.

6.7.1 ACTIVATED-SLUDGE PROPERTIES

The suspended solids in activated sludge are mostly bacteria, which contain 70–80 percent water inside the cells. In addition, even well-flocculated activated sludge traps considerable amounts of bound water outside the cells. Therefore, the activated-sludge floc is only slightly heavier than the water in which it is suspended. The difference in density is typically only about 0.0015 g/cm^3. This small difference in density means that the biomass must flocculate to a substantial size if it is to settle rapidly in a settler. For example, Stoke's law for laminar conditions gives a terminal settling velocity (v_s, m/d) of

$$v_s = \frac{g \Delta \rho d^2}{18 \mu} \cdot 86{,}400 \text{ s/d} \qquad \qquad \textbf{[6.29]}$$

in which g = gravitational constant (9.8 m/s^2), $\Delta \rho$ = density difference (g/cm^3), d = floc diameter (cm), and μ = dynamic viscosity (g/cm-s). For $T = 20 \,°$C and $\Delta \rho = 0.0015$ g/cm^3, $v_s = 12$ m/d (a typical low overflow rate) requires $d = 0.04$ cm = 0.4 mm. To have $v_s = 70$ m/d (a high overflow rate) requires $d = 0.1$ cm = 1 mm. Fortunately, well-flocculated activated sludge has floc diameters of 0.2 to 2 mm. The floc size inside the aeration basin often is smaller than 0.2 mm, but a properly designed settler allows for floc growth just after the sludge enters the settler.

Once flocculated, the sludge solids settle at a rate that depends on the sludge's physical/chemical properties and the concentration of solids, X. This relationship normally can be represented well by the empirical equation

$$v = v_0 10^{-\lambda X} \qquad \qquad \textbf{[6.30]}$$

in which v = the actual sedimentation velocity (m/d), v_0 = a maximum settling velocity (m/d) that is related to the Stoke's velocity for individual floc, λ = a sludge-compaction coefficient (m^3/kg), and X = the total suspended solids concentration (kg/m^3 = g/l). Figure 6.7 illustrates a typical settling-velocity curve. Although the sludge solids settle at near their maximum velocity for very small values of X, the velocity quickly declines due to the particle-particle interactions that occur when the solids concentration increases. In this region of *hindered settling,* the particles are close enough together that they slow each other's downward movement and the upward movement of the water that must be displaced by the settling solids. These hindering phenomena become more intense as X increases, causing the steady decrease in v.

When X becomes large enough, about 10 kg/m^3 for Figure 6.7, the sludge enters the region of compression-controlled settling. In this region, the solids particles are close enough together that the particles underneath begin to physically support part of the weight of the particles above. Then, further settling requires that the weight of the overlying solids must squeeze out water from the pores of the underlying solids. This compression (or consolidation) phenomenon greatly slows the settling velocity, which asymptotically approaches zero at high enough X: v is nearly zero (less than 0.2 m/d) for $X = 20$ kg/m^3 in Figure 6.7.

Sludge bulking by filamentous bacteria is the factor most dramatically affecting settling velocities of activated sludge. Although a bulking sludge normally has a good floc size and structure, extended filaments accentuate particle-particle interactions and the phenomena causing hindered settling. Thus, v can drop dramatically as the degree of filamentous growth increases. Sezgin (1981) correlated the settling velocity to the length of extended filaments observed microscopically in a range of bulking and nonbulking sludges. For the activated sludges he studied, Sezgin found that λ was roughly constant at 0.15 m^3/kg, but that v_0 depended strongly on the length of

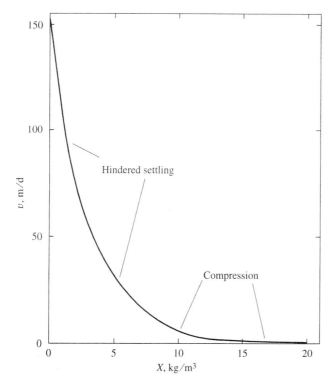

Figure 6.7   A typical profile of how the sludge's settling velocity declines with increasing solids concentration. This curve corresponds to $v_0 = 152$ m/d and $\lambda = 0.15$ m^3/kg.

extended filaments. This dependence can be expressed as

$$v_0 \text{ (in m/d)} = \frac{264}{1 + 1.17 \cdot 10^{-5}(4.57)^{\text{LF}}} \qquad \textbf{[6.31]}$$

in which LF $= \log_{10}$ of the length of extended filaments measured in μm/mg SS. Although the microscopic measurement of filament length may be too tedious for routine use in the field, Equation 6.31 demonstrates well how v_0 decreases rapidly for extended-filament lengths greater than about $10^7 \mu$m/mg, or LF $= 7$. For example, v_0 goes from 240 to 180 to 82 m/d as LF goes from 6 (no bulking) to 7 (slight bulking) to 8 (severe bulking). While the Sezgin values for λ and v_0 cannot be taken as universal, they offer good guidance on how activated-sludge settling is affected by X and bulking.

6.7.2 SETTLER COMPONENTS

Although settlers come in several different configurations, all of them must include four key components:

- an inlet zone that dissipates the momentum and kinetic energy of the inflowing sludge and allows good flocculation

- a quiescent settling zone that achieves solids separation and thickening

- an effluent outlet that prevents re-entrainment of separated solids

- a sludge collection and removal system for the thickened sludge

Settlers for activated sludge systems can have a circular or rectangular surface cross section. Because circular settlers are more common, particularly in North America, we are going to focus on them as we illustrate how the key components are incorporated.

Two types of inlet configurations are widely used: center feed and peripheral feed. These are illustrated in Figure 6.8. In the center-feed design, the influent sludge is discharged from a pipe at the center of the settler. Energy dissipation and flocculation occur in a circular center well that is defined by a continuous, tubular

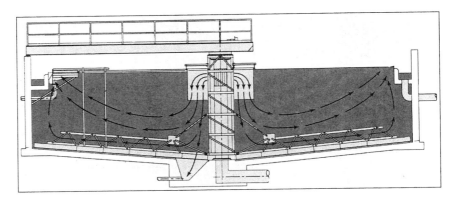

a.

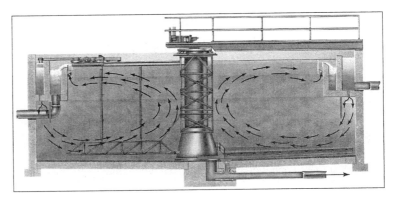

b.

Figure 6.8   Illustrations of center feed *a.* and peripheral feed *b.* for sludge inlet configurations. SOURCE: With permission from U.S. Filter.

baffle. Flocculated sludge flows slowly and evenly under the baffle to enter the quiescent settling zone. Modern center-feed clarifiers have relatively large center wells. The diameter of the well can be 25–35 percent of the total settler diameter, and the baffle can extend to a depth of one-half of the total water depth. These dimensions mean that about 5 percent of the settler volume is in the inlet well.

Peripheral feed achieves the same goals by routing the inlet sludge into an inlet trough that runs around the outer edge of the settler. Sludge flows into a baffled zone beneath the trough. Energy dissipation and flocculation occur in the trough and the baffled zone. Well-flocculated sludge then flows slowly and evenly under the baffle to enter the settling zone. In general, the volume of the trough and baffled zone for peripheral feed should be similar to the volume for the center well of center feed.

The quiescent zone must achieve a clear effluent and a thickened sludge. Most of the strategies that engineers employ for process design and operation are used to ensure that the quiescent zone is adequately sized to accomplish these two goals. The loading parameters discussed in the next section, as well as state-point analysis, are tools used to ensure proper performance of the quiescent zone. However, a properly sized quiescent zone will not yield acceptable results if the other components allow poor flocculation, turbulence, short circuiting, or sludge re-entrainment.

The effluent overflows V-notch weirs located around the periphery of the settler. V notches are used so that large variations in flow can be accommodated with small changes in head loss over the weirs. Although earlier designs put the effluent weirs and troughs flush against the outside wall, more modern designs move the weirs and troughs inboard to reduce preferential flow paths upward along the walls. Figure 6.9 is a picture of an inboard weir and trough. Scum, grease, and other floatables collect at the surface of the quiescent zone. They are removed by skimming devices that travel around the settler and push the floating materials into a scum trough. Figure 6.9 also pictures a skimmer and trough.

The thickened sludge must be collected and removed from the bottom of the settler. Because the sludge exits from the bottom of the settler, it is called the *underflow*. Two collection and removal systems are in common use: scraper with sump and hydraulic. Key aspects of both are illustrated in Figure 6.9.

Sludge scrapers physically push the sludge blanket towards a central sump, from which the underflow is pumped for wasting and recycle. Scrapers are located along trussed arms that slowly rotate. The size of the scrapers, their angle, and the rotating speed are designed to move the sludge towards the sump at a rate greater than or equal to the maximum rate at which the solids enter the settler. Failure or under-design of the scrapers can result in the buildup of an excessive sludge blanket in locations where the sludge-moving capacity is inadequate. Ironically, this situation can, in some cases, lead to a very dilute underflow sludge, because the relatively clear water is drawn into the sump when the scraper cannot move the sludge to the sump rapidly enough.

Problems that develop for scraper and sump systems can cause diminished sludge-transport capacity, even though the original design is adequate. Wearing or misalignment of the scraper shoes, an uneven floor bottom, and large debris on the floor can create excessive gaps between the floor and the scraper, thereby allowing the most

Full width skimmer and trough

a.

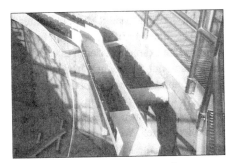

Double weir effluent trough

b.

Dual-plow scraper sweeps the floor twice for each revolution. Scrapers are attached to collector arms that are supported from the bridge shown at the top.

c.

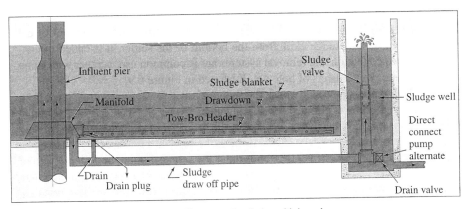

Hydraulic system for sludge withdrawal

d.

Figure 6.9   Pictures of various parts of the effluent and sludge withdrawal devices. SOURCE: With permission from Sanitaire.

thickened sludge to accumulate. Debris on the floor and (in some instances) ice buildup on exposed drives can jam and completely stop the scraper mechanism. Routine maintenance and periodic draining and inspection are key elements for ensuring that the scraper and sump system remains effective.

The hydraulic system for sludge collection and removal relies on creating a small pressure (or head) difference between the settler and a separate sludge well. Normally, a telescoping value (see Figure 6.9) is used to control the pressure difference. As the rotating sweep arms move along the bottom of the settler, sludge located at the bottom is drawn into the sweep arms through holes. The driving force is simply the pressure difference. Therefore, any liquid present at the level of the sweep arm is pulled into the sweep arm and sent to the sludge well. Thickened sludge is removed because it is located at the bottom. If the sludge blanket is very shallow, relatively clear water also is drawn into the sweep arm. Manipulating the telescoping value to give the desired pressure difference controls the rate of sludge removal. Waste and recycle sludges are pumped from the sludge well.

Figure 6.10 is a cut-away drawing of a center-feed settler that uses a scraper and sump system for underflow removal and inboard weirs for effluent removal. This figure illustrates how the different components fit together to create an integrated settler system.

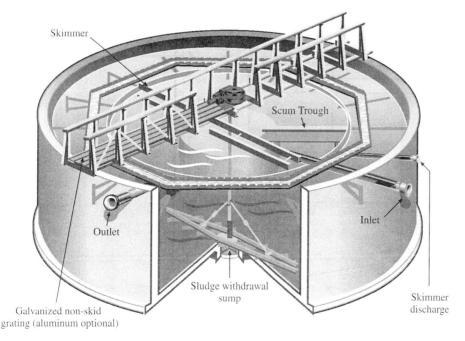

Figure 6.10   Cut-away drawing of a circular settler having center-feed, scraper and sump removal of sludge, and inboard weirs. SOURCE: With permission from Sanitaire.

6.7.3   LOADING CRITERIA

Based mainly on years of operating experience, engineers and regulators have established several loading criteria for the design and operation of settlers used with activated sludge. Although the numerical values of these criteria were empirically derived, they also are based on principles of sedimentation and thickening. The explicit connection is made through the use of flux theory and state-point analysis. In this section, we present the concepts underlying and the typical numerical value for the key loading criteria. Figure 6.11 is a schematic that defines the parameters comprising the loading criteria. This figure is for a center-feed settler, but the same parameters hold for peripheral feed.

The *overflow rate* $(O/F, LT^{-1})$ is the key criterion for ensuring that the effluent solids remain at a low concentration. The overflow rate is defined by

$$O/F = Q^e/A = (Q - Q^w)/A \qquad \textbf{[6.32]}$$

in which A = the plan-view surface area (L^2), Q^e = process effluent flow rate (L^3T^{-1}), Q = process influent flow rate (L^3T^{-1}), and Q^w = flow rate for waste sludge (L^3T^{-1}). Since the recycle flow rate (Q^r) does not exit with the effluent, it does not appear in the definition of the overflow rate. Nominally, the overflow rate

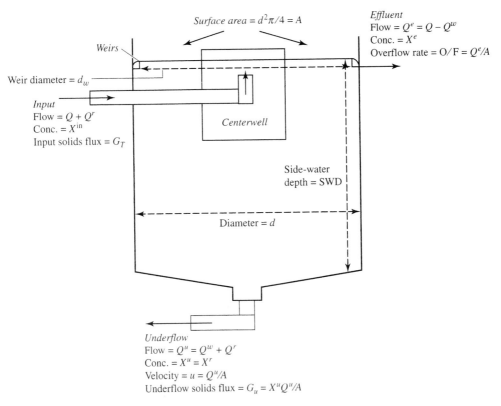

Figure 6.11   Schematic defining the parameters comprising the loading criteria for settlers.

should be smaller than the settling velocity for the slowest settling particles. However, seldom is that level of information known. Therefore, the overflow rate is set well below the velocity that would lead to a significant fraction of the solids being entrained in the overflow.

The overflow rate can be based on the average effluent flow rate or on a short-term peak flow rate. Peak-flow criteria are meant to account for the normal diurnal flow variations common to municipal sewage-treatment plants; the peak flow criteria do not ensure good performance for extended periods of high flow, such as during sustained rain events. Equalization, off-line storage, or bypassing is required to prevent long-term loading above the design average-flow overflow rate.

Due to poorer floc formation and greater risk of bulking, activated sludge processes employing an extended-aeration SRT have much lower design values for overflow rates. Whereas the average overflow rate is in the range of 12 to 40 m/d for a conventional SRT, it is only 8 to 16 m/d for extended aeration. The respective ranges for the peak flow are 40 to 70 m/d and 24 to 32 m/d.

The second key criterion is the *input solids flux*, G_T $(ML^{-2}T^{-1})$, which is defined as

$$G_T = \frac{(Q + Q^r) \cdot X^{in}}{A} \qquad [6.33]$$

Here, X^{in} is the input concentration of suspended solids, which is the same as X or the MLSS concentration. The input solids flux is tied to the ability of the settler to thicken the sludge; however, the connection is far from linear. In general, an increase in G_T caused by an increase in Q^r reduces the concentration of the underflow solids, X^u. An increase in G_T brought about by increases in Q or X^{in} may cause X^u to decrease or increase. A thorough evaluation using state-point analysis is required to ascertain how changes in G_T affect X^u. On the other hand, excessive values of G_T greatly increase the risk of thickening failure, which can lead to massive losses of solids to the effluent if the sludge blanket rises to the weirs.

Although the solids flux cannot be used alone as a control over thickening or effluent quality, experience offers valuable guidance about ranges that normally give adequate performance. Similar to the overflow rate, solids-flux criteria are given for the average flows and for short-term peak flows. For systems with a conventional SRT, the recommended ranges for solids flux are 70 to 140 kg/m²-d for average flow and < 220 kg/m²-d for peak flow. The analogous ranges for extended aeration are 24 to 120 kg/m²-d and < 170 kg/m²-d.

The third criterion is the *weir loading*, which is defined as

$$WL = Q^e / \sum d_w \pi \qquad [6.34]$$

in which WL = weir loading (L^2T^{-1}), d_w = diameter of a weir (L), and $\sum$ indicates the summation of all weirs, which may have different diameters when multiple troughs are placed away from the outer wall. Equation 6.34 is for settlers with circular weirs. Whatever the weir configuration, the denominator should equal the total length of weirs.

The concept of the weir loading is to minimize the possibility that light solids will be swept up by fast rising water near the weirs. Thus, the weir load establishes a maximum flow per unit of weir length. Whether or not the weir loading actually plays a role in preventing sludge re-entrainment is controversial. Nonetheless, empirical ranges for weir loading are available and commonly used in design. The recommended values are 100 to 150 m^2/d for average flow and < 375 m^2/d for peak flow. To keep weir loads within these bounds, especially for larger settlers, multiple rows of weirs are employed.

While not strictly loading criteria, several geometric criteria have begun to be accepted in recent years. A strong trend is towards deeper settlers. Side-water depths (SWDs) of 5 to 6 m are becoming more common today, compared to much shallower SWDs (e.g., 3 m) in older settlers. The deeper SWDs reduce negative interferences between the clear-water zone near the weirs and the sludge blanket. In parallel to greater depths is a trend towards having larger and deeper inlet zones. For example, baffles extending to 50 percent of the SWD and center-well diameter up to 35 percent of the settler diameter are now becoming more common. These larger inlet zones allow the rate of energy dissipation to be controlled for improved flocculation. Although not yet a widespread criterion for inlet design, mean-velocity-gradient values of 20–40 s^{-1} appear to be advantageous for flocculation.

6.7.4   BASICS OF FLUX THEORY

Flux theory is a practical means to combine the settling and compression characteristics of sludges with the physical characteristics of settler operation. Flux theory is well developed for sludge thickeners and can be applied through state-point analysis to activated sludge settlers, which have other goals beyond thickening. This section develops the basics of flux theory, while the next section addresses state-point analysis.

The first key to flux theory is computation of the settling flux,

$$G_s = v \cdot X^b \qquad \textbf{[6.35]}$$

in which G_s is the flux of solids past a horizontal plane and due solely to the sedimentation of the particles with respect to the water column $(ML^{-2}T^{-l})$, and X^b is the solids concentration at some location in the sludge blanket. Computing G_s requires knowing v from experimental measurements (Dick, 1980) or from Equation 6.30. Figure 6.12 is a typical settling-flux curve. A sharply rising arm occurs for very low X^b ($<$ about 1 kg/m^3 here). Intermediate values of X^b give a decreased rise rate, a maximum flux, and a gradually falling portion when the settling is more and more hindered. Finally, the curve has an asymptotic approach to zero flux as compression greatly reduces v for very high X^b ($>$ about 10 kg/m^3 here). A chord drawn from the origin to a point on the settling-flux curve has a slope equal to the settling velocity, v, for the X^b value indicated by the intersection of the chord and the settling-flux curve.

The second key to flux theory is the downward movement of the water column due to sludge withdrawal. Because settlers are continuous-flow processes, liquid volume is removed as underflow at flow rate Q^u. This underflow creates a downward velocity for the water column:

$$u = Q^u/A \qquad \textbf{[6.36]}$$

in which u is the downward flow velocity of liquid due to the underflow (LT^{-1}).

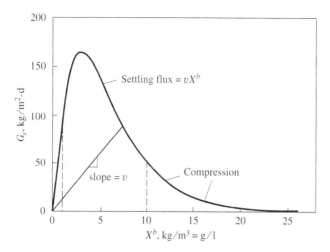

Figure 6.12 Settling-flux curve and velocity chord for a typical activated sludge ($v_0 = 152$ m/d; $\lambda = 0.15$ m^3/kg).

Thus, the flux of solids past a horizontal plane in the settler (G_b) is equal to the concentration of solids at that position (X^b) multiplied by the sum of the water column velocity (u) and the settling velocity of the solids with respect to the water column (v). For any location in the settler, the relationship is

$$G_b = X^b \cdot (u + v) \qquad [6.37]$$

Flux theory normally assumes that the system is at steady state and has essentially 100 percent solids capture. Although the steady-state assumption is not always accurate (more on that later), it allows us to write a mass balance on solids by equating the input flux to the output fluxes,

$$G_T = G_b + G_e \qquad [6.38]$$

in which G_e = the *effluent flux* $(\mathrm{ML^{-2}T^{-1}}) = X^e Q^e / A$. Furthermore, the assumption of essentially 100 percent solids capture allows us to neglect G_e, which simplifies the steady-state mass balance on solids to $G_T = G_b$.

For the underflow of the settler, the *underflow solids flux* is G_u, the solids concentration is X^u, and the settling velocity (v) is zero due to the presence of the settler's floor. Based on the flows and concentrations defined in Figure 6.11, the input and output fluxes, (G_T and G_u, respectively) are equal for steady state:

$$G_T = \frac{(Q + Q^r) \cdot X^{\mathrm{in}}}{A} = G_u = \frac{(Q^r + Q^w) \cdot X^u}{A} = u \cdot X^u \qquad [6.39]$$

When the settler is at steady state, mass balance must hold for any position in the settler. Therefore, the flux of solids across any horizontal plane also must equal G_T and G_u. This is expressed mathematically by

$$G_T = G_u = G_b = (u + v) \cdot X^b \qquad [6.40]$$

The third key to flux theory is combining the settling flux with the mass balances. In the most straightforward sense, this third step involves simultaneously solving the equations (or measurements) for settling flux (Equation 6.35), overall mass balance (Equation 6.38 or 6.39), and mass balance within the settler (Equation 6.40). While this solution can be achieved by a variety of means, a common, convenient, and intuitive method involves graphical combination of the three equations.

Figure 6.13 illustrates graphical constructions used to solve the equations for the *critical loading,* which means that all assumptions and mass balances hold exactly. Critical loading also is the situation giving the maximum solids underflow concentration for a given solids flux.

Superimposed on the settling-flux curve is the underflow line, which is a straight line that intercepts the left axis at G_T, has a slope equal to the negative of the underflow velocity $(-u)$, and is just tangent to the settling flux curve. The straight line graphically represents the overall mass balance, Equation 6.39. For this case, X^u is 15.4 kg/m^3, and u is 9.09 m/d. The point of tangency indicates what solids concentration (X_c) gives the minimum and, therefore, critical total flux. The settling velocity for the critical concentration is shown by the chord, which has slope v_c (3.04 m/d in this case). For all lower and higher values of X^b within the sludge blanket, the total solids flux could be larger; therefore, concentration X_c defines the critical condition within the sludge blanket. For X_c, the solids flux is comprised of the flux due to the water column moving downward $(uX_c = 105$ kg/m^2-d here) and the flux due to the solids settling with respect to the water columns $(v_cX_c = 35$ kg/m^2-d here). Having the total flux be dominated by the water column flux, as illustrated in Figure 6.13, is a normal situation. The solids concentration in the underflow, X^u, is larger than X_c and reflects mass balance. Thus, Figure 6.13 says that the minimum solids flux through the settler occurs at some critical concentration within the sludge blanket (X_c), not at the underflow concentration (X^u).

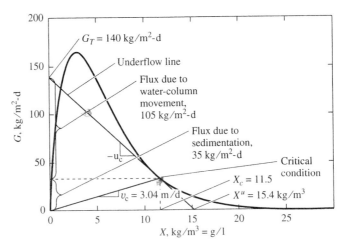

Figure 6.13   Constructions used to determine the critical-loading condition.

Figure 6.14 shows what happens if G_T is changed and the settler is maintained at a critical loading. When G_T increases from 140 to 210 kg/m²-d, the underflow velocity must be increased to 16.3 m/d in order to remove the solids from the bottom of the settler. The "price" paid for this increase in u_c is a decrease in X_u, which declines to 12.9 kg/m³. Conversely, if the total flux is decreased to 70 kg/m²-day, the underflow velocity should be decreased to maintain critical loading, and a "benefit" of a much larger X_u is obtained (almost 19 kg/m³). Figure 6.14 shows that maximum values of X_u are obtained when u_c is small, but that u_c must increase to accommodate large values of G_T.

The analysis so far has been predicated on attaining critical loading. However, fluctuations in flows, concentrations, and settling properties, as well as purposeful operating decisions, can lead to situations for which critical load is not achieved. When critical load is not achieved, the settler is either in thickening underload or overload.

Figure 6.15 illustrates two ways in which *underloading* can result. In case 2, the total flux decreases due to reductions in $(Q + Q^r)$ and/or X^{in}, but the underflow velocity remains constant at $u = 9.1$ (m/d). In case 3, the total flux remains equal to 140 kg/m²-d, but the underflow velocity increases. In both cases, the underflow line falls below the settling-flux curve for all relevant X^b. [Values of X^b below X^{in} are not relevant, since the sludge enters the settler with X^{in} much larger than these values and forms a sludge blanket with all $X^b \geq X^{in}$.] This means that, for all relevant X^b, the sum of the settlement and water-column fluxes is greater than G_T. In other words, there is no critical solids concentration that limits the downward flux of solids.

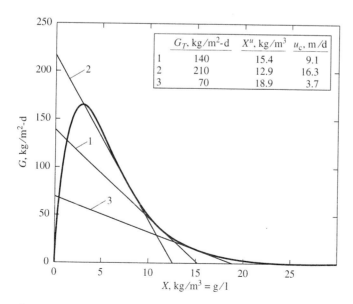

	G_T, kg/m²-d	X^u, kg/m³	u_c, m/d
1	140	15.4	9.1
2	210	12.9	16.3
3	70	18.9	3.7

Figure 6.14   Changes in G_T require that u is adjusted proportionally to maintain critical loading, while X^u varies inversely with G_T and u_c.

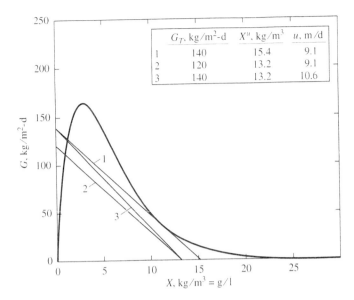

	G_T, kg/m²-d	X^u, kg/m³	u, m/d
1	140	15.4	9.1
2	120	13.2	9.1
3	140	13.2	10.6

Figure 6.15   Two ways in which underloading occurs and lowers X^u.

Therefore, the solids move directly to the bottom of the settler, where the floor stops them.

Long-term operation in an underloaded condition creates a steady state having only a minimal sludge blanket created by the floor. The underflow concentration is determined by the overall mass balance, represented by the intersection of the underflow line with the horizontal axis. As is shown clearly in Figure 6.15, the steady-state X^u for underloading is less than could be obtained by maintaining a critical load. Thus, a "cost" of underloading is more diluted underflow sludge. The corollary is that critical loading gives the maximum X^u for a given G_T.

Thickening *overload* is the converse of underloading. During thickening overload, the total input solids flux is greater than the sum of the settlement and water-column fluxes for a range of X^b values that could occur in the sludge blanket. Then, the solids cannot be transmitted to the bottom of the settler as fast as they are applied, sludge accumulates in the settler, and eventually the sludge blanket may rise to the weirs and pass to the effluent stream.

Figure 6.16 illustrates two situations of thickening overload. For case 2, the total input solids flux increases to 160 kg/m²-d, but the underflow velocity remains constant at 9.1 m/d. Thickening overload commences at a solids concentration of 7 kg/m². Therefore, solids accumulate in the sludge blanket for $X^b \geq 7$ kg/m³. Overloading stops when the underflow curve reaches tangency. For case 2, this corresponds to the same line as for case 1 for $X^b > 11.5$ kg/m³. Thus, sludge accumulates for $7 \leq X^b \leq 11.5$ kg/m³. For case 3, thickening overload is induced by a decrease in u to 8 m/day, even though G_T remains 140 kg/m³. The figure shows that sludge accumulates for $8.5 \leq X^b \leq 12.2$ kg/m³.

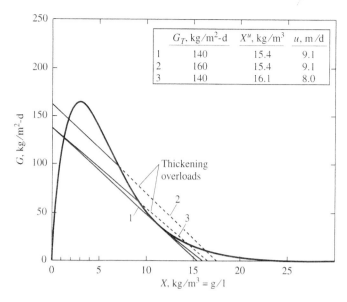

Figure 6.16 Two ways in which thickening overload occurs.

Thickening overload is an inherently nonsteady-state condition. The sludge blanket increases in mass, because the input solids flux is greater than the underflow flux. This is shown on Figure 6.16 by comparing the input flux for case 2 ($G_T = 160$ kg/m^2-d) to the underflow flux ($u X^u = 140$ kg/m^2-d). With continued operation in the overload region, the sludge blanket eventually reaches the effluent weirs and creates massive solids losses. The rate of accumulation, expressed as flux G_{acc}, can be computed from the difference between the input and underflow fluxes:

$$G_{acc} = G_T - G_u = G_T - u \cdot X^u \qquad \textbf{[6.41]}$$

For example, case 2 in Figure 6.16 gives

$$G_{acc} = 160 \text{ kg/m}^2\text{-d} - (9.1 \text{ m/d})(15.4 \text{ kg/m}^3)$$
$$= 20 \text{ kg/m}^2\text{-d}$$

The rise rate of the sludge blanket can be estimated by dividing G_{acc} by a representative solids concentration within the accumulating zone. For case 2, letting that concentration equal the low end of the accumulating zone (i.e., 7 kg/m^3) gives the fastest rise velocity, v_{acc}, for the accumulating sludge blanket:

$$v_{acc} = (19.9 \text{ kg/m}^2\text{-d})/7 \text{ kg/m}^3$$
$$= 2.8 \text{ m/d}$$

Clearly, a settler with a water depth of 5 m would experience massive solids losses within two days. The trend towards deeper settlers is partly driven by the desire to minimize the probability of massive solids losses during short-term periods of thickening overload.

Example 6.3 | **TYPICAL GOOD SETTLER PERFORMANCE**   Work by Knocke (1986) indicates that a good settling sludge has values of $v_0 = 240$ m/d and $\lambda = 0.2$ m^3/kg. Figure 6.17 shows the resultant settling-flux curve. We first find the maximum flux and corresponding underflow velocity needed to give an underflow-solids concentration of 15 kg/m^3. This maximum solids flux is achieved at critical loading (line 1) with $X^u = 15$ kg/m^3. Here, $G_T = 67$ kg/m^2-d and $u_c = 4.5$ m/d. A more typical solids flux is 120 kg/m^2-d, which yields underflow line 2, a maximum X_u of 12.5 kg/m^3, and $u_c = 9.6$ m/d for the typical flux. Activated sludge settlers more normally achieve a lower X_u for the typical solids flux. Underflow line 3 demonstrates that an underload condition gives $X_u = 10$ kg/m^3 when u is increased to 12 m/d.

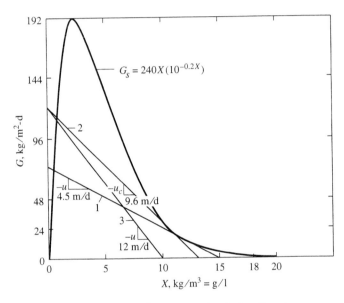

Figure 6.17   Settling-flux curve and underflow lines for Example 6.3.

6.7.5 STATE-POINT ANALYSIS

State-point analysis is an extension of flux theory (Keinath, 1985) designed to consider how activated-sludge settlers are coupled to the aeration tank. It takes into account that the recycle and effluent flow rates are independent variables and that the input concentration is controlled by bioprocess considerations that are largely independent of the settler.

Figure 6.18 illustrates the definition of the state point and how it relates to the settling flux and the underflow. The state point is the intersection of the underflow line with the chord having slope equal to the overflow rate. Its coordinates are the input solids concentration (X^{in}) and the input flux attributable to the flow that leaves

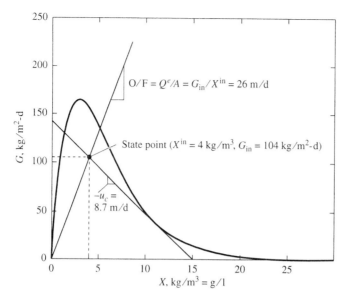

Figure 6.18 Definition of the state point in relation to the settling flux and underflow curves.

in the effluent (i.e., $G_{in} = Q^e X^{in}/A$). The difference between G_T and G_{in} is the input flux attributable to the underflow (i.e., $G_T - G_{in} = (Q_R + Q_w)X_{in}/A$).

The state point and its location are critical for two reasons. First, the overflow rate and X^{in}, which define the location of the state point, almost always are determined by factors external to the settler's operation. The overflow rate is controlled mainly by the influent flow rate, while X^{in} is controlled primarily by the influent substrate loading and the SRT. Therefore, the location of the state point usually is externally imposed and cannot be altered by operation of the settler alone.

Second, the location of the state point directly indicates whether or not the settler is experiencing *clarification overload*, which is the immediate loss of solids to the effluent. When the state point is below the settling flux curve, the possible downward flux of solids is greater than the input flux, which means that the solids initially settle to the sludge blanket and are not caught up in the effluent flow. On the other hand, a state point above the settling-flux curve indicates that the input flux is greater than the downward flux for the input concentration. Hence, some of the input solids cannot settle to the sludge blanket, but are immediately swept into the effluent flow, creating major solids loss and poor effluent quality. A settler in clarification overload automatically is in thickening overload. However, clarification overload does not require that the sludge blanket rise to the weirs to have major solids loss. Clarification overload persists for as long as the state point remains above the settling flux curve.

A number of distinctly different settler problems that commonly arise can be alleviated through appropriate strategies guided by state-point analysis. The following examples characterize these common problems and use state-point analysis to show how they can be solved. It is very important to be able to recognize what type of

problem is present, because solution strategies are quite different and even contradictory. Therefore, use of the wrong strategy often compounds the original settler
problem, instead of relieving it.

Example 6.4 | **DILUTE UNDERFLOW SLUDGE**   One common problem is dilute underflow sludge, even
though the sludge's settling properties are good. In this example (and the next 3), a good-settling
sludge has $v_0 = 152$ m/d and $\lambda = 0.15$ m^3/kg. The activated sludge system maintains a conventional θ_x of 5 d and mixed liquor suspended solids of 4,000 mg/l $= 4$ kg/m^3. The overflow
rate is 25 m/d, a normal value. Thus, $G_{in} = 4(25) = 100$ kg/m^2-d. Process performance is
good with respect to effluent BOD and suspended solids. On the other hand, the underflow
solids concentration is only 7.5 kg/m^3 (~ 0.75 percent). As is typical of some plants, little
attention is paid to the recycle flow rate.

   Figure 6.19 compiles all the known information: the settling flux curve, the overflow
rate, the underflow concentration, and the state point. Underflow line 1 is constructed from the
underflow concentration and the state point. It tells us that $u = 28$ m/d and $G_T = 210$ kg/m^2-d.
A comparison of G_T with recommended values (72–144 kg/m^2-d) suggests that a high G_T is a
possible cause of the problem. However, close inspection of Figure 6.19 shows that the settler
is very underloaded for thickening. Underload always keeps X^u below its maximum value.

   To alleviate the underload problem (i.e., dilute underflow sludge), we need to reduce G_T
by decreasing u. Because overall process performance is good with the existing θ_x, we wish to
maintain the same θ_x and X^{in}. In addition, Q^e and O/F are fixed by the plant inflow. Hence,
the state point does not move. To determine the impact of reducing u, we rotate the underflow
line counterclockwise around the state point. Figure 6.19 shows an improved new operating

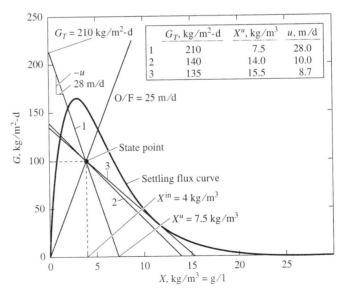

	G_T, kg/m^2-d	X^u, kg/m^3	u, m/d
1	210	7.5	28.0
2	140	14.0	10.0
3	135	15.5	8.7

Figure 6.19   Original severe underloading (underflow line 1)
gives a very dilute sludge, but reducing u greatly
improves X^u (line 2). The "best" operation is at
critical load, case 3.

condition that is only slightly underloaded (line 2). With only slight underload, this good settling sludge should give $X^u = 14$ kg/m^3 when u is reduced to 10 m/d, giving $G_T = 146$ kg/m^2-d. The reduction in u normally is achieved through a reduction in the recycle flow rate Q^r. The "best" performance (line 3) would be obtained by achieving critical load, which would have X^u is 15.5 kg/m^3. The disadvantage of trying to reach critical loading is that it increases the risk of overload. This risk is real, but can be managed as long as the sludge blanket depth is monitored regularly and kept well below the weirs. If the sludge blanket approaches the weirs, the underflow velocity can be increased to relieve thickening overload. This is illustrated in the next example.

RISING SLUDGE BLANKET   A second common problem in activated sludge treatment | **Example 6.5**
is a rising sludge blanket. Unchecked rise eventually causes massive solids losses as the blanket approaches the weirs. In this example, the same good settling sludge has $\theta_x = 5$ d, $X^{in} = 4,000$ mg/l, $G^{in} = 100$ kg/m^2-d, and satisfactory effluent quality. However, routine measurements of the sludge blanket depth show that it is continually rising and is now only about 1 m from the weir level. The total solids loading is 130 kg/m^2-d. What is the cause of the rising sludge blanket? How can it be reversed?

Line 1 in Figure 6.20 shows that the original loading condition has a slight thickening overload, which is causing a rising sludge blanket. To mitigate the thickening overload, the underflow velocity is increased to 10 m/d. Rotation about the state point (since θ_x and X^{in} should not and need not change) to line 2 shows that G_T is increased to 140 kg/m^2-d, while X^u declines from 16 to 14 kg/m^3. The second condition is in a slight underload, which will gradually draw down the sludge blanket.

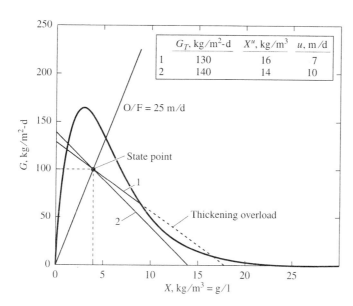

	G_T, kg/m^2-d	X^u, kg/m^3	u, m/d
1	130	16	7
2	140	14	10

Figure 6.20   A rising sludge blanket occurs with thickening overload (line 1), but increasing u (line 2) relieves the problem.

Example 6.6 | **CLARIFICATION OVERLOAD**   A sudden increase in the plant inflow, such as from a storm event, increases the overflow rate from 25 to 40 m/d. The mixed liquor suspended solids remain at 4,000 mg/l for some time, even though the flow rate increases. Assuming that u remains at 10 m/day, what is the immediate impact of the increase in overflow rate? What can be done to mitigate negative impacts?

Line 1 in Figure 6.21 shows that the higher overflow rate raises G_T to 204 kg/m^2-d and creates clarification overload. Input solids are swept into the effluent flow, creating immediate and massive solids discharges. The difference between G_T and G_u (140 kg/m^2-d) is 64 kg/m^2-d, which computes to an effluent SS concentration of 1.6 kg/m^3 ($= 1,600$ mg/l).

Clarification overload cannot be alleviated simply by increasing u, because the state point remains above the settling flux curve. The only way to stop clarification overload is to reduce the overflow rate and/or the X^{in} enough so that the state point drops below the settling flux curve. In this example, as in many real-world situations, the overflow rate is controlled solely by the plant inflow and cannot be reduced. In this case, X^{in} must decline. However, this decline generally means that sludge must be wasted and θ_x decreased! Thus, clarification overload almost always affects bioprocess operation and performance. In one sense, clarification overload is a self-correcting situation, because the massive loss of solids in the effluent is a form of sludge wasting. However, it is a highly undesirable form of sludge wasting, because it is uncontrolled and results in severe deterioration of the effluent quality. The main alternative is controlled sludge wasting (i.e., increasing Q^w) to reduce X_{in}, which also reduces θ_x.

Figure 6.21 shows two paths for eliminating the clarification and thickening overloads. The first path (line 2 in Figure 6.21) relies solely on decreasing X^{in} by wasting. In this example, u is held constant at 10 m/d, and X^{in} is decreased (by wasting) to 2.9 kg/m^3, at which point

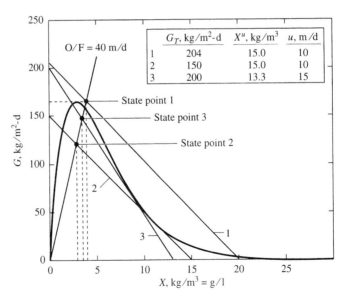

	G_T, kg/m^2-d	X^u, kg/m^3	u, m/d
1	204	15.0	10
2	150	15.0	10
3	200	13.3	15

Figure 6.21   An increase overflow rate creates immediate clarification overload (line 1 and state point 1). X^{in} must decrease in order to relieve clarification overload.

the settler is at critical loading. In the second path (line 3 in Figure 6.21), wasting is used to reduce X^{in} to 3.6 kg/m^3, a value far enough below the settling-flux curve that critical load can be achieved by increasing u to 15 m/day. Clearly, the second path results in much less wasting and smaller decrease in θ_x.

SLUDGE BULKING   The proliferation of filaments creates sludge bulking and significantly slows solids settling. Figure 6.22 shows the settling-flux curve for serious bulking: v_0 is reduced to 82 m/d, while λ remains 0.15 m^3/kg. Clearly, state point 1 gives clarification and thickening overloads for the bulking sludge. The best way to overcome the overloads is to greatly reduce filamentous growth (see discussion of bulking causes and remedies). Then, a good settling sludge gives a good X^u with slight underloading. If filaments cannot be reduced or until remedies take effect, the overload conditions can be alleviated only by some combination of reducing X^{in} and/or increasing u. Figure 6.22 shows an approach that relies on decreasing X^{in} to 3,000 mg/l (state point 2). As before, a decrease in X^{in} is brought about by increased sludge wasting and results in a smaller θ_x. The value of u is increased slightly in this case in order to maximize X^u by achieving critical loading. Although not shown on the figure, a smaller reduction in X^{in} could be accompanied by a larger increase in u to overcome overloads with a smaller effect on θ_x.

Example 6.7

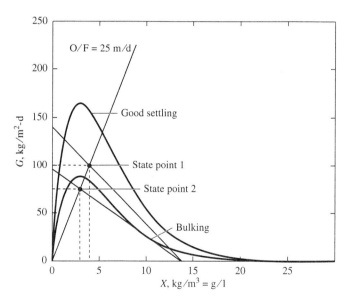

Figure 6.22   Clarification and thickening overloads are evident for the bulking sludge with state point 1. Overloads are eliminated by reducing filaments or by decreasing X^{in} when bulking persists (state point 2).

6.7.6   CONNECTING THE SETTLER AND AERATION TANK

Although each has its unique role in the activated sludge process, the settler and aeration tank are not completely independent. Earlier developments in the chapter quantified some of the connections. This section identifies and reviews which aspects are separate and which are connected.

1. Normally, settler performance has little direct control over the SRT, defined as

$$\theta_x = \frac{(V_{\text{system}})X}{(Q - Q^w)X^e + Q^w X^w}$$

In most cases, the larger amount of sludge loss comes through deliberate wasting ($Q^w X^w$), since $X^w = X^u$. The settler's performance controls X^w, and Q^w must be adjusted to compensate for changes in X^w or for effluent solids losses. On the other hand, the settler's clarification performance can control θ_x directly if $(Q - Q^w) X^e$ becomes dominant over $Q^w X^w$. This situation is most likely when the SRT is very large, which requires that the denominator be small, or when settler overload causes massive solids losses to the effluent.

2. The recycle flow rate, Q^r, is controlled strongly by settler performance. Mass balance around the settler and the entire system leads to Equation 6.14a,

$$Q^r = RQ = \frac{X\left(1 - \dfrac{\theta}{\theta_x}\right)}{X^r - X} Q$$

Because the settler's thickening performance determines X^r (i.e., $X^r = X^u$), the settler controls Q^r. Furthermore, the recycle flow usually is a major part of the solids flux to the settler and is the dominant factor in underflow velocity. Thus, the recycle flow helps determine the settler's thickening performance. If the settler is overloaded, the steady-state mass balances leading to derivation of the Q^r equation are invalid, and Q^r no longer depends on X, X^r, and θ/θ_x in the same way. For overloaded clarifiers, sludge is accumulating in the settler and, consequently, being drawn out of the aeration tank. The rates of accumulation and depletion can be computed from the difference between G_T and G_u.

6.7.7   LIMITATIONS OF STATE-POINT ANALYSIS

While a powerful tool for understanding and controlling settler performance, state-point analysis has limitations. First, it does not give direct information on the distribution of solids concentrations in the sludge blanket or on what the height of the sludge blanket should be. Second, state-point analysis does not describe the dynamics that occur immediately after the loading changes. Sophisticated finite-element representations of the settler are being developed as a research tool to address these issues not covered by state-point analysis.

6.8   CENTRIFUGAL SEPARATIONS

In a few instances, solids separation is carried out with centrifuges rather than a settling tank. Solids separation and thickening are driven by centripetal force, which is proportioned to rw^2, when r = the radius of the centrifuge, and w = the rotating speed in radians/second. (Note that rotational speed in radians/second equals (revolutions/second)/2π). For typical full-scale centrifuges, rw^2 is 1,400 to 2,100 g. Since the centripetal force is so much greater than gravity, the liquid detention time for a centrifuge is quite short, typically on the order of a minute. Thus, a main advantage of centrifuges is that they take up little space.

Despite their space advantage, centrifuges are infrequently used with activated sludge. One major disadvantage of centrifuges is that they incur significant capital and electrical costs. Furthermore, solids capture by centrifuges often is not as good as for settlers, which causes a deterioration of effluent quality. Finally, centrifuge performance responds sensitively and rapidly to changes in loading and sludge properties; therefore, major performance problems can occur rapidly and without warning.

6.9   MEMBRANE SEPARATIONS

Membranes are increasingly being used as the solids-separation device in activated sludge and other bioprocesses in environmental biotechnology. Processes using membranes are sometimes called *membrane bioreactors*. Membrane separation offers the advantage of "perfect" solids capture, because particles or macromolecules larger than the membrane's pore size, usually characterized by a nominal molecular-weight exclusion, cannot penetrate the membrane and are perfectly retained.

Figure 6.23 illustrates the key elements of an asymmetric membrane, the most common type. The pore size of the skin at the retentate side controls the cut-off size, while the membrane body, on the permeate side, provides structural strength to withstand the pressure differential used to drive the permeate flow.

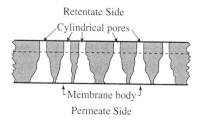

Figure 6.23   Cross-sectional view of an idealized asymmetric membrane. The pores defining the membrane's size cut-off are present in the top portion of the membrane body.

Because the membrane separator must be able to remove single bacteria of a size smaller than 1 μm, membranes used usually are microfilters. A rough definition for microfiltration is a pore size from 0.1 to 10 μm. Ultrafilters, popular in drinking-water treatment, typically remove smaller particles and macromolecules on the order of 1 nm or larger.

A pressure differential, which forces water and the nonretained solutes and colloids across the membrane, drives the membrane process. Retained materials are physically sieved at or near the surface of the membrane. The rate of permeate production (Q_{pm} in L^3T^{-1}) is measured as a permeate flux (F_{pm} in LT^{-1}) multiplied by the membrane area (A in L^2):

$$Q_{pm} = F_{pm} \cdot A \qquad \text{[6.42]}$$

The permeate flux depends on the pressure differential (ΔP in $ML^{-1}T^{-2}$), the resistance of the membrane itself (R_m in $MT^{-1}L^{-2}$), and the resistance due to polarization at the membrane surface (R_s in $MT^{-1}L^{-2}$):

$$F_{pm} = \frac{\Delta P}{R_m + R_s} \qquad \text{[6.43]}$$

When the membrane pore acts as a straight tube of radius r_p and length L_p, viscous flow according to the Hagen-Poiseuille law gives

$$R_m = \frac{8\mu L_p}{N\pi r_p^4} \qquad \text{[6.44]}$$

where μ = absolute viscosity ($ML^{-1}T^{-1}$) and N = the number of pores per unit surface area of the membrane (L^{-2}). Clearly, smaller pore size rapidly increases R_m and decreases F_{pm} for the same ΔP and N.

Concentration polarization occurs when the retained solids and large molecules accumulate at the retentate side of the membrane. A concentrated "gel" is thought to accumulate. Ultimately, the flux of permeate is controlled by the ability of the retained solids/macromolecules to be back-transported from the gel. This back-diffusion limitation can be represented mathematically by

$$F_{pm} = k_{bd} \ln\left(\frac{C_G}{C}\right) \qquad \text{[6.45]}$$

in which k_{bd} = a mass transport coefficient for back diffusion (LT^{-1}), C_G is the solid/macromolecule concentration in the gel (ML^{-3}), and C is the solid/macromolecule concentration in the bulk liquid of the retentate. In Equation 6.45, R_m is negligible compared to R_s, and F_{pm} is no longer proportional to ΔP. Instead, F_{pm} depends on k_{bd} and C_G/C according to Equation 6.45. In general, raising the tangential water velocity across the retentate skin increases k_{bd}. The ratio C_G/C depends on the desired increase in solids concentration, which is controlled mainly by the ratio of permeate flow to retentate flow. If permeate flow is relatively large, C will increase, reducing C_G/C.

When the two-resistance model of Equation 6.43 is used, R_s becomes

$$R_s = \frac{\Delta P}{k_{bd} \ln\left(\dfrac{C_G}{C}\right)}$$ [6.46]

Combining Equations 6.44, 6.45, and 6.46 yields

$$F_{pm} = \frac{\Delta P}{\dfrac{8\mu L_p}{N\pi r_p^4} + \dfrac{\Delta P}{k_{bd} \ln\left(\dfrac{C_G}{C}\right)}}$$ [6.47]

Equation 6.47 shows that F_{pm} is increased by an increase in ΔP only when the polarization resistance is not dominant. If increases in ΔP are ineffective, F_{pm} can be increased only by increasing the tangential flow velocity (to make k_{bd} larger) and/or decreasing the permeate flow in relation to the retentate flow (i.e., to make C_G/C smaller).

Probably the biggest problem of membrane separations is *fouling,* the gradual increase in resistance, which causes F_{pm} to decline while ΔP is held constant. Fouling is related to the composition and pore size of the membrane, chemical and physical parameters of the solid and dissolved components in the water, and operation of the membrane system.

"Reversible" fouling is generally associated with concentration polarization and can be reduced or reversed by increased transverse velocity and periodic flow reversals, such as back flushing or pulsing. "Irreversible" fouling is a much more difficult situation and cannot be eliminated by hydrodynamic measures. Irreversible fouling can occur by several mechanisms. When polarization is a major factor, macromolecules may concentrate in the gel to such a degree that they precipitate, blocking pore openings. This mechanism effectively decreases N in Equation 6.47. Molecules or colloids small enough to enter the pores may adsorb to the pore surfaces and occlude the pores. Pore occlusion reduces r_p in Equation 6.47. Bacterial growth on the membrane itself can block pores or be a source of soluble microbial products that absorb in the pores. Irreversible fouling, which creates a long-term decrease in permeate flux, creates economic and technical unfeasibility if the membrane must be replaced rapidly or pressure differentials become too great. Irreversible fouling can be partially overcome by washing with strong oxidants, enzyme detergents, acids, or strong chelators. The last two agents are directed mainly towards inorganic precipitates, while the first two address organic foulants. The choice of cleaning agent also depends on the type of membrane.

Membranes are made from organic polymers or ceramics. Polymers include cellulose acetate, polysulphone, polyamide, or polycarbonate. Ceramics are generally made from aluminum oxides. The polymers are more flexible and come in many configurations: bundles of hollow fibers, flat sheets, and spiral wound. Ceramic membranes are mechanically stronger, but usually only come in tube geometry. Considering the needs for preventing and reversing fouling, certain properties are key. For instance, cellulose acetate is more hydrophilic than the other polymers and seems to be much less susceptible to fouling-associated adsorption. Ceramics have by far the greatest tolerance for high temperatures ($> 100\,°C$), pH extremes (0–14), and chlorine (< 100 mg/l). Cellulose acetate is moderately tolerant to the same: temperature $< 50\,°C$, pH 2–10, and chlorine < 1 mg/l.

Membrane bioreactors introduce the mixed liquor to the retentate side of the membrane. Filtered permeate becomes the effluent, while the retentate is recycled back to the aeration basin. In some cases, the membrane unit is suspended inside the biological reactor, eliminating the need to pump retentate. Sustained operation with essentially no solids in the effluent can be achieved when permeate fluxes and pressure differentials are modest. For example, transmembrane pressure drops are less than about 3 atm (or 304 kg/m-s^2), permeate fluxes are 0.5 to 2.0 m/d, and times between cleaning are up to 150 d, although the longer cleaning periods correspond to the lower permeate fluxes and pressure drops (Manem and Sanderson, 1996).

Membrane bioreactors have been used in many applications to enhance the performance of activated-sludge systems (Manem and Sanderson, 1996; Rittmann, 1998). These include treatment of municipal wastewater, nitrogen removal, and treatment of high-strength industrial wastewaters. Advantages of using membranes include improved effluent quality, higher mixed liquor suspended solids, reduced system volume, and perfect SRT control. Disadvantages include increased capital costs for the membrane units, increased maintenance, and a more dilute waste sludge.

6.10   BIBLIOGRAPHY

Ardern, E. and W. T. Lockett (1914). "Experiments on the oxidation of sewage without the aid of filters." *J. Soc. Chem. Ind.* 33, p. 523.

Bailey, J. E. and D. F. Ollis (1986). *Biochemical Engineering Fundamentals.* New York: McGraw-Hill.

Bisogni, J. J. and A. W. Lawrence (1971). "Relationships between biological solids retention time and settling characteristics of activated sludge." *Water Research* 5, pp. 753–763.

Cheryan, M. (1986). *Ultrafiltration Handbook.* Lancaster, PA: Technomic Publishing, (1987).

Dick, R. I. (1980). "Analysis of the performance of final settling tanks." *Trib. Cebedeau* 33, pp. 359–367.

Dick, R. I. (1984). "Discussion of new activated sludge theory: Steady state. *J. Environ. Eng.* 110, pp. 1212–1214.

Eikelboom. H. A. (1975). "Filamentous organisms observed in activated sludge." *Water Research* 9, pp. 365–388

Ekama, G. A.; M. C. Wentzel; T. G. Casey; and G. Marais (1996). "Filamentous organism bulking in nutrient removal activated sludge systems." Paper 6: Review, Evaluation and Consolidation of Results. *Water SA* 22(2), pp. 147–160.

Garrett, M. T. and C. N. Sawyer (1951). "Kinetics of removal of soluble B.O.D. by activated sludge." *Proceedings, Seventh Industrial Waste Conference.* Lafayette, IN: Purdue University, pp. 51–77.

Goodman, B. L. and A. J. J. Englande (1974). "A unified model of the activated sludge process." *J. Water Pollution Control Federation* 46(2), pp. 312–332.

Gould, R. H. (1953). "Sewage aeration practice in New York City." *Proceedings, American Society of Civil Engineers* 79(Separate No. 307), pp. 307-1–307-11.

Greenberg, A. E.; L. S. Clesceri; A. D. Eaton; and M. A. H. Franson. Eds. (1992). *Standard Methods for the Examination of Water and Wastewater*, 18th ed. Washington, D.C: American Public Health Association.

Gutman, R. G. (1987). *Membrane Filtration: The Technology of Pressure-Driven Crossflow Processes*. Briston, England: Adam Hilger.

Haseltine, T. L. (1956). A rational approach to the design of activated sludge plants. In J. McCabe and W. W. J. Eckenfelder, Eds. *Biological Treatment of Sewage and Industrial Wastes, Aerobic Oxidation*. New York: Reinhold Publishing, pp. 257–270.

Henze, M., et al. (1995). *Activated Sludge Model No. 2, Scientific and Technical Report No. 3*. London: International Association on Water Quality.

Herbert, D.; R. Elsworth; and R. C. Telling. (1956). "The continuous culture of bacteria: A theoretical and experimental study." *J. General Microbiology* 14, pp. 601–622.

Hwang, H. J. and M. K. Stenstrom. (1985). "Evaluation of fine-bubble alpha factors in near full-scale equipment." *J. Water Pollution Control Federation* 57(12), pp. 1142–1151.

Jenkins, D. (1992). "Towards a comprehensive model of activated sludge bulking and foaming." *Water Science and Technology* 25(6), pp. 215–230.

Jenkins, P.; J. B. Neethling; H. Bode; and M. G. Paschard (1982). "The use of chlorination for control of activated sludge bulking." In B. Chambers and E. J. Tomlinson, Eds. *Bulking of Activated Sludge: Prevention and Remedial Methods*. Chichester, England: Ellis Horwood.

Joint Task Force (1967). *Sewage Treatment Plant Design, WPCF Manual of Practice No. 8*. Washington, D.C.: Water Pollution Control Federation.

Joint Task Force (1998). *Design of Municipal Wastewater Treatment Plants, WEF Manual of Practice No. 8*. Alexandria, VA: Water Environmental Federation.

Keinath, T. M. (1985). "Operational dynamics and control of secondary clarifiers." *J. Water Pollution Control Federation* 57, pp. 770–776.

Knocke, W. R. (1986). "Effects of floc volume variations in activated sludge thickening characteristics." *J. Water Pollution Control Federation* 58, pp. 784–791.

Laquidara, V. D. and T. M. Keinath (1983). "Mechanisms of clarification failure." *J. Water Pollution Control Federation* 55, pp. 1227–1331.

Lawrence, A. W. and P. L. McCarty (1970). "Unified basis for biological treatment design and operation." *J. Sanitary Engineering Division, ASCE* 96(SA3), pp. 757–778.

Manem, J. A. and R. Sanderson (1996). "Membrane bioreactors." In *Water Treatment Membrane Processes*. J. Mallevialle, P. E. Odendaal, and M. R. Weisner, Eds., New York: McGraw-Hill, chap. 17.

McKinney, R. E. (1962). "Mathematics of complete-mixing activated sludge." *J. Sanitary Engineering Division, ASCE* 88(SA3), pp. 87–113.

Metcalf and Eddy (1991). *Wastewater Engineering: Treatment, Disposal, Reuse*, 3rd ed. New York: McGraw-Hill.

Monod, J. (1950). "La technique of culture continue; Theorie et applications." *Annals Institute Pasteur* 79, pp. 390–410.

Novak, L.; L. Larrea; J. Wanner; and J. L. Garcia-Heras (1993). "Non-filamentous activated sludge bulking in a laboratory scale system." *Water Research* 27(8), pp. 1339–1346.

Novick, A. and L. Szilard (1950). Experiments with the chemostat on spontaneous mutations of bacteria." *Proc. National Academy of Sciences* 36, pp. 708–719.

Palm, J. C.; D. Jenkins; and D. S. Parker (1980). "Relationships between organic loading, dissolved oxygen concentration, and sludge settability in the completely mixed activated sludge process." *J. Water Pollution Control Federation* 52, pp. 2484–2506.

Pike, E. B. and C. R. Curds (1971). "The microbial ecology of the activated sludge system." In G. Sykes and F. A. Skinner, Eds. *Microbial Aspects of Pollution.* London: Academic Press, pp. 123–147.

Pitt, P. and D. Jenkins (1990). "Causes and control of *Nocardia* in activated sludge." *J. Water Pollution Control Federation* 62(2), pp. 143–150.

Rittmann, B. E. (1998). "Opportunities with membrane bioreactors." *Proc. Microfiltration Conference.* San Diego, CA: National Water Research Institute, Nov. 12–13, 1998.

Sawyer, C. N. (1965). "Milestones in the development of the activated sludge process." *J. Water Pollution Control Federation* 37(2), pp. 151–162.

Sezgin, M. (1981). "The role of filamentous microorganisms in activated sludge settling." *Prog. Water Technol.* 12, pp. 97–108.

Sezgin, M.; D. Jenkins; and D. S. Parker (1978). "A unified theory of filamentous activated sludge bulking." *J. Water Pollution Control Federation* 50, pp. 362–381.

Strom, P. F. and D. Jenkins (1984). "Identifications and significance of filamentous microorganisms in activated sludge." *J. Water Pollution Control Federation* 56, pp. 449–459.

Wanner, J. and P. Grau (1989). "Identification of filamentous microorganisms from activated sludge: A compromise between wishes, needs and possibilities." *Water Research* 23(7), pp. 883–891.

6.11   PROBLEMS

6.1.   You are to design an activated-sludge process. The influent values are:

$$Q = 4,000 \text{ m}^3/\text{d}$$
$$S^0 = 300 \text{ mg/l}$$
$$X_v^0 = 0$$

The following kinetic parameters were estimated from lab studies and stoichiometry:

$$Y = 0.4 \text{ g VS}_a/\text{g BOD}_L$$
$$\hat{q} = 22 \text{ g BOD}_L/\text{g VSS}_a\text{-d}$$
$$K = 200 \text{ mg BOD}_L/\text{l (composite waste)}$$
$$b = 0.1/\text{d}$$
$$f_d = 0.8$$

The design factors are:

Safety factor $= 60$

MLVSS $= 2,500$ mg/l; $X_v^r = X_a^r + X_i^r = 10,000$ mg VSS/l

$(X_a + X_i)^{\text{eff}} = 20$ mg VSS/l

(a) Calculate $[\theta_x^{\min}]_{\lim}$
(b) Calculate θ_x^d
(c) Calculate the concentration of substrate, S, in the effluent.
(d) Calculate U in kg BOD_L/kg VSS_a-d.
(e) Calculate the kg of substrate (BOD_L) removed per day.
(f) Calculate the kg/d of active, inactive, and total volatile solids produced in the biological reactor.
(g) Calculate the kg of active, inactive, and total VSS in the reactor. (Hint: θ_x is the same for active and inactive mass.)
(h) Calculate the hydraulic detention time, θ.
(i) Calculate the ratio $X_a/(X_a + X_i)$ (Hint: θ_x is the same for X_a and for X_i.) What is X_a in the reaction tank?
(j) Calculate the sludge wasting flow rate (Q^w) if the sludge is wasted from the return line. Calculate the kg/d of active, inactive, and total VSS that must be wasted per day. (Remember to account for the solids in the influent and effluent.)
(k) Calculate the required recycle flow rate (Q^r) and the recycle ratio, R.
(l) Calculate the volumetric loading rate in kg/m³-d.
(m) Assume that VSS_a in the effluent exerts a BOD. What is the BOD_L of the solids in the effluent?
(n) Compute the effluent SMP and the soluble COD and BOD_L.
(o) Calculate the minimum influent N and P requirements in mg/l and kg/d.

6.2. You have the following kinetic parameters:

$\hat{q} = 10$ mg BOD_L/mg VSS_a-d
$K = 20\ BOD_L$ mg/l
$b = 0.15$/d
$Y = 0.5$ mg VSS_a/mg BOD_L
$f_d = 0.9$

A completely mixed activated sludge process is operated with the following values:

$Q = 10^4$ m³/d
$\theta_x = 5$ d
$S^0 = 2000$ mg BOD_L/l
$X_i^0 = 100$ mg VSS/l
$X_a^0 = 0$
$X_v = 3,000$ mg VSS/l
$X_v^e = 30$ mg VSS/l

The design factors are:

Safety factor $= 20$

MLVSS $= X_a + X_i = 3{,}500$ mg/l;

$X_v^r = X_a^r + X_i^r = 15{,}000$ mg VSS/l

$(X_a + X_i)^{\text{eff}} = 30$ mg VSS/l

(a) Calculate $[\theta_x^{\min}]_{\lim}$
(b) Calculate θ_x^d
(c) Calculate the concentration of substrate, S, in the effluent.
(d) Calculate U in kg BOD_L/kg VSS_a-d.
(e) Calculate the kg of substrate removed (BOD_L) per day.
(f) Calculate the kg/d of active, inactive, and total volatile solids produced in the biological reactor.
(g) Calculate the kg of active, inactive, and total VSS in the reactor. (Hint: θ_x is the same for active and inactive mass.)
(h) Calculate the hydraulic detention time, θ.
(i) Calculate the ratio $X_a/(X_a + X_i)$ (Hint: θ_x is the same for X_a and for X_i.) What is X_a in the reaction tank?
(j) Calculate the sludge wasting flow rate (Q^w) if the sludge is wasted from the return line. Calculate the kg/d of active, inactive, and total VSS that must be wasted per day. (Remember to account for the solids in the influent and effluent.)
(k) Calculate the required recycle flow rate (Q^r) and the recycle ratio, R.
(l) Calculate the volumetric loading rate in kg/m³-d.
(m) Assume that VSS_a in the effluent exerts a BOD. What is the BOD_L of the solids in the effluent?
(n) Compute the effluent SMP concentration. Use standard SMP parameter values.
(o) Calculate the minimum influent N and P requirements in mg/l and kg/d.

6.3. You are to design an activated-sludge process. The influent values are:

$Q = 4{,}000$ m³/d

$S^0 = 300$ mg BOD_L/l

$X_v^0 = X_i^0 = 40$ mg VSS/l

The following kinetic parameters were estimated from lab studies and stoichiometry:

$Y = 0.6$ g VSS_a/g BOD_L

$\hat{q} = 16$ g BOD_L/g VSS_a-d

$K = 20$ mg BOD_L/l

$b = 0.2$/d

$f_d = 0.8$

The design factors are:

Safety factor $= 20$
MLVSS $= X_a + X_i = 3{,}500$ mg/l;
$X_v^r = X_a^r + X_i^r = 15{,}000$ mg VSS/l
$(X_a + X_i)^{\text{eff}} = 30$ mg VSS/l

(a) Calculate $[\theta_x^{\min}]_{\text{lim}}$
(b) Calculate θ_x^d
(c) Calculate the concentration of substrate, S in mg BOD_L/l, in the effluent.
(d) Calculate U in kg BOD_L/kg VSS_a-d.
(e) Calculate the kg of substrate removed (BOD_L) per day.
(f) Calculate the kg/d of active, inactive, and total volatile solids produced in the biological reactor.
(g) Calculate the kg of active, inactive, and total VSS in the reactor. (Hint: θ_x is the same for active and inactive mass.)
(h) Calculate the hydraulic detention time, θ.
(i) Calculate the ratio $X_a/(X_a + X_i)$ (Hint: θ_x is the same for X_a and for X_i.) What is X_a in the reaction tank?
(j) Calculate the sludge wasting flow rate (Q^w) if the sludge is wasted from the return line. Calculate the kg/d of active, inactive, and total VSS that must be wasted per day. (Remember to account for the solids in the influent and effluent.)
(k) Calculate the required recycle flow rate (Q^r) and the recycle ratio, R.
(l) Calculate the volumetric loading rate in kg/m³-d.
(m) Assume that the active VSS_a in the effluent exerts a BOD. What is the BOD_L of the solids in the effluent?
(n) Compute the effluent SMP and the soluble COD and BOD_L.
(o) Calculate the minimum influent N and P requirements in mg/l and kg/d.

6.4. A small community is receiving complaints from fishermen that their sewage-treatment effluent is creating problems in the stream to which it is discharged. The fishermen are charging that the effluent limit of 20 mg BOD_5/l must be violated, but they have no measurement to prove it. The city has very poor operations records, too. You are called in as the high-priced consultant to judge whether or not the effluent quality was in violation of the 20 mg/l standard.

Here is what you know from plant records for the activated sludge process:

Influent flow $= 4{,}000$ m³/d
Influent $BOD_L = 180$ mg/l
Total treatment volume $= 1{,}500$ m³
Influent VSS $= 50$ mg/l
D.O. in aeration basin > 3 mg/l
Mixed liquor volatile suspended solids $= 2{,}850$ mg/l
Effluent VSS $= 15$ mg/l

You cannot find any useful records on effluent BOD, sludge wasting, oxygen transfer, or sludge recycling.

Despite some big gaps in information, you must estimate the effluent BOD_5. From experience, you estimate that

$$Y = 0.45 \text{ g VSS}_a/\text{g}$$
$$\hat{q} = 16 \text{ mg BOD}_L/\text{mg VSS}_a\text{-d}$$
$$K = 20 \text{ mg BOD}_L/\text{l}$$
$$b = 0.15/\text{d}$$
$$f_d = 0.8$$

and that there is 0.68 mg BOD_5 per mg BOD_L for the influent and that $X_i^0 = 20$ mg VSS/l. Was the 20 mg BOD_5/l standard met?

6.5. You have an activated sludge reactor of volume 250 m³. The following operational parameters exist:

$$Q = 1,000 \text{ m}^3/\text{d}$$
$$S^0 = 300 \text{ mg BOD}_L/\text{l}$$
$$X_a = 3,000 \text{ mg VSS/l}$$
$$X_i = 1,000 \text{ mg VSS/l}$$
$$\text{Effl. BOD}_L = 10 \text{ mg/l}$$
$$X_a^e = 21 \text{ mg VSS/l}$$
$$X_i^e = 7 \text{ mg VSS/l}$$
$$X_a^r = 7,500 \text{ mg VSS/l}$$
$$X_i^r = 2,500 \text{ mg VSS/l}$$
$$Q^r = 667 \text{ m}^3/\text{d}$$
$$Q^w = \text{(from recycle line) } 10 \text{ m}^3/\text{d}$$

What are θ_x, U, and F/M? Would we call this process high rate, conventional rate, or extended aeration? Why?

6.6. The Muckemup Water Pollution Control Facility uses conventional activated sludge to treat its primary effluent. The wastewater is mostly domestic. The measured, average loadings to the activated sludge reactor are: volumetric loading $= 1.1 \text{ kg BOD}_L/\text{m}^3\text{-d}$ and $U = 0.34 \text{ kg BOD}_L/\text{kg VSS}_a\text{-d}$. Although effluent standards are usually met for average conditions, the plant responds poorly to short-term increases in loading. For example, soluble BOD rises above standards, and measurable D.O. is absent in much of the reactor. In addition, some neighbors have complained of odors.

Muckemup WPCF has very little money. Please suggest a process/operational change that can help improve the plant's performance. They do not have money to build new reactors or purchase large equipment. You should identify the problem, suggest a solution (or improvement), and support it with appropriate calculations.

6.7. You are asked to review the design for an industrial-wastewater treatment system. The wastewater BOD_L is 30,000 mg/l, and there are no influent solids. The design is of the complete-mix, activated sludge type. You know the following kinetic parameters:

$$\hat{q} = 15 \text{ mg } BOD_L/\text{mg } VSS_a/d$$
$$K = 20 \text{ mg } BOD_L/l$$
$$b = 0.15/d$$
$$Y = 0.45 \text{ mg } VSS_a/\text{mg } BOD_L$$
$$f_d = 0.8$$

The design SRT is 5 d. The desired effluent quality is 30 mg/l of BOD_5. In practice, the minimum effluent VSS is 10 mg/l, the maximum recycle ratio is 2, and the maximum recycle-sludge VSS is 1.5 percent. Evaluate whether or not X_v and V can have feasible values.

6.8. The step aeration and the contact stabilization modifications of the activated sludge process achieve what advantages?

6.9. An activated sludge treatment system has the following values, which you gleaned from plant records:

Influent flow rate $= 1,400 \text{ m}^3/d$

Aerator volume $= 6,000 \text{ m}^3$

Settler area $= 180 \text{ m}^2$

Settler depth $= 4.3 \text{ m}$

Recycle flow rate $= 1,100 \text{ m}^3/d$

Waste flow rate $= 20 \text{ m}^3/d$

Mixed liquor volatile suspended solids $= 4,000 \text{ mg/l}$

Effluent volatile suspended solids $= 15 \text{ mg/l}$

Recycle volatile suspended solids $= 8,800 \text{ mg/l}$

SVI $= 100 \text{ mg/g}$

(a) Compute the SRT, hydraulic detention time, recycle ratio, overflow rate, solids flux, and underflow velocity.

(b) If you wish to keep the SRT the same as it is now, what realistic change in operating strategy would be the most likely to improve process performance? How and why will it improve performance? Make a quantitative estimate of the effect.

6.10. A small community uses activated sludge to treat their sewage flow of 4,000 m^3/d. The system was sized to give conventional loadings for the average daily conditions: $\theta_x = 5$ d, $\theta = 10$ h, MLSS $= 3,000$ mg SS/l, $R = 0.30$, $X^r = 1.3$ percent. The clarifier is 5 m deep and has a surface area of 130 m^2. It has a single weir around the outside. However, the sewage flow is not even. During 8 h of the day the flow is 9,000 m^3/d, but it averages 1500 m^3/d for the other 16 h. Explain, using quantitative analysis, why this treatment plant frequently has high effluent BOD.

6.11. You are called in as the high-priced consultant to figure out how to fix a bad problem of solids carry-over in a final clarifier of an activated sludge plant. You gather quickly the following information:

> Effluent flow rate = 1,400 m^3/d (average)
>
> Mixed liquor suspended solids = 2,950 mg SS/l
>
> Recycle flow rate = 1,400 m^3/d (constant)
>
> Underflow solids concentration = 5,800 mg SS/l (average)
>
> Peak effluent flow rate = 2,800 m^3/d
>
> Effluent suspended solids = 30–500 mg/l
>
> Volume of aeration tank = 350 m^3
>
> Volume of clarifier = 230 m^3
>
> Depth of clarifier = 5 m

From the microscopic examination, you estimate that the amount of extended filament length is $10^{7.6}$ μm/mg VSS.

What is the problem and how can it be solved? Limit your answers to short-term, operational remedies. That is, do not suggest constructing any new tanks.

6.12. An activated sludge plant (complete mix) was having sludge bulking problems. A microscopic observation of the sludge revealed filamentous bacteria that appeared to be those indicative of a low-D.O.-bulking problem. To remedy the situation, the operator increased the SRT (which was calculated *correctly*) from 3.1 d (MLSS = 1,500 mg SS/l) to 7.1 d (MLSS = 3,000 mg SS/l) and increased the aeration in order to improve the aeration tank D.O. from 1 mg/l to 2 mg/l. Initially, this seemed to work: The sludge settling characteristics improved (see attached settling flux curves), and significant nitrification did not occur. However, it was not long before excessive solids were being discharged over the weirs of the clarifier again.

Your job as the high-priced consultant is to analyze quantitatively the original and altered operating conditions. Figure out what was wrong before and after the changes in operation were made. Based on this analysis, develop an improved strategy to fix the problem (be quantitative, e.g., what do Q^r and Q^w need to be, and are they feasible values?)

Operating conditions (same for initial and altered operation):

> Average influent flow rate = 4,000 m^3/d
>
> Recycle flow rate = 1,090 m^3/d
>
> Influent BOD_L = influent COD = 300 mg/l
>
> System volume (aeration basin + clarifier) = 1,332 m^3
>
> Clarifier area = 100 m^2
>
> MLVSS = 0.81 MLSS
>
> X_a/X_v = a constant at all positions

Kinetic parameters:

$\hat{q} = 20$ mg BOD_L/mg VSS_a/d

$K = 10$ mg BOD_L/l

$b = 0.06$/d

$Y = 0.5$ mg VSS_a/mg COD

$f_d = 0.8$

When you use a state-point analysis, ignore Q^w when drawing the overflow rate and recycle operating lines. Assume that, if the system is operating properly, X_v^e is negligible.

Settling-flux curve for Problem 6.12.

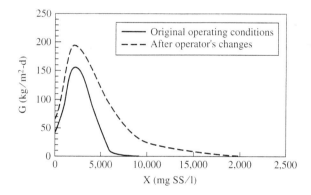

6.13. The operations staff at a sewage-treatment plant decided to reduce the SRT of their activated-sludge process from 6 d to 3 d. The goal of this change is to reduce the oxygen requirement. As the expert consultant, you must make some quick judgments about the impacts of this change. You do not need (nor do you have time) to make any sophisticated computations. You do need to identify proper trends and make semiquantitative assessments. In particular, assess each of the following issues; answer the question semiquantitatively; and *explain why* the trends you identify are correct.

(a) Will the oxygen requirement actually be reduced?

(b) What will happen to the MLVSS concentration?

(c) The settler currently is operating roughly at critical loading. What will happen to the underflow solids concentration after the change if the sludge's settling characteristics remain the same and the recycle ratio also is the same?

(d) What will happen to the underflow solids concentration if the sludge settling characteristics remain the same, but the recycle ratio is adjusted to give critical loading?

6.14. You have been hired by the EPA to evaluate if a process treating industrial wastewater is being operated properly for BOD removal. You quickly size-up the situation and note that they are using the contact-stabilization process. From the plans and plant records, you determine the following:

Dimensions:

      Contact tank: 10 m × 10 m × 5 m

      Reaeration tank: 10 m × 30 m × 5 m

      Settler: 28 m (dia.) × 5 m

Flows:

      Influent: 12,000 m^3/d

      Recycle: 6,000 m^3/d

      Waste: 240 m^3/d

VSS Concentrations:

      Influent: 200 mg/l

      Effluent: 20 mg/l

      Contact tank: 2,500 mg/l

      Reaeration tank: 7,600 mg/l

      Waste sludge: 8,000 mg/l

You determine that the SRT and the solids flux are the key parameters. What are their values? What do they tell you about the status of this plant?

6.15. You received a letter from the operator of a sewage treatment plant in Central Illinois. He is desperate, because he has a "sludge bulking" problem. From his letter and a telephone conversation, you know the following:

- The plant consists of primary sedimentation, activated-sludge aeration, secondary sedimentation, and tertiary filtration.

- From about 10:00 A.M. to about 2:00 P.M., the operator observes significant solids going over the weirs of the secondary clarifier.

- The operator does not measure the effluent SS from the secondary clarifier, but he estimates that they are about 50 mg/l from 10:00 A.M. until 2:00 P.M. Before and after those times, the SS are lower.

- The mixed liquor suspended solids are about 600 mg/l.

- Dissolved oxygen is 1.5 to 4 mg/l in the aeration basin and the secondary effluent.

This is all you now know. List at least seven pieces of information you must obtain in order to evaluate the cause of the problem and a solution. Indicate how you will use each piece of information to diagnose the problem. For example, show how key system parameters can be computed from the information.

6.16. A wastewater is found to have the following characteristics:

COD
soluble	100 mg/l
particulate	35 mg/l

BOD_L
soluble	55 mg/l
particulate	20 mg/l

Suspended solids
volatile	20 mg/l
non-volatile	10 mg/l

Estimate S^0 in mg BOD_L/l and X_i^0 in mg SS/l.

6.17. A wastewater being considered for biological treatment has the following average values:

BOD_5 (total) = 765 mg/l

BOD_5 (soluble) = 470 mg/l

$K_1 = 0.32/d$

COD (total) = 1,500 mg/l

COD (suspended) = 620 mg/l

Suspended solids (total) = 640 mg/l

Suspended solids (volatile) = 385 mg/l

From this information, estimate S^0 (BOD_L), X^0, X_v^0, X_i^0, and X_{inorg}^0 (all in mg/l) for the wastewater.

6.18. Fill out the following table to indicate what impact a small increase in each of the variables in the left hand column would have on the operating characteristics indicated in the four columns to the left for a well operating (relatively

	Operating Characteristic			
Variable	θ_x (d)	Sludge Production (kg/d)	Effluent Substrate Concentration (S^e)	Oxygen Consumption (kg/d)
Y				
$\hat{q}$				
Q^0				
S^0				
b				
K				
X^e				
X_i^0				

large SF) aerobic CSTR with cell settling and recycle, and treating a soluble wastewater. Assume all other variables in the left-hand column remain the same. Also, for this problem assume that you have an operating plant in which both the volume V and the concentration of active microorganisms in the reactor (X_a) *remain fixed.* Use: $(+) =$ increase, $(-) =$ decrease, $(0) =$ no change, and $(?) =$ undetermined.

6.19. You are asked to design an activated sludge plant to treat 10^4 m^3/d of wastewater containing 150 mg/l of phenol. The regulatory authorities are requiring that the effluent phenol concentration from this plant not exceed 0.04 mg/l. For this problem, assume the following biological coefficients apply, that phenol inhibition can be ignored, and that $X^0 = 0$:

$$Y = 0.6 \text{ g cells/g phenol}$$
$$\hat{q} = 9 \text{ g phenol/g cells-d}$$
$$b = 0.15 \text{ d}^{-1}$$
$$K = 0.8 \text{ mg/l}$$

(a) What θ_x in d would you use for design of this plant? Be sure to list all assumptions used and show calculations that are appropriate for justifying your answer.

(b) Based upon your design θ_x, what would be the reactor volume in m^3? Specify additional assumptions you may have to make.

6.20. Estimate the total biological solids production rate in kg/d from activated sludge treatment of the organic matter in a wastewater with the following characteristics (assume 98 percent biological utilization of wastewater BOD$_L$):

$$\theta_x = 6 \text{ d}$$
$$Q^0 = 1.5 \ (10)^6 \text{ m}^3/\text{d}$$
$$Y = 0.45 \text{ g cells/g BOD}_L$$
$$b = 0.2/\text{d}$$
$$S^0 = 3,300 \text{ mg/l}$$
$$X_i^0 = 600 \text{ mg/l}$$

6.21. Fill out the table on page 391 to indicate what impact an increase in each of the variables would have on other operating characteristics of activated sludge treatment. Assume that you have an operating plant with the volume V being fixed, and that all other characteristics remain the same. Use: $(+) =$ increase, $(-) =$ decrease, and $(0) =$ no change.

6.22. A wastewater has a flow rate of 5 $(10)^4$ m^3/d and S^0 of 1.2 g BOD$_L$/l (all soluble). Estimate the oxygen consumption for treatment at θ_x equal to 8 d. Assume typical values from Section 5.11.

6.23. A wastewater has a flow rate of 4 (10^4) m^3/d and BOD$_L$ of 3,000 mg/l. Estimate the nitrogen requirements in kg/d for activated sludge treatment. Assume $\theta_x = 8$ d and the efficiency of biological treatment is 95 percent. Assume typical values from Section 5.11.

Variable	Operating Characteristic		
	X_v (mg/l)	Total Biological Sludge Production (kg/d)	Oxygen Consumption (kg/d)
Y			
θ_x			
$\hat{q}$			
Q^0			
S^0			
b			
K			
X_i^0			

6.24. Determine the volume of an aerobic CSTR with recycle for BOD removal under the following conditions (assume typical values for aerobic noncarbohydrate BOD from Chapter 3):

(a) SF $= 40$
(b) $Q^0 = 10^3 \ \text{m}^3/\text{d}$
(c) $X_i^0 = 300 \ \text{mg/l}$
(d) Influent $\text{BOD}_L = 200 \ \text{mg/l}$
(e) Effluent $\text{BOD}_L = 5 \ \text{mg/l}$
(f) $X_v = 2,000 \ \text{mg/l}$

6.25. You have been approached by an industry to design an activated sludge wastewater treatment plant to meet an effluent BOD_5 of 30 mg/l and effluent suspended solids concentration of 30 mg/l. Your client wishes you to provide: (1) a first-cut estimate of the tank volume that he should use, (2) the daily quantity of all appropriate chemicals and other materials that should be added to make the system operate properly, and (3) quantitative estimates of all material handling that would be required. The industry has capable technical help to operate the treatment plant after suitable training, but they know little about wastewater treatment. Your task is to provide the industry with the information they need to help make a decision about whether or not to proceed with the design of a plant. Information available on the industrial wastewater is as follows:

Average flow rate $= 1,500 \ \text{m}^3/\text{d}$
BOD_5 (total) $= 1,330 \ \text{mg/l}$
BOD_5 (soluble) $= 1,110 \ \text{mg/l}$
COD (total) $= 2,200 \ \text{mg/l}$
COD (soluble) $= 1,800 \ \text{mg/l}$

Suspended solids
  total = 500 mg/l
  volatile = 400 mg/l
Nitrogen
  Organic = 5 mg/l
  Ammonia = 12 mg/l
  Nitrite = 1 mg/l
  Nitrate = 4 mg/l
Na^+ = 250 mg/l
K^+ = 60 mg/l
Ca^{2+} = 30 mg/l
Mg^{2+} = 19 mg/l
SO_4^{2+} = 100 mg/l
HCO_3^- = 50 mg/l
Phosphate-P = 52 mg/l
pH = 6.7

6.26. Wastewater characteristics, reaction coefficients, and treatment plant design are as follows for an activated-sludge system for aerobic treatment of organic wastes:

Wastewater characteristics:

$Q^0 = 5,000 \text{ m}^3/\text{d}$
$X_i^0 = 150 \text{ mg/l}$
$S^0 = 735 \text{ mg BOD}_L/\text{l}$
$K_1 = 0.28 \text{ d}^{-1}$

Reaction coefficients:

$Y = 0.45 \text{ mg } X_a/\text{mg BOD}_L$
$\hat{q} = 12 \text{ mg BOD}_L/\text{mg } X_a\text{-d}$
$K = 85 \text{ mg BOD}_L/\text{l}$
$b = 0.20 \text{ d}^{-1}$

Treatment plant design:

CSTR with recycle
$V = 2,600 \text{ m}^3$
$\theta_x = 6 \text{ d}$
$X_v^e = 15 \text{ mg/l}$

(*a*) Estimate the effluent soluble BOD_L concentration (S_s^e).
(*b*) What safety factor is implied by the above design?
(*c*) What will X_v be for this system?

(d) Estimate the volatile suspended solids production rate for the system in kg/d (includes effluent suspended solids plus suspended solids removed as sludge each day).

(e) If the effluent volatile suspended solids concentration were 15 mg/l, what would be the effluent BOD_5?

(f) If θ_x were increased to 12 d, what would be the *percentage* change in oxygen consumption?

(g) If θ_x were increased to 12 d, would you expect the volatile suspended solids production rate to increase or decrease? Why?

6.27. You have been hired to make an evaluation of two alternative biological treatment systems for an industrial wastewater. The first is biological treatment in an aerobic activated sludge system (CSTR with recycle), and the second is a similar system, but preceded by a primary settling tank. The question is which of the two systems is most cost effective? Assume sludge treatment costs are the same for the two systems.

(a) The wastewater characteristics are as follows:

$$S^0_{soluble} = 1,600 \text{ mg } BOD_L/l$$
$$S^0_{suspended} = 400 \text{ mg } BOD_L/l$$
$$X^0_v = 400 \text{ mg } BOD_L/l$$
$$X^0_i = 150 \text{ mg } BOD_L/l$$
$$Q^0 = 10^5 \text{ m}^3/d$$

(b) Assumed treatment plant characteristics are as follows:

1. Primary treatment:

   4 h detention time

   65 percent removal efficiency for X^0_v and X^0_i

2. Activated sludge treatment:

   $X_v = 3,000 \text{ mg/l}$

   $\theta^d_x = 8 \text{ d}$

   $Y = 0.45 \text{ g/g } BOD_L$

   $b = 0.15 \text{ d}^{-1}$

   eff. = 98 percent biological metabolism of BOD_L

3. Annual cost for construction, operation, and maintenance:

   primary treatment—$20 per year per m^3 primary tank vol.

   activated sludge—$80 per year per m^3 aeration tank vol.

chapter

7

LAGOONS

Most treatment goals of activated sludge can be achieved by lagoons, which substitute simplicity and a large volume for the mechanical complexity and compactness of activated sludge. In particular, lagoons do not employ a solids settler, solids recycling, and independently controlled solids wasting. Lagoons are popular for sewage treatment in small, rural communities and for treatment of wastewaters from industries able to dedicate large land areas to waste management.

In this chapter, we describe two types of lagoons: aerated lagoons and stabilization lagoons. They have many similarities, but differ primarily in the way in which oxygen is transferred to the liquid. Aerated lagoons use mechanical aeration devices, while stabilization lagoons rely on solar-driven oxygenation.

7.1 AERATED LAGOONS

The key features of aerated-lagoon treatment are:

- *A large liquid detention time.* Since aerated lagoons do not have sludge recycle, the solids retention time (SRT) is approximately equal to the liquid detention time, and aerated lagoons do not gain the size benefit of $\theta_x/\theta \gg 1$, as in activated sludge. Therefore, typical SRTs of 5 d for heterotrophic BOD removal or 25 d for nitrification require liquid detention times in the order of 5 and 25 d, respectively. When ample land is available, very long detention times, up to 100 d, are sometimes employed. A consequence of the large liquid detention is a small volumetric loading, often less than 0.01 kg BOD_5/m^3-d.

- *A shallow depth and a large plan-view surface area.* The typical depths of aerated lagoons range from 1 to 3 m (3 to 10 ft). These relatively shallow depths maximize the ratio of plan-view surface area to liquid volume. For example, the surface-area-to-volume ratio for a 2-m depth is 0.5 m^{-1}.

394

- *Surface aeration.* Although diffused aeration is sometimes used, most aerated lagoons use high-speed surface (mechanical) aerators. This choice is logical based on the high surface area present, the low capital costs of high-speed surface aerators, and the flexibility in locating high-speed surface aerators to maintain good solids mixing.

Although all aerated lagoons share important characteristics, their physical configurations can vary greatly. Small systems often consist of a single lagoon with uniform aeration and mixing. Large systems often are arranged with two or more cells in series, giving a plug-flow character to the system. In such staged systems, the intensity of aeration often is greatest in the first cell and diminishes in subsequent cells. Cell size often is not constant and may reflect peculiarities of site geometry.

Most aerated lagoons are shallow, earthen basins. Side slopes are generally 1:3 (vertical:horizontal). The basins should be lined with clay and/or geosynthetic membranes to prevent infiltration out of the lagoon and to the groundwater.

As a first approximation, the design of aerated lagoons follows the procedures for a chemostat (Chapter 3). For cells in series, the chemostat approach can be applied to each cell individually; however, the second and subsequent cells receive significant inputs of active biomass (X_a^0), which must be explicitly considered in the computation of the cell's θ_x value, which may differ from θ. How to incorporate X_a^o was detailed in Chapters 3 and 5 and is illustrated in the example at the end of this section.

A very important aspect of aerated lagoons is that the effluent contains substantial concentrations of active and inert biomass, in addition to residual substrate and SMP. Thus, the effluent suspended solids, BOD, and COD can be substantially elevated over those normally attained with activated sludge. Employing very long detention times minimizes the impacts, because endogenous respiration and predation oxidize much of the active biomass, which partially reduces the impact on effluent suspended solids and COD and significantly reduces the impact on effluent BOD. Again, these effects are illustrated quantitatively in the example problem.

Aerated-lagoon effluents seldom can be discharged directly to receiving water, since the suspended solids and BOD concentrations seldom meet a secondary-discharge standard. When the effluent receives further treatment—such as by being discharged into a sanitary sewer for treatment in a public-owned treatment works (POTW)—the effluent taken directly from the aerated lagoon may be acceptable. In most cases, the lagoon effluent must undergo sedimentation before discharge. Sedimentation can occur in mechanical settlers, such as are used in primary settling or in activated sludge, or in large, shallow earthen basins. In the latter case, continuous sludge removal generally is not practiced. Thus, the sedimentation basin must be designed with an adequate overflow rate to ensure solids removal and with enough volume dedicated to sludge storage that neither sludge carryover nor odors occur between the times for sludge removal (Metcalf and Eddy, 1991).

In some instances, shallow earthen basins are used in conjunction with a settler and sludge recycle. Such a system is an activated-sludge system, not an aerated lagoon. This points out that the use of an earthen basin and surface aeration is not sufficient to identify aerated-lagoon treatment. The key is the lack of solids recycled from the sedimentation basin.

The combination of large plan-view surface area and surface aeration makes aerated lagoons highly vulnerable to heat loss from evaporative and advective cooling. In cold climates, water temperatures can drop below 5 °C, which causes two problems. First, all microbial activity slows when the temperature declines. BOD and NH_4^+-N oxidation are slowed, which may jeopardize effluent quality. Furthermore, nitrifying bacteria may approach washout, since their θ_x^{min} values become very large for low temperatures (details are in Chapter 9). The second problem is ice formation, which can damage the aeration devices and preclude O_2 transfer and mixing.

A simple heat-balance approach can be used to estimate the heat-loss effects on lagoon temperature (Metcalf and Eddy, 1991). The heat content entering the lagoon is proportional to the product of the influent flow and its temperature, $Q^i T^i$. The flow is in m^3/d, and the temperature is in °C. The heat content leaving the lagoon is contained in the effluent flow ($Q^e T^e$) and in the heat transfer to the air, which is described by a heat-transfer rate proportional to the temperature-difference driving force, $k_h A_{PV}(T^e - T^a)$, in which k_h is a heat-transfer rate coefficient (m/d), A_{PV} is the plan-view surface area (m^2), and T^a is the air temperature (°C). Equating the heat input to the heat outputs (i.e., steady state) gives

$$Q^i T^i = Q^e T^e + k_h A_{PV}(T^e - T^a) \qquad \textbf{[7.1]}$$

Rearranging Equation 7.1 to solve for T^e gives

$$T^e = \frac{(k_h \theta / h)T^a + T^i}{(k_h \theta / h) + Q^e / Q^i} \qquad \textbf{[7.2]}$$

in which h = average lagoon depth (m) and $\theta = V/Q^i$ = liquid detention time (d). Equation 7.2 shows that the water temperature is lowered by decreases in T^i or T^a and by an increased aeration intensity (k_h), decreased depth (h), or reduced flow rate (Q^i). When temperatures are very cold for extended periods, lagoon depths often are increased (up to 6 m) to reduce heat loss. In such cases, draft tubes are needed to extend the mixing depth of the aerators.

Although Equation 7.2 identifies the correct trends, it should be used with caution, because of two assumptions inherent to it. First, it is based on a steady-state heat balance, which does not account for the inertia created by the specific heat of the water. Second, the k_h factor lumps together advective and evaporative heat losses. In particular, the evaporative heat loss responds directly to a humidity difference, which is only crudely represented by the temperature difference $T^e - T^a$. More meteorological data are needed to do a heat budget that separates advective and evaporative heat losses.

The aerators are used to supply oxygen and to keep the suspended solids in suspension. Oxygen requirements for aerated lagoons are computed by the same methods as are used for activated sludge. Due to the usually low volumetric BOD loading, mixing, not oxygen supply, often controls the sizing of the aeration system. The energy input required to maintain solids suspension typically is 15 to 30 kW/1,000 m^3. The relatively low suspended-solids concentrations in aerated lagoons tend to keep the mixing requirement near the lower end of the range. Good mixing also requires proper placement of the aerators in order to ensure that dead zones do not occur. The aerator manufacture should be consulted to determine the size of the mixing zone for a particular aerator under the conditions in which it will be employed.

AERATED-LAGOON PERFORMANCE An aerated-lagoon system has two cells having a total detention time of 6 d. The first and second cells are identical and have 3-d detention times. Both cells have an average depth of 2.5 m. The influent has soluble BOD_L of 600 mg/l, inert VSS of 40 mg/l, and a temperature of 30 °C. The heat-transfer coefficient is $k_h = 0.5$ m/d. The relevant kinetic and stoichiometric parameters at 20 °C are **Example 7.1**

$$\hat{q} = 10 \text{ mg BOD}_L/\text{mg VSS}_a\text{-d}$$

$$K = 10 \text{ mg BOD}_L/\text{l}$$

$$b = 0.1/\text{d}$$

$$Y = 0.45 \text{ mg VSS}_a/\text{mg BOD}_L$$

$$f_d = 0.8$$

$$k_1 = 0.12 \text{ mg COD}_P/\text{mg BOD}_L$$

$$k_2 = 0.09 \text{ mg COD}_P/\text{mg VSS}_a\text{-d}$$

$$\hat{q}_{\text{UAP}} = 1.8 \text{ mg COD}_P/\text{mg VSS}_a\text{-d}$$

$$K_{\text{UAP}} = 100 \text{ mg COD}_P/\text{l}$$

$$\hat{q}_{\text{BAP}} = 0.1 \text{ mg COD}_P/\text{mg VSS}_a\text{-d}$$

$$K_{\text{BAP}} = 85 \text{ mg COD}_P/\text{l}$$

The air temperature is 0 °C. We can assume that nitrification is not relevant.

The performance of each cell is to be analyzed for concentrations of residual substrate, SMP, active and inert VSS, soluble COD and BOD_L, and total COD and BOD_L. In addition, the aeration requirements for oxygen supply and mixing must be computed. One very key feature is that the temperature declines significantly through the system. Thus, some of the kinetic parameters vary for the different cells. In addition, cell 2 has significant inputs of active biomass and SMP.

Cell 1: We first compute the temperature using Equation 7.2. We assume that evaporation is small compared to Q^e, which makes $Q^e = Q^i$, or $Q^e/Q^i = 1$. Substituting the known values of k_h, θ, d, T^a, and T^i gives

$$T^e = \frac{\left(\dfrac{0.5 \text{ m/d} \cdot 3 \text{ d}}{2.5 \text{ m}}\right) 0 \,°\text{C} + 30 \,°\text{C}}{(0.5 \cdot 3/2.5) + 1} = 18.75 \,°\text{C}$$

The kinetic parameters are adjusted from 20 °C to 18.75 °C with a temperature-correction factor of $(1.07)^{18.75-20} = 0.92$. Thus, $\hat{q} = 9.2$ mg BOD_L/mg VSS$_a$-d, $b = 0.092$/d, $k_2 = 0.083$ mg COD$_P$/mg VSS$_a$-d, $\hat{q}_{\text{UAP}} = 1.66$ mg COD$_P$/mg VSS$_a$-d, and $\hat{q}_{\text{BAP}} = 0.092$ mg COD$_P$/mg VSS$_a$-d. We assume that all other parameters do not change with temperature. Cell 1 is treated as a chemostat with $\theta = \theta_x = 3$ d, and the effluent quality from cell 1 is computed in the same manner as for Example 3.1.

The results are:

$$S = 1.2 \text{ mg BOD}_L/\text{l}$$
$$X_a = 211 \text{ mg VSS}_a/\text{l}$$
$$X_i = 52 \text{ mg VSS}_i/\text{l}$$
$$X_v = 263 \text{ mg VSS}/\text{l}$$
$$UAP = 5.9 \text{ mg COD}_P/\text{l}$$
$$BAP = 35.4 \text{ mg COD}_P/\text{l}$$
$$SMP = 41.3 \text{ mg COD}_P/\text{l}$$
$$\text{Soluble COD} = 1.2 + 41.3 = 42.5 \text{ mg COD}/\text{l}$$
$$\text{Total COD} = 42.5 + 1.42 \cdot 263 = 416 \text{ mg COD}/\text{l}$$
$$\text{Total BOD}_L = 42.5 + 1.42 \cdot 0.8 \cdot 211 = 282 \text{ mg BOD}_L/\text{l}$$

Cell 2:   For cell 2, the liquid detention time is 3 d, but the input quality is notable for its large biomass and SMP contents: $S^0 = 1.2$ mg BOD$_L$/l, $X_i^0 = 52$ mg VSS$_i$/l, $X_a^0 = 211$ mg VSS$_a$/l, UAP$^0 = 5.85$ mg COD$_P$/l, and BAP$^0 = 35.4$ mg/l. Furthermore, the temperature is lower, due to heat loss.

The temperature in cell 2 is computed by the same approach as for cell 1.

$$T^e = \frac{\left(\dfrac{0.5 \text{ m/d} \cdot 3 \text{ d}}{2.5 \text{ m}}\right) 0\,^\circ\text{C} + 18.75\,^\circ\text{C}}{(0.5 \cdot 3/2.5) + 1} = 11.7\,^\circ\text{C}$$

Then, the kinetic parameters are adjusted with $(1.07)^{11.7-20} = 0.57$, or $\hat{q} = 5.7$ mg BOD$_L$/mg VSS$_a$-d, $b = 0.057$/d, $k_2 = 0.051$ mg COD$_P$/mg VSS$_a$-d, $\hat{q}_{\text{UAP}} = 1.03$ mg COD$_P$/mg VSS$_a$-d, and $\hat{q}_{\text{BAP}} = 0.057$ mg COD$_P$/mg VSS$_a$-d.

For the substrate and biomass components, cell 2 can be treated as a chemostat with input of active biomass. Following the procedures of Example 3.5 gives

$$\theta_x = -17.85 \text{ d (i.e., } \mu = -0.056 \text{ d}^{-1})$$
$$S = 0.0039 \text{ mg BOD}_L/\text{l}$$
$$X_a = 180.65 \text{ mg VSS}_a/\text{l}$$
$$X_i = 58 \text{ mg VSS}_i/\text{l}$$
$$X_V = 239 \text{ mg VSS}/\text{l}$$

(The result is very sensitive to X_a, and too much rounding can cause errors in the answer.) Clearly, cell 2 is a net biomass decayer, since X_a and X_v are lower in cell 2 than in cell 1.

The UAP and BAP concentrations cannot be computed with the normal steady-state solutions of Chapter 3 (e.g., Example 3.5), because UAP and BAP are input, which violates the mass balances used to obtain these solutions. Therefore, we re-derive the mass balances on each SMP component. For UAP,

$$0 = \text{UAP}^0 - \text{UAP} - k_1 r_{ut}\theta - \frac{\hat{q}_{\text{UAP}}\text{UAP}}{K_{\text{UAP}} + \text{UAP}} X_a\theta \qquad \textbf{[7.3]}$$

Substituting all known values gives UAP = 0.87 mg COD_P/l. Thus, UAP is significantly degraded in cell 2.

For BAP, the mass balance equation is

$$0 = BAP^0 - BAP - k_2 X_a \theta - \frac{\hat{q}_{BAP} BAP}{K_{BAP} + BAP} X_a \theta \qquad \textbf{[7.4]}$$

Substituting all known values gives BAP = 51.4 mg COD/l, which shows that decaying cells are net producers of BAP in this case, since the BAP-utilization kinetics are slow.

The overall effluent quality from cell 2 is:

$$\text{Soluble COD and } BOD_L = 0.0039 + 0.87 + 51.4$$
$$= 52.3 \text{ mg/l}$$
$$\text{Total COD} = 52.3 + 1.42 \cdot 239 = 391 \text{ mg/l}$$
$$\text{Total } BOD_L = 52.3 + 1.42 \cdot 0.8 \cdot 180.6 = 257 \text{ mg/l}$$

Table 7.1 summarizes the changes occurring across the two-cell system. The key trends are:

- The temperature declines significantly for this cold-weather situation.
- Without solids removal, the total BOD_L and COD removals are in the vicinity of 50 percent, since the oxygen demand of the influent substrate is being converted to biomass. The percentage removal of BOD_L is substantially greater than that of COD, since some of the suspended VSS is inert.

Table 7.1 Summary of water-quality changes across the two-cell aerated-lagoon system

Parameter	Influent	Cell-1 Effluent	Cell-2 Effluent
Temperature, °C	30	18.8	11.7
Active VSS, mg/l	0	211	181
Inert VSS, mg/l	40	52	58
Total VSS, mg/l	40	263	239
Substrate BOD_L, mg/l	600	1.15	0.0039
UAP BOD_L, mg/l	0	5.9	0.87
BAP BOD_L, mg/l	0	35	51
Soluble BOD_L, mg/l	600	42	52
VSS_a BOD_L, mg/l	0	240	205
Total BOD_L, mg/l	600	282	258
Soluble COD, mg/l	600	42	52
Total COD, mg/l	657	416	391
% BOD_L removal	-	53	57
% COD removal	-	37	40

- If all the suspended solids were removed from the effluent, the BOD_L would be 52 mg/l, corresponding to 91 percent removal.

- The majority of the removal occurs in the first cell, where the influent BOD is nearly 100 percent removed. While cell 2 net removes residual substrate, UAP, and active biomass, it is a net producer of BAP and inert biomass. Thus, the total BOD_L and COD show only small decreases in cell 2, while soluble BOD_L increases.

The last step is to compute the aerator requirements based on oxygen transfer and mixing. For oxygen supply, the required O_2 supply rate is computed from a mass-balance on oxygen demand, following the method of Example 3.4. Since the COD equivalents already are tabulated in Table 7.1, the computations are very simple.

Cell 1:

$$\frac{1}{V}\frac{\Delta O_2}{\Delta t} = \frac{(657 \text{ mg/l} - 416 \text{ mg/l})}{3 \text{ d}} = 80 \text{ mg/l-d} = 80\frac{\text{kg } O_2}{1,000 \text{ m}^3\text{-d}}$$

Cell 2:

$$\frac{1}{V}\frac{\Delta O_2}{\Delta t} = \frac{(416 \text{ mg/l} - 391 \text{ mg/l})}{3 \text{ d}} = 8.3 \text{ mg/l-d} = 8.3\frac{\text{kg } O_2}{1,000 \text{ m}^3\text{-d}}$$

Using a conservative field oxygen-transfer rate of 1 kg O_2/kWh, the aeration capacities needed to supply O_2 are 80 kWh/1,000 m^3-d and 8.3 kWh/1,000 m^3-d, respectively, for cells 1 and 2. Dividing both numbers by 24 hours per day yields the power needed for O_2 supply: 3.3 and 0.35 kW/1,000 m^3, respectively. In both cases, the aeration requirement for O_2 supply is substantially smaller than the mixing requirement of about 15 kW/1,000 m^3, which determines the size of the aerators.

7.2 STABILIZATION LAGOONS

Stabilization lagoons are widely used as low-cost alternatives for treatment of domestic and industrial wastewaters. They are known by a variety of names including stabilization ponds, oxidation lagoons, maturation lagoons, sewage lagoons, and algae lagoons. Their performance depends upon an interesting ecological mix of phototrophic and heterotrophic organisms. Their major advantages are low cost and simplicity in operation. Not only are they capable of achieving good reduction in wastewater BOD, but they can be used as well for removal of nitrogen and phosphorus, reduction in heavy metals and toxic organic compounds, and destruction of pathogens. Their disadvantages are that they require large land area, work best with moderate year-round temperature and sunlight for growth of phototrophic organisms, and generally do not achieve as low an effluent BOD as other aerobic biological treatment systems, such as activated sludge and trickling filters. If overloaded, they can create odors and thus are best not located near population centers.

Stabilization lagoons are particularly useful in developing countries, where resources for construction and operation of highly reliable treatment systems are not available. They can be designed for series operation, in which the first lagoon removes the majority of the BOD and SS, while producing plankton that are consumed by fish growing in subsequent lagoons; thus, the lagoon system can provide a local food source.

Stabilization lagoons rely on phototrophic microorganisms, including algae and cyanobacteria, to maintain an aerobic environment in the upper zones of the lagoon. Much interest has been displayed in recent years in the use of higher phototrophic organisms, such as rooted and floating aquatic plants for this purpose, thus providing a *wetlands treatment system* with a balanced ecology for the natural removal of harmful materials from wastewaters. Very recently, *phytoremediation,* in which plants are used to concentrate heavy metals for removal from wastewater or to help destroy or otherwise effect the removal of toxic organic chemicals in water, has become of interest. This chapter concentrates on stabilization lagoons, but the principles of operation are similar for wetland treatment, which is described briefly at the end of the chapter. Phytoremediation is discussed further in Chapter 15.

7.3 TYPES OF STABILIZATION LAGOONS

Stabilization lagoons are defined as wastewater treatment by any lagoon or pond system that involves the actions of phototrophic and heterotrophic microorganisms. In other words, the combined metabolic activities of algae (or cyanobacteria) and organic-degrading bacteria are required. We can further subdivide stabilization lagoons based on their characteristic mode of operation as *aerobic, facultative,* or *anaerobic. Aerobic stabilization lagoons* are generally shallow, have adequate mixing to prevent stratification, and have a sufficient balance of phototrophic over heterotrophic activity so that dissolved oxygen is present throughout the lagoon. *Facultative stabilization lagoons* tend to be somewhat deeper and have an upper aerobic zone, where phototrophs are active, and a lower zone, where anaerobic heterotrophic activity dominates. *Anaerobic stabilization lagoons* are generally deeper than facultative lagoons and depend upon anaerobic heterotrophic activity as the primary treatment method, with an upper phototrophic zone that mainly minimizes odor emissions. Stabilization lagoons also are classified as *primary stabilization lagoons,* which receive raw wastewaters, or as *secondary stabilization lagoons,* which receive wastewaters after primary sedimentation.

In order to understand the factors affecting lagoon operation, it is best to begin with a discussion of aerobic stabilization lagoons, which are somewhat less complex than the other types. Thus, we first develop the basic methods for designing aerobic lagoons. Then, we extend the methods to facultative lagoons and anaerobic lagoons. At the end, we discuss the ever-difficult issue of removing the suspended microorganisms from the effluent and give a brief introduction to wetlands treatment.

Photo 7.1   A high-rate stabilization lagoon with mixing.

7.4   AEROBIC STABILIZATION LAGOONS

Stabilization lagoons and the variety of other wastewater treatment systems that depend upon a combination of phototrophic and heterotrophic organisms are complex, and the interactions involved are not easily modeled and are dependent on site-specific conditions. Consequently, design equations are often based on rules of thumb or empirical formulations resulting from years of widespread experience rather than from use of sound scientific principles. However, we have a general understanding of the processes involved, especially for stabilization lagoons that include algae or cyanobacteria as the main phototrophs. We are wise to understand the processes and their limitations at a fundamental level so that reasonable boundaries on operation can be drawn, and factors of importance to successful operation can be better understood. This will permit rational changes in operation to improve performance or to troubleshoot the system when it is not performing as well as desired.

We begin by considering a basic aerobic stabilization lagoon with algae (or cyanobacteria) as the main phototrophs to illustrate their role and how they interact with the heterotrophs. With such knowledge, the engineer should be able to make use of the fundamentals involved for designing or analyzing any treatment system that depends upon phototrophs for successful operation. From this, the essential characteristics of any combined phototrophic/heterotrophic system can be understood, and the bounds within which successful operation occurs can be established. Empirical approaches also can be applied, but they must fit within the bounds dictated by normal mass balance and energetic considerations.

7.4.1 BASIC EQUATIONS

Figure 7.1 illustrates the completely mixed reactor of uniform depth that represents the basic aerobic stabilization lagoon. Aerobic stabilization lagoons are generally quite shallow to help keep them well mixed, which prevents anaerobic zones from developing near the bottom. Depths of only 0.3 to 1.2 m are typical. Wastewater flows into the lagoon at a rate Q and contains BOD_L with concentration S^0. Phototrophs within the lagoon grow when solar energy enters the lagoon. When carrying out photosynthesis, they reduce CO_2 carbon into phototrophic cell carbon and produce O_2 from water to gain the electrons. The oxygen so produced is used by heterotrophic bacteria to oxidize the wastewater BOD_L, thus effecting waste organic destruction. With the proper balance between algal growth to produce oxygen and organic destruction that is dependent on the oxygen, the stabilization lagoon effects good BOD removal and maintains aerobic conditions that help prevent odors associated with anaerobic processes. Such a balance depends on the quantity of sunlight available, as well as the surface area of the lagoon through which the sunlight penetrates. As for the heterotrophic organisms that are responsible for BOD_L destruction, the conditions described for aerated lagoons prevail. In short, the detention time (which relates to organism growth rate) must be sufficient to achieve the degree of BOD_L removal desired. Thus, two conditions must remain in balance: the algal growth rate for oxygen production and the heterotrophic-bacteria growth rate for BOD_L destruction.

Before we get into the details of design to meet the two basic conditions, we ask just what do we accomplish overall in the basic aerobic stabilization lagoon? We answer this through two rather simple equations in which the formula $(CH_2O)_p$ represents photosynthetic organisms and $(CH_2O)_h$ represents heterotrophic cells. We will write better empirical representations for the cells later, but, for now, this simpler approach works well for illustration.

The first equation is the balanced overall reaction for the production of phototrophs from carbon dioxide, water, and sunlight.

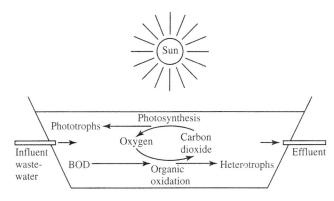

Figure 7.1   Schematic of processes of importance in an
aerobic stabilization lagoon.

Phototroph cell growth:

$$CO_2 + H_2O \xrightarrow{\text{Sunlight}} (CH_2O)_p + O_2 \qquad \textbf{[7.5]}$$

The production of one mole of oxygen also produces one empirical mole of phototrophs.

The second equation is the balanced reaction for the biodegradation of BOD_L. For this example, the wastewater organics of interest are represented as carbohydrates ($C_6H_{12}O_6$), and f_s is taken as 0.5.

Heterotroph cell growth:

$$C_6H_{12}O_6 + 3\,O_2 \rightarrow 3\,(CH_2O)_h + 3\,CO_2 + 3\,H_2O \qquad \textbf{[7.6]}$$

The oxidation of 1 mol of glucose consumes 3 mol of oxygen, which we can obtain by converting 3 mol of carbon dioxide, according to Equation 7.5, into phototrophic cells. The overall balance we thus obtain by adding three times Equation 7.5 to Equation 7.6 is

$$C_6H_{12}O_6 \longrightarrow 3\,(CH_2O)_h + 3\,(CH_2O)_p \qquad \textbf{[7.7]}$$

So, what have we accomplished by our treatment system? We have converted a wastewater constituent, glucose, into heterotrophs and phototrophs. We end up with exactly the same amount of organic matter equivalents as that with which we began. We have not destroyed organic matter at all, but have simply changed its form. Is this desirable? The answer is yes and no.

On the yes side, we have converted soluble organic matter to particulate matter, which may be removed by settling, chemical coagulation, centrifugation, or other physical means that can be designed into the system. If the treated wastewater is applied to soil for irrigation, the particulate forms may filter out in the top layers, perhaps adding organic matter and nutrients to improve soil characteristics. It may serve well as a source of food for fish in a subsequent lagoon designed for this purpose. When discharged to a stream, the organisms may not decompose as fast as the glucose, thus reducing the rate of oxygen uptake. Indeed, if the sun is shining and the stream is not too deep, the phototrophs may not decompose at all, but would keep active.

On the no side, if the lagoon effluent flows to a relatively deep slow-flowing river, the organisms may settle to the bottom and there slowly decompose, using oxygen from overlying waters. This could be worse than no treatment at all. Or, if the lagoon fills up with solids, its detention time is decreased, short-circuiting may result, and anaerobic deposits may lead to foul odors.

Thus, great caution is needed when designing an aerobic stabilization lagoon. The engineer must understand the consequences of stabilization lagoon treatment and decide whether the advantages for a given location are adequate to accomplish the goals of treatment without causing unsatisfactory outcomes. We will consider this aspect further when we consider other types of stabilization lagoons, but we proceed with our consideration of the basic system under the assumption that such a treatment lagoon may be a useful and economical approach as part of an overall treatment and disposal system.

7.4.2 SOLAR ENERGY INPUT AND UTILIZATION EFFICIENCY

We next consider the first step of growth requirements for the phototrophs. Because the energy for this step derives from sunlight, sufficient sunlight is essential to make the process work properly. Extraterrestrial solar radiation reaches the earth at a constant rate of about 84 kJ/m^2-min, or 120,000 kJ/m^2-d (Odum, 1971). Because of adsorption and reflection, only about two-thirds of that energy reaches the Earth's surface on a clear day. Because of clouds, on average only about one-half, or 60,000 kJ/m^2-d, reaches the earth's surface. The earth is a rotating sphere with its axis at an angle to that of the incoming solar radiation. At a given location in the Northern Hemisphere, the incident solar radiation varies from zero to a maximum over a 24-h period, and more reaches the surface in June than in December. As a result, the average solar radiation reaching an area with a temperate climate, such as the United States, is about 12,000 to 17,000 kJ/m^2-d. Figure 7.2 illustrates how the incidence of solar radiation varies over the period of a year for various regions.

Photosynthetic organisms capture only a small portion of the solar energy reaching the earth's surface. The visual portion of solar energy, which is the part plants can use, is only about one-half of the total, and then only a portion of the visual radiation is actually used in photosynthesis. Even the useful portion may be reflected from surfaces, used to drive evaporation, radiated back to the atmosphere, wasted as heat, or otherwise lost. Thus, under usual conditions only a small fraction of incident radiation is converted into plant growth. Odum (1971) indicates that, for photosynthetic processes in general, a maximum of 5 percent of incident radiation can be converted to primary cell production, with about 1 percent being typical under favorable average conditions. On average for the entire biosphere, the value is 0.1 percent.

Stabilization lagoons generally represent quite favorable conditions for phototrophs, with water and most nutrients in abundance. Indeed, Oswald (Oswald, Gotaas, Golueke, and Kellen, 1957), working with high-rate stabilization lagoons in California, suggested that energy efficiency from 3 to 5 percent be used.

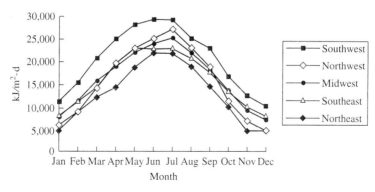

Figure 7.2 Solar radiation received on a unit horizontal surface in different regions in the United States. SOURCE: Reifsnyder and Lull (1965).

As indicated in Chapter 2, autotrophic growth requires about 133 kJ to produce an equivalent of cells (5.65 g). In other words, about 24 J are required to produce 1 mg of phototrophic cells. This is comparable to the value of 23 J suggested by Oswald (Oswald, et al., 1957) as a good average value to use for phototrophs and to the 21 J and 19 J suggested by Odum (1971) for algae and terrestrial plants, respectively.

Now, we consider a more detailed formulation for algal cell growth than indicated in Equation 7.5. We include the nitrogen and phosphorus, which can be rate-limiting in some circumstances and whose removal might be an objective in a given wastewater treatment scheme:

$$106\,CO_2 + 65\,H_2O + 16\,NH_3 + H_3PO_4 \;=\; C_{106}H_{181}O_{45}N_{16}P + 118\,O_2$$

1,272 mg CO_2-C	2.428 mg algal cells
224 mg NH_3-N	3,776 mg O_2 production
31 mg PO_4^{3-}-P	

[7.8]

This formulation indicates how much nitrogen and phosphorus are required for the growth of phototrophs. It also indicates that phototrophic growth in a stabilization lagoon results in removal of nitrogen and phosphorus from the wastewater. Phototrophs sometimes engage in "luxury uptake" of phosphorus, which means that the amount of P incorporated may be even higher than the slightly over 1 percent shown, perhaps as high as 6 percent under the right conditions. Thus, nutrient removal, as well as BOD removal, may occur in stabilization lagoons.

Example 7.2 | **PHOTOTROPHIC ACTIVITY**   What average phototroph-production and oxygen-production rates would be expected through the growth of phototrophs in a stabilization lagoon in California (southwestern region) in the summer and the winter if they were limited by solar energy? Assume an efficiency of conversion of incident solar energy of 3 percent.

From Figure 7.2, the incident radiant energy is about 30,000 kJ/m^2-d in the summer and 12,000 kJ/m^2-d in the winter. Thus, phototroph growth rates would equal:

Summer:

$$\text{Phototroph Growth Rate} = \frac{0.03(30\ 10^6\ \text{J/m}^2\ \text{d})}{24\ \text{J/mg cells}} = 3.8(10^4)\,\frac{\text{mg cells}}{\text{m}^2\text{-d}}$$

Winter:

$$\text{Phototroph Growth Rate} = \frac{0.03(12\ 10^6\ \text{J/m}^2\ \text{d})}{24\ \text{J/mg cells}} = 1.5(10^4)\,\frac{\text{mg cells}}{\text{m}^2\text{-d}}$$

The oxygen production rates would be:

Summer:

$$\text{Oxygen Production Rate} = 3.8\ 10^4\ (3{,}776/2{,}428) = 5.9\ 10^4\ \text{mg } O_2/\text{m}^2\text{-d}$$

Winter:

$$\text{Oxygen Production Rate} = 1.5\ 10^4\ (3{,}776/2{,}428) = 2.3\ 10^4\ \text{mg } O_2/\text{m}^2\text{-d}$$

7.4.3 BOD$_L$ REMOVAL

The function of phototrophic growth in the basic stabilization lagoon is to provide the oxygen required for BOD$_L$ removal. The maximum BOD$_L$ removal is thus limited by the amount of oxygen produced by phototrophic growth. Within this limitation, BOD$_L$ removal is just as it would be in an aerated lagoon, that is, in a reaction basin without recycle and acting like a CSTR. The relevant equations here come from Chapters 3 and the first part of this chapter. This connection is best illustrated with an example. For this, we will draw on the conditions of Examples 7.1 and 7.2 to illustrate the design for the stabilization lagoon as bounded by the oxygen limitations of phototrophic growth.

BOD REMOVAL Determine the depth and detention time required and the effluent characteristics for a single aerobic stabilization lagoon for treatment of the wastewater described in Example 7.1, together with the BOD$_L$ and lagoon design coefficients provided there. The influent BOD$_L$ is 600 mg/L, the inert VSS concentration is 40 mg/L, and the wastewater temperature is 30 °C. This is winter operation, since the air temperature is 0 °C. We are relying on the wastewater temperature to keep the lagoon from freezing. A short detention time is of value, since it will reduce heat losses. We assume that the lagoon is in California, so that the algal growth conditions of Example 7.2 apply. | **Example 7.3**

In order to determine the temperature within the lagoon through use of Equation 7.2, we need to know the stabilization lagoon detention time (θ) and depth. However, these are the quantities that we are seeking as answers for this problem. We can arrive at the solution through an iterative approach, using a spreadsheet. However, rather than doing that here, we begin by assuming a temperature, work through the solution for the detention time, and then check to see how far off we are from the assumed temperature.

As an initial guess, we assume 18.75 °C, which is what was estimated for cell 1 in Example 7.1; thus, the rate coefficients derived in that example from this temperature all apply. We also assume initially that the detention time of 3 d for the first cell also applies, so that the various values for influent and cell 1 effluent concentrations in Table 7.1 also apply, as does the oxygen consumption computed for cell 1 (80 kg/1,000 m^3-d).

From Example 7.2, the average oxygen production in wintertime is $2.3 \cdot 10^4$ mg O$_2$/m^2-d. Thus, with these initial assumptions, we can compute lagoon depth (h) to be:

$$h = \frac{23,000 \text{ mg O}_2/\text{m}^2\text{-d}}{80 \text{ kg O}_2/1,000 \text{ m}^3\text{-d}} \times \frac{\text{kg}}{10^6 \text{ mg}} = 0.3 \text{ m}$$

This depth is near the minimum typical depth for an aerobic stabilization lagoon. While here we have assumed the same detention time as for the aerated lagoon, the resulting stabilization lagoon depth is much less than that for the aerated lagoon (2.5 m). This means that the surface area for the stabilization lagoon is much higher by the ratio of 2.5 m/0.3 m, or by a factor of 8.3. The larger surface area for the stabilization lagoon also means that there may be a much greater loss of heat; thus, the assumed temperature of operation of 18.75 °C probably is too high. Applying Equation 7.2 to the resulting depth and detention time provides a more appropriate temperature:

$$T^e = \frac{\left(\frac{0.5 \times 3}{0.3}\right) 0 \text{ °C} + 30 \text{ °C}}{\frac{0.5 \times 3}{0.3} + 1} = 5 \text{ °C}$$

Table 7.2   BOD$_L$ removal with Example 7.3 stabilization-lagoon design

Design Characteristics		Design Coefficients		Effluent Characteristics[a]	
S^0	600 mg BOD$_L$/l	$\hat{q}$	3.62 mg BOD$_L$/mg VSS$_a$-d	S^e	2.9 mg BOD$_L$/l
X_i^0	44 mg VSS/l	K	10 mg BOD$_L$/l	X_a^e	242 mg VSS/l
		b	0.036 d^{-1}		
T_i	30 °C	Y	0.45 mg VSS$_a$/mg BOD$_L$	X_i^e	49 mg VSS/l
T_a	0 °C	f_d	0.8	UAP	14 mg COD$_P$/l
T_e	5 °C	k_1	0.12 mg COD$_P$/mg BOD$_L$	BAP	19 mg COD$_P$/l
		k_2	0.033 mg COD$_P$/mg VSS$_a$-d		
h	0.3 m	$\hat{q}_{\text{UAP}}$	0.65 mg COD$_P$/mg VSS$_a$-d	SMP	33 mg COD$_P$/L
θ	3 d	K_{UAP}	100 mg COD$_P$/l	COD$_{\text{total}}^e$	450 mg/l
		$\hat{q}_{\text{BAP}}$	0.036 COD$_P$/mg VVS$_a$-d	COD$_{\text{soluble}}^e$	36 mg/l
		K_{BAP}	85 COD$_P$/l		
		k_h	0.5 m/d	BOD$_L^e$	310 mg/l

| [a]Algal products are not considered.

We see a major drop in temperature. Thus, one would need to recalculate all the coefficients and solve the problem again, repeating this procedure until all factors come into balance.

The final results of the iterative solution using a spreadsheet are provided in Table 7.2. The changes in coefficients cause little change in the overall results, and the 0.3-m depth appears to be the appropriate solution for this problem. This corresponds to a detention time of 3 d, the same used in Example 7.1.

We also need to consider the phototrophic growth associated with the oxygen production and its impact on effluent quality. As with bacteria, phototrophs generate soluble products similar to the BAP and UAP fractions. From data on growth of phototrophs, Jewell and McCarty (1971) found that the soluble materials produced during growth represents 10 to 20 percent of the biomass production. Furthermore, a fraction of the phototrophic cells are relatively resistant to biodegradation, the fraction varying from 19 to 86 percent and averaging 44 percent for several different mixed cultures grown in a variety of waters. Such factors need to be considered when estimating organic concentration changes in a lagoon, and Example 7.4 illustrates means to do this.

Example 7.4 | **PHOTOTROPH EFFECTS ON EFFLUENT QUALITY**   Building on Examples 7.2 and 7.3, estimate the concentration of phototrophs and soluble phototroph products present in the lagoon of Example 7.3 and in the lagoon effluent.

In Example 7.2 for winter conditions, we found that phototrophic growth rate was $1.5(10)^4$ mg cells/m^2-d. A steady-state mass balance on the lagoon, assuming no phototrophs are present in the influent and assuming that 85 percent of the phototroph growth is particulate matter, gives

$$0.85 \left(1.5(10)^4 \, \frac{\text{mg cells}}{\text{m}^2 \, \text{d}} \right) \frac{V}{h} = X_{\text{pho}}^e Q$$

where V/h represents the surface area of the lagoon. Solving for the concentration of phototrophs in the lagoon effluent, X^e_{pho}, which is the same as the concentration within the lagoon, X_{pho}, we obtain:

$$X^e_{pho} = 0.85 \left(1.5(10)^4 \; \frac{mg \; cells}{m^2 \; d} \right) \frac{V}{Qh} = 1.28(10)^4 \; \frac{mg \; cells}{m^2 \; d} \frac{\theta}{h}$$

or,

$$X^e_{pho} = 1.28(10)^4 \; \frac{mg \; cells}{m^2 \; d} \frac{3 \; d}{0.3 \; m} \frac{m^3}{10^3 \; l} = 128 \; mg/l$$

The soluble products from phototrophic growth (SMP^e_{pho}) can be readily obtained from the 15 percent of photosynthetic synthesis that is not particulate and by assuming the same COD:VSS conversion factor for phototrophs as for bacteria, 1.42 mg COD per mg cells:

$$SMP^e_{pho} = 0.15(1.42) \left(1.5(10)^4 \; \frac{mg \; cells}{m^2 \; d} \right) \frac{3 \; d}{0.3 \; m} \frac{m^3}{10^3 \; l} = 32 \; mg/l$$

Finally, the organic effluent quality for an aerobic stabilization lagoon is obtained by summing up the original wastewater constituents remaining in the lagoon effluent, the concentration of bacteria and their associated soluble organic products, and the concentration of phototrophs and their soluble products. We continue the examples to illustrate this.

FINAL EFFLUENT QUALITY   For the case of Examples 7.3 and 7.4, compare the effluent organic quality from the aerobic stabilization lagoon with the influent quality. | **Example 7.5**

The comparison is made by summing the various organic products as estimated in Examples 7.3 and 7.4. The results are provided in Table 7.3. From this analysis, we find various aspects of importance with an aerobic stabilization lagoon. First, the overall concentration of COD in the lagoon effluent is similar to that in the influent, as suggested much more simply by Equation 7.7. Second, we find that most of the effluent COD (about 90 percent) is in the form of particulate matter (phototrophs plus heterotrophs). Although this was not specified in this example, if the influent BOD_L had been mostly in a soluble form, we see that treatment converts it to a particulate form that might be removable through addition of a particulate separation process on the lagoon effluent. Addition of such facilities is commonly done for aerobic stabilization lagoons. Processes added for this purpose include chemical coagulation and settling, air flotation, filtration, and centrifugation. Some of these methods are discussed in the second-to-last section of this chapter. The result is the production of heterotroph plus phototroph sludge, which may be useful as a source of single-cell protein. Aerobic stabilization lagoons have been proposed as one means of removing nitrogen and phosphorus nutrients from wastewaters. This example illustrates that soluble forms of nitrogen and phosphorus would be converted to organic forms in cells that could be removed by the physical processes just mentioned. The amount of N and P removed in this way can be computed from the sludge production rate multiplied by the N and P contents of the sludge.

We find that aerobic stabilization lagoons remove little or no COD unless a physical process of some type is added to the overall system to remove the particulate matter from the lagoon effluent. However, we have achieved some reduction in BOD_L

Table 7.3   Example 7.5 comparison between influent and effluent characteristics for stabilization pond treatment as given in Examples 7.2 through 7.4

Characteristic	Concentration (mg/l)
Influent	
BOD_L, S^0	600
Inert suspended solids, X_i^0	44
Total influent COD	662
Effluent	
Suspended solids, VSS^e	
Active heterotrophs, X_a^e	242
Inert organic solids, X_i^e	49
Phototrophs, X_{pho}^e	128
Total	419
COD	
Suspended	
Active heterotrophs, X_a^e	344
Inert organic solids, X_i^e	70
Phototrophs, X_{pho}^e	182
Subtotal	596
Soluble	
Remaining BOD_L, S^e	3
Heterotrophs, SMP^e	33
Phototrophs, UAP_{pho}^e	32
Subtotal	68
Total	664

because much of the effluent COD is in refractory heterotrophic and phototrophic cellular matter. Additionally, phototrophs generally do not decompose when in the presence of sunlight, which provides them with energy for maintenance. However, if they settle to the bottom of a river, they can form a sludge bank and there decompose, perhaps using oxygen resources from the flowing river above. The overall effect of discharge of stabilization lagoon effluent into a river is very dependent upon local conditions.

Also to be considered is the rate of BOD exertion. The decay rate and, hence, oxygen-uptake rate from decomposition of heterotrophs and phototrophs may be slower than that of the original BOD_L. Decay rates of active bacteria are typically in the range of 0.1 to 0.3 d^{-1}. The decay rates for phototrophs tend to be much slower. When first placed in the dark, mixed cultures containing phototrophs decompose and consume oxygen rapidly while storage products are consumed. Soon, the decay rates decrease to values that are, in general, quite low. Jewell and McCarty (1971) reported that decomposition rates of phototrophs measured by many different researchers varied between 0.01 and 0.08 d^{-1} at 20 °C. They also found about a two-fold decrease for a temperature decrease of 10 °C, which is a typical temperature effect for biological processes. Di Toro, O'Connor, and Thomann (1971) summarized studies by others on phototroph decay as a function of temperature and provided the following relationship:

$$b_p = K_d T \qquad [7.9]$$

where b_p is the phototrophic decay rate (d^{-1}) and K_d is a constant, which they indicated equaled 0.005 ± 0.001 °C^{-1}d^{-1}. At 20 °C, this would correspond to a decay rate of about 0.1 d^{-1}, or slightly above those mentioned above.

The fairly low decay rate for phototrophs means that the BOD$_5$ for aerobic stabilization lagoon effluent may be much less than that of the influent, even though the COD difference between the two may be negligible. Regulatory requirements are often based upon BOD$_5$ removal efficiency, and from this standpoint, aerobic stabilization lagoon treatment may be satisfactory. However, the design engineer should understand the total impacts that discharge of stabilization lagoon effluent has on receiving waters.

EFFLUENT BOD$_5$ Estimate the standard (20 °C) BOD$_5$ lagoon loading per unit surface area and removal efficiency for the aerobic stabilization lagoon design from Example 7.2. Assume a BOD exertion rate for the original wastewater of 0.3 d^{-1}. | **Example 7.6**

For the original wastewater, we find:

$$BOD_5^0 = 600 \left(1 - e^{-0.3(5)}\right) = 466 \text{ mg/L}$$

This gives the surface loading as

$$Pond\ loading = \frac{S^0 Q}{A} = \frac{S^0 Q}{V/h} = \frac{h S^0}{\theta}$$

$$= \frac{\dfrac{(0.3 \text{ m})466 \text{ mg BOD}_5}{1} \dfrac{10^3 \text{ l}}{\text{m}^3} \dfrac{\text{kg}}{10^6 \text{ mg}}}{3 \text{ d}} = 0.047 \text{ kg BOD}_5/\text{m}^2\text{-d}$$

For the stabilization lagoon effluent, we have soluble and particulate products for heterotrophs and phototrophs. We assume a 20 °C BOD exertion rate of 0.2 d^{-1} for the particulate and soluble heterotroph products and 0.04 d^{-1} for the particulate and soluble phototroph products. We thus obtain:

$$BOD_5^e = BOD_{5het}^e + BOD_{5pho}^e + BOD_{5s}^e$$

$$BOD_{5het}^e = (0.8(344) + 33) \left(1 - e^{-0.2(5)}\right) = 195 \text{ mg/l}$$

$$BOD_{5pho}^e = (0.54(182) + 32) \left(1 - e^{-0.04(5)}\right) = 24 \text{ mg/l}$$

$$BOD_{5s}^e = 3 \left(1 - e^{-0.3(5)}\right) = 2 \text{ mg/l}$$

$$BOD_5^e = 195 + 24 + 2 = 221 \text{ mg/l}$$

The BOD_5 removal efficiency is then computed to be

$$BOD_5^e \text{ Removal efficiency} = \frac{466 - 221}{466} \times 100 = 53 \text{ percent}$$

Removal of 53 percent BOD_5 might be satisfactory under certain circumstances, but it certainly is not equivalent to normal secondary treatment in an activated sludge or biofilm process.

7.4.4 KINETICS OF PHOTOTROPHIC GROWTH

One major aspect that has been ignored so far is the growth kinetics of the phototrophs. Phototroph growth kinetics is at least as important as the kinetics of heterotrophs; indeed, they may dominate in importance over heterotroph growth kinetics, because phototrophs often grow much slower. Thus, for short detention times, phototrophic growth may become the rate-limiting factor in design.

Phototroph growth kinetics is more complex than bacterial growth kinetics because of the dependence on solar radiation, which varies throughout the day and year. The amount of solar radiation received by phototrophs also depends upon water depth and the presence of other materials, especially particulate matter, which may sorb or reflect the solar radiation. The kinetics of phototrophic growth also depends on nutrient availability. Because of the many factors involved in phototroph growth kinetics, application to stabilization lagoon design is not easy. Nevertheless, it is good to understand the importance of these various factors, which are important for making informed judgments about limitations to design.

We begin with some general equations commonly used to describe phototrophic growth kinetics. Some come from efforts at modeling phytoplankton growth in lakes and estuaries (Di Toro et al., 1971), as well as from applications to design of aerobic stabilization lagoons (Shelef, Oswald, and Golueke, 1968). A general equation for growth of phototrophs is

$$\mu_p = \mu_p^m F(I) F(N) - b_p \qquad \text{[7.10]}$$

where

$$\mu_p = \text{phototroph growth rate, d}^{-1}$$
$$\mu_p^m = \text{maximum phototroph growth rate, d}^{-1}$$
$$F(I) = \text{light-intensity function, varies from 0 to 1}$$
$$F(N) = \text{nutrient function, varies from 0 to 1}$$
$$b_p = \text{phototroph decay rate, d}^{-1}$$

This equation is similar to that for heterotrophic growth (Equation 3.5), except that a light-intensity term is added. The first term on the right side of the equation represents growth and the last term represents loss through decay, whether from endogenous respiration or consumption by predators. We next evaluate each of the terms in the phototroph growth equation in order from left to right.

For a given species and temperature, the maximum growth rate (μ_p^m) occurs when all required nutrients are present in sufficient supply so that they do not limit growth and the light intensity is optimal for achieving maximum growth rate. Di Toro et al. (1971) summarized data from several sources studying different species of phototrophs over a temperature range from 2 °C to 27 °C. They summarized the results with an equation of the form

$$\mu_p^m = K_p T \qquad \text{[7.11]}$$

where T represents temperature in °C. The value for K_p was $0.10 \pm 0.025 \, \text{d}^{-1}\,°\text{C}^{-1}$, which indicates that, at 20 °C, the maximum growth rate is $2 \pm 0.5 \, \text{d}^{-1}$. The maximum growth rate also about doubled with a 10 °C temperature increase between 10 °C and 20 °C, as is generally the case with biological reactions in general. The maximum growth rates for phototrophs are much less than that for most aerobic heterotrophic bacteria by about an order of magnitude; thus, the phototrophic growth rate is likely to be more limiting in stabilization lagoon design than heterotrophic growth rate.

Maximum sunlight has an intensity that inhibits phototrophic growth. Thus, the relationship $F(I)$ is somewhat complex and must take inhibition into consideration. A commonly used expression for $F(I)$ is one attributed to Steele (Di Toro et al., 1971):

$$F(I) = \frac{I}{I_s} e^{((-I/I_s)+1)} \qquad \text{[7.12]}$$

Here, I is actual light intensity and I_s is the saturation value, or the intensity at which growth rate is maximum. For modeling studies of phytoplankton growth in Western Lake Erie, Di Toro, O'Connor, Mancini, and Thomann, (1973) used a value for I_s of 15,000 kJ/m^2-d. Using a somewhat different approach and model, Shelef et al. (1968) used a value for I_s of 2,000 kJ/m^2-d. The value of I_s is temperature and species dependent. The relationship between $F(I)$ and incident radiation based on Equation 7.12 is illustrated in Figure 7.3.

Light intensity also decreases with depth in a lagoon, largely due to sorption and reflection from particles, including phototrophs, within the lagoon. A commonly used expression for change in light intensity with depth is (Di Toro et al., 1971; Shelef et al., 1968) :

$$I_h = I_0 e^{-k_e h} \qquad \text{[7.13]}$$

where I_h is light intensity at depth h, I_0 is the intensity at the lagoon surface, and k_e is the extinction coefficient. Shelef et al. (1968) found that the extinction coefficient increased as the concentration of phototrophs increased.

$$k_e = a_p X_{\text{pho}} \qquad \text{[7.14]}$$

with a_p varying from 0.08 to 0.15 (m-mg/l)$^{-1}$ The high end of the range represents phototroph concentrations less than about 500 mg/l.

Aerobic stabilization lagoons are often mixed well enough that all the phototrophs are exposed to the whole distribution of light intensity, rather than some being exposed to intensities that are higher than optimum, while others are exposed to intensities that are lower than optimum. Many phototrophs also have an ability to store excess energy received during periods of high light exposure in the form of organic polymers,

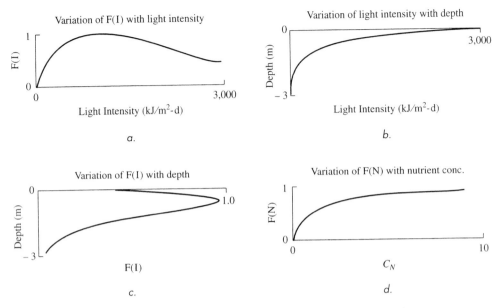

Figure 7.3   Variation of $F(I)$ with light intensity and pond depth, and of $F(N)$ with a nutrient concentration (C_N).

which they use for growth when light is absent. Thus, with good mixing, the incident radiation to which a phototroph is exposed tends to average out to some degree.

The $F(N)$ term most commonly used has the multiplicative Monod form:

$$F(N) = \frac{C_N}{K_N + C_N} \frac{C_P}{K_P + C_P} \frac{C_C}{K_C + C_C} \qquad \textbf{[7.15]}$$

where C_N, C_P, and C_C represent the concentrations of inorganic nitrogen, phosphorus, and carbon, respectively, available for phototrophic growth, and K_N, K_P, and K_C represent the concentrations of each nutrient at which the rate is one-half the maximum (affinity constant). Phototrophs may use either ammonium or nitrate as the nitrogen source, and some phototrophs can fix atmospheric nitrogen, although their growth rate is then greatly reduced. Phosphorus is generally in the form of orthophosphate. Carbon is generally taken up in the form of carbon dioxide, although bicarbonates and carbonates serve as well, due to the equilibrium among these different forms.

The values for the various affinity constants vary widely among phototrophic species. In a listing provided by Di Toro et al. (1971), K_N with ammonium nitrogen varied between 0.002 and 0.12 mg/l, with typical values between 0.01 and 0.02 mg/l. The values for nitrate nitrogen were similar. K_P values ranged between 0.006 and 0.025 mg/l of phosphate-P. Goldman, Oswald, and Jenkins (1974) reported K_C in the range of 0.1 to 0.7 mg/l of CO_2-carbon. Because of the need for nitrogen, phosphorus, and carbon for synthesis of heterotrophic and phototrophic organisms, it is possible that a deficiency in any one of these three nutrients would cause $F(N)$ to decrease below 1.0. Generally, domestic wastewaters have an excess of these nutrients, and nutrient limitations are not likely. However, this is not necessarily the case with industrial wastewaters, and mass balances on these three essential elements should be conducted to ensure they are not limiting.

Models (Fritz, Middleton, and Meredith, 1979; Llorens, Saez, and Soler, 1991; Prats and Llavador, 1994; Shelef et al., 1968) that include all these factors tend to be quite complex and difficult to apply because of the many site-specific coefficients involved. For this reason, many models used in practice (and discussed later) are quite empirical.

From the standpoint of understanding the boundary conditions for a lagoon, we can consider first the maximum growth rate of algae under conditions not limited by light or nutrient concentration ($F(I) = F(N) = 1$). Then, we can apply a safety factor, as described in Chapter 5, for design. Using this approach for estimating the minimum detention time for a stabilization lagoon, we obtain

$$\theta^d \geq \frac{S.F.}{\mu_p^m}$$

Considering aerobic stabilization lagoons to be high-rate systems, the S.F. value lies between 3 and 10.

DESIGN USING A SAFETY FACTOR   Using the safety factor approach, what should be the minimum detention time for the aerobic stabilization lagoon represented in Example 7.3? | **Example 7.7**

First, determine the maximum growth rate using Equation 7.11.

$$\mu_p^m = K_P T = 0.1(5) = 0.5 \text{d}^{-1}$$

Assuming a minimum safety factor of 3, the design detention time would be

$$\theta^d \geq \frac{S.F.}{\mu_p^m} = \frac{3}{0.5} = 6 \text{ d}$$

Using even the minimum safety factor implies that the detention time of 3 d used for design of this lagoon in Example 7.3 is much too short. The system may work, but the design is too close to the failure point to operate reliably. If the lagoon were deepened to 1 m, the detention time would increase to about 9 d, and the system would be more reliable.

Example 7.7 illustrates a case where kinetic limitations under low temperature conditions can limit stabilization lagoon operation. While solar energy was sufficient to operate the given stabilization lagoon in the winter under the conditions in Example 7.3, the kinetics of phototrophic growth, together with a desire for reliability indicated that the detention time of 3 d was not sufficient.

The many factors affecting phototrophic growth are complex, but should not be ignored in stabilization lagoon design. Rough estimates like those made in the example problems help to isolate what limits lagoon performance and where problems may result. Obviously, they should not be relied on to give exact predictions of how a given stabilization lagoon will operate.

7.4.5   FACULTATIVE STABILIZATION LAGOONS

Aerobic stabilization lagoons are not widely used because of the difficulty in maintaining aerobic, well-mixed conditions and because they produce excessive suspended solids in the effluent, unless suspended solids removal is added to treat the effluent. Instead, most stabilization lagoons are *facultative,* in which the upper portion is kept aerobic through the action of algae, while the bottom portion is allowed to go anaerobic. This is illustrated in Figure 7.4. How might this arrangement reduce the problem with high effluent suspended solids?

Several phenomena act to reduce effluent suspended solids. First, the detention times are longer, the lagoons are deeper, and the BOD_L loading rates per unit lagoon surface area are lower than with aerobic stabilization lagoons. The longer detention times allow greater decay of heterotrophic organisms. Second, a significant fraction of the biomass (phototrophs and heterotrophs) settles to the lagoon bottom, thus effecting some suspended solids removal. Third, methane fermentation of the settled solids, especially in the summertime, results in loss of organic matter in the form of methane gas, which escapes from the lagoon surface into the atmosphere. Since these phenomena are complex and highly variable from site to site, design approaches tend to be based on empirical studies and conservative design criteria. Most common are those that are based on (1) surface loading rate of BOD_5 and (2) first-order kinetics for BOD_L removal. One needs to realize that predictions of performance based on any approach are subject to great uncertainties.

7.4.6   SURFACE BOD₅ LOADING RATES

The use of surface loading is based on the concept that sufficient oxygen must be produced by phototrophs to offset the oxygen required for waste organic oxidation. Aerobic stabilization lagoon design is based on a minimum surface area for this

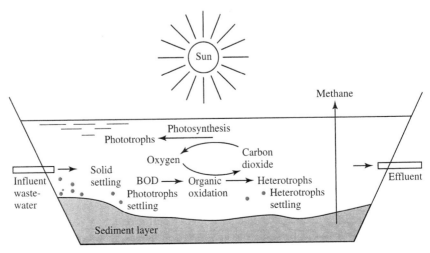

Figure 7.4   Schematic of processes of importance in a facultative stabilization pond.

purpose. With facultative lagoons, the design is much more conservative, with a much lower surface loading used. Table 7.4 indicates BOD_5 loading suggested from an evaluation of practices around the world by Gloyna (1971). Suggested loading is lower in extreme northern or southern latitudes, largely because of winter conditions when lagoon surfaces are generally frozen and phototrophic activity is absent. Long detention times are then also used, and most BOD removal undoubtedly results from settling of suspended solids. In more temperate climates, where freezing is not a significant problem, acceptable loading is higher and more in keeping with available sunlight.

Table 7.5 is a summary of incident light for various latitudes at the peak time of the year (June for northern latitudes and December for southern latitudes) and the time of minimum light (the reverse for the northern and southern latitudes). Also listed are values for maximum oxygen production based on the method of calculation given in Example 7.3. Implied safety factors for surface loading, listed in Table 7.4, were obtained by dividing the maximum computed phototrophic oxygen production, based on an energy transfer efficiency of 3 percent, and the generalized surface loading given in Table 7.4. In tropical regions near the equator, the loading is quite near the maximum phototrophic oxygen production, while the implied safety factor increases significantly away from the equator. Designs then become dependent primarily on factors other than solar radiation alone.

7.4.7 FIRST-ORDER KINETICS

As is commonly the case with complex processes, the kinetics are often simplified to first-order kinetics. This is one of the most common approaches used for design of stabilization lagoons. Here, the utilization of BOD_L is taken as in the standard 5-d

Table 7.4 Generalized BOD_5 loading for facultative stabilization ponds under various climatic conditions

Surface Loading (kg BOD_5/ha-d)	Implied Safety Factor[a]	Depth (m)	Detention Time (d)	Environmental Conditions
Less than 10	> 8–64	1.5–2.0	> 200	Frigid zones, with seasonal ice cover, uniformly low water temperatures, and variable cloud cover
10–50	4–65	1.5–2.0	100–200	Cold seasonal climate, with seasonal ice cover and temperate summer temperatures for a short season
50–150	2.5–12	1.5–2.0	33–100	Temperate to semitropical, occasional ice cover, no prolonged cloud cover
150–350	1.3–3.7	1.0	17–33	Tropical, uniformly distributed sunshine and temperature, and no seasonal cloud cover

[a] Safety Factor is based on the ratio of maximum phototrophic oxygen production from Table 7.5 to generalized surface loading.
SOURCE: Gloyna, 1971.

Table 7.5   Averaged incident radiation on earth's surface on cloudless days and estimated maximum phototrophic oxygen production assuming 3 percent energy transfer efficiency

	Incident Radiation (kJ/m²-d)		Maximum Phototrophic Oxygen Production (kg Oxygen/ha-d)	
Latitude	December	June	December	June
45°–60° N	4,300	32,800	84	640
30°–45° N	10,700	33,500	210	650
15°–30° N	19,400	31,500	380	610
0°–15° N	24,400	28,600	470	560
0°–15° S	30,700	23,600	600	460
15°–30° S	33,300	17,600	650	340
30°–45° S	35,000	11,300	680	220
45°–60° S	34,600	4,200	670	80

BOD test to be:

$$r_{ut} = -k_l S \qquad\qquad \textbf{[7.16]}$$

Introducing this expression into the steady-state mass balance equation for a CSTR yields

$$Q(S^0 - S) = -k_l S V \qquad\qquad \textbf{[7.17]}$$

Solving for S with V/Q equal to the detention time, θ, gives

$$\frac{S}{S^0} = \frac{1}{1 + k_l \theta} \qquad\qquad \textbf{[7.18]}$$

If, instead, one assumes that a stabilization lagoon acts as a plug-flow reactor, then Equation 5.16 is the mass balance on substrate. Applying it to the steady-state case ($\Delta S/\Delta t = 0$), we obtain as ΔV approaches 0 that

$$\frac{dS}{S} = -k_l \theta \qquad\qquad \textbf{[7.19]}$$

which integrates to the familiar equation for a PFR:

$$\frac{S}{S^0} = e^{-k_l \theta} \qquad\qquad \textbf{[7.20]}$$

Large reactors with long detention times and shallow depths, such as stabilization lagoons, act neither as a CSTR nor as a PFR, but as something between. Wehner and Wilhelm (1958) solved for the case that is intermediate between complete mix and plug flow.

$$\frac{S}{S^0} = \frac{4ae^{-1/2d}}{(1+a)^2 e^{a/2d} - (1 - a^2 e^{-a/2d})} \qquad\qquad \textbf{[7.21]}$$

where

$$a = \sqrt{1 + 4k_l\theta d} \quad \text{and} \quad d = \frac{D\theta}{L^2}$$

Here, D is an axial dispersion coefficient (m²/d) and L is a characteristic length (m). Solution requires some knowledge of the dispersion properties of the lagoon, which adds another complicating factor to the picture. A typical value for d, the dimensionless dispersion number, is about 1.0, with typical values ranging between 0.1 and 2 (Metcalf and Eddy, 1991). However, the question that remains is whether or not such a level of sophistication is really warranted when Equations 7.18 and 7.20 are based on such major simplifying assumptions. Figure 7.5 provides a comparison of BOD₅ removal efficiencies using Equations 7.18, 7.20, and 7.21. With a typical value for d of 1.0, the difference between predicted removals with axial dispersion and a CSTR are perhaps not great. But to view this further, information on the rate coefficient, k_l, needs to be examined.

Values suggested for k_l were reviewed by Ellis and Rodrigues (1995). Literature values ranged between 0.22 d⁻¹ to 0.5 d⁻¹, with 0.3 d⁻¹ being the most accepted value. Some indicated that k_l varies with temperature, as would be expected, and with organic loading, which would not necessarily be expected. A range of reported values was given as 0.1 to 2.0 d-1 (Metcalf and Eddy, 1991). Marais and Shaw (1961) found a mean value for stabilization lagoons in South Africa to be 0.23 d⁻¹, but suggested that a more conservative value of 0.17 d⁻¹ be used in design. Mara (1975) thought that this value is too low and suggested k_l (d⁻¹) be related to temperature by the equation, $k_l = 0.3 \cdot 1.05^{(T-20)}$, where T is °C.

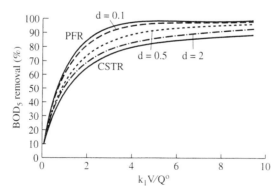

Figure 7.5   Comparison of calculated BOD₅ removal between a plug-flow reactor (PFR) model and a continuously stirred tank reactor (CSTR) model as a function of the dimensionless product of the first-order reaction rate (k_l, d⁻¹), and detention time (V/Q^0, d). Also shown are results of mixed-reaction model (Equation 7.21) for various dimensionless dispersion numbers.

These various k_l values have generally been estimated from measured total BOD_5 removal efficiencies assuming that the lagoons act as CSTRs. It is apparent from Figure 7.4 that a k_l value so obtained depends upon the model assumed. If one assumes a PFR, then lower k_l values would be obtained than if a CSTR were assumed. The lesson is that k_l values derived with one given assumption for lagoon operation should not be applied with a different model. The other feature is the wide variation in k_l values assumed. To illustrate this further, Ellis and Rodrigues (1995) estimated k_l values for two different stabilization lagoons operated near 30 °C over a year's time, assuming a CSTR model. They made monthly evaluations of k_l values, finding an average of 0.25 to 0.17 d^{-1} for the two lagoons respectively, but the monthly values varied considerably, between 0.03 and 0.40 d^{-1}. Interestingly, they also measured the soluble BOD_5 in the lagoon effluent, and they found k_{sol} values, which were computed from the total influent BOD_L, but soluble effluent BOD_5, were 0.30 to 0.38 d^{-1} for the two lagoons, with a range of 0.07 to 0.79 d^{-1}. The range of k_l values, and obviously BOD removal efficiency, is quite high even for a single location.

In order to illustrate the use of first-order models for lagoon design, we evaluated implied k_l values for the estimated BOD removal indicated for aerobic stabilization lagoons in Examples 7.3 through 7.6. The conditions indicated for those problems were maintained, except that air temperature, influent BOD_L, and lagoon depth (and detention time) were varied systematically. Table 7.6 summarizes the results. The implied values for k_l with this approach vary between 0.036 d^{-1} and 0.47 d^{-1}, with a typical value of 0.25 d^{-1}. This is in the typical range reported from empirical studies with stabilization lagoons. The k_{sol} values also were obtained in this manner, and they varied even more, between 0.19 d^{-1} and 4.3 d^{-1}, with a typical value of 2.3 d^{-1}. The implied k_l values in this example increased somewhat with temperature, but not as much as might be assumed from normal temperature relationships, and k_l did increase with increased waste organic concentration or organic loading to the lagoons, as is commonly reported from empirical studies. Thus, the procedure used to estimate lagoon performance in Examples 7.3 to 7.6 perhaps provide as accurate a picture of stabilization-lagoon operation as is given by first-order equations. The advantage of the example approach is that the processes involved in BOD removal are more easily identified.

A major and widely recognized problem in stabilization-lagoon design, as indicated in the example approach, is the large concentration of suspended solids in the lagoon effluents. With facultative lagoons, a portion of this suspended solids settles to the lagoon bottom and is removed from the effluent. The result is that the actual lagoon performance should lie somewhere between that indicated by BOD_5 removal when suspended solids are included and when they are not. If one-half of the suspended solids are removed, then the effluent BOD_5 concentration should lie just about midpoint between them. Then, what happens to the settled solids? If they are stirred up because of such things as storm conditions, then the effluent quality suffers. On the other hand, if they are subject to anaerobic decomposition with methane formation, and if the methane escapes from the lagoon into the atmosphere, then organic removal is effected and greater BOD removal will be obtained.

Another factor of importance is the growth of phototrophs. It is easily possible that conditions will be sufficiently favorable so that their growth will produce oxygen

Table 7.6 The effects of changes in air temperature, detention time, depth, and influent BOD concentration on apparent first-order k_1 values (CSTR model), pond surface loading, and percentage removals of BOD_5 considering total effluent and soluble effluent BOD_5

T_a (°C)	θ (d)	h (m)	S^0 BOD_L (mg/l)	S^0 BOD_5 (mg/l)	S^e Total BOD_5 (mg/l)	S^e Sol BOD_5 (mg/l)	k_1 Total (d^{-1})	k_1 Sol (d^{-1})	Surface Loading (kg/ha-d)	Removal (%) Total	Removal (%) Sol
Variable Air Temperature											
0	3	0.3	600	466	221	23	0.37	6.54	466	53	95
10	3	0.3	600	466	206	24	0.42	6.00	466	56	95
20	3	0.3	600	466	194	29	0.47	5.00	466	58	94
30	3	0.3	600	466	177	36	0.54	4.00	466	62	92
Variable Air Temperature (Reduced Influent BOD_L)											
0	12	1.2	200	155	74	19	0.046	0.30	155	52	88
10	12	1.2	200	155	68	23	0.053	0.24	155	56	85
20	12	1.2	200	155	69	22	0.100	0.50	155	55	86
30	12	1.2	200	155	59	28	0.068	0.19	155	62	82
Variable Detention Time (Surface Loading Constant)											
20	3	0.3	600	466	194	29	0.47	5.00	466	58	94
20	6	0.6	600	466	173	37	0.28	1.91	466	63	92
20	12	1.2	600	466	148	46	0.18	0.76	466	68	90
20	24	2.4	600	466	125	53	0.11	0.33	466	73	89
Variable Detention Time (Surface Loading Constant, Different Influent BOD_L)											
20	3	0.3	200	155	82	14	0.30	3.31	155	47	91
20	6	0.6	200	155	76	18	0.17	1.28	155	51	88
20	12	1.2	200	155	69	22	0.10	0.50	155	55	86
20	24	2.4	200	155	63	26	0.06	0.21	155	59	83
Variable Detention Time (Depth Constant)											
20	3	1.2	200	155	66	15	0.450	3.10	621	57	90
20	6	1.2	200	155	66	18	0.230	1.24	311	57	88
20	12	1.2	200	155	69	22	0.100	0.50	155	55	86
20	24	1.2	200	155	83	26	0.036	0.21	78	46	83
Variable Influent BOD_L											
20	3	0.3	600	466	194	29	0.47	5.00	466	58	94
20	3	0.3	400	311	139	22	0.41	4.30	310	55	93
20	3	0.3	200	155	82	14	0.30	3.31	155	47	91
20	3	0.3	100	78	53	9	0.16	2.59	78	32	88

in excess of that needed for BOD_L oxidation. This might occur if the lagoon surface area is larger than needed. In this case, the lagoon could become supersaturated with oxygen, and the excess might be transferred to the atmosphere and lost from the system. The excess mass of phototrophs produced increases the effluent organic content, perhaps to values higher than in the influent, resulting in increased effluent BOD! Thus, lagoon design should not have more surface area than is actually needed. Too much surface area leads to algal blooms and effluent BOD greater than influent BOD. As an example, the effluent BOD_5 from an oversized facultative lagoon in

Central Illinois was as high as 800 mg/l in the middle of the summer, even though the influent BOD_5 was only 200 mg/l. The effluent was "pea green" from an algal bloom.

7.5   ANAEROBIC STABILIZATION LAGOONS

Anaerobic stabilization lagoons are designed with a much higher surface loading than are facultative lagoons. Most of the BOD reduction that occurs is through methanogenesis in the deeper portion of the lagoon. The concept is that the surface of the lagoon is kept aerobic through phototrophic action in order to reduce odors. Anaerobic lagoons have frequently been used for treatment of concentrated industrial wastewaters and for storage and the final stages of treatment of digested sludges from municipal treatment plants. The oxygen production in the upper layers helps primarily to encourage the growth of organisms that can oxidize odorous gaseous compounds, such as hydrogen sulfide, that may otherwise pass through the surface waters. Hydrogen sulfide can be oxidized by chemolithotrophic bacteria, such as *Thiobacillus,* which oxidize sulfide for energy and produces either elemental sulfur or sulfate. Phototrophic sulfur bacteria that obtain energy from sunlight oxidize sulfide to elemental sulfur, thiosulfate, or sulfate as a means to obtain electrons for reducing carbon dioxide to cellular carbon. Odorous organic molecules, such as carbon disulfide, aldehydes, and the putrefaction products of proteins, may also be produced by anaerobic reactions and can be oxidized by aerobic heterotrophs acting in the upper aerobic layers. When the lagoon is operating well, odors are not large, but it is not appropriate to place anaerobic lagoons near population centers, where even mild odors can be very objectionable. If anaerobic lagoons mix (i.e., turn over), either because of destratification during temperature changes or strong wind action, the odors can become quite strong. Thus, while anaerobic lagoons can be an inexpensive treatment option, the engineer needs to carefully consider the odor potential.

The usual design of an anaerobic lagoon is based upon concepts of anaerobic treatment that are discussed in detail in Chapter 13. Anaerobic lagoons are generally not mixed, and temperatures are most often below the optimum range near 35 °C. Thus, designs need to be much more conservative than are used for anaerobic digestors. The detention time should be far greater than the limiting value of θ_x^{min} for the key organisms, which normally are the acetate-using methanogens. Their doubling time at temperatures of 15 °C to 20 °C is about 10 d. Thus, the detention times should not be less than about 40 d, and longer is desirable. However, when treating wastes with high suspended solids concentration, the major BOD removal may result from the settling of the suspended solids. The retention time of the suspended solids undergoing anaerobic reduction and the anaerobic bacteria that cling to the suspended solids may be much longer than that of the wastewater itself; thus, shorter liquid times may be appropriate here. Such designs then become comparable to that of a septic tank, where the liquid may have a detention time of a day or so, while the settled solids actively decompose in a lower chamber with long detention time. BOD reductions due to methanogenesis drop significantly with decrease in temperature below 15 °C. Hence, BOD reduction may be minimal in colder climates.

Empirical experience is generally used as the basis for anaerobic lagoon design, but the relative distribution of BOD between dissolved and suspended phases is crucial in order to choose an appropriate hydraulic detention time. The principles of anaerobic treatment provided in Chapter 13 need to be carefully considered. Gloyna (1971) reported on experiences in different countries with respect to domestic wastewater treatment in anaerobic lagoons. In general, he indicated that a 1-d detention time is generally sufficient to obtain 50 to 70 percent BOD reduction, with somewhat more than one-half of that removal resulting from suspended solids settling *and* decomposition. In Australia, similar BOD reductions were obtained with a 1.2-d detention time, but in the cold season, BOD reduction dropped to 45 to 60 percent, even with an increase in detention time to 5 to 7 d. It appears that most reduction then came only from settling of suspended solids. In Israel, anaerobic lagoons were designed on the basis of organic loading and depth, with 125 kg BOD_5 per 1,000 m^3-d and 1.2 to 1.5 m suggested. This then corresponds to a surface loading of 800 to 1,000 kg BOD_5 ha-d. Generally, anaerobic lagoons have been used only for preliminary treatment for domestic wastewaters in order to remove and treat suspended solids that settle. Facultative lagoons for more complete treatment then follow them.

Treatment of industrial wastes in anaerobic lagoons is often used for more concentrated wastewaters. One suggestion (Metcalf and Eddy, 1991) is that detention times of 20 to 50 d be used, with lagoon depths of 2.5 to 5 m and surface BOD_5 loading of 125 to 300 kg BOD_5/ha-d. Benefield and Randall (1980) reviewed reports on anaerobic lagoon treatment of municipal and industrial wastes and indicated surface loading between 80 and 400 kg BOD_5/ha-d. Such surface loading appears little different from those appropriate to aerobic stabilization lagoons. The main difference is in lagoon depth and detention time: The anaerobic lagoon generally is much deeper and has a longer detention time. Thus, anaerobic lagoons generally have surface areas that would permit sufficient oxygen generation through phototrophic activity to oxidize most of the BOD, but they provide volume at the bottom of the lagoon for anaerobic decomposition of suspended solids that settle and decompose. These may be suspended solids that either originate in the raw wastewater itself or are formed in the stabilization lagoon (phototrophic and heterotrophic organisms).

7.6 SERIES OPERATION

A common practice is to place stabilization lagoons in series, as this can give higher degrees of overall treatment, providing that excessive growth of phototrophic organisms is avoided. Common practice is to have two lagoons in series, but as many as four might be used. Here, the first might be an anaerobic lagoon, with suspended solids removal and anaerobic decomposition, and the second is a facultative lagoon. A third might be used for further polishing as zooplankton graze on phototrophs and heterotrophs. This is often termed a *maturation* lagoon and might have a detention time of 7 to 10 d and a depth of 1 m. A fourth lagoon might be used for growth of fish that consume the zooplankton from the maturation lagoon, and its dimensions may be similar to those of the maturation lagoon. Because of phototrophic growth, BOD removal in the second and third lagoons may be negligible, or even negative. A major feature of importance in series operation is the increased die-off of coliform organisms, as discussed in the following section.

7.7 COLIFORM REDUCTION

One of the major side benefits of stabilization lagoons is the reduction of pathogens that results without the need to add disinfectants such as chlorine. Such pathogen reduction is especially desirable when stabilization-lagoon effluents are used for crop irrigation. Stabilization lagoons thus have great advantages for developing countries, where the availability of mechanical equipment and chemicals is limited. Stabilization lagoons can provide a cost-effective means for reducing the human-health hazards associated with wastewaters.

A summary of studies on pathogen reduction was provided by Gloyna (1971). The fate of coliform organisms is often used as an indicator of the fate of pathogens in treatment systems. While there are some similarities between coliforms and human pathogens, Gloyna cautions that the fates may not always be the same. Gloyna presented an example in which 97.5 percent die-off of coliforms was obtained in a lagoon system, but the reduction in *Salmonella typhimurium*, a pathogen of concern in water, was only 92 percent. Because it is difficult to measure pathogens in wastewaters (mainly due to their low numbers), reliance on coliforms as indicators of pathogen removal will continue as a standard approach until better detection methods become available. This means that several orders of magnitude reduction in coliforms are needed in order to ensure a reasonable level of pathogen mortality.

Coliform reduction in lagoons is a result of several factors (Gloyna, 1971): dilution and mixing, aggregation and settling, toxicity from wastewater chemicals, predation, sunlight, temperature, and lack of available nutrients. High pH as a result of carbon dioxide removal by phototrophs for synthesis can also be involved. Time of exposure to environmental conditions is of importance. Mortality is often simulated as a first-order reaction, and as a conservative estimate, a completely mixed or CSTR model similar to Equation 7.14 is generally used (Marais, 1974), together with a first-order rate coefficient, K_{cm}:

$$\frac{N^e}{N^0} = \frac{1}{1 + K_{cm}\theta} \qquad [7.22]$$

where N^e is the coliform count in the lagoon effluent and N^0 is the concentration in the lagoon influent. With this model, Marais evaluated K_{cm} from lagoons operated in series and suggested that $K_{cm} = 2.6(1.19)^{(T-20)}$ d^{-1}. From this, a general value for K_{cm} of 2 d^{-1} has often been suggested (Gloyna, 1971). Some, however, have assumed a plug-flow model similar to Equation 7.20:

$$\frac{N^e}{N^0} = e^{-K_{pf}\theta} \qquad [7.23]$$

in which case the first-order decay term, K_{pf} is much lower than the equivalent K_{cm} value. For example, a K_{pf} of only about 0.1/d is required with θ of 10 to 20 d to obtain coliform removal (by Equation 7.23) equivalent to that given by Equation 7.22 with K_{cm} of 2/d. This great difference in first-order rate constants found when using the different models was noted by Mayo (1995), who underscored that coefficients must be applied only with the models from which they were estimated. Otherwise, exceptionally poor projections can result.

Studies of coliform reductions in lagoons operated in series provide some evidence that a first-order coliform decay rate with a complete-mix model may give usable estimates. For example, Gloyna (1971) reported on a field study in which 97.5 percent *E. coli* reduction was obtained with a first lagoon, and 99.2 percent reduction occurred in a second lagoon, giving an overall reduction of 99.98 percent. Here, the second lagoon acted even better than the first, even though its detention time of 15 d was lower than that of the first, which was 20 d. The implied K_{cm} values would be 2 d^{-1} and 8 d^{-1}, respectively; K_{pf} values would be 0.18 d^{-1} and 0.32 d^{-1}, respectively.

Mayo (1995) summarized the results of studies by several investigators on K_{cm} values for coliform mortality, with reported K_{cm} values at 20 °C of 0.71, 0.8, 1.94, and 9.9 d^{-1}. To further illustrate the uncertainty in K_{cm} values, Ellis and Rodrigues (1995) demonstrated the monthly variation in K_{cm} in two separate stabilization lagoon systems operated in parallel on the same waste. Each system consisted of two lagoons. K_{cm} values based on the completely mixed system model had a monthly average of 3.0 and 3.6 d^{-1} for the first lagoons in the series and 2.1 and 1.1 d^{-1} for the second lagoons in the series. However, the variations in performance throughout the year were quite significant, even though the average temperature varied by no more than 4 °C; the extreme monthly values were 0.06 to 12 d^{-1}.

It is obvious that factors other than detention time and temperature affect coliform mortality. Mayo (1995) found with stabilization lagoons in Tanzania that sunlight intensity, lagoon depth, and pH were other key factors involved. Most mortality occurred near the lagoon surface, where light intensity was the greatest. Mayo chose to use the plug-flow model, and the relationships developed should be used with Equation 7.23 rather than 7.22. Here, K_{pf} was given by

$$K_{pf} = \left[\frac{0.0135\, pH}{d} + \frac{1.36(10)^{-5}(m)^3}{kJ} \frac{I^0}{h} \right] \qquad \textbf{[7.24]}$$

The factor $1.36 \cdot 10^{-5}$ m^3/kJ was estimated from results of different studies and reported to vary between $1.18 \cdot 10^{-5}$ and $2.5 \cdot 10^{-5}$ m^3/kJ.

COLIFORM REMOVAL   For the stabilization lagoon conditions of Examples 7.2 and 7.3, | **Example 7.8**
(a) estimate the plug-flow model K_{pf} for summer and winter conditions using Equation 7.24, (b) calculate the percent coliform mortality for each case, (c) estimate the equivalent K_{cm} for the complete-mix model for the same conditions, (d) redo the calculations for parts a through

c assuming the surface loading remains the same, but the lagoon detention time is increased by a factor of 5. Assume pH $= 8.0$.

a. From Examples 7.2 and 7.3, the following conditions apply. For summer, $I_o = 30,000$ kJ/m^2-d, and for winter, $I_o = 12,000$ kJ/m^2-d. $h = 0.3$ m and $\theta = 3$ d. Using Equation 7.24:

Summer:

$$K_{pf} = 0.0135(8) + 1.36(10)^{-5}\frac{30,000}{0.3} = (0.108 + 1.36) = 1.47/\text{d}$$

Winter:

$$K_{pf} = 0.0135(8) + 1.36(10)^{-5}\frac{12,000}{0.3} = (0.108 + 0.544) = 0.65/\text{d}$$

b. Coliform removals are computed from Equation 7.23.

Summer:

$$\frac{N^e}{N^0} = e^{-1.47(3)} = 0.012, \ \text{Removal} = 100(1 - 0.012) = 98.8\%$$

Winter:

$$\frac{N^e}{N^0} = e^{-0.65(3)} = 0.14, \ \text{Removal} = 100(1 - 0.14) = 86\%$$

c. For estimating an equivalent first-order constant for the complete-mix model, rearrange Equation 7.22 to solve for K_{cm}:

$$K_{cm} = \frac{(N^0/N^e) - 1}{\theta}$$

from which,

Summer:

$$K_{cm} = \frac{(1/0.012) - 1}{3} = 27/\text{ d}$$

Winter:

$$K_{cm} = \frac{(1/0.16) - 1}{3} = 1.75/\text{ d}$$

d. New detention time is $5(3) = 15$ d. Since the surface loading remains the same, the 5 times increase in volume is achieved by increasing the depth 5 times to $5(0.3) = 1.5$ m. The changed values for parts a through c then become:

Summer:

$$K_{pf} = 0.0135(8) + 1.36(10)^{-5}\frac{30,000}{1.5} = (0.108 + 0.272) = 0.38/\text{d}$$

$$\frac{N^e}{N^0} = e^{-0.38(15)} = 0.0033, \ \text{Removal} = 100(1 - 0.0033) = 99.7\%$$

$$K_{cm} = \frac{(1/0.0033) - 1}{15} = 20/\text{d}$$

Winter:

$$K_{pf} = 0.0135(8) + 1.36(10)^{-5}\frac{12,000}{1.5} = (0.108 + 0.109) = 0.217/\text{d}$$

$$\frac{N^e}{N^0} = e^{-0.217(15)} = 0.0385, \text{ Removal} = 100(1 - 0.0385) = 96.1\%$$

$$K_{cm} = \frac{(1/0.0385) - 1}{15} = 1.67/\text{d}$$

We see from this example problem that first-order rate constants found by using the approach by Mayo, which considers the impact of solar intensity and pH, are within the very wide range noted by others, but the constants are based more directly upon factors found to affect mortality. Increasing lagoon depth or detention time, while maintaining the same surface BOD loading, results in a decrease in the rate constant, but at the same time, effects a greater net mortality to the organisms.

7.8   LAGOON DESIGN DETAILS

Once the surface area and depth of the stabilization lagoons are determined, then the lagoons can be designed. Gloyna (1971) provides details for many features of lagoon design. Generally lagoon shape is rectangular or square. The lagoons generally have flat bottoms and multiple inlets and outlets to maintain good distribution of flow. The lagoon bottoms are generally earthen, but made of compacted clay or covered with a membrane to reduce wastewater seepage from the lagoon. Side walls should be sloped at an angle of about 2.5 to 3 m horizontal to 1 m vertical, and they should be lined with stone or gravel to prevent erosion from wave action or surface runoff. Vegetation above the rock line helps reduce side erosion. Freeboard of 0.6 to 1 m above the maximum water level expected should be sufficient to prevent losses from wave action. Surface drainage should not enter the lagoon. Inlets should be submerged to reduce erosion. Outlets should be located as far from the inlet as possible to reduce short-circuiting of wastewater. Effluent is best taken below the lagoon surface to prevent the discharge of floating solids. Recirculation is sometimes used to increase mixing within a lagoon.

7.9   REMOVING SUSPENDED SOLIDS FROM THE LAGOON EFFLUENT

Often the most difficult problem of lagoons is that the effluent can have very high concentrations of suspended solids. While high suspended solids can be caused by short-circuiting of influent solids or by scouring of the sediments, the most common cause is biomass, particularly during algal blooms. In some instances, the effluent suspended solids and BOD are several times larger than they are in the influent. Even when biomass is not causing a gross violation of effluent standards, removing suspended solids from the effluent is an important strategy to help improve the effluent quality.

It always is possible to treat a lagoon effluent with a sophisticated solids-separation system, such as coagulation/flocculation, sedimentation or centrifugation, and filtration. These systems are expensive and require considerable operator attention. Alternatively, many "simple" methods have been used to remove algae and other suspended solids from lagoon effluents. Success is "spotty," but some simple approaches do achieve success. We summarize eight simple approaches that have been used.

The most effective means of suspended solids removal usually is the *intermittent sand filter.* Frequently, the final lagoon in a series has a sand bottom underlain by drains. The effluent from the previous lagoon is applied *intermittently* (say, for one day per week), and then the filter is allowed to drain, dry, and degrade. This cycling continues until the head loss becomes too great, at which time the bed is cleaned by scraping off the top layer. Key features of a properly design intermittent sand filter include:

Sand size	0.2 to 0.5-mm effective size
Uniformity coefficient	< 7, with 2 ideal
Sand depth	> 45 cm
Gravel depth	30 cm
Hydraulic loading	< 0.56 m/d for SS < 50 mg/l
	< 0.39 m/d for SS > 50 mg/l
Cleaning depth	2–5 cm

A second widely used method is *intermittent in-pond coagulation.* Ferric chloride, alum, lime, or a combination is added, usually from the back of a motor boat. Doses are determined from jar tests and are fairly large (e.g., > 100 mg/l). Effluent discharge is stopped during coagulant dosing, coagulation, and settling. Once the water is clarified, effluent discharge is resumed. The main drawback of in-pond coagulation is that it builds up significant sediment, which fills the lagoon and stores BOD, N, and P.

Third is *continuous in-pond coagulation* using metal-salt coagulants. It has the same benefits and drawbacks as intermittent coagulation. Application of the coagulant is easier, but it also may be less effective due to changing conditions.

A fourth option is to discharge effluent only when the suspended solids are low. When water inputs from the wastewater and precipitation do not greatly exceed the rate of evaporation, it is possible to retain all water during periods of high suspended solids. In addition to a high evaporation rate, this approach often requires that the lagoons be designed for variable depths for summer storage of undischarged wastewater.

Other approaches that have been used occasionally are chlorinating the water between lagoons to kill algae and promote their settling, passing the effluent through rock filters (1.5 to 2.5-cm size), treating the effluent with microstrainers before discharge, and applying the effluent to land, instead of to a receiving water.

7.10 WETLANDS TREATMENT

The use of lagoons in which photosynthetic plants, or macrophytes, are an integral part is relatively new in wastewater treatment. This form of *wetlands treatment* uses *aquatic macrophytes* (i.e., water-tolerant plants indigenous to wetlands) as part of a simple engineered lagoon system to treat wastewater. The prominent macrophytes include water hyacinths, water primrose, and cattails, although many others are possible. The type of macrophyte that is used depends solely on the local climate. For example, the water hyacinth is common in warm Southern California, while cattails are possible in the temperate Midwest. Figure 7.6 illustrates common macrophytes.

Although heterotrophic bacteria (as normal) carry out BOD removal, wetlands treatment is successful because of the benefits of the macrophytes. These benefits include:

- Leaves and stems above the water shield the water column from sunlight and reduce algal blooms.

- Leaves and stems above the water insulate the water from heat loss.

- Stems and roots in the water column are colonized by biofilms and help accumulate a large bacterial population.

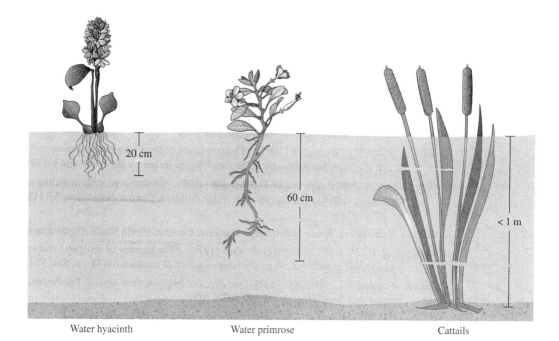

Figure 7.6 Sketches of common aquatic macrophytes used in wetlands treatment.

- Stems and roots help to capture colloids.
- Stems and roots may give off O_2 during photosynthesis and stimulate bacterial metabolism.

Wetlands treatment can be effective for nutrient removal, as well as removal of BOD and SS. Nutrient removal is a combination of microbial (Chapters 9, 10, and 11 discuss this in depth) and macrophytic. The macrophytic part is brought about by incorporation of N and P into the plant mass, which is harvested. Major removal of nitrogen, however, is more likely due to microbial nitrification and denitrification.

Design of wetlands lagoons remains rather empirical. A depth of about 1.5 m and a length-to-width ratio of 3:1 have been used successfully. Surface BOD loads of up to 220 kg BOD_5 ha-d gave good BOD and SS removals.

7.11   BIBLIOGRAPHY

Adams, C. E., Jr. and W. W. Eckenfelder, Jr. Eds. (1974). *Process Design Techniques for Industrial Waste Treatment.* Nashville, TN: Enviro.

Bartsch, E. H. and C. J. Randall (1971). "Aerated lagoons—A report on the state of the art." *J. Water Pollution Control Federation* 43(4).

Benefield, L. D. and C. W. Randall (1980). *Biological Process Design for Wastewater Treatment.* Englewood Cliffs, NJ: Prentice-Hall.

Di Toro, D. M.; D. J. O'Connor; J. L. Mancini; and R. V. Thomann (1973). "Preliminary phytoplankton-zooplankton-nutrient model of Western Lake Erie." In *Systems Analysis & Simulation in Ecology.* New York: Academic.

Di Toro, D. M.; D. J. O'Connor; and R. V. Thomann (1971). "A dynamic model of the phytoplankton population in the Sacramento–San Joaquin Delta." In *Advances in Chemistry Series No. 106, Non-Equilibrium Systems in Natural Water Chemistry.* New York: American Chemical Society, pp. 131–150.

Ellis, K. V. and P. C. Rodrigues (1995). "Developments to the first-order, complete-mix design approach for stabilisation lagoons." *Water Research* 29(5), pp. 1343–1351.

Fritz, J. J.; A. C. Middleton; and D. D. Meredith (1979). "Dynamic process modeling of wastewater stabilization lagoons." *J. Water Pollution Control Federation,* 51(11), pp. 2724–2743.

Gloyna, E. F. (1971). *Waste Stabilization Lagoons.* Geneva: World Health Organization.

Goldman, J. C.; W. J. Oswald; and D. Jenkins (1974). "The kinetics of inorganic carbon limited algal growth." *J. Water Pollution Control Federation* 46(3), pp. 554–574.

Jewell, W. J. and P. L. McCarty (1971). "Aerobic decomposition of algae." *Environmental Science and Technology* 5(10), pp. 1023–1031.

Llorens, M.; J. Saez; and A. Soler (1991). Primary productivity in sewage pond: Semi-empirical model." *J. Environmental Eng.* 117(6), pp. 771–781.

Mara, D. D. (1975). "Discussion, a note on the design of facultative sewage lagoons." *Water Research* 9(5/6), pp. 595–597.

Marais, G. v. R. (1974). "Faecal bacterial kinetics in stabilization lagoons." *J. Environmental Eng. Div., ASCE* 100(1), pp. 119–139.

Marais, G. v. R. and V. A. Shaw (1961). "A rational theory for the design of sewage stabilization lagoons in Central and South Africa." *The Civil Engineer in South Africa* 3(11), pp. 1–23.

Mayo, A. W. (1995). "Modeling coliform mortality in waste stabilization lagoons." *J. Environmental Eng.* 121(2), pp. 140–152.

Metcalf and Eddy, Inc. (1991). *Wastewater Engineering: Treatment, Disposal, Reuse*, 3rd ed. New York: McGraw-Hill.

Odum, E. P. (1971). *Fundamentals of Ecology*, 3rd ed. Philadelphia: W. B. Saunders.

Oswald, W. J.; H. B. Gotaas; C. G. Golueke; and W. R. Kellen (1957). "Algae in waste treatment." *Sewage and Industrial Wastes* 29(4), pp. 437–457.

Prats, D. and F. Llavador (1994). "Stability of kinetic models from waste stabilization lagoons." *Water Research* 28(10), pp. 2125–2132.

Reifsnyder, W. E. and H. W. Lull (1965). *Radiant Energy in Relation to Forests* (Tech. Bull. No. 1344 ed.). U.S. Department of Agriculture.

Shelef, G.; W. J. Oswald; and C. G. Golueke (1968). *Kinetics of Algal Systems in Waste Treatment* (No. SERL Report No. 68-4). University of California, Sanitary Engineering Research Laboratory.

United States Environmental Protection Agency (1977). *Wastewater Treatment Facilities for Several Small Communities*, Washington, D.C.

Water Environment Federation (1992). *Design of Municipal Wastewater Treatment Plants*, Manual of Practice No. 8, Arlington, Virginia.

Wehner, J. F. and R. H. Wilhem (1958). "Boundry conditions of flow reactor." *Chemical Eng. Science* 6, p. 89.

7.12 PROBLEMS

7.1. An industry has a soluble wastewater that contains a BOD_L of 2,000 mg/l. They wish to produce an effluent BOD_L of 1,000 mg/l. Pilot studies showed that the appropriate kinetic parameters are:

$$\hat{q} = 27 \text{ mg } BOD_L/\text{mg } VSS_a\text{-d}$$
$$K = 10 \text{ mg } BOD_L/\text{l}$$
$$b = 0.2/\text{d}$$
$$Y = 0.5 \text{ mg } VSS_a/\text{mg } BOD_L$$
$$f_d = 0.8$$

The industry wants to treat the wastewater with an aerated lagoon, which can be considered a chemostat with $\theta = 1$ d. Will they likely meet the desired effluent quality if they supply adequate O_2? Recall that the effluent BOD_L will be comprised of organized substrate, active cell mass, and products. About how much aerator capacity is needed (in kW/1,000 m^3 of tank volume), if the field oxygen transfer efficiency is 1 kg O_2/kWh?

7.2. An industry has an aerated lagoon for treatment of its wastewater. The wastewater has the characteristics:

$$Q = 10^4 \text{ m}^3/\text{d}$$
$$S^0 = 200 \text{ mg BOD}_L/\text{l}$$
$$X_a^0 = 0$$
$$X_i^0 = 30 \text{ mg VSS}_i/\text{l}$$

The lagoon is well-aerated and has a total volume of 4×10^4 m^3. Tracer studies have shown that the lagoon can be described as two CSTRs in series. Therefore, assume that the lagoon is divided into two CSTRs in series. Compute S, X_a, and X_i in the effluent from both CSTRs. Use the following parameters:

$$\hat{q} = 5 \text{ mg BOD}_L/\text{mg VSS}_a\text{-d}$$
$$K = 350 \text{ mg BOD}_L/\text{l}$$
$$b = 0.2/\text{d}$$
$$Y = 0.40 \text{ mg VSS}_a/\text{mg BOD}_L$$
$$f_d = 0.8$$

7.3. You are to design an aerated lagoon to treat 10,000 m^3/d of wastewater under the following conditions:

$$S^0 = 150 \text{ mg/l BOD}_5$$
$$T(\text{summer}) = 27 \,°\text{C (ambient)}$$
$$T(\text{winter}) = 7 \,°\text{C (ambient)}$$
$$T(\text{wastewater}) = 15 \,°\text{C}$$
First-order BOD$_5$ rate coefficient $= 2.0/\text{d}$ at 20 °C
Temperature coefficient for BOD$_5$ removal $= 1.07$
Lagoon depth $= 2.5$ m
Lagoon volume $= 100,000$ m^3
$$k_h = 0.5 \text{ m/d}$$

Use the following design steps:

(a) Calculate the surface area.
(b) Determine the summer temperature. Let the effluent flow equal the influent flow.
(c) Determine the winter temperature.
(d) Calculate the effluent concentration of the BOD$_5$ using a first-order model for the summer.
(e) Calculate the effluent BOD$_5$ concentration for the winter.
(f) If $Y = 0.7$ g VSS$_a$/g BOD$_5$ and $b = 0.1$/d at 20 °C, what are the effluent concentrations of VSS for the summer and winter? (Let $f_d = 1$ for simplicity.)

(g) What are the O_2 requirements (kg/d) for each season?

(h) Mixing controls the power requirements for the aerators. What is the power required if the mixing energy is 3 kW/1,000 m^3?

7.4. An aerated lagoon is sometimes modeled as a chemostat. However, evaporation from the large surface area can create a significant loss of water, but not suspended or dissolved components. You are to derive the expressions for the steady-state concentrations S and X_a for a completely mixed lagoon that has normal Monod kinetics and parameters, an influent soluble BOD_L concentration of S^0, no input VSS, an influent flow rate of Q, and an evaporation rate of Q'. Will S and X_a be smaller or larger when the evaporative flow rate is significant?

7.5. Design a stabilization lagoon for a cold seasonal climate when the flow rate is $3.8 \cdot 10^3$ m^3/d and the BOD_5 is 200 mg/l. Use these steps:

(a) Select a reasonable depth.

(b) Calculate the surface area based on a reasonable BOD_5 surface load.

(c) Calculate the volume and the hydraulic detention time.

(d) Calculate the volumetric loading (kg BOD_5/1,000 m^3-d).

7.6. Repeat problem 7.5 for a temperate climate.

7.7. Repeat problem 7.5 for a tropical climate.

7.8. The normal ratio of P removal to BOD_L removal in activated sludge is about 0.01 g P/g BOD_5. Wetlands treatment sometimes shows substantially greater ratios. Explain the most likely mechanism for this extra P removal in wetlands treatment.

7.9. Size a wetland-treatment lagoon. The flow is 400 m^3/d and has a BOD_5 of 180 mg/l. Give the surface area required, the depth, and the detention time.

7.10. You are to provide a wetlands lagoon for a small community in southern California. The key data are:

Population:	2,000 people
Per capita flow:	0.5 m^3/d
Per capita BOD_5:	0.09/d
Possible macrophyte:	water primrose

Your design should give the surface area, depth, length, width, and detention time.

7.11. In wetlands treatment using water hyacinths, the depth often is only about 0.5 m, instead of the more typical 1.5 m. Explain why 0.5 m might be better or worse for wetlands treatment with water hyacinths.

chapter

8

AEROBIC BIOFILM PROCESSES

Aerobic biofilm processes can accomplish the same traditional treatment goal as activated sludge—BOD oxidation. While the bacteria predominating in biofilm processes probably differ from those in activated sludge at the species and strain levels, they are very similar in terms of function and genera. This similarity is inevitable, because the microorganisms utilize the same electron donors and acceptors and are exposed to the same kind of environmental conditions, such as temperature, nutrients, and solids retention time. Furthermore, the process achieves very similar effluent quality.

The major difference between a biofilm process and activated sludge is the means of biomass retention and accumulation. Both approaches rely on natural biomass aggregation. Whereas floc formation, settling, and recycle are the mainstays of activated sludge, attachment to solid surfaces is the basis for biofilm accumulation. In some circumstances, the attachment mode offers advantages that can be exploited to give biofilm processes performance and cost benefits.

In response to these potential benefits, the late 1980s and the 1990s had major research and development activities in "new" biofilm-reactor systems and applications. On the systems side, ranges of small-granule fixed-beds, fluidized beds, and hybrid suspended/biofilm processes were developed for the traditional applications in wastewater treatment. These newer processes are reviewed throughout this chapter, which focuses on wastewater treatment. On the applications side, new biofilm processes have been developed for the treatment of drinking water and gas streams contaminated with volatile organic compounds. While design for these new applications is based on the same principles as is design for wastewater treatment, the new applications have unique aspects. Drinking water treatment is thoroughly addressed in Chapter 12, while treatment of gases is discussed in Chapter 15.

Table 8.1 summarizes the critical kinetic parameters for biofilm treatment of BOD. The values given in Table 8.1 are generic ones useful for benchmarking the design and performance of a range of processes. These generic values do not take into account site-specific effects of temperature, medium texture, presence of inhibitors, pH, or other factors that could increase or decrease substrate flux. The design and analysis techniques presented in Chapter 4 can be applied to take into account such site-specific factors. In particular, the J_R values provide good benchmarks to judge whether the performance of a system should be characterized as a low-load or high-load process. The student should review the section on Normalized Surface Loading in Chapter 4 if the meanings of J_R, low load, and high load are not familiar. The examples in Chapter 4 also provide insight into how biofilm processes behave and how to perform detailed kinetic analyses.

Table 8.1 Fundamental and derived parameters for aerobic biofilm treatment of BOD

Parameter	Value[a]
Limiting Substrate[b]	BOD_L
Y, mg VSS_a/mg BOD_L	0.45
$\hat{q}$, mg BOD_L/mg VSS_a-d	10
K, mg BOD_L/l	1.0
b, d^{-1}	0.1
b_{det}, d^{-1}	0.1
b', d^{-1}	0.2
θ_x, d	10
D, cm^2/d	0.2[c]
D_f, cm^2/d	0.16
X_f, mg VSS_a/cm^3	40
L, cm	0.0029
S_{min}, mg BOD_L/l	0.047
S_{min}^*	0.047
K^*	0.27
J_R, mg BOD_L/cm^2-d	0.033

Notes:
[a] These generic values are for $T = 15\ °C$.
[b] The limiting substrate is complex BOD_L.
[c] Assumes that the BOD_L has a formula weight averaging 1,000 g/mol.

8.1 BIOFILM PROCESS CONSIDERATIONS

The oxidation of BOD by biofilm processes dates back to the beginning of the twentieth century, when rock-media trickling filters were introduced. The advent of plastic media allowed biological towers and rotating biofilm contactors to challenge rock trickling filters beginning in the 1960s and 1970s. Then, in the late 1980s and early 1990s, a range of innovative processes was created for the purpose of significantly improving some aspect of performance over what was generally achievable with trickling filters, biological towers, and rotating biological contactors. Now, an engineer has a very wide choice of biofilm systems from which to choose for a new design. At the same time, he or she also must deal with the operation of existing facilities that run from rock trickling filters through the newest innovative system.

Despite a large degree of diversity within biofilm processes, certain features remain remarkably consistent. Paramount among these common features is the BOD

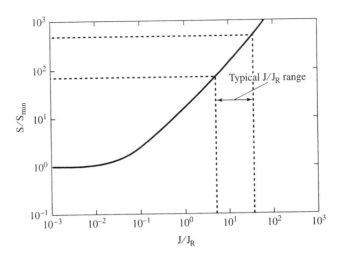

Figure 8.1   Normalized loading-curve for "generic" aerobic BOD oxidation ($S_{min} = 0.047$, $S^*_{min} = 0.047$ mg/l, $K^* = 0.27$ and $J_R = 0.033$ mg/cm^2-d) showing that the typical J/J_R values for wastewater treatment are in the high-load region.

flux employed to achieve satisfactory effluent quality for wastewater treatment. For almost all biofilm processes, the steady-state BOD flux falls in the range of 2 to 10 kg BOD$_L$/1,000 m^2-d. This range places normal operations well into the high-load region, or when $J > J_R$. Figure 8.1 illustrates this region on the normalized loading curve for the parameters of Table 8.1.

Being in the high-load range means that the steady-state effluent concentration for the original substrate is sensitive to changes in J. Figure 8.1 illustrates this sensitivity. A change in J/J_R from 8.5 to 40, roughly a five-fold range, corresponds to a three-fold increase in S/S_{min}, from about 80 to about 400. At first glance, such sensitivity may seem unacceptable. However, three factors explain why such sensitivity is acceptable in typical wastewater practice. First, S remains relatively low, since S_{min} is small: For $S_{min} = 0.047$ mg BOD$_L$/l (Table 8.1), S varies from approximately 4 to 20 mg BOD$_L$/l, values below the usual effluent standard.

Second, much of the effluent soluble BOD$_L$ and COD are comprised of SMP, which do not vary so sensitively. Third, the total effluent BOD$_L$ and COD also include the volatile suspended solids, which depend more on the biofilm detachment rate. Thus, the choice of a workable J is not controlled directly by substrate concentration. Instead, economic and practical considerations are determinative (e.g., oxygen supply, medium clogging, and biofilm detachment).

Example 8.1 | **TEMPERATURE AND BIODEGRADABILITY EFFECTS**   Some treatment systems experience low temperatures or must handle recalcitrant organic materials. This example shows how these changes affect the overall performance of a single complete-mix biofilm reactor having a surface loading of 0.45 mg BOD$_L$/cm^2-d (4.5 kg BOD$_L$/1,000 m^2-d), a common value used in wastewater treatment. The influent substrate concentration is 200 mg BOD_L/l, and $aV/Q = 45$ m^{-1}. To account for slowly biodegradable organic material, K is increased from 1 mg BOD_L/l to 100 mg BOD_L/l. The lower temperature is 5 °C, for which all biological rate coefficients are reduced by 50 percent (i.e., $\hat{q}$ from 10 to 5 mg BOD_L/mg VSS_a-d, b from 0.1 to 0.05/d, $\hat{q}_{UAP}$ from 1.8 to 0.9 mg COD_P/mg VSS_a-d, $\hat{q}_{BAP}$ from 0.1 to 0.05 mg COD_P/mg VSS_a-d, and k_2 from 0.09 to 0.045 mg COD_P/mg VSS_a-d), and the diffusion coefficients are reduced to $D = 0.16$ cm^2/d and $D_f = 0.125$ cm^2/d. The other SMP parameters remain the same: $K_{UAP} = 100$ mg COD_P/l, $K_{BAP} = 85$ mg COD_P/l, and $k_1 = 0.12$ mg COD_P/mg BOD_L.

Table 8.2 summarizes the key derived kinetic parameters and the process performance determined by the methods of Examples 4.4 and 4.5. The main effect of treating a much more slowly biodegradable substrate is that S_{min} increases 100-fold; however, this is partly compensated by a decrease in relative mass-transport resistance (K^* increases 10-fold). The net effect is a major increase in effluent substrate concentration (S). Because SMP constitutes the majority of the effluent soluble organic material ($S + SMP$) and is relatively stable, the percentage increase in $S + SMP$ is only 28.5 percent, even though S increases by 154 percent. The amount of biomass attached and in the effluent declines somewhat, since J is lower.

The significant decrease in temperature increases S_{min} and S^*_{min}, but both remain low values. Soluble effluent quality slightly improves, even though the effluent substrate increases. This occurs because BAP production, which dominates the soluble organic material, is produced more slowly at the lower temperature. The lower decay rate allows greater biofilm accumulation, which gives a larger effluent VSS concentration. This result is for steady state only and does not reflect transient effects on effluent quality as the attached biomass ($X_f L_f$) gradually increases.

Table 8.2   Summary of process performance for Example 8.1

Parameter	Generic Case	Slowly Biodegradable	Low Temperature
T, °C	15	5	5
S_{min}, mg BOD_L/l	0.047	4.7	0.071
S^*_{min}	0.047	0.047	0.071
K^*	0.27	2.7	0.35
J_R, mg BOD_L/cm^2-d	0.033	0.33	0.03
J, mg BOD_L/cm^2-d	0.43	0.42	0.40
S, mg BOD_L/l	9.0	23	13
UAP, mg COD_P/l	2.8	2.7	4
BAP, mg COD_P/l	31	29	22
SMP, mg COD_P/l	34	32	26
$S + SMP$, mg COD/l	43	55	39
$X_f L_f$, mg VS_a/cm^2	0.97	0.9	1.4
$X_f L_f a$ mg VS_a/l	1950	1800	2800
X^e_v, mg VSS/l	49	45	70
$S + SMP + 1.42 \cdot X^e_v$, mg COD/l	112	119	139

8.2   TRICKLING FILTERS AND BIOLOGICAL TOWERS

Rock-medium trickling filters, first used in England in 1893, became widespread in the United States in the 1920s and remain in common use today. The filter rocks generally are 25 to 100 mm (1 to 4 inches) in major dimension and quite irregular in shape. Figure 8.2 gives close-up and panoramic views of the top of rock trickling filters. The depth of the rock bed typically is 1 to 2 m and is controlled by weight-bearing limitations of the supporting underdrain system. The specific surface of a rock filter is approximately 40 m^{-1}.

By the 1970s, plastic media began to replace rocks for new designs. The plastic media are much lighter, which allows filter depths up to 12 m. They also have much greater porosity (95 percent versus about 40 percent for rocks) and specific surface area (ca. 200 m^{-1}). Figure 8.3 shows a close-up of the most widely used type of plastic media, corrugated modules that are stacked inside a round or square tower. Due to the greater heights, trickling filters using plastic media often are termed *biological towers*.

Whether in the rock or tower form, all modern (or high-rate) trickling filters share these characteristics, which are shown schematically in Figure 8.4:

- The wastewater is applied in a "trickling" or three-phase condition in which the water (first phase) moves downward along the surfaces of an unsaturated porous medium (second phase). Air moves upward or downward as a third phase for the purpose of supplying oxygen.

- The water exiting the bottom of the trickling filter is routed to a settler, which is needed to reduce the concentration of effluent suspended solids and total BOD_L. Sludge, frequently called *humus,* is wasted through the settler underflow.

- Effluent (or occasionally filter effluent prior to clarification) is recycled to establish control over the hydraulic loading to the trickling filter. Recycle flow rates (Q^r) typically are 50 to 400 percent of the influent flow rate (Q).

In high-rate trickling filters, the *hydraulic loading* is defined as

$$\text{H.L.} = \frac{Q + Q^r}{A_{pv}} \qquad \textbf{[8.1]}$$

in which A_{pv} is the plan-view surface area of the trickling filter; for a circular filter, $A_{pv} = r^2\pi$, where r is the radius of the filter. The normal range of hydraulic load for a high-rate trickling filter is 10–40 m/d. In order to maintain allowable BOD surface loads (discussed below), the influent flow must be mixed with recycled effluent to achieve a hydraulic load in this range.

Although using effluent recycle to control the hydraulic loading has been applied successfully with high-rate trickling filters for decades, why it works is a subject of debate. Elevating the hydraulic loading to this normal range does four things that should enhance process performance. One, increased hydraulic loading causes the water-layer thickness to be larger, which increases the liquid hold-up and detention time and may also increase the amount of surface area that is wetted and active in biodegradation. Two, the greater hydraulic load tends to distribute the biofilm deeper

a.

b.

Figure 8.2 Close-up and panoramic view of the top of rock trickling filters.

into the bed, which should allow more surface area to be active in biodegradation and should reduce the clogging potential. Three, the higher shear stress may cause increased detachment of excess biomass, further reducing the potential for clogging. Four, an increased hydraulic loading probably increases the rate of oxygen mass transfer from the air phase.

The use of recycled effluent (instead of just more influent) to elevate the hydraulic loading creates two other effects that can improve process performance. First, recycling effluent dilutes the influent BOD at the filter inlet and may also increase the D.O. concentration. Therefore, the BOD/D.O. ratio in the filter is reduced, which makes oxygen limitation less likely. Second, the mixing of the influent and recycle flows dampens BOD-loading fluctuations and the resulting problems of BOD overload and D.O. depletion.

Along with the hydraulic load, the most important design parameter is the *BOD surface loading*, which is roughly equal to the substrate flux. It is computed from

$$\text{S.L.} = \frac{Q S^0}{A_{pv} h a} \qquad \textbf{[8.2]}$$

where h is the filter depth, S^0 is the influent BOD concentration, and a is the medium's specific surface area. Most trickling filters operate in the range of S.L. = 2–7 kg BOD_L/1,000 m^2-d (0.2–0.7 mg/cm^2-d). As described earlier in this chapter, this loading range puts the process into the high-load category. Lower loads generally

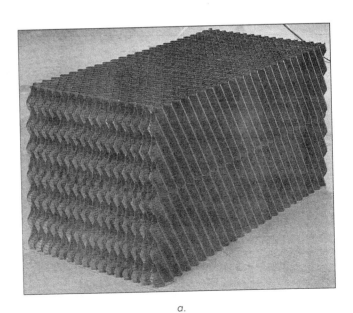

a.

Figure 8.3   Close-up of a module of corrugated plastic media.
SOURCE: With permission from Brentwood Industries.

b.

c.

Figure 8.3 Continued

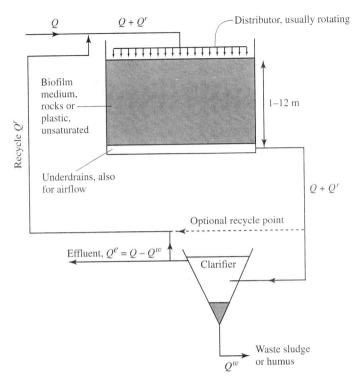

Figure 8.4   Schematic of a high-rate trickling filter system. Of particular importance are the definitions of the different flow rates.

are uneconomical. Higher loads appear to be infeasible due to practical limits on the oxygen transfer rate or biofilm detachment. Parker and coworkers estimate that the maximum practical oxygen-transfer rate is about 28 kg O_2/1,000 m^2-d; thus, the normal design range in BOD flux is safely below the maximum oxygen flux. Roughing filters take advantage of this "extra" oxygen-transfer capacity by increasing the surface loading up to 45 kg BOD_L/1,000 m^2-d and the hydraulic loading up to 200 m/d. They achieve much less BOD percentage removal and can be plagued by odors and excessive suspended solids in the effluent.

Of great economic importance is the *volumetric loading*,

$$\text{V.L.} = \frac{QS^0}{A_{pv}h} \qquad \text{[8.3]}$$

Typical volumetric loadings for high-rate trickling filters range from 0.3 to 1.0 kg BOD_5/m^3-d. Roughing filters can have much higher volumetric loads, up to 6 kg BOD_5/m^3-d. It is interesting that the 0.3–1.0 kg/m^3-d range is essentially the same as the volumetric loading for conventional activated sludge and explains how both systems coexisted as competing treatment alternatives for many decades.

The settler's function is to produce an effluent low in suspended solids. The settled solids are wasted, but not recycled to the trickling filter, because biofilm attachment provides the necessary biomass retention. The main design parameter is the *overflow rate,*

$$O/F = \frac{Q + Q^r - Q^w}{A_s} \qquad \text{[8.4]}$$

where Q^w is the waste-sludge flow rate and A_s is the plan-view surface area of the settler. The overflow rate should remain less than 48 m/d (1,200 gpd/ft^2) for peak flow. This value is similar to that used as a peak-flow criterion for conventional activated sludge. Due to the high effluent recycle rate, the ratio of peak to average overflow rates is substantially lower for trickling filters, compared to activated sludge.

The underdrain system for a trickling filter usually is constructed of vitrified-clay blocks that have slotted tops to simultaneously support the media weight and admit the water. The underdrains, which are laid directly on the filter floor, have a 1 to 2 percent gradient for water drainage to a central collection trench and are open around the filter's circumference in order to allow air ventilation.

Natural-draft trickling filters rely on a temperature difference between the ambient air and the air in the filter pores. The pressure head resulting from a temperature difference is

$$\Delta P_{\text{draft}} = 0.353 \left(\frac{1}{T_{\text{am}}} - \frac{1}{T_{\text{pore}}} \right) h \qquad \text{[8.5]}$$

in which ΔP_{draft} is the air-draft pressure head in centimeters of H_2O, T_{am} is the ambient air temperature in K, T_{pore} is the pore-air temperature in K, and h is the depth of the filter in meters. If $T_{\text{am}} > T_{\text{pore}}$, then $\Delta P_{\text{draft}} < O$, and the air flows downward. Conversely, the air flows upward when $T_{\text{pore}} > T_{\text{am}}$ and $\Delta P_{\text{draft}} > O$. The volumetric flow rate that results is the one for which the head losses at that rate just balance ΔP_{draft}. Guidelines for minimizing head losses in the design of the underdrains are given by the Water Environment Federation (1992).

Despite proper underdrain design, some trickling filters experience times when T_{am} and T_{pore} are so close that virtually no airflow occurs. Air stagnation tends to occur most in spring, when air and wastewater temperatures are similar. Lack of airflow can cause serious problems, including loss of treatment efficiency, massive losses of biofilm (i.e., sloughing), and odors. Forced-air ventilation sometimes is used to overcome stagnation or when the trickling filter is heavily loaded. A fan system to generate an air velocity of at least 0.3 m/min (1 ft/min) is required.

During periods of cold weather, large airflow (natural or forced) can be deleterious if evaporative cooling lowers the water temperature too much. Low temperatures reduce rates of biological activity and can lead to ice formation.

Trickling filters are susceptible to sudden, massive detachment of biofilm, called *sloughing* (recall Chapter 4). A major sloughing event causes a serious deterioration of effluent quality directly through the suspended solids discharge, and indirect impacts occur with the loss of biomass active in substrate removal. While the details of sloughing are far from well understood, sloughing events typically occur during periods of air stagnation, such as during the late spring, when air and water temperatures are similar. Presumably, anaerobic conditions in the biofilm cause a structural weakening (perhaps through local acid generation) that leads to fault planes that allow large "chunks" of biofilm to break free. Thus, maintaining adequate air ventilation is a key to preventing sloughing.

Albertson (1989) suggests that the typical way in which water is applied to the top of the trickling filter exacerbates sloughing problems. Based on work in Germany, Albertson recommends that the rotary distribution arms be slowed, which creates a strong "pulsed" hydraulic loading. This pulsing prevents excessive build-up of biofilm, which he believes is a key to sloughing. To achieve the proper degree of pulsing, Albertson recommends that the *flushing intensity* be in the range of 0.1 to 0.5 m/arm-revolution, with larger values needed for higher BOD loads. The flushing intensity, denoted SK for its German name *Spulkraft*, is computed by

$$SK = \frac{Q + Q^r}{A_{pv}n\omega} \cdot (d/1{,}440 \text{ min})$$
[8.6]

where

SK = flushing intensity, m/arm-revolution

n = numbers of distributor arms

ω = rotation speed in revolutions per minute (rpm).

Typical North American practice has SK values of 0.002 to 0.01 m/arm-revolution. To increase the SK value, the rotating speed of the distributor must be slowed. For example, a two-arm distributor ($n = 2$) with a hydraulic loading $[(Q + Q^R)/A_{pv}]$ of 30 m/d must have a rotating speed (ω) of 0.1 rpm to reach a SK of 0.1 m/arm-revolution.

Example 8.2 | **PERFORMANCE OF HIGH-RATE AND ROUGHING BIOLOGICAL TOWERS** The steady-state biofilm model is applied (following the steps in Examples 4.4 and 4.5) to compare the performance of typical high-rate and roughing biological towers. Both towers have a 10-m depth and are divided into three equal segments. The high-rate filter has a surface loading of 4.5 kg $BOD_L/1{,}000$ m^2-d, while the surface loading of the roughing tower is 45 kg/1,000 m^2-d. Each filter has a hydraulic load of 40 m/d or greater without recycle. The kinetic parameters of Table 8.1 and Example 8.1 (15 °C, easily biodegraded BOD) are used in both cases.

Tables 8.3 and 8.4 summarize the performance for each segment of the two towers. For a typical high-rate tower, the concentration of original substrate drops rapidly in the first segment and is negligible in the effluent, which is dominated by BAP. The biofilm accumulation and substrate flux have sharp gradients, which reinforce that most of the substrate removal occurs in the first segment. The oxygen flux also declines strongly, and its maximum value, 5.4 kg $O_2/1{,}000$ m^2-d, is far below the practical limit of 28 kg $O_2/1{,}000$ m^2-d.

The roughing tower has a quite different situation. The concentration of original substrate is reduced steadily, but remains high in the effluent. Although SMP concentrations are higher in the roughing tower than in the high-rate tower, the effluent quality is dominated by original substrate. Biofilm accumulation and substrate flux are high and relatively uniform throughout.

Table 8.3 Performance of the high-rate biological tower for Example 8.2

Parameter	Segment 1	Segment 2	Segment 3 = Effluent
S.L., kg BOD_L 1,000 m^2-d	------------------------4.5------------------------------		
H.L., m/day	------------------------40------------------------------		
V.L., kg BOD/m^3-d	------------------------0.9------------------------------		
S^0, mg BOD_L/l	------------------------225------------------------------		
S, mg BOD_L/l	30	3.2	0.3
UAP, mg COD_P/l	2.7	2.9	3.0
BAP, mg COD_P/l	28	33	33
SMP, mg COD_P/l	31	36	36
S+ SMP, mg COD_P/l	62	39	37
L_f, μm	790	110	12
$X_f L_f$ total, mg VS/cm^2	3.2	0.44	0.05
$X_f L_f$ active, mg VS_a/cm^2	2.6	0.37	0.04
J_S, kg BOD_L/1,000 m^2-d	11.7	1.6	0.17
J_{O2}, kg O_2/1,000 m^2-d	5.4	0.69	0.06
VSS, mg/l*	53	60	61

The values were computed using the approach of Examples 4.4 and 4.5 and with the parameters of Table 8.1.

*Assumes that biofilm detached in one segment does not reattach in a later segment.

Table 8.4 Performance of the roughing biological tower for Example 8.2

Parameter	Segment 1	Segment 2	Segment 3 = Effluent
S.L., kg BOD_L 1,000 m^2-d	------------------------45------------------------------		
H.L., m/d	------------------------100------------------------------		
V.L., kg BOD/m^3-d	------------------------9------------------------------		
S^0, mg BOD_L/l	------------------------900------------------------------		
S, mg BOD_L/l	580	340	180
UAP, mg COD_P/l	2.9	3.0	3.1
BAP, mg COD_P/l	41	66	81
SMP, mg COD_P/l	44	69	84
S+ SMP, mg COD_P/l	625	412	264
L_f, μm	5,200	3,900	2,600
$X_f L_f$ total, mg VS/cm^2	21	15	9.5
$X_f L_f$ active, mg VS_a/cm^2	16	13	8.8
J_S, kg BOD_L/1,000 m^2-d	48	36	24
J_{O2}, kg O_2/1,000 m^2-d	23	18	13
VSS, mg/l*	140	240	310

The values were computed using the approach of Examples 4.4 and 4.5 and with the parameters of Table 8.1.

*Assumes that biofilm that detaches in one segment does not reattach in a later segment.

The oxygen flux in the first segment is nearly 23 kg O_2/1,000 m^2-d, which approaches the practical maximum and signals the potential for oxygen limitation, odor development, and sloughing.

One of the biggest disadvantages of trickling filters, compared to activated sludge, has been their relatively higher concentration of effluent suspended solids. The cause seems to be that the eroding solids detaching from the biofilm on a continuous basis (not sloughing, which is periodic) are poorly flocculated. An approach to alleviate the flocculation problem is *solids contact*. A small tank for flocculation is included in the line from the trickling-filter underflow to the settler. This solids-contact tank can be a separate vessel, an open channel, or a center well in the settler. The most fundamental aspect is that it is gently aerated to provide time and particle contacts for the biological solids to form flocs. Liquid detention times range from 2 to 60 min, although 20 min is typical. In some cases, underflow solids from the settler are recycled to and mixed with the trickling-filter underflow to "seed" flocculation. Given such a floc-promoting environment, the detached bacteria and other organic colloids form aggregates large enough to settle well in the settler.

Rotating distributor arms are designed according to their pipe diameter and length. Increasing the arm length and diameter increases the flow rate obtainable. Table 8.5 summarizes typical capacities. Details of the flow capacity of a distributor can be obtained from the manufacturer. Actual flows obtained depend on the headloss that can be allowed. In some cases, *fixed-nozzle distributors* are employed. Special flat-spray nozzles are employed to ensure that the water is distributed uniformly across the bed for a range of flow rates per nozzle.

Table 8.5   Typical flow capacities for distributor arms according to diameter and length

Length of One Arm, m	Diameter of Arm, inches (cm)	Total Capacity (m^3/min) for Two Arms	Four Arms
13	3 (7.6)	0.7	1.4
13	4 (9.2)	1.2	2.4
13	5 (13)	1.9	3.8
13	6 (15)	2.7	5.5
25	8 (20)	4.7	9.5
25	10 (25)	7.6	15
30	12 (31)	11	21
30	14 (36)	13	26
30	16 (41)	17	34
30	18 (46)	22	44
30	20 (51)	27	54
38	24 (61)	42	84

SOURCE: Walker Process Corporation.

Low-rate or *intermittent trickling filters* are mainly of historic interest only, although some small installations remain in service. Low-rate trickling filters utilize low hydraulic loading (1–4 m/d), no effluent recycle, and intermittent dosing. These features lead to low volumetric loading (0.08–0.4 kg BOD_5/m^3-d). A dosing tank, which operates as a siphon, is used to ensure that the distributor arms rotate properly during application. Dosing tanks are small, resulting in a roughly 4-min dosing cycle for average flow conditions. Between dosings, no wastewater is applied. During low flows, the filter remains "dry" for extended periods until the dosing tank refills. Long dry periods—longer than an hour—can cause a deterioration of process performance, as the biofilm desiccates.

Since the 1950s, a series of mathematical design "formulas" have been derived. Some are strictly empirical, while others attempt to fit empirical results to a theoretically based model. The strictly empirical models can be useful when the design falls within the constraints of the original data. Two classic empirical formulas are NRC and Galler-Gotaas.

After World War II, a committee of the National Research Council (NRC) statistically analyzed operating data from 34 rock filters on military bases, which generated essentially domestic sewage. They derived Equations 8.7 and 8.8 to represent how the percent removal efficiency for BOD_5 varied with volumetric loading, recycle, and staging:

$$E_1 = \frac{100}{1 + 0.505(W_1/VF)^{1/2}} \qquad \textbf{[8.7]}$$

$$E_2 = \frac{100}{1 + \dfrac{0.505}{1 - E_1}(W_2/VF)^{1/2}} \qquad \textbf{[8.8]}$$

in which

E_1 = efficiency of first-stage BOD_5 removal
E_2 = efficiency of second-stage BOD_5 removal
W_1 = first-stage BOD_5 load, kg/d
W_2 = second-stage BOD_5 load, kg/d
V = filter volume, m^3
$F = 1 + Q^r/Q$ (F can be different for each stage)

In the 1960s, Galler and Gotaas statistically analyzed a much larger database (322 observations) of plant data for rock filters treating domestic sewage. The larger database and a more sophisticated multiple-regression analysis allowed Galler and Gotaas to include the effects of temperature, BOD concentration, and filter geometry, as well as the loading. The Galler-Gotaas formula describes the effluent BOD_5 concentration by

$$BOD_5 = \frac{K(vBOD_5^0 + v^r BOD_5)^{1.19}}{(v + v^r)^{0.78}(0.305 + h)^{0.67} r^{0.25}} \qquad \textbf{[8.9]}$$

in which

$$K = \frac{0.57}{v^{0.28} T^{0.15}}$$ [8.10]

and

v = influent hydraulic loading rate, m^3/m^2-d = m/d

v^r = recycle loading rate, m/d

BOD_5^0 = influent concentration of BOD$_5$, mg/l

BOD_5 = effluent concentration of BOD$_5$, mg/l

h = depth of filter, m

r = radius of filter, m

T = temperature, °C

Note that BOD$_5$ appears on both sides of Equation 8.9, which must be solved iteratively.

Velz, Holland, Eckenfelder, and many others have used a *first-order model* to provide a semi-empirical means of interpreting and extrapolating plant data for design. All of these first-order approaches are built upon the assumption that the removal of BOD is a first-order function,

$$\frac{d\,BOD_5}{dz} = -k_1\,BOD_5$$ [8.11]

in which z is the depth dimension in the filter and k_1 is a first-order rate parameter (T^{-1}). The various first-order formulas reflect different ways to explain how k_1 depends on hydraulic loading, temperature, wastewater characteristics, and medium characteristics. A common form is the Eckenfelder formula, which is given in English units in Equation 8.12 and converted to metric units for a typical situation in Equation 8.13:

$$BOD_5 = BOD_5^{in} \exp\left\{ \frac{-k_1 h^{(1-m)}}{\left(Q^i / A_{pv}\right)^n} \right\}$$ [8.12]

where

BOD_5^{in} = actual inlet concentration, mg BOD$_5$/l

$$= \frac{Q\,BOD_5^0 + Q^r\,BOD_5}{Q + Q^r}$$

BOD_5^0 = influent BOD$_5$ concentration, mg/l

A_{pv} = plan-view surface area, ft^2

$Q^i = Q + Q^r$, in gallons per min

h = depth, ft

k_1 = treatability factor, typically 0.088

m = slime distribution factor, $0 = m$ if evenly distributed

n = filter medium exponent, typically 1/3 to 2/3.

If $m = 0$, $n = 2/3$, and $k_1 = 0.088$, the Eckenfelder formula can be converted to metric units:

$$BOD_5 = BOD_5^{in} \exp \left\{ \frac{-4.36 \, (m^{-1/3} \, d^{-2/3}) \, h \, (m)}{\left(\dfrac{Q^i \, (m^3/d)}{A_{pv} \, (m^2)} \right)^{2/3}} \right\} \qquad \textbf{[8.13]}$$

These trickling-filter formulas can be useful for design and analysis, but they should be used with caution. The strictly empirical formulas are valid only for rock filters and sewage; they should not be used for plastic media or industrial wastewaters. The Eckenfelder formula can be an excellent tool for fitting plant data, especially since it has several calibration parameters (m, n, and k_1). However, extrapolating outside the tested operating range should be avoided. Finally, the BOD_5 values computed are composites that include, but do not distinguish among, all organic forms that contribute to BOD_5.

DESIGN WITH EMPIRICAL FORMULAS   For rock filters treating sewage, the NRC and Galler-Gotaas formulas can provide guidance for design and operation. This example illustrates use of the Galler-Gotaas formula, Equations 8.9 and 8.10. In particular, it shows how effluent recirculation affects effluent BOD. The sewage has $BOD_5^0 = 150$ mg/l and a flow rate of 1,000 m^3/d. The plan-view surface area is 750 m^2 (i.e., $r = 15.5$ m or 51 ft), giving an influent hydraulic loading of $v = (1,000 \, m^3/d)/750 \, m^2 = 1.33$ m/d. The recycle flow rates considered are 0, 1,000, 2,000, 3,000, and 6,500 m^3/d, which have v^r values of 0, 1.33, 2.67, 4.0, and 8.7 m/d, respectively. The temperature is $T = 15\,°C$. The filter depth is $h = 1.5$ m (5 ft).

From Equation 8.10,

$$K = \frac{0.57}{(1.33)^{0.28}(15)^{0.15}} = 0.35$$

The effluent S is then found from Equation 8.9 and the parameters in the proper units:

$$BOD_5 = \frac{0.35 \cdot (1.33 \cdot 150 + v^r \cdot BOD_5)^{1.19}}{(1.33 + v^r)^{0.78}(0.305 + 1.5)^{0.67}(15.5)^{0.25}}$$

This equation is solved iteratively for BOD_5 for each value of v^r. Table 8.6 summarizes

Example 8.3

Table 8.6   Summary of how the recirculation flow rate affects effluent BOD_5 according to the Galler-Gotaas formula

Recycle Ratio	v^r m/d	H.L., m/d	BOD_5, mg BOD_5/l
0	0	1.33	52
1	1.33	2.67	40
2	2.67	4.0	34
3	4.0	5.3	32
6.5	8.7	10	28

Notes:
S.L. = 3.3 kg BOD_5/1,000 m^2-d with $a = 40$ m^{-1}.
V.L. = 0.13 kg BOD_5/m^3-d

results, which show that increasing the recycle flow rate reduces effluent BOD_5. This result may appear to contradict the pattern that decreasing the plug-flow nature of a process increases the effluent concentration of substrate. The Galler-Gotaas equation, which was based on empirical results, probably reflects that effluent recycle improves the biomass distribution and/or mass transport rates.

Example 8.4

COMPREHENSIVE TRICKLING-FILTER SYSTEM DESIGN   In order to maximize the performance of a trickling-filter system, we wish to optimize the design of the filter itself and the settler. To do this, we select the following design criteria:

S.L. = 4 kg BOD_L/1,000 m^2-d

Medium: cross-flow plastic modules with $a = 200$ m^{-1}

Medium depth = 10 m

H.L. = 40 m/d at average flow

Solids-contact tank included with a detention time of 20 min

Settler O/F = 20 m/d

SK = 0.1 m/arm-revolution

These criteria are combined to provide a design for a wastewater having an influent BOD_L of 250 mg/l and a flow rate of 5,000 m^3/d (1.3 MGD) per filter system.

The surface load determines the total biofilm area and volume required.

$$\text{S.L.} = 4 \text{ kg BOD}_L/1{,}000 \text{ m}^2 \text{ d} = \frac{5{,}000 \text{ m}^3/\text{d} \cdot 0.25 \text{ kg BOD}_L/\text{m}^3}{aV}$$

or

$$aV = 3.125 \cdot 10^5 \text{ m}^2$$

and

$$V = 3.125 \cdot 10^5 \text{ m}^2/200 \text{ m}^{-1} = 1{,}560 \text{ m}^3$$

This gives an empty bed detention time of 1,560/5,000 = 0.31 d, or 7.5 h. The volumetric loading is then 0.8 kg BOD_L/m^3-d, a typical value (see text under Equation 8.3). The depth of 10 m gives a plan-view surface area of 156 m^2 and a radius of approximately 7 m.

To obtain a hydraulic loading of 40 m/d, the effluent recirculation flow (Q^r) must be large enough that

$$\text{H.L.} = 40 \text{ m/d} = \frac{5{,}000 \text{ m}^3/\text{d} + Q^r}{156 \text{ m}^2}$$

Solving for Q^r yields

$$Q^r = 1{,}240 \text{ m}^3/\text{d}$$

or a recirculation ratio of 25 percent. The solids contact tank receives the combined flow and should have a detention time of 20 min, or 0.0139 d. Thus,

$$V_{sc} = (5{,}000 + 1{,}240 \text{ m}^3/\text{d}) \cdot 0.0139 \text{ d}$$
$$= 87 \text{ m}^3$$

The settler also receives the combined flow. Its surface area is determined from the O/F of 20 m/d.

$$A_{settler} = \frac{(5{,}000 + 1{,}240 \text{ m}^3/\text{d})}{20 \text{ m/d}} = 312 \text{ m}^2$$

A circular settler with a radius of 10 m is satisfactory.

The rotating speed of the distributor in the trickling filter must be slow enough to an give $SK = 0.1$ m/arm revolution. From Equation 8.6,

$$SK = 0.1 \text{ m/arm–rev} = \frac{(5{,}000 + 1{,}240 \text{ m}^3/\text{d})}{156 \text{ m}^2 \cdot n \cdot \omega} \cdot \frac{d}{1{,}440 \text{ min}}$$

This gives that $(n \cdot \omega)^{-1} = 3.6$ min/rev. If n is selected as 4, then $\omega = 0.07$ rpm, or 4.2 revolutions per hour.

The total flow rate in m^3/min is 6,240 m^3/d/(24 h/d $\cdot$ 60 min/h) = 4.33 m^3/min. Manufacturers of distributor systems should be contacted to find a proper combination of distributor number, diameter, and length to give at least 4.33 m^3/min.

8.3 ROTATING BIOLOGICAL CONTACTORS

The development of lightweight plastic media led to an alternative to biological towers, the rotating biological contactor (RBC). RBCs were introduced in the 1960s, enjoyed substantial popularity in the 1970s, and lost favor in the 1980s as problems of early designs became evident. Today, RBCs are important biofilm processes in existing treatment facilities and are an option for new design.

As illustrated in Figure 8.5, an RBC has its biofilm attached to plastic media that rotate into and out of a trough of wastewater. The rotation causes turbulent mixing, circulation, and aeration of the liquid contents in the trough. Probably even more important is that the biofilm and its adhering water layer are exposed to O_2 when they are above the water level. The majority of oxygen transfer to the biofilm is attributable to the above-water exposure.

Figure 8.5 illustrates corrugated plastic media in a spiral format. Many other formats are used and are patented features for a given manufacturer. Despite different media formats, the specific surface area of the medium itself ranges from about 110 m^{-1} (standard density) to about 170 m^{-1} (high density). The high-density media, which have smaller channels for liquid and gas penetration, are more susceptible to clogging and should be used only with low loading (described below). For full-scale RBCs, the diameter of the medium is approximately 3.6 m (12 ft).

The medium is attached to a steel shaft (or axle) that is supported on bearings and rotated by a direct mechanical drive in most cases. (In a few cases, the trough is bubble aerated, and the bubbles are trapped in "cups" to create air-driven rotation.) Modules of medium are attached to the shaft to give a total medium length of approximately 8 m (26 ft). In fact, RBCs are generally sold in units of "shafts" of approximately 8-m length. A "shaft" of standard-density medium has approximately 9,300 m^2 (100,000 ft^2) of surface area, while a high-density shaft has approximately 14,000 m^2 (150,000 ft^2). Generally, the high-density media are used only in the later stages of an RBC system, when the BOD concentration has been reduced enough that the smaller pore openings are not clogged.

The medium is partly submerged in the water. A key operating parameter is the *percent submergence,* which is the percentage fraction of the medium diameter

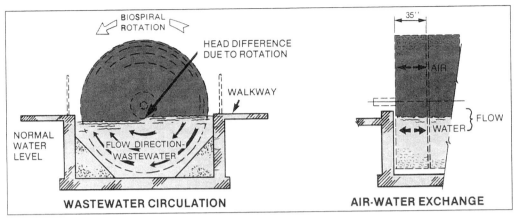

a.

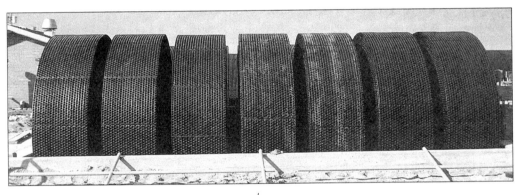

b.

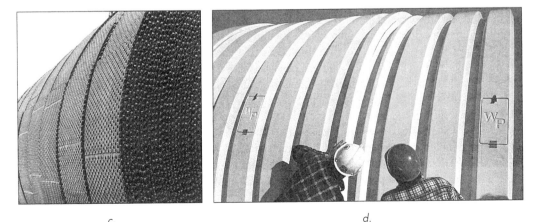

c.   *d.*

Figure 8.5   Key components of an RBC. *a.* Schematic of medium rotation and water circulation. *b.* A "shaft" of media modules. *c.* Left: Close-up of spiral-type media. Right: The fiberglass enclosure over the operating media. SOURCE: Walker Process Corporation.

(measured at the centerline) that is submerged. Typical percent submergence is 25–40 percent. An increased submergence increases the oxygen-transfer rate, but also requires greater energy to maintain a given rotating speed. The rotating speed for a full-scale unit normally is around 2 rpm. However, the most widely used criterion for setting the rotating speed is the *tip velocity*, (rpm) πD_m, in which D_m is the diameter of the medium. A common guideline for tip velocity is 20 m/min for a full-scale unit. Increasing the tip velocity increases the oxygen-transfer rate (the proportionality is roughly linear), but the energy requirement also increases.

One key to the long-term success of an RBC system is protecting it from the elements. Most RBC media are black in color due to the inclusion of carbon black with the plastic. Carbon black absorbs UV sunlight and helps protect the plastic from UV-induced deterioration. In most cases, the medium modules are enclosed in a fiberglass cover that shields the medium and biofilm from sunlight, protects the drive mechanism from the weather, reduces heat loss and icing during cold weather, and helps to control odors. Of course, the enclosures must be well ventilated to allow oxygen supply, structurally designed to withstand wind and snow loads, and equipped with access doors and inspection ports for maintenance.

RBC systems are almost always operated in a series mode with three to five stages. A settler follows to reduce effluent suspended solids. The settler normally is designed with an overflow rate of 16 to 32 m/d. In some cases, settlers do not provide a satisfactory effluent quality; chemical coagulation before sedimentation and the use of wire-screens ($< 30\text{-}\mu$m mesh) as an adjunct to sedimentation have proven successful for removing poorly flocculated solids.

RBC design easily can be based on biofilm kinetics, for which each stage is treated as a completely mixed biofilm reactor. The biofilm-kinetics approach should be supplemented with empirically derived design criteria from operating experience. Based on experience and consistent with biofilm fundamentals, an overall surface loading is in the range 3 to 15 kg soluble $BOD_5/1{,}000$ m^2-d, with a typical range of 5 to 8 in the same units. In order to avoid excessive growth, oxygen depletion, and odors in the first stage, a first-stage-loading limit of 45 kg total $BOD_5/1{,}000$ m^2-d is often employed as an auxiliary design criterion. Early designs frequently used a hydraulic loading (Q/A, where A is the medium surface area) as the key design loading. Typical values were 0.04 to 0.16 m/d. Hydraulic loading does not take into account the BOD flux and (correctly) has been abandoned.

Deficiencies in early RBC designs led to catastrophic failures that damaged the reputation of the RBC option. Fortunately, modern designs have overcome these early deficiencies. The early deficiencies included:

1. *Shaft and Bearing Failure.* Early designs were insufficient in their structural strength, because the designers did not realize that the accumulation of biofilm greatly increased the weight of the media. Therefore, the shafts cracked, the bearings wore out, and the media ripped off the shaft. These components are now designed to accommodate the added weight of the biofilm.

2. *Eccentric Growth.* If the RBC stopped its rotation, the submerged portion would continue to have growth, while the exposed portion would stop growth,

drain, and desiccate. Then, the eccentric weight distribution required too much torque to restart the rotation. Higher capacity motors and drives are used to overcome this problem.

3. *Nuisance Organisms.* Sulfate-reducing bacteria, fostered by too high loads and insufficient aeration, caused odors, heavy precipitates (S^0, FeS), filamentous growths, and other serious nuisances. Appropriate first-stage and overall surface loads, as well as the tip-velocity criterion, generally prevent nuisance organisms.

Example 8.5 | **RBC DESIGN**   An RBC system is to be designed for the same wastewater as in Example 8.4: $Q = 5,000$ m^3/d and $S^0 = 250$ mg BOD$_L$/l, where one-half of the BOD$_L$ is soluble. A three-stage system is to be employed with standard media in the first and second stages, but high-density media in the third stage. The first-stage maximum and overall soluble-BOD loading criteria are 66 and 9 kg BOD$_L$/1,000 m^{2}d, respectively. (We are assuming a BOD$_5$/BOD$_L$ ratio of 0.68). The tip speed is assumed to be 20 m/min. The settler has an overflow rate of 20 m/d for average flow. We are to determine how many "shafts" are needed for each stage, the rotating speed of the shaft, and the settler surface area.

The BOD$_L$ mass loading is 250 mg BOD$_L$/l · 5,000 m^3/d · 10^{-6} kg/mg · 10^3 l/m^3 = 1,250 kg BOD$_L$/d. The soluble BOD$_L$ loading is 625 kg BOD$_L$/d. To satisfy the first-stage maximum surface load, the first-stage should have an area of at least

$$A_1 = (1,250 \text{ kg BOD}_L/\text{d})/(66 \text{ kg BOD}_L/1,000 \text{ m}^2\text{-d}) = 19 \cdot 10^3 \text{ m}^2$$

To satisfy the overall surface loading, the total area must be

$$A_{\text{tot}} = 625 \text{ kg BOD}_{L(\text{sol})}/\text{d})/(9 \text{ kg BOD}_{L(\text{sol})}/1,000 \text{ m}^2\text{-d}) = 69 \cdot 10^3 \text{ m}^2$$

Dividing the total area by three gives an average of $23 \cdot 10^3$ m^2 per segment. In this case, the overall loading controls the area for the first stage.

At 9,300 m^2/shaft for standard media, the first and second stages require 2.5 shafts each. One means to allocate the shafts is to use three standard shafts for the first stage and two standard shafts for the second stage. Alternately, both stages could have three shafts.

The third stage uses high-density media with 14,000 m^2/shaft. This requires 1.64 shafts. The actual design then requires two high-density shafts for the third stage.

To maintain a tip speed of 20 m/min, the 3.6-m-diameter shafts require a rotating speed of

$$\omega = (20 \text{ m/min})/(\pi \cdot 3.6 \text{ m}) = 1.8 \text{ rpm}$$

The settler requires an overflow rate of 20 m/day, which gives a surface area of

$$A_{\text{settler}} = (5,000 \text{ m}^3/\text{d})/(20 \text{ m/d}) = 250 \text{ m}^2$$

This requires a 9-m radius for a circular settler.

Example 8.6 | **PREDICTED RBC PERFORMANCE**   The steady-state biofilm model, including SMP formation, is used to predict the performance of the RBC design in Example 8.5 ($Q = 5,000$ m^3/d and $S^0 = 250$ mg BOD$_L$/l). As in Example 8.2, the computational methods (following Examples 4.4 and 4.5) and parameters (Table 8.1) of Example 8.2 are used.

Table 8.7 summarizes how the water quality changes through the three stages. Whereas the original BOD dominates the soluble effluent quality in stage 1, BAP is dominant by stage 3.

Table 8.7 Performance summary for the RBC stages of Example 8.6

Parameter	Stage 1	Stage 2	Stage 3
No. Shafts	3	2	2
Surface area, m^2	27,900	18,600	28,000
S, mg BOD$_L$/l	100	44	12
UAP, mg COD$_p$/l	2.6	2.9	3.0
BAP, mg COD$_p$/l	23	31	36
SMP, mg COD$_p$/l	26	34	39
S + SMP, mg COD/l	126	78	51
L_f, μm	1,800	1,030	380
$X_f L_f$ total, mg VS/cm^2	7.25	4.1	1.5
J_S, kg BOD$_L$/1,000 m^2-d	27	15	5.6
J_{O2} kg O$_2$/1,000 m^2-d	12	7.1	2.7
VSS, mg/l	41	56	64
S + SMP + 1.42 VSS mg COD/l	184	157	142

The parameters are from Table 8.1, and the methods of Examples 4.4 and 4.5 were used to compute the values shown.

The first-stage BOD$_L$ and O$_2$ fluxes do not exceed the O$_2$ supply capacity. The effluent VSS concentrations illustrate why a settler is required.

APPARENT REMOVAL KINETICS FOR SOLUBLE BOD$_L$ Figure 8.6 plots | **Example 8.7**
ln ($S^0/(S+$SMP)) versus the cumulative relative surface area for the segments of the trickling filters in Example 8.2 and the RBC in Example 8.6. If the overall soluble effluent quality

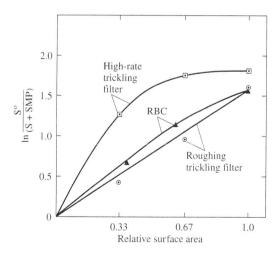

Figure 8.6 Comparison of the performance by stages for trickling filters and an RBC. A straight line indicates apparent first-order kinetics for soluble BOD$_L$.

behaved as though there were first-order removal, the data would be linear and extrapolate through the origin. Only the data for the roughing trickling filter approximate this pseudo first-order behavior. The other systems show a significant decrease in slope, which signifies a decrease in the rate of net-COD removal. The nonlinear responses occur because of the accumulation of SMP, which dominates effluent quality. On the other hand, the roughing filter's COD remains dominated by the original substrate. A close examination of the data for the roughing filter shows a slight increase in the slope (and apparent first-order rate) in the later segments. This increase in slope occurs because the substrate concentration is in or near the $\frac{1}{2}$-order range in the first segment and approaches first order in later segments.

8.4 GRANULAR-MEDIA FILTERS

The late 1980s and the 1990s saw the development of a set of biological filters that utilize a granular medium of gravel-sized clay particles. Called *Biolite,* the vitrified clay particles have an average dimension of approximately 4 mm, but are irregularly shaped. Although several process configurations are employed (described below), these granular-media filters share two important characteristics. First, the bed depth of approximately 3 m and the 4-mm medium size allow good particle filtration, as well as excellent biofilm accumulation. Thus, tertiary quality effluent (e.g., < 10 mg BOD$_5$/l and 5 mg SS/l) can be achieved without final clarification or tertiary filtration. Second, accumulations of biofilm and filtered particles require backwashing to preclude excessive headloss. Roughly daily backwashing is needed when the process is operating well.

Biolite filters are marketed in three configurations, which are shown schematically in Figure 8.7. The *Biocarbone* process, which is marketed as the Biological Aerated Filter (BAF) in North America, was developed by OTV in France and uses countercurrent flows of wastewater and air. Primary-treated wastewater is applied downflow. Air is sparged at approximately the two/thirds depth. Backwashing is upflow and with air scour. The average COD surface loading is 5–10 kg COD/1,000 m^2-d, yielding volumetric loads of 5 to 10 kg COD/m^3-d. Although the surface loads for *Biocarbone* are only slightly higher than for trickling filters, the volumetric loading are about 10 times greater, due to the much higher specific surface area of the *Biolite* medium (ca. 1,000 m^{-1}).

The second configuration, marketed by Dégremont of France, is called *Biofor.* Like *Biocarbone, Biofor* is a fully submerged filter with in-bed air sparging. Surface and volumetric loading also are similar. *Biofor* differs in that the wastewater and airflow are co-current. The co-current configuration is reported to give superior oxygen-transfer efficiency, since the oxygen is supplied at the location of maximum COD concentration, giving fewer problems with medium clogging.

Dégremont also markets *Biodrof,* which operates in a downflow, unsubmerged mode. A vacuum pump draws the air through the bed in the co-current direction. Thus, *Biodrof* effectively is a trickling filter with *Biolite* media. Biodrof is reported to have problems with medium clogging.

A newer form of granular-medium reactor, also developed by Dégremont, is *Biostyr. Biostyr* is an upflow, porous bed of expanded polystyrene beads, which are

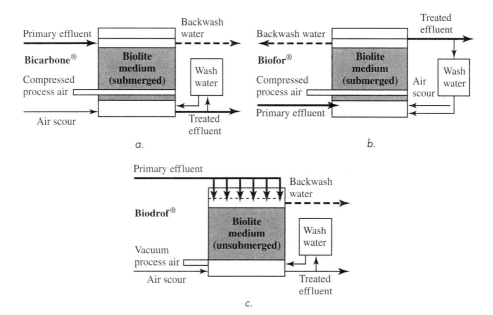

Figure 8.7 Schematics of three *Biolite* process configurations.

slightly lighter than water and float. They are retained in the reactor by a screen. The *Biostyr* beads are about 3.5-mm in size, and the bed height is around 3 m. Backwashing is required. Hydraulic loading is around 10 m/h, and the bed is aerated. COD surface and volumetric loadings are similar to *Biocarbone* and *Biofor*.

8.5 FLUIDIZED-BED AND CIRCULATING-BED BIOFILM REACTORS

The size benefit of biofilm reactors can be maximized by making the media as small as possible, which increases the specific surface area and the volumetric loading for a given substrate flux. However, sand-sized media (i.e., smaller than 1 mm) clog far too quickly when operated in a rapid-filtration mode.

A means to gain the surface-area advantages of small media, while avoiding the clogging drawback, is medium expansion, which greatly increases the liquid pore size. Concomitant with the pore-size enlargement are increases in the bed volume and porosity, which reduces the operating specific surface area. Recall that $a = 6(1 - \varepsilon)/d_p \Psi$, where ε is the porosity, d_p is the medium diameter, and Ψ is the shape factor. Thus, fluidized beds maximize a by making d_p small, but incur a countereffect as ε increases. In most cases, the former effect is dominant, resulting in substantial increases in a. For example, rock trickling filters have $a \approx 40$ m^{-1}, plastic-medium biological towers have $a = 100$–200 m^{-1}, *Biolite* granular filters have $a \approx 1,000$ m^{-1}, and a fluidized bed of 0.5-mm sand has $a = 2,000$–$10,000$ m^{-1}, depending on the bed expansion and particle shape. The gain in specific surface area is evident.

Expanded beds of small media can be achieved with two main process configurations: fluidized beds and circulating-bed reactors. A main difference between the two approaches is the density of the biofilm-carrier particles.

Fluidized beds work when the density of the carrier particles is greater than that of water. Carrier particles can range from sand or glass beads, which are much heavier than water (specific gravity of 2.5 to 2.65), to coal or activated-carbon particles, which are only slightly heavier than water (with specific gravity of 1.1 to 1.3). When friction from the upward water flow velocity first equals the negative buoyancy of the carrier particles, the bed expands. This is called incipient fluidization. Increasing liquid flow velocity causes the bed to expand further, although the friction loss on the carrier particles remains the same. This latter situation was used in Chapter 4 to develop the shear-stress relationship (Equation 4.34) for fluidized particles; it is repeated here as Equation 8.14 for convenience:

$$\sigma = \frac{(\rho_p - \rho_w)(1 - \varepsilon)g}{a} \qquad \textbf{[8.14]}$$

Clearly, heavier particles require a larger σ and a larger water velocity to create the frictionally induced buoyancy. As the bed expands and ε increases, the actual or interstitial upward flow velocity (v_{1i}) of the water approaches the superficial flow velocity, Q/A_{cs}:

$$v_{1i} = (Q/A_{cs})/\varepsilon_1 \qquad \textbf{[8.15]}$$

in which v_{1i} is the interstitial flow velocity of the liquid [LT^{-1}], Q is the total volumetric flow rate [L^{3}T^{-1}], A_{cs} is the cross-sectional area of the reactor [L^2], and $\varepsilon_1 =$ the liquid hold-up, or the liquid volume divided by the total volume in the fluidized bed. When ε_1 increases, Q/A_{cs} must increase proportionally in order to maintain the σ needed for particle fluidization. The relationship between ε_1 and Q/A_{cs} is generally given by the Richardson and Zaki (1954) correlation:

$$\varepsilon_1 = [(Q/A_{cs})/v_t]^{1/n} \qquad \textbf{[8.16]}$$

in which v_t is the terminal settling velocity for the carrier particles and n is the expansion index, which generally takes a value of approximately 4.5 for two-phase systems. The carrier terminal velocity can be computed from Stokes Law,

$$v_t = \left[\frac{4g(\rho_g - \rho_w)\, d_p}{3C_D \rho_w} \right]^{0.5} \qquad \textbf{[8.17]}$$

in which C_D is the drag coefficient computed iteratively from

$$C_D = \frac{24}{\text{Re}_t} + \frac{3}{\text{Re}_t^{0.5}} + 0.34 \qquad \textbf{[8.18]}$$

and

$$Re_t = \rho_w v_t d_p / \mu \qquad \text{[8.19]}$$

and μ is viscosity of water $[ML^{-1}T^{-1}]$. For further information on fluidization dynamics in general, the reader should consult Fair, Geyer, and Okun (1971), Darton (1985), or Yu and Rittmann (1997).

In a two-phase system, the liquid hold-up is easily computed from the bed height,

$$\varepsilon = 1 - \frac{H_{un}}{H}(1 - \varepsilon_{un}) \qquad \text{[8.20]}$$

in which ε_{un} is the liquid hold-up of the unexpanded bed, H_{un} is the height of the unexpanded bed, and H is the height of the expanded (or fluidized) bed. When the fractional bed expansion (FBE) is expressed by

$$FBE = 100\% \cdot (H - H_{un})/H_{un} \qquad \text{[8.21]}$$

the liquid hold-up is expressed as

$$\varepsilon = \frac{H_{un}}{H}\left(\frac{FBE}{100\%} + \varepsilon_{un}\right) \qquad \text{[8.22]}$$

The most common form of the fluidized-bed biofilm reactor is the *two-phase system*, which is shown schematically in Figure 8.8a. An effluent recycle line serves two crucial purposes. First, it controls the upward flow velocity (Q/A_{cs}) to maintain the proper degree of bed fluidization independently of the influent flow rate, which may be variable or too low to allow fluidization. Second, oxygen is transferred to the system via an oxygenator system built into the recycle line. These oxygenator systems often are patented devices.

An alternative form of fluidized-bed biofilm treatment is the *three-phase system*, which is illustrated schematically in Figure 8.8b. While three-phase systems usually maintain effluent recycle to control bed expansion, oxygen is supplied by direct air

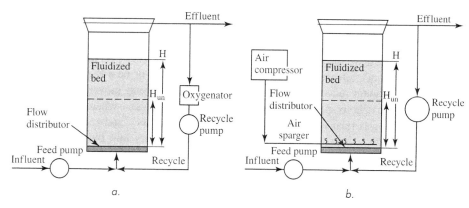

Figure 8.8   Fluidized-bed biofilm reactors. *a.* Two-phase system. *b.* Three-phase system.

sparging into the fluidized bed. Air sparging alters at least three aspects of fluidized bed operation. First, the presence of the gas phase reduces the liquid hold-up (ε), which decreases the liquid detention time and increases the water's interstitial velocity. Second, these effects of the gas phase modify how the solid carriers are expanded in response to the upward water flow. Although the response is complicated, bed aeration generally causes a decrease in the FBE for the same Q/A_{cs}. Chang and Rittmann (1994) and Yu and Rittmann (1997) discuss these interacting factors. Third, the input of the energy from aeration increases the bed turbulence, which can result in a significant increase in the biofilm detachment rate. Since aerobic heterotrophic systems have high biomass yields and growth potential (i.e., S_{min}^* is very low), the added detachment rate is not necessarily an impediment in terms of BOD removal and process stability, although effluent suspended solids may be increased.

Fluidized beds offer reduced volumes, due to their high specific surface area. Liquid detention times can be as low as a few minutes. The size advantage of fluidized beds generally is limited by the ability to transfer oxygen to the water and the biofilm. Thus, surface loads for fluidized beds may be somewhat lower than for the other biofilm systems. As a consequence, volumetric loads probably cannot be increased in direct proportion to the increase in specific surface area. However, the short liquid detention times of fluidized beds are particularly advantageous for aerobic treatment of low concentrations of contaminants, for which the oxygen demand is relatively low.

One operating problem that arises in some situations is *bed stratification,* which usually arises when the carrier particles are not sufficiently uniform in size. The smaller particles accumulate near the top and also experience a lower biofilm-detachment rate. Over time, these smaller particles accumulate more biofilm than do the larger particles, making them less dense, which increases their degree of fluidization. The problem is that the continued expansion of the stratified bed leads ultimately to entrainment of the carrier particles in the effluent and recycle flows. Using a highly uniform medium most effectively prevents bed stratification. Other control measures include designing a conical section at the top of the reactor to allow light particles to settle; installing a mechanical shear device, such as a propeller mixer, to detach the excess biofilm from the small particles; or withdrawing carriers from the top of the bed for cleaning.

When the medium is uniform, the carrier particles do not stratify, but instead circulate throughout the bed height. This medium mixing can provide a performance advantage when the substrate flux is in or near the low-load region and the effluent recycle ratio is not large. Then, medium movement allows each biofilm particle to spend some time near the column inlet—where the substrate concentration is relatively high and biofilm growth occurs—and some time near the column outlet—where the substrate concentration is very low, but the already accumulated biofilm can continue removing substrate. Movement of the biofilm particles disconnects substrate utilization and biofilm accumulation at any particular location. Therefore, biofilm grown at a concentration well above S_{min} near the inlet can remove substrate to well below S_{min} near the outlet (Rittmann, 1982). In this way, a steady-state-biofilm process can sustain effluent concentrations substantially below S_{min}, as long as the medium mixes, while the substrate concentration changes across the reactor.

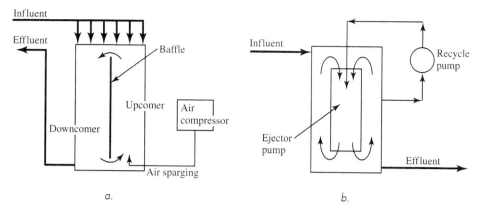

Figure 8.9 Circulating-bed reactors of the *a.* air-lift and *b.* ejector-mixed types.

Flow distribution is one of the most important engineering features of full-scale fluidized beds. Poor flow distribution allows short-circuiting and uneven bed expansion. Nozzles, lances, and underdrain systems typically are patented devices utilized to provide distribution for technical sized units.

In order to enhance oxygen transfer and to reduce the process footprint, full-scale fluidized beds generally are tall columns, up to 10 m in height. The performance implications of using such tall columns have not been fully explored. One effect of a large height-to-cross-section ratio is that the effluent recycle ratio can be low and still have a high enough flow velocity for bed expansion. The lower recycle ratio and the greater column distance work together to create larger substrate gradients along the column length.

An alternative to fluidized beds is the *circulating-bed biofilm reactor*. Figure 8.9 illustrates two types of circulating-bed reactor: an air lift reactor and an ejector-mixed reactor. Airlift reactors are subdivided into upflow and downflow sections. The upflow section receives aeration, which makes its effective density lower and creates the airlift pumping that drives liquid circulation. In an airlift biofilm reactor, light carrier particles are carried with the circulating water, creating a well-mixed biofilm distribution and a degree of complete mixing of the water contents. It is equally possible to create a strictly liquid-driven circulation pattern by using an external recycle to create an ejector pump inside the reactor.

In circulating beds, the carrier particles are smaller than 5 mm and have a density near that of water, although their specific gravity can be slightly greater than or less than 1. In some cases, the carriers can be porous with bacteria attached inside the pores, as well as on the outside. The solids hold-up is up to 40 percent, which is similar to a fluidized bed having an FBE of 50 to 100 percent.

CAPACITY OF A FLUIDIZED-BED BIOFILM REACTOR   We are designing a fluidized-bed biofilm reactor to treat a wastewater. Oxygenation of the recycle flow is used to provide oxygen without causing extra detachment from in-bed aeration. Previous pilot studies have

Example 8.8

yielded these substrate and microorganism parameters:

$$\hat{q} = 8 \text{ g/g VS}_a\text{-d}$$
$$K = 15 \text{ mg/l}$$
$$Y = 0.4 \text{ g VS}_a/\text{g}$$
$$b = 0.08/\text{d}$$
$$X_f = 30 \text{ mg VS}_a/\text{cm}^3$$
$$L = 60 \ \mu\text{m}$$
$$D = 1.6 \text{ cm}^2/\text{d}$$
$$D_f = 1.28 \text{ cm}^2/\text{d}$$

The medium particles are spherical and have a diameter of 1 mm, a density of 1.04 g/cm^3, and an unexpanded porosity of 0.3. We wish to operate the bed with an expanded porosity substantially greater than 0.3 in order to minimize detachment from abrasion (recall the discussion on detachment in the section on Estimating Parameter Values in Chapter 4). We use $\varepsilon = 0.46$.

The treatment goal is to have an effluent substrate concentration of 1 mg/l when treating a flow of 100 m^3/d with a bed volume (expanded) of 10 m^3. What is the maximum influent concentration that can be treated? We will assume that effluent recycle is large enough that we can treat the fluidized bed as a completely mixed biofilm reactor (CMBR). We need to draw extensively from information in Chapter 4.

First, we must compute the percent expansion. To do so, we need to know the original volume of the medium, to which we give the symbol V_{un}. Conservation of medium mass requires that

$$10 \text{ m}^3 (1 - \varepsilon) = V_{\text{un}}(1 - \varepsilon_{\text{un}})$$

or

$$V_{\text{un}} = 10 \cdot 0.54/0.7 = 7.7 \text{ m}^3$$

Therefore, the bed expansion is $((10 - 7.7)/7.7) \times 100\% = 30\%$.

Second, the specific surface area at 30 percent expansion is

$$a = \frac{6(1 - \varepsilon)}{d_p \psi} = \frac{6(1 - 0.46)}{1 \text{ mm (1)}} = 3.24/\text{mm} = 32.4/\text{cm} = 3,240/\text{m}$$

Third, the detachment is controlled by the shear stress when abrasion is unimportant. The shear stress in the fluidized bed is determined from Equation 4.34:

$$\sigma = \frac{(\rho_p - \rho_w)(1 - \varepsilon)g}{a} = \frac{(1.04 - 1.00 \text{ g/cm}^3)(1.00 - 0.46)980\text{cm/s}^2}{32.4/\text{cm}}$$

$$= 0.653 \text{ dynes/cm}^2$$

Thus, the specific detachment rate can be estimated as

$$b_{\text{det}} = 8.42 \times 10^{-2}(\sigma)^{0.58} = 8.42 \times 10^{-2}(0.653)^{0.58}$$

$$= 0.066/\text{d}$$

when $L_f \leq 30 \ \mu\text{m}$. (The actual biofilm thickness must be checked.)

The parameters needed to utilize the steady-state biofilm model are

$$b' = b + b_{det} = 0.08 + 0.066 = 0.146/\text{d}$$

$$S_{min} = Kb'/(Y\hat{q} - b') = 15 \cdot 0.146/(0.4 \cdot 8 - 0.146) = 0.72 \text{ mg/l } (< 1 \text{ mg/l})$$

$$S_{min}^* = S_{min}/K = 0.72/15 = 0.05 \text{ (high-growth potential)}$$

$$K^* = \frac{D}{L}\sqrt{\frac{K}{\hat{q}X_f D_f}} = \frac{1.6}{0.006}\sqrt{\frac{0.015}{8 \cdot 30 \cdot 1.28}} = 1.86$$

$$J_R = 0.284 \text{ mg/cm}^2\text{-d} \quad \text{(using Fig. 4.11)}$$

When we set $S = 1$ mg/l, then $S/S_{min} = 1.39$. J/J_R is estimated from the normalized loading curve with $S_{min}^* = 0.05$ and $K^* = 2$. $J/J_R = 0.12$, which gives $J = 0.034$ mg/cm²-d. A mass balance around the reactor (Equation 4.38) is

$$0 = Q(S^0 - S) - JaV$$

which yields S^0 when the other values are substituted: $S^0 = 12.0$ mg/l.

With a flux of 0.034 mg/cm²-d, the biofilm thickness (Equation 4.16b) is

$$L_f = YJ/b'X_f = 0.4 \cdot 0.034/0.146 \cdot 30 = 3.1 \cdot 10^{-3} \text{ cm} = 31 \ \mu\text{m} \approx 30 \ \mu\text{m}.$$

Thus, the computation of σ was adequate.

8.6 HYBRID BIOFILM/SUSPENDED-GROWTH PROCESSES

Activated sludge processes can be enhanced in terms of capacity and reliability through the introduction of biofilm media that create hybrid biofilm/suspended-growth systems. Existing activated sludge processes can be upgraded by adding biofilm surface area inside the aeration basin. Immobile biofilms include plastic screens, ribbons, or lace strings that are held inside the aeration basin by fixed frames immersed in the mixed liquor. Mobile biofilm carriers are mixed into and travel with the mixed liquor. They include sponges, plastic mesh-cubes or cylinders, porous cellulosic pellets, or polyethylene glycol pellets. The latter can be imbedded with bacteria, although this is not necessary to achieve the mobile-biofilm effect. The mobile biofilm carriers must be easily separated from the mixed liquor and held in the aeration basin. Screens or wire wedges at the aeration basin outlet are effective. When mobile carriers are used, the system can be called a *moving-bed biofilm reactor*.

Whether immobile or mobile biofilm carriers are employed, the goal is to increase the total mass of active bacteria in the system and to increase the SRT. The suspended and biofilm bacteria easily can have quite different SRTs. Normally, the biofilm SRT is longer, which makes it especially important for accumulating slow-growing species, such as nitrifiers. An example of the analysis of a hybrid process is given in Chapter 9.

Another hybrid process is the *Activated Biofilter (ABF) process* that was developed in the 1970s by Neptune-Microfloc. Figure 8.10 is a schematic of how the ABF process combines a biofilm reactor with suspended growth. The biofilm reactor is

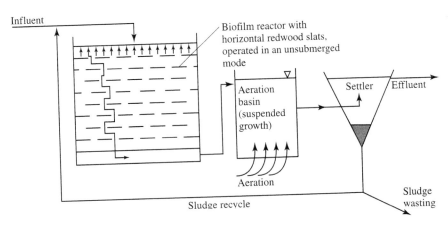

Figure 8.10   Schematic of the activated biofilter, a hybrid biofilm/activated-sludge process.

comprised of horizontal redwood slats. The applied water "trickles" down over the slats in an unsubmerged mode that allows oxygen supply from the air. This design accommodates the applied liquor, which contains suspended biomass recycled from the settler underflow. Between the biofilm reactor and the settler is an activated sludge aeration basin in which the MLVSS is maintained at around 2,500 mg/l. ABF practice generally fixes the BOD loading to the biofilm process at about 9 kg BOD_5/m^3-d, which is approximately 14 kg $BOD_L/1,000$ m^2-d. Although this surface loading is relatively large for a biofilm process, the subsequent aeration basin is used to ensure adequate BOD removal and nitrification, if desired. The size of the aeration basin is adjusted to achieve the desired treatment. With no nitrification, an aeration basin detention time of 1 to 2 h is typical, while nitrification generally requires 3 to 6 h.

A final hybrid process is the *PACT process,* which is an acronym for Powdered Activated Carbon Treatment. Originally developed by Zimpro to be accompanied by its wet-air oxidation of spent activated carbon, PACT is used most often for the biological treatment of industrial wastewater, which often contains organic chemicals inhibitory to bacteria. Powdered activated carbon (PAC) is added to the process influent at a dosage of 10 to 150 mg/l. Of course, the PAC builds up in the process with a θ_x/θ concentration factor. The primary purpose of the PAC is to adsorb inhibitory organic components. Experience also shows that the PAC can improve sludge flocculation and settling and adsorb slowly biodegraded components, such as BAP, that otherwise add to effluent BOD, COD, color, and toxicity to microorganisms. The main disadvantage of PACT is the high cost of the PAC.

8.7   BIBLIOGRAPHY

Albertson, O. E. (1989). "Slow down that trickling filter." *Operations Forum* 6(1), pp. 15–20.

Chang, H. T. and B. E. Rittmann (1994). "Predicting bed dynamics in three-phase, fluidized-bed biofilm reactors." *Water Sci. Technol.* 29(10–11), pp. 231–241.

Chang, H.T.; B.E. Rittmann; D. Amar; O. Ehlinger; and Y. Lesty (1991). "Biofilm detachment mechanisms in a liquid fluidized bed." *Biotechnol. Bioengr.* 38, pp. 499–506.

Darton, R. C. (1985). "The physical behavior of three-phase fluidized beds." In *Fluidization,* 2nd ed., J. F. Davidson, R. Clift, and D. Harrison, Eds. London: Academic Press, pp. 495–528.

Eckenfelder, W. W., Jr. (1966). *Industrial Water Pollution Control.* New York: McGraw-Hill.

Fair, G. M; J. C. Geyer; and D. A. Okun (1971). *Water and Wastewater Engineering.* New York: John Wiley.

Heijnen, J. J.; M. C. M. van Loosdrecht; R. Mulder; R. Weltevrede; and A. Mulder (1993). "Development and scale up of an aerobic biofilm air-lift suspension reactor." *Water Sci. Technol.* 27(6), pp. 253–261.

Lazarova, V. and J. Manem (1994). "Advances in biofilm aerobic reactors ensuring effective biofilm control." *Water Sci. Technol.* 29(10–11), pp. 345–354.

Metcalf and Eddy, Inc. (1979). *Wastewater Engineering: Treatment. Disposal. Reuse.* New York: McGraw-Hill.

National Research Council (1946). "Trickling filters in sewage treatment at military installations. " *Sewage Works J.,* Vol. 18, No. 5.

Richardson, J. F. and W. N. Zaki (1954). *Trans. Inst. Chem. Engr.* 32, p. 35.

Rittmann, B. E. (1982). "Comparative performance of biofilm reactor types." *Biotechnol. Bioengr.* 24, pp. 1341–1370

Rittmann, B. E. (1987). "Aerobic biological treatment." *Environ. Sci. Technol.* 21, pp. 128–136.

Rittmann, B. E.; R. Suozo; and B. Romero (1983). "Effect of temperature on oxygen transfer to rotating biological contactors." *J. Water Pollution Control Federation* 55, pp. 270–277.

Schroeder, E. D. and G. Tchobanoglous (1976). "Mass transfer limitations in trickling filter design." *J. Water Pollution Control Federation* 48(4), pp. 771–775.

Tanaka, K.; A. Oshima; and B. E. Rittmann (1987). "Performance evaluation of rotating biological contactor process." *Water Sci. Technol.* 19, pp. 483–494.

Trinet, F.; R. Heim; D. Mar; H. T. Chang; and B. E. Rittmann (1991). "Study of biofilm and fluidization of bioparticles in a three-phase liquid-fluidized-bed reactor." *Water Sci. Technol.* 23, pp. 1347–1354.

Water Environment Federation (1992). *Design of Municipal Wastewater Treatment Plants* Manual of Practice No. 8, Arlington, VA.

Yu, H. and B. E. Rittmann (1997). "Predicting bed expansion and phase holdups for three-phase, fluidized-bed reactors with and without biofilm." *Water Research* 31, pp. 2604–2616.

8.8 PROBLEMS

8.1. A rotating biological contactor can be considered to be a series of completely mixed biofilm reactors. In this case, assume that you have a system of three reactors in series. Each reactor has a total surface area of 10^4 m^2. The total flow is $1,500 \text{ m}^3/\text{d}$, and it contains 400 mg/l of soluble BOD_L. The necessary kinetic parameters are

$$K = 10 \frac{\text{mg BOD}_L}{1} \qquad D_f = 0.8 \frac{\text{cm}^2}{\text{d}} \qquad Y = 0.4 \frac{\text{mg VS}_a}{\text{mg BOD}_L}$$

$$L = 100 \ \mu\text{m} \qquad b_{\text{det}} = 0.10 \ \text{d}^{-1} \qquad \hat{q} = 16 \frac{\text{mg BOD}_L}{\text{mg VS}_a \cdot \text{d}}$$

$$D = 1.0 \frac{\text{cm}^2}{\text{d}} \qquad b = 0.10 \ \text{d}^{-1} \qquad X_f = 25 \frac{\text{mg VS}_a}{\text{cm}^3}$$

Estimate the steady-state effluent substrate concentration from each complete mix segment in the series. Assume no suspended reactions. This is an example requiring the steady-state-biofilm model.

8.2. You wish to design a fixed-film treatment process that will achieve an effluent concentration of 100 mg/l of BOD_L when the influent concentration is 5,000 mg BOD_L/l. You know that your process will be a fixed-bed process that will have mixing characteristics intermediate between plug flow and complete-mix. You use four reactor segments in series to describe the mixing. Reasonable kinetic parameters for the BOD utilization are

$$K = 200 \frac{\text{mg BOD}_L}{1} \qquad D_f = 0.4 \frac{\text{cm}^2}{\text{d}} \qquad Y = 0.2 \frac{\text{mg VS}_a}{\text{mg BOD}_L}$$

$$L = 150 \ \mu\text{m} \qquad b = 0.04 \ \text{d}^{-1} \qquad \hat{q} = 10 \frac{\text{mg BOD}_L}{\text{mg VS}_a \cdot \text{d}}$$

$$D = 0.5 \frac{\text{cm}^2}{\text{d}} \qquad X_f = 20 \frac{\text{mg VS}_a}{\text{cm}^3} \qquad b_{\text{det}} = 0.04 \ \text{d}^{-1}$$

$$f_d = 0.9$$

If you have a total flow of $1,000 \text{ m}^3/\text{d}$, approximately how much fixed-film surface area (m^2) is needed to achieve your goal? Use the normalized loading curves (Chapter 4) to solve this problem.

8.3. You are called as the high-priced expert because an engineering consultant cannot figure out if it is possible to design and operate a fluidized-bed biofilm reactor (with effluent recycle) to achieve his client's objectives. Here are the facts:

(a) The industry must treat $1,000 \text{ m}^3/\text{d}$ of a wastewater that contains 5,000 mg/l of BOD_L.

(b) The goal is to reduce the original substrate BOD_L to 100 mg/l.

(c) The consultant has used chemostats to estimate kinetic parameters for the wastewater BOD_L at 20 °C:

$$\hat{q} = 3.5 \frac{\text{mg BOD}_L}{\text{mg VS}_a \cdot \text{d}} \qquad K = 30 \frac{\text{mg BOD}_L}{1}$$

$$b = 0.04 \ \text{d}^{-1} \qquad Y = 0.1 \frac{\text{mgVS}_a}{\text{mg BOD}_L}$$

(d) The actual wastewater BOD_L is a polymer and has a diffusion coefficient of $0.3 \text{ cm}^2/\text{d}$.

(e) The consultant found that the monomer of the wastewater polymer has exactly the same values of $\hat{q}$, K, b, and Y, but its diffusion coefficient is $1.2 \text{ cm}^2/\text{d}$.

(f) The consultant ran a pilot-scale fluidized-bed reactor. The reactor used 0.2 mm sand, $\rho_p = 2.65 \text{ g/cm}^3$, with an expansion of 25 percent (i.e., the height of the bed was 1.25 times its unexpanded height). The expanded porosity was 0.5. To effect fluidization to a 25-percent expansion, effluent recycle was practiced, such that the total superficial flow velocity for the column reactor was 95,000 cm/d. The pilot reactor had an unexpanded bed volume of 250 cm^3 and an unexpanded bed height of 50 cm. It treated a steady flow of 100 l/d of monomer-containing synthetic wastewater (5,000 mg/l of BOD_L). The measured effluent quality, 200 mg/l of BOD_L, did not meet the given goal.

(g) Your goal is to determine whether or not the same reactor configuration can meet the 100 mg/l goal for a full-scale design. You must decide if a design is feasible (even though the pilot unit did not meet the goal for the synthetic wastewater). If a design is feasible, you are to provide the unexpanded bed volume. If it is not feasible, you must explain to the client why the design will not work well enough. Certain things must remain the same in the full-scale design as they were in the pilot study:

- Same medium, type, and size (0.2-mm sand)
- Same bed expansion (25 percent)
- Same total superficial flow velocity (95,000 cm/d)
- Same temperature (20 °C)
- Same influent substrate concentration (5,000 mg/l)

However, you will be treating the real wastewater BOD_L, not the synthesized monomer wastewater. The higher-priced consultant that you hired suggested that you determine the following:

(a) For both reactor systems, estimate S_{min}. Note that you need to estimate b'. Assume $L_f < 30 \ \mu m$.

(b) Estimate $L(\rho_w = 1 \text{ g/cm}^3$, $\mu = 864 \text{ g} \cdot \text{cm}^{-1} \cdot \text{d}^{-1}$, and let $D_f^* = 0.8$ and $X_f = 50 \text{ mg/cm}^3$).

(c) Use the loading criteria concepts to estimate the expected performance for the pilot-scale reactor. Would you have expected the pilot unit to have the 100 mg/l effluent criterion? Did it perform the way you would have expected?

(d) Based upon your analysis so far and the loading criteria, is it feasible for a full-scale design to achieve the 100 mg/l goal?

(e) If the answer to d is yes, estimate the unexpanded volume needed to achieve the treatment goal. If the answer to d is no, explain why the analysis of the pilot study demonstrates that the design is not feasible.

8.4. An aerobic biofilm reactor may be limited in the maximum BOD_L flux possible by the maximum flux of oxygen that can be transferred to the biofilm. You wish to estimate what these upper oxygen and BOD_L fluxes may be when considering the design of a roughing filter. The problem here is to estimate the following.

(a) The maximum oxygen flux to a trickling filter biofilm (mg/cm^2-d) when the D.O. concentration at the biofilm surface is near the air-saturation value of 8 mg/l and the biofilm is deep for oxygen. The molecular diffusion coefficient for dissolved oxygen is 1.28 cm^2/d. Assume D_f is 80 percent of this value, and that K for dissolved oxygen is 0.3 mg/l. Obtain other values that you need to solve this problem from Table 8.1. You will need to determine $\hat{q}$ for dissolved oxygen by multiplying the $\hat{q}$ for BOD_L by f_e, which you may assume is 0.5.

(b) Compute the maximum BOD_L flux (mg/cm^2-d) that can be handled by the biofilm before D.O. flux becomes rate limiting. (Hint: This is related to D.O. flux by stoichiometry in the same way that the $\hat{q}$ values are related.)

(c) Compare your results in (b) with the statement about the maximum BOD_L loading for roughing filters.

8.5. You have a wastewater with a flow rate of 1,000 m^3/d and a BOD_5 concentration of 400 mg/l. Using typical hydraulic loading rates, how much surface area would be required for a high-rate trickling filter and an RBC? Why are the values so different? Calculate the required volumes based on typical volumetric loading. Why are the volumes different?

8.6. A single-stage trickling filter treats a wastewater with these design conditions:

$$Q = 4,000 \text{ m}^3/\text{d}$$
$$A \text{ (plan view)} = 100 \text{ m}^2$$
$$\text{Radius} = 5.6 \text{ m}$$
$$\text{Depth} = 5 \text{ m}$$
$$\text{Specific surface area of medium} = 100 \text{ m}^2/\text{m}^3$$
$$BOD_L^0 = 1,000 \text{ mg } BOD_L/\text{l}$$
$$BOD_5/BOD_L \text{ ratio} = 0.68 \text{ g } BOD_5/\text{g } BOD_L$$
$$Q' = 8,000 \text{ m}^3/\text{d}$$
$$T = 15\,°\text{C}$$

(a) Calculate the effluent BOD_5 and estimate the BOD_L using the NRC and Galler-Gotaas formulas.

(b) The maximum oxygen flux to the biofilm is 2.85 mg O_2/cm^2-d, and the oxygen flux can be estimated from

$$J_{O_2} = J_{BOD_L}(1 - 0.3 \text{ mg VSS/mg } BOD_L \cdot 1.42 \text{ mg } BOD_L/\text{mgVSS})$$

where 0.3 is the net yield (mg VSS/mg BOD_L). If you assume that the total removal rate of BOD_L is based on J_{O_2}, then what is the effluent BOD_L concentration?

(c) Compute the surface loading (kg $BOD_L/1{,}000$ m^2-d), the hydraulic load (m/d), and the volumetric load (kg BOD_L/m^3-d) for the situation in part (b). Are these "typical" values? Characterize the process.

8.7. You are to predict the effluent concentration of soluble BOD_L from a *Biocarbone* (BAF) process. The input is $Q = 100$ m^3/h and $S^0 = 200$ mg BOD_L (soluble)/l. The *Biocarbone* medium has a porosity of 0.4, a depth of 3 m, a size equivalent to 4.5-mm spheres. The filter has a total plan-view surface area of 20 m^2. You may assume that aeration is sufficient so that BOD_L is rate-limiting.

The bed had a degree of backmixing. Therefore, consider that it behaves as 3 1-m complete-mix sections in series. For kinetic analysis, you may assume that $S_{min} = 1$ mg BOD_L/l, $J_R = 0.001$ kg BOD_L/m^2-d $= 0.1$ mg/cm^2-d, $S^*_{min} = 0.05$, and $K^* = 1.0$. What will be the effluent concentration of the input BOD_L?

8.8. The solids retention time in an activated-sludge process is easily computed from a solids mass balance. However, this is not so easy in a fixed-film process. Another way to estimate θ_x is by the ratio of BOD_L consumption to D.O. consumption. For a submerged biofilter, you have these pieces of information:

Input $BOD_L = 22$ mg/l

Input D.O. $= 9$ mg/l

Output $BOD_L = 9$ mg/l

Output D.O. $= 0$ mg/l

Let $Y = 0.54$ mg VSS_a/mg BOD_L, $b = 0.05$/d, and $f_d = 1.0$. Also, no oxygen transfer occurs within the process. Estimate an average θ_x from the BOD_L and D.O. data.

8.9. Do problem 8.6, but with these conditions:

$Q = 6{,}000$ m^3/d

Plan-view surface area $= 100$ m^2

Radius $= 5.6$ m

Depth $= 2$ m

Specific surface area $= 40$ m^{-1}

$BOD^0_5 = 250$ mg BOD_5/l

BOD_5:BOD_L ratio $= 0.68$

$Q^r = 6{,}000$ m^3/d

$T = 18\,°C$.

8.10. Estimate how much RBC surface area is needed to treat a waste flow of 10,000 m^3/d when the influent BOD_5 is 400 mg/l and the desired effluent BOD_5 is 20 mg/l, 50 percent of which is suspended solids. How many *shafts* are needed when conventional RBC media are used? How many first-stage shafts are needed?

chapter

9

NITRIFICATION

Nitrification is the microbiological oxidation of NH_4^+-N to NO_2^--N and NO_3^--N. Because of its oxygen demand (up to 4.57 g O_2/g NH_4^+-N) and toxicity to aquatic macroorganisms, NH_4^+-N removal is a mandated process for some wastewaters. In addition, wastewater treatment that involves denitrification of NO_3^--N frequently requires nitrification to convert the input NH_4^+-N to NO_3^--N. Nitrification to remove NH_4^+-N from drinking-water supplies also is practiced in order to make the water *biologically stable* (discussed in Chapter 12) and to eliminate the free-chlorine demand that produces chloramines when free chlorine is desired.

This chapter first reviews the key biochemical and physiological characteristics of the unique bacteria that perform nitrification. These characteristics are responsible for the process design features that are essential for successful nitrification. Then, activated sludge and biofilm nitrification are described.

9.1 BIOCHEMISTRY AND PHYSIOLOGY OF NITRIFYING BACTERIA

The nitrifying bacteria are autotrophs, chemolithotrophs, and obligate aerobes. Each factor is crucial for understanding when nitrifiers can be selected and accumulated in a biological process. Being *autotrophs,* the nitrifiers must fix and reduce inorganic carbon. This is an energy-expensive process that is primarily responsible for nitrifiers having much smaller values of f_s^0 and Y than do the aerobic heterotrophs that always populate activated sludge and biofilm systems. Their *chemolithotrophic* nature makes f_s^0 and Y still smaller, because their nitrogen electron donors release less energy per electron equivalent than do organic electron donors, H_2, or reduced sulfur. Of course, the low Y value translates into a small maximum specific growth rate (μ_m) and a large θ_x^{min}. Therefore, nitrifiers are slow growers.

470

Nitrifiers are *obligate aerobes,* and they use O_2 for respiration and as a direct reactant for the initial monooxygenation of NH_4^+ to NH_2OH (hydroxylamine). The latter use of oxygen may be the reason why nitrifiers are relatively intolerant of low dissolved-oxygen concentrations; nitrifier catabolism is slowed by oxygen limitation at concentrations that have no effect on many heterotrophs.

Nitrification is a two-step process. In the first step, NH_4^+ is oxidized to NO_2^- according to the following energy-yielding reaction, which is normalized to one electron equivalent.

$$\frac{1}{6} NH_4^+ + \frac{1}{4} O_2 = \frac{1}{6} NO_2^- + \frac{1}{3} H^+ + \frac{1}{6} H_2O$$

$$\Delta G^{0\prime} = -45.79 \text{ kJ per e}^- \text{ eq}$$

[9.1]

The most commonly recognized genus of bacteria that carries out the first step is *Nitrosomonas;* however, *Nitrosococcus, Nitrosopira, Nitrosovibrio,* and *Nitrosolobus* are also able to oxidize NH_4^+ to NO_2^-. The ammonium-oxidizing nitrifiers, which all have the genus prefix *Nitroso,* are genetically diverse, but related to each other in the beta subdivision of the proteobacteria (Teske et al., 1994). This diversity suggests that neither the *Nitrosomonas* genus nor any particular species within it (e.g., *N. europa*) necessarily is dominant in a given system.

The second stage of the nitrification reaction is the oxidation of NO_2^- to NO_3^-:

$$\frac{1}{2} NO_2^- + \frac{1}{4} O_2 = \frac{1}{2} NO_3^-$$

$$\Delta G^{0\prime} = -37.07 \text{ kj per e}^- \text{ eq}$$

[9.2]

Although *Nitrospira, Nitrospina, Nitrococcus,* and *Nitrocystis* are known to sustain themselves from the second-stage reaction, *Nitrobacter* is the most famous genus of the NO_2^- oxidizers. Within the *Nitrobacter* genus, several subspecies are distinct, but closely related genetically within the alpha subdivision of the proteobacteria (Teske et al., 1994). Recent findings using oligonucleotide probes targeted to the 16S rRNA of *Nitrobacter* indicate that *Nitrobacter* is not the most important nitrite-oxidizing genus in most wastewater-treatment processes. *Nitrospira* more often is identified as the dominant nitrite oxidizer.

Since nitrifiers exist in environments in which organic compounds are present, such as in wastewater treatment plants, it might seem curious that they have not evolved to use organic molecules as their carbon source. While the biochemical reason that organic-carbon sources are excluded is not known, the persistence of their autotrophic dependence probably is related to their evolutionary link to photosynthetic microorganisms (Teske et al., 1994).

Tables 9.1 and 9.2 summarize the basic and derived parameters that describe the stoichiometry and kinetics of ammonium oxidizers and nitrite oxidizers, respectively. Rittmann and Snoeyink (1984) critically reviewed the available parameters and selected the most reliable values for the basic parameters. These parameters succinctly summarize the key features of nitrifiers that must be taken into account in design and

Table 9.1   Basic and derived parameter values for ammonium oxidizers at 5 to 25 °C

	Parameter Values at T =				
Parameter	5 °C	10 °C	15 °C	20 °C	25 °C
f_s^0	0.14	0.14	0.14	0.14	0.14
Y, mg VSS_a/mg NH_4^+-N	0.33	0.33	0.33	0.33	0.33
$\hat{q}_n$, mg NH_4^+-N/mg VSS_a-d	0.96	1.3	1.7	2.3	3.1
$\hat{q}_{02}$, mg O_2/mg VSS_a-d	2.9	3.8	5.1	6.8	9.2
$\hat{\mu}$, d^{-1}	0.32	0.42	0.58	0.76	1.02
K_N, mg NH_4^+-N/l	0.18	0.32	0.57	1.0	1.50
K_O, mg O_2/l	0.50	0.50	0.50	0.50	0.50
b, d^{-1}	0.045	0.060	0.082	0.11	0.15
$[\theta_x^{min}]_{lim}$, d	3.6	2.8	2.1	1.5	1.2
$S_{min\,N}$, mg NH_4^+-N/l	0.029	0.053	0.094	0.17	0.26
$S_{min\,O}$, mg O_2/l	0.081	0.083	0.084	0.085	0.085

Table 9.2   Basic and derived parameter values for nitrite oxidizers at 5 to 25 °C

	Parameter Values at T =				
Parameter	5 °C	10 °C	15 °C	20 °C	25 °C
f_s^0	0.10	0.10	0.10	0.10	0.10
Y_N, mg VSS_a/mg NO_2^--N	0.083	0.083	0.083	0.083	0.083
$\hat{q}_n$, mg NO_2^--N/mg VSS_a-d	4.1	5.5	7.3	9.8	13.
$\hat{q}_{02}$, mg O_2/mg VSS_a-d	4.2	5.6	7.5	10.1	13.5
$\hat{\mu}$, d^{-1}	0.34	0.45	0.61	0.81	1.1
K_N mg NO_2^--N/l	0.15	0.30	0.62	1.3	2.7
K_O mg O_2/l	0.68	0.68	· 0.68	0.68	0.68
b, d^{-1}	0.045	0.060	0.082	0.11	0.15
$[\theta_x^{min}]_{lim}$, d	3.5	2.6	1.9	1.4	1.1
$S_{min\,N}$, mg NO_2^--N/l	0.024	0.047	0.10	0.20	0.42
$S_{min\,O}$, mg O_2/l	0.11	0.11	0.12	0.11	0.11

operation of a nitrifying process. These parameters are consistent with autotrophic growth.

The first observation is that f_s^0 is very low for each group. Compared to the typical f_s^0 value of 0.6–0.7 for aerobic heterotrophs, nitrifiers conserve very few electrons in biomass. The low f_s^0 values translate directly to low Y values. While the numerical value of Y for the ammonium oxidizers might not appear to be much lower than for aerobic heterotrophs, the apparent closeness is an illusion caused by using different units in the denominator. In units of g VSS_a/g OD, Y for the ammonium oxidizers is approximately 0.1, compared to about 0.45 for heterotrophs.

The maximum specific growth rates of both organisms are low, with both of them less than 1/d at 20 °C. With such small values of $\hat{\mu}$, the limiting value of θ_x^{min} must be large: All values are greater than 1 d. On the other hand, nitrifiers are able to drive effluent NH_4^+ or NO_2^- concentrations to very low levels, since S_{min} values are well below 1 mg N/l. Thus, nitrification can be highly efficient, as long as the SRT is maintained well above θ_x^{min} and sufficient dissolved oxygen is present. The relatively high values of K_O quantify that nitrifiers are not tolerant of low D.O. concentrations. Continued operation with a D.O. below K_O will increase θ_x^{min} for the nitrifiers and can lead to washout, as well as a high NH_4^+-N concentration in the effluent.

The temperature effects shown in Tables 9.1 and 9.2 are quite important, because nitrification is sometimes considered impossible for low-water temperatures. In fact, stable nitrification can be maintained at 5 °C or lower, as long as the SRT remains high enough. For 5 °C, a safety factor of only 5 requires that θ_x be $3.6 \cdot 5 = 18$ d. One problem with low-temperature nitrification is that $\hat{\mu}$ becomes quite small, making recovery of nitrification after a washout a very slow process. Thus, avoiding nitrifier washout due to excess sludge wasting, low D.O., or inhibition must be an absolute priority, particularly for low temperatures.

The directly comparable parameters—such as f_s^0, $\hat{\mu}$, $[\theta_x^{min}]_{lim}$, and S_{min}—are very similar for the two types of nitrifiers. This circumstance is completely logical, since both are aerobic chemolithoautotrophs oxidizing N. Furthermore, they almost always coexist in the same habitats, experiencing the same SRT and oxygen concentrations. The correspondence of limiting parameters reflects their biochemical and ecological similarities.

The following equation is an overall, balanced reaction for the complete oxidation of NH_4^+ to NO_3^--N by nitrifiers having $\theta_x = 15$ d and $f_s = 0.067$. It represents a typical situation for both nitrifier types together.

$$NH_4^+ + 1.815\ O_2 + 0.1304\ CO_2 = 0.0261\ C_5H_7O_2N + 0.973\ NO_3^-$$
$$+ 0.921\ H_2O + 1.973\ H^+ \qquad \textbf{[9.3]}$$

Besides the low net formation of nitrifier biomass ($Y_{net} = 0.21$ g VSS_a/g N), this stoichiometric equation illustrates the two other important features of nitrification. First, nitrification creates a major oxygen demand. For this example, it is $1.815 \cdot 32/14 = 4.14$ g O_2/g NH_4^+-N consumed. Second, nitrification produces almost two strong-acid equivalents per mole of NH_4^+ removed. In common mass units, the alkalinity consumption is $1.973 \cdot 50/14 = 7.05$ g as $CaCO_3$/g NH_4^+-N. The first step, ammonium oxidation, is responsible for the acid production.

Nitrifiers produce soluble microbial products, which can be consumed by heterotrophic bacteria (Rittmann, Regan, and Stahl, 1994; deSilva and Rittmann, 2000a,b). It appears that most of the nitrifier-produced SMP are BAP. SMP generally are important in two ways. First, they are part of the decay process of the nitrifiers and reduce the net synthesis of the nitrifiers. Second, they are a way in which nitrifiers create electron donors for heterotrophs and increase the heterotrophic biomass.

Nitrifiers are reputed to be highly sensitive to chemical inhibition. This belief is partly accurate. The very slow growth rate of nitrifiers magnifies the negative impacts of inhibition and, in part, makes it appear that nitrifiers are more sensitive than are faster growing bacteria. Furthermore, some apparent inhibitors are electron donors whose oxidation depletes the D.O. and may cause oxygen limitation. However, nitrifiers are sensitive to inhibition from a range of organic and inorganic compounds. Among the most relevant ones are: unionized NH_3 (at higher pH), undissociated HNO_2 (usually at low pH), anionic surfactants, heavy metals, chlorinated organic chemicals, and low pH.

A controversial issue is *mixotrophy,* in which nitrifiers use an organic carbon source and an inorganic electron donor. Mixotrophy should significantly reduce the energy cost of cell synthesis, thereby increasing Y and $\hat{\mu}$. Some claim that almost all nitrifier strains are capable of mixotropic growth. Cases of mixotrophic growth show increased values of Y_{net} and $\hat{\mu}$, but not all results are consistent. Furthermore, culturing nitrifiers without heterotrophic contamination is extremely difficult. Although mixotrophy cannot be discounted, when studied carefully, nitrification processes behave in a manner consistent with strictly autotrophic anabolism.

9.2   COMMON PROCESS CONSIDERATIONS

A successful nitrification process—suspended growth or biofilm—must account for the reality that heterotrophic bacteria always are present and competing with the nitrifiers for dissolved oxygen and space. The nitrifiers' relatively high K_O value puts them at a disadvantage in the competition for oxygen. Their slow growth rate is a disadvantage when competing for any space that requires a high growth rate.

These two disadvantages are overcome by ensuring that the nitrifiers have a long SRT, typically greater than 15 d, although larger values may be needed in the presence of toxic materials, a low D.O. concentration, or low temperature. In activated sludge, maintaining a SRT of 15 d or greater corresponds to the loading condition termed *extended aeration.* Thus, maintaining an extended aeration loading usually is synonymous with having a nitrifying process. For biofilm processes, the BOD flux and the detachment rate indirectly control the nitrifiers' SRT.

Even with the relatively long SRT of extended aeration, nitrifying processes often have relatively small safety factors. The low safety factor occurs for economic reasons. When $[\theta_x^{min}]_{lim}$ is 1 to 3 d, the reactor volume is too great when the safety factor is greater than about 10, because the θ_x/θ ratio cannot be increased indefinitely to compensate. Of course, operating with a small safety factor increases the risk of washout due to solids loss or inhibition and increases the needs for operator attention. Unfortunately, the risk is high, and instability in nitrification is a common problem in treatment operations.

9.3   ACTIVATED SLUDGE NITRIFICATION: ONE-SLUDGE VERSUS TWO-SLUDGE

One-sludge nitrification is the process configuration in which heterotrophic and nitrifying bacteria coexist in a single mixed liquor that simultaneously oxidizes ammonium and organic BOD. In other words, *one sludge* contains the nitrifiers and the

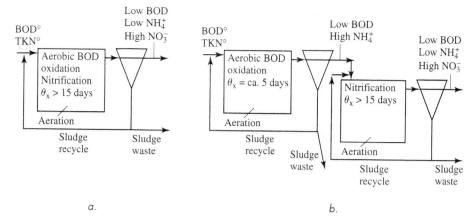

Figure 9.1 Schematics of the one-sludge *a.* and two-sludge *b.* approaches to nitrification with activated sludge.

heterotrophs, which carry out their metabolisms simultaneously. Figure 9.1*a* illustrates schematically the one-sludge approach, which has one reactor and one settler for all types of biomass. One-sludge nitrification also is called *one-stage nitrification,* because it has only one reactor stage. We use the term *one sludge* to emphasize the ecological relationship between the nitrifiers and heterotrophs: They are present together in one sludge, or one microbial community.

Two-sludge nitrification is an attempt to reduce the competition between heterotrophs and nitrifiers by oxidizing most of the organic BOD in a first stage, while the ammonium is oxidized in a second stage. Figure 9.1*b* shows how the two-sludge process is comprised of two complete activated-sludge systems in series. Because the biomass in each stage is captured and recycled solely within that stage, each stage develops its own biomass. From an ecological perspective, we have two different communities, one for each "sludge." Thus, we call it a *two-sludge process.* The first sludge is essentially free of nitrifiers, while the second sludge has a major fraction of nitrifiers, since all the nitrification and only a small amount of the BOD oxidation occur there. Two-sludge nitrification also is called *two-stage nitrification.*

One-sludge nitrification can be carried out in sequencing batch reactors, which involve sequential periods of filling, aerobic reaction, settling, and effluent draw-off in one tank. As for any SBR system, multiple units are almost always needed, in order that storage requirements for incoming wastewater are not excessive. In most ways, SBR nitrification resembles any other one-sludge system. However, the fill-and-draw-feeding scheme can create conditions resembling plug flow in a continuous feed system.

ONE-SLUDGE DESIGN   We are to make a preliminary design of a one-sludge nitrification system for a wastewater that has the following characteristics:

Example 9.1

$$Q = 500 \text{ m}^3/\text{d}$$
$$BOD_L^0 = 500 \text{ mg/l}$$
$$TKN^0 = 60 \text{ mg/l}$$
$$\text{Inert VSS}^0 = 25 \text{ mg/l}$$
$$T = 15\,°\text{C}$$

Note that the influent concentration is expressed as total Kjeldahl nitrogen (TKN), instead of NH_4^+-N. Frequently, wastewaters contain a significant fraction of their reduced nitrogen as organic nitrogen. In most systems active in nitrification, the organic nitrogen is hydrolyzed to form NH_4^+-N, which will be available for nitrification. Thus, the proper measure of influent reduced nitrogen is the TKN, not NH_4^+-N.

First, we must select a design θ_x. Because $[\theta_x^{min}]_{lim}$ is much larger for nitrification, we use an appropriate value for that process. Tables 9.1 and 9.2 indicate that $[\theta_x^{min}]_{lim}$ is approximately 2 d for $T = 15\,°\text{C}$. Applying a safety factor of 10 gives:

$$\theta_x = 10 \cdot 2 \text{ d} = 20 \text{ d}$$

Second, we compute the steady-state concentrations of BOD_L, NH_4^+-N, and NO_2^--N from typical kinetic parameters:

$$S = K \frac{1 + b\theta_x}{Y\hat{q}\theta_x - (1 + b\theta_x)}$$

$$BOD_L = 10 \frac{\text{mg BOD}_L}{1} \frac{1 + 0.1/\text{d} \cdot 20 \text{ d}}{0.45 \frac{\text{mg VSS}_a}{\text{mg BOD}_L} \cdot 10 \frac{\text{mg BOD}_L}{\text{mg VSS}_a - \text{day}} \cdot 20 \text{ d} - (1 + 0.1 \cdot 20)}$$

$$= 0.34 \text{ mg BOD}_L/1$$

$$NH_4^+\text{-N} = 0.57 \frac{\text{mg N}}{1} \frac{1 + 0.082/\text{d} \cdot 20 \text{ d}}{0.33 \frac{\text{mg VSS}_a}{\text{mg N}} \cdot 1.7 \frac{\text{mg N}}{\text{mg VSS}_a\text{-d}} \cdot 20 \text{ d} - (1 + 0.082 \cdot 20)}$$

$$= 0.18 \text{ mg NH}_4^+\text{-N}/1$$

$$NO_2^-\text{-N} = 0.62 \frac{\text{mg N}}{1} \frac{1 + 0.082/\text{d} \cdot 20 \text{ d}}{0.083 \frac{\text{mg VSS}_a}{\text{mg N}} \cdot 7.3 \frac{\text{mg N}}{\text{mg VSS}_a\text{-d}} \cdot 20 \text{ d} - (1 + 0.082 \cdot 20)}$$

$$= 0.17 \text{ mg NO}_2^-\text{-N}/1$$

Third, we compute the VSS production rates for all types of biomass. We must start with the *aerobic heterotrophs:*

$$\frac{\Delta X_v}{\Delta t} = Q(S^0 - S)Y \frac{1 + (1 - f_d)b\theta_x}{1 + b\theta_x}$$

$$= 500 \frac{\text{m}^3}{\text{d}} \left(500 - 0.34 \frac{\text{mg BOD}}{1}\right) 0.45 \frac{\text{mg VSS}_a}{\text{mg BOD}_L} \frac{1 + (1 - 0.8) \cdot 0.1/\text{d} \cdot 20 \text{ d}}{1 + 0.1 \cdot 20}$$

$$\cdot 10^{-3} \frac{\text{kg} - 1}{\text{mg-m}^3}$$

$$= 52.5 \text{ kg VSS/d}$$

For the ammonia-oxidizing nitrifiers, we need to adjust their S^0 to account for the NH_4^+-N assimilated into the aerobic heterotrophs and not available for nitrification.

$$N_1^0 = 60 \frac{\text{mg NH}_4^+\text{-N}}{\ell} - 52.5 \frac{\text{kg VSS}}{\text{d}} \cdot 0.124 \frac{\text{kg N}}{\text{kg VSS}} \cdot \frac{10^3 (\text{mg-m}^3)/(\text{kg-1})}{500 \text{ m}^3/\text{d}}$$

$$= 60 - 13 = 47 \text{ mg NH}_4^+\text{-N/1}$$

where N_1^0 is the available NH_4^+-level nitrogen for nitrification (mg N/l). Thus, a significant part of the input TKN is assimilated into heterotrophic biomass, leaving only about 80 percent of the input TKN to be nitrified. (Of course, this fraction depends on the SRT and the ratio of influent BOD_L to TKN.) The VSS production rate for the *ammonium oxidizers* is

$$\frac{\Delta X_v}{\Delta t} = 500 \cdot (47 - 0.18) \cdot 0.33 \frac{1 + (0.2) \cdot 0.082 \cdot 20}{1 + 0.082 \cdot 20} \cdot 10^{-3}$$

$$= 3.9 \text{ kg VSS/d}$$

Based on the lower yield and lower TKN available, production of ammonium-oxidizers is very much less than for heterotrophs. This is a typical situation.

For the *nitrite oxidizers,* we repeat the same pattern as for the ammonium oxidizers, where N_2^0 is the available NO_2^--N for the second step of nitrification.

$$N_2^0 = 60 - 13 - 3.9 \cdot 0.124 \cdot 10^3 / 500 - 0.18$$

$$= 45.8 \text{ mg NO}_2^-\text{-N/1}$$

$$\frac{\Delta X_v}{\Delta t} = 500 \cdot (45.8 - 0.17) \cdot 0.083 \frac{1 + (0.2) \cdot 0.082 \cdot 20}{1 + 0.082 \cdot 20} \cdot 10^{-3}$$

$$= 0.95 \text{ kg VSS/d}$$

Inert VSS passes through:

$$\frac{\Delta X_i^0}{\Delta t} = 500 \cdot 25 \cdot 10^{-3} = 12.5 \text{ kg VSS}_i/\text{d}$$

The total amount of VSS that must leave the system is the sum, or

$$\left(\frac{\Delta X_v}{\Delta t} \right)_{\text{total}} = 52.5 + 3.9 + 0.95 + 12.5 = 69.8 \text{ kg VSS/d}$$

Fourth, we compute the oxygen requirement from a steady-state mass balance on O_2 equivalents. We ignore SMP for this example in order to simplify the equations and mathematics. SMP contain oxygen demand and would reduce the oxygen requirement from what is computed here.

$$\text{Input O}_2 \text{ equivalents} = 500 \text{ m}^3/\text{d} \cdot \left[\begin{array}{l} 500 \frac{\text{mg BOD}_L}{1} \\[6pt] + 60 \frac{\text{mg NH}_4^+\text{-N}}{1} \cdot 4.57 \frac{\text{mg OD}}{\text{mg NH}_4^+\text{-N}} \\[6pt] + 25 \frac{\text{mg VSS}_i}{1} \cdot 1.42 \frac{\text{mg OD}}{\text{mg VSS}} \end{array} \right] \cdot 10^{-3} \frac{\text{kg-1}}{\text{mg-m}^3}$$

$$= 404.9 \text{ kg OD/d}$$

Soluble output O_2 equivalents $= 500 \text{ m}^3/\text{d}$

$$\cdot \begin{bmatrix} 0.34 \text{ mg BOD}_L/\text{l} + (4.57 \text{ mg OD/NH}_4^+\text{-N}) \cdot (0.18 \text{ NH}_4^+\text{-N/l}) \\ + (1.14 \text{ mg OD/NO}_2^-\text{-N}) \cdot (0.17 \text{ mg NO}_2^-\text{-N}) \end{bmatrix} \cdot 10^{-3} \frac{\text{kg-l}}{\text{mg-m}^3}$$

$= 0.68 \text{ kg OD/d}$

Solid output O_2 equivalents $= 1.42 \dfrac{\text{mg OD}}{\text{mgVSS}_i} \cdot 12.5 \dfrac{\text{kg VSS}_i}{\text{d}}$

$$+ 1.98 \frac{\text{mg OD}}{\text{mg VSS}} \left(52.5 + 3.9 + 0.95 \frac{\text{kg VSS}}{\text{d}} \right)$$

$= 131 \text{ kg OD/d}$

For the computation of the output solid oxygen equivalent, the factor 1.98 mg OD/mg VSS is used to account for the 20 electron equivalents per mole of $C_5H_7O_2N$ in the C and the 8 electron equivalents in the N. (This also can be seen from Equation C-2 in Table 2.4.) Counting the N in this case is necessary, because the input oxygen equivalents explicitly include the reduced N. The inert VSS just passes through and must have the same multiplier in the input and output. We used 1.42 mg OD/mg VSS here, but we could have used 1.98 mg OD/mg VSS, as long as it was used for the input and output.

Therefore, the total oxygen required is

$$\frac{\Delta O_2}{\Delta t} = 404.9 - 0.68 - 131 = 273 \text{ kg O}_2/\text{d}$$

Fifth, we must provide a total system volume. As usual, we must select either a VSS concentration or a hydraulic detention time. In this case, let's select $X_v = 3,000 \text{ mg/l}$. Then,

$$V = \frac{\theta_x}{X_v} \left(\frac{\Delta X_v}{\Delta t} \right)_{\text{total}}$$

$$= \frac{20 \text{ d}}{3,000 \text{ mg VSS/l}} \left(69.8 \frac{\text{kg VSS}}{\text{d}} \right) \left(10^3 \frac{\text{mg-m}^3}{\text{kg-l}} \right)$$

$$= 465 \text{ m}^3$$

$$\theta = \frac{465 \text{ m}^3}{500 \text{ m}^3/\text{d}} = 0.93 \text{ d or 22 h}$$

This is a fairly typical situation for extended aeration with domestic wastewater (i.e., θ near 1 d), which almost always supports one-sludge nitrification.

Finally, we provide a preliminary settler design based on the overflow rate and the solids flux. For extended aeration systems, the overflow rate typically takes a conservative value, 8 to 16 m/d (recall the section on Loading Criteria in Chapter 6). We use 12 m/d. Based on that criterion, the required surface area for the settler is approximately (we will ignore the small flow rate of waste sludge):

$$A_s = \frac{500 \text{ m}^3/\text{d}}{12 \text{ m/d}} = 41.7 \text{ m}^2$$

The solids flux used for extended aeration also is conservative, typically around 70 kg/m²-d (see Loading Criteria in Chapter 6). To compute the solids flux, we need an estimate of the recycle flow rate, which requires a judgment about the concentrations of the underflow solids,

X_v^r. We assume that $X_v^r = 8,000$ mg VSS/l. Then, using Equation 6.14a:

$$Q^r = Q\frac{X_v(1 - \theta/\theta_x)}{X_v^r - X_v}$$

$$= 500 \text{ m}^3/\text{d}\frac{3,000 \text{ mg/l} \left(1 - \dfrac{0.93 \text{ d}}{20 \text{ d}}\right)}{8,000 \text{ mg/l} - 3,000 \text{ mg/l}}$$

$$= 286 \text{ m}^3/\text{d (or } R = 0.57)$$

Having Q^r, we can determine the surface area based on the solids flux.

$$A_s = \frac{(Q + Q^r)X_v}{G_t}$$

$$= \frac{(500 + 286 \text{ m}^3/\text{d})3,000 \text{ mg/l} \times 10^{-3}\dfrac{\text{kg-1}}{\text{m}^3\text{-mg}}}{70 \text{ kg/m}^2\text{-d}}$$

$$= 33.7 \text{ m}^2$$

The overflow rate controls in this case, making $A_s = 42 \text{ m}^2$.

TWO-SLUDGE DESIGN   For the same conditions as in the previous example, we are to make a preliminary design of a two-sludge nitrification system. At the end, we will compare the one-sludge and two-sludge designs.

Example 9.2

For the first stage, we wish to suppress nitrification by using a suitably low θ_x. We select 4 d, which normally does not allow nitrification. Then, we compute the sludge production, oxygen required, and unit sizes following the same pattern as in the previous example. (Again, we ignore SMP for simplicity.)

$$\text{BOD}_L = 10\frac{1 + 0.1 \cdot 4}{0.45 \cdot 10 \cdot 4 - (1 + 0.1 \cdot 4)}$$

$$= 0.84 \text{ mg BOD}_L/1$$

$$\frac{\Delta X_v}{\Delta t} = 500(500 - 0.84)0.45\frac{1 + 0.2 \cdot 0.1 \cdot 4}{1 + 0.1 \cdot 4} \times 10^{-3}$$

$$= 86.6 \text{ kg VSS/d}$$

$$\frac{\Delta X_i}{\Delta t} = 500 \cdot 25 \cdot 10^{-3} = 12.5 \text{ kg VSS}_i/\text{d}$$

$$\left(\frac{\Delta X_v}{\Delta t}\right)_{\text{total}} = 86.6 + 12.5 = 99.1 \text{ kg VSS/d}$$

$$\text{input O}_2 \text{ equivalents} = 500 \times \left[500\frac{\text{mg BOD}_L}{1} + 1.42\frac{\text{mg OD}}{\text{mg VSS}_i} \cdot 25\frac{\text{mg VSS}_i}{1}\right]10^{-3}$$

$$= 267.8 \text{ kg OD/d}$$

$$\text{Output O}_2 \text{ equivalents} = 500 \cdot 0.84 \cdot 10^{-3} + 99.1 \cdot 1.42$$

$$= 141.1 \text{ kg OD/d}$$

$$\frac{\Delta O_2}{\Delta t} = 267.8 - 141.1 = 127 \text{ kg O}_2/\text{d}$$

Again, we assume $X_v = 3,000$ mg VSS/l to estimate V and θ:

$$V = \frac{4\ \text{d}}{3,000\ \text{mg/l}} (99.1\ \text{kg VSS/d}) 10^3 \frac{\text{mg-m}^3}{\text{kg-1}}$$

$$= 132.1\ \text{m}^3$$

$$\theta = 132.1/500 = 0.26\ \text{d}\ (6.3\ \text{h})$$

For a conventional loading, we can use a settler-overflow rate of 40 m/d. Then, based on the overflow,

$$A_s = \frac{500}{40} = 12.5\ \text{m}^2$$

For the solids-flux computation, we assume $X_v^r = 10,000$ mg/l and $G_T = 140$ kg/m^2-d.

$$Q^r = 500 \frac{3,000\left(1 - \dfrac{0.26}{4}\right)}{10,000 - 3,000} = 200\ \text{m}^3/\text{d}$$

$$R = 0.4$$

$$A_s = \frac{(500 + 200)3,000 \times 10^{-3}}{140} = 15\ \text{m}^2$$

In this case, the solids flux controls, and $A_s = 15$ m^2.

For the second stage, we assume that BOD oxidation and heterotrophic growth are negligible. This is a significant oversimplification, because (1) the second stage receives active heterotrophs and SMP from the first stage effluent, and (2) the nitrifiers in the second stage produce SMP that allow more heterotrophic growth. However, these complications require making several key assumptions about uncertain factors, particularly carry over of volatile solids and SMP formation by nitrifiers. Ignoring these factors may cause an underestimation of X_v, which affects the system's volume and the settler area.

We select $\theta_x = 20$ d for the second-stage SRT. Then, effluent NH$_4^+$-N and NO$_2^-$-N remain 0.18 and 0.17 mg/l, respectively. As before, we must adjust the TKN input to take into account N assimilated into heterotrophic biomass in the first stage and released from decay of heterotrophic biomass input from the first stage as 0.52 kg NH$_4^+$-N/d (calculated later and used here).

For the *ammonia oxidizers*,

$$N_1^0 = 60 - 86.6 \cdot 0.124 \cdot 10^3/500 + 0.52 \cdot 10^3/500$$

$$= 60 - 21.5 + 1.1 = 39.6\ \text{mg NH}_4^+\text{-N/l}$$

$$\frac{\Delta X_v}{\Delta t} = 500 \cdot (39.6 - 0.18) \cdot 0.33 \cdot \frac{1 + 0.2 \cdot 0.082 \cdot 20}{1 + 0.082 \cdot 20} \cdot 10^{-3}$$

$$= 3.3\ \text{kg VSS/d}$$

For the *nitrite oxidizers*,

$$N_2^0 = 60 - 20.4 - 0.124 \cdot 3.3 \cdot 10^3/500 - 0.18 = 38.6\ \text{mg NO}_2\text{-N/l}$$

$$\frac{\Delta X_v}{\Delta t} = 500 \cdot (38.6 - 0.17) \cdot 0.083 \frac{1 + 0.2 \cdot 0.082 \cdot 20}{1 + 0.082 \cdot 20} \cdot 10^{-3}$$

$$= 0.80\ \text{kg VSS/d}$$

We also assume that 20 mg VSS/l is not removed by the first-stage settler and enters the second stage, it contains 80 percent active biomass, and this active biomass decays with a rate of 0.1/d. The input rates are

$$\left(\frac{\Delta X_i^0}{\Delta t}\right) = 0.2 \times 20 \times 500/1{,}000 = 2 \text{ kg VSS}_i/\text{d}$$

$$\left(\frac{\Delta X_a^0}{\Delta t}\right) = 0.8 \times 20 \times 500/1{,}000 = 8 \text{ kg VSS}_a/\text{d}$$

With a 20-d SRT, the input active biomass decays to leave a residual X_a, produce more X_i, and release NH_4^+-N.

$$\left(\frac{\Delta X_a}{\Delta t}\right)_{\text{residual}} = \frac{8 \text{ kg VSS}_a/\text{d}}{1 + 0.1 \cdot 20} = 2.7 \text{ kg VSS}_a/\text{d}$$

$$\left(\frac{\Delta X_i}{\Delta t}\right)_{\text{increase}} = 0.2(8 - 2.7) = 1.1 \text{ kg VSS}_i/\text{d}.$$

Therefore,

$$\left(\frac{\Delta X_v}{\Delta t}\right)_{\text{residual}} = 2 + 2.7 + 1.1 = 5.8 \text{ kg VSS}/\text{d}.$$

The total VSS that must exit the second stage is then

$$\frac{\Delta X_v}{\Delta t} = 5.8 + 3.3 + 0.80 = 9.9 \text{ kg VSS}/\text{d}.$$

The increase in NH_4^+-N due to decay of heterotrophic biomass is

$$\left(\frac{\Delta N_1}{\Delta t}\right)_{\text{increase}} = 0.124(8 - 2.7 - 1.1) = 0.52 \text{ kg } NH_4^+\text{-N}/\text{d}$$

Note here that the active nitrifiers are about one-third of the total VSS, even though we have ignored the input of any BOD from substrate or SMP. This shows that second-stage processes often are far from being completely dominated by nitrifiers.

The oxygen required is computed from:

$$\text{Input } O_2 \text{ equivalents} = 500[4.57 \cdot 39.6 + 1.42 \cdot 4 + 1.98 \cdot 16] \cdot 10^{-3}$$

$$= 109 \text{ kg OD}/\text{d}$$

$$\text{Soluble output } O_2 \text{ equivalents} = 500[4.57 \cdot 0.18 + 1.14 \cdot 0.17] \cdot 10^{-3}$$

$$= 0.5 \text{ kg OD}/\text{d}$$

$$\text{Solid output } O_2 \text{ equivalents} = 1.42 \cdot 2 + 1.98 \cdot (2.7 + 1.1 + 3.3 + 0.8)$$

$$= 18.5 \text{ kg OD}/\text{d}$$

Hence, the total O_2 requirement for the second stage is

$$\frac{\Delta O_2}{\Delta t} = 109 - 0.5 - 18.5 = 90 \text{ kg } O_2/\text{d}$$

To compute the total system volume for the second stage, we must assume a reasonable MLVSS concentration. The low VSS production generally prevents obtaining the typical values seen when heterotrophic bacteria are dominant. Therefore, we assume here that $X_v = 1{,}000$ mg/l. Then,

$$V = \frac{20}{1{,}000}(9.9) \cdot 10^3 = 198 \text{ m}^3$$

and

$$\theta = \frac{198}{500} = 0.40 \text{ d (or 9.5 h)}$$

For the settler, we will use the extended aeration values of overflow rate and solids flux. With the overflow rate of 12 m/d,

$$A_S = \frac{500 \text{ m}^3/\text{d}}{12 \text{ m/d}} = 41.7 \text{ m}^2$$

The solids flux can be used after assuming an X_v^r value of 5,000 mg/l:

$$Q^r = 500\frac{1{,}000\left(1 - \dfrac{0.35}{20}\right)}{5{,}000 - 1{,}000} = 123 \text{ m}^3/\text{d}$$

and

$$R = 0.25$$

Then,

$$A_S = \frac{(500 + 123)1{,}000 \times 10^{-3}}{70} = 8.9 \text{ m}^2.$$

Again, the overflow rate controls, giving $A_S = 42$ m^2.

Table 9.3 compares the one-sludge and two-sludge designs from Examples 9.1 and 9.2. It demonstrates the general trade-off between the two approaches.

• The one-sludge system has less sludge production, since its SRT for BOD oxidation is large, but its oxygen requirement is greater.

Table 9.3 Comparison of the one-sludge and two-sludge designs for Examples 9.1 and 9.2

		Two Sludge		
Parameter	One Sludge	Stage 1	Stage 2	Total
θ_x, d	20	4	20	
Solids wasting, kg VSS/d	69.8	99.1	9.9	99*
Oxygen required, kg O$_2$/d	273	127	90	217
System volume, m^3	465	132	198	330
Settler Area, m^2	42	15	42	57

*The total solids wasting for the two-sludge systems is less than the sum of the two stages, because 10 kg/d from stage 1 is transferred to stage 2, where most of it decays. Thus, 10 kg/d is subtracted from the total to eliminate the double counting.

- The total system volume is greater for the one-sludge system, again due to the long SRT for the heterotrophs, but the two-sludge system must operate two separate reactor-settler systems and has more settler area.

9.4 BIOFILM NITRIFICATION

All of the biofilm processes useful for aerobic BOD oxidation also can be employed to perform nitrification, provided that the process design appropriately considers the slow specific growth rate inherent to nitrifiers and the competition between nitrifiers and heterotrophs for oxygen and space. The typical nitrification parameters are listed in Table 9.4.

Figure 9.2 presents the "generic" normalized loading curve for the nitrifiers and points out that most successful nitrification biofilm processes (discussed below) operate in the transition region between low load and high load, that is, J/J_R from about

Table 9.4 Fundamental and derived parameters for NH_4^+-N oxidation[1]

Parameters	Value
Limiting substrate	NH_4^+-N[2]
Y, mg VSS_a/mg NH_4^+-N	0.33
$\hat{q}$, mg NH_4^+-N/mg VSS_a-d	1.7
K, mg NH_4^+-N/l	0.57
b, 1 d	0.08
b_{det}, 1 d	0.05[3]
b', 1 d	0.13
θ_x, d	20
D, cm^2/d	1.3
D_f, cm^2/d	1.04
X_f, mg VSS_a/cm^3	10
L, cm	0.004
S_{min}, mg NH_4^+-N/l	0.17
S_{min}^*	0.30
K^*	1.85
J_R, mg NH_4^+-N/cm^2-d	0.072

[1] Notes:
  1. Generic values for $T = 15\,°C$.
  2. The limiting substrate is NH_4^+-N, which assumes that the first step of nitrification is limiting.
  3. Assumes that nitrifiers exist mainly deep within a multispecies biofilm and are partially protected from detachment.

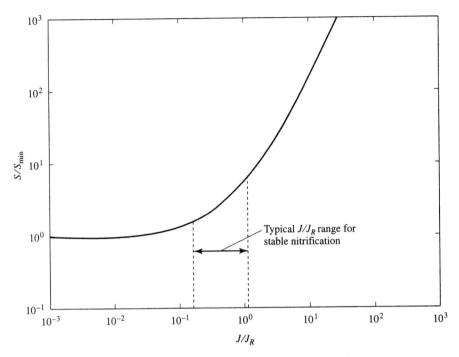

Figure 9.2   Normalized loading curve for "generic" nitrification (S^*_{min} = 0.3, S_{min} = 0.17 mg NH_4^+-N/l, K^* = 1.85, J_R = 0.072 mg NH_4^+-N/cm^2-d) shows that practical surface loading is in the transition range between low load and high load.

0.2 to 1.1 (recall the section on Normalized Surface Loading in Chapter 4). Because S_{min} is small (0.17 mg NH_4^+-N in Table 9.4), the generic normalized loading range yields S/S_{min} values of 1.6 to 6, which keeps S less than about 1 mg NH_4^+-N/l.

Because BOD almost always is present in wastewaters and because nitrifiers generate SMP that further supports heterotrophic growth, coexistence with heterotrophs is a normal situation for nitrifying bacteria in all nitrification processes. Experimental and modeling results indicate that the nitrifying bacteria tend to accumulate closer to the attachment surface, while heterotrophs dominate near the outer surface of the biofilm. This "layering" is a natural consequence of the heterotrophs' much faster specific growth rate, which allows them to exist stably in the region of the biofilm experiencing the greatest detachment and predation rates. On the other hand, the nitrifiers accumulate most successfully deeper inside the biofilm, where they are at least partly protected from detachment and predation. This phenomenon is discussed further in the website chapter, "Complex Systems."

While having the nitrifiers deep inside the biofilm helps stabilize the slow-growing nitrifiers against washout, it also increases their susceptibility to oxygen limitation, since dissolved oxygen must diffuse through the heterotroph layer before it reaches the nitrifiers. The added mass-transport resistance and oxygen consumption by the heterotrophs lower biofilm D.O. concentrations available to the nitrifiers. When coupled with the relatively high oxygen sensitivity of nitrifiers (e.g., $K_O \approx 0.1$ mg/l for many heterotrophs while it is around 0.5 mg/l for nitrifiers), these low D.O. concentrations inside the biofilm can negate the advantages of protection, unless the bulk-liquid D.O. concentration is maintained high enough. For example, Furumai and Rittmann (1994) described declines in nitrifier populations and increases in effluent concentrations of NH_4^+-N and NO_2^--N once the bulk D.O. concentration dropped below approximately 3 mg/l, even though the nitrifiers were well-protected deep inside the biofilm. Keeping surface loading low can preclude the adverse effects of D.O. depletion.

In practice, imposing a maximum BOD or COD surface loading, which is in the range of 2 to 6 kg BOD_L/1,000 m^2-d, controls the competition by heterotrophs. When this design criterion for the organic loading augments a surface-load criterion for NH_4^+-N, nitrification performance usually is stable. Keeping the organic loading below 2 to 6 kg BOD_L/1,000 m^2-d makes it possible to maintain a sufficient bulk-liquid D.O. concentration and also prevents medium clogging, short-circuiting, sloughing, and/or the need for excessive backwashing caused by too much heterotrophic growth. The last two items can lead to such large detachment rates that the nitrifiers never are able to establish protected layers, and they wash out.

Table 9.5 summarizes successfully used N and BOD_L surface loads for various biofilm processes which were described in Chapter 8. The ranges of N and BOD_L loads are relatively narrow. The BOD_L loading for nitrification overlaps the BOD_L loading used for strictly aerobic BOD oxidation, which explains why biofilm processes used primarily for BOD oxidation often nitrify, as long as the BOD surface loading is in the lower part of the typical range. Because organic N generally is hydrolyzed to release NH_4^+-N, the N loading should be in terms of TKN, not just NH_4^+-N.

When the wastewater's BOD_L:TKN ratio is too large to allow both surface-loading criteria to be met simultaneously, *staged treatment* can be used to reduce the BOD loading to the nitrification process. Two-stage trickling filters are one good example. The normal staging in RBCs is another example. Normally, nitrification occurs in the later stages of RBCs, after the BOD loading is reduced in earlier stages. The reduced growth of heterotrophs in nitrifying stages of RBCs allows high-density media to be used in order to maximize the specific surface area.

Table 9.5 Nitrogen and BOD loadings used successfully in various nitrifying biofilm processes

Process Type	N Surface Loading* kg N/1,000 m^2-d	BOD Surface Loading kg BOD_L/1,000 m^2-d
Trickling filters	0.5–0.8	< 4.4
Rotating biological contactors	0.2–0.6	< 6
Biolite granular filters	< 0.7	< 6
Fluidized beds	0.5	not given
Circulating beds	< 1	not given

*The N surface loading should be computed from the TKN concentration, not only the NH_4^+-N concentration, since organic N can be hydrolyzed to NH_4^+-N.

Example 9.3 | **BASIC NITRIFICATION BIOFILM DESIGN**   A wastewater has a volumetric flow of 1,000 m^3/d and influent concentrations of 300 mg/l of BOD_L and 50 mg/l of TKN. We are to design a one-stage and then a two-stage system that achieves full nitrification.

First, the total mass loading rates of each electron donor are 300 kg BOD_L/d and 50 kg TKN/d. The 6:1 ratio is typical for domestic wastewaters.

Second, we find the required surface area of a one-stage system by computing the areas for BOD_L oxidation and TKN nitrification. We then use the larger value. For BOD_L, we use a maximum surface load of 4 kg BOD_L/1,000 m^2-d (Table 9.5). That gives a biofilm surface area of

$$A = (300 \text{ kg } BOD_L/\text{d})/(4 \text{ kg } BOD_L/1{,}000 \text{ } m^2\text{-d}) = 7.5 \cdot 10^4 \text{ } m^2$$

The similar computation for TKN and with a conservative TKN load of 0.5 kg N/1,000 m^2-d (Table 9.5) gives

$$A = (50 \text{ kg N/d})/(0.5 \text{ kg N/1,000 } m^2\text{-d}) = 10 \cdot 10^4 \text{ } m^2$$

We use the larger value, or $10 \cdot 10^4$ m^2.

Second, we do a two-stage design. We begin with BOD_L oxidation. From Chapter 8, we select a fairly high surface loading of 7 kg BOD_L/1,000 m^2-d. This gives the first stage area of

$$A_1 = (300 \text{ kg } BOD_L/\text{d})/(7 \text{ kg } BOD_L/1{,}000 \text{ } m^2\text{-d}) = 4.3 \cdot 10^4 \text{ } m^2$$

We then compute the area for the second, nitrifying stage. Before we compute the area, we must adjust the TKN loading to account for N used in heterotroph synthesis in the first stage. An exact computation can be achieved following the methods in Examples 9.1 and 9.2, as long as a detachment rate is estimated. For this simple example, we assume that the TKN load is reduced from 50 to 40 kg N/d by heterotroph synthesis. Then, we assume a less conservative TKN surface load of 0.8 kg N/1,000 m^2-d (Table 9.5) and compute the second-stage area:

$$A_2 = (40 \text{ kg N/d})/(0.8 \text{ kg N/1,000 } m^2\text{-d}) = 5 \cdot 10^4 \text{ } m^2$$

The total surface area for both stages is $9.3 \cdot 10^4$ m^2, which gives a small area savings over the one-stage design. On the other hand, we must design, build, and operate two separate systems.

9.5 HYBRID PROCESSES

Nitrification is one of the applications for which hybrid suspended-growth/biofilm processes are used to increase the volumetric loading rate. Hybrid processes were mentioned in Chapter 8. Hybrid processes are becoming popular because of the widespread need to oxidize ammonium in wastewaters and because of the strict requirement that the SRT be high enough to sustain the slow-growing nitrifiers. Example 9.4 illustrates how the capacity of an overloaded activated sludge process can be expanded through adding biofilm surface area.

Example 9.4 | **HYBRID ONE-SLUDGE NITRIFICATION**   Example 9.1 provided a design for a one-sludge nitrification process treating $Q = 500$ m^3/d, $BOD_L = 500$ mg/l, TKN = 60 mg/l, $X_i^0 = 25$ mg/l, and $T = 15\,°C$. An SRT of 20 d gave adequate nitrification and BOD removal and required a total system volume of 465 m^3 and a MLVSS of 3,000 mg/l. However, structural defects forced shutdown of one-half of the system's volume. The settler now is severely overloaded when the MLVSS is increased enough to maintain the same θ_x. In fact, the practical maximum MLVSS is only about 3,000 mg/l, which makes the maximum SRT 10 days. Nitrification and effluent VSS are unsatisfactory.

To improve the performance, we propose to add to the remaining reactor volume biofilm carriers that are cubes 3 mm on a side and having a wet density of 1.02 g/cm^3. The outside area per unit cube mass is 0.0196 cm^2/mg. The maximum density of the cubes in the basin is 10 percent by weight, or 100,000 mg/l. Thus, the biofilm surface area is up to 100,000 mg/l · 0.0196 cm^2/mg = 1,960 cm^2/l.

To achieve the same performance as before, the biofilm should have an average SRT of 20 d or greater. For the biofilm, that means $b_{det} \leq 0.05/d$, since $b_{det} = 1/\theta_x$ for a biofilm (Chapter 4, section on Average Solids Retention Time). Having $b_{det} = 0.05/d$ corresponds to the parameters for nitrification in Table 9.4. Mass-transfer limitation will cause the NH$_4^+$-N concentration to be higher than the 0.18 mg/l computed in Example 9.1. We assume that we are willing to tolerate a higher value of 0.25 mg/l. Using the steady-state biofilm model and $S/S_{min} = 0.25/0.17 = 1.47$, we get that $J/J_R \approx 0.25$, and $J \approx 0.018$ mg N/cm^2-d.

A mass balance assuming that all nitrifiers are in the biofilm yields

$$0 = Q(N_1^0 - N_1) - aVJ$$
$$0 = 500 \text{ m}^3/\text{d} \cdot (60 - 0.25 \text{ mg/l}) \cdot 1{,}000 \text{ l/m}^3$$
$$\quad - 1{,}960 \text{ cm}^2/\text{l} \cdot V \cdot 1{,}000 \text{ l/m}^3 \cdot 0.018 \text{ mg/cm}^2\text{-d}$$

giving

$$V = 8.47 \cdot 10^5 \text{ l} = 850 \text{ m}^3$$

Clearly, the required volume is far greater than the volume available. Operating a hybrid process does not work when we force the effluent ammonium concentration to be only 0.25 mg N/l.

If we are willing to tolerate an NH$_4^+$-N increase to 1 mg/l, J goes up to 0.072 mg/cm^2-d. Then,

$$0 = 500 \times (60 - 1) - 1960 \cdot V \cdot 0.072$$
$$V = 209 \text{ m}^3$$

which is approximately the aeration-tank volume available.

For the BOD removal and heterotrophs, it is not necessary that all the biomass be in the biofilm or that the biofilm SRT be 20 d. For this situation, we assume that $b_{det} = 0.1/d$ for the heterotrophs, and the parameters of Table 8.1 can be used. In Example 9.1, the effluent substrate concentration was 0.34 mg/l, which is a reasonable target. It gives $S/S_{min} = 7.23$, $J/J_R = 0.6$, and $J = 0.02$ mg/cm^2-d. The volume required to have all the heterotrophic biomass in the biofilm would be

$$0 = 500 \times (500 - 0.34) - 1960 \cdot V \cdot 0.02$$
$$V = 6{,}400 \text{ m}^3$$

This computation shows that almost all of the heterotrophic biomass would need to be suspended, an unworkable situation. If the substrate BOD were allowed to rise to 6 mg/l, then $S/S_{min} = 128$, $J/J_R \cdot 10$, and $J \approx 0.33$ mg/cm^2-d. This gives

$$0 = 500 \times (500 - 6) - 1960 \cdot V \cdot 0.33$$

and

$$V = 380 \text{ m}^3$$

This volume, compared to the available system volume of 232 m^3 indicates that a substantial portion of the heterotrophic biomass would be in the biofilm, which reduces the suspended MLVSS and eliminates the settler overloading. Of course, the effluent substrate concentration rises and is a trade-off.

This example shows that the treatment capacity of a suspended-growth process can be significantly increased *if* a large enough amount of surface area is added and *if* the effluent substrate concentrations are allowed to rise. The success of such a scheme depends upon having sufficient aeration capacity to keep the biofilm cubes suspended and, most importantly, to supply oxygen at a volumetric rate approximately twice that of the system in Example 9.1.

9.6   THE ROLE OF THE INPUT BOD$_L$:TKN RATIO

Nitrifying systems are affected by the influent BOD$_L$:TKN ratio in three ways. The first way, illustrated in Examples 9.1 and 9.2, is that synthesis of heterotrophic biomass sequesters nitrogen and reduces the flow of nitrogen from ammonium, to nitrite, and to nitrate. If the influent BOD$_L$:TKN ratio is large enough—say, greater than about 25 g BOD$_L$/g TKN—little or no reduced nitrogen is available for nitrification.

Second, the BOD$_L$:TKN ratio determines what fraction of the active biomass is comprised of nitrifiers. Due to the low f_s^0 for the nitrifiers, their fraction normally is low. For typical BOD$_L$:TKN ratios in municipal sewage (5 to 10 g BOD$_L$/g N), the nitrifiers normally constitute less than 20 percent of the active biomass and are a smaller fraction of the volatile suspended solids.

Finally, the BOD$_L$:TKN ratio exerts some control over how heterotrophs and nitrifiers compete for common resources: dissolved oxygen and space in flocs or biofilms. In the long term, a higher BOD$_L$:TKN ratio tends to force the nitrifiers deeper into the floc or biofilm. This incurs greater mass-transport resistance for the nitrifiers' substrates, particularly NH$_4^+$ and O$_2$ (Rittmann and Manem, 1992). In the short term, a high growth rate for heterotrophs could create negative impacts on nitrifiers by sequestering nitrogen, consuming oxygen, or physically sweeping nitrifiers out of the floc or biofilm.

9.7   THE ANAMMOX PROCESS

Recently, a novel bacterium in the planctomycetes group has been discovered for its ability to anaerobically oxidize NH$_4^+$-N to N$_2$, not to NO$_2^-$ (Strous, Kuenen, and Jetten, 1999; Strous et al., 1999). It is called the ANAMMOX microorganism because it does ANaerobic AMMonium OXidation.

The ANAMMOX bacterium uses ammonium as its electron donor and nitrite as its electron acceptor. The energy reaction is

$$NH_4^+ + NO_2^- = N_2 + 2\,H_2O$$

The cells are autotrophs, and the reduction of inorganic carbon to the oxidation state of cellular carbon is via oxidation of nitrite to nitrate. Nitrite also is the nitrogen source:

$$5\,CO_2 + 14\,NO_2^- + 3\,H_2O + H^+ = C_5H_7O_2N + 13\,NO_3^-$$

The yield and specific growth rate reported for ANAMMOX are low, about 0.14 g VSS_a/g NH_4^+-N and 0.065/d, respectively (Strous, Heijnen, Kuenen, Jetten, 1998). This gives an overall stoichiometry of approximately:

$$NH_4^+ + 1.26\,NO_2^- + 0.085\,CO_2 + 0.02\,H^+ = N_2 + 0.017\,C_5H_7O_2N$$
$$+ 0.24\,NO_3^- + 1.95\,H_2O$$

The discovery of ANAMMOX is one of the most startling ones in environmental biotechnology. However, the importance of ANAMMOX bacteria in environmental-biotechnology practice is not yet known. Conditions favoring their accumulation included exceptional biomass retention (to give a very long SRT), stable operation, the presence of nitrite, lack of oxygen, and lack of donors that could cause the reduction of nitrite via denitrification (Chapter 10).

9.8   **BIBLIOGRAPHY**

de Silva, D. G. V. and B. E. Rittmann (2000a). "Nonsteady-state modeling of multispecies activated sludge processes." *Water Environment Research,* in press.

de Silva, D. G. V. and B. E. Rittmann (2000b). "Interpreting the response to loading changes in a mixed-culture CSTR." *Water Environment Research,* in press.

Furumai, H. and B. E. Rittmann (1994). "Evaluation of multi-species biofilm and floc processes using a simplified aggregate model." *Water Sci. Technol.* 29(10–11), pp. 439–446.

Lazarova, V. and J. Manem (1994). "Advances in biofilm aerobic reactors ensuring effective biofilm control." *Water Sci. Technol.* 29(10–11), pp. 345–354.

Metcalf & Eddy, Inc. (1979). *Wastewater Engineering: Treatment, Disposal, Reuse.* New York: McGraw-Hill.

Mobarry, B. K.; M. Wagner; V. Urbain; B. E. Rittmann; and D. A. Stahl (1996). "Phylogenetic probes for analyzing abundance and spatial organization of nitrifying bacteria." *Applied Environ. Microb.* 62, pp. 2156–2162.

Rittmann, B. E. (1987). "Aerobic biological treatment." *Environ. Sci. Technol.* 21, pp. 128–136.

Rittmann, B. E. and J. A. Manem (1992). "Development and experimental evaluation of a steady-state, multi-species biofilm model." *Biotechnol. Bioengr.* 39, pp. 914–922.

Rittmann, B. E.; J. M. Regan; and D. A. Stahl (1994). "Nitrification as a source of soluble organic substrate in biological treatment." *Water Sci. Technol.* 30(6), pp. 1–8.

Rittmann, B. E. and V. L. Snoeyink (1984). "Achieving biologically stable drinking water." *J. Amer. Water Works Assn.* 76(10), pp. 106–114.

Schroeder, E. D. and G. Tchobanoglous (1976). "Mass transfer limitations in trickling filter design." *J. Water Pollution Control Federation* 48(4).

Strous, M.; J. A. Fuerst; E. H. M. Kramer; S. Logeman; G. Muyzer; K. T. van de Pas-Schoonen; R. Webb; J. G. Kuennen; and M. S. M. Jetten (1999). "Missing lithotroph identified as a new planctomycete." *Nature* 400, 446–449.

Strous, M.; J. G. Kuenen; and M. S. M. Jetten (1999). "Key physiology of anaerobic ammonium oxidation." *Appl. Environ. Microb.* 65, pp. 4248–3250.

Strous, M.; J. J. Heijnen; J. G. Kuenen; and M. S. M. Jetten (1998). "The sequencing batch reactor as a powerful tool for the study of slowly growing ammonium-oxidizing microorganisms." *Appl. Environ. Microb.* 50, pp. 589–596.

Teske, A.; E. Alm; J. M. Regan; S. Toze; B. E. Rittmann; and D. A. Stahl (1994). "Evolutionary relationships among ammonia- and nitrite-oxidizing bacteria." *J. Bacteriology* 176, pp. 6623–6630.

9.9 PROBLEMS

9.1. A nitrification process can easily be operated in a fixed-film mode. The typical kinetic parameters

$$K = 0.5 \frac{\text{mg NH}_4^+\text{-N}}{1} \quad D_f = 1.2 \frac{\text{cm}^2}{\text{d}} \quad Y = 0.26 \frac{\text{mg VS}_a}{\text{mg NH}_4^+\text{-N}} \quad L = 65 \ \mu\text{m}$$

$$\hat{q} = 2.0 \frac{\text{mg NH}_4^+\text{-N}}{\text{mg VS}_a \cdot \text{d}} \quad D = 1.5 \frac{\text{cm}^2}{\text{d}} \quad b' = 0.1 \ \text{d}^{-1} \quad X_f = 5 \frac{\text{mg VS}_a}{\text{cm}^3}$$

You are to determine the response of a complete-mix biofilm reactor to changes in detention time. The reactor has specific surface area, a, of 1 cm^{-1}. The influent feed strength (S^0) is 30 mg N/l of NH$_4^+$-N. You can assume that NH$_4^+$-N completely limits the kinetics.

The steps for the solution are

(a) Calculate $S_{\min}$, $S_{\min}^*$, and J_R.

(b) Generate a curve of J/J_R versus $S/S_{\min}$. You may need to interpolate between given normalized curves.

(c) Set up the mass balance for substrate in the reactor.

(d) For each detention time (1, 2, 4, 8 and 24 h), solve for S. You may wish to use an iterative technique. Guess S, solve for J (from curve from part b), calculate S from the mass balance, check for agreement of S, repeat until convergence. Ignore suspended reactions. Alternatively, you can superimpose the mass-balance curve on the loading curve.

(e) Plot S versus θ. Interpret the curve.

(f) Plot S^0/S versus θ and see if an apparent first-order reaction seems like a good description of the process. If so, what is the apparent first-order coefficient? If not, why is the reaction not first-order?

9.2. You are to design a fixed-film biological process to oxidize ammonia nitrogen. The wastewater contains 50 mg NH$_4^+$-N/l in a flow of 1,000 m^3/d. Available are modules of complete-mix biofilm reactors; each module has 5,000 m^2 of surface area for biofilm colonization. How many modules, operated in series, are needed to achieve an effluent concentration of 1 mg NH_4^+-N/l or less? You may use the following parameters and may assume that NH_4^+-N is the rate limiting substrate.

$$K = 1.0 \frac{\text{mg } NH_4^+\text{-N}}{l} \quad D_f = 1.3 \frac{\text{cm}^2}{\text{d}} \quad Y = 0.33 \frac{\text{mg VS}_a}{\text{mg } NH_4^+\text{-N}} \quad b_{det} = 0.1 \text{ d}^{-1}$$

$$\hat{q} = 2.3 \frac{\text{mg } NH_4^+\text{-N}}{\text{mg VS}_a \cdot \text{d}} \quad D = 1.5 \frac{\text{cm}^2}{\text{d}} \quad b = 0.11 \text{ d}^{-1} \quad X_f = 40 \frac{\text{mg VS}_a}{\text{cm}^3}$$

L = a value so small that S_s approaches S.

What do you expect the actual effluent concentration to be? Use the normalized loading-criterion approach.

9.3. A set of kinetic coefficients for nitrification is:

Species	$\hat{q}$ [kg NH_4^+-N/kg $VSS_a \cdot$ d]	K [mg NH_4^+-N/l]	b [1/d]	Y [kg VSS_a/kg NH_4^+-N]	f_d
Amonium-oxidizers	1.4	0.7	0.05	0.5	0.8
Nitrite-oxidizers	9	2.2	0.05	0.1	0.8

If nitrification is to work reliably, the apparent safety factor (SF) for activated sludge will be at least 20 for the lowest growing species. If SF = 20, what will be the sludge production rate of active, inactive, and total VSS (kg/d) if the NH_4^+-N removal is 367 kg N/d? Note that you must consider each species separately. You also may ignore SMP and heterotrophs.

9.4. Use the kinetic parameters in Chapter 9 to calculate N_1 and θ for an activated-sludge nitrification process (complete-mix reactor with settler and recycle) treating 500 mg/l of NH_4^+-N with $\theta_x = 15$ d and MLVSS = 2,500 mg/l. Assume no influent solids or BOD. Calculate the N_1 and N_2 values for 5, 10, 15, and 20 °C. Hint: First calculate S values for NH_4^+-N and NO_2^-, then calculate X_v for both species of nitrifiers. Ignore SMP. Do not ignore N used for synthesis.

9.5. A treatment plant was designed to carry out nitrification, as well as BOD removal, in a one-sludge system. In order to assure that nitrification occurred, the designers used a low loading rate of 0.2 kg BOD_5/m^3 d. Nonetheless, nitrification performance has been poor. You are the high-priced consultant and have obtained the following operating data: water temperature: 15 °C, aeration basin D.O. = 2.5 mg/l, aeration basin regime: essentially completely mixed, MLVSS = 1,200 mg/l, $Q = 10^4$ m^3/d, reactor volume = 1.25×10^4 m^3.

- Influent quality: BOD_5 = 250 mg/l, total Kjeldahl N = 23 mg/l
- Effluent quality: BOD_5 = 25 mg/l, VSS = 25 mg/l, NH_4^+-N = 20 mg N/l

- Recycle sludge quality: recycle ratio = 0.32, volatile solids content = 5×10^3 mg/l
- Waste sludge flow rate: 9×10^2 m³/d

What is the problem and how can it be solved?

9.6. Your client wants to treat his ammonium-containing wastewater by a trickling filter nitrification process to eliminate the nitrogenous oxygen demand. The water is at 20 °C, and you have estimated the limiting kinetic coefficients to be:

$$\hat{q} = 2.6 \frac{\text{mg NH}_4^+\text{-N}}{\text{mg VSS}_a \cdot \text{d}} \quad K = 3.6 \frac{\text{mg NH}_4^+\text{-N}}{\text{l}} \quad Y = 0.26 \frac{\text{mg VSS}_a}{\text{mg NH}_4^+\text{-N}}$$

$$Y_{\text{obs}} = 0.2 \frac{\text{mg VSS}_a}{\text{mg NH}_4^+\text{-N}} \quad b = 0.05 \frac{1}{\text{d}}$$

The process configuration that your client wants includes $Q = 10{,}000$ m³/d, $N_1^0 = 50$ mg NH$_4^+$-N/l, $a = 50$ m²/m³, depth of the filter = 2 m, and volume of filter medium = 500 m³. Does it seem feasible that the client's design can achieve the desired 95 percent removal with these conditions?. You'll need to consider $N_{1\ \text{min}}^0$ and O$_2$ flux first. Compare necessary N_1 and J_{O_2} with $N_{1\ \text{min}}^0$ and the practical J_{O_2} (about 2.85 mg O$_2$/cm²-d). Are they reasonable? If yes, is the volume reasonable? You may ignore heterotrophs here.

9.7. RBCs are designed for nitrification with two loading criteria. First, the BOD load must be low enough to allow nitrifiers to compete. Second, the NH$_4^+$-N flux must be sufficient to give adequate N oxidation. Estimate the required surface area if the BOD load is 10 kg BOD$_5$/1,000 m²-d for the carbonaceous stages, but less than 4 kg BOD$_5$/1,000 m²-d for the whole process, and the NH$_4^+$-N load is 0.4 kgN/1,000 m²-d in the nitrifying stages. The flow is 0.1 m³/s, influent BOD$_5$ = 250 mg/l, and influent TKN = 50 mg N/l.

9.8. An existing sewage treatment plant has a reactor volume of 2,000 m³. The clarifier has a volume of 750 m³ and a surface area of 300 m². Aeration is supplied by brush aerators having a capacity of 110 kW. The brush aerators have a field OTE of 1.2 kg O$_2$/kWh. The elevation is sea level. The desire is to operate the plant at an extended aeration loading, and the design θ_x is 30 d. You may assume the following kinetic parameters:

Process	$\hat{q}$	K	b	Y	f_d
BOD oxidation	$16 \dfrac{\text{g BOD}_L}{\text{g VSS}_a \cdot \text{d}}$	$20 \dfrac{\text{mg BOD}_L}{\text{l}}$	$0.2 \dfrac{1}{\text{d}}$	$0.6 \dfrac{\text{g VSS}_a}{\text{g BOD}_L}$	0.8
Nitrification	$1.7 \dfrac{\text{g NH}_4^+\text{-N}}{\text{g VSS}_a \cdot \text{d}}$	$0.5 \dfrac{\text{mg NH}_4^+\text{-N}}{\text{l}}$	$0.05 \dfrac{1}{\text{d}}$	$0.413 \dfrac{\text{g VSS}_a}{\text{g NH}_4^+\text{-N}}$ (total oxidation)	0.8

Let T = 15 °C. Let the ratio of TSS/VSS be 1.1 g TSS/g VSS. Let the D.O. concentration be 2 mg/l. For simplicity, you can assume 100 percent oxidation of NH_4^+-N and BOD_L. Estimate the maximum practical BOD_L load (kg BOD_L/d) and the maximum flow rate (m³/d) when the maximum BOD_L load is taken if the BOD_L:TKN ratio is 10 g BOD_L-/g TKN in the influent. Let $X_V^0 = 0$.

9.9. Operation of the nitrification plant as a one-sludge or two-sludge system can affect the total O_2 consumption and total sludge production. In this problem you will compare a one-sludge system having a sludge age of 15 d with a two-sludge system that has sludge ages of 3 and 15 d, respectively.
   Kinetic parameters are:

Process	$\hat{q}$	K	b	Y	f_d
BOD oxidation	$16 \dfrac{g\ BOD_L}{g\ VSS_a \cdot d}$	$20 \dfrac{mg\ BOD_L}{1}$	$0.2\dfrac{1}{d}$	$0.6 \dfrac{g\ VSS_a}{g\ BOD_L}$	0.8
Nitrification (complete reaction)	$1.7 \dfrac{g\ NH_4^+\text{-}N}{g\ VSS_a \cdot d}$	$0.5 \dfrac{mg\ NH_4^+\text{-}N}{1}$	$0.05\dfrac{1}{d}$	$0.4 \dfrac{g\ VSS_a}{g\ NH_4^+\text{-}N}$	0.8

The flow rate is 1,000 m³/d, and influent concentrations of BOD_L and TKN are 400 and 40 mg/l, respectively. Assume no input solids, and ignore products. You will take a stepwise approach:

A. For the first stage of the two-sludge system,

   1. Calculate effluent BOD_L concentration
   2. Calculate total kg BOD_L removed per day
   3. Calculate the VSS production rate per day
   4. Calculate the kg N/d incorporated into cells
   5. Calculate the kg O_2/d consumed for heterotrophic reactions

B. For the second stage of the two-sludge system,
   1. Calculate the effluent TKN concentration
   2. Calculate the kg/d of TKN oxidized
   3. Calculate the production rate of VSS from nitrification
   4. Calculate the kg O_2/d consumed in nitrification

C. For the one-sludge system,
   1. Calculate the effluent BOD_L concentration
   2. Calculate the kg BOD_L/d removed
   3. Calculate the VSS production rate from heterotrophic reactions

4. Calculate the rate of incorporation of N into heterotrophic cells

5. Calculate the O_2 consumption rate for heterotrophic reactions

6. The effluent TKN concentration is the same as before. Calculate the total TKN removed by nitrification.

7. Calculate the VSS produced per day via nitrification.

8. Calculate the kg O_2/d consumed via nitrification

D. Prepare a table with separate columns for the one-sludge and two-sludge systems. Each column should have entries of VSS production rate and O_2 consumed for BOD oxidation, nitrification, and total.

9.10. Estimate the required volume of a Biocarbone (also called BAF) process to treat wastewater that has a flow rate of 500 m^3/d, and influent NH_4^+-N concentration of 100 mg N/l, and a temperature of 25 °C. You can use the ammonium-oxidizers kinetic parameters from Table 9.1. Assume $b_{det} = 0.1$ d^{-1}, that you wish to achieve an effluent concentration of 1 mg NH_4^+-N/l, diameter, a porosity of 0.4, $L = 50$ μm, and $X_f = 40$ mg/cm^3. You should assume that the reactor is moderately well mixed by aeration, such that it has two segments in series (Hint: the normalized curves should be a good approach).

9.11. A reasonable estimate for θ_x is $1/b_{det}$ for a biofilm process. A plug-flow, fixed-bed nitrification process is to have an effluent concentration of 0.2 mg NH_4^+-N/l when it is operated in the low load region and with a temperature of 15 °C. At approximately what sludge age is the biofilm? Does this value make sense? Use the 15 °C kinetic parameters for ammonium-oxidizers in Table 9.1 to make your analysis.

9.12. You are attending a meeting with your client who operates a two-stage trickling-filter system. The following sketch is shown to you:

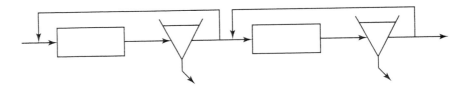

The two trickling filters are of the biological-tower type. They each have a depth of 8 m, a diameter of 16 m, and plastic media having specific surface area of 200 m^{-1}. Each clarifier is 3 m deep and has a diameter of 32 m. In addition, you are given the following information: influent flow rate (Q) = 10,000 m^3/d, influent $BOD_5 = 600$ mg/l, and influent TKN = 100 mg/l. You have only a few minutes to give your expert opinion concerning whether or not this system ought to be able to successfully achieve BOD oxidation and nitrification. Make a rapid (but relevant and expert) analysis of the system. Explain why the system should or should not achieve the goal. Use the first-stage and overall BOD and TKN loads.

9.13. Bioaugmentation is a relatively recent concept for biological processes. The idea is to add a biological material, such as special bacteria or enzymes, to enhance the performance of the process. One application is nitrification. Here, nitrifying bacteria are added to the influent of a reactor to "bioaugment" nitrification. You are to analyze how well bioaugmentation will work to upgrade an aerated lagoon. The lagoon's current parameters are:

> Flow rate: 1,000 m^3/d
>> Influent BOD$_L$: 20 mg/l
>> Volume: 3,000 m^3
>> Temperature: 10 °C
> Influent NH$_4^+$-N: 200 mg/l
>> D.O.: 4 mg/l
>> Effluent NH$_4^+$-N: variable, but mostly near 200 mg/l

The bioaugmentation materials consist of freeze-dried *Nitrosomonas*. The cost is $100/kg VSS, and you can assume that the VSS are 100 percent active. How much money will it cost to bioaugment with the *Nitrosomonas* materials if you are to have a stable effluent concentration of 0.1 mg N/l of NH$_4^+$-N? You can use the kinetic parameters in Table 9.1 for 10 °C. You may ignore any heterotroph growth and SMP. Assume that the oxygen supply is adequate.

9.14. You wish to design a one-sludge activated sludge process to nitrify. The influent has these characteristics: BOD$_L$ = 250 mg/l, TKN = 100 mg/l, NO$_3^-$ = 40 mg N/l, inert VSS$_i$ = 35 mg/l, and alkalinity = 1,000 mg/l as CaCO$_3$. Determine the O$_2$ demand (in mg O$_2$/l of wastewater) if the following parameters are valid:

Process	θ_x	$\hat{q}$	K	b	Y	f_d
Heterotrophs	15 d	$\dfrac{10 \text{ mg VSS}_a}{\text{mg BOD}_L}$	10 mg BOD$_L$/l	0.15 d^{-1}	0.45 mg VSS$_a$/mg BOD$_L$	0.8
Nitrifiers (total)	15 d	$\dfrac{2 \text{ mg NH}_4^+\text{-N}}{\text{mg VSS}_a\text{-d}}$	1.5 mg NH$_4^+$-N/l	0.15 d^{-1}	0.41 g VSS$_a$/g NH$_4^+$-N (total oxidation)	0.8

9.15. A biological tower has a height of 10 m, a square cross-section of 5 m × 5 m, and a modular media with a = 200 m^{-1}. It receives influent having a flow rate of 350 m^3/d, 600 mg BOD$_L$/l, and 50 mg TKN/l. Effluent recycle is at 350 m^3/d. Evaluate whether or not this process ought to be successful at nitrification and BOD removal.

9.16. You have a wastewater with the following characteristics: Q = 2,000 m^3/d, BOD$_L$ = 400 mg/l, TKN = 50 mg/l, and NO$_3^-$ + NO$_2^-$-N = 0 mg/l. You decide to use a multistage RBC to achieve excellent BOD and NH$_4^+$ removals. You design the first stage just for BOD oxidation. Its surface loading is the "traditional" first-stage one: 45 kg $BOD_5/1,000$ m^2 d, which converts to 66 kg $BOD_L/1,000$ m^2-d. You then follow that first stage with three more stages of exactly the same "size." Using the normalized loading curves, you estimate that the BOD_L concentration will decline along the RBC stages according to:

Location	Influent	Stage 1 effluent	Stage 2 effluent	Stage 3 effluent	Stage 4 effluent
BOD_L concentration (mg/l)	400	135	50	30	20

Will this RBC system achieve excellent removal of NH_4^+-N? If yes, in which stages will it occur? If not, why is NH_4^+-N removal a failure?

chapter

10

DENITRIFICATION

Denitrification is the dissimilatory reduction of NO_3^- or NO_2^- to (mainly) N_2 gas. In other words, NO_3^- or NO_2^- is the electron acceptor used in energy generation. Denitrification is widespread among heterotrophic and autotrophic bacteria, many of which can shift between oxygen respiration and nitrogen respiration.

In environmental biotechnology, denitrification is applied when the complete removal of N is required. Key examples include advanced treatment of wastewater discharged to watersheds that must be protected against eutrophication; treatment of high-N wastes, such as agricultural runoff and wastewater from feedlots; and treatment of drinking waters that contain elevated $NO_3^- + NO_2^-$ levels, thereby reducing risks to infants. In each of these applications, soluble N is converted to N_2 gas, which evolves to the atmosphere. In order to have denitrification, the nitrogen must be in one of its oxidized forms, NO_3^- or NO_2^-. Because many wastewaters contain reduced nitrogen, denitrification frequently is coupled to nitrification, which is needed to create the oxidized nitrogen.

This chapter is divided into three sections. The first section reviews the fundamentals of denitrification and denitrifying bacteria. Keys are the stoichiometric and kinetic characteristics of the denitrification reactions. The second section describes so-called *tertiary denitrification*, in which the water to be treated by denitrification does not contain the necessary electron donor; thus, an exogenous electron donor must be supplied. The third section describes the so-called *one-sludge denitrification*, in which the water contains an electron donor that can drive denitrification.

10.1 PHYSIOLOGY OF DENITRIFYING BACTERIA

Denitrification is widespread in nature. Denitrifiers are common among the Gram-negative *Proteobacteria*, such as *Pseudomonas*, *Alcaligenes*, *Paracoccus*, and *Thiobacillus*. Some Gram-positive bacteria, including *Bacillus*, can denitrify. Even a few halophilic Archaea, such as *Halobacterium*, are able to denitrify. All the denitrifiers are facultative aerobes, which means that they shift to NO_3^- or NO_2^- respiration when O_2 becomes limiting.

The denitrifiers used in environmental biotechnology are chemotrophs that can use organic or inorganic electron donors. Those that utilize organic electron donors are heterotrophs and are widespread among the *Proteobacteria*. A more limited group of autotrophs can utilize H_2 and reduced sulfur. Because of their great metabolic diversity, denitrifiers are commonly found in soils, sediments, surface waters, groundwaters, and wastewater treatment plants.

Denitrification proceeds in a stepwise manner in which nitrate (NO_3^-) is sequentially reduced to nitrite (NO_2^-), nitric oxide (NO), nitrous oxide (N_2O), and N_2 gas. Each half-reaction and the enzyme catalyzing it are shown below.

$$NO_3^- + 2\,e^- + 2\,H^+ = NO_2^- + H_2O \qquad \text{Nitrate Reductase}$$
$$NO_2^- + e^- + 2\,H^+ = NO + H_2O \qquad \text{Nitrite Reductase}$$
$$2\,NO + 2\,e^- + 2\,H^+ = N_2O + H_2O \qquad \text{Nitric Oxide Reductase}$$
$$N_2O + 2\,e^- + 2\,H^+ = N_2(g) + H_2O \qquad \text{Nitrous Oxide Reductase}$$

The overall reaction from NO_3^- to N_2 reduces the N by 5 electron equivalents per N. The first step is a two-electron reduction, while the following three steps are one-electron reductions for each N.

The oxygen concentration controls whether or not the facultative aerobes respire nitrogen. Oxygen can control denitrification in two ways. The first is repression of the several nitrogen-reductase genes. Research with *Pseudomonas stutzeri* indicates that these genes are repressed by D.O. concentrations greater than 2.5 to 5 mg O_2/l (Körner and Zumft, 1989). The second control mechanism is inhibition of the activity of the reductase by D.O. concentrations greater than a few tenths of a mg O_2/l (Tiedje, 1988; Rittmann and Langeland, 1985). The fact that the D.O. concentrations that repress the reductase genes are much higher than the concentrations that inhibit their activity means that denitrification can occur when D.O. concentrations are well above zero (Rittmann and Langeland, 1985; deSilva, 1997). This situation is enhanced when the denitrifying bacteria are located inside flocs or biofilms, where the oxygen concentration is lower than in the bulk liquid (Rittmann and Langeland, 1985).

Very low concentrations of the electron donor or too high concentrations of D.O. concentration can lead to accumulation of the denitrification intermediates: NO_2^-, NO_2, and N_2O. The latter two are greenhouse gases whose release should be avoided. A low concentration of the donor limits the supply of electrons to drive the reductive half-reactions. A high O_2 concentration tends to repress the nitrite and nitrous oxide reductases before the nitrate reductase is repressed.

Although denitrifiers are not especially pH sensitive, pH values outside the optimal range of 7 to 8 can lead to accumulation of intermediates. In low alkalinity waters, pH control can become an issue, because denitrification produces strong base. Base production is illustrated by the balanced reactions in which acetate and H_2 are electron donors for the heterotrophic and autotrophic denitrifiers, respectively:

$$CH_3COOH + \frac{8}{5}\,NO_3^- + \frac{4}{5}\,H_2O \rightarrow \frac{4}{5}\,N_2 + 2\,H_2CO_3 + \frac{8}{5}\,OH^- \qquad \textbf{[10.1]}$$

$$4\,H_2 + \frac{8}{5}\,NO_3^- \rightarrow \frac{4}{5}\,N_2 + \frac{8}{5}\,OH^- + \frac{16}{5}\,H_2O \qquad \textbf{[10.2]}$$

In both cases, the water's alkalinity is increased by the $\frac{8}{5}$ equivalents of strong base produced when $\frac{8}{5}$ mol of NO_3^--N is reduced. In mass terms, that is an alkalinity increase of $(50)/(14) = 3.57$ g as $CaCO_3$/g NO_3^--N consumed. For the acetate case, the effect is altered slightly, because 2 mol of a weaker acid (H_2CO_3, $pK_a = 6.3$) replace 1 mol of a weak acid (CH_3COOH, $pK_a \approx 4.3$).

Although heterotrophic denitrifiers exhibit a nearly infinite range for their organic substrates, a few simple organic substrates have been intensively studied. This intensive study came about because several early applications were for systems in which high NO_3^--N levels occurred in waters that had little or no BOD (McCarty, Beck, and St. Amant, 1969). Agricultural runoff and advanced treatment of secondary effluents were the critical applications. Thus, research addressed exogenous electron donors and carbon sources. Simple compounds that can be purchased in bulk quantity were evaluated: methanol, acetate, glucose, ethanol, and a few others. Because methanol (CH_3OH) was relatively inexpensive, it gained widespread use, and a very large database on methanol has been developed. Being a one-carbon compound, methanol has some unusual characteristics as an organic electron donor. Although the large database on methanol is a great resource for analyzing situations when methanol is the exogenous donor, it cannot be applied directly for situations in which another organic molecule is the donor.

Table 10.1 summarizes representative stoichiometric and kinetic parameters (at 20 °C) for heterotrophs using methanol, other organic material, H_2, and elemental sulfur (S) as the electron donor for denitrification. Table 10.1 utilizes the very useful concept of *oxygen demand (OD)*. OD represents the mass of electron donor expressed

Table 10.1 Representative stoichiometric and kinetic parameters for denitrifiers $(T = 20 °C)$

Electron Donor	Methanol	BOD	H_2	S^0
C-source	methanol	BOD	CO_2	CO_2
f_s^0	0.36	0.52	0.21	0.13
Y, g VSS_a/g donor	0.27	0.26	0.85	0.10
g VSS_a/g OD	0.18	0.26	0.11	0.07
$\hat{q}$, g donor/g VSS_a-d	6.9	12	1.6	8.1
g OD/g VSS_a-d	10.4	12	11.8	11.2
K, mg donor/l	9.1	1	1	?
mg OD/l	13.7	1	0.13	?
b, d^{-1}	0.05	0.05	0.05	0.05
$[\theta_x^{min}]_{lim}$, d	0.55	0.33	0.76	1.3
S_{min}, mg donor/l	0.25	0.017	0.04	?
mg OD/l	0.38	0.017	0.005	?
D, cm^2/d	1.3	1.0	0.9	—
J_R, kg OD/1,000 m^2-d	1.5	0.5	1.2	?
S_{min}^* (no detachment)	0.027	0.017	0.040	0.066
($b_{det} = 0.2$/d)	0.15	0.087	0.23	0.45
K^*	1.8	0.4	2.2	?

Notes: For K^*, $L = 40$ μm, $D_f/D = 0.8$, and $X_f = 40$ mg VSS_a/cm^3. ? = not yet determined—not applicable.

in oxygen equivalents. This is the same as the mass of O_2 required for complete oxidation of the donor. Eight grams OD equals one e^- eq. For organic donors, the OD equals the BOD_L.

The values in Table 10.1 lead to the following observations about denitrifiers and how processes involving denitrification should perform.

1. While the f_s^0 value for heterotrophs using general BOD_L is only slightly smaller than f_s^0 for aerobic heterotrophs (around 0.6 e^- eq synthesis/e^- eq donor), the f_s^0 values for the two autotrophs are much smaller, similar to nitrifiers. The f_s^0 value for the one-carbon oxidizers that consume methanol is lower than for other heterotrophs. True yield values parallel the f_s^0 values.

2. Since $\hat{q}$ and b values are roughly similar (cf. 12 g OD/g VSS_a-d and 0.05/d), $[\theta_x^{min}]_{lim}$ is controlled mainly by Y.

3. S_{min} values are less than 1 mg OD/l, which means that high residuals of BOD in the effluent are not a special problem.

4. For biofilm processes, all microbial types show high growth potential (low S_{min}^*) when detachment is negligible. On the other hand, the autotrophic processes are subject to growth limitations as b_{det} increases.

5. The K^* value for biofilm processes is not large. This means that external mass transport controls the substrate flux when the biofilm loading is in the medium- or high-load region.

In summary, the heterotrophic denitrifiers have kinetic characteristics similar to aerobic heterotrophs. Because they are facultative aerobes, the shifts from O_2 respiration to NO_3^- or NO_2^- respiration causes only a small decrease in f_s^0 and Y, which gives only modest increases in $[\theta_x^{min}]_{lim}$. Thus, denitrification processes should perform similarly to aerobic processes used for BOD_L removal.

On the other hand, the kinetic characteristics of heterotrophic denitrifiers and autotrophic nitrifiers are very different. The nitrifiers have lower f_s^0 values, are much slower growers, and require substantially longer solids retention times. Furthermore, maximum nitrification rates require a high D.O. concentration, while high D.O. concentration slows or stops denitrification. Because nitrification often is necessary to provide the NO_3^- or NO_2^- for denitrification, process design and operation must reconcile these conflicting physiological characteristics. It is the reconciliation of the needs of the nitrifiers and heterotrophic denitrifiers that distinguishes the different approaches to denitrification in environmental biotechnology.

One important feature of the denitrifying bacteria is that they often use NO_3^- (or NO_2^-) as the N source for cell synthesis. The added electron cost of reducing the N source to the -3 oxidation state reduces f_s^0 and the true yield. For example, using NO_3^- as the N source requires 8 extra electron equivalents per mole of biomass, represented as $C_5H_7O_2N$. As shown in Table 2.4, $C_5H_7O_2N$ requires 20 electron equivalents to reduce the C to oxidation state zero in all cells (reaction C-1), but it requires 28 electron equivalents to reduce the C and N when the N source is NO_3^- (reaction C-2). Thus when NO_3^- is the N source, the oxygen demand of biomass is (28 e^- eq/mol cells) (1 mol cells/113 g cells) (8 g O_2/e^- eq) = 1.98 g OD/g cells, not 1.42 g OD/g cells.

10.2 TERTIARY DENITRIFICATION

Denitrification processes can be divided into two major classes: tertiary versus one-sludge. The distinction is based on whether or not an exogenous electron donor is added. Tertiary denitrification requires the addition of an exogenous donor, while one-sludge denitrification uses a donor already present in the wastewater. This section describes tertiary denitrification used in wastewater treatment. Its application for drinking water is described in Chapter 12.

Tertiary denitrification is appropriate whenever the water to be treated contains NO_3^- or NO_2^-, but little or no electron donor. This situation occurs naturally with agricultural runoff contaminated with nitrogen fertilizers. Drinking-water supplies in agricultural regions also contain high NO_3^- levels, but little electron donor. Tertiary denitrification also follows aerobic biological processes (i.e., secondary treatment) of wastewater. When the secondary treatment oxidizes all electron donors originally present, reduced nitrogen is oxidized to NO_3^-, while reduced carbon is mineralized. The outcome of secondary treatment is N converted to NO_3^-, but almost no donors. In fact, the terminology tertiary denitrification derives from its following directly after secondary treatment.

Organic electron donors are most commonly supplied, and they promote the accumulation of heterotrophic denitrifiers. In terms of physiology and kinetics, the heterotrophic denitrifiers are very similar to the aerobic heterotrophs used for BOD oxidation. Thus, many design criteria for heterotrophic denitrification processes are close to those for aerobic BOD removal. One major difference, however, is that denitrification requires no oxygen supply. This difference relieves several design constraints that rule how "high rate" an aerobic process can be.

Almost any organic compound could be used as an exogenous electron donor. Historically, methanol was chosen for its economic benefits, not because it is a "better" exogenous electron donor than any other choice. When available, concentrated organic wastes can be used as an inexpensive (or even free) electron donor. Waste streams from the food-processing and beverage industries are most often employed because they have high BOD concentrations (greater than 10,000 mg BOD_L/l) and very high C-to-N ratios. A high C-to-N ratio, characteristic of a carbohydrate source, provides a potent supply of electrons, but releases little reduced nitrogen. Since the goal of denitrification is total removal of N from the water, addition of reduced N would defeat the goal.

Inorganic electron donors also can be used and are gaining popularity. Hydrogen gas (H_2) is an excellent electron donor for autotrophic denitrification. Its advantages include lower cost per electron equivalent compared to organic compounds, less biomass production than with heterotrophs, and absolutely no reduced nitrogen added. The main disadvantage of H_2 in the past has been lack of a safe and efficient H_2-transfer system. The recent development of membrane-dissolution devices overcomes the explosion hazard of conventional gas transfer and makes H_2 a viable alternative (Lee and Rittmann, 2000).

Reduced sulfur also can drive autotrophic denitrification. The most common source of reduced S is elemental sulfur, $S(s)$, which is oxidized to SO_4^{2-}. The S normally is embedded in a solid matrix that includes a solid base, such as $CaCO_3$, because the oxidation of S(s) generates strong acid.

$$S(s) + \frac{6}{5} NO_3^- + \frac{2}{5} H_2O \rightarrow SO_4^{2-} + \frac{3}{5} N_2 + \frac{4}{5} H^+ \qquad \textbf{[10.3]}$$

In almost all circumstances, the electron donor is the rate-limiting substrate. Even though NO_3^- may be the target pollutant, the electron donor controls the kinetics of substrate utilization and biomass growth. Thus, design normally should be based on the donor as the rate-limiting component. Example 10.1 illustrates how key stoichiometric features are computed for different donors.

Example 10.1 | **STOICHIOMETRY OF DENITRIFICATION REACTIONS**   Three different denitrification schemes are being tested: heterotrophic with methanol as the donor, heterotrophic with acetate as the donor, and autotrophic with H_2 as the donor. From Table 10.1, we use f_s^0 values of 0.36, 0.52, and 0.21, respectively. Each reactor is operated with an SRT of 15 d. We are to compute the overall stoichiometric reaction for each system under the operating conditions.

First, each system has its own f_s and f_e values computed from

$$f_s = f_s^0 \frac{1 + (1 - f_d)b\theta_x}{1 + b\theta_x}$$

When $b = 0.05/d$ and $f_d = 0.8$ for all systems, we compute:

Reaction type	f_s	f_e
Heterotrophic with methanol	0.267	0.733
Heterotrophic with acetate	0.342	0.658
Autotrophic with H_2	0.138	0.862

Second, the relevant half reactions are:

Cell synthesis with NO_3^- as N source (Equation C-2, Table 2.4, same for all)

$$\frac{1}{28} NO_3^- + \frac{5}{28} CO_2 + \frac{29}{28} H^+ + e^- = \frac{1}{28} C_5H_7O_2N + \frac{11}{28} H_2O$$

(Note here that 1 e$^-$ eq of biomass is $\frac{1}{28}$ mol of $C_5H_7O_2N$. If NH_4^+ were available for synthesis, Equation C-1, Table 2.4, should be used instead.).

Acceptor (same for all)

$$\frac{1}{5} NO_3^- + \frac{6}{5} H^+ + e^- = \frac{1}{10} N_2 + \frac{3}{5} H_2O$$

Donors

Methanol

$$\frac{1}{6} CO_2 + H^+ + e^- = \frac{1}{6} CH_3OH + \frac{1}{6} H_2O$$

Acetate

$$\frac{1}{8} CO_2 + \frac{1}{8} HCO_3^- + H^+ + e^- = \frac{1}{8} CH_3COO^- + \frac{3}{8} H_2O$$

H_2 Gas

$$H^+ + e^- = \frac{1}{2} H_2$$

Third, the half-reactions are combined according to

$$R = -R_{\text{donor}} + f_e R_{\text{acceptor}} + f_s R_{\text{cell}}$$

Heterotrophic with Methanol

$$0.1667 \, CH_3OH + 0.1561 \, NO_3^- + 0.1561 \, H^+$$

$$= 0.00954 \, C_5H_7O_2N + 0.0733 \, N_2 + 0.3781 \, H_2O$$

$$+ 0.119 \, CO_2$$

Heterotrophic with Acetate

$$0.125 \, CH_3COO^- + 0.1438 \, NO_3^- + 0.1438 \, H^+$$

$$= 0.0122 \, C_5H_7O_2N + 0.0658 \, N_2 + 0.125 \, HCO_3^-$$

$$+ 0.0639 \, CO_2 + 0.1542 \, H_2O$$

Autotrophic with H_2 Gas

$$0.5 \, H_2 + 0.1773 \, NO_3^- + 0.0246 \, CO_2 + 0.1773 \, H^+$$

$$= 0.00493 \, C_5H_7O_2N + 0.0862 \, N_2$$

$$+ 0.5714 \, H_2O$$

Table 10.2 summarizes key stoichiometric features of the reactions. The table documents that the observed yield of denitrifiers—expressed as f_s, the ratio of net VSS produced per g NO_3^- consumed, or the ratio of net VSS produced per g OD consumed—declines significantly from acetate to methanol to H_2 as the donor. Another key trend is that 4.3 to 13 percent of the e^- eq of NO_3^--N consumed is used as the N source in synthesis; again, the greatest percentage is associated with the greatest f_s. The table also points out that 28.6 percent ($8 \cdot 100/28$) of the oxygen demand of the biomass is invested in reducing the N source to the -3 oxidation state.

The key ratios in Table 10.2 give a quick guide to practical features of the design. For example, heterotrophic denitrification with acetate requires an input of almost 4 g BOD_L for each g NO_3^--N removed. At the same time, it produces about 0.69 g VSS of biomass and 3.6 g as $CaCO_3$ of alkalinity. The former determines the sludge-wasting rate, while the latter can be used to decide if the water is sufficiently buffered.

10.2.1 ACTIVATED SLUDGE

Tertiary denitrification with activated sludge is a common approach for treating nitrate-bearing waters. The basic configuration is the same as for aerobic treatment using activated sludge: a mixed reactor, a quiescent settler, sludge recycle, and sludge wasting to control the solids retention time. The design SRT generally is around 5 d when a heterotrophic electron donor is added. It is longer (e.g., 15 d) when denitrification is autotrophic.

Although denitrification by activated sludge resembles aerobic activated sludge in many ways, several crucial aspects are distinctly different. First, the reactor is

Table 10.2   Summary of stoichiometry for various denitrification reactions at $T = 20\ °C$ (Example 10.1)

Reaction Type	Heterotrophic with Methanol	Heterotrophic with Acetate	Autotrophic with H_2
f_s	0.267	0.342	0.138
Electron equivalents in donor	1	1	1
Electron equivalents in biomass			
Total $(= f_s)$	0.267	0.342	0.138
in C $(= \frac{20}{28} \cdot f_s)$	0.191	0.244	0.099
in N $(= \frac{8}{28} \cdot f_s)$	0.076	0.098	0.039
NO_3^- consumed			
mol	0.1561	0.1438	0.1773
e^- eq as acceptor $(= f_e)$	0.733	0.658	0.862
e^- eq as N source	0.076	0.098	0.039
e^- eq total	0.809	0.756	0.901
Net H^+ consumed			
H^+ equivalents	0.1561	0.1438	0.1773
Key ratios			
g OD/g NO_3^--N	3.66	3.97	3.22
g alk as $CaCO_3$/g NO_3^--N	3.57	3.57	3.57
g VSS/g NO_3^--N	0.490	0.685	0.224
g VSS/g OD $(= Y_n)$	0.135	0.172	0.0696

designed to minimize aeration. Although the liquid contents need to be well mixed, contact with the air needs to be minimized. This is achieved by subsurface mixing, usually with a submerged turbine. In some cases, the reactor is covered, although it is not the usual practice.

Second, supplementation with electron donor is required. Although the exact dose of donor can be computed with stoichiometry (e.g., Example 10.1), a rule of thumb is 4 g BOD_L/g NO_3^--N removed through denitrification. Extra electron donor must be supplied if O_2 enters the system. Organic donors can be supplied in concentrated liquid streams. Hydrogen is best added with bubbleless membrane dissolution. Elemental sulfur is not easily added in a suspended-growth system, since it is supplied by dissolution of a solid. In principle, sulfide (HS^-), sulfite (SO_3^{2-}), or thiosulfate ($S_2O_3^{2-}$) can be added in liquid form, but this practice is likely to lead to the proliferation of filamentous bacteria and sludge bulking.

Settler design is based on the same principles as for aerobic activated sludge. It might seem that rising sludge should be a problem; however, successful denitrification drives the NO_3^- concentration to a very low level in the reactor, which means that N_2 generation in the settler is minimal.

Sludge production can be computed from the SRT and stoichiometry. Again, Example 10.1 shows how to do the estimation. A rule of thumb is about 0.75 g VSS/g NO_3^--N removed by heterotrophic denitrification and an SRT near 5 d, but it is lower for autotrophs or if the SRT is increased.

BASIC DESIGN OF TERTIARY DENITRIFICATION USING ACTIVATED SLUDGE | **Example 10.2**

A wastewater flow of 1,000 m^3/d contains 50 mg/l of NO_3^--N, 2 mg/l of O_2, and essentially no electron donor. We are to complete a preliminary design that supplements with acetic acid and has an SRT of 5 d. The mixed-liquor volatile suspended solids is to have a concentration of 2,500 mg/l. Although aeration is minimized, O_2 is added at a rate equal to a concentration of 6 mg O_2/l in the influent flow. (Note that full aeration, as in aerobic activated sludge, would add hundreds of mg/l of oxygen to the water based on the influent flow.)

First, we must compute the dose rate of acetic acid, which must be added to remove the nitrate and the O_2. When $f_s^0 = 0.52$ for denitrification with acetic acid, $\theta_x = 5$ d, $b = 0.05$/d, and $f_d = 0.8$, then $f_s = 0.4368$, making the overall stoichiometry of the denitrification reaction

$$0.125\ CH_3COOH + 0.1282\ NO_3^- + 0.1282\ H^+$$

$$= 0.05632\ N_2 + 0.0156\ C_5H_7O_2N$$

$$+ 0.1720\ H_2CO_3 + 0.0875\ H_2O$$

From this stoichiometry, we see that the ratio of acetic acid consumption to NO_3^- consumption is

$$\frac{0.125\ \text{mol}\ CH_3COOH}{0.1282\ \text{mol}\ NO_3^-\text{-N}} = 0.975\ \frac{\text{mol}\ CH_3COOH}{\text{mol}\ NO_3^-\text{-N}}$$

Converting to mass units gives

$$0.975\ \frac{\text{mol}\ CH_3COOH}{\text{mol}\ NO_3^-\text{-N}} \cdot \frac{64\ \text{g}\ BOD_L}{\text{mol}\ CH_3COOH} \cdot \frac{\text{mol}\ NO_3^-\text{-N}}{14\ \text{g}\ N} = 4.46\ \frac{\text{g}\ BOD_L}{\text{g}\ N}$$

Thus, removal of all 50 mg/l of NO_3^--N requires $4.46 \times 50 = 223$ mg BOD_L/l of acetic acid.

The O_2 also must be reduced. Stoichiometry for the aerobic reaction can be obtained similarly from $f_s^0 = 0.6$, $b = 0.1$/d, $f_d = 0.8$, and $\theta_x = 5$ d.

$$0.125\ CH_3COOH + 0.14\ O_2 + 0.0157\ NO_3^- + 0.0157\ H^+$$

$$= 0.0157\ C_5H_7O_2N + 0.1714\ H_2CO_3 + 0.0315\ H_2O$$

The stoichiometry gives

$$\frac{0.125\ \text{mol}\ CH_3COOH}{0.14\ \text{mol}\ O_2} \cdot \frac{64\ \text{g}\ BOD_L}{\text{mol}\ CH_3COOH} \cdot \frac{\text{mol}\ O_2}{32\ \text{g}\ O_2} = 1.79\ \frac{\text{g}\ BOD_L}{\text{g}\ O_2}$$

This stoichiometry means that the acetic acid supplement to deplete the O_2 is

$$1.79\ \frac{\text{g}\ BOD_L}{\text{g}\ O_2} \cdot (2 + 6\ \text{mg}\ O_2/\text{l}) = 14\ \text{mg}\ BOD_L/\text{l}$$

In total, the acetic acid must be added to give $223 + 14 = 237$ mg BOD_L/l in the influent.

Second, the sludge production rate also is obtained from stoichiometry and assuming 100 percent removal of the acetic acid. The observed yields are

$$\frac{0.0156\ \text{mol}\ C_5H_7O_2N}{0.125\ \text{mol}\ CH_3COOH} \cdot \frac{113\ \text{g}\ VSS}{\text{mol}\ C_5H_7O_2N} \cdot \frac{\text{mol}\ CH_3COOH}{64\ \text{g}\ BOD_L} = 0.22\ \frac{\text{g}\ VSS}{\text{g}\ BOD_L}$$

for denitrification and

$$\frac{0.0157}{0.125} \cdot \frac{113}{64} = 0.22\ \text{g}\ VSS/\text{g}\ BOD_L$$

for aerobic oxidation. That the numbers are exactly the same is coincidental. Thus, the overall VSS production is

$$\frac{237 \text{ mg BOD}_L}{1} \cdot 0.22 \frac{\text{g VSS}}{\text{BOD}_L} \cdot 1{,}000 \text{ m}^3/\text{d} \cdot \frac{10^{-3} \text{ kg-l}}{\text{mg-m}^3} = 52 \text{ kg VSS/d}$$

Third, the volume of the system can be computed from θ_x, the VSS production rate, and the desired MLVSS:

$$V = \frac{\theta_x (\Delta X_v / \Delta t)}{X_v} = \frac{5 \text{ d}(52 \text{ kg VSS/d}) \cdot 10^6 \text{ mg/kg}}{(2{,}500 \text{ mg VSS/l}) \cdot (10^3 \text{ 1/m}^3)}$$

$$= 104 \text{ m}^3$$

The volumetric loading (QS^0/V) is 2.3 kg $\text{BOD}_L/1{,}000$ m^3-d, a value higher than normally used in aerobic activated sludge. A higher volumetric loading is feasible, because oxygen-transfer limitation is not an issue. However, settler design still must be considered and may force the system volume to increase, which might require that the MLVSS be lower for the same θ_x.

As a final note, supplementation with acetic acid adds acid to the system. However, denitrification consumes H^+ (or generates alkalinity) at a rate almost equal to the rate at which the acetic acid is removed. Therefore, the supplementation with acetic acid will not depress the pH in the reactor. The exact impact on the pH can be computed using principles of acid/base chemistry and considering all relevant weak acid/base systems.

10.2.2 BIOFILM PROCESSES

Denitrification is one of the easiest applications for a wide range of biofilm processes. Eliminating oxygen transfer relieves the biggest limitation on volumetric loading. In addition, accumulation by biofilm attachment eliminates volume constraints imposed by a settler. Thus, volumetric loading can be much higher than for aerobic biofilm processes.

Although nitrate is the target pollutant, the electron donor almost always is rate-limiting. Therefore, design of biofilm systems for denitrification should be based on the flux of the donor. Table 10.1 indicated that J_R for electron donors is near 1 kg OD/1,000 m^2-d. Of course, the exact value depends on the donor used and operating conditions, such as the detachment rate and temperature. Operation of different biofilm systems with J values for organic donors ranging from 15 to 22 kg $\text{BOD}_L/1{,}000$ m^2-d generally gives essentially complete N removal and BOD concentrations in the effluent acceptable for wastewater. Using H_2 as an autotrophic electron donor allowed nearly complete NO_3^- removal for fluxes up to 25 kg OD/1,000 m^2-d, which is the same as 3.1 kg H_2/1,000 m^2-d. By stoichiometry, NO_3^- fluxes (kg N/1,000 m^2-d) are one-third to one-fourth the value of the donor fluxes.

Almost any biofilm system works well for denitrification, as long as oxygen transfer can be kept to a minimum and plugging does not occur. Successful systems include:

- RBCs in which air ventilation is restricted
- Submerged fixed beds of rocks, sand, limestone, or plastic media
- Fluidized beds of sand, activated carbon, and pellets of ion-exchange resin
- Circulating beds of a range of lightweight particles
- Membrane bioreactors in which the membrane supplies H_2 and is the biofilm attachment surface.

Hydraulic detention times can be extremely low when support media of high specific surface area are utilized in a nonclogging configuration. For example, fluidized-bed treatment can give hydraulic detention times less than 10 min (Jeris, Beer, and Mueller, 1974). The volumetric loading in such cases can exceed 20,000 kg $BOD_L/1,000$ m^3-d.

FLUIDIZED-BED DENITRIFICATION USING H$_2$ The wastewater containing 50 mg NO_3^--N/l in a flow of 1,000 m^3/d is to be treated with autotrophic denitrification in a fluidized bed. The biofilm detachment rate will be controlled such that the average biofilm SRT is 15 d; this means that we can use the stoichiometry of Example 10.1. No O_2 transfer will occur in the system, and the influent D.O. is 4 mg/l. We provide a preliminary design for a fluidized bed in which we use 0.75-mm particles and a bed expansion that makes the specific surface area 4,000 m^2/m^3 of bed. We select an H_2 flux of 20 kg OD/1,000 m^2-d, which gives a high-load process ($J \gg J_R$, Table 10.1) | **Example 10.3**

Based on the stoichiometry in Example 10.1, the H_2 supply rate for denitrification is determined from

$$\left(\frac{\Delta H_2}{\Delta t}\right)_{den} = \frac{50 \text{ mg } NO_3^-\text{-N}}{l} \cdot \frac{1,000 \text{ m}^3}{d} \cdot \frac{0.5(2) \text{ g } H_2}{0.1773(14) \text{ g } NO_3^-\text{-N}} \cdot \frac{10^3 \text{ kg-1}}{\text{mg-m}^3} = 20.1 \text{ kg } H_2/d$$

This converts to

$$\frac{20.1 \text{ kg } H_2}{d} \cdot \frac{8 \text{ g OD}}{0.5(2) \text{ g } H_2} = 161 \text{ kg OD}/d$$

With a specific surface area of 4,000 m^2/m^3, the volume becomes

$$V = \frac{161 \text{ kg OD}}{d} \cdot \frac{1,000 \text{ m}^2\text{-d}}{20 \text{ kg OD}} \cdot \frac{\text{m}^3}{4,000 \text{ m}^2} = 2.01 \text{ m}^3$$

Then, the hydraulic detention time is

$$\theta_{calc} = \frac{V}{Q} = \frac{2.01 \text{ m}^3}{1,000 \text{ m}^3/d} = 0.002 \text{ d (2.9 min)}$$

This short detention time could result in a fairly high upward flow velocity, depending upon the cross-sectional area selected. As discussed in Chapter 8, the upward flow will need to be consistent with the settling velocity of the particles selected for the fluidized bed. The volumetric loading is

Volumetric Loading

$$= \frac{S^0 Q}{V}$$

$$= \frac{(50 \text{ mg NO}_3^- \text{-N/l})(1{,}000 \text{ m}^3/\text{d})}{2.01 \text{ m}^3} \cdot \frac{8 \text{ mg OD}}{0.1773(14) \text{ mg NO}_3^- \text{-N}} \cdot \frac{1{,}000 \text{ l}}{\text{m}^3} \cdot \frac{10^{-6} \text{ kg}}{\text{mg}}$$

$$= 80 \text{ kg OD/m}^3\text{-d} = 80{,}000 \text{ kg OD/1,000 m}^3\text{-d}$$

The H_2 demand for O_2 reduction must be added to the denitrification demand in order to know the total H_2 feed rate. We assume that $f_s = 0.25$ for the aerobic reaction.

$$\left(\frac{\Delta H_2}{\Delta t}\right)_{\text{ox}} = \frac{4 \text{ mg O}_2}{1} \cdot \frac{1}{1 - 0.25} \cdot \frac{1{,}000 \text{ m}^3}{\text{d}} \cdot \frac{1 \text{ g H}_2}{8 \text{ g O}_2} \cdot \frac{10^{-6} \text{ kg}}{\text{mg}} = 0.7 \text{ kg H}_2/\text{d}$$

Thus, the total H_2 feed rate is

$$(H_2)_{\text{total}} = 20.1 + 0.7 = 20.8 \text{ kg H}_2/\text{d}$$

Similar to oxygen, hydrogen is a sparingly soluble gas with solubility of about 2 mg/l at one atmosphere of hydrogen and room temperature. The 20.8 kg H_2/d represents a concentration in the 1,000 m^3/d of flow of about 21 mg/l, or ten times the H_2 solubility. Thus, a hydrogen-transfer process will be required and may affect the actual reactor detention time required. For the economical and safe use of hydrogen for denitrification, a novel approach may be required for its transfer to the wastewater, such as gas-permeable membranes (Lee and Rittmann, 2000).

10.3 ONE-SLUDGE DENITRIFICATION

One-sludge denitrification, sometimes called single-sludge or combined denitrification, involves using the BOD in the influent of a wastewater to drive denitrification. Thus, one-sludge denitrification cannot work as an add-on process after secondary treatment, because secondary treatment removes the organic electron donors. Instead, denitrification must be fully integrated with the aerobic processes, BOD oxidation and nitrification. This integration must reserve organic electron donor for anoxic denitrification, while at the same time providing aerobic conditions that allow full nitrification, which generates the nitrate for denitrification. At first glance, the twin goals seem to conflict with each other. In reality, they can be reconciled.

Attaining the twin goals of one-sludge denitrification provides many benefits for treating wastewater:

- Because no exogenous electron donor needs to be added, chemical costs are reduced over tertiary denitrification.

- Because some of the influent BOD is oxidized with nitrate as the electron acceptor, not O_2, aeration costs are reduced compared to alternative systems that nitrify the reduced-nitrogen forms in the influent.

- Full or nearly full N removal is achieved, thereby protecting receiving waters at risk from cultural eutrophication.

In recent years, engineers have developed many clever approaches for achieving the apparently conflicting goals of one-sludge denitrification. Despite a wide range of engineering configurations, all one-sludge processes rely on one or more of three basic strategies, which are described in the next section. How these strategies are combined to create more efficient processes is reviewed in the section that follows. Finally, a quantitative approach for design and analysis of one-sludge processes is presented.

10.3.1 BASIC ONE-SLUDGE STRATEGIES

Influent wastewater normally contains organic BOD and reduced nitrogen, often called total Kjeldahl nitrogen, or TKN. The TKN must be oxidized to NO_3^--N without oxidizing all the BOD before denitrification takes place. The three basic strategies for reserving organic electron donor while nitrification takes place are:

- Biomass storage and decay
- Classical predenitrification
- Simultaneous nitrification with denitrification

Figure 10.1 illustrates each of these strategies, which are described below.

Biomass storage and decay preserves some of the electrons from the original donor in biomass. The synthesis of biomass stores electron equivalents that originally came from the BOD and can be released through endogenous respiration to drive denitrification. This biomass approach often is associated with the famous Swiss engineer, K. Wuhrmann (Wuhrmann, 1964), and the approach can be called a *Wuhrmann biomass decayer.*

Figure 10.1 shows that biomass storage and decay involves an initial aerobic, activated sludge tank in which TKN is nitrified to NO_3^--N, while BOD is partly oxidized and partly stored through biomass synthesis. The mixed liquor then flows to an anoxic tank, where the NO_3^--N is respired to N_2 through endogenous respiration. The settler, sludge recycle, and sludge wasting are used for their usual functions of solids capture, solids recycle, and SRT control, respectively.

Although biomass storage and decay is a simple and effective way to reserve electrons for denitrification, it is not often employed as a stand-alone process. Two shortcomings explain why biomass storage and decay has limited applicability by itself. First, endogenous respiration has slow kinetics. A typical b value (Table 10.1) is 0.05/d for denitrification. The slow kinetics of endogenous respiration means that a high concentration of mixed-liquor volatile suspended solids and a long hydraulic detention time in the anoxic tank are necessary. Both lead to high capital costs, while the former can create operating problems with the settler and recycle. Second, the decay of biomass always releases NH_4^+-N. Thus, the anoxic step returns N and nitrogenous BOD to the water, albeit at concentrations lower than in the process influent.

Classical predenitrification, shown in Figure 10.1b, directly utilizes the influent BOD for denitrification. To do so, the first tank is anoxic. Influent BOD is the electron

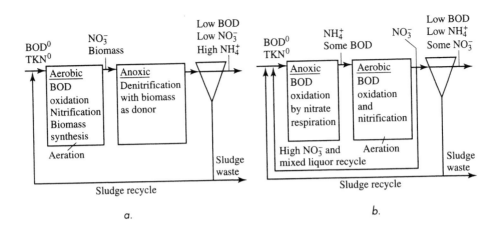

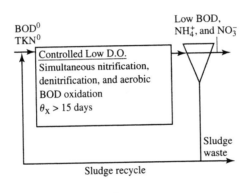

Figure 10.1   Schematic representations of one-sludge denitrification by *a.* biomass storage and decay, *b.* classical predenitrification, and *c.* simultaneous nitrification with denitrification.

donor for denitrification, which occurs in the first tank. The second tank, which is fully aerobic, is where the influent TKN is nitrified to NO_3^-, and any BOD not utilized in the anoxic tank is aerobically consumed. The nitrate formed in the aerobic tank is supplied to the anoxic tank by a large mixed-liquor recycle flow. The system is called *predenitrification,* because denitrification occurs in the first tank and precedes the aerobic reactions. This prepositioning of denitrification in the plant flow is how electron equivalents in the influent BOD are preserved for denitrification. The key to being able to use the influent BOD is the large recycle flow of nitrate from the aerobic tank. NO_3^- not recycled leaves in the effluent, and the fractional removal of N by denitrification is roughly equal to $Q^{r2}/(Q + Q^{r2})$, where Q is the plant flow rate and Q^{r2} is the mixed-liquor recycle flow rate. (This relationship is demonstrated in the last part of the section on one-sludge denitrification.) Therefore, recycle ratios (100 percent $\cdot$ Q^{r2}/Q) of 400 percent or more are employed to bring enough NO_3^- back to the anoxic tank so that total N removals are substantial.

Classical predenitrification has come into widespread use worldwide. Its advantages include the direct use of influent BOD for denitrification, which reduces aeration costs compared with strictly aerobic removal of BOD; faster kinetics than with biomass storage and decay; and no release of NH_4^+-N, as with biomass storage and decay. The main disadvantage of predenitrification is the large mixed-liquor recycle rate, which can substantially increase costs of piping and pumping.

The third approach to one-sludge denitrification is *simultaneous nitrification with denitrification,* illustrated in Figure 10.1c. When the dissolved oxygen concentration is poised at a suitably low level—typically less than about 1 mg/l—anoxic denitrification can occur in parallel to the aerobic reactions of nitrification and aerobic BOD oxidation. Three factors allow all reactions to occur simultaneously. First, the various nitrogen reductases are repressed only when the D.O. concentration is well above 1 mg/l. Second, inhibition of the nitrogen reductase is not severe when the D.O. concentration is less than 1 mg/l. Third, the D.O. concentration is depressed inside the aggregates that normally form in treatment systems; thus, denitrification can occur even more vigorously inside the floc (or biofilm), as long as the electron donor penetrates inside.

Essentially 100 percent N removal by simultaneous nitrification with denitrification has been documented (Rittmann and Langeland, 1985), and small amounts of denitrification probably occur in most activated sludge systems that nitrify and have D.O. concentrations below saturation (deSilva, 1997). When it is exploited reliably, simultaneous nitrification with denitrification offers all the advantages of predenitrification, but overcomes the main disadvantage, the high recycle rate. In effect, maintaining a low D.O. concentration throughout one reactor creates an infinitely high recycle ratio and allows essentially 100 percent N removal. The main drawback of implementing simultaneous nitrification with denitrification today is that we do not yet know the combinations of SRT, hydraulic retention time, and D.O. concentration that guarantee reliability. Past successes were documented for systems with conservative values of θ_x and θ.

The common feature of all one-sludge processes is that one community of microorganisms carries out all the reactions. *One community* is a synonym for *one sludge.* The nitrifying bacteria oxidize ammonia whenever D.O. is present. The heterotrophs switch back and forth between aerobic and anoxic respiration, or they do both simultaneously. Because the nitrifiers are autotrophs, their growth rate controls the SRT needed, but all microbial types have the same SRT. SRTs greater than 15 d are required in most cases, and sometimes much longer SRTs are used. The longer SRTs provide an added safety factor for the nitrifiers, who experience periods of low or zero D.O. The long SRTs mean that accumulation of inert suspended solids is important.

A practical outcome of a long SRT is that the hydraulic detention time (θ) needs to increase in order to keep the mixed-liquor suspended solids concentrations within reasonable limits dictated by settler performance. Settler parameters for one-sludge denitrification are similar to those used for extended-aeration activated sludge (Chapter 8). Hydraulic detention times in the reactors for predenitrification and simultaneous nitrification with denitrification normally are at least 10 h for typical sewage, and

24 h or greater are used in some instances. Very high strength wastewaters require even longer hydraulic detention times.

10.3.2 VARIATIONS ON THE BASIC ONE-SLUDGE PROCESSES

The disadvantages of the basic strategies for one-sludge denitrification served as impetus to develop schemes that combine the basic strategies in ways that overcome the limitations. Three variations are the Barnard process, sequencing batch reactors, and biofilm systems.

Probably the most famous combined process is named for its inventor, Dr. J. Barnard of South Africa (Barnard, 1975). The *Barnard process,* shown schematically in Figure 10.2, begins with classical predenitrification in reactors 1 and 2. The typical hydraulic retention time, based on the plant flow Q, is around 14 h for the

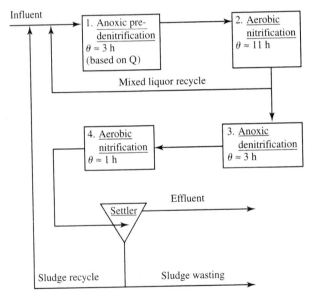

Reactions in each stage
- <u>Anoxic Denitrification1</u> — Influent BOD_L is the main e^- donor, while NO_3^--N recycled from (2) is the electron acceptor. In (1), NH_4^+ passes through unreacted, while most organic N from BOD_L is released as NH_4^+. Virtually, all recycled NO_3^- is removed as N_2 as long as sufficient BOD_L is available.
- <u>Aerobic Reactor 2</u> — The principal reaction is nitrification of the NH_4^+ entering (2). Also, any BOD_L not removed in (1) is oxidized in (2), and some cell decay also occurs.
- <u>Anoxic Denitrification 3</u> — The electron donor is cells from (2), and the electron acceptor is NO_3^- produced in (2).
- <u>Aerobic Reactor 4</u> — The main reaction is nitrification of NH_4^+ released by cell decay in (3). So, effluent from (4) has NO_3^- as the N species.
- <u>Settler</u> — Suspended solids are captured to produce a clear effluent and to provide a concentrated sludge underflow for recycling and wasting.

Figure 10.2 Schematic of the Barnard process (top) and the reactions occurring in each stage (bottom).

predenitrification stage, with a range from 7 to 24 h. With a recycle ratio for Q^{R2} of 400 percent, approximately 80 percent of the influent TKN is denitrified to N_2, while about 20 percent leaves reactor 2 as NO_3^--N. If the influent TKN were 50 mg/l, the NO_3^--N concentration leaving reactor 2 would be about 10 mg/l.

The effluent from reactor 2 flows to reactor 3, in which endogenous respiration fuels denitrification of the 10 mg NO_3^--N/l to N_2 gas. A typical hydraulic detention time is 3 h. Cell decay releases NH_4^+-N to the water with a ratio of roughly 0.3 mg NH_4^+-N/mg NO_3^--N. Thus, the effluent from the biomass decayer (reactor 3) contains about 3 mg NH_4^+-N/l.

Reactor 4 provides about 1 h of aeration in order to oxidize the NH_4^+-N to NO_3^--N. Hence, the effluent from the Barnard process contains NO_3^--N at about 6 percent of the influent TKN. The settler performs its usual functions and is designed with extended-aeration criteria.

The Barnard process is well established worldwide as a means to use one-sludge denitrification to achieve greater than 90 percent total N removal. Its main liabilities are the many tanks, the relatively long hydraulic detention time, and the significant mixed-liquor recycle between reactors 2 and 1. A closely related, but different process is the Bardenpho process, which is used for phosphorus removal and is discussed in Chapter 11.

The *sequencing batch reactor (SBR)* is an alternate means for achieving the same goals as predenitrification and the Barnard process. Figure 10.3 shows a typical 9-h cycle for achieving greater than 90 percent N removal, as with the Barnard process. During the anoxic fill stage, the influent BOD is denitrified with the NO_3^- carried over with the settled sludge from the previous cycle. The first aerobic react period oxidizes the influent TKN to NO_3^-, while residual BOD is utilized aerobically. The anoxic react period denitrifies mainly through endogenous respiration. The second aerobic react period (React 3) converts NH_4^+-N, released in anoxic react, to NO_3^-, which is discharged in the effluent (after settling) and carried over for the anoxic fill

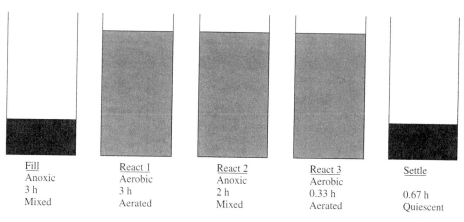

Fill	React 1	React 2	React 3	Settle
Anoxic	Aerobic	Anoxic	Aerobic	
3 h	3 h	2 h	0.33 h	0.67 h
Mixed	Aerated	Mixed	Aerated	Quiescent

Figure 10.3   A typical cycle (9 h) for achieving 90 percent N removal with a sequencing batch reactor.

stage of the next cycle. The five-step SBR has nearly the same functions in time as the 5-stage Barnard process has in space. The only significant difference is that the SBR only uses NO_3^- carryover from settle and draw to drive predenitrification, while the Barnard process mainly uses mixed-liquor recycle from the first aerobic stage.

Although the SBR cycle normally lasts for 8 to 12 h, the SRT and hydraulic detention time are similar to those of the Barnard process. The SRT is 15 d or greater, while the hydraulic detention time is 8 to 24 h. The SBR operation can be truncated to mimic only predenitrification. In this case, the anoxic fill, one aerobic react, and settling-plus-draw periods are retained.

Simultaneous nitrification with denitrification can be achieved with *biofilm processes,* as well as suspended growth. Due to the aggregated nature of biofilms, aerobic BOD oxidation, nitrification, and denitrification can take place in parallel within different layers of biofilms. Thus, simultaneous nitrification with denitrification by biofilms can be achieved in one reactor in a manner very much analogous to what occurs in activated sludge. The bulk D.O. concentration, the biofilm detachment rate, and the surface loading of BOD are the critical design parameters. Future research and development are needed to define the reliable ranges for these design parameters.

Biofilm predenitrification also is feasible. Figure 10.4 shows that an anoxic biofilm reactor precedes an aerobic biofilm reactor, and the aerobic effluent is recycled to provide NO_3^- to the anoxic biofilm. In biofilm predenitrification, two distinct microbial communities develop. Thus, it is not strictly "one sludge," although it achieves all the goals of suspended one-sludge processes. Because the biomass does not move with the water, the biofilm system shown in Figure 10.4 does not exploit biomass storage and decay. If further denitrification is needed, a tertiary process could be added to what is shown in Figure 10.4. Alternately, a third stage could be operated in the biomass storage-and-decay mode if it were operated in a sequencing mode with the second stage. The biomass would grow and store OD during the aerobic period, while it would perform endogenous denitrification when in the anoxic mode.

Each of the two biofilm reactors can be designed using principles of biofilm process design. The aerobic reactor is an example in which nitrification and aerobic

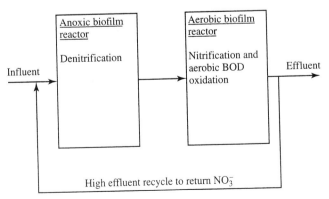

Figure 10.4   Schematic of biofilm predenitrification. A settler may be added to the effluent stream after the recycle flow is taken out.

BOD oxidation must coexist; this was described in Chapter 9. The anoxic reactor normally has a significant residual of BOD, while the NO_3^--N can be driven to a very low concentration. Thus, the rate-limiting substrate may be NO_3^-, not the BOD. Further research is needed before we know the importance of NO_3^- limitation in the anoxic biofilm reactor.

10.3.3 QUANTITATIVE ANALYSIS OF ONE-SLUDGE DENITRIFICATION

Design of a process to carry out one-sludge denitrification must take into account the important interactions between N species and BOD. This section develops and applies quantitative tools that make these connections, but are not excessively complicated. This balancing is achieved by exploiting the stoichiometric relationships that exist for the three key reactions: nitrification, aerobic BOD oxidation, and denitrifying BOD oxidation. Because most one-sludge denitrification processes use conservative designs in order to achieve stable nitrification, kinetics do not control the overall performance in terms of N transformations and BOD oxidation. Thus, a detailed kinetic analysis is not necessary, as long as the SRT is sufficiently long. Similarly, we do not explicitly include SMP, since the focus is on N removal more than BOD removal.

This section first presents a comprehensive design tool for predenitrification. It then applies that tool for distinctly different situations in which predenitrification can be applied. Finally, a design tool is developed for a biomass-decay system.

Modeling Predenitrification   Figure 10.5 identifies the tanks, flows, and concentrations for a classic predenitrification process. The recycle and waste flows are identified by recycle ratios: R^2 for the mixed-liquor recycle normally employed to return NO_3^--N to the (first) anoxic reactor, R^1 for the sludge recycling from the settler underflow, and R^w for the waste sludge from the settler underflow. The total system volume, V, is comprised of the volumes of the anoxic tank (V_{an}), aerobic tank (V_{aer}), and settler (V_{set}). As we did before, we assume that all volumes contain the same average volatile suspended solids concentration, X_v.

When the design is conservative, stoichiometry, not kinetics, controls the concentrations of substrates. Thus, we assume that NO_3^--N is completely denitrified in the anoxic reactor, or $(NO_3^-)^1 = 0$. Likewise, BOD_L and TKN are fully oxidized in the aerobic reactor, making $BOD_L^2 = TKN^2 = 0$, while the maximum amount of NO_3^--N is generated in the aerobic reactor, $(NO_3)^2$.

This analysis does not compute the volumes required for the anoxic reactor, aerobic reactor, and settler. The distribution of volume between the anoxic reactor and aerobic reactor can be estimated from successful practice, in which $V_{aer} = 1.5 V_{an}$. The settler volume should be based on the solids flux and overflow rate.

The main objective of the analysis procedure is to compute the relationships among the SRT, effluent NO_3^--N concentration, the influent BOD and TKN concentrations, the recycle flow ratios, and the volume of the system. First, the stoichiometric relationships are derived for each reactor. Second, these relationships are used to compute effluent NO_3^--N, the biomass production rates and concentrations, the system volume, the oxygen supply rate, and the BOD supply rate (if needed).

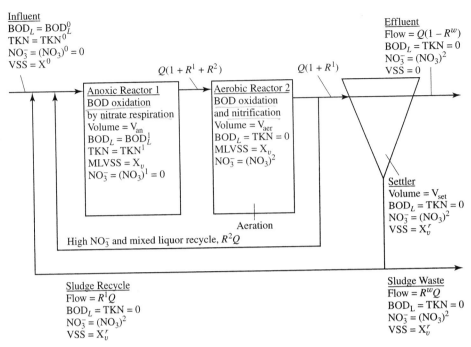

Influent
$BOD_L = BOD_L^0$
$TKN = TKN^0$
$NO_3^- = (NO_3)^0 = 0$
$VSS = X^0$

$Q(1 + R^1 + R^2)$

Anoxic Reactor 1
BOD oxidation
by nitrate respiration
Volume = V_{an}
$BOD_L = BOD_L^1$
$TKN = TKN^1$
$MLVSS = X_v$
$NO_3^- = (NO_3)^1 = 0$

$Q(1 + R^1)$

Aerobic Reactor 2
BOD oxidation
and nitrification
Volume = V_{aer}
$BOD_L = TKN = 0$
$MLVSS = X_v$
$NO_3^- = (NO_3)^2$

Aeration

High NO_3^- and mixed liquor recycle, R^2Q

Effluent
Flow = $Q(1 - R^w)$
$BOD_L = TKN = 0$
$NO_3^- = (NO_3)^2$
$VSS = 0$

Settler
Volume = V_{set}
$BOD_L = TKN = 0$
$NO_3^- = (NO_3)^2$
$VSS = X_v^r$

Sludge Recycle
Flow = R^1Q
$BOD_L = TKN = 0$
$NO_3^- = (NO_3)^2$
$VSS = X_v^r$

Sludge Waste
Flow = R^wQ
$BOD_L = TKN = 0$
$NO_3^- = (NO_3)^2$
$VSS = X_v^r$

Figure 10.5   Schematic of the tanks, volumes, flows, and concentrations used in the analysis of a classic predenitrification system.

In the anoxic reactor, denitrification occurs to the maximum degree possible. In other words, NO_3^--N is driven to zero when BOD is in excess, or BOD_L is driven to zero when NO_3^--N is in excess. The electron donor is the input BOD, which we represent as complex organic matter, $C_{10}H_{19}O_3N$. When $C_{10}H_{19}O_3N$ is oxidized as the electron donor for anoxic heterotrophs (NH_4^+-N is available as the N source), the donor half reaction is (Equation O-18, Table 2.3)

$$\frac{1}{50} C_{10}H_{19}O_3N + \frac{9}{25} H_2O = \frac{9}{50} CO_2 + \frac{1}{50} NH_4^+ + \frac{1}{50} HCO_3^- + H^+ + e^-$$

which is written for 1 electron equivalent, or 8 g BOD_L. The acceptor half-reaction is that for reduction of NO_3^--N to N_2 (Equation I-7, Table 2.2):

$$f_e \left(\frac{1}{5} NO_3^- + \frac{6}{5} H^+ + e^- = \frac{1}{10} N_2 + \frac{3}{5} H_2O \right)$$

The acceptor half-reaction is multiplied by f_e to account for the actual flow of electrons for energy generation under the operating SRT. The synthesis half-reaction (Equation C-1, Table 2.4), multiplied by f_s, is

$$f_s \left(\frac{1}{20}NH_4^+ + \frac{1}{5} CO_2 + \frac{1}{20} HCO_3^- + H^+ + e^- = \frac{1}{20} C_5H_7O_2N + \frac{9}{20} H_2O \right)$$

The three half-reactions can be combined in the usual manner ($R = -R_d + f_e R_a + f_s R_c$) if f_s and f_e are known. We can compute f_s and f_e if we choose an SRT for the predenitrification system using Equation 3.33:

$$f_s = f_s^0 \frac{1 + (1 - f_d)b\theta_x}{1 + b\theta_x}$$

$$f_e = 1 - f_s$$

Table 10.3 summarizes the results of the stoichiometric computations for the common case in which $f_s^0 = 0.52$, $b = 0.05/d$, and $f_d = 0.8$. The table describes how θ_x systematically alters f_s, f_e, the observed yield, and the ratios of N and BOD transformations. The computations of the ratios are given below the table. Key trends in the table are:

- As expected, f_s and $Y_{n(den)}$ for denitrification decrease with increasing θ_x, while f_e increases. For an SRT of 16 d, $Y_{n(den)} = 0.24$ g VSS/g BOD_L.

- The ratio of NO_3^--N consumed (as acceptor and N source), g NO_3^--N/g BOD_L, systematically increases as θ_x goes up, since more biomass is oxidized by endogenous respiration.

- The ratio g BOD_L/g NO_3^--N is the reciprocal of g NO_3^--N/g BOD_L and is very useful for determining whether BOD_L or NO_3^--N is limiting denitrification. For example, an SRT of 16 d has 4.3 g BOD_L/g NO_3^--N. This means that NO_3^--N can be limiting (and driven to zero in the anoxic reactor) only when the ratio of influent BOD_L to TKN is greater than approximately 4.3 g BOD_L/g NO_3^--N. We need less BOD_L per gram of input N when the SRT increases.

Table 10.3 Summary of stoichiometric ratios for denitrification of complex organic matter

θ_x, d	f_s	f_e	$Y_{n(den)}$	$\frac{g\ VSS}{g\ BOD_L}$	$\frac{g\ NO_3^-\text{-}N}{g\ BOD_L}$	$\frac{g\ BOD_L}{g\ NO_3^-\text{-}N}$	$\frac{g\ N_2}{g\ BOD_L}$	$\frac{g\ NH_4^+\text{-}N}{g\ BOD_L}$
5	0.44	0.56	0.31		0.20	5.07	0.20	-0.003
10	0.38	0.62	0.27		0.22	4.62	0.22	0.002
12	0.36	0.64	0.26		0.22	4.49	0.22	0.003
14	0.35	0.65	0.25		0.23	4.39	0.23	0.004
16	0.34	0.66	0.24		0.23	4.30	0.23	0.006
20	0.31	0.69	0.22		0.24	4.15	0.24	0.008
25	0.29	0.71	0.20		0.25	4.02	0.25	0.010
30	0.27	0.73	0.19		0.26	3.92	0.26	0.011
40	0.24	0.76	0.17		0.27	3.77	0.27	0.014
50	0.22	0.78	0.16		0.27	3.68	0.27	0.016
100	0.17	0.83	0.12		0.29	3.46	0.29	0.020

Notes:
1. $f_d = 0.8$, $f_s^0 = 0.52$, $Y = 0.37$ g VSS_a/g BOD_L, $b = 0.05/d$
2. $Y_{n(den)} = f_s(5.65$ g $VSS_a/8$ g $BOD_L)$, which assumes that NH_4^+ is the N source.
3. g NO_3^--N)/g $BOD_L = f_e(14/5$ g NO_3^--N)/(8 g $BOD_L)$
4. g N_2/g $BOD_L = f_e(14/5$ g $N_2)/(8$ g $BOD_L)$
5. g NH_4^+-N/g $BOD_L = [(14/50$ g Org-N)- $f_s(14/20$ g cell-N)]/(8 g $BOD_L)$

- The ratio g N_2/g BOD_L tells what portion of the N is denitrified, versus synthesized into heterotrophic biomass. For $\theta_x = 16$ d, 0.23 g N/g BOD_L goes to N_2, and this is 100 percent of the total N removal.

- The last ratio, g NH_4^+-N/g BOD_L, indicates whether or not the organic N present in the influent organic matter ($C_{10}H_{19}O_3N$) satisfies the cells' need for nitrogen in net synthesis. For all SRTs greater than 5 days, the N in the organic matter supplies enough NH_4^+-level N for synthesis. This explains why 100% of the NO_3^--N can go to N_2.

Nitrification and aerobic BOD oxidation occur simultaneously in the aerobic reactor. TKN that passes through the anoxic reactor is fully oxidized to NO_3^-. Any BOD_L not utilized in the anoxic reactor is fully oxidized aerobically. The stoichiometries of both reactions are important and have been fully developed in prior chapters. Thus, the presentations are focused on the most critical factors, the observed yields.

For nitrification, the donor, acceptor, and synthesis half-reactions are

$$\frac{1}{8} NH_4^+ + \frac{3}{8} H_2O = \frac{1}{8} NO_3^- + \frac{5}{4} H^+ + e^-$$

$$f_e \left(\frac{1}{4} O_2 + H^+ + e^- = \frac{1}{2} H_2O \right)$$

$$f_s \left(\frac{1}{20} HCO_3^- + \frac{1}{5} CO_2 + \frac{1}{20} NH_4^+ + H^+ + e^- = \frac{1}{20} C_5H_7O_2N + \frac{9}{20} H_2O \right)$$

These lead to the computation of the observed yield from

$$Y_{n(\text{nit})} = f_s \cdot \frac{113 \text{ g cells}/20 \text{ e}^- \text{ eq cells}}{\left(\frac{14}{8} + \frac{14}{20} f_s \right) (\text{g } NH_4^+\text{-N/e}^- \text{ eq } NH_4^+\text{-N})} = \frac{8.07 f_s}{2.5 + f_s}$$

f_s can be computed in the usual manner, and typical f_s^0 and b values are 0.127 and 0.05/d, respectively.

For aerobic BOD oxidation, a parallel development yields the following, in which typical f_s^0 and b values are 0.6 and 0.15/d, respectively:

$$\frac{1}{50} C_{10}H_{19}O_3N + \frac{9}{25} H_2O = \frac{9}{25} CO_2 + \frac{1}{50} NH_4^+ + \frac{1}{50} HCO_3^- + H^+ + e^-$$

$$f_e \left(\frac{1}{4} O_2 + H^+ + e^- = \frac{1}{2} H_2O \right)$$

$$f_s \left(\frac{1}{20} HCO_3^- + \frac{1}{5} CO_2 + \frac{1}{20} NH_4^+ + H^+ + e^- = \frac{1}{20} C_5H_7O_2N + \frac{9}{20} H_2O \right)$$

$$Y_{n(\text{aer})} = 0.706 f_s$$

in which the donor is BOD_L and units on $Y_{n(\text{aer})}$ are g cells/g BOD_L.

The analysis continues by computing the synthesis rates of the different types of biomass. The nitrifiers in the aerobic tank consume all of the input TKN that passed through the anoxic reactor, except for the N synthesized into heterotrophs.

$$\left(\frac{\Delta X_v}{\Delta t}\right)_{nit} = Y_{n(nit)} \left[Q \cdot TKN^0 - \left[\left(\frac{\Delta X_v}{\Delta t}\right)_{aer} + \left(\frac{\Delta X_v}{\Delta t}\right)_{den} \right] \cdot 0.124 \frac{g\ N}{g\ VSS} \right] \quad \textbf{[10.4]}$$

in which

$$\left(\frac{\Delta X_v}{\Delta t}\right)_{den} = \text{net mass production rate for denitrifiers in the anoxic reactor } [M_x T^{-1}]$$

$$\left(\frac{\Delta X_v}{\Delta t}\right)_{nit} = \text{net mass production rate for nitrifiers } [M_x T^{-1}]$$

$$\left(\frac{\Delta X_v}{\Delta t}\right)_{aer} = \text{net mass production rate for aerobic heterotrophs } [M_x T^{-1}]$$

The aerobic heterotrophs grow by consuming the entire BOD that escapes the anoxic reactor

$$\left(\frac{\Delta X_v}{\Delta t}\right)_{aer} = Y_{n(aer)} Q(1 + R^1 + R^2)BOD_L^1 \quad \textbf{[10.5]}$$

The denitrifying heterotrophs grow by consuming the influent BOD to the degree allowed by the available NO_3^--N.

$$\left(\frac{\Delta X_v}{\Delta t}\right)_{den} = Y_{n(den)} Q[BOD_L^0 - (1 + R^1 + R^2)BOD_L^1] \quad \textbf{[10.6]}$$

The amount of NO_3^--N that exits the aerobic reactor is the difference between the TKN entering it ($Q \cdot TKN^0$ on a mass/time basis) and the amount of NH_4^+-N synthesized into biomass $((\Delta X_v/\Delta t)_{nit} + (\Delta X_v/\Delta t)_{aer} + (\Delta X_v/\Delta t)_{den}) \cdot 0.124$ g N/g cells). Thus, the concentration of NO_3^--N leaving the aerobic reactor is computed by mass balance on N in the aerobic reactor:

$$\left(NO_3^-\right)^2 = \frac{1}{Q + R^2 Q + R^1 Q}$$
$$\cdot \left\{ Q \cdot TKN^0 - \left[\left(\frac{\Delta X_v}{\Delta t}\right)_{aer} + \left(\frac{\Delta X_v}{\Delta t}\right)_{nit} + \left(\frac{\Delta X_v}{\Delta t}\right)_{den} \right] \cdot 0.124 \frac{g\ N}{g\ VSS} \right\} \quad \textbf{[10.7]}$$

This critical equation shows that the effluent NO_3^--N concentration is controlled most strongly by the mixed-liquor recycle ratio R^2. To achieve a lower $(NO_3)^2$, R^2 must be increased. A typical range of R^2 values, 4 to 6, gives roughly 80 to 86 percent N removal.

The NO_3^--N recycled back to the anoxic reactor controls how much BOD is removed by denitrification. The BOD removed by denitrification is, on a mass/time basis, $Q(R^1 + R^2)(NO_3^-)^2/(g\ NO_3^-$-N/g $BOD_L)$. By mass balance, the concentration of BOD_L leaving the anoxic reactor is

$$BOD_L^1 = \frac{BOD_L^0 - (R^2 + R^1)(NO_3^-)^2 \cdot (g\ BOD_L/g\ NO_3^-\text{-N})}{1 + R^2 + R^1} \quad \textbf{[10.8]}$$

If BOD_L^1 is negative, the influent BOD is insufficient to drive full denitrification. To get full denitrification, supplemental BOD must be added, increasing BOD_L^0 so that BOD_L^1 is greater than or equal to zero. The minimum supplement of BOD occurs where $BOD_L^1 = 0$. In that case, the supplemental BOD_L concentration to add to the influent flow $Q(BOD_L^{sup})$ is

$$BOD_L{}^{sup} = \frac{(R^2 + R^1) \cdot (NO_3^-)^2}{\left(\dfrac{g\ NO_3^--N}{g\ BOD_L}\right)} - BOD_L^0 \qquad \textbf{[10.9]}$$

If supplemental BOD is not provided, $(NO_3)^1$ will not be zero, and $(NO_3)^1$ must increase. Solution for $(NO_3)^2$ when $BOD_L^1 = 0$ gives the highest value of $(NO_3)^2$:

$$(NO_3^-)_{max}^2 = \frac{BOD_L^0 \left(\dfrac{g\ NO_3^--N}{g\ BOD_L}\right)}{R^2 + R^1} \qquad \textbf{[10.10]}$$

Equations 10.4 to 10.8 must be solved simultaneously for the influent quality (BOD_L^0, TKN^0), influent flow (Q), selected SRT (θ_x), and selected recycle ratios ($R^2 + R^1$). The total sludge loss rate is then computed from

$$\left(\frac{\Delta X_v}{\Delta t}\right)_{tot} = \left(\frac{\Delta X_v}{\Delta t}\right)_{nit} + \left(\frac{\Delta X_v}{\Delta t}\right)_{aer} + \left(\frac{\Delta X_v}{\Delta t}\right)_{den} + QX_i^0 \qquad \textbf{[10.11]}$$

The hydraulic detention times (θ) or mixed-liquor volatile suspended solids concentration (X_v) is then computed from

$$\theta = \frac{\theta_x}{X_v Q} \cdot \left(\frac{\Delta X_v}{\Delta t}\right)_{tot} \qquad \textbf{[10.12]}$$

Either X_v or θ must be selected as a design judgment in order to compute the other parameter. The underflow sludge concentration (X_v^r) and the waste flow ratio (R^w) are determined from a mass balance around the settler:

$$X_v^r = \frac{\left(1 + R^1 - \dfrac{\theta}{\theta_x}\right)}{R^1} \cdot X_v \qquad \textbf{[10.13]}$$

$$R^w = \frac{\theta}{\theta_x} \cdot \frac{X_v}{X_v^r} \qquad \textbf{[10.14]}$$

The amount of N denitrified to N_2 gas ($\Delta N_2/\Delta t$ in MT^{-1}) can be computed from

$$\frac{\Delta N_2}{\Delta t} = Q \frac{(R^2 + R^1)(NO_3^-)^2}{\left(\dfrac{g\ NO_3^--N}{g\ BOD_L}\right)} \cdot \left(\frac{g\ N_2}{g\ BOD_L}\right) \qquad \textbf{[10.15]}$$

The required oxygen supply rate to satisfy aerobic BOD oxidation and nitrification

$(\Delta O_2/\Delta t$ in $MT^{-1})$ is

$$\left(\frac{\Delta O_2}{\Delta t}\right) = (Q + Q^1 + Q^2) \cdot BOD_L^1 + QX_i^0 \cdot 1.98\frac{g\ OD}{g\ VSS} + 4.57\frac{g\ OD}{g\ NH_4^+ \text{-} N}$$

$$\cdot \left[Q \cdot TKN^0 - \left(\left(\frac{\Delta X_v}{\Delta t}\right)_{aer} + \left(\frac{\Delta X_v}{\Delta t}\right)_{nit} + \left(\frac{\Delta X_v}{\Delta t}\right)_{den}\right) \cdot 0.124\frac{g\ N}{g\ VSS}\right]$$

$$- 1.98\frac{g\ OD}{g\ VSS} \cdot \left[\left(\frac{\Delta X_v}{\Delta t}\right)_{aer} + \left(\frac{\Delta X_v}{\Delta t}\right)_{nitr} + QX_i^0\right]$$

[10.16]

DESIGN OF A CLASSIC PREDENITRIFICATION SYSTEM FOR TYPICAL SEWAGE | Example 10.4

The design method is applied to domestic sewage containing $BOD_L^0 = 300$ mg/l; $TKN = 50$ mg/l; and $X_i^0 = 30$ mg/l. The f_s^0 and b values are:

	f_s^0	b, d^{-1}
denitrification	0.52	0.05
aerobic BOD oxidation	0.60	0.15
nitrification	0.11	0.11

Typical R^2 and R^1 values are selected as 6 and 0.25, respectively. The mixed-liquor volatile suspended solids are fixed at 2,000 mg/l.

Table 10.4 summarizes the design outputs for three sludge ages: 15, 30, and 50 d. It shows that the effluent NO_3^--N $((NO_3)^2)$ is only slightly affected by SRT. Greater biomass synthesis for lower θ_x increases N removal marginally. To increase the N removal, R^2 must be increased. For example, increasing R^2 to 10 reduces $(NO_3)^2$ to 3.8 mg/l for a θ_x of 30 d. The SRT shows more dramatic effects for sludge wasting, oxygen required, and hydraulic detention time. Increasing θ_x from 15 d to 50 d increases the hydraulic detention time 2.4 times (from 19 to 46 h), increases the O_2 requirement by 46 percent, and decreases the sludge wasting by 26 percent.

Table 10.4 Summary of key design outputs for classic predenitrification of a domestic sewage (BOD_L = 300 mg/l, TKN^0 = 50 mg/l, X_i^0 = 30 mg/l)

θ_x d	$Y_{n(den)}$ $\frac{g\ VSS}{g\ BOD_L}$	$\left(\frac{\Delta X_v}{Q\Delta t}\right)_{den}$ $\frac{mg\ VSS}{l}$	BOD_L rem by denitr $\frac{g\ NO_3^-\text{-}N}{g\ BOD_L}$	BOD_L rem by denitr $\frac{mg\ BOD_L}{l}$	$Y_{n(nit)}$ $\frac{g\ VSS}{g\ NH_4^+\text{-}N}$	$\left(\frac{\Delta X_v}{Q\Delta t}\right)_{nit}$ $\frac{mg\ VSS}{l}$	$Y_{n(aer)}$ $\frac{g\ VSS}{g\ BOD}$	$\left(\frac{\Delta X_v}{Q\Delta t}\right)_{aer}$ $\frac{mg\ VSS}{l}$	$\left(\frac{\Delta X_v}{Q\Delta t}\right)_{tot}$ $\frac{mg\ VSS}{l}$	$(NO_3)^2$ $\frac{mg\ N}{l}$	$\left(\frac{\Delta O_2}{Q\Delta t}\right)_{tot}$ $\frac{mg\ O_2}{l}$	θ h
15	0.24	37	0.24	155	0.18	8.8	0.19	37	104	5.7	185	19
30	0.19	27	0.26	141	0.14	6.8	0.15	30	88	6.0	240	32
50	0.16	22	0.27	140	0.11	5.7	0.12	26	77	6.1	269	46

Notes:
MLVSS = 2,000 mg/l, R^1 = 6, R^2 = 0.25
$X_v^r \approx$ 10,000 mg/l for each case

Table 10.5   Summary of key design outputs for the high-TKN wastewater treated by predenitrification in a membrane bioreactor ($\theta_x = 30$ d, $BOD_L^0 = 3{,}000$ mg/l, $TKN^0 = 1{,}105$ mg/l, MLVSS = 7,000 mg/l, and $X_i^0 = 0$)

	θ, d				$\dfrac{(\Delta X_v/\Delta t)_{tot}}{Q}$	$\dfrac{\Delta O_2/\Delta t}{Q}$			Net Alkalinity
R^2	Total	Anoxic	Aerobic	$(NO_3)^2$ (mg NO_3^--N/l)	(mg VSS/l)	(mg O_2/l)	BOD_L^{sup} (mg BOD_L/l)	R^w	Consumed mg/l $CaCO_3$
20	2.7	1.1	1.6	49	638	2.250	554	0.09	3,100
50	2.8	1.1	1.7	20	652	2,190	624	0.09	2,970
100	2.8	1.1	1.7	10	656	2,160	648	0.09	2,930

Example 10.5 | **PREDENITRIFICATION OF A HIGH-TKN INDUSTRIAL WASTEWATER BY A MEMBRANE BIOREACTOR**   An industrial wastewater has high concentrations of BOD_L (2,000 mg/l, all soluble) and TKN (1,105 mg/l). Even though the ratio of BOD_L^0:TKN^0 is less than the minimum required for full predenitrification (3.5 to 5.1, depending on θ_x, Table 10.3), the client desires that predenitrification be used to achieve the maximum N removal. To improve effluent quality and process stability, a membrane separator is to be used instead of a settler. Thus, no separate sludge recycle is needed. Sludge wasting to control the SRT occurs directly from the aerated reactor.

The design of predenitrification for this unusual wastewater requires special consideration of the amount of supplemental BOD in the influent and the loss of alkalinity. An SRT of 30 d is selected, and the volume of the aerated reactor is set equal to 1.5 times the volume of the anoxic reactor. The mixed-liquor volatile suspended solids are set at 7,000 mg/l. Then, all equations (as in Example 10.4) are solved simultaneously, subject to the criterion that $BOD_L^1 = 0$, which minimizes the required BOD supplement. The mixed-liquor recycle ratio (R^2) is varied from 20 to 100.

Table 10.5 summarizes the key results of the design. The change in R^2 greatly affects the effluent NO_3^--N, which can be driven to 10 mg/l when the mixed-liquor recycle ratio is 100. Changes in R^2 have only small effects on all other design results. Wasting sludge from the aerobic reactor gives a relatively dilute waste sludge and requires that 9 percent of the plant flow be removed for sludge wasting, even though the SRT is 30 d.

The high strength nature of the wastewater causes the hydraulic residence time (2.7 to 2.8 d) and the oxygen requirement per unit of flow (c. 2,200 mg O_2/l) to be large. The high ratio TKN^0:BOD_L^0 requires that BOD_L be supplemented by 554 to 648 mg/l in the influent, and the larger supplement for the larger R^2 reflects increased denitrification in the anoxic reactor. The high influent TKN means that over 3,000 mg/l as $CaCO_3$ of alkalinity is destroyed; alkalinity must be supplemented to the influent if the wastewater's alkalinity is not substantially greater than 3,000 mg/l as $CaCO_3$.

Modeling Denitrification That Uses Biomass Decay   Biomass can be used as the electron donor for denitrification. The main issues involved in using biomass decay are the release of NH_4^+-N and the slow kinetics of biomass oxidation. This section links stoichiometry and kinetics to provide a rational design basis for one-sludge denitrification that relies on biomass decay.

The electron donor is biomass, while the electron acceptor is NO_3^--N. The half-reactions are

$$\frac{1}{20} C_5H_7O_2N + \frac{9}{20} H_2O = \frac{1}{5} CO_2 + \frac{1}{20} HCO_3^- + \frac{1}{20} NH_4^+ + H^+ + e^-$$

$$\frac{1}{5} NO_3^- + \frac{6}{5} H^+ + e^- = \frac{1}{10} N_2 + \frac{3}{5} H_2O$$

Key ratios are

$$\frac{g\ cells_{ox}}{g\ NO_3^-N_{rem}} = \frac{0.05\ mol\ cells}{0.2\ mol\ NO_3^-} \cdot \frac{113\ g\ VSS}{mol\ cells} \cdot \frac{1\ mol\ NO_3^-}{14\ g\ NO_3^--N}$$

$$= 2.02\ g\ VSS_{ox}/g\ NO_3^--N_{rem}$$

$$\frac{g\ NH_4^+-N}{g\ NO_3^+-N_{rem}} = \frac{0.05\ mol\ NH_4^+-N}{0.2\ mol\ NO_3^--N} = 0.25\ g\ NH_4^+-N/g\ NO_3^--N$$

Thus, it takes oxidation of approximately 2 g VSS to denitrify 1 g NO_3^--N, while about 0.25 g NH_4^+-N is released. Thus, the net soluble N removal is 75 percent.

The appropriate θ is determined by the kinetics of decay, which are relatively slow. The rate of denitrification in a decay reactor is equal to the active biomass oxidation rate divided by g $cells_{ox}$/g NO_3^--N_{rem}:

$$\left(\frac{\Delta NO_3^--N}{\Delta t} \right) = \frac{X_a f_d bV}{2.02\ g\ VSS_{ox}/g\ NO_3^--N_{rem}} \qquad \textbf{[10.17]}$$

A steady-state mass balance on NO_3^--N for the reactor is

$$0 = Q[(NO_3^-)^0 - (NO_3^-)] - \frac{X_a f_d bV}{2.02} \qquad \textbf{[10.18]}$$

which can be solved to give θ:

$$\theta = \frac{[(NO_3^-)^0 - (NO_3^-)] \cdot 2.02}{f_d b X_a} \qquad \textbf{[10.19]}$$

The effluent NH_4^+-N, computed from its mass balance, is

$$(NH_4^+) = (NH_4^+)^0 + 0.25[(NO_3^-)^0 - (NO_3^-)] \qquad \textbf{[10.20]}$$

A simple mass balance on X_a can determine the output concentration of X_a.

$$0 = QX_a^0 - QX_a - bX_aV \qquad \textbf{[10.21]}$$

or

$$X_a = X_a^0/(1 + b\theta) \qquad \textbf{[10.22]}$$

Example 10.6

DESIGN OF A BIOMASS DECAYER   Example 10.3 showed that a predenitrification scheme with an SRT of 30 d and $R^2 = 6$ transformed 50 mg/l of TKN to 6 mg/l of NO_3^--N. If the effluent were routed to a biomass decayer, what hydraulic residence time would be needed for full denitrification if the decayer has $X_a^0 = 1,000$ mg VSS_a/l? What would be the effluent NH_4^+-N and X_a in the reactor?

The hydraulic detention time requires the assumption that $(NO_3^-) = 0$ for full denitrification:

$$\theta = \frac{6 \text{ mg } NO_3^-\text{-N/l} \cdot 2.02 \text{ g } VSS_{ox}/\text{g } NO_3^-\text{-N}}{0.8 \cdot 0.05/\text{d} \cdot 1,000 \text{ mg } VSS_a/\text{l}} = 0.3 \text{ d} = 7.3 \text{ h}$$

The residual NH_4^+-N is determined from the assumption that $(NH_4^+)^0 = O$:

$$(NH_4^+) = 0 + 0.25\frac{\text{g } NH_4^+\text{-N}}{\text{g } NO_3^-\text{-N}} \cdot [(6 - 0) \text{ mg } NO_3^-\text{-N/l}] = 1.5 \text{ mg } NH_4^+\text{-N/l}$$

If the nitrogenous oxygen demand of the 1.5 mg NH_4^+-N/l (6.8 mg NOD/l) cannot be tolerated, then an aerobic polishing step must follow. In either case, the final effluent contains about 1.5 mg N/l, giving a total N removal efficiency of 97 percent.

The output value of X_a is computed from

$$X_a = \frac{1,000 \text{ mg } VSS_a/\text{l}}{1 + 0.05/\text{d} \cdot 0.3 \text{ d}} = 985 \text{ mg } VSS_a/\text{l}$$

We see that the reactor is a cell decayer, as X_a declines from 1,000 to 985 mg VSS_a/l.

10.4   BIBLIOGRAPHY

Barnard, J. L. (1975). "Biological nutrient removal without the addition of chemicals." *Water Res.* 9, p. 485.

Davies, T. R. and W. A. Pretorious (1975). "Denitrification with a bacterial disc unit." *Water Res.* 9, p. 459.

deSilva, D. G. V. (1997). "Theoretical and experimental studies on multispecies bioreactors involving nitrifying bacteria." Ph.D. dissertation, Dept. Civil Engineering, Northwestern University, Evanston, IL.

Furumai, H. and B. E. Rittmann (1994). "Evaluation of multispecies biofilm and floc processes using a simplified aggregate model." *Water Sci. Technol.* 29(10–11), pp. 439–446.

Jeris, J. S.; C. Beer; and J. A. Mueller (1974). "High rate biological denitrification using a fluidized bed." *J. Water Pollution Control Fedn.* 46, p. 2118.

Kissel, J. C.; P. L. McCarty; and R. L. Street (1984). "Numerical simulation of mixed-culture biofilm." *J. Environ. Engr.* 110, p. 393.

Knowles, R. (1982). "Denitrification." *Microbio. Rev.* 46, pp. 43–70.

Körner, H. and W. G. Zumft (1989). "Expression of denitrification enzymes in response to the dissolved oxygen level and respiratory substrate in continuous culture of *Pseudonomas stutzeri. Appl. Environ. Microb.* 55, pp. 1670–1676.

Lee, K. C. and B. E. Rittmann (2000). "A novel hollow-fiber biofilm reactor for autohydrogenotrophic denitrification of drinking water." *Water Sci. Technol.*, in press.

McCarty, P. L.; L. Beck; and P. St. Amant (1969). "Biological denitrification of wastewater by addition of organic materials." *Proc. 24th Annual Purdue Industrial Waste Conf.*, p. 1271.

Paepcke, B. H. (1983). "Performance and operational aspects of biological phosphorous removal plants in Johannesburg" *Water Sci. Technol.* 15 (3/4), p. 219.

Payne, W. J. (1973). "Reduction of nitrogenous oxides by microorganisms." *Bacteriol. Rev.* 37, pp. 409–452.

Rittmann, B. E. and W. E. Langeland (1985). "Simultaneous denitrification with nitrification in single-channel oxidation ditches." *J. Water Pollution Control Fedn.* 57, pp. 300–308.

Schulthess, R. V. and W. Gujer (1996). "Release of nitrous oxides from denitrifying activated sludge: Verification and application of a mathematical model." *Water Res.* 30, pp. 521–530.

Schulthess, R. V.; D. Wild; and W. Gujer (1994). "Nitric and nitrous oxides from denitrifying activated sludge at low oxygen concentration." *Water Sci. Technol.* 30(6), pp. 123–132.

Tiedje, J. M. (1988). "Ecology of denitrification and dissimilarity nitrate reduction to ammonium." *Biology of Anaerobic Microorganisms*, A. J. B. Zehnder, Ed., New York: John Wiley, pp. 179–244.

Van Haandel, A. C.; G. A. Ekama; and G. V. R. Marais (1981). "The activated sludge process. III. Single-sludge denitrification." *Water Res.* 15, p. 1135.

Wanner, O. and W. Gujer (1986). "A multispecies biofilm model." *Biotechnol. Bioengr.* 28, pp. 314–328.

Wuhrman, K. (1964). "Nitrogen removed in sewage treatment processes." *Verh. Internat. Verein Limnol.* XV: 580.

Zumft, W. G. (1992). "The denitrifying prokaryotes." In *The Prokaryotes: A Handbook on the Biology of Bacteria*, A. Balows, H. G. Trüper, M. Dworkin, W. H. Harder, and K. Schleifer, Eds., Berlin: Springer-Verlag, pp. 554–582.

10.5 PROBLEMS

10.1. RBCs have been used for denitrification, as well as for aerobic processes. Estimate the surface area needed to remove essentially all of 30 mg NO_3^--N/l from a wastewater by a complete-mix RBC having a temperature of 20 °C, methanol is the carbon source and limiting substrate, the effluent methanol concentration is 1 mg/l, and $Q = 1,000$ l/d. You may assume that the average SRT is 33 d to calculate the stoichiometry, which you may assume is constant. Also, you may assume steady-state operation, and the following parameters:

$$K = 9.1 \frac{\text{mg } CH_3OH}{l} \qquad D_f = 1.04 \frac{cm^2}{d} \qquad b_{det} = 0.03 \text{ d}^{-1}$$

$$b = 0.05 \text{ d}^{-1} \qquad L = 60 \ \mu\text{m} \qquad \hat{q} = 6.9 \frac{\text{mg } CH_3OH}{\text{mg } VS_a \cdot d}$$

$$D = 1.3 \frac{cm^2}{d} \qquad X_f = 20 \frac{\text{mg } VS_a}{cm^3} \qquad Y = 0.27 \frac{\text{mg } VS_a}{\text{mg } CH_3OH}$$

Use the steady-state-biofilm model directly.

10.2. A set of kinetic coefficients for denitrification in a CSTR is:

Substrate	$\hat{q}$	K	b	Y
methanol	$8.3 \dfrac{\text{g methanol}}{\text{g VSS}_a \cdot \text{d}}$	$15 \dfrac{\text{mg methanol}}{\text{l}}$	$0.05 \dfrac{1}{\text{d}}$	$0.27 \dfrac{\text{g VSS}_a}{\text{g methanol}}$
Nitrate	$2.8 \dfrac{\text{g N}}{\text{g VSS}_a \cdot \text{d}}$	$0.1 \dfrac{\text{mg N}}{\text{l}}$	$0.05 \dfrac{1}{\text{d}}$	$0.81 \dfrac{\text{g VSS}_a}{\text{g N}}$

(a) Calculate the steady-state concentrations of methanol and nitrate-N for $\theta_x = 1, 2, 3, 4,$ and 5 d, assuming the respective substrate is limiting. In other words, there are two answers for each θ_x.

(b) If you have an initial concentration of 20 mg/l of NO_3^--N, what is the maximum concentration of methanol that may be present in the feed to preclude nitrate-limitation for each θ_x.

10.3. An industrial wastewater is characterized as follows: $BOD_L = 5,000$ mg/l, $TKN = 150$ mg/l, $NO_3^- + NO_2^-$-N $= 0$ mg/l, $SS = 250$ mg/l, pH = 8.4, alkalinity $= 2,000$ mg/l, and $TDS = 4,000$ mg/l. Evaluate the technical and economic suitability of using a predenitrification scheme to treat this wastewater. In other words, is it possible to do predenitrification? Compared to just aerobic oxidation, what are the effects of predenitrification on O_2 usage, sludge production, and pH?

10.4. Predenitrification is being used for BOD removal and N removal. The denitrification reactor is first, and the flow then goes to the nitrification reactor; a clarifier follows the nitrification reactor, and sludge recycle is to the head of the denitrification reactor. The θ_x value is 15 d. The coefficients are as presented in Table 10.3 for predenitrification. Your ultimate goals are to determine the sludge production rate and the oxygen-transfer rate. The influent BOD_L is 300 mg/l. The influent TKN is 25 mg/l. NO_3^--N is negligible in the feed. The influent flow rate is 770 m³/d. The mixed-liquor recycle rate is 7,700 m³/d, and the sludge recycle rate is 385 m³/d. To get your final result, take the following steps:

1. Calculate the amount of BOD_L removed by denitrification. The amount of BOD_L removed by denitrification is proportional to the NO_3^--N removed ultimately as N_2. Assume that all NH_4^+-N is nitrified to NO_3^-.

2. Calculate the amount of BOD_L removed in the aerobic nitrification tank. The BOD_L removal aerobically is that BOD_L remaining in the influent to the aerobic tank minus the soluble effluent BOD_L, which can be assumed to be 2.5 mg/l.

3. Calculate the cells produced (kg VSS/d) via denitrification (use $Y_{n(\text{den})}$).

4. Calculate the cells produced via aerobic BOD_L oxidation (let $Y = 0.5$ g VSS_a/g BOD_L, $f_d = 0.8$, and $b = 0.05$ d^{-1}).

5. Calculate the cells produced during nitrification (let effluent NH_4^+-N$=0$, $Y = 0.4$ g VSS_a/g BOD_L, $f_d = 0.8$, and $b = 0.05$ d^{-1}).

6. Calculate the total cell production rate.

7. Calculate the O_2 required (kg/d) for nitrification, BOD_L oxidation, and total.

10.5. List six advantages and/or disadvantages of using denitrification before nitrification (with recycle of NO_3^-), instead of the three-sludge approach for sequential BOD oxidation, nitrification, and denitrification.

10.6. An industrial waste contains nitric acid (HNO_3) at 7 mM. An aspiring young environmental engineer, eager to impress her new employer, suggests they use waste syrup as the carbon source. Syrup has a BOD_L of 100,000 mg/l and is basically pure carbohydrate ($C_6H_{12}O_6$). The engineer says to operate the system with $\theta_x = 8$ d. If $f_d = 0.8$, $f_s^0 = 0.6$, $b = 0.08$ d^{-1}, $\hat{q} = 18$ g BOD_L/g VSS_a-d, and $K = 20$ mg BOD_L/l, how much syrup (volume) must be added to remove essentially all the nitric acid? What is the absolute minimum amount of alkalinity needed to neutralize the acid in the treated wastewater?

10.7. A wastewater has the following characteristics:

Parameter	BOD_L	TKN	NO_3^--N	PO_4^{3-}-P	K$^+$	Na$^+$	Ca^{2+}	SO_4^{2-}	Cl$^-$	Mg^{2+}	Alkalinity (as $CaCO_3$)
Concentration (mg/l)	2,000	59	1	10	78	170	200	500	300	49	200

Select a biological process or series of biological processes to reduce the BOD_L to 20 mg/l and the total N to about 1 mg/l. What should θ_x or θ (as appropriate) be? Must there be any chemical additions? If so, what and how much? What is the O_2 utilization? How much sludge is produced?

10.8. You wish to design a denitrification process that will remove the nitrate-N in a wastewater to less than 1 mgN/l. Your source of BOD is spent-grains liquor from a brewery. Not only must you achieve virtually 100 percent N removal, but you must achieve an effluent BOD_L of 5 mg/l or less from the influent BOD. The influent has 50 mg NO_3^--N/l; no nitrite, ammonia, or BOD; and a flow rate of 100 m^3/d. You can use the following kinetic parameters for $T = 20$ °C.

Substrate	$\hat{q}$	K	b	Y	f_s^0	D	D_f	X_f
BOD_L	$10.4 \dfrac{\text{g } BOD_L}{\text{g } VS_a \cdot \text{d}}$	$50 \dfrac{\text{mg } BOD_L}{\text{l}}$	$0.05\dfrac{1}{\text{d}}$	$0.18\dfrac{\text{g } VS_a}{\text{g } BOD_L}$	0.36	$0.60\dfrac{\text{cm}^2}{\text{d}}$	$0.48\dfrac{\text{cm}^2}{\text{d}}$	$20\dfrac{\text{mg } VS_a}{\text{cm}^3}$
Nitrate	$2.3\dfrac{\text{g N}}{\text{g } VS_a \cdot \text{d}}$	$0.1\dfrac{\text{mg } NO_3^-\text{-N}}{\text{l}}$	$0.05\dfrac{1}{\text{d}}$	$0.81\dfrac{\text{g } VS_a}{\text{g } NO_3^-\text{-N}}$	0.36	$1.00\dfrac{\text{cm}^2}{\text{d}}$	$0.80\dfrac{\text{cm}^2}{\text{d}}$	$20\dfrac{\text{mg } VS_a}{\text{cm}^3}$

You chose to use a fluidized-bed biofilm reactor having a once-through liquid regime, 2 mm sand for media, and fluidization to 1.5 times the unfluidized bed height. Your overall goal is to determine a feasible design size (i.e., empty-bed contact time), if it is possible to achieve your treatment goals.

(a) Determine ε and σ for fluidized conditions, if the sand has a specific gravity of 2.65, a nonfluidized porosity of 0.35, and can be considered as a sphere.

(b) Estimate $S_{\min}$ for NO_3^--N and BOD_L. Choose the process-limiting substrate.

(c) Select an appropriate loading criterion for the process-limiting substrate. If the design is not feasible, state the reasons for infeasibility at this point and then select an appropriate load to come as close as possible to the desired effluent quality. (You may assume $L = 50\ \mu$m.)

(d) Estimate the biofilm thickness and check your estimate of $S_{\min}$. If you must adjust $S_{\min}$, describe qualitatively how the change would affect your criterion or performance, but do not rework the problem. Continue on to part e using the same design value.

(e) Calculate the required influent BOD_L concentration.

(f) Calculate the fluidized-bed empty-bed detention time for your design.

10.9. Consider a fluidized-bed denitrification reactor that has an empty-bed detention time of 10 min. The original medium is 0.2 mm sand particles having a packed porosity of 0.25. In operation, the bed is expanded to 1.6 times its packed height. The feed strength is 100 mg/l of methanol and 25 mg NO_3^--N/l. Calculate the following:

(a) Fluidized-bed porosity.

(b) Specific surface area of fluidized-bed (assume 0.2 mm spheres).

(c) Shear stress in reactor (based on fluidization of sand at a density of 2.65 g/ml).

(d) Approximate biofilm loss coefficient due to shearing and decay, b'.

(e) $S_{\min}$ for methanol. Use b' and kinetic parameters in Table 10.1.

(f) If methanol is limiting and its effluent concentration approaches $S_{\min}$, what is the "average" methanol flux in the reactor? What is the "average" biofilm thickness if $X_f = 40$ mg/cm^3? What is the effluent NO_3^--N concentration?

10.10. The operator of an activated sludge treatment plant was puzzled when an engineer from the state EPA indicated that the activated-sludge plant was improperly designed and could not possibly provide adequate treatment of the BOD_L load. The EPA engineer stated that the operator must install twice as much aeration capacity to make the plant work. The operator was puzzled because effluent standards have been met for several years. The effluent standards are: 20 mg BOD_5/l, 20 mg SS/l, and 1.5 mg NH_4^+-N/l. In addition, the engineer noted that D.O. was present in the mixed liquor.

The activated sludge plant has the following specifications: aeration tank volume = 2,500 m^3, settler area = 330 m^2, aerator capacity = 65 kW. The influent characteristics to the activated-sludge unit are: flow rate $= 10,000 \text{ m}^3/\text{d}$, $BOD_5 = 200$ mg/l, SS $= 75$ mg/l, TKN $=15$ mg/l, NO_3^--N $= 40$ mg/l, and P $= 12$ mg/l. The mixed liquor has 4,000 mg SS/l. Waste sludge has 9,000 mg SS/l and is wasted at rate of 75 m^3/d. An aeration test gave an aerator FOTE of 1 kg O_2/kWh under actual conditions.

You are called in as the high-priced consultant to help the operator defend himself. You have only a few minutes to analyze the situation and present a lucid and correct explanation of why the plant works properly when the EPA engineer says that it is underdesigned with respect to oxygen transfer capacity. You need not make a highly detailed analysis, but you must be able to quantify your explanation. Therefore, demonstrate if and why the EPA engineer has made an incorrect analysis.

10.11. Denitrification is sometimes used to remove nitrate from drinking water. Provide a preliminary design for a fixed-film process to remove 10 mg NO_3^-/l to less than 0.1 mg/l. The preliminary design involves estimating the total reactor volume and the amount of added acetate needed as the electron donor. The known information is:

Substrate	$\hat{q}$	K	b	b_{det}	Y	f_d	D	D_f	X_f
Acetate	$10\,\dfrac{\text{mg Ac}^-}{\text{mg VS}_a \cdot \text{d}}$	$15\,\dfrac{\text{mg Ac}^-}{1}$	$0.1\dfrac{1}{\text{d}}$	$0.05\dfrac{1}{\text{d}}$	$0.18\,\dfrac{\text{mg VS}_a}{\text{mg Ac}^-}$	0.8	$1.2\dfrac{\text{cm}^2}{\text{d}}$	$1.0\dfrac{\text{cm}^2}{\text{d}}$	$40\dfrac{\text{mg VS}_a}{\text{cm}^3}$
Nitrate	$2.3\,\dfrac{\text{mg NO}_3^- \text{-N}}{\text{mg VS}_a \cdot \text{d}}$	$0.1\,\dfrac{\text{mg NO}_3^- \text{-N}}{1}$			$0.81\,\dfrac{\text{mg VS}_a}{\text{mg NO}_3^- \text{-N}}$				

You also know that $Q = 1,000 \text{ m}^3/\text{d}$, $a = 300 \text{ m}^{-1}$, $L = 90 \ \mu\text{m}$ (for acetate). You may assume that the added acetate is the rate-limiting substrate, as long as it is not fully depleted. You will need to estimate $S_{\min}$ for each substrate and the amount of acetate needed to remove the 10 mg NO_3^--N/l. Then, determine the needed J and volume to keep residual acetate to as low a level as practical.

10.12. You need to design a biofilm process to be used as the first step in a pre-denitrification scheme that uses only biofilm processes. The wastewater to be treated has the following characteristics: $Q = 4,000 \text{ m}^3/\text{d}$, $BOD_L = 300$ mg/l, TKN $= 65$ mgN/l, $NO_3^- + NO_2^-$-N $= 0$ mg N/l. Your design calls for 85 percent removal of N, and you have correctly figured out that the recycle flow rate from a second (aerobic) biofilm reactor is 22,700 m^3/d. Also, you are able to assume that N uptake in cell synthesis is not important. Furthermore, pilot studies with denitrifying biofilm reactor of a similar type gave values of $b_{\text{det}} = 0.04 \text{ d}^{-1}$, $X_f = 40 \text{ mg/cm}^3$, and $L = 40 \ \mu\text{m}$.

Your goal is to determine the total reactor volume of a completely mixed biofilm reactor when the specific surface area is 200 m^{-1}. Some potentially important kinetic parameters for the rate-limiting substrate for denitrification are:

$$\hat{q} = 12\frac{\text{mg BOD}_L}{\text{mg VSS}_a \cdot \text{d}} \quad K = 20\frac{\text{mg BOD}_L}{\text{l}} \quad b = 0.05 \text{ d}^{-1}$$

$$Y = 0.27\frac{\text{mg VSS}_a}{\text{mg BOD}_L} \quad D = 0.6\frac{\text{cm}^2}{\text{d}} \quad D_f = 0.48\frac{\text{cm}^2}{\text{d}}$$

10.13. Your firm needs to design a process to treat a wastewater with somewhat unusual composition. The key features are: $BOD_L = 250$ mg/l, $TKN = 50$ mg/l, $NO_3^--N = 50$ mg/l, and $SS = 0$ mg/l. Your boss wishes to use an oxidation ditch to treat this wastewater by the technique of simultaneous nitrification with denitrification in a single reactor. The boss assigns to you the task of assessing the feasibility of such an approach. In particular, you are to assess whether or not you can accomplish nearly complete BOD and N removals with this wastewater in such a process. You brilliantly realize that the one-reactor system can be represented as a predenitrification system with a very high mixed-liquor recycle ratio, say 1,000. Can this process be successful?

10.14. One potential benefit of biofilm processes is that they can achieve a very high volumetric loading, which makes the process compact. To achieve a high volumetric loading, you need a high specific surface area, such as with a fluidized-bed biofilm reactor. Some say that *denitrification* is the *ideal* reaction for a compact fluidized-bed process. Give two aspects of denitrification that make it a good choice for a compact biofilm process.

10.15. You are going to treat a wastewater having the following characteristics: $BOD_L = 500$ mg/l, $TKN = 50$ mg/l, inert $SS = 50$ mg/l, $NO_3^--N = 0$ mg/l. You have five options for treatment, and they are shown schematically at the top of the next page. You are to indicate for each comparison how sludge production (in kg VSS/d) and the oxygen required (in kg O_2/d) change by making the changes indicated by the comparisons. For example, how do the sludge production and oxygen required change when you go from process (a) to process (b), which is comparison 1? Your choices in all cases are: increases, decreases, or no change.

10.16. Predenitrification ought to be achievable using biofilm reactors, instead of suspended-growth reactors. Draw a schematic diagram of such a system. Label the five most important parts of the system. Describe what each part does and why it is necessary.

10.17. You are the lead engineer for the design of a predenitrification system for a somewhat unusual wastewater: $BOD_L = 2,000$ mg/l, $TKN = 300$ mg/l, and $NO_3^--N = 150$ mg/l. Your first design decision is to set the SRT at 20 d. At this SRT, you can assume that the anoxic reactor will drive the NH_4^+-N and BOD_L concentrations so close to zero that you can take them as zero.

  (*a*) You need first to assure yourself that the influent BOD is sufficient to drive complete denitrification. Is it?

Figure for Problem 10.15.

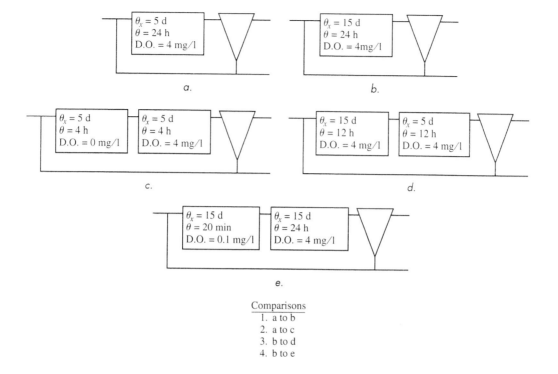

Comparisons
1. a to b
2. a to c
3. b to d
4. b to e

(*b*) What mixed-liquor recycle ratio is required to achieve a total N concentration in the influent of 30 mg/l if the underflow sludge recycle ratio is 0.5?

(*c*) How much BOD_L (expressed in mg/l in the influent flow) will be removed by denitrification and by aerobic oxidation?

10.18. You wish to consider the addition of methanol to groundwater for anaerobic biological removal of nitrate by denitrification. If the nitrate nitrogen concentration is 84 mg/l, what minimum concentration of methanol should be added to achieve complete nitrate reduction to nitrogen gas? Assume no ammonia is present and that f_s for the reaction is 0.30.

10.19. You wish to design a plug-flow biofilm reactor without recycle for denitrification to N_2 of a wastewater containing 65 mg NO_3^-/l and no ammonia, and you have selected acetate to add as the electron donor for the reaction.

(*a*) What concentration of acetate should be added to the wastewater to achieve complete removal of nitrate by denitrification?

(*b*) Given the above concentration of acetate added to the wastewater prior to treatment, is the reaction at the entrance to the reactor rate-limited by acetate, nitrate, or neither? Show appropriate calculations to support your conclusion.

Assume that nitrate is used for cell synthesis, and that the following characteristics apply for reaction kinetics and biofilm characteristics:

	Acetate	Nitrate	Biofilm
K	10 mg/l	1 mg/l	
$\hat{q}$	15 mg/mg VS_a-d		
D	0.9 cm^2/d	0.7 cm^2/d	
D_f	0.8 D_w	0.8 D_w	
L			0.150 cm
f_s			0.55
X_f			12 mg/cm^3
b			0.15/d
Y			0.4 g VS_a/g acetate

10.20. A deep biofilm reactor is used for denitrification of a water supply, and methanol is being considered as an electron donor for the reaction. For the following conditions, estimate the flux rate for nitrate into the biofilm (mg/cm^2-d) Assume NH_3-N is available for cell synthesis:

	Methanol	Nitrate	Biofilm
S	40 mg/l	60 mg/l	
K	15 mg/l	3 mg/l	
$\hat{q}$	10 mg/mg VS_a-d		
D	1.3 cm^2/d	0.7 cm^2/d	
D_f	0.8 D	0.8 D	
L			0.090 cm
f_s			0.40
X_f			15 mg/cm^3

10.21. You are designing a CSTR with recycle for anoxic denitrification of an industrial wastewater containing nitrate, and are considering the addition of thiosulfate ($S_2O_3^{2-}$), a waste chemical at the plant, to be used as the electron donor for the biological reaction. As such the thiosulfate would be oxidized to sulfate. You are conducting a sensitivity analysis for the design to determine the influence of different variables on reactor size, nitrate concentration in the reactor effluent, and thiosulfate addition required. Since nitrate is the contaminant of interest, you desire to keep it, rather than thiosulfate, as rate-limiting for the reaction occurring, while at the same time, maintaining thiosulfate additions as low as possible. That is, S represents nitrate concentration, not thiosulfate concentration. Fill out the following table to indicate how an increase in each of the variables in the left-hand column will affect the three process variables. Assume all other factors listed in the left-hand column remain constant and also that X, the reactor total suspended solids concentration remains constant. Use: (+) = increase, (−) = decrease, (0) = no change, and (i) = need more information to tell.

Variable	Process Variables		
	Reactor Size V m^3	Effluent Nitrate Concentrate (mg/l)	Thiosulfate Addition Required (kg/d)
K			
$\hat{q}$			
Q^0			
S^0			
b			
θ_x			
X_i^0			
Y			

10.22. A wastewater has a BOD_L of 1,200 mg/l and a NO_3^--N concentration of 400 mg/l. For denitrification in a CSTR with recycle, operating at $\theta_x = 10$ d, will the electron donor or electron acceptor be rate limiting to the overall reaction? Assume ammonia nitrogen is available and used for cell synthesis. Provide appropriate calculations to support your answer.

10.23. Draw a flow diagram for a two-stage organic oxidation, nitrification plant, designed in such a way as to reduce overall oxygen requirements.

10.24. In a study of denitrification of an industrial wastewater, 120 mg/l acetate was used in the removal of 30 mg/l nitrate-nitrogen. Estimate f_s and f_e.

10.25. What concentration of lactate would be required for denitrification of 60 mg/L nitrate-nitrogen in a CSTR with settling and recycle in which θ_x was maintained at 6 d?

10.26. Draw schematic diagrams of two economical biological treatment systems that you might design to achieve removal of ammonia, nitrite, and nitrate nitrogen from a municipal wastewater. For the first system (low nitrogen removal), assume the requirement is simply to remove about 50 percent of the nitrogen, and in the second treatment system (high nitrogen removal), assume the requirement is to remove 95 percent of the nitrogen. Indicate all fluid and gaseous inputs and outputs from each reactor, indicate what chemicals, if any, would be added to each reactor in the system, and indicate the dominate forms of nitrogen in streams entering and exiting each reactor. Indicate what are the dominant electron donors and acceptors for the biological reactions occurring in each reactor in the system.

10.27. A $10\,m^3$ fluidized bed biofilm reactor containing sand as the attachment surface is used for denitrification of a wastewater containing 50 mg NO_3^--N/l, and the total surface area within the reactor for the sand grains coated with a growth of bacteria is $3(10^8)$ cm^2. Methanol is added as the primary substrate for bacterial growth, leading to the following stoichiometric equation:

$$0.1667\ CH_3OH + 0.1343\ NO_3^- + 0.1343\ H^+$$
$$= 0.0143\ C_5H_7O_2N + 0.06\ N_2 + 0.3505\ H_2O + 0.0952\ CO_2$$

(a) If the reactor is assumed to be completely mixed, what minimum concentration of methanol in solution would be required in order to insure that an effluent nitrate-nitrogen concentration of 1.0 mg/l or less can be obtained?

(b) Under such conditions, what would be the removal rate of nitrate-nitrogen in kg/d, and the detention time for the reactor under these operating conditions (assume V equals the total reactor volume of $10\ m^3$)?

Assume deep biofilm kinetics apply ($S_w = 0$). Pertinent coefficients are as follows:

Coefficient	Units	Methanol	Nitrate-Nitrogen
D	cm^2/d	1.3	0.7
D_f	cm^2/d	0.9	0.5
$\hat{q}$	mg/mg VS_a-d	9	
K	mg/cm^3	0.008	0.002
L	cm	0.004	0.004
X_a	mg VS_a/cm^3	15	15
b	d^{-1}	0.1	0.1

11

PHORPHORUS REMOVAL

Phosphorus is an essential macronutrient that spurs the growth of photosynthetic algae and cyanobacteria, leading to accelerated eutrophication of lakes. Wastewater discharges that reach lakes sensitive to eutrophication often require phosphorus removal over and above that normally taking place in primary and secondary treatment. For example, a typical municipal wastewater in the United States has a BOD_5 of 250 mg/l, a BOD_L of 370 mg/l, a COD of 500 mg/l, a TKN of 60 mg/l, and total P of 12 mg/l. Conventional primary sedimentation and secondary activated sludge reduce the effluent total P to about 6 mg/l. However, a typical effluent standard for a protected watershed is 1 mg P/l. Therefore, additional phosphorus removal is necessary.

Phosphorus can be removed from wastewater prior to biological treatment, as part of biological treatment, or following biological treatment. The first and third approaches almost always are carried out by chemical precipitation of the phosphate anion (PO_4^{3-}) with Ca^{2+}, Al^{3+}, or Fe^{3+} cations. The second approach is the subject of this chapter, which shows how we can extend the capabilities of biotechnology processes already discussed for BOD and N removal.

Three phenomena can be exploited to remove phosphorus as part of a microbiological treatment process:

- Normal phosphorous uptake into biomass
- Precipitation by metal-salts addition to a microbiological process
- Enhanced biological phosphorus uptake into biomass

Each phenomenon and its application are described below. The latter two phenomena are the bases for so-called "advanced treatment" for phosphorus.

11.1   NORMAL PHOSPHORUS UPTAKE INTO BIOMASS

The biomass that develops normally in an aerobic biological process, such as activated sludge, contains 2 to 3% P in its dry weight. The stoichiometric formula for biomass can be modified to include this amount of P. The formula $C_5H_7O_2NP_{0.1}$ has a formula weight of 116 g/mol, of which P is 2.67 percent. Sludge wasted from the process removes P in proportion to the mass rate of sludge VSS wasted ($Q^w X_v^w$). A steady-state mass balance on total P is

$$0 = Q P^0 - Q P - Q^w X_v^w (0.0267 \text{ g P/g VSS}) \qquad \textbf{[11.1]}$$

in which P^0 and P are the influent and effluent total P concentrations, and Q is the influent flow rate. Equation 11.1 can be solved for the effluent total P concentration,

$$P = \frac{Q P^0 - Q^w X_v^w (0.0267)}{Q} = P^0 - \frac{Q^w X_v^w (0.0267)}{Q} \qquad \textbf{[11.2]}$$

The rate of sludge wasting is proportional to the BOD removal (ΔBOD_L) and the observed yield (Y_n):

$$Q^w X_v^w = Y_n Q (\Delta\text{BOD}_L) \qquad \textbf{[11.3]}$$

The net yield depends on the SRT (θ_x), the true yield (Y), the endogenous decay rate (b), and the biodegradable fraction of the new biomass (f_d) (same as Equation 3.32):

$$Y_n = Y \frac{1 + (1 - f_d) b \theta_x}{1 + b \theta_x} \qquad \textbf{[11.4]}$$

Combining Equations 11.2 to 11.4 gives us the relationship describing how the effluent P concentration depends on the influent P concentration, the BOD_L removal, and the SRT.

$$P = P^0 - \frac{(0.0267) Y (1 + (1 - f_d) b \theta_x)(\Delta\text{BOD}_L)}{1 + b \theta_x} \qquad \textbf{[11.5]}$$

Table 11.1 compiles effluent P concentrations for a range of SRTs and BOD_L removals when $Y = 0.46$ mg VSS_a/mg BOD_L, $b = 0.1$/d, $f_d = 0.8$, and $P^0 = 10$ mg P/l. Only one scenario yields an effluent P concentration less than 1 mg/l: an SRT of 3 d and BOD_L removal of 1,000 mg/l. Increasing SRT and decreasing BOD removal result is less P removal and failure to meet the effluent standard of 1 mg/l.

Table 11.1   Effects of SRT and BOD_L removal on the effluent P concentration when the influent P concentration is 10 mg P/l

BOD_L removal, mg BOD_L/l	Effluent PO_4-P Concentration, mg/l			
	3 d	6 d	15 d	30 d
100	9.0	9.1	9.4	9.5
300	7.0	7.4	8.1	8.5
500	5.0	5.7	6.8	7.5
1,000	0	1.4	3.6	5.1

Note: $Y = 0.46$ mg VSS_a/mg BOD_L, $f_d = 0.8$, $b = 0.1$/d, and biomass P content = 2.67 percent.

For domestic sewage, which is best represented by the BOD_L removal of 300 mg/l, the effluent P concentration is well above 1 mg P/l, and the need for advanced treatment is obvious. In many cases, N removal must accompany P removal. Thus, the large SRTs needed for nitrification reduce sludge wasting and normal P removal. For example, a conservative SRT of 30 d leaves 8.5 mg P/l in the effluent, whereas a conventional SRT of 6 d leaves 7.4 mg P/l.

These simple calculations for the normal removal of phosphorus are important to perform when phosphorus is regulated. In most cases, they define the amount of additional P removal required by one of the advanced techniques. On the other hand, a wastewater having a low $P:BOD_L$ ratio may not require advanced treatment if the SRT is carefully controlled to keep a low, but positive P concentration.

11.2 PRECIPITATION BY METAL-SALTS ADDITION TO A BIOLOGICAL PROCESS

Aluminum and ferric cations precipitate with the orthophosphate anion at pH values that are compatible with microbiological treatment. Therefore, salts of Al^{3+} or Fe^{3+} can be added directly to the wastewater as it enters or leaves the bioreactor. Precipitates form, are incorporated into sludge, and are removed from the system by sludge wasting.

The key precipitates are $AlPO_{4(s)}$ and $FePO_{4(s)}$. Their standard dissolution reactions and solubility products (pK_{so}) are

$$AlPO_{4(s)} = Al^{3+} + PO_4^{3-} \qquad pK_{so} = 21$$

$$FePO_{4(s)} = Fe^{3+} + PO_4^{3-} \qquad pK_{so} = 21.9 \text{ to } 23$$

Although the solubility products are very small, the conditional solubility of phosphates is not necessarily miniscule, because Al^{3+}, Fe^{3+}, and PO_4^{3-} undergo competing acid/base and complexation reactions. For phosphate, the key competing reactions are acid/base ones that result in the formation of protonated species. The acid/base reactions and pK_a values are

$$HPO_4^{2-} = H^+ + PO_4^{3-} \qquad pK_{a,3} = 12.3$$

$$H_2PO_4^- = H^+ + HPO_4^{2-} \qquad pK_{a,2} = 7.2$$

$$H_3PO_4 = H^+ + H_2PO_4^- \qquad pK_{a,1} = 2.1$$

At near neutral pH, HPO_4^{2-} and $H_2PO_4^-$, not PO_4^{3-}, are the dominant species. For example, at pH = 7.0, PO_4^{3-} comprises about 0.00025 percent of the total dissolved orthophosphates. Higher pH gives a greater fraction as PO_4^{3-}.

The aluminum and ferric cations form numerous complexes, and the hydroxy complexes always are very important. Key complex-formation reactions and stability constants (pK_{1-4}) include:

$$Fe^{3+} + OH^- = FeOH^{2+} \qquad pK_1 = 11.8$$

$$FeOH^{2+} + OH^- = Fe(OH)_2^+ \qquad pK_2 = 10.5$$

$$Fe(OH)_2^+ + OH^- = Fe(OH)_3^0 \qquad pK_3 = 7.7$$

$$Fe(OH)_3 + OH^- = Fe(OH)_4^- \qquad pK_4 = 4.4$$

$$Al^{3+} + OH^- = AlOH^{2+} \qquad pK_1 = 9.0$$

$$AlOH^{2+} + OH^- = Al(OH)_2^+ \qquad pK_2 = 9.7$$

$$Al(OH)_2^+ + OH^- = Al(OH)_3^0 \qquad pK_3 = 8.3$$

$$Al(OH)_3 + OH^- = Al(OH)_4^- \qquad pK_4 = 6.0$$

The acid/base nature of the hydroxy complexes means that the iron or aluminum speciation is pH dependent. At neutral pH, the dominant species are $Fe(OH)_2^+$, $Fe(OH)_3^0$, and $Al(OH)_3^0$. Al^{3+} and Fe^{3+} are tiny fractions of the total aluminum and iron. Lower pH is required to give greater fractions of Al^{3+} or Fe^{3+}.

Due to the opposing pH trends for the precipitating anion (PO_4^{3-}) versus the precipitating cations (Al^{3+} or Fe^{3+}), an optimal pH exists. Based on the reactions shown above, the pHs of minimum solubility are near 6 for $AlPO_{4(s)}$ and near 5 for $FePO_{4(s)}$. In principle, stoichiometric additions of the cations can drive the total phosphate concentration to well below 1 mg P/l near these optimal pH values.

In reality, the theoretical predictions only define a rough location for a "window" in which precipitation through metal-salts addition can work within a biological-treatment process. Empirical testing is used to define the best dosage in practice, and the metal-salts dosage always exceeds the stoichiometric amount defined by the solubility products for $AlPO_{4(s)}$ and $FePO_{4(s)}$. Several factors act to complicate the chemistry and increase the metal-salts addition.

1. Phosphate forms competing complexes, such as $CaHPO_4$, $MgHPO_4$, and $FeHPO_4^+$. Thus, the fraction of total phosphate that is present as PO_4^{3-} is less than predicted by acid-base chemistry alone.

2. Aluminum and iron form other complexes, particularly with organic ligands, or precipitate as $Al(OH)_{3(s)}$. These reactions reduce the available Al^{3+} and Fe^{3+}.

3. Some of the total phosphorus is not orthophosphate, but is tied up in organic compounds.

4. The optimal pH for precipitation may not be compatible with the optimal microbiological activity. The pH cannot be changed so much that metabolic activity is significantly inhibited.

5. The precipitation reaction may be kinetically controlled and not reach its maximum extent, which occurs at equilibrium.

Typically, the metal-salts dosage is 1.5 to 2.5 times the stoichiometric amount. Since the stoichiometric mole ratio is 1 mol metal to 1 mol P, the practical ratios, determined empirically, are 1.5 to 2.5 mol metal:mol P. This translates to 1.3 to 2.2 g Al/g P and 2.7 to 4.5 g Fe/g P. For example, a practical mole ratio of 2:1 requires an Fe dose of 36 mg Fe/l to treat a wastewater with 10 mg P/l and provide an effluent with 1 mg P/l.

An important consideration for metal-salts addition is that the added metals behave as acids and consume alkalinity as they precipitate with PO_4^{3-} or complex with hydroxide. For example, when Fe^{3+} is added (such as in $FeCl_3$) and reacts to form $FePO_{4(s)}$ or $Fe(OH)_3$, it consumes three base equivalents per mole. The same occurs for Al^{3+}, such as from alum ($Al_2(SO_4)_3 \cdot 14H_2O$). A dosage at a 2:1 mol ratio consumes 9.7 mg as $CaCO_3$ of alkalinity per mg P. For low alkalinity waters, pH depression is a major risk. In such cases, a base, such as lime (CaO), can be added to supplement the alkalinity. Another alternative is to dose sodium aluminate ($NaAlO_2$), which dissolves in water to give

$$AlNaO_2 + 2H_2O = Na^+ + Al(OH)_3 + OH^-$$

This dissolution reaction is a net generator of alkalinity: one base equivalent/mol Al. Another alternative is to dose ferrous iron (Fe^{2+}), which is oxidized to Fe^{3+} in a reaction that produces 1 base equivalent per mole of Fe^{2+} oxidized:

$$Fe^{2+} + 0.25\, O_2 + 0.5\, H_2O = Fe^{3+} + OH^-$$

Precipitation of $AlPO_{4(s)}$ or $FePO_{4(s)}$ creates significant quantities of inorganic sludge that is retained within the biological process and ultimately wasted. The stoichiometric production ratios are 3.9 g $AlPO_{4(s)}$/g P and 4.9 g $FePO_{4(s)}$/g P. For example, removal of 8 mg P/l produces 39 mg $FePO_{4(s)}$/l of plant flow. If an activated sludge process has an SRT (θ_x) of 6 d and a hydraulic reaction time of 0.25 d (θ), the concentration factor (θ_x/θ) is 24, which means the concentration of $FePO_{4(s)}$ built up in the mixed liquor is 935 mg/l. Clearly, phosphate precipitation increases the sludge to be wasted and significantly enriches it in inorganic solids.

Most experience with metal-salts addition is with activated sludge treatment, in which the metal salt can be added with the influent wastewater, directly to the aeration basin, or to the mixed liquor before it goes to the settler. Phosphorus removal by metal-salt addition also is possible with biofilm processes that have excellent capability for capturing the precipitated solids and wasting them regularly. Large-granule, fixed-bed filters seem most amenable and have been used successfully (Clark, Stephensen, and Pearce, 1997).

11.3 ENHANCED BIOLOGICAL PHOSPHORUS REMOVAL

Certain heterotrophic bacteria are capable of sequestering high levels of phosphorus as *intracellular polyphosphate* (poly P), which is an energy storage material. If such microorganisms are selected, induced to store poly P, and wasted when rich in poly P, the net removal of P through biomass uptake can be increased significantly. The goal of *enhanced biological phosphorus removal* is to create a process environment that attains each of those three goals.

When enhanced biological phosphorus removal is successful, the biomass contains 2 to 5 times the P content of normal biomass. Equation 11.5 can be revised to account for the enrichment in poly P. For example, a three-fold enrichment gives

$$P = P^0 - \frac{(0.0801)Y(1 + (1 - f_d)\, b\theta_x)(\Delta BOD_L)}{1 + b\theta_x} \qquad \textbf{[11.6]}$$

The effluent P concentration for a BOD_L removal of 300 mg/l and an influent P concentration of 10 mg/l then becomes 4.2 mg/l for $\theta_x = 15$ d and 2.2 mg/l for $\theta_x = 6$ d. These values approach a typical P standard of 1 mg/l. Increased poly-P enrichment or a higher BOD_L:P ratio in the influent make the effluent goal achievable.

Figure 11.1 sketches the key components of an activated sludge system active in enhanced biological phosphorus removal. Many technological innovations are used to optimize the performance, and they are summarized later. Figure 11.1 presents the essential items and facilitates an explanation of how enhanced biological phosphorous removal works.

The four essential components for enhanced biological phosphorus removal are:

1. The influent wastewater and recycle sludge must mix and first enter an *anaerobic bioreactor*. Electron acceptors—particularly O_2 and NO_3^-—must be excluded to the maximum degree possible so that BOD oxidation is insignificant in this reactor. Due to the high mixed-liquor biomass level, hydrolysis and fermentation steps occur, but the electron equivalents in the influent BOD are not transferred to a terminal electron acceptor.

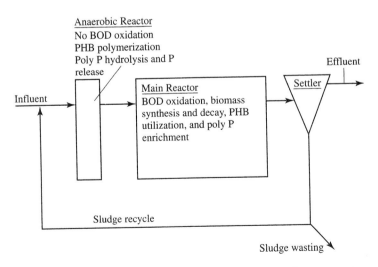

Figure 11.1   Schematic of the required components of an activated sludge process active for enhanced biological phosphorus removal.

2. The mixed liquor flows from the anaerobic tank to the *main activated sludge bioreactor or bioreactors*. Depending on the system's SRT, nitrification and denitrification may or may not occur. Ample electron acceptors are available through aeration, which directly supplies O_2 and allows generation of NO_3^- if nitrification occurs. These electron acceptors make it possible for the heterotrophic bacteria to oxidize electron donors, gain energy, and grow.

3. The mixed liquor exiting the main bioreactor is settled, and most of it is *recycled* back to the head of the process, where it mixes with the influent and enters the anaerobic tank. This recycling ensures that all of the biomass *experiences alternating anaerobic and respiring conditions.*

4. The sludge that was most recently in the main bioreactor is *wasted* to control the SRT and to remove the biomass when it is enriched in poly P.

Each of the four essential components must be present in order that the biochemical and ecological mechanisms that drive enhanced biological phosphorus removal act. Figure 11.2 summarizes the biochemical mechanisms acting in the anaerobic and main bioreactors.

In the anaerobic phase (Figure 11.2*a*), certain heterotrophic bacteria are able to take up simple organic molecules produced by hydrolysis and fermentation. Because no electron acceptors are available, they sequester the electrons and carbon in insoluble intracellular solids, such as *polyhydroxybutyrate (PHB)*. To do the polymerization, the cells require an activated chemical form, acetyl coenzyme A (HSCoA). Formation of HSCoA is an energy-consuming step, and the energy (ultimately transported as ATP) comes from the hydrolysis of poly P, which these microorganisms also contain and use as an energy-storage material. The hydrolysis of poly P releases phosphate

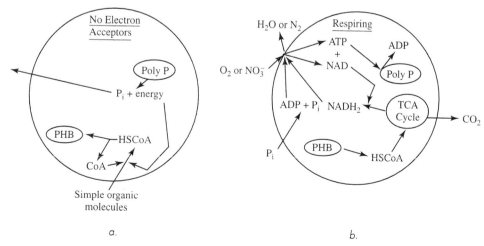

a.   b.

Figure 11.2   Biochemical mechanisms operating in the anaerobic *a.* and main *b.* bioreactors of enhanced biological phosphorus removal. HSCoA = Acetyl coenzyme A, PHB = polyhydroxybutyrate.

(P_i) to the cellular P_i pool, and much of this P_i is released to the environment during the anaerobic phase.

When these heterotrophic bacteria move to the main reactor, they have an ample supply of electron acceptors, and the biochemical machinery essentially works in the opposite direction. The electron storage material (PHB) is hydrolyzed to HSCoA, which is then oxidized in the TCA cycle. The released electrons, carried on $NADH_2$, are used for ATP synthesis through respiration with O_2 or NO_3^- as the electron acceptor. Some of the ATP generated is "invested" in the synthesis of poly P, the energy-storage material. Inorganic phosphate, P_i, must be imported for poly P synthesis. Thus, these bacteria are significant sinks for environmental P_i when they are synthesizing poly P. They must be harvested and wasted when they are enriched in poly P and before they cycle to the anaerobic phase.

Not all heterotrophic bacteria are capable of synthesizing poly P and PHB. In order to select for these bacteria, the so-called *Bio-P bacteria*, the process must exert a strong ecological pressure to favor them. The cycling between the anaerobic and respiratory phases creates that ecological pressure. During the anaerobic phase, the Bio-P bacteria take-up and sequester the available electron donors and carbon sources into PHB. They require poly P to fuel this "investment." During the respiratory phase, the Bio-P bacteria have rich reserves of electrons and carbon in PHB. Hydrolysis and oxidation of PHB gives them a major energy source for growth, even though the aqueous environment in the main bioreactor is oligotrophic. Generating energy from storage reserves allows the Bio-P bacteria to outgrow other heterotrophic bacteria in the main bioreactor and gradually establish themselves as a major fraction of the biomass.

In the past, Bio-P bacteria were identified as belonging to the genus *Acinetobacter*. Further research has shown that *Acinetobacter* is only one genus of Bio-P bacteria and often is not the dominant genus. Bio-P bacteria also are found among the *Pseudomonas, Arthrobacter, Nocardia, Beyerinkia, Ozotobacter, Aeromonas, Microlunatus, Rhodocyclus,* and others.

Although the biochemical and ecological foundation for enhanced biological phosphorus removal by Bio-P bacteria now seems to be firmly established, some evidence supports the concept that chemical precipitation plays a role, at least in some instances. The main candidate for the solid phase is $Ca_5(OH)(PO_4)_{3(s)}$, or hydroxyapatite. Its solubility product is $pK_{so} = 55.9$, which suggests that moderately hard waters could drive the orthophosphate below 1 mg P/l for pH values greater than about 7.5. Struvite ($MgNH_4PO_4$) is another possibility when Mg^{2+} and NH_4^+ concentrations are elevated. Strong denitrification produces base that can increase the pH, particularly inside a floc or biofilm, and foster precipitation of hydroxyapatite or struvite. Very high aeration rates, which strip CO_2 from solution, also raise the pH and can foster precipitation of hydroxyapatite. Chemical precipitation as part of enhanced biological phosphorous removal is accentuated by low alkalinity, which minimizes the buffering against these pH increases.

Figure 11.3 summarizes schematically two approaches that link enhanced biological phosphorous removal to well-known predenitrification processes. The approach in Figure 11.3*a*, sometimes called *Phoredox*, adds the initial anaerobic tank to a

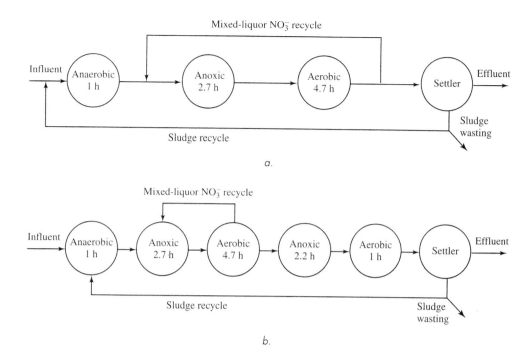

Figure 11.3   Two approaches to incorporate enhanced biological phosphorus removal into classical predenitrification. The top a. links an anaerobic tank to classical predenitrification, while the bottom b. links to the Barnard process. Hydraulic detention times are typical and for general guidance only.

classical predenitrification system. The process in Figure 11.3b, called *Bardenpho,* adds the anaerobic tank to the Barnard process, which has anoxic cell decay and aerobic polishing steps to effect added N removal. Burdick, Refling, and Stensel (1982) reported 93 percent removal of total N and 65 percent removal of total P with the Bardenpho process shown in Figure 11.3. The sludge contained 4 to 4.5% P, which is about twice that of normal biomass. The effluent total P was 1.9 to 2.3 mg/l. Meganck and Faup (1988) summarized results from several other Bardenpho processes. In general, the effluent total ranged from about 0.5 to 6.3 mg/l, with the majority being close to 1 mg/l.

One possible drawback of the modification of predenitrification is that the sludge recycle returns NO_3^-; thus, some electron acceptor is applied to the anaerobic tank. Workers at the University of Cape Town (South Africa) developed the configurations shown in Figure 11.4 to eliminate NO_3^- from the return sludge. The sludge recycle goes directly to the anoxic predenitrification tank. Then, biomass is recycled to the anaerobic tank from the anoxic tank, where active denitrification should keep the NO_3^- concentration very low. The modified UCT process further reduces NO_3^- transfer to the anaerobic tank by partitioning the anoxic tanking into two zones. The second zone receives the main mixed-liquor NO_3^- recycle, while the first zone receives only

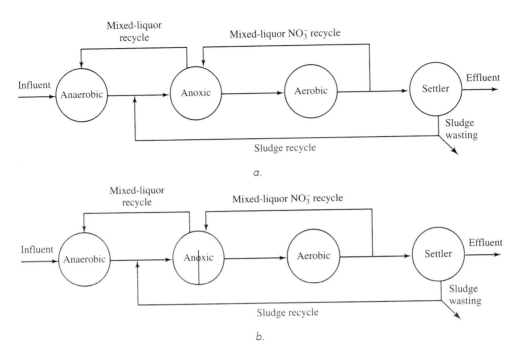

Figure 11.4 Two University of Cape Town (UCT) processes minimize the recycling of NO_3^- to the anaerobic tank. In the modified UCT process b., the anoxic tank is divided into two compartments. The mixed-liquor NO_3^- recycle enters in the downstream part of the anoxic tank, while the mixed-liquor recycle to the anaerobic tank leaves from the upstream part of the anoxic tank. Mixing between the two parts is restricted by a physical barrier.

the small NO_3^- input from sludge recycle and is the source of mixed liquor sent to the anaerobic tank.

Sequencing batch reactors (SBRs) also can be used for enhanced biological phosphorous removal, as well as nitrogen removal. Carryover of settled sludge from the previous cycle always contains some NO_3^-. Once the NO_3^- is denitrified during an unaerated fill period, anaerobic conditions can be established and lead to poly P and PHB cycling.

Biofilm processes can be used for enhanced biological phosphorous removal if the biofilm is exposed to alternating anaerobic and respiratory periods (Shanableh, Abeysinghe, Higazi, 1997). Characteristic responses of phosphate release during the anaerobic phase and phosphate uptake during the respirator phase occur. Oxygen or NO_3^- could be used as the terminal electron acceptor. Cycle times of around 6 h are workable, but further research is needed to optimize these processes for N and P removal.

Quite a unique process for enhanced biological phosphorus removal is *PhoStrip*, which combines Bio-P bacteria with chemical precipitation. Figure 11.5 sketches the key features of PhoStrip. The aerobic bioreactor and settler form a typical activated

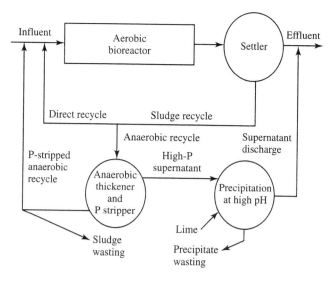

Figure 11.5   Schematic of the PhoStrip process.

sludge system, which usually is operated with a conventional SRT, does not nitrify, and has a "plug flow" configuration. The ecological and biochemical pressures for the Bio-P bacteria come from routing part of the sludge recycle through a thickener, which becomes completely anaerobic. The thickened sludge is returned to the aeration basin after having hydrolyzed and released its poly P to the supernatant, which contains 30–70 mg P/l and is sent to high-lime precipitation. Addition of lime raises the pH to around 11 and precipitates $Ca_3(PO_4)_{2(s)}$. The $Ca_3(PO_4)_2$ sludge is wasted, while the clarified supernatant can be discharged to the effluent. Field results show that PhoStrip can attain an effluent with less than 1 mg P/l in most cases (Meganck and Faup, 1988). PhoStrip is unique in that it does not rely on direct wasting of biomass enriched in poly P. Instead, it strips the poly P to soluble phosphates, which are then precipitated separately from the biomass. Despite this important difference from the other processes used for enhanced biological phosphorus removal, PhoStrip still relies on the same biochemical and ecological forces to select for and activate Bio-P bacteria.

11.4   BIBLIOGRAPHY

Arvin, E. (1985). "Observations supporting phosphate removal by biologically mediated precipitation: A review." *Water Sci. Technol.* 15(3/4), p. 43.

Barnard, J. L. (1976). "A review of biological phosphorus removal in the activated sludge process." *Water S.A.* 2(3), p. 136.

Battistoni, P.; G. Fava; P. Pavan; A. Musacco; and F. Cecci (1997). "Phosphate removal in anaerobic liquors by struvite crystallization without addition of chemicals: Preliminary results." *Water. Res.* 31(1), pp. 2925–2929.

Burdick, C. R.; D. R. Refling; and D. H. Stensel (1982). "Advanced biological treatment to achieve nutrient removal." *J. Water Pollution Control Federation* 54, pp. 1078–1086.

Clark, T.; T. Stephensen; and P. A. Pearce (1997). "Phosphorous removal by chemical precipitation in a biological aerated filter." *Water Res.* 31(10), pp. 2557–2563.

Comeau, Y.; K. J. Hall; R. E. W. Hancock; and W. K. Oldham (1986). "Biochemical model for enhanced biological phosphorus removal." *Water Res.* 20(12), pp. 1511–1522.

Dainema, M. H.; M. vanLoosdrecht; and A. Scholten (1985). "Some physiological characteristics of *Acinetobacter* spp. accumulating large amounts of phosphate." *Water Sci. Technol.* 17, pp. 119–125.

Ekama, G. A.; I. P. Siebritz; and G. v. R. Marais (1983). "Considerations in the process design of nutrient removal activated sludge processes." *Water Sci. Technol.* 15(3/4), p. 283.

Henze, M. (1996). "Biological phosphorus removal from wastewaters: Processes and technology." *Water Quality Intl.*, July/August, 1996, pp. 32–36.

Kerrn-Jespersen, J.; M. Henze; and R. Strube (1994). "Biological phosphorus release and uptake under alternating anaerobic and anoxic conditions in a fixed-film reactor." *Water Res.* 28(5), pp. 1253–1256.

Levin, G. V. and J. Shapiro (1965). "Metabolic uptake of phosphorus by wastewater organisms." *J. Water Pollution Control Federation* 37(6), p. 800.

Levin, G. V., G. J. Topol; A. G. Tarnay, and R. B. Samworth (1972). "Pilot test of a phosphate removal process." *J. Water Pollution Control Federation* 44(10), p. 1940.

Marais, G. v. R.; R. E. Loewenthal; and I. P. Siebritz (1984). "Observations supporting phosphate removal by biological uptake—a review." *Water Sci. Technol.* 15 (3/4), p. 15.

Meganck, M. I. C. and G. M. Faup (1988). "Enhanced biological phosphorus removal from wastewaters." In D. L. Wise, Ed. *Biotreatment Systems, Vol. III*. Boca Raton, FL: CRC Press.

Mino, T.; T. Kawakami; and T. Matsuo (1985). "Behavior of intracellular polyphosphate in a biological phosphate removal process." *Water Sci. Technol.* 17(11/12), pp. 11–21.

Nakamura, K.; A. Hiraishi; Y. Yoshimi; M. Kawaharasaki; K. Masuda; and Y. Kamogata (1995). "*Microlunatus phosphorus* gen. nov., sp. nov., a new gram-positive polyphosphate-accumulating bacterium isolated from activated sludge." *Intl. J. Syst. Bact.* 45(1), pp. 17–22.

Peirano, L. E.; D. P. Henderson; J. G. M. Gonzales; and E. F. Davis (1983). "Full-scale experiences with the PhoStrip process." *Water Sci. Technol.* 15 (3/4), p. 181.

Shanableh, A.; D. Abeysinghe; and A. Higazi (1997). "Effect of cycle duration on phosphorus and nitrogen transformations in biofilters." *Water Res.* 31(1), pp. 149–153.

Shoda, M.; T. Oshima; and S. Udaka (1980). "Screening for high phosphate accumulating bacteria." *Agric. Biol. Chem.* 44, pp. 319–324.

Snoeyink, V. L. and D. Jenkins (1980). *Water Chemistry*. New York: John Wiley.

Stumm, W. and J. J. Morgan (1981). *Aquatic Chemistry*, 2nd ed. New York: John Wiley.

Suresh, N.; R. Warburg; M. Timmerman; J. Wells; M. Coccia; M. F. Roberts; and H.O. Halvorsen (1985). "New strategies for the isolation of microorganisms responsible for phosphate accumulation." *Water Sci. Technol.* 17(11/12), pp. 43–56.

11.5   PROBLEMS

11.1. To remove most of 10 mg P/l from a wastewater, you add alum $[Al_2(SO_4)_3 \cdot 6\,H_2O$, FW $= 450$ g/mol] at a 2X stoichiometric dose. How much $AlPO_{4(s)}$ is built up in the reactor if $\theta_x = 6$ d, $\theta = 6$ h, and the total P removal by precipitation is 9 mg P/l? How much Al(III) remains in the water if Al(III) reacts with no other species?

11.2. Explain the purpose of each reactor and sludge flow in the Bardenpho process.

11.3. A predenitrification process is operated for enhanced biological P removal by adding an anaerobic tank before the denitrification tank. The influent has 250 mg/l of BOD_L, 20 mg/l of TKN, and 7.5 mg/l of phosphate P. You may assume that the overall SRT is 20 d and that the denitrification parameters for Table 10.4 are appropriate. Estimate the fraction of biomass that must be P if the effluent P concentration is to be 1 mg/l. The mixed-liquor recycle ratio is 6, while the return-sludge recycle ratio is 0.3. For comparison, you may assume that the P content of "normal" biomass is 2.5 percent.

11.4. Denitrification has been implicated in phosphorus removal in biological processes. The two mechanisms for P removal are biomass synthesis and production of alkalinity, which raises the pH. For a wastewater having the following characteristics, show quantitatively how denitrification might bring about P removal. Specifically, how much biomass is produced and how much alkalinity is produced? You may assume that $f_s = 0.2$. Which mechanism seems more important?

$$T = 20\,°C \qquad\qquad BOD_L^0 = 100 \text{ mg/l, added as methanol}$$
$$(NO_3^- \text{-N})^0 = 30 \text{ mg/l} \qquad \text{alkalinity} = 100 \text{ mg/l as } CaCO_3$$
$$P^0 = 12 \text{ mgP/l}$$

11.5. You have a wastewater with the following characteristics:

$$Q = 2{,}500 \text{ m}^3/\text{d} \qquad BOD_L = 450 \text{ mg/l} \qquad TKN = 50 \text{ mg/l}$$
$$PO_4^{3-}\text{P} = 12 \text{ mg/l} \qquad SS = 300 \text{ mg/l}$$

You wish to treat this wastewater directly by a Pho-Strip process for enhanced biological phosphorus removal. If you maintain an SRT of 4 d, can you meet a 1-mg/l standard for phosphate-P? Why is the design feasible or infeasible? You may assume that $\hat{q} = 12$ mg BOD_L/mg VSS_a-d, $K = 10$ mg BOD_L/l, $b = 0.15$/d, $f_d = 0.8$, $Y = 0.45$ mg VSS_a/mg BOD_L.

11.6. A treatment facility treats domestic sewage with the contact stabilization process. Aluminum salts are added to the contact tank for phosphorus removal. All the collected solids are then sent to thickening and aerobic digestion. The typical influent and effluent quality is:

Parameter	Influent	Effluent
Flow rate, m^3/d	20,000	20,000
BOD_L, mg/l		
Total	350	6
Suspended	50	1
Suspended solids, mg/l		
Total	100	2
Volatile	70	1
Biodegradable	35	0.7
TKNitrogen, mg/l	35	1
Total phosphorus, mg/l	8	1

Appropriate parameters are:

$$\hat{q} = 12 \text{ mg BOD}_L/\text{mg VSS}_a\text{-d}$$
$$K = 20 \text{ mg BOD}_L/\text{l}$$
$$Y = 0.5 \text{ mg VSS}_a/\text{mg BOD}_L$$
$$b = 0.1/\text{d}$$
$$f_d = 0.8$$
$$\theta_x = 5\text{d}$$

You will do a stepwise evaluation.

(a) Calculate the waste rate (kg/d) of sludge solids that originally are in the influent, but are not biodegraded. Distinguish inorganic solids from inert volatile solids.

(b) Calculate the production rate of active and inert volatile solids due to net biomass synthesis.

(c) Calculate the P incorporated into the biomass (kg P/d).

(d) Calculate total P removal rate through the process (kg P/d).

(e) If all the organic P originally in the influent is released as phosphates, how much P (kg/d) is removed by precipitation?

(f) How much inert solids (kg/d) are produced by P precipitation as $AlPO_{4(s)}$ if the practical mole ratio for Al^{3+} addition is 1.5:1 and any excess Al^{3+} precipitates as $Al(OH)_{3(s)}$?

(g) Tabulate the waste rates (kg/d) of all active VSS, inert VSS, inorganic SS, and total SS. Express each category as a percent of the total.

11.7. Describe the purpose of each reactor and each recycle flow in the UCT process.

11.8. The Bardenpho process has a small unaerated tank at its beginning, but it is not called a selector tank. What are the similarities and the differences (at least two of each) between a selector tank and the first tank in the Bardenpho process?

11.9. Evaluate the following process for its ability to carry out total N removal and enhanced biological P removal. First, describe the role of each numbered tank and flow. If the process is not going to be successful in N or P removal, explain why. You may assume that the process receives normal domestic sewage.

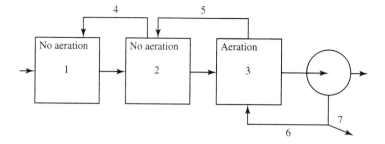

chapter

12

DRINKING-WATER TREATMENT

Biological treatment to produce safe and pleasing drinking water is beginning to play a key role in North America, and it already is common throughout much of Europe (Rittmann and Huck, 1989; Rittmann and Snoeyink, 1984). Biological treatment has its most widespread use as a means to remove biodegradable electron donors from the water. Called *biological instability,* these electron donors include biodegradable organic matter (BOM), ammonium, nitrite, ferrous iron, manganese (II), and sulfides. The presence of even small concentrations of these instability components can foster the growth of bacteria in the distribution system. Although the bacteria grown through oxidation of instability components seldom are directly harmful to humans, they create several undesirable water-quality changes as a result of their presence and metabolic activity:

- Increases in heterotrophic plate counts, regrowth of coliform bacteria, and turbidity

- Formation of tastes, odors, and nitrite

- Consumption of dissolved oxygen

- Accelerated corrosion

In the absence of biological treatment to eliminate biological instability, the common approach to control its adverse effects during distribution is maintaining a substantial chlorine residual in all parts of the distribution system. Since the mid-1970s, evidence continues to grow that the presence of the chlorine residuals needed for biologically unstable waters generate significant concentrations of disinfection by-products (DBPs), including chloroform and other trihalomethanes (THMs), haloacetic acids, and chlorinated aromatics. Because of the known or suspected health risks of the DBPs, regulations are limiting their concentrations in distributed waters. These regulations are forcing water utilities to reduce chlorine doses and/or shift from free chlorine to combined chlorine, which reduces (but does not eliminate) DBP formation. Reductions in chlorine use also are being implemented in order to minimize the formation of chlorinous tastes and odors, corrosion, and chemical costs. In some case, such as at Amsterdam, chlorination has been totally abandoned.

If bacterial growth cannot be suppressed with high chlorine residuals, the distributed water must be made *biologically stable* by eliminating the biodegradable electron donors during treatment. Fortunately, a range of biofilm processes is available to achieve this treatment goal. These processes are the main subjects of Chapter 12.

A complicating factor for the inorganic electron donors is that they can be generated during water distribution. The decay of monochloramine (NH_2Cl) produces NH_4^+, which can spur nitrification. Corrosion of elemental iron (Fe^0) in unlined pipes may directly or indirectly provide an inorganic electron donor for autotrophs, and Fe^{2+} released by corrosion certainly is an electron donor for iron-oxidizing autotrophs. Growth and metabolism by autotrophs always results in release of SMP, which increases the potential for growth of heterotrophs.

A second use for biological treatment in drinking-water treatment is denitrification of NO_3^- (or occasionally NO_2^-) to N_2 gas. Nitrate is seasonally present in water supplies subject to runoff or infiltration of water that has been in contact with agricultural fertilizers. In most parts of the world, NO_3^- has a maximum standard of 10 mg NO_3^--N/l, while Europe also maintains a 0.1 mg/l standard for NO_2^--N. These standards are set to prevent *methemoglobinemia* in infants, in whose guts NO_3^- is reduced to NO_2^-, which binds to hemoglobin and causes suffocation. When biofilm processes are used for denitrification, NO_3^- is the electron-acceptor substrate.

This chapter is divided into two major sections. The first section address elimination of biological instability in aerobic biofilm processes. It begins by discussing some general characteristics of these processes. Then, it describes the three main approaches to biofilm treatment: biological pretreatment, hybrid biofiltration, and slow biofiltration. The first section concludes with information on the biodegradation of organic components of special health and aesthetic concerns. The second major section discusses anoxic processes used for denitrification.

12.1 AEROBIC BIOFILM PROCESSES TO ELIMINATE BIOLOGICAL INSTABILITY

All biological processes used to create a biologically stable drinking water share three key characteristics.

- Their environment is *oligotrophic*, which means that the electron-donor substrates are present in very low concentrations. This occurs because influent concentrations of electron donors usually are quite low, and effluent concentrations must be driven to extremely low concentrations (i.e., near S_{min}) in order to severely limit bacterial growth during distribution.

- They rely on *biofilms* for excellent biomass retention without requiring long liquid detention times. This biofilm dependence is a direct consequence of the very oligotrophic environment.

- They are *aerobic*. In many (but not all) cases, the relatively low concentrations of instability components do not create an oxygen demand capable of depleting the dissolved oxygen (D.O.) present in the water. Even more important is the need to keep the processes aerobic in order to avoid adverse quality impacts of anaerobisis: taste and odor formation, color, biofilm detachment, and metals solubilization.

Virtually all water supplies contain some BOM, but two aspects of the BOM make its concentration and instability impact difficult to quantify: very low concentration and heterogeneous nature. Measured BOM concentrations in most water supplies range from less than 100 μg/l as C up to 1 mg/l as C; these numbers convert approximately to less than 260 μg as BOD_L up to 2.6 mg/l as BOD_L (Woolschlager and Rittmann, 1995). Although these measured BOM values may underestimate the true BOM (Woolschlager and Rittmann, 1995), it is clear that BOM concentrations entering a drinking-water process are lower than the effluent concentrations exiting most wastewater processes. Most of the BOM in water supplies is natural organic matter (NOM) derived from the decay of the complex polymers comprising vegetation and microorganisms, and much of it is humiclike. The humiclike BOM appears to have relatively slow biodegradation kinetics and moderate molecular size, typically 1,000 to 5,000 daltons effective formula weight. A fraction of the BOM contains small, easily degradable molecules, such as organic acids, aldehydes, alcohols, ketones, and carbohydrates. While water supplies naturally contain small concentrations of this labile BOM, significantly larger concentrations are produced during chemical disinfection, especially with ozone, which is well known for producing aldehydes and other rapidly biodegradable by-products from NOM (Huck, 1990). Whereas the measured BOM values are only a few percent of the water source's total organic carbon (TOC), ozonation can make 30 to 40 percent of the TOC biodegradable (Owen, Amy, and Chowdbury, 1993).

Table 12.1 summarizes "generic" parameters that can be used to represent the biodegradation kinetics of the easily degraded and humiclike BOM. The main dif-

Table 12.1   Generic kinetic parameters for the biodegradation of easily degraded and humiclike natural organic matter

Parameter	Easily Degraded	Humiclike
$\hat{q}$, mg BOD_L/mg VSS_a-d	10	10
K, mg BOD_L/l	1	7.5
Y, mg VSS_a/mg BOD_L	0.42	0.42
b, d^{-1}	0.1	0.1
b_{det}, d^{-1}	0.05	0.05
b', d^{-1}	0.15	0.15
D, cm^2/d	0.85	0.053
D_f, cm^2/d	0.68	0.042
X_f mg VSS_a/cm^3	40	40
L, cm	0.0040	0.0016
S_{min} mg BOD_L/l	0.037	0.28
S_{min}^*	0.037	0.037
K^*	0.41	0.70
J_R, mg BOD_L/cm^2-d	0.054	0.037

Parameters are complied from Woolschlager and Rittmann (1995) and Rittmann(1990) and are for 15 °C.

ferences are the larger K and smaller D and D_f values for the humiclike material. While both substrates show very high growth potential ($S_{min}^* = 0.037$), the humiclike BOM has a higher S_{min} and a lower J_R. Because biologically stable waters should have BOM concentrations approaching S_{min}, the expected J/J_R values should be substantially less than 1.0 (see Normalized Surface Loading in Chapter 4). This suggests that BOM fluxes should be much less than 0.034 to 0.054 mg BOD_L/cm^2-d, which corresponds to 0.34 to 0.54 kg $BOD_L/1,000$ m^2-d. Rittmann (1990) tabulated observed flux values and found that they ranged from 0.0005 to 0.72 kg COD/1,000 m^2-d for a variety of process configurations. These empirical results are consistent with the goal of driving S to values very close to S_{min}, which is around 0.3 mg BOD_L/l (in this generic example) and comprised mostly of humiclike substances having slow biodegradation kinetics.

Because BOM typically is present at quite low concentrations in drinking waters, its measurement becomes a severe challenge. Gross measures like COD and TOC are inappropriate, since much of the NOM is not biodegradable at all or with kinetics fast enough to make it relevant. The BOD test is not sensitive at sub-mg/l concentrations. Therefore, a suite of tests has been designed specifically for the measurement of low-levels of BOM in drinking water. Most of the methods were reviewed by Huck (1990) and Rittmann and Huck (1989), where details and additional primary sources can be located. A brief summary is provided here.

12.1.1 BOM MEASUREMENT TECHNIQUES

Most of the BOM measurement techniques involve batch incubations of the water with a controlled bacterial inoculum. These batch tests can be divided into three groups.

1. The methods of van der Kooij (1982) use pure strains of bacteria selected for their ability to grow on different types of organic compounds. The maximum growth of the pure strain, most commonly assayed by plate counts, is converted to a concentration of *assimilable organic carbon (AOC)* by conversion factors derived from pure-substrate tests. Due to the metabolic specificities of the organisms tested and the generally low concentrations measured, AOC most probably assays only the easily biodegraded BOM. van der Kooij (1982) suggests that water is biologically stable if its AOC is less than about 10 μg C/l, or about 26 μg/l as BOD_L, a value in line with S_{min} for easily biodegradable organic molecules. By biologically stable, van der Kooij means that it can be distributed with no chlorine residual, as is common in The Netherlands. When a chlorine residual can be maintained throughout the distribution system, higher AOC levels can be tolerated.

2. Some techniques employ a small inoculum of a mixed bacterial culture and measure the maximum decrease in dissolved organic carbon to estimate the *biodegradable dissolved organic carbon (BDOC)*. BDOC tests are not as sensitive as tests that monitor the growth of bacteria, and minimum detection levels are around 100 μg C/l, or approximately 260 μg/l as BOD_L. The use of a mixed culture, normally acclimated to the water being tested, as well as the

normally higher values compared to the AOC test, suggests that BDOC assays a wider range of biodegradable organic matter than does the AOC test.

3. Large-inoculum BDOC tests are used to overcome one limitation of the first two types of tests: a long incubation time, often 30 d. All high-inoculum tests are of the BDOC type. The original tests were of the batch type (Joret, Levi, and Volk, 1991) and took 1 to 7 d. They used biofilm-coated sand taken from a biologically active filter. Emerging is a suite of high-inoculum tests that pass water through biofilm columns and measure the change in BDOC (e.g., Kaplan and Newbold, 1995).

Woolschlager and Rittmann (1995) systematically analyzed the low- and high-inoculum BOM methods. They found that formation of soluble microbial products (SMP) and the endogenous decay of the bacteria caused both types of tests to underestimate the true BOM concentration. The effects are greatest when the true BOM is near S_{min}. For example, 300 μg/l (as BOD_L) of humiclike BOM gives essentially zero BDOC or AOC, since net growth is impossible. At high BOM concentrations, BDOC still underestimates the true BOM, because the cells continually produce BAP, which is measured as organic carbon. In summary, current BOM methods provide good trends about changes in the biological stability of a water, but are not perfectly quantitative for the true BOM concentration.

The treatment goal for AOC or BDOC depends on disinfection practices for the distributed water. In Europe, where many countries use small or zero residuals of chlorine, a treatment goal to achieve a biologically stable water is often taken as 10 μg/l as AOC or 50 μg/l as BDOC. In North America, where large chlorine residuals are common, acceptable AOC or BDOC concentrations may be higher. No exact values can be given, because the loss of disinfectant residual is affected by temperature, pipe size and materials, corrosion, and water-distribution time.

12.1.2 REMOVING INORGANIC SOURCES OF BIOLOGICAL INSTABILITY

The next sections describe the techniques used to treat a biologically unstable water. The focus is on removal of BOM. However, the inorganic sources of biological instability (mainly NH_4^+, NO_2^-, Fe^{2+}, HS^-, and Mn^{2+}) also can be removed by similar processes. NH_4^+-N is the most prevalent inorganic electron donor, and its oxygen demand of 4.57 mg O_2/mg N gives a very potent impact to a small concentration. Nitrification produces NO_2^-, which is known to react with chloramines and destroys the combined-chlorine residual.

Nitrification in biofilm processes is reviewed in Chapter 9, and most of the information provided there is directly applicable to drinking-water treatment. Key features of biofilm nitrification are that it has a poor growth potential ($S_{min}^* > 0.3$), but S_{min} is quite low (< 0.2 mg N/l). Thus, if the substrate flux can be kept substantially below J_R (which is 0.4–0.8 kg NH_4^+-N/1,000 m^2-d), the NH_4^+ concentration can be driven quite low. Rittmann (1990) reviewed nitrification practice in drinking-water treatment and found that successful treatment occurred for NH_4^+-N fluxes from 0.003 to

0.5 kg/1,000 m^2-d. Ironically, NH_4^+-N fluxes for wastewater treatment (Chapter 9) are about the same (0.2–0.8 kg/1,000 m^2-d) and reflect that stable nitrification usually results in S values approaching S_{min}. Stable nitrification also oxidizes the NO_2^- to NO_3^-.

The oxidations of Fe^{2+} to Fe^{3+} and Mn^{2+} to Mn^{4+} are catalyzed by aerobic bacteria that can gain energy from these reactions. Once in the oxidized form, the metals usually precipitate as oxides or carbonates. Bacterially catalyzed oxidations seem to occur in biofiltration processes, but the documentation is sketchy (Rittmann and Snoeyink, 1984). The physiology of the prevalent strains is not well understood. The Fe^{2+} and Mn^{2+} oxidizers probably are autotrophs and should flourish in biofilters that promote nitrification. Bacterial oxidation of HS^- is well established, although it has not been studied much in the drinking-water context.

12.1.3 BIOFILM PRETREATMENT

When concentrations of instability components are high, the most logical approach to biological treatment is to employ a biofilm process for the sole purpose of pretreating the water to make it biologically stable for subsequent physical/chemical treatment and distribution. High-BOM water applied directly to a particle filter, such as a rapid sand filter, can disrupt the performance of the particle filter. For instance, high BOM loading can cause excessive backwashing, particle breakthrough, taste and odor formation, or other operating problems. The dividing line for when the BOM is large enough to warrant biofilm pretreatment is not clearly defined. The boundary probably lies around 2 mg/l of oxygen demand, and the next section explores the boundary. Biofilm pretreatment also can be used when the BOM is low, for which quite high hydraulic loading is possible.

The guiding concept for biofilm pretreatment processes is to create relatively large pores, so that the medium does not encounter clogging, which causes excessive head losses and/or requires bed cleaning. Large pores are attained by using a fixed-bed of gravel-sized medium (4 to 14 mm) or by fluidizing sand-sized media (< 1 mm). In either case, the large pores minimize the need for backwashing and also allow for bed aeration to meet oxygen demands greater than the D.O. content of the input water. Even though the pores are relatively large, these processes still have high specific surface areas that allow short hydraulic detention times, typically 5 to 30 min (Rittmann and Huck, 1989, Rittmann and Snoeyink, 1984).

Most of the *fixed-bed biofilm processes* used in wastewater treatment (Chapter 8) have been used for drinking-water applications (Rittmann and Huck, 1989; Rittmann, 1990; Rittmann and Snoeyink, 1984). Most notable are the processes using the *Biolite* medium (4-mm expanded clay) and pozzolana of approximately 14 mm size. Pozzolana is a volcanic rock of relatively low density, but very irregular shape. When needed, bed aeration is provided. Hydraulic loading can be quite high—50 to 110 m/d—and gives liquid detention times measured in minutes, BOM fluxes up to 0.14 kg BOD_L/1,000 m^2-d, and ammonium fluxes up to 0.38 kg N/1,000 m^2-d. Periodic backwashing may be required for the *Biolite* system. Unsubmerged gravel and plastic-medium filters also have been used, but cold-weather performance was poor.

Fluidized-bed biofilm processes offer great potential for having very short liquid detention times, perhaps as low as one minute. Pilot studies carried out in Great Britain (Rittmann and Huck, 1989; Rittmann and Snoeyink, 1984) gave excellent performance, including full nitrification at temperatures as low as 4 °C . Hydraulic loading was 140–720 m/d, giving an empty-bed detention time of 5 to 10 min, a BOM flux greater than 0.01 kg BOD_L/1,000 m^2-d, and an ammonium flux up to 0.032 kg N/1,000 m^2-d. Given the high potential of fluidized-bed treatment, especially for retrofits into existing facilities, further research and development are warranted to answer questions about substrate loading, medium selection, oxygen supply, control of biofilm accumulation, and the role of pre-ozonation to increase the biodegradability of the NOM.

Example 12.1

DESIGN OF A *BIOLITE* PRETREATMENT PROCESS   After ozonation, a rather poor-quality surface water contains BDOC = 1,500 μg/l and NH_4^+-N = 2 mg/l in a flow of 1,000 m^3/d. The treatment goal is to create a biologically stable water with BOM and NH_4^+-N concentrations near their S_{min} values. A *Biolite* medium is selected. We must determine the biofilm surface area, the process volume and dimensions, and the empty-bed detention time. We also must estimate the D.O. supply rate.

First, we calculate the mass loading rates of BOD_L and NH_4^+-N.

$$\frac{\Delta S}{\Delta t} = 1,000 \text{ m}^3/\text{d} \cdot 1.5 \text{ mg BODC/l} \cdot 2.6 \text{ mg BOD}_L/\text{mg BDOC} \cdot \frac{10^{-3} \text{ kg-l}}{\text{mg-m}^3}$$

$$= 3.9 \text{ kg BOD}_L/\text{d}$$

$$\frac{\Delta N}{\Delta t} = 1,000 \text{ m}^3/\text{d} \cdot 2.0 \text{ mg N/l} \cdot \frac{10^{-3} \text{ kg-l}}{\text{mg-m}^3} = 2.0 \text{ kg N/d}$$

Second, we select design surface loads. For nitrification, we select an average value of 0.25 kg N/1,000 m^2-d. This gives a biofilm surface area of (aV)

$$aV = \frac{2 \text{ kg N/d}}{0.25 \text{ kg N/1,000 m}^2\text{-d}} = 8,000 \text{ m}^2$$

For BOM removal, we choose a conservative value of the BOD_L surface load to reduce competition with the nitrifiers. We choose 0.10 kg BOD_L/1,000 m^2-d. The area is then

$$aV = \frac{3.9 \text{ kg BOD}_L/\text{d}}{0.10 \text{ kg BOD}_L/1,000 \text{ m}^2\text{-d}} = 39,000 \text{ m}^2$$

The BOM area controls the biofilm area, 39,000 m^2.

Third, we compute the volume, empty-bed detention time, and the dimensions. The *Biolite* medium has a specific surface area of about $a = 1,000$ m^{-1} (Chapter 8). Thus,

$$V = 39,000 \text{ m}^2/1,000 \text{ m}^{-1} = 39 \text{ m}^3$$

The empty-bed detention time is simply EBDT = V/Q = 39 m^3/(1,000 m^2/d) = 0.039 d, or 56 min. The typical depth of a *Biolite* filter is 3 m, which makes the plan-view surface area (A) to be

$$A = 39 \text{ m}^3/3 \text{ m} = 13 \text{ m}^2$$

which gives a hydraulic loading (H.L.) of

$$\text{H.L.} = (1{,}000 \ m^3/d)/13 \ m^2 = 77 \ m/d \ (or \ 3.2 \ m/h)$$

Finally, we estimate the oxygen-supply rate by assuming that f_e is 0.7 for the heterotrophs and 0.9 for the nitrifiers:

$$\frac{\Delta O_2}{\Delta t} = 0.7 \cdot 3.9 \ kg \ BOD_L/d + 0.9 \cdot 4.57 \ kg \ O_2/kg \ N \cdot 2 \ kg \ N/d$$

$$= 2.7 + 8.2 = 10.9 \ kg \ O_2/d$$

This is the same as delivering 10.9 mg/l to the influent flow. Clearly, aeration is necessary to sustain the reactions and prevent D.O. depletion, but delivering the equivalent of 10.9 mg/l to the influent flow is easily achieved by bed sparging.

DESIGN OF A FLUIDIZED-BED PRETREATMENT PROCESS   A poor-quality ground- | **Example 12.2**
water has a BDOC of 4 mg/l after ozonation to remove color. The treatment goal is to use a very compact process to reduce the BOM before membrane ultrafiltration. We will use a fluidized-bed biofilm process with GAC medium of 0.5 mm. We must compute the surface area, bed volume, empty-bed detention time, and oxygen supply rate. The flow rate is 10,000 m^3/d.

The BOD_L mass load is

$$\frac{\Delta S}{\Delta t} = 10{,}000 \ m^3/d \cdot 4 \ mg \ BDOC/l \cdot 2.6 \ mg \ BOD_L/mg \ BDOC \cdot \frac{10^{-3} \ kg\text{-}l}{mg\text{-}m^3}$$

$$= 104 \ kg \ BOD_L/d$$

We select a fairly high surface loading to minimize the reactor size: 0.2 kg $BOD_L/1{,}000$ m^2-d. This leads to the biofilm surface area:

$$aV = (104 \ kg \ BOD_L/d)/(0.2 \ kg \ BOD_L/1{,}000 \ m^2\text{-}d) = 520{,}000 \ m^2$$

The specific surface area depends on the medium size ($d_p = 0.5$ mm $= 5 \cdot 10^{-4}$ m), the expanded-bed porosity (we assume it to be $\varepsilon_{ex} = 0.46$), and the shape factor for the GAC (we assume $\Psi = 0.8$) as discussed in Chapter 8:

$$a = \frac{6(1 - \varepsilon_{ex})}{\Psi d_p} = \frac{6(1 - 0.46)}{0.8 \cdot 5 \cdot 10^{-4}} = 8{,}100 \ m^{-1}$$

This gives us the bed volume and the empty-bed detention time:

$$V_{bed} = 520{,}000 \ m^2/8{,}100 \ m^{-1} = 64 \ m^3$$

$$EBDT = 64 \ m^3/10{,}000 \ m^3/d = 0.0064 \ d \ (or \ 9.2 \ min)$$

We note that the total reactor volume is 1.5 to 2 times the bed volume of the expanded bed. The oxygen-supply rate is based on $f_e = 0.7$:

$$\frac{\Delta O_2}{\Delta t} = 0.7 \cdot 104 = 73 \ kg \ O_2/d$$

This is an oxygen-addition rate of 7.3 mg/l in the influent flow. If the influent does not contain D.O. well in excess of 7.3 mg/l, aeration is needed. The oxygen is most easily added with an oxygen-transfer device placed in the effluent-recycle line, although bed aeration or influent oxygenation can be used.

12.1.4   HYBRID BIOFILTRATION

When the biological instability is low in the raw water, traditional filtration processes can be enhanced to create a hybrid biofiltration process that eliminates significant amounts of BOM and inorganic electron donors, while maintaining its traditional function. Examples include rapid filters of sand, anthracite, granular activated carbon (GAC), and their dual-media combinations; GAC absorbers; and floc-blanket reactors (Rittmann and Huck, 1989; Urfer et al., 1997). Although not required, ozone often is applied before these processes to make the NOM more biodegradable, as well as for primary disinfection, taste and odor removal, or enhanced coagulation. Other strategies to optimize the biodegradation aspect include eliminating bed chlorination, using a larger medium size (e.g., effective sizes of 1 mm versus 0.5 mm for rapid filtration), and controlling the length and severity of backwashing to alleviate head loss while maintaining a good biofilm base.

Unchlorinated *rapid filters* always are active in BOM biodegradation, and oxidation of NH_4^+, Fe^{2+}, HS^-, and Mn^{2+} also can be obtained. Hydraulic loading for hybrid biofilters generally is similar to other rapid filters, 5–10 m/h, and is based more on particle removal than on BOM surface loading. Due to the accumulation of biofilm, backwash intervals may be more frequent than when the bed is chlorinated continuously. Biofilm control strategies require further systematic research, because biofilm control is critical. On the one hand, excessive removal of biofilm causes an undesirable "ripening" period before excellent BOM removal commences after backwash, and removal of inorganic electron donors surely will be prevented. On the other hand, inadequate biofilm removal during backwash can lead to rapid buildup of excessive headloss, breakthrough of turbidity, and growth of undesirable macroinvertebrates and nematode worms. Chlorination of the backwash water is a compromise strategy that reduces the amount of biofilm accumulation without eliminating biological activity. Air scour during backwashing improves the removal of particles in biofilters without causing a problem of inadequate biofilm retention. Controlling bed expansion during backwash also should affect the balance between particle removal and biofilm loss.

Biological activity in *GAC adsorbers* was well documented beginning in the 1970s (Rittmann and Huck, 1989). The original interest was in the role of biodegradation for extending the adsorptive life of adsorbers by "bioregenerating" the activated carbon. While that objective remains an important benefit, the long-term, steady-state removal of BOM and other instability components is now recognized. Frequently, ozone is applied to the influent water at doses of 0.5 to 1.0 mg O_3/mg DOC. Ozonation increases the biodegradability of the NOM, thereby allowing a larger steady-state DOC removal across the adsorber. Sometimes the ozone-GAC combination is called the *biological activated carbon (BAC) process*; although ozone increases the biomass and biodegradation in a GAC column, GAC columns are biologically active when pre-ozonation is not practiced. The required empty-bed contact time needed for adsorption (15–30 min) and the high specific surface area of a GAC adsorber combine to give relatively low BOM surface loading, less than 0.03 kg BOD_L/1,000 m^2-d in most cases. Periodic backwashing often is required.

A very important, but poorly defined issue in hybrid biofiltration is how high the BOM concentration can be before excessive biofilm accumulation causes operating problems, such as excessive headloss. When the influent BOM is too high, biofilm pretreatment should be used. No direct work on BOM loading is available, but Rittmann (1993) evaluated organic-substrate surface loading causing rapid and severe clogging of porous media. He concluded that a surface loading (QS^0/aV) of 0.04 kg BOD_L/1,000 m²-d gave discontinuous biofilms having little clogging. This value is approximately 10 percent of J_R (Table 12.1). On the other hand, a surface loading of 0.12 kg BOD/1,000 m²-d or greater definitely gave continuous biofilms with clogging potential. Clearly, higher surface loading increases the clogging severity.

Table 12.2 presents the surface loading for a range of relevant influent BOM concentration, filter hydraulic loading, and medium diameter. The surface loading (S.L.) is computed as

$$\text{S.L.} = QS^0/aAh \qquad \textbf{[12.1]}$$

in which

Q = influent flow rate $[L^3 T^{-1}]$
A = plan-view area $[L^2]$
(Q/A) = hydraulic loading $[LT^{-1}]$
S^0 = influent BOM concentration $[M_s L^{-3}]$
a = specific surface area of the medium $[L^{-1}]$
h = depth of the filter bed $[L]$

The value of a depends on the medium diameter (d_p) according to

$$a = 6(1 - \varepsilon)/d_p \Psi \qquad \textbf{[12.2]}$$

For the computations, filter-depth h is fixed at a typical value of 1 m, Q/A spans the usual range of 5 to 10 m/h (2–4 gpm/ft²), porosity ε is set at 0.4, shape factor $\Psi = 1$, and particle-diameter d_p is 1.0 mm.

Table 12.2 Surface BOM loading (in kg BOD_L/1,000 m²-d) for a range of influent BOM concentrations and filter-operating conditions

Operation Conditions		Influent BOM Concentration, mgBOD_L/l				
Medium Diameter, mm	Hydraulic Loading, m/h	0.25	1.0	2.0	4.0	10
1	5	**0.0083**	**0.033**	0.066	*0.13*	*0.33*
1	10	**0.017**	0.066	*0.13*	*0.27*	*0.66*

Conditions held constant: Filter depth = 1 m, medium porosity = 0.4, shape factor = 1.

Relevant Conversions: 1 m/h = 0.4 gpm/ft²; 1mg BOD_L/l ≈ 0.4 mg BDOC/l.

Boldface indicates a surface loading less than 0.04 kg BOD_L/1,000 m²-d, or that not likely to cause serious clogging.

Italic indicates more than 0.12 kg BOD_L/1,000 m²-d, or that very likely to cause serious clogging.

The results in Table 12.2 suggest that rapid filtration probably does not cause excessive clogging (**boldface entries**) when the influent BOM has a concentration less than about 1 mg BOD_L/l, which corresponds roughly to a BDOC of 0.4 mg/l. For concentrations greater than about 2 mg BOD_L/l (ca. 0.8 mg/l of BDOC), operation of a hybrid filter may lead to excessive headloss (*italic entries*) and frequent backwashing, and problems seem very likely for 4 mg BOD_L/l (ca. 1.6 mg/l of BDOC).

Montgomery (1985) states that backwashing is needed when the accumulation of solids in the rapid filter is more than 550 g/m^3, or 550 $kg/1,000 \, m^3$. If a biofilm were to reach steady state, the accumulation is computed as $X_f L_f a = Y Ja/b'$. We assume that J can be taken as the surface loading, $Y = 0.42$ kg VSS_a/kg BOD_L, and b' can be estimated as 0.5/d, which corresponds to one-half of the biofilm being removed by backwashing and detachment each day. The surface load (J) that corresponds to $X_f L_f a = 550 \, kg/1,000 \, m^3$ and backwashing to remove *only* biofilm accumulation is 0.18 kg $BOD_L/1,000 \, m^2$-d for $d_p = 1$ mm. Of course, filtered influent solids also build up and contribute to the clogging. If biofilm accumulation accounts for one-half of the total solids accumulation, then the BOD_L flux to give excessive clogging is about 0.09 kg $BOD_L/1,000 \, m^2$-d, which corresponds well to the bold values in Table 12.2. This lends further support to the idea that hybrid biofiltration may cause problems when the influent BOM is greater than roughly 2 mg BOD_L/l, or about 0.8 mg/l of BDOC.

Example 12.3 | **ANALYSIS OF A HYBRID RAPID FILTER**   A water supply has influent suspended solids of 2 mg/l and a BDOC of 500 μg/l. A rapid filter is designed with a hydraulic loading of 8 m/d (32 gpm/ft^2), 1-mm sand, and a 1-m depth. It is supposed to operate as a hybrid biofilter. We are to evaluate the BOD_L loading and the clogging potential.

The BOM surface loading (S.L. in kg $BOD_L/1,000 \, m^2$-d) is computed from

$$\text{S.L.} = \text{H.L.} \cdot \frac{S^0}{a \cdot h} \cdot (24 \text{ h/d})$$

where H.L. is the hydraulic loading ($Q/A = 8$ m/h), S^0 is the influent BOD concentration (0.5 mg BDOC/l · 2.6 mg BOD_L/mg BDOC = 1.3 mg BOD_L/l = 1.3 kg $BOD_L/1,000 \, m^3$), a is the specific surface area (5,600 m^{-1} for sand of 1-mm diameter), and h is the bed depth (1 m). Substituting yields

$$\text{S.L.} = 8 \text{ m/h} \cdot \frac{1.3 \text{ kg } BOD_L/1,000 \text{ m}^3}{5,600 \text{ m}^{-1} \cdot 1 \text{ m}} \cdot (24 \text{ h/d}) = 0.045 \text{ kg } BOD_L/1,000 \text{ m}^2\text{-d}$$

This value of S.L. is less than J_R, which suggests that BOM removal will be good. It also is approximately equal to 0.04 kg $BOD_L/1,000 \, m^2$-d, which should preclude excess backwashing from biomass accumulation.

The accumulation of filtered influent suspended solids is

$$\frac{\Delta SS}{\Delta t} = \frac{8 \text{ m}}{h} \cdot \frac{24 \text{ h}}{d} \cdot \frac{1}{1 \text{ m}} \cdot \frac{2 \text{ kg SS}}{1,000 \text{ m}^3} = 384 \text{ kg SS}/1,000 \text{ m}^3\text{-d}$$

This value gives less than 550 kg $SS/1,000 \, m^3$ each day. Thus, the hybrid biofilter should perform well for BOM removal and should not be subject to premature clogging from biofilm or influent suspended solids.

12.1.5 SLOW BIOFILTRATION

When ample and appropriate land area is available, slow biofiltration can be used. Here, the infiltration rate is much slower than in rapid filtration. The slow infiltration rates have three important effects. First, they require substantial land surface areas. Second, most of the microbiological activity is concentrated in a small bed depth. Third, the slow infiltration rates make headloss buildup a much less serious drawback than in rapid filtration.

The *slow sand filter* is a traditional treatment process that is still in use in non-urban areas. A bed of sand having an effective size of 0.25 to 0.35 mm is slowly and continuously loaded at 3 to 15 m/d, with 5 m/d being typical (Rittmann and Huck, 1989; Collins, Eighmy, and Malley, 1991). A layer of microorganisms and captured particles (the *schmutzdecke,* which is German for "dirty layer") develops at the top of the sand bed and is responsible for removing instability and turbidity. When the headloss becomes too great, the top 50–75 mm of sand plus *schmutzdecke* is scraped away. The slow sand filter is a very simple and effective process when land area is available (note that the hydraulic loading is more than 20 times smaller than for rapid filters and most pretreatment biofilters) and when BOM and turbidity are modest in the raw water.

In situ ground passage is used in several European countries as a form of slow filtration. *Bank filtration,* widespread in Germany, pulls river water through the riverbank and a groundwater aquifer before it is extracted from wells (Rittmann and Huck, 1989). In The Netherlands, river water is applied to *sand dunes* and travels 60–100 m during a 5- to 30-week period. In several countries, raw or pretreated river water is infiltrated to *groundwater aquifers* before it is withdrawn for treatment. Each system takes advantage of naturally present "filters" and uses slow passage through the ground to remove BOM, iron, phosphorous, metals, and halogenated organic chemicals. The extracted water is substantially upgraded in terms of its biological stability and is easier to treat by subsequent filtration and adsorption steps.

SIZING A SLOW SAND FILTER How much surface area is needed for a slow sand filter | **Example 12.4**
for a town 1,000 people if the water usage is 100 gallons per capita per day?
The total water use rate is (100 gallons/person-day) · (3.8 liters/gallon) · (0.001 m³/liter) · (1,000 people) = 380 m³/d. With a design hydraulic loading of 5 m/d, the plan-view surface area is $A = (380 \text{ m}^3/\text{d})/(5 \text{ m/d}) = 76 \text{ m}^2$. This equates to 0.076 m² per capita.

12.2 RELEASE OF MICROORGANISMS

All biofilm processes release bacteria to the effluent due to detachment. The predominant microorganisms in the effluent are the heterotrophic and autotrophic bacteria that oxidize substrates and grow in the reactors. These microorganisms generally are harmless to humans, although their release in high numbers can cause problematic levels of turbidity and heterotrophic plate counts. Furthermore, they are themselves a source of biological instability whose presence during distribution could create the instability problems mentioned earlier in the chapter.

Pathogenic microorganisms can be input to a biofilter. Their fate in a biofilter is not well established. Without question, they can be filtered out and captured on or in the biofilm. Competition and predation probably reduces the numbers of invading species, although some pathogenic bacteria are able to grow under oligotrophic conditions. Detachment of biofilm or release of GAC fines allows invading species to be carried to the effluent, along with the more prevalent heterotrophs and autotrophs.

Prudent design suggests that biofilm processes should be located before the main particle-removal processes whenever possible. Biofilm pretreatment and slow filtration through natural media accomplish this objective. Because hybrid biofilters and slow sand filters are themselves particle filters, they seldom are followed by another particle-removal process. While this practice appears to be adequate in most cases, further investigation is justified, particularly on the transient release of microorganisms just after backwashing or scraping.

12.3 BIODEGRADATION OF SPECIFIC ORGANIC COMPOUNDS

In addition to making waters biologically stable, biofilm processes are effective at biodegrading several types of organic compounds that create chronic health risks and/or tastes and odors. Rittmann (1995) and Rittmann, Gantzer, and Montiel (1995) provide in-depth reviews of the capabilities of biofilm processes for biodegrading these kinds of organic compounds in the drinking-water setting. A brief summary that focuses on drinking water is given here, and Chapter 14 provides more information on the ways in which microorganisms detoxify these kind of chemicals.

Petroleum hydrocarbons are widely present as micropollutants in groundwater and surface-water supplies. The most prevalent are the soluble aromatic components of gasoline: benzene, toluene, ethyl benzene, and xylenes (BTEX). In some cases, polynuclear aromatic hydrocarbons, such as naphthalene and phenanthrene, are present. All of the common petroleum hydrocarbons are biodegradable by aerobic bacteria that initiate biodegradation by oxygenase reactions. These initial oxygenation reactions are sensitive to low D.O. concentrations, which emphasizes the need to maintain aerobic conditions. A recent development for gasoline contamination is the addition of a large fraction of methyl-*tert*-butyl ether (MTBE) to gasoline in order to reduce the formation of smog. MTBE is very mobile in groundwater, as are the BTEX compounds (more on this in "Chapter 15"). The biodegradability of MTBE is a current research issue, but it is certain to be much more slowly biodegraded than BTEX. The fate of MTBE in a biofilter for drinking-water treatment cannot be predicted well at this time.

Widely used industrial solvents, plasticizers, and paints include oxygenated derivatives of petroleum hydrocarbons. The high water solubility of these alcohols, phenols, ethers, and ketones makes them common contaminants in waters subject to industrial inputs. These compounds usually are fully mineralizable by a series of hydroxylation and dehydrogenation reactions common to many aerobic bacteria.

The halogenated one- and two-carbon aliphatics are common industrial solvents (e.g., 1, 1, 1-trichloroethane, trichloroethene, tetrachloroethane, and carbon tetrachloride), and some are formed as disinfection by-products from chlorination (e.g., chloroform and dibromochloromethane). Most of these compounds are considered resistant to biodegradation in aerobic systems. However, lightly chlorinated methanes are susceptible to hydrolytic dechlorination, and dichloromethane and trichloroethene can be oxidatively dechlorinated by several microorganisms containing monooxygenase enzymes that normally are induced by phenol, methane, or ammonium. Because they are highly volatile, halogenated aliphatics are much more likely to be stripped to the air phase than be biodegraded in aerated systems.

Chlorinated benzenes, phenols, and cyclohexanes are used in pesticides, as well as in industry. These compounds can be biodegraded aerobically, usually after initial monooxygenation of the ring, followed by ring cleavage via a dioxygenation. Halogenated aliphatics and aromatics also are susceptible to reductive dehalogenation reactions. Recent evidence shows that reductive dehalogenations can occur in aerobic systems, especially when electron-donor substrates are present in fairly high concentrations. However, the practical applicability of these recent findings is unknown and seems highly speculative, due to the low concentrations of electron donors in drinking water.

Key taste-and-odor compounds are known to be biodegradable by bacteria present in biofilm processes. Most notable are the proven biodegradabilities of *geosmin, methylisoborneol,* and *trichloroanisole*—the causes of earthy and musty odors—by biofilms grown on a humic primary substrate. Likewise, the causes of fishy odors (amines and aliphatic aldehydes), marshy-swamp-septic odors (reduced sulfur compounds), and medicinal-antiseptic odors (chlorinated phenols and benzenes) are degradable under the aerobic conditions of drinking-water treatment (Rittmann, 1995; Rittmann et al., 1995).

12.4 DENITRIFICATION

The basics of denitrification, which uses NO_3^- or NO_2^- as the primary electron acceptor, were discussed in Chapter 10. In many ways, denitrification for drinking water resembles the tertiary biofilm processes described in Chapter 10. Key common features are:

- An exogenous electron donor must be added to drive NO_3^- respiration, as well as to eliminate D.O.

- A low D.O. concentration, usually less than 0.2 mg/l, must be maintained by minimizing contact with the atmosphere.

- A wide range of fixed-bed or fluidized-bed processes is appropriate.

The drinking-water application has two unique features that distinguish it from tertiary wastewater treatment. The first special feature is the effluent quality criterion for NO_3^--N or NO_2^--N. Almost all standards for NO_3^--N are 10 mg N/l, which means that partial denitrification is acceptable. The second feature is that residual electron donor should be minimal in order to keep the water biologically stable. These two features are compatible as long as the electron-donor substrate is kinetically limiting and driven to its S_{min} value. Thus, the J/J_R ratio for the electron donor should be below 1.0, and the amount of donor added may be less than the stoichiometric requirement for full denitrification.

A third unique feature is the choice of the added electron donor. Although methanol is widely used in wastewater treatment and has been used with drinking water, it may be a poor choice since it is acutely toxic to humans, is a relatively costly source of electrons, and is regulated. Ethanol and acetate are alternative organic electron donors; ethanol is substantially less expensive than acetate or methanol, but may have political or regulatory liabilities as an additive in drinking-water treatment because it is an alcohol.

Autotrophic denitrification using elemental sulfur or H_2 gas as the electron donor is highly promising. The elemental sulfur usually is impregnated in the solid medium and is supplied "on demand" by dissolution. The best means to supply H_2 gas is by bubbleless membrane dissolution. Autotrophic denitrification offers these advantages: Residual electron donor is negligible in the effluent, biomass yields are low (low sludge production), and both are less expensive than the organic electron donors.

The fourth unique feature is the need to keep NO_2^--N levels very low. Although only the European Community has an NO_2^- standard at this time (0.1 mg NO_2^--N/l), NO_2^- is the direct cause of methemoglobinemia and should be kept as low as possible. NO_2^--N can be formed as a partial reduction intermediate when the electron donor is depleted. Therefore, achieving the first and second features may create some risk of NO_2^- formation, an area deserving additional research.

Heterotrophic systems have used methanol, ethanol, and acetate as the supplementary electron donor. Fixed beds, fluidized beds, upflow sludge blanket reactors, and membrane bioreactors have been used for biomass retention. Most published results have used high donor-substrate fluxes, typically 4,000 to 18,000 kg COD/1,000 m^2-d. The advantages of high loads are small reactor volumes (detention times generally only a few minutes) and suppression of NO_2^- buildup (generally less than 0.02 mg NO_2^--N/l). On the other hand, leakage of the supplemental electron donor, as well as SMP, to the effluent is significant and requires post treatment.

Autotrophic denitrification using H_2 as the supplemental donor (i.e., autohydrogenotrophic denitrification) has been practiced in fixed-bed and fluidized-bed systems. The most common way to feed H_2 has been via gas sparging, although bubbleless membrane transfer is a very promising new alternative (Lee and Rittmann, 2000). Gas sparging tends to keep higher-than-needed H_2 concentrations, which "wastes" donor, can create an explosive atmosphere in the off-gas, and promotes stripping of gaseous intermediates NO and N_2O. Bubbleless gas transfer precludes these disadvantages. In general, H_2 substrate fluxes have been kept low, around 1 kg as COD/1,000 m^2-d, which gives low NO_3^- concentrations. Higher fluxes could be used when partial NO_3^- removal is the goal.

Although heterotrophic and autotrophic denitrification of drinking water has been demonstrated, substantial research and development are needed to ensure economical designs that address each of the four unique features of the drinking-water application. Rational bases for selecting donor fluxes and concentrations need to be established. Balancing is needed between minimizing effluent biological instability and intermediates (particularly NO_2^-) formation. Once these rational design bases are established, the analysis/design procedure given for tertiary denitrification in Chapter 10 can be used directly for drinking-water scenarios. Of the greatest importance is the stoichiometry between the desired removal of NO_3^--N and the addition of the organic or inorganic electron donor. Example 10.3 is especially relevant, and a drinking-water variation on it is given below.

An interesting application for denitrification is the microbiological regeneration of anion-exchange resins used to remove the NO_3^- from the water supply. The anion-exchange resin can be directly bioregenerated, or it can first be chemically regenerated with an HCO_3^- or Cl^- brine, which is subsequently treated by denitrification. This regeneration mode eliminates concerns over residual donor, NO_2^- intermediates, or the type of added donor.

A final application of denitrification in drinking water is the reduction of perchlorate, which is a component of rocket fuel that has contaminated water supplies in the western United States. Many denitrifying bacteria are able to use perchlorate as an alternate electron acceptor, reducing perchlorate to chloride ion. Perchlorate reduction is an area of active research today.

DRINKING-WATER DENITRIFICATION BY HYDROGEN AUTOTROPHY | **Example 12.5**

A drinking-water supply of 1,000 m^3/d has 15 mg/l of NO_3^--N. In order to treat it, an autohydrogenotrophic biofilm system is to be used. In parallel to Example 10.3, we use a fluidized bed of specific surface area 4,000 m^{-1} and average biofilm SRT of 15 d. This means that the stoichiometry between H_2 and NO_3^--N is 3.2/8 = 0.4 g H_2/g N. The treatment goal is to reduce the nitrate-N from 15 to 5 mg/l, since the standard is 10 mg N/l. A substrate flux of 10 kg COD/1,000 m^2-d is to be used for partial denitrification, and this converts to 1.25 kg H_2/1,000 m^2-d.

The N load is

$$\frac{\Delta N}{\Delta t} = 1,000 \ m^3/d \cdot 10 \ mg \ N/l \cdot 10^{-3} \ kg\text{-}1/mg\text{-}m^3 = 10 \ kg \ N/d$$

The H_2-supply rate is then 0.4 g H_2/g N times the N removal rate: 4 kg H_2/d. The volume of the fluidized bed is then

$$V_{bed} = \frac{4 \ kg \ H_2}{d} \cdot \frac{1,000 \ m^2\text{-}d}{1.25 \ kg \ H_2} \cdot \frac{1}{4,000 \ m^{-1}} = 0.8 \ m^3$$

which gives a hydraulic detention time for the fluidized bed of 0.8 m^3/(1,000 m^3/d) = 0.0008 d, or 1.2 min. The hydrogen supply rate is equivalent to 4 mg/l in the influent flow, which is about double the H_2 solubility. Thus, H_2-supply considerations similar to those in Example 10.3 must be evaluated in selecting the size and configuration of the treatment system.

12.5 BIBLIOGRAPHY

Bouwer, E. J. and P. B. Crowe (1988). "Assessment of biological processes in drinking water treatment." *J. Amer. Water Works Assn.* 80(9), pp. 82–93.

Collins, M. R.; T. T. Eighmy; and M. P. Malley (1991). "Evaluating modifications to slow sand filters." *J. Amer. Water Works Assn.* 83(9), pp. 62–70.

Fiessinger, F.; J. Mallevialle; and A. Benedek (1983). "Interaction of adsorption and bioactivity in full-scale activated carbon filters. The Mont Valerien experiment." *Adv. Chem. Ser.* 202, p. 319.

Huck, P. M. (1990). "Measurement of biodegradable organic matter and bacterial growth potential in drinking water." *J. Amer. Water Works Assn.* 82(7), pp. 78–86.

Izaguire, G.; R. L. Wolfe; and E. G. Means (1988). "Degradation of 2-methylisoborneol by aquatic bacteria." *Appl. Environ. Microb.* 52, p. 2424.

Joret, J. C. and Y. Levi (1986). "Method rapide d'evaluation du carbone eliminable des eaux par voie biologique." *Trib. Cebedeau.* 519(39), p. 3.

Joret, J.; Y. Levi; and C. Volk (1991). "Biodegradable dissolved organic carbon (BDOC) content in drinking water and potential regrowth of bacteria." *Water Sci. Technol.* 24(2), pp. 95–100.

Kaplan, L. A. and J. D. Newbold (1995). "Measurement of streamwater biodegradable dissolved organic carbon with a plug-flow bioreactor." *Water Research* 29, pp. 2696–2706.

Kaplan, L.; T. Bott; and D. Reasoner (1993). "Evaluation and simplification of the assimilable organic carbon nutrient assay for bacterial growth in drinking water." *Appl. Environ. Microb.* 59, pp. 1532–1539.

LeChevallier, M. W.; N. E. Shaw; L. A. Kaplan; and T. L. Bott (1993). "Development of a rapid assimilable organic carbon method for water." *Appl. Environ. Microb.* 59, pp. 1526–1531.

Lee, K. C. and B. E. Rittmann (2000). "A novel hollow-fiber membrane biofilm reactor for autohydrogenotrophic denitrification of drinking water." *Water Sci. Technol.* 40(4/5), pp. 219–226.

Lundgren, B. V.; A. Grimvall; and R. Sävenhed (1988). "Formation and removal of off-flavour compounds during ozonation and filtration through biologically active sand filters." *Water Sci. Technol.* 20(8/9), p. 245.

Manem, J. A. and B. E. Rittmann (1992). "Removing trace-level pollutants in a biological filter." *J. Amer. Water Works Assoc.* 84(4), pp. 152–157.

Manen, J. A. and B. E. Rittmann (1992). "The effects of fluctuations in biodegradable organic matter on nitrification filters." *J. Amer. Water Works Assn.* 84(4), pp. 147–151.

Montgomery, J. M. (1985). *Water Treatment Principles and Design.* New York: Wiley-Interscience.

Namkung, E. and B. E. Rittmann (1987). "Removal of taste and odor compounds by humic-substances-grown biofilms." *J. Amer. Water Works Assn.* 59, pp. 670–678.

Owen, D. M.; G. L. Amy; and Z. K. Chowdhury (1993). *Characterization of Natural Organic Matter and Its Relationship to Treatability.* Denver, CO: Amer. Water Works Assoc. Research Foundation.

Rittmann, B. E. (1990). "Analyzing biofilm processes in biological filtration." *J. Amer. Water Works Assn.* 82(12), pp. 62–66.

Rittmann, B. E. (1993). "The significance of biofilms in porous media." *Water Research* 29, pp. 2195–2202.

Rittmann, B. E. (1995). "Fundamentals and applications of biological processes in drinking water treatment." In. J. Hrubec, Ed. *Handbook of Environmental Chemistry*, Vol. 5B, Verlag, NY: Springer, pp. 61–87.

Rittmann, B. E. (1995). "Transformation of organic micropollutants by biological processes." In J. Hrubec, Ed. *Handbook of Environmental Chemistry*, Vol. 5B, Verlag, NY: Springer, pp. 31–60.

Rittmann, B. E. (1996). "Back to bacteria: A more natural filtration." *Civil Engineering* 66(7), Verlag, NY: Springer, pp. 50–52.

Rittmann, B. E. and V. L. Snoeyink (1984). "Achieving biologically stable drinking water." *J. Amer. Water Works Assn.* 76(10), pp. 106–114.

Rittmann, B. E. and P. M. Huck (1989). "Biological treatment of public water supplies." *CRC Critical Reviews in Environmental Control.* Vol. 19, Issue 2, pp. 119–184.

Rittmann, B. E.; C. J. Gantzer; and A. Montiel (1995). "Biological treatment to control taste and odor compounds in drinking-water treatment." In M. Suffett and J. Mallevialle, Eds., *Advances in the Control of Tastes and Odors in Drinking Water.* Denver, CO: Amer. Water Works Assoc., pp. 203–240.

Rogalla, F.; G. Larminat; J. Contelle; and H. Godart (1990). "Experience with nitrate removal methods from drinking water." *Proc. NATO Advanced Research Workshop on Nitrate Contamination: Exposure, Consequences, and Control.* Lincoln, NB, Sept. 9–14, 1990.

Servais, P.; G. Billen; and M. C. Hascoet (1987). "Determination of the biodegradable fraction of dissolved organic matter in waters." *Water Research* 21, pp. 445–450.

Urfer, D.; P. M. Huck; S. D. J. Booth; and B. M. Coffery (1997). "Biological filtration for BOM and particle removal: A critical review." *J. Amer. Water Works Assn.* 89(12), pp. 83–98

van der Kooij, D.; A. Visser; and W. A. M. Hignen (1982). "Determining the concentration of easily assimilable organic carbon in drinking water." *J. Amer. Water Works Assn.* 74, pp. 540–550.

Woolschlager, J., and B. E. Rittman (1995). "Evaluating what is measured by BDOC and AOC tests. *Revue Science de l'Eau* 8, pp. 372–385

12.6 PROBLEMS

12.1. Denitrification is sometimes used to remove nitrate from drinking water. Provide a preliminary design for a fixed-film process to remove 10 mg NO_3^-/l to less than 0.1 mg/l. The preliminary design involves estimating the total reactor volume and the amount of added acetate needed as the electron donor. The known information is:

Substrate	$\hat{q}$	K	b	b_{det}	Y	f_d	D	D_f	X_f
Acetate	$10\,\dfrac{\text{mg Ac}^-}{\text{mg VS}_a \cdot \text{d}}$	$15\,\dfrac{\text{mg Ac}^-}{1}$	$0.1\,\dfrac{1}{\text{d}}$	$0.05\,\dfrac{1}{\text{d}}$	$0.18\,\dfrac{\text{mg VS}_a}{\text{mg Ac}^-}$	0.8	$1.2\,\dfrac{\text{cm}^2}{\text{d}}$	$1.0\,\dfrac{\text{cm}^2}{\text{d}}$	$40\,\dfrac{\text{mg VS}_a}{\text{cm}^3}$
Nitrate	$2.3\,\dfrac{\text{mg NO}_3^- \text{-N}}{\text{mg VS}_a \cdot \text{d}}$	$0.1\,\dfrac{\text{mg NO}_3^- \text{-N}}{1}$			$0.81\,\dfrac{\text{mg VS}_a}{\text{mg NO}_3^- \text{-N}}$				

You also know that $Q = 1,000$ m^3/d, $a = 300$ m^{-1}, and $L = 90$ μm (for acetate). You may assume that the added acetate is the rate-limiting substrate, as long as it is not fully depleted. You will need to estimate S_{min} for each substrate and the amount of acetate needed to remove the 10 mg NO$_3^-$-N/l. Then, determine the needed J and volume to keep residual acetate to as low a level as practical.

12.2. Design a fixed-bed pretreatment biofilter to treat a surface water with a flow rate of 2,000 m^3/d and a BDOC of 1,000 μg/l. Estimate the biofilm surface area and the O$_2$ supply rate. Is aeration needed?

12.3. Design a fixed-bed pretreatment biofilter to treat a surface water with a flow rate of 5,000 m^3/d and a BDOC of 3,000 μg/l after ozonation. Estimate the biofilm surface area and the O$_2$ supply rate. Is aeration needed?

12.4. The water in Problem 12.3 also has 1 mg/l of NH$_4^+$-N. How does this change the answers?

12.5. The water in Problem 12.2 also has 0.4 mg/l of NH$_4^+$-N. How does this change the answers?

12.6. A rapid filter has media with an average diameter of 1 mm, porosity of 0.4, depth of 1 m, and hydraulic loading of 10 m/h (4 gpm/ft^2). The influent BDOC is 300 μg/l. Compute the BOD$_L$ surface loading and assess whether or not excessive backwashing is likely. Is this a good situation for hybrid rapid biofiltration?

12.7. A rapid filter has media with an average diameter of 1 mm, porosity of 0.4, depth of 1 m, hydraulic loading of 10 m/h (4 gpm/ft^2). The influent BDOC is 2,000 μg/l. Compute the BOD$_L$ surface loading and assess whether or not excessive backwashing is likely. Is this a good situation for hybrid rapid biofiltration?

12.8. Size a slow sand filter for a flow of 1,000 m^3/d.

12.9. A slow sand filter had plan-view dimensions of 10 m wide by 10 m long. It treats a flow of 2,000 m^3/d. Is its size adequate? How many people can it serve if the per capita demand is 250 l/d?

12.10. Design a biofilm reactor to denitrify 10,000 m^3/d of surface water having a NO$_3^-$ content of 15 mg N/l. The effluent goal is 5 mg N/l of NO$_3^-$. What is the H$_2$ supply rate (kg H$_2$ per day) if $f_s = 0.1$? What is the required biofilm surface area if the design flux is 2 kg as COD/1,000 m^2-d? What limitations must be considered in the design?

chapter

13

ANAEROBIC TREATMENT
BY METHANOGENESIS

Methanogenesis refers to an anaerobic process in which the electron equivalents in organic matter (BOD_L) are used to reduce carbon to its most reduced oxidation state, -4, in CH_4, or methane. Methane is a poorly soluble gas that evolves from the water. Thus, the BOD_L is removed from the water by directing electron equivalents to CH_4, a result that we call *waste stabilization*. Each mole of CH_4 contains 8 electron equivalents, or 64 g of BOD_L or COD. At standard temperature and pressure (i.e., STP $= 0$ °C and 1 atm), each mole of CH_4 has a volume of 22.4 l. Thus, each g of BOD_L stabilized generates 0.35 l of CH_4 gas at STP.

Although the microbial community in a methanogenic process usually is quite complex (see the Microbial Ecology section of Chapter 1), it always contains the *methanogens,* the unique group of Archaea that produces methane. Since the methanogens must be present to have a methanogenic process, their physiological characteristics are critical. Listed here are generic parameters that characterize the two distinct groups of methanogens: acetate fermenters and hydrogen oxidizers.

	Acetate Fermenters	Hydrogen Oxidizers
Electron Donors	Acetate	H_2 and Formate
Electron Acceptors	Acetate	CO_2
Carbon Sources	Acetate	CO_2
f_s^0	0.05	0.08
Y	0.04 g VSS_a/g Ac	0.45 g VSS_a/g H_2
$\hat{q}$ (at 35 °C)	7 g Ac/g VSS_a-d	3 g H_2/g VSS_a-d
K	400 mg Ac/l	?
b	0.03/d	0.03/d
$[\theta_x^{min}]_{lim}$	4 d	0.76 d
S_{min}	48 mg Ac/l	?

These generic values point out that methanogens—particularly the acetate fermenters—are slow-growing microorganisms that require a relatively long solids retention time (θ_x) just to avoid washout. For example, a minimal safety factor of 5 requires a design θ_x of 20 d for the acetate-fermenting methanogens. The listing also emphasizes that all the electrons originally present in the input BOD must ultimately be funneled into acetate, H_2, or formate. Other microorganisms must be present to carry out the funneling reactions. The others can vary widely, but the slow-growing methanogens are the foundation of the process.

13.1 USES FOR METHANOGENIC TREATMENT

Anaerobic treatment by methanogenesis is widely used for the stabilization of municipal wastewater sludges and municipal solid wastes. Methane fermentation of high-strength industrial wastewaters is also widely practiced, but, as yet, the applications are far below the potential that actually exists. The reasons for the underuse for industrial wastewaters include lack of experience with use of the process, lack of adequate understanding of process chemistry and microbiology, the presence of toxic compounds, and concern over the need for high process reliability in a strong regulatory atmosphere. Anaerobic treatment also can be used to treat more dilute wastewaters, such as domestic sewage. This application is gaining popularity in the developing world, particularly where the climate is warm for most of the year. Anaerobic treatment of sewage is not practiced in the developed world, where other technologies (e.g., activated sludge) already are in place and effluent standards are strict.

The chemistry and microbiology of anaerobic treatment are, indeed, more complex than for aerobic systems, and, for this reason, the operator of an anaerobic treatment system must have a solid understanding of the factors of importance in operation and control. With large municipal treatment systems—where wastewaters and sludges that contain essentially all the nutrients required for good biological growth, pH problems and the presence of toxic materials are relatively low, and operators are generally well trained—anaerobic treatment can be quite reliable. In an industrial setting, one might expect that highly skilled personnel also would be present. Unfortunately, wastewater operation is not a profit-making center of industrial plants, where the operating staff may or may not be skilled. When proper attention is paid to the process, and especially when the wastes are most suitable for anaerobic treatment, such as in the food-processing and fermentation industries, anaerobic treatment can offer highly significant advantages over aerobic treatment for industrial organic wastes.

Table 13.1 contains a summary of the advantages and disadvantages of the anaerobic treatment process, compared with aerobic treatment. First, the fraction of wastewater BOD_L synthesized into biological solids is much less in anaerobic treatment, as the energy yield from transferring electrons from BOD_L to methane is low. With aerobic treatment, up to half of the waste BOD_L is converted into bacterial cells, which then represent a significant and costly disposal problem. In anaerobic treatment, only 5 to 15 percent of the BOD_L is converted into biological solids, thus reducing this disposal problem considerably. Second, synthesis of the biomass requires essential nutrients, such as nitrogen and phosphorus. With some industrial wastewaters, such nutrients are in limited supply and must be added. With the smaller biomass production in anaerobic systems, the nutrient requirements are proportionally less.

A third major advantage of anaerobic treatment is in the methane gas produced, which constitutes a readily available energy source that can be used for heating or to generate electrical power. The energy value of CH_4 is 35.8 kJ/l at STP. Anaerobic

Table 13.1   Advantages and disadvantages of anaerobic treatment

Advantages

1. Low production of waste biological solids
2. Low nutrient requirements
3. Methane is a useful end product
4. Generally, a net energy producer
5. High organic loading is possible

Disadvantages

1. Low growth rate of microorganisms
2. Odor production
3. High buffer requirement for pH control
4. Poor removal efficiency with dilute wastes

treatment is generally a net producer of energy. In contrast, the energy requirements to operate aeration systems are substantial and constitute the major operating cost for such systems. In many municipal treatment plants employing aerobic secondary treatment, methane gas generated from anaerobic treatment of the sludges often satisfies all the energy requirements for running the plant. The net savings can be substantial. If anaerobic treatment of the wastewater could be substituted for most or all of the wastewater treatment, the treatment facility could become an exporter of energy.

Another advantage of anaerobic treatment is that organic loading per unit reactor volume can be very high compared with aerobic treatment. A typical loading for an anaerobic system may be 5 to 10 kg COD per day per m^3 of reactor volume, whereas it is less than 1 kg COD per day per m^3 for aerobic systems, in which transferring O_2 is a major limitation. For this reason, anaerobic treatment is especially useful with more concentrated wastewaters, those with a COD of 5,000 mg/l or higher.

ENERGY VALUE OF METHANE   We are treating an industrial wastewater with a flow of 10^4 m^3/d and a BOD_L of 20,000 mg/l. We are to compute the volume of methane gas generated if the waste stabilization is 90 percent. Also, we are going to compute the energy value of that methane.

Example 13.1

$$BOD_L \text{ Loading Rate} = 20,000 \text{ mg } BOD_L/\text{l} \cdot 10^4 \text{ m}^3/\text{d} \cdot 10^3 \text{ l/m}^3 \cdot 10^{-6} \text{ kg/mg}$$
$$= 2 \cdot 10^5 \text{ kg } BOD_L/\text{d}$$
$$BOD_L \text{ Stabilization Rate} = 0.9 \cdot 2 \cdot 10^5 = 1.8 \cdot 10^5 \text{ kg } BOD_L/\text{d}$$
$$\text{Methane Generation Rate} = 1.8 \cdot 10^5 \text{ kg } BOD_L/\text{d} \cdot 0.35 \text{ m}^3 \text{ CH}_4(\text{at STP})/\text{kg } BOD_L$$
$$= 6.3 \cdot 10^4 \text{ m}^3 \text{ CH}_4 \text{ (at STP)/d}$$
$$\text{Energy Rate} = 6.3 \cdot 10^4 \text{ m}^3/\text{d} \cdot 35,800 \text{ kJ/m}^3 \text{ CH}_4 \text{ (at STP)}$$
$$= 2.26 \cdot 10^9 \text{ kJ/d}$$

(The energy rate also can be expressed as $5.4 \cdot 10^8$ kcal/d, $2.1 \cdot 10^9$ BTU/d, or $2.6 \cdot 10^4$ kW, which equals $6.3 \cdot 10^5$ kWh/d.)

Disadvantages of the anaerobic process are somewhat related to some of the advantages. The low energy available for biological synthesis translates into the smaller bacterial yields and slower growth rates, particularly for some of the key methane producers, which have doubling times measured in days, rather than the hours characteristic of typical aerobic microorganisms. This long doubling time means that it takes longer to start the process if one does not have a very large mass of seed sludge. It also means that if an upset occurs, say because of the introduction of toxic materials into the treatment process, a long time will be required for recovery. This slow response time is, perhaps, the major disadvantage, because it leaves little margin for error in operation.

Another disadvantage is that anaerobic processes often produce sulfides from sulfate reduction and from decomposition of protein-containing wastewaters. Sulfides are toxic and corrosive, and gaseous H_2S has the strong, unpleasant odor of rotten eggs. The anaerobic process must be protected from the negative effects from sulfides, and H_2S odors emanating from upset anaerobic treatment systems are particularly troublesome and always a cause for concern. Intermediate products from decomposition of some wastewaters, such as that from wood-pulping processes, can also at times be odorous. Combustion of the methane gas normally destroys the odor-causing intermediate compounds and oxidizes the H_2S to SO_2. While combustion solves the odor problem, SO_2 is an air pollutant that may need to be scrubbed in order to meet air-pollution standards and to protect equipment from corrosion. Therefore, the best strategy is to guard against the formation of H_2S and other odor-causing agents in anaerobic treatment. Sulfate reducers also divert electron equivalents away from CH_4 when they produce sulfides. Each gram of sulfide S is equal to 2 g BOD_L or 0.7 l of CH_4 lost.

The control of pH in anaerobic treatment is critical, as the desired range for methanogens is rather narrow, generally between 6.5 and 7.6. Organic acids produced as intermediates in the process, as well as carbonic acid associated with the high concentrations of carbon dioxide gas produced, tend to depress the reactor pH. Buffer requirements in anaerobic treatment tend to be relatively high compared with aerobic treatment, and supplying this requirement can be expensive with industrial wastewaters that do not contain sufficient natural buffer.

A final disadvantage of anaerobic treatment is the poor efficiency of treatment that generally results when attempting to treat relatively dilute wastewaters with COD of 1,000 mg/l or less. Using anaerobic processes to treat municipal wastewaters offers all the advantages listed above, but generally it needs to be followed by aerobic treatment to reach effluent BOD standards common in the developed world. Applications for municipal treatment in developing countries are increasing because of the low energy requirements, and new reactor technologies may make anaerobic treatment more efficient in terms of effluent BOD.

The engineer needs to weigh the various advantages and disadvantages of anaerobic treatment for a given application. Unfortunately, the real benefits of resource conservation and reduced energy usage with anaerobic treatment are often not given the weight they deserve. Greater attention to anaerobic treatment surely will occur if energy prices escalate in the future.

13.2 REACTOR CONFIGURATIONS

Several excellent summaries are available for those wishing an expanded knowledge of the methanogenic processes and their applications. One of the most extensive, with emphasis on industrial wastewater treatment, is a textbook on the subject by Speece (1996). A less extensive, but nevertheless thorough overview of the process, with emphasis on municipal sludge treatment, is Parkin and Owen (1986). Other valuable summaries include: Jewell (1987), McCarty (1964a, 1964b, 1964c, 1964d, 1981), McCarty and Smith (1986), and Speece (1983).

Many reactor configurations are used for treatment of municipal and industrial wastewaters or sludges. Several described by Speece (1983) are illustrated in Figure 13.1 and discussed here.

13.2.1 COMPLETELY MIXED

The *completely mixed process* represents the basic anaerobic treatment system that has been used for treatment of municipal sludges since the first separate, heated anaerobic

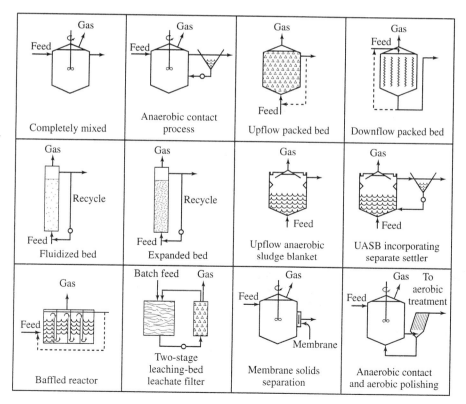

Figure 13.1 Typical anaerobic reactor configurations. SOURCE: Speece, 1983.

digester was built in 1927 at the Essen-Rellinghausen plant in Germany. Heating of digesters with the methane gas produced to an optimum mesophilic temperature, about 35 °C, is used at most municipal wastewater treatment plants, since heat recovery can be done economically and results in the more stable operation. Detention times in such tanks today are commonly 15–25 d, which is well above $[\theta_x^{min}]_{lim}$ of about 4 d at that temperature for the critical acetate-using methanogens. Thus, washout of the critical organisms is avoided by using an implied safety factor, based on $\theta_x/[\theta_x^{min}]_{lim}$, of about 4 to 6.

Earlier designs did not have mixing, which caused two problems. First, the fresh sludge and fermenting microorganisms are not brought together efficiently. Second, the denser solids, such as grit and sand, tend to settle in the reactor, thereby lessening the volume available for treatment. To overcome these problems, unmixed reactors often are operated at detention times, based on the ratio of tank volume to influent sludge flow rate, on the order of 60 or more days.

Municipal primary and secondary sludges entering an anaerobic sludge digester normally have concentrations from 2.5 to 15 percent total solids. With a detention time of about 20 d, the organic loading to a well-mixed digester of the continuous-flow stirred-tank reactor (CSTR) type is 1 to 4 kg biodegradable COD per d per m^3 of digester volume. (This calculation assumes that 65 percent of the influent suspended solids are volatile and that 50 percent of the volatile solids are biodegradable.) This is a loading comparable to, but higher than the volumetric loading achieved in aerobic systems. In addition, the digester produces energy, rather than consuming it. It is easy to see why so many municipal treatment plants employ anaerobic treatment for waste sludges.

A disadvantage of the CSTR for anaerobic treatment is that a high loading per unit volume is obtained only with quite concentrated waste streams, similar to municipal

Photo 13.1   Mesophilic anaerobic sludge digester with a floating cover.

sludge, which has a biodegradable COD content of 8,000 to 50,000 mg/l. However, a great many waste streams are much more dilute. If they were treated in a CSTR with a detention time of 15 to 25 d, the COD loading per unit volume would be very low, reducing or eliminating the cost advantage of anaerobic treatment. The secret to treating such wastes economically is to use reactor systems that separate the detention time of the liquid passing through the reactor from that of the biomass, as is accomplished in the aerobic activated sludge and biofilm systems. In other words, θ_x/θ must be made substantially greater than 1. Figure 13.1 illustrates many other systems that separate biomass and fluid retention times in anaerobic systems, a strategy that began in the 1950s and continues today.

13.2.2 ANAEROBIC CONTACT

The *anaerobic contact process* is an analogy to the aerobic activated sludge system. The first such system was developed and described in 1955 by Schroepfer and his co-workers (Schroepfer et al., 1955) for the treatment of relatively dilute packing house wastes with COD of about 1,300 mg/l. By adding a settling tank and recycle of the biomass back to the reactor, they separated θ_x from θ and achieved a hydraulic detention time of about 0.5 d, which is significantly less than the 4-d $[\theta_x^{min}]_{lim}$ of the acetoclastic methanogens. They obtained 91 to 95 percent BOD removals at loading rates of 2 to 2.5 kg/m^3-d.

Although the anaerobic contact process has seen many applications, one recurring problem is a tendency for the biosolids in the settling tank to rise due to bubble generation and attachment in the settling tank. Biosolids loss to the effluent is a more severe problem than in aerobic systems, because the quantity of microorganisms

Photo 13.2   Anaerobic contact process. Anaerobic reactor is to the left and the settler to the right.

produced is so much less; hence, a small biosolids loss can significantly reduce θ_x and adversely affect process stability, as well as effluent quality. This problem is present in all anaerobic treatment systems and has been addressed in different ways. One solution is applying a vacuum to the water passing to the settling tank in order to remove some of the gas supersaturation.

13.2.3   UPFLOW AND DOWNFLOW PACKED BEDS

The *upflow packed bed process,* also commonly called the *anaerobic filter,* was developed through laboratory studies in the late 1960s (Young and McCarty, 1969). This system is similar to a trickling filter system in that, originally, a rock medium was used for attaching the biosolids. The anaerobic filter was used for treating soluble substrates with COD from 375 to 12,000 mg/l and had detention times of 4 to 36 h. The first full-scale application, in the early 1970s, also used rock media for the treatment of a starch-gluten wastewater stream having a COD of 8,800 mg/l. Because of low void volume and specific surface area with rock media, most of the subsequent applications used plastic media, just as the aerobic biological towers do today.

The anaerobic filter is excellent for the retention of biosolids and has seen wide application. The main concern with this system is clogging by biosolids, influent suspended solids, and precipitated minerals. Because of this potential problem, packed-bed systems work best for wastewaters containing few suspended solids, as these are likely to be removed by the process and clog pore spaces. Various innovations have been made to reduce the clogging problem, such as occasional high-rate application into the bottom of the system of the gases produced. The turbulent fluid motion accompanying the rapid rise of the gas bubbles through the reactor tends to dislodge solids that might otherwise clog the system.

An alternative to the upflow packed-bed system is the *downflow packed bed.* Reasons why the downflow system might be superior are quite subtle. With the downflow system, the solids tend to accumulate more near the top surface, where substrate concentration and biological growth are higher. This may make it easier to achieve solids removal from the top by gas recirculation. Another possible advantage of the downflow system is that sulfide produced through sulfate reduction may be stripped from the liquid in the upper part of the column. Normally, the sulfate-reducing population resides in the upper levels of the reactor, while the methanogenic population is at lower levels. Hydrogen sulfide can be toxic to the methanogens, and stripping H_2S before it reaches the methanogenic part of the reactor can reduce toxicity to the methanogens. This separation appeared to be advantageous in a Puerto Rico plant treating rum distillery wastewater that had a high sulfate content. On the other hand, the downflow system, especially with plastic media, may have a greater tendency to lose biosolids to the effluent.

Young and Yang (1989) presented an excellent discussion of the various applications, the relative advantages and disadvantages of upflow versus downflow, and general operational criteria for packed-bed anaerobic treatment. Applications have been mostly in the food-processing, beverage, and pharmaceutical industries, although the chemical industry has used anaerobic filters as well. Wastewater CODs

Photo 13.3   Packed-bed anaerobic filter with plastic media used to treat rum wastewater.

generally are in the 2,500 to 10,000 mg/l range, although some applications are for CODs over 100,000 mg/l. Design loading often is in the 10 to 16 kg/m^3-d range, more than tenfold higher than for normal aerobic processes.

13.2.4   FLUIDIZED AND EXPANDED BEDS

The *fluidized-bed* anaerobic reactor is a unique system that was originally conceived for the removal of nitrate from wastewater through biological denitrification (Chapter 10). However, it also is well suited for methanogenic treatment of wastewaters. The fluidized-bed reactor contains small media, such as sand or granular activated carbon, to which bacteria attach. A relatively high upflow velocity of wastewater causes the biofilm carriers to rise to a point where the carrier's negative buoyancy is just countered by the upward force of friction from the water. Then, the height of the fluidized bed stabilizes. Normally, a portion of the effluent is recycled back to the influent to maintain the high upward velocity, even when wastewater flow rates are low.

    The high flow rate around the particles creates good mass transfer of dissolved organic matter from the bulk liquid to the particle surface. Bed expansion creates relatively large pore spaces, even though the carriers are small. The large pores mean that clogging and short-circuiting of flow through the reactor are much less than in the packed-bed systems. The small carrier size gives a very high specific surface area for biofilm, even when the bed is fluidized to open up the pores. All these features make fluidized beds highly efficient in terms of loading per unit volume.

Among possible disadvantages of fluidized beds is the difficulty in developing strongly attached biosolids containing the correct blend of methanogens. Abrasion between particles and fluid shear stress can increase detachment rates for microorganisms exposed to the outer environment of the carrier. Other possible disadvantages come when a high recycle flow is needed to maintain bed fluidization. On the one hand, the recycle flow dilutes the substrate concentration near the inlet and reduces benefits of a plug-flow liquid regime. On the other hand, high recycle rates can incur energy costs, mainly due to head loss in the recycle piping. Finally, defluidization for an extended period, such as from a power failure, can cause process instability, as the bed may not refluidize properly when the liquid flow is returned to its normal rate. Loss of the carriers and/or the biofilm can occur if the refluidization occurs suddenly.

A significant advantage, however, has been found in the use of granular activated carbon as the support media for the treatment of wastewaters containing toxic, but sorbable materials, such as phenolic compounds. Compounds such as phenol are toxic at concentrations of a few hundred mg/l, but are biodegradable under anaerobic conditions after acclimation by methanogenic consortia. Suidan, Najm, Pfeffer, and Wang (1988) found that the granular activated carbon acted as a buffer to reduce the concentration of such chemicals during process start-up or following a process overload. The adsorbed compounds then desorb and are biodegraded once conditions return to normal. Another major advantage of using granular activated carbon as a biofilm carrier is that the macropores at the outer surface provide sheltered niches for initial colonization by methanogens. Even when toxic materials are not present, the sheltering mechanism of the macropores greatly accelerates start-up or recovery of an anaerobic fluidized bed.

There have now been several laboratory studies of the anaerobic fluidized bed reactor and some full-scale applications (Speece, 1996). Volumetric loading to fluidized beds can be extraordinarily high, some approaching 100 kg COD per m^3/d.

The *expanded-bed* anaerobic reactor is a variation on the fluidized-bed reactor and contains similar support media. The difference is that fluid's upward flow velocity through the expanded-bed reactor is not maintained as high as in the fluidized bed; thus, full bed fluidization does not result. Possible advantages to this partial-fluidization mode are that solids may be captured better and that a high recycle rate does not need to be maintained. Possible disadvantages are that mass transfer is not as good as in the fluidized bed systems, clogging or short-circuiting are more likely, and detachment might be greater due to abrasion. Full-scale methanogenic applications with these systems are not as well developed as with other reactors.

13.2.5   UPFLOW ANAEROBIC SLUDGE BLANKET

Lettinga, van Velsen, de Zeeuw, and Hobma (1979) developed an important new anaerobic reactor, the *upflow anaerobic sludge blanket (UASB)*, which has had wide application for the treatment of industrial wastewaters and has been used to some extent for the treatment of relatively dilute municipal wastewaters as well. This system was first described in the late 1970s and saw extensive full-scale application in the 1980s. The UASB reactor is similar in many ways to the "clarigester" reactor used in the full-scale treatment of winery wastewaters as described by Stander (1966). The main difference between Stander's and Lettinga's reactors is in the method for separating biosolids from the effluent stream. Stander had a funnel-shaped settling tank on the top of the reactor. Exiting wastewater and some suspended solids passed upward through the funnel hole in the middle of the settling tank; escaping solids settled in the settling tank and returned to the reactor through the same hole. The gas produced in the process passed outward around the sides of the funnel. The return of solids worked in many cases, but represented the bottleneck in this reactor. With the Lettinga system, shown in Figure 13.1, a funnellike top is also used, but here, the funnel is inverted and wastewater passes out around the edges of the funnel rather than through the middle, and the gases pass out through the hole in the middle of the funnel. This edge passage of liquid provides a much greater area for the effluent, with the result that upward velocities are greatly reduced, and solids retention in the reactor is enhanced. Therefore, solids separation from the outward flowing wastewater tends to be much more efficient in the Lettinga approach.

Stander found that his system tended to improve its performance over time, an effect that he described as "maturing." Lettinga much later found a phenomenon that is perhaps related to this maturing: With time, the biosolids form what Lettinga called "granules." These granules, which naturally form after several weeks of reactor operation, are compact spherical gray-white particles about 0.5 mm in diameter. They have a small ash content, about 10 percent, and consist primarily of a dense mixed population of bacteria that are required to carry out the overall methane fermentation of substrates. Microorganisms found to dominate in granules are acetate-utilizing methanogens, especially *Methanothrix* and *Methanosarcina*. Settled granules can attain concentrations between 1 and 2 percent, and the specific activity of the particles can be on the order of 1 to 2 g COD/g VSS-d (Lettinga et al., 1988; Speece, 1996). The formation of granules depends upon characteristics of the waste stream, the substrate loading, and operational details, such as the upward fluid velocity. Serious problems that occur at times are the formation of granules that float and the lack of granule formation, both of which result in loss of biomass from the system. Thus, a knowledge of factors affecting granulation is key to successful use of the UASB system. For more on granulation, the interested reader is directed to Lettinga et al. (1988) and Speece (1996).

Many UASB systems are being used with a great deal of success on many food-processing industry wastewaters, as well as on wastewaters from the paper and chemical industries (Lettinga et al., 1988). Design loading typically is in the range of 4 to 15 kg COD/m^3-d. Because the UASB system at times forms granules or biosolids that do not settle well within the reactor, a separate settler can be provided as a safeguard against excessive loss of biosolids from the reactor.

13.2.6 MISCELLANEOUS ANAEROBIC REACTORS

A simple approach is the *baffled reactor*, which is a series of reactors connected so that the wastewater alternately moves upward and downward. Each time the wastewater moves upward, it passes through a sludge-blanket chamber, similar to the UASB reactor. As wastewater leaves each sludge-blanket chamber at the top, it is directed by a baffle to the bottom of the next chamber. A recycle line can be added to remove biosolids from the last chamber and return them to the first. Biosolids moving out of one chamber simply move into the next chamber, rather than being lost. Thus, the baffled reactor can offer some advantages to help overcome the biosolids loss from the UASB system. A few full-scale baffled reactor systems have been built to date, and so experience with them is rather limited.

A related process is the *horizontal sludge-blanket reactor,* which is not pictured in Figure 13.1. The wastewater flows into one side of a long, flexible bladder made of rubberlike material. Solids settling and biomass growth lead to a sludge blanket at the inlet end of the bladder. All the wastewater must flow through the sludge blanket, which ensures excellent contact between the wastewater and the microorganisms. When fully operational, the bottom half or so of the bladder is filled with liquid, while the top half contains the gas phase. The bladder normally is placed into an excavated depression in the soil. The volumetric loading of the horizontal sludge-blanket reactor is similar to that of the UASB. However, its horizontal nature means that its "footprint" is much larger. Countering this in terms of costs is the simplicity and lack of structures. The main operating problem with the horizontal sludge blanket is solids accumulation, which eventually leads to short circuiting and poor performance. Thus, solids need to be wasted on a regular basis.

The *two-stage leaching-bed leachate filter* evolved from an interest in converting biomass (sawdust, wheat straw, corn stover, and other lignocellulosic materials) into methane to serve as a readily usable energy source. The limiting step in biomass conversion to methane is the hydrolysis of cellulose contained in a lignocellulosic matrix. Here, the organic solid materials are placed in a pile or chamber, and liquid is passed over them to extract organic acids formed from the initial steps of cellulose hydrolysis and fermentation. The organic acids are passed to a methanogenic reactor that is optimized for methane production. By separating acid formation from methanogenesis, the more sensitive methanogenic system can be operated as a high-rate reactor. A system similar in concept is a municipal refuse pile or sanitary landfill through which water is passed. The resulting leachate is then fed to an anaerobic reactor, where the organic acids are efficiently converted to methane gas. The effluent from the reactor is then recirculated through the refuse pile or landfill. Such a system may help speed up the conversion of refuse to methane, a process that otherwise takes decades.

Another interesting variation on two-stage anaerobic treatment is the so-called *two-phase digestion.* In two-phase treatment (not shown in Fig. 13.1), the sludge or other complex organic substrate is first treated in a CSTR that has a short detention time, such as 1 or 2 d. Although the methanogens should be washed out of the first stage, anaerobic bacteria able to hydrolyze and ferment the complex organic material can survive. The concept is that the first stage has significant acidification, and the low pH results in accelerated hydrolysis. The contents of the first reactor move to the second reactor, where fermentation is completed and methanogenesis takes place. The liquid retention time of the second reactor normally is significantly less than the conventional 15 to 25 d needed for a CSTR digester. A detention time of 8 to

10 d is normal for the second stage. The overall volume of the two-phase system is less than for a conventional one-reactor CSTR. Extensive laboratory and pilot-scale research has shown that the two-phase system allows higher volumetric loading than a conventional CSTR, but evidence for accelerated hydrolysis as the cause is not clear. Further fundamental research into the mechanisms acting in two-phase digestion is warranted.

A *membrane bioreactor* system that allows passage of treated liquid, but retains the solids within the reactor, has been used successfully in France for the treatment of leachate from a sanitary landfill, and it is being used for other wastewaters. The membrane bioreactor offers three major advantages that stem from the fact that the membrane is a perfect separator for solids. First, the membrane eliminates the possibility of uncontrolled biomass loss to the effluent and, therefore, a sudden washout of slow-growing methanogens. Second, effluent quality is improved, since it contains no suspended BOD, and the membrane also removes some fraction of the SMP and other macromolecules. Third, the volumetric loading can be increased to very high levels, since loss of biomass is impossible. The main disadvantage of the membrane bioreactor is added costs: capital costs to install the membrane, energy costs to pump the water to and through the membrane, and replacement or cleaning costs to overcome membrane fouling.

An anaerobic treatment system may produce a waste stream that has a BOD that exceeds regulatory requirements, contains some degree of odor, and is deficient in oxygen. Thus, a common addition to an anaerobic system is an *aerobic polishing* step. The excess biosolids from the aerobic process can be treated within the anaerobic system. By using aerobic treatment for polishing, the advantages of anaerobic treatment can be maintained (minimal waste biosolids for disposal and minimal, or even negative, energy costs), while the effluent is acceptable for discharge to rivers or streams.

13.3 PROCESS CHEMISTRY AND MICROBIOLOGY

The anaerobic treatment process is more complex in chemistry and microbiology than aerobic treatment. A thorough understanding of both is essential by the designer and operator of anaerobic systems that are to be used successfully. This section provides the basics of anaerobic treatment microbiology and chemistry. More thorough treatments can be found elsewhere (McCarty, 1964a, b, c, d; Parkin and Owen, 1986; Speece, 1996). Chapter 1, which uses the anaerobic system as an example in the section on Microbial Ecology, is a good reference for the microbiology.

13.3.1 PROCESS MICROBIOLOGY

The consortia of microorganisms involved in the overall conversion of complex organic matter to methane begins with bacteria that hydrolyze complex organic matter—such as carbohydrates, proteins, and fats—into simple carbohydrates, amino acids, and fatty acids. The simple carbohydrates and acids are then utilized to obtain energy for growth by fermenting bacteria, producing organic acids and hydrogen as the dominant intermediate products. The organic acids are then partially oxidized by other fermenting bacteria, which produce additional hydrogen and acetic acid. Hydrogen and acetic acid are the main substrates used by archael methanogens, which convert them into methane. Hydrogen (H_2) is used as an electron donor, with carbon dioxide as an electron acceptor to form methane, while acetate is cleaved (the acetoclastic reaction) to form methane from the methyl group and carbon dioxide from the carboxyl group in a fermentation reaction. The complex and close community interactions of many prokaryotic organisms from two entirely different biological kingdoms—Bacteria and Archaea—in this widely prevalent natural process are truly amazing.

Thermodynamics and kinetics are crucial to the mixed microbial community involved in methane fermentation. Both must be understood if one is to appreciate this process in its various complexities. From an operational standpoint, the overall complexity of the process can be simplified into a few principles that can be applied readily. For instance, the process can be broken into two basic steps, as illustrated in Figure 13.2: (1) hydrolysis and fermentation of complex organic matter into simple

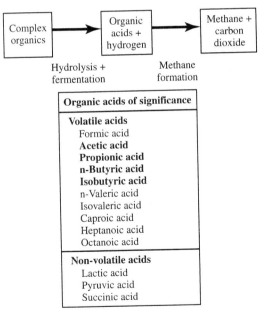

Figure 13.2   Simplified two-step view of the overall conversion of complex organics into methane in anaerobic wastewater treatment. Volatile acids shown in bold are the most prevalent intermediates found in the process.

organic acids and hydrogen, and (2) the conversion of the organic acids and hydrogen into methane.

The microorganisms involved in the first step grow relatively rapidly, because the fermentation reactions give a greater energy yield than the reactions that lead to methane formation. For this reason, the methanogens are more slowly growing and tend to be rate-limiting in the process. This generalization is true with domestic wastewater organic matter, municipal sludges, and most industrial wastewaters. However, with certain organic materials, for example the anaerobic decomposition of lignocellulosic materials such as grasses, agricultural crop residues, or newsprint, the hydrolysis step may be very slow and rate-limiting.

The successful start-up and operation of an anaerobic system requires that a proper balance be maintained between the hydrolytic and fermentative organisms involved in the first step and the methanogenic organisms responsible for the second step. This balance is accomplished through proper seeding, as well as through control of organic-acid production and pH during the start-up, when the microbial populations are establishing themselves. Ideally, an anaerobic reactor is seeded with digested sludge or biosolids from an active anaerobic treatment system. This kind of balanced, active seeding is necessary because of the slow doubling time (ca. 4 d at 35 °C) of the critical microorganisms involved in the second step. If the seed contains only a small number of methanogens, the start-up time may be long. For example, about 10^8 to 10^9 of the critical microorganisms are required per ml of reactor volume to ensure successful operation of an anaerobic treatment system. If a seed with only 10^3 per ml of the critical organisms is available, the population would need to be increased by a factor of about 10^6. This requires about 20 doubling times, or about 80 d at 35 °C. At lower temperatures, the doubling time increases by a factor of about two for each 10 °C drop in temperature.

During reactor start-up, the operator must maintain a sufficiently small loading on the reactor so that organic acids produced by the much faster growing fermentative bacteria do not exceed the buffering capacity of the system. If this occurs, the pH will drop, and the methanogenic population can be killed. The crucial steps during start-up are: (1) begin with as much good anaerobic seed as possible, (2) fill the digester with this seed and water, (3) bring the system to temperature, (4) add buffering material in the form of a chemical such as sodium bicarbonate to protect against pH drop, and (5) add a small amount of organic waste sufficient to let the organic acid content from fermentation reach no more than about 2,000 to 4,000 mg/l, while keeping pH between 6.8 and 7.6. These organic acids are the food source required for the methanogenic population to grow. The time when sufficient doublings have occurred will be evidenced through a drop in the organic acid concentration. Feeding with additional waste can then be initiated, slowly at first, until a balance is reached between the step 1 and step 2 reactions in the system. At such balance, the organic acid concentration will generally remain below 100 to 200 mg/l, depending upon the loading on the system.

Organic acid concentration and reactor pH should be determined on a daily basis to ensure that the operation of the anaerobic system remains in balance. Inhibition of the biological reactions or an overload on the system with organic wastes often can be evidenced through a sudden increase in the organic acid concentration. If the buffering capacity is becoming depleted, chemical base (like bicarbonate) must then be added quickly to prevent a drop in pH from occurring, which would kill the critical methanogenic population. Thus, monitoring the organic acids and the buffer capacity provides the first line of defense for control of anaerobic systems so that the acid producers and the acid consumers achieve and maintain a proper balance.

The organic acid concentration is a key indicator of system performance. The question then is what organic acids should be measured on a routine basis, and how can this be accomplished? The key organic acids are the series of short chain fatty acids listed in Figure 13.2 and which vary in chain length from formic acid with one carbon per mole to octanoic acid with eight carbon atoms per mole. These acids have been termed *volatile acids* because, in their unionized form, they can be distilled from boiling water. This meaning of the term *volatile* is different from its meaning in *volatile organic compounds (VOCs),* a term generally used to describe organic compounds that are readily removed from water by simple air stripping. The short-chain fatty acids cannot be removed from water by air stripping.

The volatile acids that are generally found present in highest concentrations as intermediates during start-up of an anaerobic system or during organic overload are acetic, propionic, butyric, and isobutyric acids. These comprise the bulk of the organic acids found in anaerobic systems. Other nonvolatile organic acids also are formed as intermediates of waste organic degradation (e.g., lactic, pyruvic, and succinic acids) but their concentrations generally are much below those of the volatile acids and thus are of less general concern for control. The volatile acids are all quite soluble in their ionized and unionized forms and are present as dissolved species. At normal pH of operating systems, they all are present for the most part in the ionized (or deprotonated) form. Typical values at 35 °C for the negative logarithm of the acidity constant, or pK_a (the pH at which the acids are 50 percent in the acid form and 50 percent in the ionized form), vary from a low of 3.8 for formic acid to more typical values of 4.8 for acetic and n-butyric acids and 4.9 for propionic acid.

The routine measurement of volatile acid concentration is of the greatest importance in the operation and control of anaerobic systems. Various analytical procedures can be used to measure volatile acids. Methods vary from those that require expensive instrumentation, such as gas chromatography and high performance liquid chromatography, to relatively inexpensive wet chemical procedures involving distillation, column chromatography, or acid/base titration. The instrumental approaches allow one to differentiate among the various organic acids present, which can at times be used to help diagnose the cause of digester failure. The wet chemical procedures generally provide information only on the total organic acid or total volatile acid concentration present, but this is often sufficient for routine control.

A desirable analysis would be that for the active population of microorganisms present in the system, especially of the populations responsible for the critical steps of the overall process, including the methanogens. At this point, tools for such analyses are not available for routine use. However, new methods adapted from molecular biology research and reviewed in Chapter 1 eventually will become available for routine operation. So far, oligonucleotide probes have been used successfully to determine the relative abundances of methanogenic and sulfidogenic populations, as well as to differentiate among the different methanogenic and sulfidogenic populations. Analyses for the more complex fatty acids that comprise bacterial cell walls has also been used to characterize bacterial populations in complex systems.

13.3.2 PROCESS CHEMISTRY

Along with process microbiology, one needs to understand basic chemical characteristics of the anaerobic treatment process. Aspects of importance here include reaction stoichiometry, pH and alkalinity requirements, nutrient requirements, and the effect of inhibitory materials on the process.

Stoichiometry The microbiology of the anaerobic process is quite complex, and organic compounds are generally converted in many intermediate steps before the basic end product, methane, is formed. Mass balances are maintained at every step on carbon, nitrogen, hydrogen, oxygen, and other elements. Most importantly, an electron balance must be maintained, because most of the electron equivalents, or BOD_L, entering the anaerobic process in organic matter are conserved in CH_4, which evolves to the gas phase. Therefore, the removal of BOD_L, or waste stabilization, depends totally on the formation of methane.

While some intermediate products remain after treatment, most of the organic matter consumed by the microorganism is converted to the main end products: carbon dioxide, methane, water, and biomass. If other elements, such as nitrogen and sulfur, are part of the consumed organic matter, then they are converted to inorganic form, generally ammonium and sulfides. On this basis, the end products of methanogenic treatment of an organic waste can be determined readily using the procedures for writing stoichiometric equations given in Chapter 2. For example, we consider the empirical molecular formula for the organic matter (and electron donor, or BOD_L) to be $C_nH_aO_bN_c$. A certain portion of its electron equivalents, f_s, is (net) synthesized into biomass, and ammonium is the source of cell nitrogen.

Since we know that carbon dioxide is the true electron acceptor for the portion of the methane formed from oxidation of H_2, then it is a good choice for the electron acceptor for the methane formed from H_2. However, what do we do about the portion of methane that comes from acetic acid? We can assume for the basis of writing the stoichiometric equation that carbon dioxide serves here as the electron acceptor as well. While CO_2 is not the true acceptor for the acetoclastic methanogens, the exact pathway by which compounds are converted to end products is not important for maintaining a mass balance. This can be illustrated by writing out the conversion of acetate to methane, which we know takes place through the acetoclastic reaction:

$$CH_3 \vdots COO^- + H_2O \rightarrow CH_4 + HCO_3^- \qquad \textbf{[13.1]}$$

The three dots in a vertical line identify the cleavage location, with the CH_3 going to form CH_4. However, if we did not know this was a fermentation pathway and, instead, assumed acetate to be an electron donor and carbon dioxide as electron acceptor for the reaction, then using half-reactions, we would find:

$$-R_d : \quad \frac{1}{8} CH_3CO^- + \frac{3}{8} H_2O \rightarrow \frac{1}{8} CO_2 + \frac{1}{8} HCO_3^- + H^+ + e^- \quad \textbf{[13.2]}$$

$$R_a : \quad \frac{1}{8} CO_2 + H^+ + e^- \rightarrow \frac{1}{8} CH_4 + \frac{1}{4} H_2O \quad \textbf{[13.3]}$$

$$R : \quad \frac{1}{8} CH_3COO^- + \frac{1}{8} H_2O \rightarrow \frac{1}{8} CH_4 + \frac{1}{8} HCO_3^- \quad \textbf{[13.4]}$$

It is obvious that Equation 13.4 is identical to Equation 13.1 divided by 8. Thus, the assumption that CO_2 is the electron acceptor for the reaction makes no difference in the overall balanced reaction.

Applying this same assumption of carbon dioxide being the electron acceptor and using the principles from Chapter 2 for writing balanced overall stoichiometric equations, we obtain the following equation for our generalized organic waste:

$$C_n H_a O_b N_c + \left(2n + c - b - \frac{9df_s}{20} - \frac{df_e}{4} \right) H_2O \rightarrow$$

$$\frac{df_e}{8} CH_4 + \left(n - c - \frac{df_s}{5} - \frac{df_e}{8} \right) CO_2 + \frac{df_s}{20} C_5 H_7 O_2 N \quad \textbf{[13.5]}$$

$$+ \left(c - \frac{df_s}{20} \right) NH_4^+ + \left(c - \frac{df_s}{20} \right) HCO_3^-$$

where

$$d = 4n + a - 2b - 3c$$

The value f_s represents the fraction of waste organic matter synthesized or converted to cells, while f_e represents the portion converted for energy, such that $f_s + f_e = 1$ (recall Chapter 2). The value for f_s depends on the energetics of the cell's energy-generation and synthesis reactions, as well as the decay rate (b) and θ_x. For a reactor operating at steady state, f_s can be estimated from Equation 3.33:

$$f_s = f_s^0 \left[\frac{1 + (1 - f_d)b\theta_x}{1 + b\theta_x} \right]$$

Typical values for f_s^0 and b for methane fermentation of common organic compounds are summarized in Table 13.2. The f_s^0 values include the methanogens *and* all bacteria needed to convert the original organic matter to acetate and H_2. Values of f_s^0 for a mixture of waste materials can be estimated from a weighted average of values from Table 13.2 based on relative electron equivalents (as COD or BOD_L) for the different electron donors present.

Table 13.2 Coefficients for stoichiometric equations for anaerobic treatment of various organic materials

Waste Component	Typical Chemical Formula	f_s^0	Y g VSS_a per g BOD_L removed	b d^{-1}
Carbohydrates	$C_6H_{10}O_5$	0.28	0.20	0.05
Proteins	$C_{16}H_{24}O_5N_4$	0.08	0.056	0.02
Fatty acids	$C_{16}H_{32}O_2$	0.06	0.042	0.03
Municipal sludge	$C_{10}H_{19}O_3N$	0.11	0.077	0.05
Ethanol	CH_3CH_2OH	0.11	0.077	0.05
Methanol	CH_3OH	0.15	0.11	0.05
Benzoic acid	C_6H_5COOH	0.11	0.077	0.05

STOICHIOMETRY OF GLUCOSE FERMENTATION TO METHANE A wastewater **Example 13.2** from food processing contains 1.0 M glucose. For operation of an anaerobic treatment process for this wastewater, estimate the methane production, the mass of biological cells produced, and the concentration of ammonia-N required for cell growth per m^3 of wastewater treated. Assume that f_s for the operating conditions used is 0.20, and that the glucose is essentially 100 percent consumed.

The molecular formula for glucose is $C_6H_{12}O_6$. Then, for Equation 13.5, $n = 6$, $a = 12$, $b = 6$, $c = 0$, $d = (4 \times 6 + 12 - 2 \times 6) = 24$, and $f_e = 1 - 0.2 = 0.8$. The resulting equation is

$$C_6H_{12}O_6 + 0.24\,NH_4^+ + 0.24\,HCO_3^- \rightarrow 2.4\,CH_4 + 2.64\,CO_2 + 0.24\,C_5H_7O_2N + 0.96\,H_2O$$

For each liter of wastewater, 2.4 mol methane is produced, along with 2.64 mol carbon dioxide. The cell production is 0.24 times the empirical cell formula weight of 113, or 27.2 g, and the nitrogen requirement is for 0.24 mol, or 3.36 g/l based on the atomic weight of nitrogen of 14. Thus, for each m^3 of wastewater, 2.4 kmol methane and 27.2 kg cells are produced. This requires that 3.36 kg ammonia nitrogen be present in each m^3 of wastewater to satisfy the needs for biological growth.

STOICHIOMETRY OF THE METHANOGENESIS OF AN ORGANIC MIXTURE An **Example 13.3** industrial wastewater has a flow rate of 100 m^3/d and a COD of 5,000 mg/l. Its COD consists of 50 percent fatty acids and 50 percent protein based on COD. The wastewater is treated in an anaerobic reactor with θ_x of 20 d and at 35 °C. Estimate the methane production in m^3/d, biological cell production rate in kg/d, and the relative percentages of carbon dioxide and methane produced. Also, estimate the bicarbonate alkalinity formed from the biological reaction. Assume 80 percent conversion of waste COD to end products. A typical empirical formula for protein is $C_{16}H_{24}O_5N_4$ and for fatty acids is $C_{16}H_{32}O_2$.

First, one needs to consider how to handle the mixture of organic matter. Calculations could be made separately for the fatty acid and protein components, and the results could then be added together, or one might devise an empirical wastewater organic matter formula that represents a composite of the different waste types. The latter here appears simpler and so is used. Since the wastewater is 50 percent fatty acids and 50 percent protein on a COD or electron equivalent basis, we first need to devise an electron-donor equation that is consistent with this fact. From the generalized Reaction O-19 in Table 2.3, we obtain:

Fatty Acids:

$$\frac{4}{23} CO_2 + H^+ + e^- = \frac{1}{92} C_{16}H_{32}O_2 + \frac{15}{46} H_2O$$

Protein:

$$\frac{2}{11} CO_2 + \frac{2}{33} NH_4^+ + \frac{2}{33} HCO_3^- + H^+ + e^- = \frac{1}{66} C_{16}H_{24}O_5N_4 + \frac{31}{66} H_2O$$

From this, we can generate a donor half-reaction by taking 50 percent of each equation and adding the results together:

$$0.1779\ CO_2 + 0.0303\ NH_4^+ + 0.0303\ HCO_3^- + H^+ + e^-$$
$$= 0.013\ C_{16}H_{27.3}O_{3.75}N_{2.33} + 0.398\ H_2O$$

For the organic compound, the coefficient 0.013 results from $(1/66 + 1/92)/2$. For CO_2, the 0.1779 comes from $(4/23 + 2/11)/2$. Since the values for f_s^0 in Table 13.2 are on an electron equivalent basis, we can take 50 percent of each for fatty acids and proteins, respectively, and add the results together to give the composite $f_s^0 = 0.07$. The value b is the same for both, $0.05\ \mathrm{d}^{-1}$. Then,

$$f_s = 0.07\left[\frac{1 + 0.2(0.05)(20)}{1 + 0.05(20)}\right] = 0.042 \quad \text{and} \quad f_e = 1 - 0.042 = 0.958$$

Substituting these results into Equation 13.5 gives us the overall reaction for this case:

$$C_{16}H_{27.3}O_{3.75}N_{2.33} + 10.73\ H_2O \rightarrow$$
$$9.20\ CH_4 + 3.83\ CO_2 + 0.161\ C_5H_7O_2N + 2.17\ NH_4^+ + 2.17\ HCO_3^-$$

From the donor half-reaction above, 0.013 empirical mol of substrate represents one electron equivalent. Because one electron equivalent of any organic donor has a COD of 8 g, the COD represented by an empirical mol of waste organic matter is 8/0.013 or 615 g COD/mol. We now have all the information needed to answer the questions raised:

$$\text{COD removal rate} = (S^0 - S)Q$$

$$= [5,000 - 0.2(5,000)]\frac{mg}{1} \cdot 100\frac{m^3}{d} \cdot \frac{10^3}{m^3}\frac{1}{} \cdot \frac{g}{10^3\ mg} = 4(10^5)\ \text{g/d}$$

One mol CH_4 at 35 °C occupies $22.4\frac{1}{mol} \cdot \frac{273 + 35}{273} = 25.3$ l; thus, from the stoichiometric equation:

$$\text{CH}_4\ \text{production} = 25.3\frac{1}{mol} \cdot 9.20\frac{mol}{mol} \cdot \frac{4(10^5)\ \text{g COD/d}}{615\ \text{g COD/mol}} \cdot \frac{m^3}{10^3\ 1} = 151\ m^3/\text{d}$$

One empirical mole of cells has a formula weight of 113 g; thus,

$$\text{Cell production} = 0.161\frac{mol}{mol} \cdot 113\frac{g}{mol} \cdot \frac{4(10^5)\ \text{g COD/d}}{615\ \text{g COD/mol}} \cdot \frac{kg}{10^3\ g} = 11.8\ \text{kg/d}$$

Since the gas from the reaction is essentially all methane and carbon dioxide, the two must equal 100 percent of the mixture. Also, gas volumes are proportional to the molar fractions of each gas; thus,

$$CH_4 = \frac{9.20}{9.20 + 3.83}(100) = 71 \text{ percent}$$

and

$$CO_2 = 100 - 71 \text{ percent} = 29 \text{ percent}$$

We can determine the alkalinity formed from the stoichiometric equation, which indicates 2.17 mol HCO_3^- alkalinity is formed per mol of substrate consumed. Generally, alkalinity is reported as $CaCO_3$, which has an equivalent weight of 50 g. Thus,

$$\text{Alk (as } CaCO_3) = \frac{0.8 \times 5 \text{ g COD}}{1} \cdot \frac{2.17 \text{ mol } HCO_3^-}{\text{mol}} \cdot \frac{\text{mol}}{615 \text{ g COD}} \cdot \frac{50,000 \text{ mg alk}}{\text{mol } HCO_3^-}$$

$$= 706 \text{ mg/l as } CaCO_3$$

We have answered all the questions, but we might go one step further. Let us assume that there was no alkalinity present in the original wastewater; that is, the only alkalinity available to buffer the wastewater is that produced during the biological reaction. We can then ask, what will be the resulting pH if no additional buffer were added? We find how to approach the answer to this rather simple question in the next section.

---

pH and Alkalinity Requirements   The desired pH for anaerobic treatment is between 6.6 and 7.6. Values outside this range can be quite detrimental to the process, particularly to methanogenesis. The biggest problem generally is to maintain the pH above 6.6, because organic acids produced as intermediates in the process during start-up, overload, or other unbalance can cause a rapid pH drop and cessation of the methane production. Start-up after such an event can be very slow, on the order of weeks or months; thus it is crucial that low pH be avoided. The relationships between the various factors that affect reactor pH must be well understood by those in charge of anaerobic treatment systems.

The main chemical species controlling pH in anaerobic treatment are those related to the carbonic acid system, as governed by the following reactions:

$$CO_2(aq) = CO_2(g) \qquad \textbf{[13.6]}$$

$$CO_2(aq) + H_2O = H_2CO_3 \qquad \textbf{[13.7]}$$

$$H_2CO_3 = H^+ + HCO_3^- \qquad \textbf{[13.8]}$$

$$HCO_3^- = H^+ + CO_3^{2-} \qquad \textbf{[13.9]}$$

$$H_2O = H^+ + OH^- \qquad \textbf{[13.10]}$$

The equilibrium relationships among the various species are given by

$$\frac{[CO_2(g)]}{[H_2CO_3^*]} = K_H = 38 \text{ atm/mol } (35 \text{ °C}),$$

where

$$H_2CO_3^* = CO_2(aq) + H_2CO_3 \qquad \textbf{[13.11]}$$

$$\frac{[H^+][HCO_3^-]}{[H_2CO_3^*]} = K_{a,1} = 5 \cdot 10^{-7}(35\,°C) \qquad \textbf{[13.12]}$$

$$\frac{[H^+][CO_3^{2-}]}{[HCO_3^-]} = K_{a,2} = 6 \cdot 10^{-11}(35\,°C) \qquad \textbf{[13.13]}$$

$$[H^+][OH^-] = K_w = 2 \times 10^{-14}(35\,°C) \qquad \textbf{[13.14]}$$

At the normal pH of anaerobic treatment, carbonate (CO_3^{2-}) is not important, and Equations 13.9 and 13.13 need not be considered.

Alkalinity is defined as the acid-neutralizing capacity of water. When the carbonic-acid system dominates the buffering (as it does in most anaerobic treatment processes), alkalinity can be quantified from a proton condition on the species of interest (Sawyer, McCarty, and Parkin, 1994):

$$[H^+] + [\text{Alkalinity}] = [HCO_3^-] + 2[CO_3^{2-}] + [OH^-] \qquad \textbf{[13.15]}$$

in which all species are in mol/l. With respect to the usual pH and conditions of anaerobic treatment, the concentrations of $[H^+]$, $[CO_3^{2-}]$, and $[OH^-]$ are quite small compared with $[HCO_3^-]$. Also, alkalinity can be expressed in the conventional units of mg/l as $CaCO_3$. Making those two changes converts Equation 13.15 to the following good approximation:

$$\frac{\text{Alkalinity(bicarb)}}{50,000} = [HCO_3^-] \qquad \textbf{[13.16]}$$

Equation 13.16 shows that the total alkalinity in an anaerobic process is effectively equal to the bicarbonate concentration, or the bicarbonate alkalinity.

By taking the log of both sides of Equation 13.12 and using the relationships that $pH = -\log[H^+]$ and $pK_{a,1} = -\log K_{a,1}$, we obtain:

$$pH = pK_{a,1} + \log \frac{[HCO_3^-]}{[H_2CO_3^*]} \qquad \textbf{[13.17]}$$

Finally, substituting Equations 13.11 and 13.16 into Equation 13.17, we obtain:

$$pH = pK_{a,1} + \log \frac{\dfrac{\text{Alkalinity(bicarb)}}{50,000}}{\dfrac{[CO_2(g)]}{K_H}} \qquad \textbf{[13.18]}$$

Equation 13.18 indicates that the pH is controlled by the concentrations of alkalinity in the reactor liquid and carbon dioxide in the reactor gas phase, assuming that CO_2 equilibrium exists between the gas phase and the reactor liquid phase. Generally, CO_2 equilibrium is closely approached in anaerobic treatment systems. The customary units for $[CO_2(g)]$ are atmospheres, and $[CO_2(g)]$ is obtained by multiplying the fraction of CO_2 in the gas phase times the total pressure in atm. For example, if there

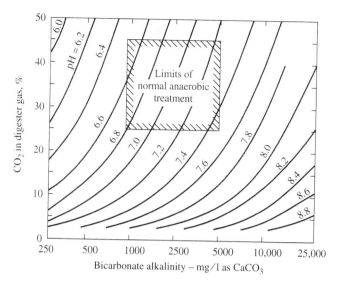

Figure 13.3 Relationship among bicarbonate alkalinity, the percentage of carbon dioxide in the gas phase (at 1 atm total pressure), and reactor pH in anaerobic treatment.

is 35 percent CO_2 in the digester gas and the gas is under one atmosphere of pressure, then $[CO_2(g)] = 0.35$ atm.

Figure 13.3 gives a helpful pictorial representation of Equation 13.18 for 35 °C and 1 atm total pressure. It is apparent from Figure 13.3 that, at the normal percentages of carbon dioxide in digester gas, 25 to 45 percent, a bicarbonate alkalinity of at least 500 to 900 mg/l as $CaCO_3$ is required to keep the pH above 6.5. A higher carbon dioxide partial pressure makes the alkalinity requirement larger. Figure 13.3 also shows two other important and generalizable trends: (1) a quite high alkalinity of 5,000 mg/l with normal carbon dioxide content does not lead to an excessively high pH for anaerobic treatment, and (2) the pH is not sensitive to increases in alkalinity once the pH and alkalinity are about 7.4 and 5,000 mg/l as $CaCO_3$, respectively. In practical terms, increasing the alkalinity above about 5,000 mg/l gives little benefit and incurs little risk.

CALCULATING THE pH   Calculate the pH for the anaerobic treatment system of Example 13.3. The temperature was 35 °C; the carbon dioxide partial pressure was computed as 29 percent, which corresponds at sea level to a partial pressure of 0.29 atm; and the alkalinity was calculated as 706 mg/l. Using Equation 13.18:

| **Example 13.4** |

$$pH = -\log(5 \times 10^{-7}) + \log \frac{706/50,000}{0.29/38} = 6.6$$

This pH is on the borderline for satisfactory anaerobic treatment. The 706 mg/l of alkalinity was produced solely as a result of the release of ammonia from the organic matter. If alkalinity were naturally present in the water before treatment, then this could help to increase the amount available. Also, alkalinity could be added to supplement that initially present and that produced in the reaction to maintain a more satisfactory level.

The overall stoichiometric equation in Example 13.2 shows that the moles of bicarbonate alkalinity just equal the moles of NH_4^+. This correspondence underscores that release of organic N forms the base NH_3, which then accepts a proton from water and releases OH^- (which is base, or alkalinity) to the water: $H_2O + NH_3 = NH_4^+ + OH^-$. Release of ammonium is the main way in which alkalinity is added to the water through biodegradation of organic matter.

Biodegradation of organic matter also can destroy alkalinity, and this is one of the dangers in anaerobic treatment. Fermentation of complex organic molecules to organic acids is the way in which alkalinity is destroyed, since an acid neutralizes base. In reaction form, the production of an organic acid (represented as HA) *destroys bicarbonate alkalinity* according to

$$HA + HCO_3^- \rightarrow H_2CO_3^* + A^- \qquad \textbf{[13.19]}$$

Strong acids, such as HCl, also reduce bicarbonates and alkalinity. However, weak acids such as the volatile acids will destroy bicarbonate species, but do not change the total alkalinity very much. This dichotomy occurs because volatile acids are weak acids ($pK_a > 2$), and their neutralized forms (for example, acetate) titrate as alkalinity in the standard acid titration, which has an end point around $pH = 4.3$. Thus, an increase in volatile acid concentration does not cause a commensurate decrease in the measured total alkalinity. The lack of response of total alkalinity is very dangerous, and total alkalinity alone must not be used to monitor the buffering status of an anaerobic process. As shown by Equation 13.18, the bicarbonate alkalinity is the right measure.

In the case when weak acids and bicarbonate are present together, Equation 13.15 for alkalinity needs to be expanded as follows:

$$[H^+] + [Alkalinity] = [A^-] + [HCO_3^-] + 2[CO_3^{2-}] + [OH^-] \qquad \textbf{[13.20]}$$

where $[A^-]$ represents the summation of the molar concentrations of all the weak acid salts present (except for bicarbonate and carbonate). Thus, when volatile acid increase occurs, $[A^-]$ simply substitutes for some of the $[HCO_3^-]$, but the total alkalinity is not affected. However, the bicarbonate alkalinity (along with the partial pressure of CO_2) is what controls the pH within the normal pH range of anaerobic treatment.

The important point is that, in the absence of volatile acids, the total alkalinity measurement is generally a good measure for $[HCO_3^-]$, as represented in Equation 13.16. When volatile acids are present, the bicarbonate alkalinity is reduced in proportion according to Equation 13.19. If the volatile acid concentration is known, then the new $[HCO_3^-]$ can be determined by appropriate adjustments to the measured total alkalinity concentration to determine the bicarbonate alkalinity to use in Equation 13.18. This adjustment is shown in Example 13.5.

CALCULATING THE EFFECT OF VOLATILE ACIDS ON THE pH An anaerobic treat- | **Example 13.5**
ment system is operating at 35 °C, at sea level, with 25 percent carbon dioxide in the gas phase, with a very low volatile acid concentration, and with a total alkalinity of 2,800 mg/l measured as $CaCO_3$. Due to an overload, the volatile acid concentration rapidly increases to 2,500 mg/l, measured as acetic acid (molecular weight = 60). Estimate the pH before the volatile acid increase and after it.

Initially, alkalinity(bicarb) is equal to the total alkalinity, 2,800 mg/l. After the increase in volatile acids, alkalinity(bicarb) is equal to the same total alkalinity minus that represented by the volatile acids:

$$\text{Alkalinity(bicarb)} = 2{,}800 \text{ mg/l} - 2{,}500(50/60) = 717 \text{ mg/l}$$

in which 50 is the equivalent weight of $CaCO_3$ and 60 is the equivalent weight of acetic acid. The equivalent-weight ratio (50/60) converts the 2,500 mg as HAc/l to 2,083 mg as $CaCO_3$/l. In other words, 2,083 mg/l of the total alkalinity of 2,800 mg/l is acetate, leaving only 717 mg/l in bicarbonate. (Note that the measured total alkalinity would have decreased slightly if titration to pH $\approx$ 4.3 had been performed for the second case; this occurs because the pK_a values of the volatile acids are lower than $pK_{a,1}$ for carbonic acid (Sawyer, McCarty, and Parkin, 1994).)

The pH values are computed from Equation 13.18:

$$\text{pH initial} = 6.3 + \log \frac{2{,}800/50{,}000}{0.25/38} = 7.2$$

$$\text{pH final} = 6.3 + \log \frac{717/50{,}000}{0.25/38} = 6.6$$

We see that the buildup in volatile acids causes a significant drop in pH, and, indeed, the remaining bicarbonate buffer of 717 mg/l is exceedingly low. A small additional increase in volatile acids could be disastrous to system operation. As an additional problem, we have assumed that the carbon dioxide concentration here remained the same during volatile acid buildup, but according to Equation 13.19, the destruction of bicarbonate alkalinity results in the production of carbonic acid, and with it carbon dioxide would be released to the gas phase. An increase in the gas-phase CO_2 fraction would depress the pH even lower. For example, a CO_2 partial pressure of 0.43 atm would drop the pH to 6.4 for the situation in which the volatile acids were 2,500 mg/l as HAc.

Alkaline materials often are added to provide adequate buffer when it is not present in a wastewater or to prevent an excessive drop in pH during unbalanced conditions. Common materials used for this purpose are lime ($Ca(OH)_2$), sodium bicarbonate ($NaHCO_3$), soda ash (Na_2CO_3), sodium hydroxide ($NaOH$), ammonia (NH_3), or ammonium bicarbonate (NH_4HCO_3). Generally, lime, sodium hydroxide, and ammonia are the cheapest of these chemicals and thus the ones selected. However, each has its own potential problem, which needs to be carefully understood before use.

Lime usually is the cheapest chemical, but is the one that should be used with most caution, because of the high potential to form $CaCO_{3(s)}$ within the reactor. The reactions of interest here are:

Formation of Ca^{2+} and bicarbonate from the base in the added $Ca(OH)_2$ and CO_2 in the gas phase:

$$Ca(OH)_2 + CO_2 = Ca^{2+} + 2\,HCO_3^-$$ [13.21]

Formation of carbonate from the bicarbonate:

$$2\,HCO_3^- = 2\,H^+ + 2\,CO_3^{2-}$$ [13.22]

Precipitation of calcium carbonate solid from the Ca^{2+} and the CO_3^{2-}:

$$Ca^{2+} + CO_3^{2-} = CaCO_{3(s)}$$

$$[Ca^{2+}][CO_3^{2-}] = K_{sp} = 2.9 \times 10^{-9}(35\ ^\circ C)$$ [13.23]

Equation 13.21 represents the desired reaction, the formation of bicarbonate alkalinity. However, it comes with several costs.

One cost is the consumption of some carbon dioxide. The potential danger here is that, when lime is directly added to a digester, the carbon dioxide equilibrium is disturbed, and carbon dioxide may leave the gas phase and enter the liquid phase within a digester. This results in a reduction in the total pressure of the gas, and a vacuum is created. If nothing is done to relieve this vacuum, such as the introduction of water or a provision to allow digester gas to flow back into the digester from a storage tank, then the digester could collapse, an event which is not all that uncommon.

A second cost is the formation of $CaCO_{3(s)}$. As the bicarbonate concentration increases (Equation 13.21), the carbonate concentration also increases according to Equation 13.22, especially since H^+ also is decreasing due to the resulting increase in pH that $Ca(OH)_2$ addition causes. As indicated by Equations 13.21 and 13.22, the addition of lime results in the increase in the Ca^{2+} *and* CO_3^{2-} concentrations. When the concentrations reached a sufficient point, then $CaCO_{3(s)}$ precipitation takes place according to Equation 13.23. This precipitate may simply mix with the suspended solids already present in the reactor, or it may precipitate to form a tough scale on the reactor parts or attachment media in a biofilm reactor. This precipitation also results in the removal of bicarbonate alkalinity from the reactor (Equation 13.22) and thus defeats the original purpose of the lime addition! Theoretically, the $CaCO_3$ would serve as a buffer as well, but it is so insoluble and unreactive that the reverse reactions generally occur too slowly to be beneficial to the system. Thus, lime addition must be carried out very carefully and with full knowledge of the chemistry involved. Generally, the undesirable reactions start to occur once the pH has been increased just above 6.8. If one attempts to increase the pH to 7.5 by lime addition, the likely result is a treatment reactor full of concretelike precipitate.

The above set of problems is less likely if the other alkaline materials are added. However, adding hydroxides and carbonates can cause a consumption of carbon dioxide, and so the potential problem from vacuum formation needs to be addressed before they are added. Sodium bicarbonate addition avoids all these problems, but has a higher dollar cost associated with it, and the sodium can be inhibitory. Ammonia may be added, but has a potential problem of toxicity if too much is added, a problem addressed later under Inhibitory Materials.

Nutrient Requirements   As with all biological treatment systems, trace nutrients must be present to satisfy the growth requirements of the microorganisms involved. With municipal wastewater and treatment plant sludges, essentially all nutrients required for growth are present. This is true of many food-processing wastes as well, especially those resulting from processing of animals, fish, and fowl. However, many industrial wastewaters, especially those from the chemical industry, may have deficiencies in some required nutrients. Among the inorganic nutrients required for growth are the major ones, nitrogen and phosphorus. The quantity needed can be determined from estimates of net biological growth, such as obtained from stoichiometric equations in Chapter 2. Nitrogen represents about 12 percent by weight of the cell, while phosphorus represents about 2 percent. Nitrogen should be in the reduced form (NH_3 or organic amino-nitrogen) for anaerobic treatment, as nitrate and nitrite are likely to be lost by denitrification in the anaerobic environment. Also, nitrogen over and above that simply needed for growth should be present in the reactor to ensure that it does not become rate-limiting; an excess concentration of about 50 mg/l should be sufficient for this purpose. Methanogens also have a requirement for sulfur of about the same order of magnitude as that for phosphorus, or perhaps even a little more. This can generally be satisfied by sulfate in the original wastewater. If sulfur is deficient, sulfate can be added easily, but it should not be overdosed, as sulfate reduction will reduce methane generation and produces sulfides that can have numerous adverse effects (described below).

An additional requirement in anaerobic systems is for trace metals, which are needed for activation of key enzymes for methanogenesis. Table 13.3 is a listing adapted from Speece (1996) of trace metals that have been found to stimulate the anaerobic treatment process. Iron, cobalt, and nickel are known requirements for key enzymes within methane-producing species and must always be present for effective anaerobic treatment. Lack of sufficient trace nutrients may be a cause of failure of anaerobic treatment for many industrial wastewaters. The required concentration of each differs considerably. Iron often needs to be present in concentrations as high as 40 mg/l, while 1 mg/l or less of the others is generally sufficient. Speece (1996) provides an excellent summary of the stimulatory benefit of trace metal additions to reactors treating various wastewaters.

One major difficulty in anaerobic treatment is the interaction between metals and sulfide, both of which are required for biological growth. Sulfide combines with many metals to form highly insoluble complexes that are not readily available for use by microorganisms. This has always presented a bit of a dilemma for the engineer to know how to add the nutrients. With complex municipal sludges, it is likely that the many organic complexing ligands present help keep sufficient concentrations of the metals present in a soluble form that is available for microorganisms. In addition, soluble microbial products contain carboxylate groups that complex metal cations. However, addition of strong chelating agents, such as EDTA (ethylene diamine tetraacetic acid) may create complexes that are so strong that the metals are unavailable, even though they are soluble. Because of the slow bioavailability of metal sulfides, the addition of higher metal concentrations than might be needed just to satisfy the physiological needs of the microorganisms often stimulates the process.

Table 13.3   Nutrient requirements for anaerobic treatment

Element	Requirement mg/g COD	Desired Excess Concentration mg/l	Typical Form for Addition
Macronutrients			
Nitrogen	5–15	50	NH_3, NH_4Cl, NH_4HCO_3
Phosphorus	0.8–2.5	10	NaH_2PO_4
Sulfur	1–3	5	$MgSO_4 \cdot 7\,H_2O$
Micronutrients			
Iron	0.03	10	$FeCl_2 \cdot 4\,H_2O$
Cobalt	0.003	0.02	$CoCl_2 \cdot 2\,H_2O$
Nickel	0.004	0.02	$NiCl_2 \cdot 6\,H_2O$
Zinc	0.02	0.02	$ZnCl_2$
Copper	0.004	0.02	$CuCl_2 \cdot 2\,H_2O$
Manganese	0.004	0.02	$MnCl_2 \cdot 4\,H_2O$
Molybdenum	0.004	0.05	$NaMoO_4 \cdot 2\,H_2O$
Selenium	0.004	0.08	Na_2SeO_3
Tungsten	0.004	0.02	$NaWO_4 \cdot 2\,H_2O$
Boron	0.004	0.02	H_3BO_3
Common Cations			
Sodium		100–200	$NaCl$, $NaHCO_3$
Potassium		200–400	KCl
Calcium		100–200	$CaCl_2 \cdot 2\,H_2O$
Magnesium		75–250	$MgCl_2$

| SOURCE: Speece, 1996.

Finally, microorganisms need an aqueous environment that has a balance of the common cations: sodium, potassium, calcium, and magnesium. These are generally present in a balanced quantity in most water supplies and sewage. Minimum concentrations needed are generally in the 40 to 60 mg/l range. However, if one of these cations is present in exceptionally high concentration compared with the others—such as sodium when added in the form of sodium bicarbonate, sodium carbonate, or sodium hydroxide as alkalinity supplements—then the cations may become unbalanced. Such unbalance may be relieved by increasing the concentration of the others, especially that of potassium.

The question of adequate trace nutrient availability for an industrial wastewater can be addressed in simple laboratory pilot studies with the wastewater of interest. The potential stimulation of anaerobic treatment by the addition of trace nutrient mixtures can aid greatly in this determination. This aspect is discussed well by Speece (1996).

Inhibitory Materials   Many materials can cause toxicity in all types of biological treatment. For two reasons, toxicity is, perhaps, a larger problem with anaerobic treatment than aerobic treatment. First, the concentrations of organic material treated are often so much higher, and with higher concentration of wastewater organic matter, the concentration of other materials, including those that are inhibitory, are likely to be higher as well. Second, the specific growth rates of the anaerobic microorganisms are much lower. Low specific growth rates mean that the biological safety factor that is economically acceptable is lower, thereby putting the anaerobic process at greater risk in general. In addition, recovery times are lengthened by the slower growth rates.

Toxicity is a relative term, as indicated in Figure 13.4. A great many materials are stimulatory at low concentration, have no impact at intermediate concentration, but cause inhibition to performance when present at high concentrations. This is true of most materials present in normal wastewaters, including simple salts. Microorganisms have some ability to adapt to inhibitory materials given time to adjust. Some microorganisms are less susceptible to some inhibitory materials than others, and so this problem can at times be addressed by finding a better culture to use. Because microbial communities differ and can adapt, it is difficult to draw firm conclusions about toxic concentrations of given materials.

Controlling Toxicity in General From a control standpoint, inhibitory materials need to be reduced in concentration in some manner to below a toxic threshold. Table 13.4 summarizes methods that may be used to control inhibitory materials. Elimination of the toxic material from the waste stream and dilution of the waste stream (e.g., with another waste stream) are approaches that might be used in any situation. Dilution may be quite expensive, as it could result in a larger reactor to achieve a given treatment efficiency. Other methods to control toxicity are specific to the given toxicant.

Toxicity normally comes from a soluble material, one that is available to the microorganisms. If the available concentration of an inhibitory species can be decreased by some procedure, such as precipitation or strong-complex formation, then inhibition can be prevented within the reactor. In this manner, toxicity of some heavy metals, such as copper or zinc, can be removed by addition of sulfide, which forms an insoluble precipitate with these metals. The easiest way to put sulfides into the

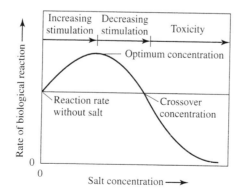

Figure 13.4 The effect of concentration of a typical inhibitory compound on reaction rate in biological processes. SOURCE: McCarty, 1964c.

Table 13.4   Possible methods to control
toxicity in anaerobic treatment

1. Remove toxic material from waste stream
2. Dilute waste so toxicant is below toxic threshold
3. Form insoluble complex or precipitate with toxicant
4. Change form of toxicant through pH control
5. Add material that is antagonistic to toxicant

process is by the addition of sulfate, which is biologically reduced to sulfide in the reactor. The addition rate must be carefully controlled, because too much sulfate reduction reduces methane formation and can lead to an excess buildup of sulfides, which are themselves inhibitory. In fact, the simplest way to eliminate sulfide toxicity is to precipitate it with a metal, such as through addition of ferrous iron.

Inhibition by synthetic detergents, such as the anionic linear alkyl benzene sulfonates, can be eliminated through the addition of cationic quaternary ammonia compounds, which form a complex with the anionic detergents. The presence of calcium can complex and reduce the toxicity of long-chain fatty acids, such as oleic acid, which can then be eliminated through biodegradation. A change in pH can alter the toxicity of materials. For example, ammonia is toxic primarily in the NH_3 form, not its more prevalent conjugate acid, NH_4^+. Maintaining a lower pH can reduce ammonia toxicity by shifting the acid-base equilibrium towards NH_4^+. On the other hand, volatile acids, such as propionic acid, are more inhibitory in their unionized acid form (CH_3CH_2COOH) than in their deprotonated base form ($CH_3CH_2COO^-$). Thus, a higher pH is more favorable here. Toxicity from sulfide (H_2S) can be reduced by gas stripping from the reactor. Thus, higher gas production and a lower pH (to shift the acid-base equilibrium to the acid $H_2S(g)$ form) can be beneficial in reducing its concentration below an inhibitory level.

Precipitation, complex formation, and stripping occur during treatment of many wastes, good examples being primary and secondary sludges from municipal wastes, which contain very high concentrations of synthetic detergents, fatty acids, and heavy metals. However, some industrial wastewaters do not have the same breadth of materials able to complex or precipitate inhibitory materials. The potential toxicity of materials in wastewaters to anaerobic treatment can be determined best through laboratory bioassay procedures, such as the anaerobic toxicity assay (Owen et al., 1979).

Salts Toxicity   Some industrial wastewaters have relatively high concentrations of normal alkali and alkaline earth salts, and this can cause inhibition to the anaerobic process. Indeed, if one attempts to control very high volatile acid concentrations through the addition of sodium bicarbonate or other sodium-containing base, inhibition can result from the resulting high salt concentration. Such inhibition is related to the cation part of the salt, rather than the anion. Table 13.5 is a summary of concentrations of various common cations that may cause inhibition. Also listed in the table are concentrations that are stimulatory. One phenomena associated with salt toxicity

Table 13.5   Common cations, stimulatory and inhibitory concentration ranges, mg/l

Cation	Stimulatory	Moderately Inhibitory	Strongly Inhibitory
Sodium	100–200	3,500–5,500	8,000
Potassium	200–400	2,500–4,500	12,000
Calcium	100–200	2,500–4,500	8,000
Magnesium	75–150	1,000–1,500	3,000

| SOURCE: McCarty, 1964c.

is the antagonistic effect. Here, if a cation such as sodium is present in an inhibitory concentration, this inhibition might be relieved if another cation, such as potassium, is added. If the stimulatory concentrations of the various cations listed in the table are present, then they will help reduce the extent of inhibition caused by any of the other cations present at a moderately inhibitory concentration.

Ammonia Toxicity   Ammonia is produced anaerobically from the degradation of proteinaceous wastes, as indicated in Example 13.3. Ammonia, which is a base, combines with carbon dioxide and water to form ammonium bicarbonate, the bicarbonate being the natural pH buffer. However, if the protein concentration is too high, such as it might be with treatment of slaughterhouse or piggery waste that contain much urine, the ammonia concentration resulting from their treatment may be excessively high and cause ammonia toxicity.

It is ammonia, not the ionized ammonium ion, that more often causes inhibition. An ammonia concentration of about 100 mg/l caused inhibition in acetate-fed anaerobic systems (McCarty and McKinney, 1961). The NH_4^+ nitrogen concentration found to cause inhibition was much higher, about 3,000 mg/l. At high ammonia nitrogen concentrations, whether NH_3 or NH_4^+ is more inhibitory depends upon system pH. The normal equilibrium between these two species is (K_a is for 35 °C):

$$NH_4^+ = H^+ + NH_3 \qquad K_a = 5.56(10^{-10}) \qquad \textbf{[13.24]}$$

The distribution of the two species is related to pH by

$$pH = 9.26 + \log \frac{[NH_3]}{[NH_4^+]} \qquad \textbf{[13.25]}$$

For a pH of 7.0, $[NH_3] = 0.0055[NH_4^+]$, and ammonium toxicity is generally more important. However, if the pH were 8.0, ammonia toxicity is more severe. Here, $[NH_3] = 0.055[NH_4^+]$.

Of interest is the fact that the release of ammonia nitrogen from protein increases the bicarbonate alkalinity concentration (see Equation 13.5). As a result, the pH can increase, too. It is not uncommon with high concentration proteinaceous wastes that treatment pH can be quite high, approaching 8.0. If the total ammonia nitrogen concentration ($NH_3 + NH_4^+$) were 2,000 mgN/l and the pH were 8.0, the NH_3 nitrogen concentration would be 110 mgN/l, which is in the inhibitory range. The response of the anaerobic treatment system to such inhibition is to reduce the rate of consumption of the volatile acid intermediates, which in turn drives the pH lower so that inhibition is reduced. One piece of evidence for NH_3 inhibition is an increase in the volatile acid concentration as the total ammonia nitrogen concentration increases. If enough ammonia is released, the NH_4^+ concentration builds up and reaches its inhibitory level, which cannot be relieved by a lower pH. The best control of ammonia toxicity is to reduce the waste's N concentration through dilution. Hydrochloric acid addition to reduce pH somewhat could also be used if NH_3 toxicity was the problem, but it does not work if NH_4^+ is the cause of toxicity.

Sulfide Toxicity   Sulfide toxicity is a common problem with wastewaters containing high concentrations of sulfate, which is used preferentially as an electron acceptor in anaerobic wastewater treatment and converted to sulfide. Sulfide complexed with heavy metals, such as iron, zinc, or copper, is not toxic. It is the soluble form—primarily the unionized H_2S form—that is most inhibitory. Sulfide toxicity tends to become a problem when the soluble sulfide concentration reaches about 200 mg/l (McCarty, 1964c). Theoretically, 600 mg/l of sulfate produces 200 mg/l of sulfide. In reality, more sulfate must be reduced, because two of the chemical forms of sulfide are removed from solution.

H_2S, one of the sulfide species formed, is a relatively insoluble gas and is partially stripped from solution through normal gas production. For a normal pH of anaerobic treatment, almost all soluble sulfide is either H_2S or HS^-. Although never in a high concentration in solution, S^{2-} also is important, as it forms precipitates with many metals. The various reactions of sulfide are:

$$M^{2+} + S^{2-} = MS_{(s)}, \quad [M^{2+}][S^{2-}] = K_{sp} \quad K_{sp}(Fe^{2+}) = 6 \times 10^{-18} \quad \textbf{[13.26]}$$

$$H_2S = H^+ + HS^-, \quad \frac{[H^+][HS^-]}{[H_2S]} = K_{a,1}, \quad pK_{a,1} = 7.04 \quad \textbf{[13.27]}$$

$$HS^- = H^+ + S^{2-}, \quad \frac{[H^+][S^{2-}]}{[HS^-]} = K_{a,2} \quad pK_{a,2} = 12.9 \quad \textbf{[13.28]}$$

$$H_2S(aq) = H_2S(g), \quad \frac{[H_2S(g)]}{[H_2S(aq)]} = K_H = 13 \text{ atm/mol (35 °C)} \quad \textbf{[13.29]}$$

M^{2+} represents a divalent heavy metal. The value for K_{sp} shown here is for ferrous iron, which is more soluble than the sulfides of zinc, copper, nickel, mercury, or most other of the divalent heavy metals. However, ferrous sulfide is still sufficiently insoluble that it complexes and precipitates essentially all sulfide in the system if ferrous iron is present in excess. Once all of the heavy metals are complexed with sulfide according to Equation 13.26, then concentrations beyond this point remain in solution and partition between the gas phase as indicated by the series of Equations 13.27 to 13.29.

The distribution of sulfide between various phases is similar to that of the carbon dioxide/bicarbonate/carbonate interactions. Equation 13.27 indicates that a neutral pH gives about an even distribution between H_2S and HS^- ($pK_{a,1} = 7.04$). H_2S is a gas of intermediate solubility as indicated by Equation 13.29, but can be stripped from solution. Higher gas production per unit of wastewater volume and a lower pH favor stripping from solution. Figure 13.5 indicates the portion of total soluble sulfides that remain in solution as a function of pH and gas production.

Besides being toxic to the anaerobic microorganisms, H_2S also is a toxic and odorous gas that poses health and aesthetic problems for workers and those who live around anaerobic systems. Hydrogen sulfide in anaerobically produced gases not only causes odors, but it is also corrosive and quite detrimental to the operation of combustion engines used in energy recovery. Furthermore, hydrogen sulfide is oxidized to sulfur dioxide during combustion, creating an air pollution problem.

Sulfide production in anaerobic systems is not all bad. Sulfide serves as an essential nutrient for biological growth, and a certain amount is required for successful operation. It also helps maintain a low oxidation-reduction potential, which is required for successful treatment operation. In addition and as described in the next section, it helps prevent toxicity that might result from excessive concentrations of heavy metals. A balance needs to be obtained such that the benefits from sulfide production are realized, while the problems are kept to a minimum.

Heavy Metals Toxicity and Iron Protection   Heavy metals are often blamed for the failure of anaerobic systems, and in many systems such blame may be well justified. Copper, nickel, zinc, cadmium, and mercury can be inhibitory to anaerobic microflora at concentrations of less than 1 mg/l. Obviously, the best method for keeping heavy-metal toxicity from becoming a problem is to prohibit their introduction into the wastewater stream. However, if this cannot be adequately controlled, then the formation of an iron sulfide buffer system may be an approach to consider.

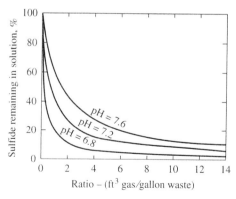

Figure 13.5   Fraction of produced soluble sulfide remaining in a CSTR liquid phase as a function of pH and gas production during anaerobic treatment. 1 ft³/gallon = 7.47 m³ gas/m³ waste. SOURCE: McCarty, 1964c.

Iron, which is commonly present in quite high total concentration, is generally nontoxic, which is fortunate, as it can serve a useful role when complexed with sulfide to reduce the toxicity of the other heavy metals. The main protector against heavy metal toxicity in anaerobic systems is sulfide, which forms highly insoluble nontoxic complexes. As noted above, however, sulfides can themselves be harmful if too high in concentration. The beneficial effect that ferrous iron can produce in the system is that it too forms an insoluble precipitate with sulfide. Iron addition can help keep sulfide under control in order to reduce the problem aspects that it may cause.

Since iron does not form as insoluble a precipitate as do most other metals, iron sulfide can serve as a buffer to reduce the toxicity caused by the sudden introduction of a toxic heavy metal. The other metal displaces the iron from the sulfide and, in this manner, is converted to an insoluble, noninhibitory form. The degree to which such an FeS buffer works in an anaerobic system depends on the total sulfides and the relative concentrations of heavy metals in the treatment system.

Organic Toxicants   Organic compounds that can be toxic to anaerobic systems are common in some industrial wastes, such as that from the chemical industry. Many organic chemicals that are toxic to anaerobic systems at higher concentrations can serve as a source of food for anaerobic microorganisms at lower concentrations. By proper design, such chemicals can be treated readily in an anaerobic system. Phenol is one such chemical. There are also other potentially toxic organic chemicals that can be biotransformed in the anaerobic system to nonharmful chemicals. An example here is chloroform. Dealing with a waste containing a potentially toxic organic chemical requires an understanding of its potential for biotransformation in the process. An excellent discussion of the toxicity of organic compounds to anaerobic system is given by Speece (1996), who provides much more detail on this problem than space allows here.

Blum and Speece (1991) conducted a comparative analysis of the toxicity of a large number of organic compounds to unacclimated mixed cultures. They used an acetate-using methanogenic culture for this analysis and found concentrations that resulted in a 50 percent reduction in gas production. Their results for selected compounds are summarized in Table 13.6. These results should be used with caution, as the organisms were not given time to adapt to inhibition. Since the cultures were unacclimated, the compounds probably were not degraded following addition. Despite this cautionary note, the study indicates concentrations that could be a problem in a treatment system. It is apparent that toxicity thresholds vary widely, but frequently they are less than 100 mg/l and in some cases are in the low mg/l range.

Many models have been used to describe the effect of the inhibitors on methane fermentation (Speece, 1996). The Haldane model (Chapter 3) has commonly been used to describe reaction kinetics for toxic organic compounds that can also be used as a source of energy anaerobically. Other models may be more appropriate when the toxicant is not a growth substrate, but affects the kinetics for microorganisms using the normal substrates in an anaerobic system. The reader should consult Chapter 3 for more information on how to incorporate inhibition in modeling a microbial system. One lesson is that a table of toxicity concentrations, like Table 13.6, cannot necessarily be used for developing mathematical models of relationships between concentration and rate, even though it is very useful for indicating concentration ranges where inhibition should be considered.

Table 13.6 Concentrations of organic compounds that reduce gas production by 50 percent (IC50) with nonacclimated acetate-utilizing methanogens

Toxicant	mg/L	Toxicant	mg/L
Hydrocarbons		**Halogenated Alkanes**	
Alkanes		Chloromethane	50
Cyclohexane	150	Methylene chloride	7
Octane	2	Chloroform	1
Decane	0.35	Carbon tetrachloride	6
Undecane	0.61	1,1-Dichloroethane	6
Dodecane	0.23	1,2-Dichloroethane	25
Pentadecane	0.09	1,1,1-Trichloroethane	0.5
Heptadecane	0.03	1,1,2-Trichloroethane	1
Nonadecane	0.01	1,1,1,2-Tetrachloroethane	2
Aromatics		1,1,2,2-Tetrachloroethane	4
Benzene	1,200	Pentachloroethane	11
Toluene	580	Hexachloroethane	22
Xylene	250	1-Chloropropane	60
Ethylbenzene	160	2-Chloropropane	620
Phenols		1,2-Dichloropropane	180
Phenol	2,100	1,2,3-Trichloropropane	0.6
o-Cresol	890	1-Chlorobutane	110
p-Cresol	91	1-Chloropentane	150
2,4-Dimethylphenol	71	Bromomethane	4
4-Ethylphenol	240	Bromodichloromethane	2
Alcohols		1,1,2-Trichlorotrifluoroethane	4
Methanol	22,000	**Halogenated Alkenes**	
Ethanol	43,000	1,1-Dichloroethene	8
1-Propanol	34,000	1,2-Dichloroethene	19
1-Butanol	11,000	t-1,2-Dichloroethene	48
1-Pentanol	4,700	Trichloroethene	13
1-Hexanol	1,500	Tetrachloroethene	22
1-Octanol	370	1,3-Dichloropropene	0.6
1-Decanol	41	5-Chloro-1-pentyne	44
1-Dodecanol	22	**Halogenated Aromatics**	
Ketones		Chlorobenzene	270
Acetone	50,000	1,2-Dichlorobenzene	150
2-Butanone	28,000	1,3-Dichlorobenzene	260
2-Hexanone	6,100	1,4-Dichlorobenzene	86
Miscellaneous		1,2,3-Trichlorobenzene	24
Catechol	1,400	1,2,3,4-Tetrachlorobenzene	20
Resorcinol	1,600	2-Chlorotoluene	53
Hydroquinone	2,800	2-Chloro-p-xylene	89
2-Aminophenol	6	2-Chlorophenol	160
Isopropylether	4,200	3-Chlorophenol	230
Ethylacrylate	130	4-Chlorophenol	270
Butylacrylate	150	2,3-Dichlorophenol	58
Acetonitrile	28,000	3,5-Dichlorophenol	14
Acrylonitrile	90	2,3,4-Trichlorophenol	8
Carbon disulfide	340	2,3,5,6-Tetrachlorophenol	0.1
2-Aminophenol	6	Pentachlorophenol	0.04
4-Aminophenol	25	2,2-Dichloroethanol	18
2-Nitrophenol	12	2,2,2-Trichloroethanol	0.3
3-Nitrophenol	18	3-Chloro-1,2-propanediol	630
4-Nitrophenol	4	2-Chloropropionic Acid	0.01
2,4-Dinitrophenol	0.01	Trichloroacetic Acid	<0.001

SOURCE: Blum and Speece, 1991.

13.4   PROCESS KINETICS

Anaerobic treatment requires the action of many different groups of microorganisms to bring about the conversion of complex organic materials into methane gas. Chapter 1 provides an overview of the various steps involved in anaerobic treatment of complex materials, which include hydrolysis of complex proteins, carbohydrates, and fats to simpler molecules such as amino acids, sugars, and fatty acids. These simpler molecules then are fermented to form fatty acids and hydrogen. The fatty acids are oxidized further to acetate and hydrogen. Finally, two different methanogenic groups convert acetate and hydrogen to methane. Mathematical models that describe the rate of each step in the overall process can be highly complex and difficult to apply in a practical setting.

However, if the rate-controlling step in the overall process can be determined, then simpler models can be developed and are useful for design and operation. Figure 13.2 simplifies for practical purposes the complex series of steps into two basic steps, the hydrolysis and fermentation step and the methane-formation step. Practical mathematical models also can be simplified into two rate-limiting steps. One is the hydrolysis of complex organic materials, and the other is the overall methane fermentation of fatty acids and hydrogen to methane. For some organic materials, hydrolysis tends to be rate-limiting: for example, lignocellulosic materials like newsprint, grasses, corn stover, or straw. However, with most industrial and municipal wastes, the last step—the conversion of fatty acids to methane—tends to be rate-limiting. Factors important to these two potential rate-controlling steps are considered in this section on Process Kinetics.

13.4.1   TEMPERATURE EFFECTS

Temperature affects reaction rates considerably. In anaerobic treatment, the slow growth rate of microorganisms most critical to the process makes temperature all the more important for reactor design. Growth rates in general roughly double for each 10 °C rise in temperature within the usual mesophilic operational range from 10 to 35 °C. Growth rates generally do not change between 35 and 40 °C, but denaturation of proteins at higher temperatures slows growth rates for mesophiles. However, different mixed cultures adapted to thermophilic temperatures have optimum temperatures in the 55 to 65 °C range. Thermophiles do not function as well at the intermediate temperature of 40 to 45 °C as do mesophilic organisms. Thus, one must make the decision to operate at the lower mesophilic range with an optimum temperature of around 35 °C or in the thermophilic range with a temperature optimum of 55 to 60 °C.

With dilute wastewaters at ambient temperature, the methane produced may be insufficient to raise the wastewater temperature, and operation at ambient wastewater temperature may be the economical option. With more concentrated wastewaters that produce larger volumes of methane per unit volume of reactor or with high-temperature wastewaters, operation at the mesophilic optimum of 35 °C or at thermophilic temperatures is the best option. With the latter, the rates are typically 50 to

100 percent higher than at the optimum mesophilic temperature. Thus, the advantage of the higher temperature is faster reactions and smaller required tank volumes. The disadvantages are a greater energy cost to maintain the higher temperature and the risk of a rapid loss in treatment capacity due to failure of the reactor heating system.

The change in rate of a chemical reaction with temperature is generally expressed with the Arrhenius equation:

$$\frac{d \ln k}{dT} = \frac{E_a}{RT^2} \qquad \textbf{[13.30]}$$

which indicates that the change in the natural log of the rate constant with change in temperature is equal to some activation energy for the reaction, E_a, divided by the gas constant R and the square of the absolute temperature. If the Arrhenius equation is integrated over the temperature range from T_1 to T_2, then the following results:

$$\ln \frac{k_2}{k_1} = \frac{E_a(T_2 - T_1)}{RT_2 T_1} \qquad \textbf{[13.31]}$$

Equation 13.31 is generally simplified for use. Since E_a is generally not known, the ratio of E_a/R is taken to be a constant. Over the temperature range of interest, whether mesophilic or thermophilic, the product $T_2 T_1$ will not vary greatly and may be taken to be constant. Thus, Equation 13.31 can be simplified to either of the forms:

$$k_2 = k_1 e^{\phi(T_2 - T_1)} \qquad \textbf{[13.32]}$$

or

$$k_2 = k_1 \phi'^{(T_2 - T_1)} \qquad \textbf{[13.33]}$$

from which

$$\phi' = e^{\phi} \quad \text{or} \quad \phi' \approx 1 + \phi \qquad \textbf{[13.34]}$$

Table 13.7 summarizes the different ϕ values obtained by several researchers for the kinetic parameter of anaerobic systems. A ϕ of 0.07 corresponds to ϕ' of 1.07 and gives a doubling of the parameter for each 10 °C rise in temperature. The values provided are for the utilization of volatile fatty acids, which is often the rate-limiting step in the process. The coefficients affected by temperature are the maximum rate of substrate utilization ($\hat{q}$), the rate of organism decay (b), the affinity constant (K), and the maximum growth rate ($\hat{\mu}$).

Buhr and Andrews (1977) reviewed temperature impacts on volatile-acid-utilization rates over the range of mesophilic and thermophilic temperatures. They concluded that the change in growth rate with temperature for organisms involved in methane production from volatile acids could be approximated by a single equation that includes separate terms for growth and decay:

$$\hat{\mu}_{net} = \hat{\mu}_{T_1} e^{\phi_\mu(T_2 - T_1)} - b_{T_1} e^{\phi_b(T_2 - T_1)} \qquad \textbf{[13.35]}$$

Based on the literature, they assigned appropriate coefficients for the methanogens:

$$\hat{\mu}_{net} = 0.324 e^{0.06(T_2 - 35)} - 0.02 e^{0.14(T_2 - 35)} \qquad \textbf{[13.36]}$$

Table 13.7   Temperature coefficient ϕ for determining the effect of temperature on various rate constants for anaerobic treatment

Rate Constant	Substrate	ϕ	Temperature Range °C	Reference
$\hat{\mu}$	volatile acids	0.06	15–70	(Buhr and Andrews, 1977)
$\hat{q}$	volatile acids	0.077	15–35	(Lin et al., 1987)
	acetate	0.11	37–70	(van Lier et al., 1996)
	primary sludge	0.035	20–35	(O'Rourke, 1968)
b	volatile acids	0.14	15–70	(Buhr and Andrews, 1977)
	acetate	0.30	37–70	(van Lier et al., 1996)
	primary sludge	0.035	20–35	(O'Rourke, 1968)
K	volatile acids	−0.077	25–35	(Lawrence and McCarty, 1969)
	volatile acids	−0.061	15–35	(Lin et al., 1987)
	primary sludge	−0.112	20–35	(O'Rourke, 1968)

The graph in Figure 13.6a illustrates how the maximum net growth rate varies with temperature according to this relationship. The decay rate overtakes the growth rate at temperatures greater than 60 °C. The growth rate at the optimum thermophilic temperature of 60 °C is about 2.5 times that at 35 °C. Although the curve in Figure 13.6 is useful for comparing the growth rates within the mesophilic and thermophilic ranges, the smooth transition between them is inaccurate. The mesophilic and thermophilic microorganisms are completely different from each other. The mesophiles die off above about 40 °C, and the thermophiles pick up above about 50 °C.

The graph in Figure 13.6b presents the reciprocal of the maximum net growth rate, which is $[\theta_x^{\min}]_{\lim}$, the solids retention time in a reactor at which washout occurs. Except for the curve being continuous from the mesophilic range to the thermophilic range, the information in the bottom graph is important for the design of treatment systems and will be discussed in the next section.

13.4.2   REACTION KINETICS FOR CSTR

Modeling the anaerobic treatment process can be greatly simplified if the rate-controlling step is first determined, and then the model is built around this step. The two most likely limiting steps in the anaerobic process are conversion of volatile acid to methane or hydrolysis of complex substrate. These two cases are handled separately in the following. First, we begin with simple soluble substrates, and then we proceed to more complex substrates, all within the framework of the CSTR, the easiest of the processes for making mass balances and illustrating the effect of process variables. Finally, we consider complex substrates to illustrate how process kinetics developed from simple substrates can be applied to the more complex materials.

Simple Soluble Substrates   Figure 13.7 illustrates a CSTR around which the mass balances for anaerobic treatment of simple substrates are applied. Here, the reaction kinetics developed in Chapter 3 apply. The master variable is the solids

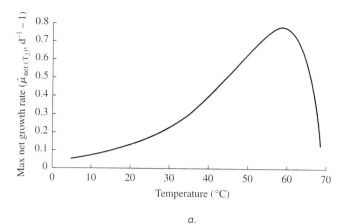

a.

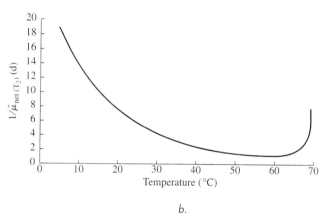

b.

Figure 13.6   Effect of temperature on the maximum growth rate and its reciprocal for volatile acid using methanogenic mixed cultures. SOURCE: Using formulation after Buhr and Andrews, 1977b.

retention time or θ_x. Lawrence and McCarty (1969) used a CSTR to evaluate the kinetics of methanogenesis from acetate, propionate, and butyrate, the major volatile acids intermediates formed in anaerobic treatment. Figure 13.8 presents the results obtained for 35 °C and the rate coefficients estimated from the results. These results indicate that washout or system failure occur at $\theta_x = 2.5$ to 4 d with each of these substrates. Experimental data in each case indicate that some substrate reduction occurred at θ_x less than the minimum value, but subsequent studies indicated this was the result of wall growth that prevented complete hydraulic washout of the organisms.

With these three important substrates, acetate is converted to methane in a single step, while propionate and butyrate are converted in three separate steps: The first is their fermentation to acetate and hydrogen, and the second and third are the independent conversion of these two substrates to methane. Nevertheless, propionate and butyrate conversion can be modeled adequately as if they were single substrates,

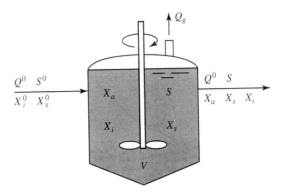

Figure 13.7   Schematic of anaerobic CSTR used for conducting mass balances and deriving equations for simulating performance.

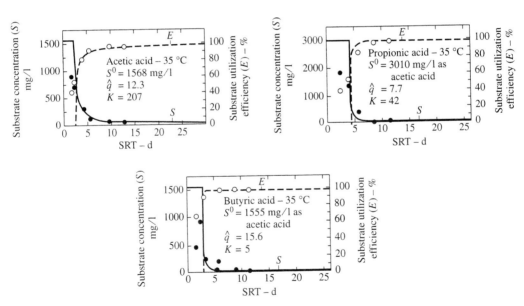

Figure 13.8   Effect of θ_x on reactor concentration at 35 °C for the methane formation stage with a. acetate, b. propionate, and c. butyrate. SOURCE: Lawrence and McCarty, 1969.

like acetate, because the three steps involved with propionate and butyrate are closely coupled through thermodynamics. When multiple steps can be handled as though they behave as one reaction, the modeling of the overall process is nearly as simple as it is for one simple soluble substrate.

Table 13.8   Rate coefficients reported for anaerobic treatment of volatile acids

Substrate	T °C	$\hat{\mu}$ d^{-1}	$\hat{q}$ mg/mgVSS$_a$ -d	K mg/l	b d^{-1}	Y mg/mg	Reference
Acetate	25	0.23	4.7	869	0.011	0.050	(Lawrence and McCarty, 1969)
	30	0.26	4.8	333	0.037	0.054	(Lawrence and McCarty, 1969)
	35	0.32	8.1	154	0.019	0.040	(Lawrence and McCarty, 1969)
	35	0.39	9.8	168	0.033	0.041	(Kugelman and Chin, 1971)
	35			220			(van Lier et al., 1996)
	40			560			(van Lier et al., 1996)
	45			320			(van Lier et al., 1996)
	50			220			(van Lier et al., 1996)
	55			820			(van Lier et al., 1996)
	60			1,150			(van Lier et al., 1996)
Propionate	25	0.50	9.8	613	0.040	0.051	(Lawrence and McCarty, 1969)
	35	0.40	9.6	32	0.010	0.042	(Lawrence and McCarty, 1969)
	40			120			(van Lier et al., 1996)
	45			60			(van Lier et al., 1996)
	50			60			(van Lier et al., 1996)
	55			86			(van Lier et al., 1996)
	60			140			(van Lier et al., 1996)
Butyrate	35	0.64	15.6	5	0.010	0.042	(Lawrence and McCarty, 1969)
	35			240			(van Lier et al., 1996)
	40			140			(van Lier et al., 1996)
	45			160			(van Lier et al., 1996)
	55			16			(van Lier et al., 1996)
	60			11			(van Lier et al., 1996)

1 Biomass is mg of VSS. Substrate mass is mg of the noted substrate.

Several studies have evaluated the kinetics of volatile-acids conversion. Coefficients determined from these various studies are summarized in Table 13.8. As noted in the last section, temperature has an important effect on the coefficients and needs to be considered in any models developed.

13.4.3 COMPLEX SUBSTRATES

O'Rourke (1968) produced within his Ph.D. dissertation one of the most widely referenced sets of data on the kinetics of degradation of complex organic materials. He never published these results in a journal, but many others have done so. O'Rourke used primary municipal wastewater sludge and a series of chemostats to study the complex of reactions involved. Primary sludge contains a mixture that is dominated by proteins, carbohydrates, and fats. With this mixture, most of which was in particulate form, it was not possible for O'Rourke to measure biomass production. However, he could evaluate the effect of θ_x on overall performance.

Figure 13.9 illustrates, for 25 °C, the effect of θ_x on the overall performance as documented by COD, volatile solids, and methane production, which are the

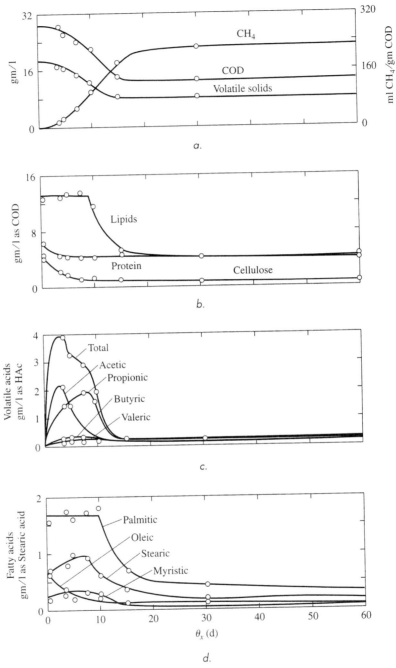

Figure 13.9   Effect of θ_x on a. overall *COD* and volatile solids removal and gas production, b. protein, cellulose, and lipid removal, c. volatile acid concentration, and d. fatty acid concentration for anaerobic treatment of primary municipal sludge in a *CSTR* at 25 °C. SOURCE: O'Rourke, 1968.

parameters of general interest for design and operation. Also shown are details for the categories of proteins, carbohydrates, lipids, and the individual volatile acids, as well as the longer chain fatty acids of which the lipids are mostly comprised. The lines connecting the data points is not model fits, but simply lines drawn through the points to help visualize the results. With θ_x greater than about 20 d, little change occurs in the concentration of any component. The residual organic matter persisting beyond this point is largely refractory and not usable by anaerobic organisms, at least within the time frame of normal interest in wastewater treatment. Some of this residue is the bacteria produced during treatment, but most represents organic matter originally present in the primary sludge. While much organic material remains for a large θ_x, it is quite stable and useful as a soil conditioner or as a slow release organic fertilizer.

The most interesting aspect is what occurs with θ_x less than 20 d. For lipid degraders, washout occurs at about 10 d, and this is generally true for the fermenters of volatile acids as well. Oleic acid tends to decrease in concentration at a lower θ_x, but this is a result of the hydrogenation of this unsaturated fatty acid to produce the saturated stearic acid. The decomposable portion of cellulose and protein degrades at θ_x much less than 10 d, suggesting that hydrolysis to simpler compounds occurs readily for this wastewater and is not a limiting factor in the overall conversion to methane. Fermentation of the resulting hydrolysis products to form volatile acids is indicated by their rise in concentration at θ_x of 3 to 5 d.

O'Rourke based his modeling development for overall COD reduction and methane production on the assumption that volatile-acid utilization was the rate-limiting step as far as washout of the critical methane-producing bacteria were concerned. However, he assumed that the affinity coefficient, K, could be represented by a composite K_c value for all the organic matter. This is based on the supposition that, for a given θ_x, a predictable concentration of each of the original organic compounds in the wastewater and each intermediate that may be formed remains. The residual concentration of each intermediate can be predicted based on the assumption that it is consumed by a single bacterial species. Then, its concentration is predictable from the normal solution of the CSTR mass balances, the various coefficients for that substrate, and θ_x. The COD is then the sum of the CODs of all the original and intermediate substrates:

$$S = \sum_{i=1}^{i=n} S_i = \sum_{i=1}^{i=n} \frac{K_i(1 + b_i\theta_x)}{\theta_x(Y_i\hat{q}_i - b_i) - 1} \qquad \textbf{[13.37]}$$

where S represents the total COD concentration of all the biodegradable organic matter remaining, and S_i represents the remaining or effluent concentration for substrate i out of the n different original substrates and intermediate compounds formed in the process. He also assumed, based upon literature data on kinetics of individual substrates, that S_i would be much smaller for most substrates compared with that for the volatile acids and the longer chain fatty acids, all of which tend to be the rate-limiting substrates for the process. He noted, based on the data within Figure 13.9, that these substrates tended to have similar washout times, which means that their values for Y_i, $\hat{q}_i$, and b_i must be similar. With these assumptions, Equation 13.37 was simplified to:

$$S = \frac{(1 + b\theta_x)}{\theta_x(Y\hat{q} - b) - 1} \sum_{i=1}^{i=n} K_i = \frac{K_c(1 + b\theta_x)}{\theta_x(Y\hat{q} - b) - 1} \qquad \textbf{[13.38]}$$

In this way, the values of K_i for the individual volatile and fatty acids are added together to give the composite value K_c. Equation 13.38 then appears with the same form as that for a single microorganism using a single substrate in a CSTR. The difference is that K_c tends to be much higher than values for K for single substrates, as it is a composite of K values.

The applicable equations and values for the coefficients evaluated by O'Rourke are summarized in Figure 13.10. A differentiation is made between growth coefficients related to substrate removal related to the rate-limiting step versus growth from consumption of the complete mixed waste. Equations for substrate removal include values for Y_a and b_a, which are the yield and decay coefficient for the microorganisms using the volatile acids. Equations relating to growth of total biomass use Y and b, which relate to the net overall growth and decay of the first-step and second-step organisms.

The distribution of COD between refractory organic matter in the sludge, X_i^0, and the microorganisms formed in the process, X_a and X_i, was not considered by O'Rourke, but it is included in Figure 13.10 for completeness and to be consistent with the theory of Chapter 3. With sludge, the majority of the COD is associated with suspended solids, as are the long chain fatty acids, which represent the major portion of K_c. Therefore, effluent COD is strongly controlled by the effluent volatile solids concentration, for which the COD/VSS ratio for the primary sludge was 1.61, a number close to that ratio of 1.42 for microorganisms.

Figure 13.11 compares experimental results to model calculations for effluent COD and reactor methane production. θ_x and temperature varied within the range of 20 to 35 °C. O'Rourke also evaluated performance at 15 °C, but decomposition of lipids was negligible, making treatment ineffective. Treatment at this low temperature may be possible for more readily degraded organic matter, but not for the primary municipal sludge.

The simplified model for steady-state operation of a well-mixed CSTR for the treatment of primary sludge captured the key trends quite satisfactorily. Figure 13.11 indicates that, with operation at long θ_x, on the order of 40 to 60 d, the difference in operation between 20 and 35 °C is not great. However, with operation at θ_x of 20 d or less, highly different results are produced. The time for washout is also quite different for the different temperatures. Operation at 35 °C provides for a higher degree of treatment at the lower θ_x values, and washout does not become significant until about 4 d.

High-rate anaerobic treatment of municipal wastewaters is typically conducted with θ_x of 15 to 20 d at 35 °C. Figure 13.11 indicates that operational results over this range are not much different. However, results do change significantly with operation below 10 d. Those advocating design at such short θ_x values risk system failure from

Primary Sludge Characteristics

$X_v^0 = X_i^0 + X_d^0$ (X_d represents biodegradable VSS)

$X_v^0 = 17.7$ g/l

$X_d^0 = 12.4$ g/l

$S^0 = \gamma X_d^0 = 20.0$ g COD/l

$S_T^0 = \gamma X_v^0 = 28.5$ g COD/l

$\gamma = 1.61$ g COD/g X_v

Reaction Coefficients

$\hat{q}_T = 6.67 \text{g/g} X_a \text{-d} \left(e^{0.035(T-35)}\right)$

$K_{cT} = 1.8 \text{ g/COD/l} \left(e^{0.12(35-T)}\right)$

$b_{aT} = 0.03 \text{ d}^{-1} \left(e^{0.035(T-35)}\right)$

$b_T = 0.05 \text{ d}^{-1} \left(e^{0.035(T-35)}\right)$

$Y_a = 0.04$ gX_a/g COD

$Y = 0.10$ gX_a/g COD

$f_d = 0.8$

Reactor Equations

$$S_T \text{ (g COD/l)} = K_{cT} \frac{1 + b_{aT}\theta_x}{Y_a\hat{q}_T - (1 + b_{aT}\theta_x)}$$

$$X_a \text{ (g/l)} = \frac{Y(S^0 - S)}{1 + b_T\theta_x}$$

$$X_d \text{ (g/l)} = \frac{S}{\gamma}$$

$$X_v \text{ (g/l)} = X_i^0 + X_d + X_a[1 + (1 - f_d)b_T\theta_x]$$

$$S_T \text{ (g COD/l)} = \gamma X_i^0 + \gamma X_d + 1.42 X_a[1 + (1 - f_d)b_T\theta_x]$$

$$r_{COD} \text{(g COD/l reactor-d)} = \frac{Q}{V}([S^0 - S] - 1.42X_a[1 + (1 - f_d)b_T\theta_x])$$

$$r_{CH_4} \text{(l methane/l reactor-d)} = \frac{22.4 \text{l}}{64 \text{ g COD}} r_{COD}$$

$$CH_4/COD \text{(l/g COD)} = \frac{\frac{22.4 \text{l}}{64 \text{ g COD}}(S_T^0 - S_T)}{S_T^0}$$

Figure 13.10 Summary of primary municipal sludge characteristics and relevant equations for modeling sludge anaerobic treatment in a CSTR between the temperature range of 20 to 35 °C. SOURCE: O'Rourke, 1968.

small perturbations to the system, such as organic or hydraulic overload, introduction of toxic materials, or loss of temperature control. The designer and operator should be cautious about operating so close to the brink of failure.

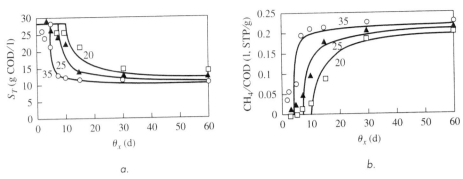

Figure 13.11   Comparison between experimental results (data points) and model simulations (lines) using data from O'Rourke (1968) and Figure 13.10 equations for a CSTR treating municipal wastewater sludge over the temperature range of 20 to 35 °C.

13.4.4   PROCESS OPTIMIZATION

Another objective that can be evaluated with our CSTR model is the COD throughput per unit volume of reactor, which is illustrated in Figure 13.12 for the primary municipal sludge studied by O'Rourke. Shown are model fits using Figure 13.10 equations and the data resulting from the laboratory studies. One θ_x for each temperature gives the peak in the stabilization rate or COD conversion to methane per day per unit

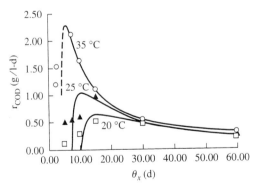

Figure 13.12   Relationship among θ_x, temperature, and COD destruction rate per unit volume of reactor for a CSTR treating municipal wastewater sludge after O'Rourke (1968). Lines represent simulations using Figure 13.10 equations and symbols the experimental data developed by O'Rourke.

volume of reactor. This peak is more pronounced and occurs at a lower θ_x for 35 °C than for 25 °C and 20 °C.

A question for design is what should be the design θ_x for anaerobic treatment of primary municipal-wastewater treatment sludge? A typical design used 20 d at an operating temperature of 35 °C, but Figures 13.9 and 13.11 tell us that a shorter θ_x could be used at this temperature. We consider the 20 d and determine what safety factors are implied by this design. The concept of safety factor, described in Chapter 5, is defined in terms of the limiting $\theta_x^{\min}$:

$$\theta_x^d = \text{SF}[\theta_x^{\min}]_{\lim} \qquad \textbf{[5.61]}$$

We now expand this concept to consider safety factors with respect to other θ_x criteria, for example, with respect to organism washout for a given waste ($\theta_x^{\min}$), to the maximum utilization rate per unit volume ($\theta_x^{\max}$), or to some desired removal efficiency at steady-state (θ_x^{eff}). These various operational θ_x values can be represented as follows:

$$[\theta_x^{\min}]_{\lim} = \frac{1}{Y_a \hat{q} - b} \qquad \textbf{[13.39]}$$

$$\theta_x^{\min} = \frac{1}{\dfrac{Y_a \hat{q} S^0}{K + S^0} - b} \qquad \textbf{[13.40]}$$

$$\theta_x^{\max} = \frac{[\theta_x^{\min}]_{\lim}}{1 - \dfrac{K(1 + b[\theta_x^{\min}]_{\lim})}{K + S^0}} \qquad \text{(after O'Rourke, 1968)} \qquad \textbf{[13.41]}$$

$$\theta_x^{\text{eff}} = \frac{1}{\dfrac{Y_a \hat{q}_a S_{\text{eff}}}{K + S_{\text{eff}}}}, \quad \text{where} \quad S_{\text{eff}} = \left(1 - \frac{\% \text{ eff}}{100}\right) S^0 \qquad \textbf{[13.42]}$$

We can then use the equations developed from O'Rourke's data for primary sludge treatment and calculate implied safety factors with respect to the various values of θ_x defined by Equations 13.39 to 13.42.

To illustrate, we assume design θ_x of 20 d and a desired removal efficiency for biodegradable organic matter of 90 percent. Calculated values for the respective θ_x values from the above equations, together with the resulting implied safety factors, are summarized in Table 13.9. First, with respect to the desired treatment efficiency of 90 percent, the S.F. with operation at 35 °C is 2.2. If the operating temperature dropped to 25 °C, we could no longer maintain this removal efficiency with 20 d, although the system would not fail. Indeed, we can see that if the temperature decreased to 20 °C, we would be operating near a point where maximum rate of waste treatment per unit volume occurs. Returning to the 35 °C case, operation at 20 d provides a safety factor of 3.3 with respect to $\theta_x^{\max}$. From a review of Figure 13.12, it would appear obvious that we would never wish to operate at a θ_x less than $\theta_x^{\max}$, as the rate of waste decomposition per unit volume would decrease significantly. We would not achieve as much with our reactor as we could, say by simply not treating

Table 13.9   Implied safety factors for treatment of primary municipal wastewater sludge at a 20-d θ_x^d and removal efficiency for degradable organic material of 90 percent*

Temperature of Operation °C	$[\theta_x^{min}]_{lim}$ d	θ_x^{min} d	θ_x^{max} d	θ_x^{eff} d	Implied Safety Factors for $\theta_x^d = 20$ d			
					$\theta_x^d/[\theta_x^{min}]_{lim}$	$\theta_x^d/\theta_x^{min}$	$\theta_x^d/\theta_x^{max}$	$\theta_x^d/\theta_x^{eff}$
20	7.1	11	17	91	2.8	1.8	1.1	0.22
25	6.0	7.8	12	33	3.3	2.6	1.7	0.61
35	4.2	4.7	6.1	9.2	4.8	4.3	3.3	2.2

*Based on Data Obtained by O'Rourke (1968).

some of the waste so that we could operate at a longer θ_x. From the standpoint of optimizing the throughput of the reactor, θ_x^{max} would be a better reference point for applying the safety factor concept for a CSTR without organism recycle than would be $[\theta_x^{min}]_{lim}$, which is related to the washout.

13.4.5   REACTION KINETICS FOR BIOFILM PROCESSES

While reactors to which media are added for biofilm attachment are easily recognized as being "biofilm" processes, categorization of others as biofilm reactors may not be as obvious. For example, is the *UASB* process a suspended-growth or biofilm process? Because the size and density of *UASB* granules are large, the kinetics probably obeys biofilm kinetics more closely than suspended-growth kinetics. Likewise, the flow rates around particles is relatively slow, making mass transfer from the bulk liquid to the granule surface relatively slow, another characteristic of biofilm processes.

Reaction kinetics for biofilm anaerobic processes are more difficult to apply than for suspended-floc systems when the surface area of the biofilm systems is not easily determined and mixing is inadequate to prevent short-circuiting or dead zones. Another complication of biofilm kinetics with anaerobic systems is the numerous reaction stages and microorganisms involved. A waste stream organic may diffuse into a biofilm and there be converted into intermediates within the biofilm, and these intermediates then are subject to diffusion in two directions, one deeper into the biofilm and the other outward toward the bulk water. At the same time, microorganisms within the biofilm consume the intermediate. While all microorganisms responsible for the overall biodegradation of the complex substrates present to methane reside together within the biofilm, they are not necessarily located homogeneously throughout the biofilm.

As with suspended-growth systems, such as the CSTR, reaction kinetics is best simplified by focusing on the rate-limiting step in the reaction. For biofilm systems, this is often the rate of diffusion of the substrate into the biofilm. However, with relatively dilute wastewaters where desired detention times are on the order of a few hours to a day, hydrolysis of suspended particles can be rate-limiting when suspended organic substrates are input. For such materials, a hydrolysis model, such as the Contois model (Chapter 3), may be appropriate. Use with these more complex models, however, is quite limited. Designs for biofilm systems for industrial wastewaters are generally based on empirical pilot-scale systems, for which much information is available in the literature. The value of biofilm models is to understand better the variables of importance and the type of relationships that may be most useful for analysis of data obtained through empirical studies.

An example of the application of a biofilm model to empirical data is that by Bachmann, Beard, and McCarty (1985), who applied biofilm kinetics to a baffled reactor with recycle. Such a reactor is in essence a series of UASBs. They assumed that, because of rapid gas production within each chamber, a given chamber acted as a completely mixed reactor, and thus they used an equation equivalent to Equation 4.42:

$$S_i = S_{i-1} - \frac{J_i a_i V_i}{Q_i} \qquad \textbf{[13.43]}$$

where S_i represents the substrate concentration in chamber i, and S_{i-1} represents the substrate concentration in the waste stream entering chamber i. They applied this model to each chamber in a manner similar to the procedure outlined in Example 4.6. Assuming that the biofilm was deep, they used the variable-order model from Equation 4.64:

$$J_i = C_i S_i^{q_i} \qquad \textbf{[13.44]}$$

which is coupled to Equations 4.65 through 4.68.

Equation 13.43 was solved iteratively to obtain the solution for the first chamber, and then this approach was applied to each subsequent chamber in order to obtain the overall solution for a given case. As with O'Rourke's models for sludge, methanogenesis of volatile acids was rate limiting, and the coefficients given in Table 13.8 were used. The mass transfer coefficient for movement of substrate from bulk liquid to the biofilm surface was estimated from a dimensionless correlation similar to that of Equation 4.30. Additional key assumptions were the fraction of biomass representing active microorganisms (1.8 to 5.7 percent), the average diameter of a biofilm particle (2.0 mm), and the biomass concentration within the particle (100 mg/cm^3). The model was applied for volumetric organic loading rates for the total reactor varying from 10.4 to 36 kg COD/m^3-d and for detention times from 4.8 to 18.7 h. A constant influent COD to the reactor of 8 g/l was used. The synthetic wastewater studied was entirely soluble and represented a mixture of a proteinaceous substrate (nutrient broth) and a simple carbohydrate (sucrose). A comparison of model fits for the series of 6 chambers used and at 4 organic loading rates is illustrated in Figure 13.13. The figure shows that the removal of COD was rapid in the first chamber and slowed for the downstream chambers. Increasing loading gave a higher effluent COD concentration.

The single-reaction model can be very useful for describing performance over the range of loading, although its applicability for full-scale application has not been well evaluated. The results in Figure 13.14 suggest that COD removal efficiency depends on the organic loading and the influent COD concentration. For the same volumetric loading, a higher influent concentration allows a larger removal percentage. This is highly important in design and is consistent with empirical findings that removal efficiencies with relatively dilute wastewaters (COD < 4,000 mg/l) are not generally very high except at very low organic loading.

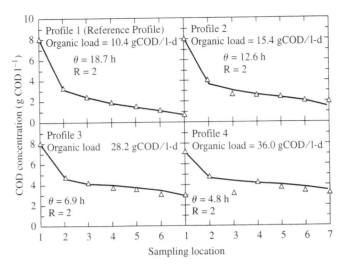

Figure 13.13   Comparison between biofilm model simulations (lines) and experimental data (symbols) for an anaerobic baffled reactor treating a soluble wastewater at different loading rates with a constant influent *COD* of 8 g/l and *L* = 0.05 cm. SOURCE: Bachmann and McCarty, 1985.

13.4.6   KINETICS WITH HYDROLYSIS AS LIMITING FACTOR

Many wastes to be treated anaerobically contain organic materials for which the rate-controlling step is hydrolysis. This is the case with many lignocellulosic materials, such as grasses, agricultural crop residues, and most wood products, including newsprint. Enzymes cannot easily access cellulose when it is present in a lignocellulosic matrix. Cellulose can be hydrolyzed readily when in a free form, such as in high-quality paper or cotton fibers (see Photo 13.4, page 620). Lignin, a complex material consisting of interconnected aromatic groups, is not easily degraded anaerobically, if at all, and acts as a barrier to hydrolysis of cellulose and hemicellulose. Although the cellulose and hemicellulose in materials having a low lignin content, like grasses, is close to 100 percent biodegradable, the reaction is quite slow. With softwoods, which have a high lignin content, only a small fraction of the cellulose and hemicellulose is available anaerobically in reasonable time periods. Some wastes, such as municipal refuse, are mixtures of lignocellulosic materials and more readily biodegradable organic matter; so, for one portion of the wastes, hydrolysis is rate-limiting, while for the other, volatile-acid utilization is rate-limiting. A mixed model that is appropriate to each fraction may be used for such cases.

Eastman and Ferguson (1981) included hydrolysis of complex organic material and methane fermentation from volatile acids in their overall model for anaerobic treatment of primary municipal sludges using a CSTR. Thus, their model was more comprehensive than that by O'Rourke and was capable of indicating acid production for θ_x values causing washout of methanogens. The hydrolysis model they used

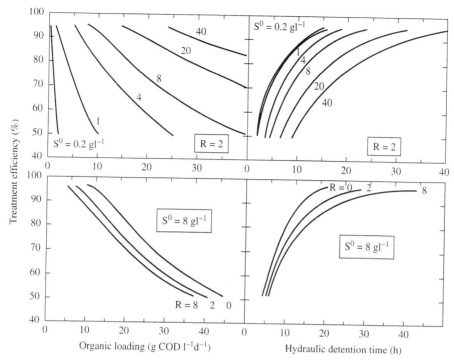

Figure 13.14   Anaerobic baffled reactor performance from biofilm model simulations showing the predicted effect of influent wastewater COD concentration and hydraulic detention time on *COD* removal efficiency. SOURCE: Bachmann and McCarty, 1985.

was a simplification of the Contois model in which the concentration of hydrolyzing bacteria is very high such that $BX_a \gg S$, for which the Contois Equation 3.73 reverts to a simple first-order reaction for hydrolysis:

$$-r_{hyd} = k_{hyd} S_{hyd} \qquad\qquad [13.45]$$

where S_{hyd} represents the concentration of substrate that is subject to hydrolysis and k_{hyd} represents the first-order rate constant for such hydrolysis. The value they found for primary sludge was $k_{hyd} = 3/d$. Gossett and Belser (1982) studied the anaerobic digestion of waste activated sludge and found that hydrolysis was rate-limited. Thus, the model O'Rourke had found to be satisfactory for primary sludge was inadequate for waste activated sludge. Indeed, they found that k_{hyd} was almost an order of magnitude lower than Eastman and Ferguson had found for primary sludge (Table 13.10). The important practical finding is that anaerobic digestion of waste activated sludge is much different from digestion of primary sludge.

Tong, Smith, and McCarty (1990) studied the anaerobic treatment of a variety of complex lignocellulosic materials and also found that the O'Rourke model was not satisfactory for this material. They found an excellent agreement between the first-order hydrolysis model and degradation rates for particulate cellulosic materials.

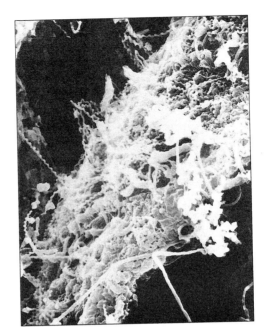

Photo 13.4   Scanning electron photomicrograph of mixed methanogenic culture digesting a cellulosic fiber (Courtesy of Xingyang Tong).

Table 13.10   First-order anaerobic hydrolysis rate constants for primary municipal sludge, activated sludge, and various cellulosic and lignocellulosic materials at 35 °C

Material	k_{hyd} d^{-1}	Reference
Primary sludge	3.0	Eastman and Ferguson, 1981
Activated sludge	0.22	Gossett and Belser, 1982
Filter paper	0.25	Tong et al., 1990
BW 200	0.20	Tong et al., 1990
Corn stover	0.14	Tong et al., 1990
Wheat straw no. 1	0.086	Tong et al., 1990
Wheat straw no. 2	0.088	Tong et al., 1990
Napier grass	0.090	Tong et al., 1990
Wood grass	0.079	Tong et al., 1990
Newspaper	0.049	Tong et al., 1990
White fir	0.039	Tong et al., 1990

They found hydrolysis rates for pure cellulose (filter paper and BW 200) were about the same as found by Gossett and Belser (1982) for activated sludge. However, the hydrolysis rates for lignocellulosic materials (all materials below BW 200 in Table 13.10)

were very low. Indeed, these hydrolysis rates were so low that models for overall degradation of such materials did not need to consider volatile acid rates at all. From a simple mass balance with a CSTR, the effect of θ_x on degradation can readily be obtained from mass balance:

$$\frac{S_{hyd}}{S_{hyd}^0} = \frac{1}{1 + k_{hyd}\theta_x}$$ [13.46]

Anaerobic treatment of a slurry of lignocellulosic materials was studied in CSTR and batch reactors (Tong and McCarty, 1991). In the batch reactors, seed from a CSTR reactor treating the respective lignocellulosic material was added along with the lignocellulosic material slurry, and the decomposition with time was monitored. Beyond a small minimum of seed, the rate of reaction was independent of seed (i.e., microorganism) concentration. Thus, Equation 13.45 was appropriate for the cases studied. With the batch reactor, which acts like an ideal plug-flow reactor, the appropriate mass balance equation is integrated to give:

$$\frac{S_{hyd}}{S_{hyd}^0} = 1 - e^{-k_{hyd}t}$$ [13.47]

where t represents the time of reaction. They found that the rate of hydrolysis was temperature dependent (Figure 13.15). Here, a culture adapted at 35 °C was used. The rate at 20 °C was much slower than at higher temperatures, although with time the rate improved. This suggests that adapted cultures are better at hydrolysis. The optimum hydrolysis was at 40 °C. At a higher temperature, hydrolysis was initially similar, but it decreased significantly after a few days, perhaps through some adverse affect of high temperature on the hydrolytic enzymes or their production.

Of particular interest in this study was a comparison between treatment effectiveness with a CSTR, versus that with a batch reactor. The input material was wheat straw, for which k_{hyd} was 0.1/d. The results of this comparison are given in

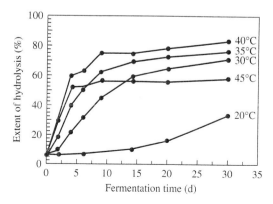

Figure 13.15   Effect of temperature and time on the hydrolysis of cellulose and hemicellulose by an anaerobic mixed culture. SOURCE: Tong et al., 1990.

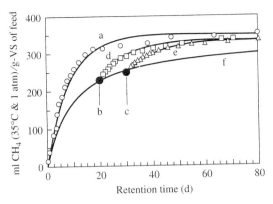

Figure 13.16   Comparison of the methane conversion fractions for CSTR and batch reactions with wheat straw, which has $K_{hyd} = 0.1/d$. The text explains the points and lines. SOURCE: Tong and McCarty (1991).

Figure 13.16. Two lines passing through zero are shown; the upper (a) represents the model for batch treatment (Equation 13.47) together with batch experimental data, and the lower one (f) is for a CSTR (Equation 13.46), with experimental data for θ_x of 20 and 30 d as points b and c, respectively. The great difference between the CSTR and batch kinetics is obvious. Batch reaction gave much more hydrolysis and methane formation. The comparisons between experimental results and the simple models of Equations 13.46 and 13.47 are excellent. In addition, lines d and e show the results of removing a portion of the mixed fluid slurry from the two CSTRs and placing this mixture in batch reactors. The resulting calculated result from Equation 13.47 and experimental data also show an excellent fit. The conclusion from this study is that with slow kinetics and hydrolysis limitation, batch reactors are much better than CSTRs. Here, sequencing batch reactors would be very appropriate if one wished to achieve the maximum rate of biodegradation per unit reactor volume.

13.5   SPECIAL FACTORS FOR THE DESIGN OF ANAEROBIC SLUDGE DIGESTERS

The most common application of methanogenic treatment is anaerobic sludge digestion. A digester must accomplish two goals. First, the biodegradable volatile suspended solids must be hydrolyzed. This first step reduces the mass of solids that need to be dewatered and disposed of. The hydrolysis step also is the precursor for achieving the second goal, which is stabilization of the BOD_L to methane. Extensive experience is available to guide design, and this is very valuable, since sludges are complicated mixtures that are difficult to handle and treat. In this section, we review five key factors of digester design: process-loading criteria, mixing, heating, gas collection, and performance.

13.5.1 LOADING CRITERIA

High-rate digesters are heated and well mixed. Thus, they are to behave as a CSTR. For high-rate mesophilic digestion, a temperature of 35 °C normally is used, because it provides the optimal balance between increasing reaction rates and incurring added costs. The primary design criterion must be the *detention time*, which equals θ_x for a CSTR. An absolute minimum value is 10 d to allow stable methanogenesis, and 15 to 25 d are typical. A longer θ_x improves process stability, lowers the net sludge production (which must be disposed of), and increases the methane-gas yield. A second criterion is the *volatile-solids loading rate*, which is computed as $QX_v^\circ/V = X_v^\circ/\theta$. The normal range for high-rate digestion is 1.6 to 4.8 kg VSS/m^3-d. Although the volatile-solids loading rate was established empirically, it has a sound basis. First, it recognizes that the hydrolysis step is the first step towards stabilization. In some cases, hydrolysis is rate-limiting. If the volatile-solids loading rate is too high, hydrolysis and, ultimately, methane stabilization are incomplete. Second, an excessive concentration of solids in the digester can hinder mixing, and the volatile-solids loading rate precludes such a high buildup for most applications. For design purposes, the digester volume is computed from the detention time and the volatile-solids loading rate; the larger volume controls the design.

HIGH-RATE DIGESTER SIZING   The size of a digester is to be computed based on both criteria: $\theta_x = 20$ d and a volatile-solids loading rate of 3 kg VSS/m^3-d. The sludge has a VSS concentration of 5 percent, which is approximately 50 kg/m^3. The flow rate is 100 m^3/d. | **Example 13.6**

The θ_x criterion gives $\theta = 20$ d. The volatile-solids loading rate (VSLR) gives

$$\theta = X_v^0/\text{VSLR} = (50 \text{ kg VSS/m}^3)/(3 \text{ kg VSS/m}^3\text{-d}) = 16.7 \text{ d}$$

Thus, the detention time controls, and $\theta = 20$ d. The volume is then

$$V = Q \cdot \theta = 100 \text{ m}^3/\text{d} \cdot 20 \text{ d} = 2{,}000 \text{ m}^3$$

It is proposed to thicken the sludge to 100,000 kg VSS/m^3, which also reduces the flow rate to 50 m^3/d. How would that affect the size?

The only change is to X_v^0 for the VSLR. Recomputation of θ from the VSLR gives

$$\theta = 100/3 = 33.3 \text{ d}$$

Now, the θ from the VSLR controls, and $\theta = 33.3$ d. The volume is

$$V = 50 \text{ m}^3/\text{d} \cdot 33.3 \text{ d} = 1{,}670 \text{ m}^3$$

Thus, the thickening reduces the digester volume by 17 percent.

Historically, *"standard rate" digesters* were designed without mixing or heating. Sludge settles at the bottom to form a thickened (and stabilized) sludge layer. A somewhat clarified supernatant is removed from the top of the digester. The most common design criterion is the volatile-solids loading rate. It is in the range of 0.5 to 1.6 kg VSS/m^3-d. Another historical design criterion that is completely restricted to municipal sewage-treatment sludges is the per capita loading, or the digester volume required per person connected to the sewerage system. Typical per capita volumes are 65 m^3/1,000 persons for primary sludge alone and 150 m^3/1,000 persons for primary plus waste activated sludge.

The solids retention time of a standard-rate digester can be approximated from measurements of the incoming and outgoing flows. The incoming sludge flow rate is Q, the outgoing digested-sludge removal rate is Q^d, and the digester volume is V. Then, a semi-empirical relationship for θ_x is (Metcalf & Eddy, 1979)

$$\theta_x \approx \frac{V}{Q - \frac{2}{3}(Q - Q^d)} \qquad \textbf{[13.48]}$$

If the digested sludge is thickened to T times the concentration of the incoming sludge and the total solids mass is reduced by a fraction D by digestion, then Q^d is approximately equal to $Q \cdot (1 - D)/T$. This gives a relationship among the desired θ_x, the detention time ($\theta = V/Q$), the thickening (T), and the solids destruction (D) of

$$\theta_x \approx \theta \cdot \left(\frac{T}{\frac{2}{3} + \frac{1}{3} \cdot T - \frac{2}{3} \cdot D} \right) \qquad \textbf{[13.49]}$$

If the desired mean cell residence time is 30 d, the thickened sludge is 2 times more concentrated than the input sludge, and the total solids reduction is 33.3 percent, then θ is computed from

$$30 \text{ d} \approx \theta \cdot \left(\frac{2}{\frac{2}{3} + \frac{1}{3} \cdot 2 - \frac{2}{3} \cdot 0.333} \right)$$

which gives θ of about 17 d.

13.5.2   MIXING

High-rate digesters are mixed in order to improve mass transfer between the microorganisms and their substrates and to prevent formation of scum at the water level and sediments at the bottom. Digester mixing is accomplished by one or a combination of three ways.

1. The liquid contents are recirculated by pumping them from one location to another.

2. The digester gas is compressed and injected into the liquid through lances or diffusers.

3. Mechanical mixers, usually slow speed turbines, create liquid motion inside the digester.

No matter how the mixing is effected, the goal is to maintain a high enough liquid velocity so that all the solids remain in suspension. Popular in Europe are digesters that have an "egg shape" (see Photo 13.5), which is supposed to enhance mixing and minimize scum and sediment formation. More details can be found in Speece (1996).

13.5.3 HEATING

High-rate digesters are heated in order to increase hydrolysis and methanogenesis rates, as well as to hold digester temperature steady despite fluctuations in the temperature of the incoming sludge. The amount of energy required depends on the temperature differential with the incoming sludge, the heat losses through the digester's surfaces, and the degree to which the heat in the effluent can be recovered. In principle, the energy cost to heat a digester can be quite small, if the digester is well insulated and the effluent heat is recovered with heat exchangers.

The contents of the digester can be heated with three techniques.

1. The sludge is pumped through external heat exchangers and returned warm to the digester. This method is well suited for situations in which liquid recirculation is used for mixing.

2. Heat exchangers are located inside the digester.

3. Steam is injected through lances or diffusers. This approach works best when gas recirculation is used for mixing.

When heat exchangers are used, the temperature on the sludge side of the exchanger must not be too much greater than the temperature inside the digester. Very

Photo 13.5   Egg-shaped anaerobic sludge digester.

high temperature causes the proteinaceous material in the sludge to coagulate and precipitate on the exchanger surface, significantly reducing the heat exchange capacity. Thus, the surface area of a heat exchanger must be large to compensate for the modest temperature.

13.5.4 GAS COLLECTION

Most high-rate digesters in the United States have floating covers that move up and down as the gas production and removal rates fluctuate. The covers must be sealed so that air does not enter the digester, nor does digester gas escape. While loss of a valuable energy resource is one reason to avoid leakage around the covers, the more important reason is to avoid creating an explosive atmosphere where the CH_4 and O_2 mix. Whether the cover floats or is fixed, gas storage must be provided in low-pressure tanks with floating covers or in pressurized tanks.

The collected gas is cleaned of H_2S and H_2O before it is combusted to generate steam or electricity. Typical digester gas is about 70 percent CH_4, with most of the rest of the gas being CO_2. Since the energy content of pure methane is 35,800 kJ/m^3 (at STP), the digester gas has a net heat value of about 25,000 kJ/m^3. The heat value of methane is roughly the same as natural gas, which has a heat content of approximately 37,000 kJ/m^3.

13.5.5 PERFORMANCE

The performance of an anaerobic sludge digester can be measured against its two goals: destroying biodegradable solids and stabilizing the BOD to methane. The normal performance measure for solids destruction is the percent volatile-solids reduction. The values can vary greatly and depend on the quality of the input sludge and the solids retention time. A high value of volatile solids reduction is 67 percent, and it would be associated with a large θ_x and an input sludge that has not undergone prior biological treatment (e.g., the sludge is from primary sedimentation). However, the volatile-solids reduction normally is substantially lower, from 30 to 50 percent. These lower reductions are due in part to employing modest values of θ_x. Even more important is when the input sludge already is partially digested, as is the case when waste activated sludge is sent to anaerobic digestion. The volatile-solids reduction is lowest for waste activated sludge from extended-aeration processes, in which the large solids retention time enhances endogenous decay of the biomass within the activated sludge process. Thus, volatile solids entering digestion are already partially depleted of biodegradable solids.

The stabilization of the BOD_L is measured by the methane gas production rate, as illustrated in Example 13.1. If the digester is operating stably (i.e., with little buildup of volatile acids), the volatile-solids removal should be directly proportional to the methane generation rate. Volatile solids normally have a COD content of 1.4 to 1.6 g COD/g VSS. Thus, the mass rate of VSS destruction is easily related to the rate of BOD_L stabilization and methane generation. Using a conversion factor of

1.5 g BOD_L/g VSS, we obtain

BOD_L Stabilization Rate (kg BOD_L/d) = 1.5 g BOD_L/g VSS destroyed

$$\cdot \text{ volatile-solids reduction rate (kg VSS/d)}$$

Methane Generation Rate (m^3 at STP/d) = 0.525 m^3 CH_4/kg VSS destroyed

$$\cdot \text{ volatile-solids reduction rate (kg VSS/d)}$$

13.6 BIBLIOGRAPHY

Bachmann, A.; V. L. Beard; and P. L. McCarty (1985). "Performance Characteristics of the Anaerobic Baffled Reactor." *Water Research* 19(1), pp. 99–106.

Blum, D. J. W. and R. E. Speece (1991). "A Database of Chemical Toxicity to Environmental Bacteria and Its Use in Interspecies Comparisons and Correlations." *Research J. Water Pollution Control Federation* 63(3), pp. 198–207.

Buhr, H. O. and J. F. Andrews (1977). "Review Paper, The Thermophilic Anaerobic Digestion Process." *Water Research* 11(2), pp. 129–143.

Eastman, J. A. and J. F. Ferguson (1981). "Solubilization of Particulate Organic Carbon During the Acid Phase of Anaerobic Digestion." *J. Water Pollution Control Federation* 53(3), pp. 352–366.

Gossett, J. M. and R. L. Belser (1982). "Anaerobic Digestion of Waste Activated Sludge." *J. Environmental Eng. Division, ASCE* 108(EE6), pp. 1101–1120.

Jewell, W. J. (1987). "Anaerobic Sewage Treatment." *Environmental Sci. & Tech.* 21(1), pp. 14–21.

Kugelman, I. J. and K. K. Chin (1971). "Toxicity, Synergism, and Antagonism in Anaerobic Waste Treatment Processes." In F. G. Pohland, Ed., *Anaerobic Biological Treatment Processes,* American Chemical Society, Washington, D.C., pp. 55–90.

Lawrence, A. W. and P. L. McCarty (1969). "Kinetics of Methane Fermentation in Anaerobic Treatment." *J. Water Pollution Control Federation* 41(2), pp. R1–R17.

Lettinga, G.; A. F. M. van Velsen; W. de Zeeuw; and S. W. Hobma (1979). Feasibility of the Upflow Anaerobic Sludge Blanket (UASB)—Process. In *National Conference on Environmental Eng.,* American Society of Civil Engineers, New York, p. 35.

Lettinga, G.; A. J. B. Zehnder; J. T. C. Grotenhuis; and L. W. Hulshoff Pol (1988). *Granular Anaerobic Sludge; Microbiology and Technology.* Wageningen, The Netherlands: Centre for Agricultural Publishing and Documentation.

Lin, C. Y.; T. Noike; K. Sato; and J. Matsumoto (1987). "Temperature Characteristics of the Methanogenesis Process in Anaerobic Digestion." *Water Sci. & Technology* 19, p. 299.

McCarty, P. L. (1964a). "Anaerobic Waste Treatment Fundamentals, Part I, Chemistry and Microbiology." *Public Works* 95(September): pp. 107–112.

McCarty, P. L. (1964b). "Anaerobic Waste Treatment Fundamentals, Part II, Environmental Requirements and Control." *Public Works* 95(October): pp. 123–126.

McCarty, P. L. (1964c). "Anaerobic Waste Treatment Fundamentals, Part III, Toxic Materials and Their Control." *Public Works* 95(November): pp. 91–94.

McCarty, P. L. (1964d). "Anaerobic Waste Treatment Fundamentals, Part IV, Process Design." *Public Works* 95(December): pp. 95–99.

McCarty, P. L. (1981). "One Hundred Years of Anaerobic Treatment." In D. E. Hughes, Ed., *Anaerobic Digestion 1981,* Amsterdam: Elsevier Biomedical, pp. 3–22.

McCarty, P. L. and R. E. McKinney (1961). "Salt Toxicity in Anaerobic Digestion." *J. Water Pollution Control Federation* 33, pp. 399–414.

McCarty, P. L. and D. P. Smith (1986). "Anaerobic Wastewater Treatment." *Environmental Sci. & Technology* 20(12), pp. 1200–1206.

Metcalf & Eddy, Inc. (1979). *Wastewater Engineering: Treatment/Disposal/Reuse,* 2nd ed. New York: McGraw-Hill.

O'Rourke, J. T. (1968). *Kinetics of Anaerobic Treatment at Reduced Temperatures.* Stanford University.

Owen, W. F., et al. (1979). "Bioassay for Monitoring Biochemical Methane Potential and Anaerobic Toxicity." *Water Research* 13, pp. 485–492.

Parkin, G. F. and W. F. Owen (1986). "Fundamentals of Anaerobic Digestion of Wastewater Sludges." *J. Environmental Eng.,* 112(5), pp. 867–920.

Sawyer, C. N.; P. L. McCarty; and G. F. Parkin (1994). *Chemistry for Environmental Engineering,* 4th ed. New York: McGraw-Hill.

Schroepfer, G. J., et al. (1955). "The Anaerobic Contact Process as Applied to Packinghouse Waste." *Sewage and Industrial Wastes,* 27, pp. 460–486.

Speece, R. E. (1983). "Anaerobic Biotechnology for Industrial Wastewater Treatment." *Environmental Sci. & Tech.* 17(9), pp. 416A–427A.

Speece, R. E. (1996). *Anaerobic Biotechnology for Industrial Wastewaters.* Nashville: Archae Press.

Stander, G. J. (1966). "Water Pollution Research—A Key to Wastewater Management." *J. Water Pollution Control Federation* 38, p. 774.

Suidan, M. T.; I. N. Najm; J. T. Pfeffer; and Y. T. Wang. (1988). "Anaerobic Biodegradation of Phenol: Inhibition Kinetics and System Stability." *J. Environmental Eng.* 114(6), pp. 1359–1376.

Tong, X. and P. L. McCarty (1991). "Microbial Hydrolysis of Lignocellulosic Materials." In R. Isascson, Ed. *Methane from Community Wastes.* London: Elsevier Applied Science, pp. 61–100.

Tong, X.; L. H. Smith; and P. L. McCarty (1990). "Methane Fermentation of Selected Lignocellulosic Materials." *Biomass* 21, pp. 239–255.

van Lier, J. B.; J. L. S. Martin; and G. Lettinga (1996). "Effect of Temperature on the Anaerobic Thermophilic Conversion of Volatile Fatty Acids by Dispersed and Granular Sludge." *Water Res.* 30(1), pp. 199–207.

Young, J. C. and B. S. Yang (1989). "Design Considerations for Full-Scale Anaerobic Filters." *J. Water Pollution Control Federation* 61(9), pp. 1576–1587.

Young, J. C. and P. L. McCarty (1969). "The Anaerobic Filter for Waste Treatment." *J. Water Pollution Control Federation* 41, pp. R160–R173.

13.7 PROBLEMS

13.1. You are to design an anaerobic contact process. The influent values are:

$$Q = 1,000 \text{ m}^3/\text{d}$$
$$S^0 = 1,500 \text{ mg BOD}_L/\text{l}$$
$$X_v^0 = X_i^0 = 50 \text{ mg VSS/l}$$

The following kinetic parameters were estimated from lab studies and stoichiometry:

$$\hat{q} = 10 \text{ g BOD}_L/\text{g VSS}_a\text{-d}$$
$$K = 100 \text{ mg BOD}_L/\text{l}$$
$$b = 0.05\text{d}^{-1}$$
$$Y = 0.2 \text{ g VSS}_a/\text{g BOD}_L$$
$$f_d = 0.8$$

The design factors are:

Safety factor $= 20$

MLVSS $= 2,500$ mg/l

$$X_v^r = 10,000 \text{ mg VSS/l}$$
$$X_v^{\text{eff}} = 50 \text{ mg VSS/l}$$

(a) Calculate $[\theta_x^{\min}]_{\lim}$
(b) Calculate θ_x^d
(c) Calculate the concentration of substrate, S, in the effluent.
(d) Calculate U (Equation 6.9) in kg/kg VSS$_a$-d.
(e) Calculate the kg of substrate removed per day.
(f) Calculate the kg/day of active, inactive, and total volatile solids produced in the biological reactor.
(g) Calculate the kg of active, inactive, and total VSS in the reactor. (Hint: θ_x is the same for active and inactive mass.)
(h) Calculate the hydraulic detention time, θ.
(i) Calculate the ratio X_a/X_v (Hint: θ_x is the same for X_a and for X_v.) What is X_a in the reaction tank?
(j) Calculate the sludge wasting flow rate (Q^w) if the sludge is wasted from the return line. Calculate the kg/day of active, inactive, and total VSS that must be wasted per day. (Remember to account for the solids in the influent and effluent.)
(k) Calculate the required recycle flow rate (Q^r) and the recycle ratio, R.
(l) Calculate the volumetric loading rate in kg/m^3-d.
(m) Assume that the active VSS in the effluent exerts a BOD. What is the BOD$_L$ of the solids in the effluent?
(n) Compute the effluent SMP concentration.
(o) Calculate the minimum influent N and P requirements in mg/l and kg/d.

13.2. You are to treat a wastewater that is contaminated with 100 mg/l of SO_4^{2-} as S. You will do the treatment biologically by sulfate reduction, in which SO_4^{2-}-S is reduced to H_2S-S, which is stripped to a gas phase. You do not need to be concerned about the stripping. The electron donor will be acetate (CH_3COO^-), which you will need to add, say in the form of vinegar. You know the following about the wastewater:

flow rate = 1,000 m^3/d

sulfate = 100 mg/l as S

volatile suspended solids = 0

From previous tests and theory, you have estimated the following parameters for the rate-limiting substrate, which will be acetate (expressed as COD):

$f_s^0 = 0.08$

$\hat{q} = 8.6$ g COD/g VSS_a-d

$b = 0.04/d$

$K = 10$ mg COD/l

$f_d = 0.8$

N-source = NH_4^+-N

You also have selected these design criteria:

- use a solids-settling and recycle system
- have $\theta_x = 10$ d and $\theta = 1$ d
- have an effluent sulfate concentration of 1 mg S/l

You must provide the following crucial design information.

(*a*) Compute the effluent concentration of acetate (in mg/l of COD).

(*b*) Compute (using stoichiometric principles) the input concentration of acetate (in mg/l as COD) needed to take the effluent sulfate concentration from 100 to 1 mg S/l. [Note that the formula weight of S is 32 g/mol.]

(*c*) Compute the concentrations (in mg VSS/l) of active, inert, and total volatile suspended solids.

(*d*) Compute the waste-sludge mass flow (in kg VSS/d) and volumetric flow rate (in m^3/d) if the effluent VSS is 5 mg/l and the recycle VSS is 5,000 mg/l.

(*e*) Compute the recycle ratio, R.

(*f*) Based on the (normal) multisubstrate approach, determine the effluent concentration of SMP.

(*g*) Do you see any problems with the design? If yes, explain the cause of any problems and ways to overcome them.

13.3. Submerged filters have been used for methane fermentation. In this problem, the volume of a reactor required to achieve 90 percent removal of acetate, the major degradable organic in a wastewater, using a submerged filter is to be determined.

The submerged filter is to be considered as a CMBR, which is a reasonable assumption for methane fermentation because of gas mixing by gas evolution. Thus, the concentration of the rate-limiting substrate is the same throughout the reactor and equal to the effluent concentration. It will be assumed to equal 10 percent of the influent concentration. Assume the following conditions apply:

$$\hat{q} = 8 \text{ mg BOD}_L/\text{mg VS}_a\text{-d} \qquad X_f = 20{,}000 \text{ mg VS}_a/\text{l}$$
$$K = 50\text{mg BOD}_L/\text{l} \qquad L = 0.015 \text{ cm}$$
$$Y = 0.06 \text{ mg VS}_a/\text{mg BOD}_L \qquad D = 0.9 \text{ cm}^2/\text{d}$$
$$b' = 0.02/\text{d} \qquad D_f = 0.8 \text{ D}$$
$$b_{\text{det}} = 0.01/\text{d}$$
$$Q = 15 \text{ m}^3/\text{d} \qquad h = 0.8 \text{ (liquid hold-up)}$$
$$S^0 = 2{,}000 \text{ mg BOD}_L/\text{l} \qquad a = 1.2 \text{ cm}^2/\text{cm}^3$$
$$X_i^0 = 0$$

Using the steady-state-biofilm model:

(a) Determine the required reactor volume, V, of the reactor in m^3.
(b) Determine the hydraulic detention time (V/Q) in hours.
(c) Estimate the effluent active suspended solids concentration, X_a, in mg/l.
(d) Estimated the effluent inert organic suspended solids concentration, X_i, in mg/l.
(e) Estimate the total effluent volatile suspended solids concentration, X_v, in mg/l.

13.4. A CSTR has a volume of 20 m^3 and is used to treat a fatty-acid wastewater with a flow rate 2 m^3/d and a BOD$_L$ of 11,000 mg/l by methane fermentation. What is the methane production rate in m^3/d at standard temperature and pressure (STP)?

13.5. A wastewater with flow rate of 5,000 l/d and containing soluble organic material only with a BOD$_L$ of 6,000 mg/l is being treated by methane fermentation at 35 °C in a fixed-film reactor with a volume of 2,500 l. The reactor effluent BOD$_L$ is found to be 150 mg/l. Estimate the total biofilm surface area in the reactor. The wastewater organic matter is primarily acetic acid (a fatty acid). Assume the reactor acts like a deep-biofilm $(S_w = 0)$ completely mixed system. Also, for acetate assume D equals 0.9 cm^2/d, K is 50 mg BOD$_L$/l, $D_f = 0.8$ D, L equals 0.01 cm, $\hat{q}$ is 8.4 g BOD$_L$/g VS$_a$-d, $b = 0.1$/d, and X_f equals 20 mg VS$_a$/cm^3.

13.6. Fill out the following table to indicate what impact a small increase in each of the variables in the left-hand column would have on the operating characteristics indicated in the four columns to the right for a good operating (relatively large SF) anaerobic CSTR without recycle (methane fermentation) treating an industrial waste at 30 °C. Assume all other variables in the left-hand column remain the same. Also, for this problem, assume that you have an operating plant in which the reactor volume V remains fixed. Use: $(+)$ = increase, $(-)$ = decrease, and (0) = no change, and $(?)$ as undetermined.

Variable	θ_x (d)	Operating Characteristic		
		Total Effluent Solids Concentration (X^e, mg/l)	Substrate Removal Efficiency (%)	Methane Production (m³/d)
T (°C)				
$\hat{q}$				
Q^0				
S^0				
b				
K				
X_i^0				
Y				

13.7. What volume of methane in liters at STP (0 °C and 1 atm) should be expected per kg BOD_L removed by methane fermentation in a CSTR with detention time of 20 d if $f_s^0 = 0.11$ and $b = 0.05$/d?

13.8. In a laboratory study of the anaerobic biological treatment of a wastewater with a biodegradable COD of 12,000 mg/l, 98 percent removal of the COD was obtained, and 800 mg/l of biological cells were formed. Estimate the methane production resulting from this treatment in grams of methane per liter of wastewater.

13.9. You have been asked to design a treatment system to treat by methane fermentation a wastewater with flow rate of 10,000 m³/d and containing only soluble components including 4,000 mg/L BOD_L. The wastewater consists primarily of proteinaceous materials. One treatment system being considered is a CSTR with settling and recycle.

  (a) Make appropriate assumptions and calculations to determine a suitable reactor volume (m³) for this treatment system using a safety factor that is in the conventional range.

  (b) Estimate the sludge production rate from this treatment system.

  (c) Estimate the methane production from this system in m³/d at standard temperature (0 °C) and pressure (1 atm).

  Be sure to indicate all assumptions made and give a reason for using the values chosen.

13.10. A wastewater contains 0.2 M pyruvate. Estimate the volume of methane in liters (STP) that would be produced per liter of wastewater from anaerobic treatment in a CSTR with a detention time of 15 d, assuming an efficiency of wastewater treatment of 97 percent. In order to solve this problem use thermodynamics to estimate f_s^0 with an energy transfer efficiency of 60 percent, and assume the organism decay rate is 0.08 d^{-1}.

13.11. What tends to be the two rate-limiting steps around which design of an anaerobic treatment system should normally be based?

13.12. When an anaerobic reactor treating wastes by methane fermentation is temporarily overloaded or else is stressed by some perturbation, an increase in the concentrations of short-chain fatty acids, such as acetic acid and propionic acid, often results to the point where the pH may drop significantly. What is the microbiological reason for this occurrence?

13.13. Fill out the following table to indicate what impact a small increase in each of the variables in the left-hand column would have on the operating characteristics indicated in the four columns to the right for a well operating (relatively large SF) anaerobic CSTR without recycle (methane fermentation) treating an industrial waste at 30 °C. Assume all other variables in the left-hand column remain the same. Also, for this problem, assume that you have an operating plant in which the reactor volume V *remains fixed.* Use: $(+) =$ increase, $(-) =$ decrease, and $(0) =$ no change.

Variable	θ_x, (d)	Total Effluent Solids Concentration (X^e, mg/l)	BOD_L Removal Efficiency (%)	Methane Production (m^3/d)
		Operating Characteristic		
$\hat{q}$				
Q^0				
S^0				
b				
K_c				
X_i^0				
Y				

13.14. Consider a CSTR (without recycle) anaerobic digester for the following wastewater:

$Q^0 = 160 \ m^3$/d

$S^0 = 32,000 \ mg \ BOD_L$/l (mostly soluble)

Carbohydrates $= 50$ percent of degradable organics

Fatty acids $= 50$ percent of degradable organics

SF $= 5$

Temperature $= 35 \ °C$

K_c for fatty acids $= 1,000 \ mg$/l

Estimate the methane production from the reactor in m^3/d (at standard conditions of $0\,°C$ and 1 atm).

13.15. Anaerobic treatment of a waste containing primarily soluble carbohydrates is desired. The wastewater flow rate is $10^3\ m^3/d$, and the BOD_L is $10,000$ mg/l. What quantity of methane in m^3/d at STP (standard temperature of $0\,°C$ and pressure of 1 atm—thus, 1 mol of methane equals 22.4 l) would you estimate would be produced from a CSTR with θ_x of 15 d? Assume K_c is 600 mg BOD_L/l.

13.16. An anaerobic digester (CSTR without recycle and with methanogenesis) is being designed for sludge treatment at $35\,°C$ using typical reaction coefficients and with a safety factor of 8. Estimate the methane production in m^3/d at standard temperature and pressure (1 atm pressure and $0\,°C$) for a sludge having the following characteristics:

$$Q^0 = 20\ m^3/d$$
$$X_i^0 = 21,000\ mg\ VSS_i/l$$
$$X_v^0 = 33,000\ mg\ VSS/l$$
$$\gamma = 1.6\ g\ BOD_L/g\ \text{degradable volatile suspended solids}$$
$$S_{(sol)}^0 = 18,000\ mg\ BOD_L/l$$
$$k_{hyd} = 0.15d^{-1}$$
$$f_s = 0.14\ \text{(for the overall reaction)}$$
$$K_c = 4,000\ mg\ BOD_L/l\ \text{(for volatile fatty acids)}$$
$$b = 0.08\ d^{-1}\ \text{(for the overall reaction)}$$
$$b = 0.02\ d^{-1}\ \text{(for the volatile fatty acid organisms)}$$

13.17. If a methane fermentation reactor were suddenly overloaded by a slug introduction of a carbohydrate such as glucose, would each of the following likely to (1) increase, (2) decrease, (3) stay the same, or (4) be as likely to increase as to decrease?

(a) pH
(b) Volatile organic acids
(c) Alkalinity
(d) Total gas production rate
(e) Percentage methane
(f) Percentage carbon dioxide
(g) Percentage hydrogen
(h) Effluent COD
(i) Reactor temperature

13.18. A waste sludge with $X_v^0 = 30$ g/l, $X_i^0 = 10$ g/l, ratio of suspended solids COD to VSS mass $= 1.8$, and $Q^0 = 10^3\ m^3/d$, is treated in a CSTR without recycle by methane fermentation. $S_s^0 = 0$. If the hydrolysis rate $k_{hyd} = 0.15\ d^{-1}$, the rate coefficients for methane fermentation of fatty acids are $Y = 0.04$ g

VSS_a/gCOD, $\hat{q} = 8$ g COD/VSS_a-d, and $K_c = 2$ g COD/l, $\theta_x^d = 20$ d, and $f_e = 0.9$, what is the methane production rate from the reactor in m^3/d at standard conditions of 1 atm and 0 °C?

13.19. Determine the volume for a sludge digester if the design is based on (a) a volatile-solids loading factor of 3.8 kg VSS/m^3-d or (b) the volume-reduction method with a sludge residence time of 30 d. In both cases, the sludge has an SS concentration of 50 kg/m^3, a flow rate of 100 m^3/d, and a ratio of VSS to SS of 75 percent. You estimate that the VSS destruction will be 60 percent, and the output thickened sludge has SS = 80 kg/m^3.

13.20. Compute the pH for a digester liquor that has a total alkalinity of 2,400 mg/l as $CaCO_3$ and volatile fatty acids of 1,800 mg/l as HAc, if it is exposed to a CO_2 partial pressure of 0.6 atm. Is this digester "in trouble?" If yes, what must be done immediately to preclude a sudden drop in pH?

13.21. Two anaerobic digesters are treating the same sludge, but with different organic loading. Digesters A and B have total alkalinity of 2,750 mg/l as $CaCO_3$. Digester A has volatile fatty acids of 200 mg/l as HAc, while Digester B has 2,000 mg/l as HAc. If both have a partial pressure of CO_2 of 0.3 atm, which digester has the higher organic loading? Is it too high?

13.22. If the partial pressure of CH_4(g) is 0.6 atm, what is the molar concentration of CH_4(aq) in the reactor liquid? The Henry's constant is $1.5 \cdot 10^{-3}$ mol/l atm. What is the COD (mg/l) of the liquid-phase CH_4? What COD removal (mg/l) is needed for the effluent CH_4(aq) to equal 50 percent of the total CH_4 generated? 10 percent? 1 percent?

13.23. Use the steady-state biofilm model and a normalized loading curve to design an anaerobic biofilm reactor to treat a wastewater containing acetate. Due to the slow growth rate of the acetate-cleaving methanogens, you select a medium that has highly protected areas and gives a specific detachment rate of $b_{det} = 0.01$/d and a moderate specific surface area of 200 m^{-1}. The other relevant parameters are:

$$\hat{q} = 10 \text{ mg Ac/mg } VS_a\text{-d}$$
$$K = 400 \text{ mg Ac/l}$$
$$Y = 0.04 \text{ mg } VS_a/\text{mg Ac}$$
$$b = 0.02/\text{d}$$
$$X_f = 50 \text{ mg } VS_a/\text{cm}^3$$
$$L = 50 \ \mu\text{m} = 5 \cdot 10^{-3} \text{ cm}$$
$$D = 1.3 \text{ cm}^2/\text{d}$$
$$D_f = 1.04 \text{ cm}^2/\text{d}$$

Determine the design acetate flux, surface area, volume, and methane production rate (m^3/d) if you obtain 95 percent removal of acetate originally present at 46,900 mg/l (or 50,000 mg/l as BOD_L) in a flow of 400 m^3/d.

13.24. The approach of using primary and secondary sludge digesters is largely predicated upon the desire to thicken the digested sludge that exits the primary digester. Here, you analyze the solids concentration in a primary digester. What is the effluent VSS concentration and the percent VSS destruction for a primary, anaerobic digester if:

> Influent flow rate = 100 m^3/d
> Working liquid volume = 1,500 m^3
> Total influent VSS = 5 percent by weight
> Inert influent VSS = 1.75 percent by weight
> Hydrolysis rate coefficient = 0.3/d

Do not consider synthesis of new biomass.

13.25. You determine that the pH must be increased to 7.2 in order to alleviate pH inhibition. The current pH is 6.5, and you measure total alkalinity = 2,400 mg/l as $CaCO_3$, VFAs = 2,000 mg/l as HAc, and P_{CO2} = 0.5 atm. You can increase the pH by making the CO_2 partial pressure lower or by increasing the bicarbonate alkalinity. What partial pressure is needed to get pH = 7.2 if the alkalinity does not change? How much base must be added to get pH = 7.2 if the partial pressure of CO_2 stays at 0.5 atm? Which approach seems more feasible?

DETOXIFICATION OF HAZARDOUS CHEMICALS

Over 70,000 synthetic organic chemicals are in general use today (Schwarzenbach, Gschwend, and Imboden, 1993), and few have been tested for their effects on human health and the environment. We now know that several with widespread use in the past have adverse effects on the environment or human health; for this reason, they are banned from commercial production. Every year, we find additional nonnatural chemicals that produce harmful impacts, and new efforts are made to determine safe concentrations or to find alternatives with less adverse impacts.

New organic chemicals that are readily biodegradable generally do not have as many adverse effects as those that are highly resistant to biodegradation, as the latter tend to persist in the environment and spread throughout the world's ecosystems. Biodegradation generally leads to detoxification of a chemical. Thus, the biodegradation potential of a new compound is of great interest when judging its potential for environmental harm.

Among the first synthetic organic chemicals to create environmental problems were synthetic detergents, developed in Germany during World War II because of a lack of fats, the normal materials from which soap is made. The synthetic detergents have excellent properties for cleaning, particularly because they do not react with calcium and magnesium in hard water to cause unwanted precipitates. However, it soon became apparent that the introduction of synthetic detergents was creating a problem, because foam began to form on the tops of aeration tanks at activated sludge treatment plants (Figure 14.1). Similar foam formed in tap water drawn from groundwaters near septic tanks, and streams receiving treated wastewater discharges also began to produce suds. While these synthetic detergents were not toxic to humans, the aesthetic problems with foam led to bans on synthetic detergents in Germany and England, and banning in the United States looked imminent.

Researchers soon found the cause of the foaming problem. The *alkyl benzene sulfonate* (ABS) detergent is not readily biodegradable in the environment (Sawyer, 1958). Poor biodegradability is due primarily to the quaternary carbon that attaches the aromatic structure to the alkyl part of the ABS molecule. Other properties that led to resistance are the sulfonate group and the high degree of branching in the alkyl chain. The solution was to straighten the alkyl chain, which removed the branching and the quaternary carbon. The result was *linear alkylbenzene sulfonate* (LAS). When this alternative detergent was introduced in the 1960s, the foaming problem disappeared, and we continue to use LAS today.

Figure 14.1   Foaming at an activated sludge plant due to ABS detergents in the early 1960s (courtesy of Bruce M. Wyckoff).

While synthetic detergents perhaps were the first recognized problem of this type, the public's awareness of the potential for environmental problems from synthetic chemicals was greatly stimulated by the publication of *Silent Spring* by Rachel Carson (1962). She noted the growing incidences of problems to wildlife from the widespread use of pesticides. Her book created much controversy, but was one of the major forces that led to the environmental movement in the early 1970s. As a result of the concerns raised about chlorinated pesticides, most of those then in use are now either banned or greatly restricted in application. Since then, we have found problems with many other synthetic organic chemicals and have restricted their use. Among the problem chemicals are polychlorinated biphenyls (PCBs), halogenated solvents, and chlorofluorocarbons (CFCs). Indeed, halogenated organic compounds probably cause about half of the environmental problems attributable to organic pollution in the world today (Tiedje et al., 1993). One problem is that the halogen atoms block the sites normally attacked by enzymes.

Many naturally occurring compounds also are difficult to biodegrade and can be of health or environmental concern. Included here are the aromatic components of gasoline, such as benzene, ethylbenzene, toluene, and xylenes (BETX); polycyclic aromatic hydrocarbons (PAHs); and dioxins. Thus, while our focus is primarily on the synthetic organic chemicals used in industry and commerce, we also need to consider some naturally occurring compounds that human activities have introduced to the biosphere.

Before the problems with synthetic detergents and pesticides arose, the general philosophy among microbiologists was that all organic compounds (generally formed through biological activity) can and will be destroyed in the environment through biological activity. The concept (Alexander, 1965) was termed "microbial infallibility." The difficulties that arose in biodegradation of synthetic detergents and pesticides appeared to be exceptions to this concept, exceptions for which humans received the blame. The fact that many naturally formed organics also are resistant to biodegradation suggests that microbial infallibility is inaccurate as a general concept. Lignin, which is the second most abundant natural organic chemical class (second only to cellulosic materials), is highly resistant to biodegradation. Only fungi have enzyme systems that can destroy lignin, and this occurs quite slowly, as one can readily see from the longevity of fallen trees in the forest. Indeed, the coal and oil deposits and the organic material in shale indicate that many naturally produced organic materials are not rapidly mineralized back to carbon dioxide and water. This nonreactivity is indeed fortunate for human life, because the long-term burial of poorly degradable natural organic materials has led to the high oxygen content in our current atmosphere.

At the same time that we were realizing that many organic compounds are resistant to biodegradation, we also learned that most, if not all, synthetic organic compounds are biodegradable under some conditions. For example, the peroxidases that many fungi produce are powerful enzymes that can effect the degradation of most synthetic organic compounds in use today. Anaerobic microorganisms can attack carbon-halogen bonds not accessible by aerobic microorganisms.

The fact that biodegradation can occur under some conditions is not sufficient to protect humans and the environment from a hazardous organic compound. The compound must be biodegradable under most natural conditions, including aerobic and anaerobic environments. For instance, conditions necessary for fungal growth are not prevalent everywhere; indeed, fungi require oxygen and an ample supply of organic matter to thrive. And, the anaerobic attack on halogenated organics requires strongly reducing conditions.

Many natural and synthetic organic chemicals are readily biodegradable in natural environments and need not be considered environmental threats, providing their toxicity is also relatively low. Organic chemicals meeting a standard of ready biodegradability and low toxicity can enter the environment in significant amounts.

On the other hand, many synthetic and natural organic compounds present in soils, groundwaters, and surface waters are harmful and persist over time. Persistence generally means that conditions necessary for their biodegradation are not present. The question then is whether conditions necessary for biodegradation can be brought about so that the compound can be destroyed in a sufficiently short time that human health and the environment are protected. In that case, the compound is *detoxified*.

A corollary question of great importance is whether biological transformations reduce the hazard. In some cases, a more hazardous daughter product is formed. Common examples of this are the anaerobic biological transformation of the solvent tetrachloroethene (PCE) to vinyl chloride (VC), a known human carcinogen; the microbial conversion of secondary amines to produce highly toxic N-nitroso compounds; and the epoxidation of the pesticide aldrin to produce the more toxic dieldrin (Alexander, 1994). Thus, we need to know the pathway and completeness of the biotransformation. The ultimate goal is to mineralize organic compounds to inorganic elements, particularly carbon dioxide and water. Determining whether or not a compound can be mineralized is not an easy task, as some of the organic carbon of the compound is likely to be synthesized into cellular material. We must distinguish the innocuous organic compounds of cellular material from a hazardous organic by-product.

This chapter addresses the *potential for detoxification* of several classes of synthetic organic chemicals, as well as some that are naturally produced. It indicates the range of conditions required for biodegradation to occur and the pathways for biodegradation. A section on inorganic hazardous compounds also is included, as their biotransformations are of increasing interest.

14.1 FACTORS CAUSING MOLECULAR RECALCITRANCE

Alexander (1965) described factors causing organic compounds to resist biodegradation in the environment, which he termed "molecular recalcitrance:"

1. A structural characteristic of the molecule prevents an enzyme from acting.

2. The compound is inaccessible, or unavailable.

3. Some factor essential for growth is absent.

4. The environment is toxic.

5. Requisite enzymes are inactivated.

6. The community of microorganisms present is unable to metabolize the compound because of some physiological inadequacy.

The first factor suggests that, for biodegradation to occur, the *molecular structure* must be inherently biodegradable. The next four factors are related to *environmental conditions* that are necessary for biodegradation to take place. The last indicates that *appropriate organisms* must be present. If biodegradation does not occur naturally, then at least one of the factors listed is likely to be the cause. In order to bring about biodegradation, the factors imparting recalcitrance must be overcome. The objective of engineering strategies for enhancing biodegradation is to overcome the factor effecting recalcitrance.

The following sections address the three general areas of recalcitrance: molecular structure, environmental conditions, and microorganism presence.

14.1.1 MOLECULAR STRUCTURE

Most natural and synthetic organic compounds are biodegradable by microorganisms as part of their normal metabolism for energy and growth. A portion of the organic material, serving as a primary electron and energy source, is converted to oxidized end products through oxidation/reduction reactions. The other portion of the organic carbon is synthesized into cellular material. Such conversions can take place in aerobic environments, in which oxygen serves as the terminal electron acceptor. They also occur in anaerobic environments, in which nitrate, sulfate, carbon dioxide, other oxidized inorganic elements, or the organic compounds themselves serve as electron acceptors. When bacteria use the compounds of concern as a primary electron donor, the pathways, end products, and rates of reaction usually are well known. Then, engineered systems can be designed and operated according to the principles described in the earlier chapters of this book.

On the other hand, some natural and synthetic organic compounds have structures that impart resistance to biodegradation. In many cases, the compound can be biotransformed, but its structure prevents the biotransformation from becoming part of the microorganism's primary metabolic system for energy generation and growth. In these cases, the principles presented so far must be expanded.

14.1.2 ENVIRONMENTAL CONDITIONS

The top of Figure 14.2 illustrates contamination of soil and groundwater from a leaking storage tank, a common way by which contamination with liquids occurs (McCarty, 1990). As the liquid is pulled downward by gravity, residuals left behind contaminate the surface soil, the unsaturated (vadose) zone, and finally the aquifer

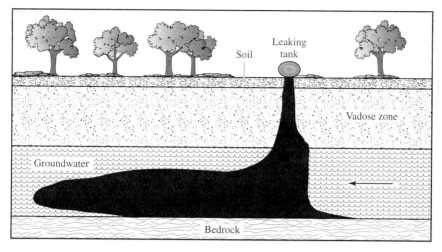

a.

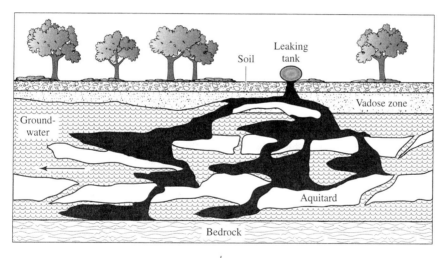

b.

Figure 14.2   Groundwater contamination resulting from a leaking tank containing denser than water nonaqueous-phase liquids (DNAPL). *a.* Illustrates contamination in an idealized homogeneous subsurface hydrogeological system, while *b.* illustrates the much more complex result of contamination in a highly heterogeneous system. SOURCE: McCarty, 1990.

containing the groundwater itself. After the leakage is stopped and the most highly contaminated soil around the tank is excavated, one must deal with a lower concentration residual in the soil, the vadose zone, and the groundwater. If the contaminating liquid is a mixture of many different compounds, each may move and be transformed at different rates.

Subsurface environments usually are much more complex, resembling the situation illustrated in Figure 14.2b (McCarty, 1990). Layering of permeable (sands and gravels) and impermeable (silts, clays, and rock) strata are common; discontinuities can result from faults or large-scale stratigraphic features. Conductivity of water and contaminants through rocks and other such barriers may result from joints and fractures that are difficult to locate and to describe. The mixture of gravel, sand, silt, clay, and organic matter of which the subsurface environment consists can vary widely from location to location, as can the grain-size distribution and mineral composition within each broad class of subsurface strata. In addition, abandoned wells can often provide passages between separated aquifers.

Recalcitrance of contaminants in systems like that shown in Figure 14.2 may result from the contaminant remaining in the nonaqueous phase liquid, penetrating into fissures or small pores in minerals, and strongly sorbing to particle surfaces. All of these render the contaminants *unavailable* to microorganisms and their enzymes.

The absence of suitable *electron acceptors* is a key factor that can affect biodegradability in an environmental setting. For aromatic hydrocarbons, degradation rates are generally enhanced through aerobic decomposition. Thus, introduction of oxygen may be essential. Generally, the quantity of oxygen required is similar to the mass of contaminants present. In complex subsurface systems, getting the oxygen to the areas of need can prove difficult.

Recalcitrance also may result from *insufficient nutrients,* such as nitrogen and phosphorus, for bacterial growth. For aerobic treatment, optimal concentrations of ammonium or nitrate nitrogen are in the range of 2 to 8 grams per 100 grams of organic material, while inorganic phosphorus requirements are about one-fifth of this. When these nutrients are below optimum levels, the amount of biomass is limited, and rates of biodegradation slow considerably and may depend on rates of nitrogen and phosphorus regeneration. With anaerobic degradation, nutrient needs are generally less.

When environmental conditions are not appropriate for biodegradation to occur, the solution often involves adding chemicals that overcome the limiting factor. This approach is called *engineered bioremediation* and is discussed in more detail in Chapter 15. Adding chemicals is simple when the contaminated water or soil is transferred to an engineered reactor, such as those described earlier in the book. In some cases, they can be added fairly easily to contamination near the soil surface. However, adding materials is more difficult and sometimes impossible with some subsurface contamination, depending upon the hydrogeology. The conditions that make it difficult to pump out the contaminants also make it difficult to add material. If it is difficult to pump contaminants out of the ground, then it is also difficult to pump chemicals into the ground to reach the contaminants.

Where environmental conditions are suitable (and the proper microbial populations are present), complete mineralization of organic contaminants can occur, even within the most complex hydrogeological environments. Where environmental conditions are not ideal, degradation of many organic chemicals still may take place at reduced rates. When the environment is suitable, one option is to leave the contaminants alone, allowing then to be biodegraded by strictly natural processes, an approach termed *natural attenuation* or *intrinsic bioremediation*. The challenge is obtaining evidence that convinces the technical experts, the regulatory authorities, and the public that such natural processes are indeed occurring and protecting human health and the environment now and for the future. Collecting sufficient evidence may be difficult and expensive. Intrinsic bioremediation is discussed more in Chapter 15.

14.1.3 MICROORGANISM PRESENCE

Even though a contaminant is known to be readily biodegradable, the absence of a suitable microbial population may be a limiting factor. Methodologies for determining microorganism presence are under development. Some include the simple exposure of aseptically obtained soil or water to the contaminants of concern under ideal conditions for biodegradation. If the microorganisms are naturally present, then degradation of the contaminant occurs. Others attempt to isolate species known to biodegrade the compounds of interest. Molecular techniques can identify the presence of specific microorganisms, nucleic acid sequences, or enzymes that are key to compound degradation. These more sophisticated techniques are not yet fully developed, but offer promise for the future.

If appropriate organisms are not present, they may be introduced into the environment, a technique called *bioaugmentation* (Omenn et al., 1988; Rittmann and Whiteman, 1994). The bioaugmented organisms may be natural, but not ubiquitous in nature. Their introduction into a new system may thus be legally and politically acceptable. Genetically engineered microorganisms can be developed to degrade organic compounds that are inherently recalcitrant. The use of such organisms raises legal and political concerns, as well as questions about the fate of these organisms after their release.

An important question is whether specialized organisms (natural or engineered) can survive in the new environment, and, if so, can they be transported to the place where they are needed. Survival is an especially important issue when the biotransformation of the contaminant is not part of the microorganism's energy and growth metabolism. However, survival of introduced strains into a totally new environment is always a question. If the hydrogeology is complex, then transport may be most difficult. If the bacteria do not encounter the contaminants, then they cannot bring about bioremediation.

14.2 SYNTHETIC ORGANIC CHEMICAL CLASSES

Table 14.1 is a summary of various classes of synthetic chemicals of environmental significance. Also included are some compounds of natural origin that have become of increased environmental significance through human activity. Physical properties for some of the synthetic organic chemicals of concern are listed in Table 14.2. Chemicals that are common groundwater contaminants have moderate to high water solubility and moderate to low octanol/water partition coefficients. These properties permit the chemicals to migrate through the soil so that they reach the groundwater

Table 14.1   Categories of environmental contaminants, their frequency, and sources

Chemical Classes	Example Compounds[a]	Occurrence Frequency[b]	Examples of Industrial Sources or Applications
Organic			
Hydrocarbons			
low FW	Benzene, xylenes, toluene (BTEX); alkanes	F	Crude oil, refined fuels, dyestuffs, solvents
high FW	Polycyclic aromatic hydrocarbons (PAHs), creosotes	C	Creosote, coal tar, crude oil, dyestuffs
Oxygenated Hydrocarbons			
low FW	Alcohols, ketones, esters, ethers, phenols methyltertiarybutylether (MTBE)	F	Fuel oxygenates, solvents, paints, pesticides, adhesives, pharmaceuticals, fermentation products, detergents
Halogenated Aliphatics			
Highly chlorinated	Tetrachloroethene (PCE), trichloroethene (TCE), 1,1,1-trichloroethane (1,1,1-TCA), carbon tetrachloride (CT)	F	Dry cleaning, degreasing solvents
Less chlorinated	Dichloroethane (DCAs), dichloroethene (DCE), vinyl chloride (VC), dichloromethane (DCM)	F	Solvents, pesticides, landfills, biodegradation by-products
Halogenated Aromatics			
Highly chlorinated	Pentachlorophenol, polychlorinated biphenyls (PCBs), polychlorinated dioxins (e.g., TCDD), polychlorinated dibenzofurans (e.g., TCDF)	C	Wood treatment, insulators, heat exchangers, by-products of chemical synthesis and combustion
Less chlorinated	Dichlorobenzene, PCBs	C	Solvents, pesticides
Nitroaromatics	Trinitrotoluene (TNT), RDX, HMX	C	Explosives
Inorganic			
Metals	Cr, Cu, Ni, Pb, Hg, Cd, Zn, etc.	F	Mining, gasoline additive, batteries, paints, fungicides
Nonmetals	As, Se	F	Mining, pesticides; irrigation drainage
Oxyanions	Nitrate, (per)chlorate, phosphate	F	Fertilizers, paper manufacturing, disinfection, aerospace
Radionuclides	Tritium (^{3}H), Pu, U, Ra, Ce, Tc	I	Nuclear reactors, weaponry, medicine, food irradiation

[a] Compound abbreviations: FW = formula weight; TCDD = 2,3,7,8-tetrachlorodibenzo-*p*-dioxin, TCDF = 2,3,7,8-tetrachlorodibenzoforan; RDX = 1,3,5-trinitrohexahydro-*s*-triazine; HMX = octahydro-1,3,5,7-tetranitro-1,3,5,7-tetrazocine

[b] Based on survey of groundwater contaminants, F = Very Frequent; C = Common; I = Infrequent. National Research Council. (2000). *Natural Attenuation for Ground Water Remediation*, Washington, D.C.: National Academy of Sciences Press.

Table 14.2   Physical characteristics of common organic contaminants[a]

Compound	Density (g/cm^3 at 20 °C)	Henry's Constant, H (atm-m^3/mol 20 °C)	Water Solubility (mg/l at 20 °C)	Octanol/Water Partition Coefficient (log K_{OW})
Aromatic Hydrocarbons				
Anthracene	1.24[b]	2.3×10^{-5c}	0.075[b]	4.54[c]
Benzene	0.88[b]	0.0055	1,780	2.13
Benzo(a)pyrene	1.35[b]	4.9×10^{-7}	0.0038	6.06
Biphenyl	0.866[e]		7.48[b]	4.09[c]
Ethylbenzene	0.88[b]	0.0088	152	3.34
Fluorene	1.203(0 °C)[e]	7.24×10^{-5c}	1.8[c]	4.18[c]
Naphthalene	1.14[b]	0.00046	31	3.29
Phenanthrene	0.980(0 °C)[e]	2.57×10^{-5c}	1.1[c]	4.57[c]
Pyrene	1.27[b]	8.9×10^{-6c}	0.15[b]	5.13
Toluene	0.87[b]	0.0066	535	2.69
o-Xylene	0.88		175	
m-Xylene	0.864[e]			
p-Xylene	0.861[e]			
Oxygenated Hydrocarbons				
Phenol	1.06[b]	4.57×10^{-7}	93,000	1.48
Halogenated Aliphatics				
Bromoform	2.890[e]	0.0005	3,100[b]	2.30
Carbon Tetrachloride	1.594[e]	0.023	800	2.64
Chloroethane	0.898[e]	0.0093[d]	5,740	1.49
Chloroform	1.483[e]	0.0032	8,200	1.97
Chloromethane	0.916[e]	0.0066[d]	6,450	0.95
Dichlorodifluoromethane		2.98	280 (25 °C)	2.16
		0.4[c]		
1,1-Dichloroethane	1.176[e]	0.0043	400	1.80
1,1-Dichloroethene	1.218[e]	0.021[d]	400	1.48
1,2-Dichloroethane	1.235[e]	0.0011	8,000	1.48
				1.79[b]
1,2-trans-Dichloroethene	1.257[e]	0.0072[d]	600	2.09
1,2-cis-Dichloroethene	1.284[e]	0.0029[d]		
Dichloromethane	1.33	0.002	20,000	1.26
Hexachloroethane	2.091[e]	0.0025	50	3.34
Tetrachloroethene	1.623[e]	0.012	200	2.88
1,1,1-Trichloroethane	1.339[e]	0.018	4,400	2.51
			1,240[b]	
1,1,2-Trichloroethane	1.440[e]	0.00074	4,500	2.07
Trichloroethene	1.464[e]	0.0088	1,100	2.29

[a] Unless otherwise noted, data from J. L. Schnoor et al., (1987) "Process Coefficients, and Models for Simulating Toxic Organics and Heavy Metals in Surface Waters," U.S. Environmental Protection Agency, EPA/600/3-87/015.

[b] Thibodeaux, L. J., (1996) *Environmental Chemodynamics*, 2nd ed. New York: John Wiley.

[c] R. P. Schwarzenbach, P. M. Gschwend, and D. M. Imboden. (1993). *Environmental Organic Chemistry*, New York: John Wiley.

[d] Gossett, J. M. (1987). "Measurement of Henry's Law Constants for C$_1$ and C$_2$ Chlorinated Hydrocarbons," *Environmental Sci. Technology.* 21(2), pp. 202–208.

[e] Weast, R. C. and M. J. Astle. (1980). *CRC Handbook of Chemistry and Physics*, 61 ed. Boca Raton, FL: CRC Press.

Table 14.2   *Continued*

Compound	Density (g/cm^3 at 20 °C)	Henry's Constant, H (atm-m^3/mol 20 °C)	Water Solubility (mg/l at 20 °C)	Octanol/Water Partition Coefficient (log K_{OW})
Trichlorofluoromethane	1.48[b]	0.126[c]	1,240[b]	2.53[b]
Vinyl Chloride	0.911[e]	0.022[c]	90	0.60
		0.022[d]	1.1[b]	
			2,800[c]	
Halogenated Aromatics				
Arochlor 1242	1.39[b]		0.24[b]	3.71[b]
Arochlor 1254	1.51[b]		0.057[b]	5.61[b]
Chlorobenzene	1.11[b]	0.0037	500	2.84
Hexachlorobenzene	1.569[e]	0.00068	0.006	6.41
				3.4[b]
				5.5[c]
Pentachlorophenol	1.98[b]	3.4×10^{-6}	14	5.04
1,2-Dichlorobenzene	1.305[e]	0.0018	100	3.56
1,3-Dichlorobenzene	1.288[e]	0.0036	123	3.56
1,4-Dichlorobenzene	1.248[e]	0.0031	79	3.56
2,3,7,8-TCDD	1.83[b]	5×10^{-5c}	0.0193[b]	6.66[b]
			1.6×10^{-5c}	
1,2,4-Trichlorobenzene	1.574[b]	0.0023	30	4.28
2-Chlorophenol	1.26[b]	1.03×10^{-5}	28,500	2.17
Nitroaromatics				
Nitrobenzene	1.204[e]	2.2×10^{-5}	1,900	1.87
2-Nitrophenol	1.485(14 °C)[e]	7.56×10^{-6}	2,100	1.75
2,4,6-Trinitrotoluene	1.654[e]			

[a]Unless otherwise noted, data from J. L. Schnoor et al., (1987) "Process Coefficients, and Models for Simulating Toxic Organics and Heavy Metals in Surface Waters," U.S. Environmental Protection Agency, EPA/600/3-87/015.

[b]Thibodeaux, L. J., (1996) *Environmental Chemodynamics*, 2nd ed. New York: John Wiley.

[c]R. P. Schwarzenbach, P. M. Gschwend, and D. M. Imboden. (1993). *Environmental Organic Chemistry*, New York: John Wiley.

[d]Gossett, J. M. (1987). "Measurement of Henry's Law Constants for C$_1$ and C$_2$ Chlorinated Hydrocarbons," *Environmental Sci. Technology*. 21(2), pp. 202–208.

[e]Weast, R. C. and M. J. Astle. (1980). *CRC Handbook of Chemistry and Physics*, 61 ed. Boca Raton, FL: CRC Press.

without excessive partitioning onto soils. This group includes chlorinated solvents, many of the oxygenated aliphatic hydrocarbons, BTEX, and chlorinated benzenes. Compounds with low water solubility or high octanol/water partition coefficients tend to partition from water to soil and thus are more likely to be contaminants in surface soils or stream sediments. Included here are most pesticides, PAHs, and PCBs. Some compounds have high vapor pressure, and thus are transferred readily to the atmosphere. Included here are chlorinated solvents, aliphatic hydrocarbons, and BTEX compounds. Chapter 15 provides more quantitative information on the partitioning effects for the organic contaminants.

It is also useful to separate the compounds into classes because of the difference in biodegradation potential exhibited. Some compounds can be used as primary electron donors for energy and growth, some are transformed as electron acceptors, and some are biodegraded through cometabolism, which does not benefit the cells. These characteristics are discussed below, followed by a discussion of the factors necessary for biodegradation of each class of compounds listed in Table 14.1.

14.3 ENERGY METABOLISM VERSUS COMETABOLISM

Organic compounds that can be used by organisms to obtain energy for growth are the ones most susceptible to biodegradation in the natural environment and in engineered systems. Microorganisms that obtain electrons and energy from the transformations carry out most degradation that occurs in biological wastewater treatment plants. Some compounds, however, are not used as a source of energy for microorganisms, primarily because no organism has the enzymes necessary for their complete biodegradation. Some such compounds may be transformed through *cometabolism*, which is the fortuitous transformation of a compound by enzymes or cofactors designed for other purposes.

Most enzymes are quite specific in the transformations they catalyze, but some are not very specific and transform other compounds. Excellent examples of nonspecific enzymes are oxygenases that initiate the oxidation of hydrocarbons such as methane or toluene. These oxygenases often *fortuitously oxidize* a compound such as trichloroethene (TCE), forming an epoxide that is chemically unstable and degrades to simpler organic compounds that are readily biodegraded (McCarty, 1997a). In order to have cometabolic degradation of TCE, the normal electron donor for the organisms (methane or toluene) must also be present to induce production of the oxygenase and also to provide the electrons and energy needed for the organism to grow and the enzymes to act. Without the presence of the inducing and energy-supplying compounds, transformation by cometabolism does not occur.

Peroxidases used by fungi to degrade lignin are also often not very specific and can bring about biodegradation of many different organic chemicals. However, the presence of peroxidases requires the presence of lignins and oxygen, which are needed for activation of peroxidases.

Why are some enzymes not very specific? Does this nonspecificity serve a useful purpose for the organisms? The answer is not clear. One hypothesis is that an organism whose growth substrate is contained in a mixture of many materials may find it advantageous to oxidize compounds that other organisms utilize. By making the substrates more available for other organisms, the first organism gains better access to its growth substrate. Complex ecological systems often involve many different microorganisms working together in a symbiotic fashion so that they all can benefit. Whether or not nonspecific reactions benefit the microorganisms, we take advantage of cometabolism to biodegrade compounds that otherwise would remain persistent in the environment.

14.4   ELECTRON DONOR VERSUS ELECTRON ACCEPTOR

Organic compounds normally are electron donors in energy metabolism. However, that is not always the case. The most common exception is in fermentations, where one part of an organic compound is oxidized and another part is reduced. For example, some of the carbons in glucose can be oxidized to CO_2, while the other carbons are reduced to ethanol:

$$C_6H_{12}O_6 = 2\,CH_3CH_2OH + 2\,CO_2 \qquad \textbf{[14.1]}$$

Here, four of the carbon atoms in glucose are partially reduced to become the carbon atoms in ethanol, while two of the carbon atoms are completely oxidized to form carbon dioxide. The part of the glucose that becomes ethanol is an electron acceptor, while the part becoming CO_2 is an electron donor.

In recent years, halogenated organic compounds were recognized as potential electron acceptors in energy metabolism. The first observation was with chlorobenzoate (Dolfing and Tiedje, 1987). Using a three-organism system, they demonstrated reductive dehalogenation of chlorobenzoate (Figure 14.3). In the first step of the reaction, chlorobenzoate acts as an acceptor for species DCB-1, which gains electrons from molecular hydrogen. This results in the substitution of a hydrogen atom for the chlorine atom, releasing hydrogen chloride and forming benzoate. A second organism (B2-2) ferments benzoate to form 3 mol of acetate, 3 mol of hydrogen, and 1 mol of carbon dioxide. Two mol of the hydrogen formed are used by organism PM-1 to reduce the 1 mol of carbon dioxide to methane. The other mole of hydrogen is used by DCB-1 to reduce another mole of chlorobenzoate so that the process can continue. If another methanogen were present to convert the 3 mol of acetate to 3 mol of carbon dioxide and 3 mol of methane, the anaerobic biodegradation of chlorobenzoate would be complete.

DCB-1 obtains energy from hydrogen as an electron donor and chlorobenzoate as an electron acceptor. This is an example of *dehalorespiration*. We now know that several other chlorinated organic chemicals serve as electron acceptors in energy metabolism. In most cases, H_2, formed as an intermediate in anaerobic biodegradation of organic matter (Chapters 1 and 13), is the electron donor. Some exceptions are known, and organics such as acetate or pyruvate may also be donors.

It is obvious from Figure 14.3 that, if the hydrogen formed from benzoate fermentation were not shared by PM-1 with DCB-1, chlorobenzoate fermentation would stop, and none of the organisms would benefit. Thus, a key to this system is the sharing of the hydrogen between PM-1 and DCB-1. What is the control mechanism for this sharing? The answer is thermodynamics. More energy is released from the dehalogenation of chlorobenzoate than from methanogenesis. For this reason, dehalogenation has a lower energy threshold than methanogenesis. The thermodynamic threshold for methanogenesis is a hydrogen concentration of about 10 nM (Lovley and Goodwin, 1988), while the hydrogen threshold for dehalogenation is on the order of 2 nM or lower (Yang and McCarty, 1998). Thus, DCB-1 can lower

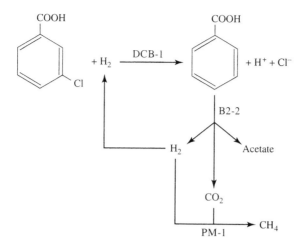

Figure 14.3 Anaerobic degradation of m-chlorobenzoate, which involves three bacterial strains (Dolfing and Tiedje, 1986). The first (DCB-1) uses hydrogen as an electron donor and chlorobenzoate as an electron acceptor, thus removing the chlorine atom on benzoate and replacing it with hydrogen. The subsequent oxidation of benzoate by a second strain (B2-2) produces the hydrogen needed for dehalogenation. The third strain (PM-1) removes excess hydrogen in methane production.

the hydrogen concentration below 10 nM and still capture energy. By keeping the hydrogen concentration low, DCB-1 limits the growth of PM-1 and ensures that it obtains hydrogen.

Dehalorespiration does not require that the dehalogenated compound also serve as the ultimate source of H_2. If a different organic compound is the initial electron donor that supplies hydrogen for dehalogenation, then competition for the available hydrogen can be quite keen. This is illustrated in Figure 14.4 (McCarty, 1997b). Here, complex organic matter is fermented to produce hydrogen and acetate. The hydrogen serves as an electron donor for a variety of different organisms using different electron acceptors, including sulfate, Fe(III), Mn(IV), chlorinated solvents, and CO_2 (for methanogens and homoacetogens, microorganisms that form acetate from hydrogen). Homoacetogens have a hydrogen threshold of about 200 nM and thus compete for hydrogen only when the concentration is relatively high (Lovley and Goodwin, 1988). The thresholds for iron, manganese, and sulfate reducers are in the same range as or somewhat lower than for the dehalogenators. Thus, reductive dehalogenation may not occur or may be greatly limited if these other electron

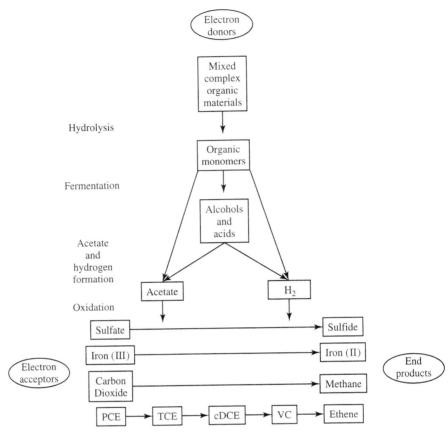

Figure 14.4   Electron flow from electron donors to electron acceptors in the anaerobic oxidation of mixed and complex organic materials. Microorganisms that can use chlorinated compounds (PCE, TCE, cDCE, and VC) as electron acceptors in *dehalorespiration* compete for the electrons in the acetate and hydrogen intermediates with microorganisms that can use sulfate, iron (III), and carbon dioxide. SOURCE: P. L. McCarty, 1997b.

acceptors are present. This illustrates that, while reductive dehalogenation can be very important in the transformation of halogenated organics, it is not something that can be assumed to occur automatically.

A key aspect of dehalorespiration is that each organism tends to have a very limited range of halogenated compounds that it can transform. Chlorobenzoate degraders that dehalogenate *m*-chlorobenzoate may be different from those that dehalogenate *o*-chlorobenzoate. Organisms that dehalogenate tetrachloroethene (with one known exception) generally are different from those that dehalogenate dichloroethene. This means that a great variety of different organisms may be necessary for dehalogenation of mixtures of chlorinated solvents or of the many different congeners of PCBs.

While DCB-1 gains energy from reducing chlorobenzoate, not all reductive de-halogenation reactions are part of energy metabolism. The relative frequencies with which microorganisms obtain energy (dehalorespiration) versus carry out fortuitous (cometabolic) reductive dehalogenation are not yet known. Early observations of reductive dehalogenation suggested that it was strictly cometabolic. However, as more organisms are isolated and studied in greater detail, we find that many of them obtain energy from the process. The difference is critically important, because organisms that can obtain energy from a process can grow and multiply from it. An energy-yielding process can be continued, and, with increased growth, the rate can increase. In addition, electron and energy sources external to the dehalogenation may not be required if the microbial community can use the product of dehalorespiration as an electron donor, as in the case of chlorobenzoate.

14.5   MINIMUM SUBSTRATE CONCENTRATION (S_{min})

Biodegradation of synthetic organic compounds is strongly affected by S_{min}, the lower threshold concentration below which reaction speed is insufficient to supply the organism with sufficient energy for net growth. For a single contaminant and a single organism, S_{min} was defined in Chapter 3 (Equation 3.28) as

$$S_{min} = K \frac{b}{Y\hat{q} - b}$$

In a CSTR, S_{min} is the lowest concentration attained for steady state in this case.

Many hazardous synthetic organic compounds have drinking-water standards that are very low, on the order of a few micrograms per liter. Such concentrations are frequently less than S_{min}. While a given compound may be easily biodegradable at higher concentrations, the question is whether or not it can be biodegraded at concentrations that meet regulatory and health standards.

Concentrations less than S_{min} can be obtained in two circumstances. The first is when the contaminant in used as a *secondary substrate* (McCarty, Reinhard, and Rittmann, 1981; Rittmann et al., 1994; Namkung and Rittmann, 1987a,b). Here, a microorganism obtains energy by utilizing more than one compound. The rates of electron and energy gain are the sum of the rates for all compounds. An example is growth of one organism on acetate and dichloromethane (methylene chloride). If acetate is present at a concentration higher than its S_{min}, it can be the main substrate supporting biological growth. Then, the organism may use dichloromethane at the same time in order to obtain additional energy, even when dichloromethane is present at a concentration below its S_{min}. A *Pseudomonas* sp. uses either acetate or dichloromethane for energy and growth. When grown on 1 mg/l of acetate, it was able to reduce methylene chloride from an initial concentration of 10 μg/l to below 1 μg/l, much lower than in the absence of the acetate (LaPat-Polasko, McCarty, and Zehnder, 1984). Here, feeding a nonhazardous compound at a concentration above its S_{min} threshold provided the organism with energy for growth, allowing it to remove the secondary hazardous compound to well below the hazardous compound's S_{min} threshold.

Secondary utilization also works well when many compounds present at concentrations less than S_{min} are utilized simultaneously (Namkung and Rittmann, 1987a,b). No one compound is the main primary substrate, but the primary substrate is an aggregate of many secondary substrates. This phenomenon of degradation was illustrated for a mixture of chlorinated benzenes under aerobic conditions and for many halogenated methanes under anaerobic conditions (Bouwer and McCarty, 1984). Several similar examples for other compounds have been given as well by Alexander (1994).

The second method by which a contaminant can be biodegraded below S_{min} is through proper design of a treatment system. For example, if the contaminant concentration in a wastewater is well above its S_{min}, then a plug-flow reactor (PFR) is theoretically capable of reducing the concentration to below S_{min}, while a continuous-flow stirred-tank reactor (CSTR) is not. The key is that the biomass must be retained in the plug-flow system by attachment as a biofilm (Rittmann, 1982) or recycling of suspended growth (Chapter 5). Then, the biomass at the effluent end of the process can remove the contaminant to a very low ($\ll S_{min}$) concentration, because it was grown at the head end of the process, where the concentration was well above S_{min}. Of course, the influent concentration must be greater than S_{min}.

The advantage of plug flow with biomass recycling is illustrated in Figure 14.5. For the example shown, the contaminant is toluene, a regulated chemical that is

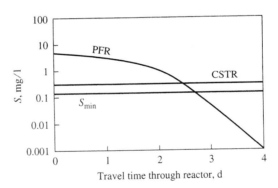

Figure 14.5   Comparison between the ability of a plug-flow reactor (PFR) and a continuous-flow stirred-tank reactor (CSTR) to reduce the concentration of toluene to below S_{min}. The reactor detention time (V/Q^0) is 4 d in both cases. The recycle ratio for the PFR is 1.0. With a PFR, the concentration is high at the inlet and low at the outlet. With a CSTR in contrast, the toluene concentration is the same throughout the reactor, which is the same as the effluent concentration.

a common groundwater contaminant resulting from gasoline spills. Toluene also is used as a growth substrate and inducer of toluene oxygenases that can degrade chlorinated ethenes, such as trichloroethene, through cometabolism. Coefficients that apply for toluene are $Y = 0.77$ g VSS_a/g toluene, $K = 1.0$ mg/l, $\hat{q} = 1.5$ g toluene/g VSS_a-d, and $b = 0.15$/d (McCarty et al., 1998). These values yield an S_{min} of 0.15 mg/l. This is the lowest concentration that can theoretically be achieved in a CSTR, even with infinite detention time. However, with an inlet toluene concentration of 10 mg/l and a PFR with a recycle ratio (R) of 2 and a detention time of 2 d, an effluent concentration of 0.001 mg/l theoretically can be achieved (Figure 14.5). For the same size of reactor, the CSTR would produce an effluent concentration of 0.53 mg/l. The advantage of plug flow with biomass retention is striking.

14.6 BIODEGRADATION OF PROBLEM ENVIRONMENTAL CONTAMINANTS

14.6.1 SYNTHETIC DETERGENTS

Figure 14.6 shows the three basic types of synthetic detergents: anionic, cationic, and nonionic. The biodegradation pathways of each have been reviewed (van Ginkel, 1996). The anionic detergents still are the major detergents used in laundry soaps. Cationic detergents are toxic to microorganisms in the concentrations normally used for cleaning. They commonly are used when good disinfection, as well as cleaning action, is desired, such as in hand detergents for medical cleaning, restaurants, and home cleaning of diapers. Nonionic detergents are not ionized and depend for solubility on polymers of ethylene oxide connected to a hydrophobic end that dissolves in grease. Of particular concern in recent years is the discovery that the alkyl-type of nonionic detergents tends to persist in natural waters.

The relative persistence of akylphenol polyethoxylates is related to the branched alkyl hydrophobic group connected to the aromatic ring. They generally contain 8 (octyl) or 9 (nonyl) carbons. As with the older ABS anionic detergents, the high degree of branching of these alkyl groups imparts resistance to biodegradation. The general pathway for biodegradation under aerobic and anaerobic conditions begins with degradation of the ethylene oxide side chain. Cleavage of the side chain from the aromatic ring leaves an alkyl phenol residue (Ball, Reinhard, McCarty, 1989). Also of interest is the finding that, upon chlorination of water supplies and wastewaters, the alkyl group can become chlorinated or brominated, leading to halogenated organic compounds similar in structure to the 2,4-D pesticides and may have toxic properties (Ball et al., 1989). Even without halogenation, the residues of alkyl phenol polyethoxylates show toxicity to fish and are suspected of being *endocrine disrupters* (Rudel et al., 1998). Endocrine disrupters mimic hormones and may alter reproductive patterns of fish and other aquatic organisms. Because of such concerns, the use of alkylphenol polyethoxylates as part of detergent formulations is decreasing.

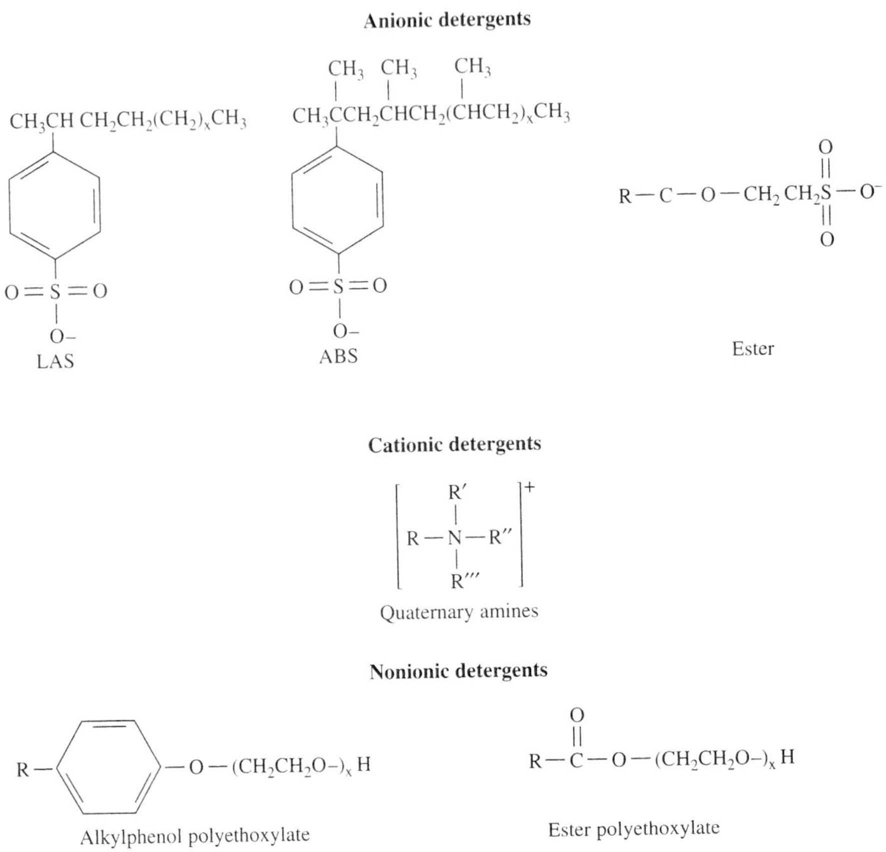

Anionic detergents

$$CH_3CH\ CH_2CH_2(CH_2)_xCH_3$$

$$CH_3CCH_2CHCH_2(CHCH_2)_xCH_3$$
(with three CH_3 groups labeled above)

$$R—C—O—CH_2\ CH_2S—O^-$$
(with O double bonds on the carbonyl and on the sulfur)

LAS

ABS

Ester

Cationic detergents

$$\left[R—\overset{R'}{\underset{R'''}{N}}—R'' \right]^+$$

Quaternary amines

Nonionic detergents

$$R—\!\!\!\bigcirc\!\!\!—O—(CH_2CH_2O-)_x\ H$$

Alkylphenol polyethoxylate

$$R—C—O—(CH_2CH_2O-)_x\ H$$
(with O double bond on the carbonyl)

Ester polyethoxylate

Figure 14.6   Different general classes of synthetic detergents.

14.6.2 PESTICIDES

Rachael Carson's 1962 book, *Silent Spring*, was a major force behind the change in thinking of the public about the introduction of new chemicals into the environment. The main problem chemicals in 1962 were the *chlorinated pesticides* (Figure 14.7), the first major one of great importance being DDT. DDT was of great benefit to humans in their fight against malaria, as it was highly effective in controlling the *Anopheles* mosquito, the carrier and transmitter of the *Plasmodium* protozoan to humans. While malaria was not a disease of significance in the United States, DDT found other uses for fighting crop insects. Its use became very widespread, and because of its great resistance to biodegradation and hydrophobic character, DDT's presence in the fatty tissues of birds and mammals, including humans, increased rapidly. DDT causes weakening of egg shells of birds, which led to dramatic decreases in the populations of brown pelicans, osprey, bald eagles, and peregrine falcons. DDT was one of the first of the widely used chlorinated pesticides banned in the United States. All of

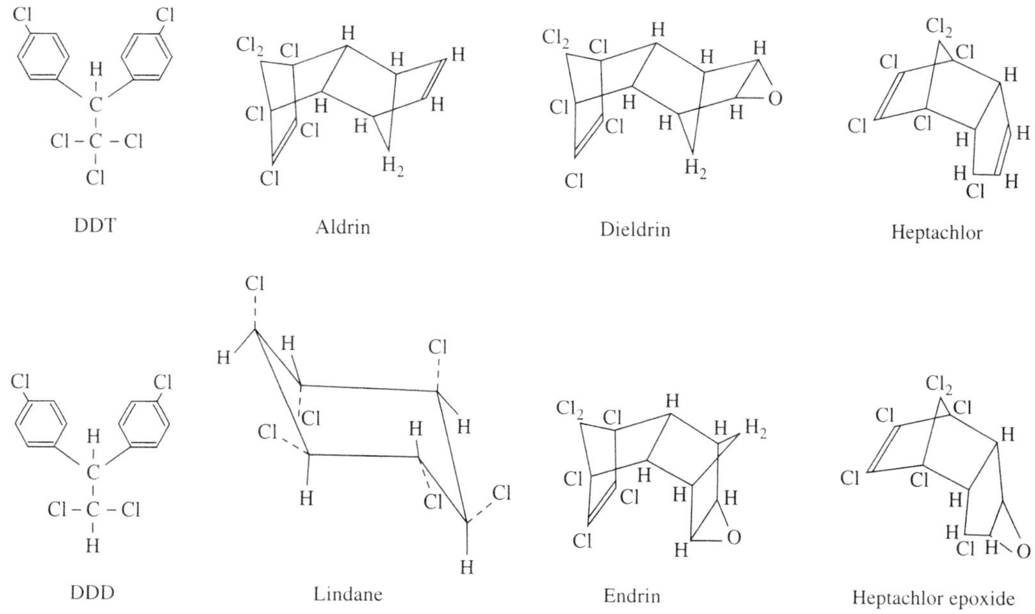

Figure 14.7   Typical chlorinated pesticides used widely in the past.

those illustrated in Figure 14.7 are relatively resistant to biodegradation and are now greatly restricted in use.

The experience with DDT was one of the first examples that target organisms develop a resistance to the toxic chemicals, leading to the need for ever-increasing concentrations for control of the pests. Because DDT-resistant strains of the *Anopheles* mosquito have now arisen, DDT's value for malaria control has greatly diminished. The same observation of increasing resistance by target organisms is being observed with other pesticides and antibiotics.

Two characteristics of chlorinated pesticides that led to great harm are their resistance to biodegradation and their hydrophobicity. The latter is quantified by a high octanol/water partition coefficient, which signifies a strong affinity for soil organic matter and fatty tissue. Chlorinated pesticides are taken up by plankton and concentrated. They are further concentrated in the fatty tissue of fish that eat plankton, and then further concentrated by birds and other carnivores such as humans. This tendency to *bioaccumulate* magnifies their negative impacts on higher organisms.

The tendency of chlorinated pesticides to partition strongly into soils and fatty tissue was recognized early as one aspect of their resistance to biodegradation by microorganisms (Alexander, 1994). Partitioning out of the water greatly reduces the solution concentration for such compounds, possibly below the S_{min} concentration required for organism growth.

The chemical structure of the chlorinated pesticides also makes them poorly susceptible to biotransformation reactions, especially under aerobic conditions. The chlorine substituents and the heavy branching block the normal sites for enzyme attack. Half-lives in active aerobic systems usually are measured in weeks to months.

While chlorinated pesticides are highly resistant to transformation under aerobic conditions, anaerobic conditions are more favorable. Hill and McCarty (1967) found that lindane added to digested sludge from methanogenic fermentation disappeared rapidly, within days. Indeed, all of the pesticides in Figure 14.7 except heptachlor epoxide were transformed under anaerobic methanogenic conditions. Also of interest is that heptachlor disappeared rapidly, even though the epoxide did not. An interesting observation with DDT was that it was reductively dehalogenated rapidly to DDD (shown in Figure 14.7), and DDD later disappeared as well, but much more slowly. Commonly recognized now is that reductive dehalogenation occurs with strongly reducing anaerobic conditions for essentially all chlorinated aromatic and aliphatic compounds (more on this later). While this observation is interesting and at times of importance, strong reducing conditions do not occur everywhere in the environment and so cannot be counted on to bring about detoxification of chlorinated compounds in general. The other aspect is that reductive dehalogenation does not necessarily result in the complete destruction of the chemical. Indeed, it may create products that are hazardous to humans and the environment. The formation of DDD from DDT is an example.

Several new pesticides were developed in order to capture the advantages of pesticides and yet reduce their adverse environmental impacts. Representative members of three important classes are indicated in Figure 14.8. The *phosphorus-based pesticides* are, in general, highly toxic, but upon exposure to water are enzymatically or chemically hydrolyzed quite rapidly. Hydrolysis occurs at the ester linkage in malathion, yielding ethanol and the carboxylated pesticide residue, or between the carbon/oxygen bond in the carbon/oxygen/phosphorous linkages. Other transforma-

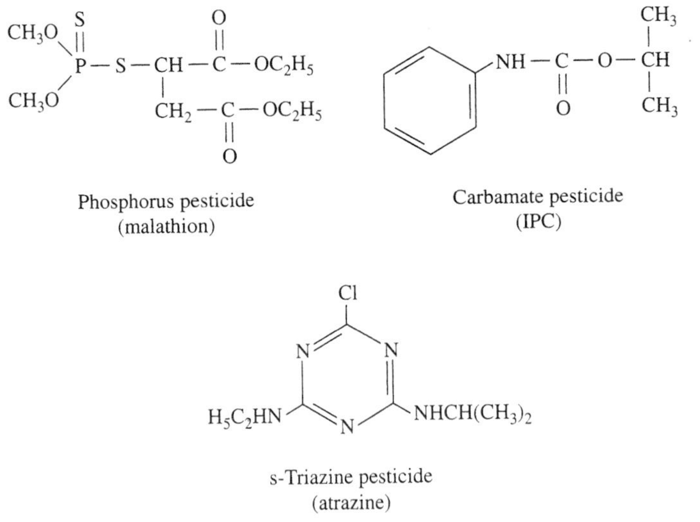

Phosphorus pesticide
(malathion)

Carbamate pesticide
(IPC)

s-Triazine pesticide
(atrazine)

Figure 14.8   Other pesticide forms that are in common use.

tions also occur. One is the conversion of the sulfur atom connected to the phosphorus, forming a double bond to oxygen. A typical example is the conversion of parathion to paraoxon in soil, a compound that is much more potent as a chlolinesterase inhibitor than the parent compound (Alexander, 1994). Alexander gives several other examples where daughter products resulting from partial degradation of pesticides lead to the formation of more highly toxic compounds. This further indicates that disappearance of a compound in itself does not mean protection to human health and the environment has been achieved.

The *carbamate pesticides* are transformed rapidly in the environment through hydrolysis. Although the *s-triazine pesticides* are biodegradable, they are frequently found in groundwaters throughout the United States. Atrazine is one of the most commonly found members of the s-triazines, which are among the most widely used herbicides. Because triazines have relatively low octanol/water partition coefficients, they do not sorb readily on soils, but become groundwater contaminants. Atrazine biodegradation involves the sequential removal of the alkyl side chains followed by deamination, dehalogenation, and ring cleavage (Radosevich, Traina, and Tuovinen, 1996). The alkyl carbon can serve as a carbon and energy source to degrading microorganisms (Cook, 1987), but the ring carbon is fully oxidized and does not serve for catabolic or biosynthetic purposes. Atrazine biodegradation is spatially variable and usually slow, perhaps because of low temperature in the subsurface or lack of degrading organisms (Radosevich et al., 1996).

14.6.3 HYDROCARBONS

Organic compounds composed solely of carbon and hydrogen are termed hydrocarbons. They are widely distributed natural products having many aliphatic and aromatic representatives. They may be rather simple compounds, such as methane or benzene, or highly complex and large molecules, as represented by petroleum and coal. Many are saturated completely, while others contain unsaturated carbon bonds.

A common characteristic of the hydrocarbons is that they are biodegradable aerobically. However, the rate of aerobic biodegradation depends on the complexity of the molecule. Large hydrocarbons with much branching or containing many aromatic rings are difficult to degrade. One aspect of their poor biodegradability is their low water solubility. The other is that the complex structure makes it difficult for microorganisms to find a location for an initial enzymatic attack.

The initial step in hydrocarbon degradation by bacteria is generally the introduction of oxygen into the molecule with an *oxygenase,* which requires an energy investment in the form of NAD(P)H and molecular oxygen. Oxygenation reactions, which introduce one or two −OH groups into the hydrocarbon structure, are described in Chapter 1. The hydrocarbon is oxidized by two electrons for each −OH group added, but the organisms do not recover the electrons as NADH. In fact, monooxygenation reactions, which insert one −OH, consume an NADH. Thus, the cells do not gain the energy that could be gained if NADH were produced. The advantage to the microorganisms is that the products of the oxygenation reactions are more available. Specifically, the products are easily attacked in subsequent dehydrogenation and hydroxylation reactions, which yield NADH, and are more water-soluble.

Only about a decade ago, the common understanding was that hydrocarbons could not be biologically degraded under anaerobic conditions. Without the presence of molecular oxygen, there was then no known process for introducing an oxygen atom into the molecule. This thinking has changed radically. Many aromatic hydrocarbons are now known to be biodegradable in the absence of molecular oxygen, organisms effecting the biodegradation of many aromatic hydrocarbons have been isolated, and biochemical pathways through which oxygen from water or carboxylate is introduced into the aromatic ring are being discovered. In general, the anaerobic reactions are quite slow compared with their aerobic counterparts, but they do occur and are of importance to the fate of aromatic hydrocarbons. Evidence that aliphatic hydrocarbons may also be degraded under anoxic conditions is also beginning to surface.

These new discoveries call for a major revision in our thinking about the capabilities of microorganisms. They are much more versatile than had ever before been imagined. The aerobic and anaerobic biodegradation of two important classes of hazardous hydrocarbons, the BTEX compounds and the polycyclic aromatic hydrocarbons (PAHs), are summarized here as examples of hydrocarbon biodegradation. Closely associated with BTEX is the fate of methyl-*tertiary*-butyl ether (MTBE), which also is discussed in the context of BTEX.

BTEX and MTBE   Gasoline is a complex mixture of hundreds of different hydrocarbons, most of which are saturated and contain 4 to 12 carbon atoms per molecule. From 22 to 54 percent of gasoline is composed of aromatic hydrocarbons, the most common being benzene, toluene, ethylbenzene, and the three different isomers of xylene (Figure 14.9), which comprise the BTEX compounds. Other aromatic compounds in gasoline include methylethyl benzenes and trimethyl benzenes. Toluene tends to be the dominant aromatic hydrocarbon, followed by the isomers of xylene and then ethylbenzene and benzene in about equal amounts. Chapter 15 gives more information on the composition of gasoline. Number 2 petroleum fuels are similar in aromatic content to gasoline, while jet fuels and lubrication oils have somewhat lower aromatic content.

The BTEX compounds are of special significance because they are toxic and more water-soluble than the other hydrocarbons in gasoline. Benzene is a known human carcinogen. When a gasoline spill occurs, the gasoline seeps into the soil

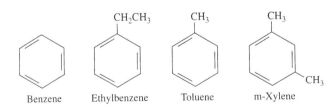

Figure 14.9   The major aromatic hydrocarbon constituents of gasoline, generally referred to as the BTEX compounds.

and reaches the groundwater table, where the BTEX compounds partition into the groundwater. The individual BTEX compounds also are of commercial interest and are used as solvents and in the production of other chemicals. The frequency with which BTEX leaks occur caused the U.S. Environmental Protection Agency to set drinking-water standards for BTEX compounds. They also cause taste and odor at low concentrations in drinking water (see Chapter 12 for more information).

BTEX are natural products, but their concentrations in natural waters are low and generally of little concern. The widespread human use of gasoline and other products containing BTEX has caused their wide environmental distribution. Being natural products, they are biodegraded under aerobic conditions. The biochemical pathways and enzymes involved in aerobic oxidation are well established, and microorganisms that can oxidize BTEX are numerous and ubiquitous in the environment.

As with all hydrocarbons, the initial oxidation step is difficult and requires an oxidase, along with molecular oxygen and energy in the form of NADH. Many different oxidases can initiate BTEX biodegradation. Figure 14.10 illustrates four monooxygenation reactions and one dioxygenation reaction that can occur with toluene (Mikesell, Kukor, and Olsen, 1993). The dioxygenase adds two oxygen atoms at a time (the second pathway from the top), while the monooxygenases add one atom of oxygen at a time to the ring (bottom three reactions) or the methyl group (top reaction).

The pathways for aromatic hydrocarbons essentially all involve oxidation reactions up to the point at which two hydroxyl groups are adjacent to each other on the ring, as illustrated in the last column of Figure 14.10. The ring is then cleaved by addition of more oxygen either between the two carbons connected to the hydroxyl groups (*ortho* fission) or at one of the two adjacent double bonds (*meta* fission). This results in a carboxylated aliphatic compound that can be oxidized further by the usual dehydrogenation and hydroxylation reactions (Chapter 1).

The new finding is that BTEX compounds can be biodegraded under anoxic conditions. Reinhard, Goodman, and Barker (1984) first reported that xylene had disappeared in an anaerobic plume and hypothesized that this was due to biodegradation. This possibility was confirmed just two years later by Vogel and Grbic-Galic (1986). Since then, numerous microorganisms capable of degrading BTEX compounds under anoxic conditions have been isolated, and the biochemical pathways by which this occurs are being determined. One such pathway for toluene is illustrated in Figure 14.11. The novel occurrence is the introduction of oxygen by addition of fumarate to the molecule. Fumarate is formed as an intermediate in toluene anaerobic degradation (Beller and Spormann, 1997), as shown in the figure. Fumarate addition is an example of the general strategy of adding carboxylate groups (fumarate has two carboxylate groups) as a means to introduce oxygen to the aromatic molecule when O_2 is absent.

Many isolated microorganisms can biodegrade toluene, ethylbenzene, and xylenes through denitrification. Isolates that can degrade BTEX compounds under more reducing conditions are more limited, except for toluene, for which several organisms are known. Nevertheless, degradation of BTEX compounds under sulfate-reducing or methanogenic conditions in complex mixture of organisms has been observed frequently, and use of Fe(III) or Mn(IV) as electron acceptors also is

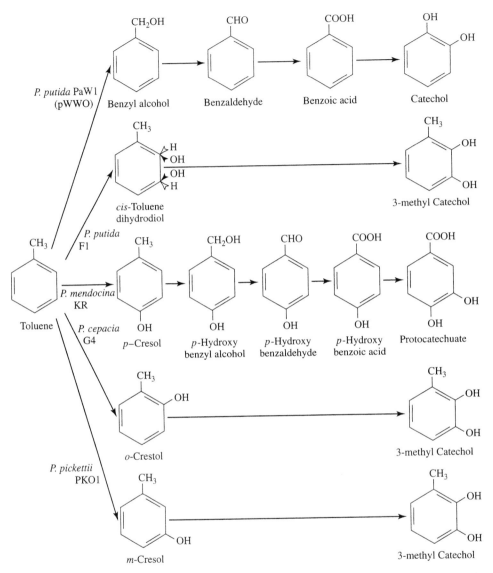

Figure 14.10   Five different aerobic biodegradation pathways for toluene, each initiated through the activity of a mono- or di-oxygenase together with molecular oxygen. SOURCE: Mikesell et al., 1993.

possible. The rates of degradation tend to be slower than under aerobic conditions, and the organisms that can carry out the reactions do not appear to be as prevalent in the environment as their aerobic counterparts. Nevertheless, disappearance of BTEX plumes with time is very commonly observed, and regulators consider the *natural attenuation* or *intrinsic bioremediation* of BTEX in groundwater to be an acceptable cleanup alternative for many gasoline spills (Chapter 15). Except when the gaso-

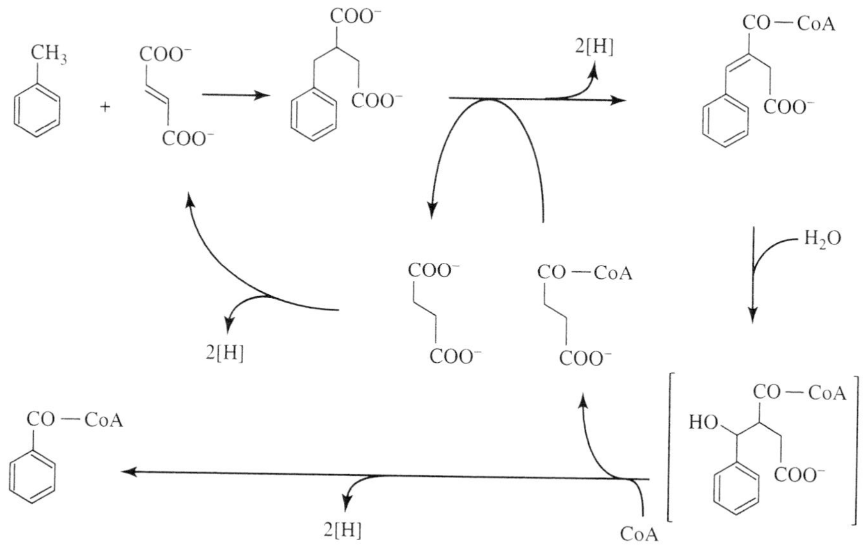

Figure 14.11 Anaerobic pathway for toluene degradation. SOURCE: Beller and Spormann, 1997; Spormann and Widdel, 2000.

line spill is very small, anaerobic reactions are very important, in some cases more important than the aerobic reactions.

Although BTEX compounds are generally biodegraded naturally under either aerobic or anaerobic conditions, other compounds that may be present in gasoline are not. This led to a problem in California and other states that required that methyl *tertiary* butyl ether (*MTBE*) be added to gasoline in order to reduce smog-producing emissions from automobile exhausts. For a while, MTBE comprised 10–12 percent of the gasoline mixture in California. While the BTEX compounds in gasoline-contaminated groundwater biodegrade, MTBE often does not. For example, at Vandenberg Air Force Base in California, the BTEX plume from a gasoline spill migrated only 17 to 35 m from the source before being biodegraded, while the MTBE plume extended about 550 m (National Research Council, 2000). Since MTBE has a low microgram-per-liter threshold for taste and odor and may be a health hazard, its presence in groundwaters and also in surface waters where gasoline-driven recreational vehicles were being used led to a reversal of California's earlier decision to require that it be added to gasoline. This is another of the many examples where attempts to solve one environmental problem have led to another. While microorganisms appear to biodegrade MTBE under certain conditions, the needed conditions or the appropriate organisms are not present with sufficient frequency at contaminated sites for natural attenuation to be relied upon to protect human health and the environment.

Polycyclic Aromatic Hydrocarbons (PAHs)

PAHs comprise another important group of naturally formed aromatic hydrocarbons that are of general concern because of toxicity to humans and other species, including their potential for causing cancer. PAHs are formed from incomplete combustion of organic matter—naturally and through human activities—and thus are found widely in the environment. One of the first observations of human cancer was in the late eighteenth century by Potts, who noted a high incidence of skin cancer in chimney sweeps, who came into extensive contact with soots and tars, which have high levels of PAHs. Major concerns today are the high concentrations of PAHs surrounding commercial processes that involve incomplete combustion of coal, such as at coking plants and manufactured-gas plants. The coal tars obtained when making coke for use in steel production are commonly used as creosote wood preservatives, a use in itself that suggests coal tars are toxic to living things. Manufactured-gas plants (MGPs) converted coal or petroleum to gaseous carbon-monoxide fuel used at one time in homes and industry. The coal-tar residue, which contains PAHs, contaminates soils surrounding such plants. While such locations have high PAH concentrations, any combustion process, including cigarette smoking, can cause exposure to PAHs.

Figure 14.12 illustrates the predominant PAHs formed from combustion (Cerniglia, 1992). Benzo(a)pyrene is the known carcinogen in this group. As in-

PAH	FW	Sol (mg/l)	Genotoxicity
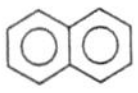 Naphthalene	128.2	31.7	—
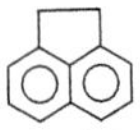 Acenaphthene	154.2	3.9	+ Ames
 Anthracene	178.2	0.07	—
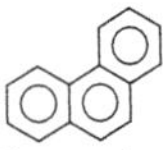 Phenanthrene	178.2	1.3	—
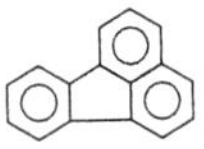 Fluoranthene	202.3	0.26	Weak Carcinogen
 Pyrene	202.3	0.14	± Ames + UDS + SCE
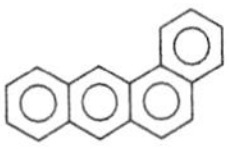 Benz[a]anthracene	228.3	0.002	+ Ames + CA + SCE + Carcinogen
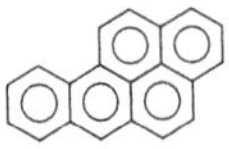 Benzo[a]pyrene	252.3	0.003	+ Ames + CA + UDS + DA + SCE + Carcinogen

(Recalcitrance — arrow pointing downward along left side)

Figure 14.12   Various polycyclic aromatic hydrocarbons of importance. Solubility is in water at room temperature. SOURCE: Cerniglia, 1992.

dicated in Table 14.1, the solubility of the higher-ringed compounds is very low. They tend to remain in surface soils and do not migrate in groundwaters significantly. However, the smallest member of this group, naphthalene, is fairly water-soluble and frequently is found as a groundwater contaminant. In general, PAHs are assumed to be primarily soil, rather than water, contaminants. However, they can concentrate in the sediments of rivers, lakes, and streams.

The PAHs are biodegradable under aerobic conditions (Cerniglia, 1992). The two- to four-ring compounds are quite readily biodegraded aerobically by many organisms. As with BTEX compounds, oxidation is initiated by oxygenases that require molecular oxygen. The higher ringed PAHs, however, are quite resistant to biodegradation, due again to their complexity and their very low solubility. Because of the very low solubility of the larger PAHs, their *bioavailability* to microorganisms that might degrade them is a significant issue. Many-ringed PAHs can be biodegraded aerobically, but then biodegradation may be through cometabolism by organisms living on PAHs with a few rings. If cometabolic oxygenation occurs, then the compound is no longer a PAH, but whether or not this is sufficient to reduce toxicity is not clear.

As with BTEX and other hydrocarbons, anaerobic degradation of PAHs was formerly thought impossible. However, naphthalene biodegradation through denitrification was documented several years ago (Mihelcic and Luthy, 1988). More recently, the biodegradation of phenanthrene under similar conditions was reported (Strand, 1998). Rates are generally quite low compared with aerobic biodegradation, and the bioavailability issue for the higher-ringed PAHs remains very significant.

Some question whether or not many-ringed PAHs that are tightly bound to soils should be of environmental concern. If they are not bioavailable to microorganisms, are they bioavailable to cause harm if the soil is inadvertently ingested, such as by a child playing in the soil? Perhaps biodegradation only of the PAHs that are bioavailable to the microorganisms may be sufficient to be protective of human health and the environment if the PAHs left behind are only those that are biounavailable to humans and other higher life forms. This issue has not been adequately resolved. This bioavailability issue is of practical significance, because the alternatives to biodegradation for remediation of PAH contaminated soils are removal and burial in a secured landfill or incineration, both of which can be very expensive. Hopefully, this issue will be resolved in the near future.

14.6.4 CHLORINATED SOLVENTS AND OTHER HALOGENATED ALIPHATIC HYDROCARBONS

Chlorinated solvents are among the most prevalent organic contaminants in groundwater. The chlorinated aliphatic hydrocarbons (CAHs) and related chlorinated compounds generally contain one or two carbon atoms and one to six chlorine atoms (Table 14.3). CAHs with two carbon atoms may be saturated (ethanes) or unsaturated (ethenes). The chlorinated solvents have been and are widely used for cleaning clothes, engines, electronic parts, and other items contaminated with grease. Because the chlorinated solvents are not explosive or readily combustible, they are superior to hydrocarbon alternatives. However, most commonly used CAHs are not natural compounds, and, perhaps for this reason, enzyme systems capable of their degradation

Table 14.3 Major chlorinated aliphatic hydrocarbon (CAH) contaminants found in groundwater and biotransformation pathways known to exist (indicated with an X)

CAH	Formula	Acronym	Primary Substrate			Cometabolism	
			Aerobic Donor	Anaerobic Donor	Anaerobic Acceptor	Aerobic	Anaerobic
Methanes							
Carbon Tetrachloride	CCl_4	CT					X
Chloroform	$CHCl_3$	CF				X	X
Dichloromethane	CH_2Cl_2	DCM	X	X		X	X
Chloromethane	CH_3Cl	CM	X			X	X
Ethanes							
1,1,1-Trichloroethane	CH_3CCl_3	TCA				X	X
1,1,2-Trichloroethane	$CH_2ClCHCl_2$	1,1,2-TCA				X	X
1,1-Dichloroethane	CH_3CHCl_2	1,1-DCA				X	X
1,2-Dichloroethane	CH_2ClCH_2Cl	1,2-DCA	X	X		X	X
Chloroethane	$CH3CH_2Cl$	CA	X			X	X
Ethenes							
Tetrachloroethene	$CCl_2=CCl_2$	PCE			X		X
Trichloroethene	$CHCl=CCl_2$	TCE			X	X	X
cis-1,2-Dichloroethene	$CHCl=CHCl$	c-DCE		X	X	X	X
trans-1,2-Dichloroethene	$CHCl=CHCl$	t-DCE		X		X	X
1,1-Dichloroethene	$CH_2=CCl_2$	1,1-DCE				X	X
Vinyl Chloride	$CH_2=CHCl$	VC	X	X	X	X	X

have not evolved sufficiently to make them widely biodegradable. This refractory nature also is related to the fact that they are highly resistant to chemical breakdown, which is a major reason they have come to enjoy such widespread commercial use. The order of their resistance to biological degradation, which is the same as to chemical degradation, is closely related to the halogen-carbon bond strength tabulated in Table 14.4 (Vogel, Criddle, and McCarty, 1987). The relative resistance to biodegradation of halogenated organics increases as one goes from bromide to chloride to fluoride. Brominated compounds are much easier to biodegrade than chlorinated

Table 14.4 Strengths of carbon-halogen bonds

Diatomic Molecule	Bond Strength kJ/mol
C-F	536
C-Cl	397
C-Br	280
C-I	209

SOURCE: *Handbook of Chemistry and Physics.* Boca Raton, FL: CRC Press.

compounds, but fluoridated compounds are highly resistant to biodegradation (Key, Howell, and Criddle, 1997).

Although CAHs are recalcitrant, many can be transformed biologically under suitable conditions. First, some can be used as an electron donor for energy generation and growth, either under aerobic or anaerobic conditions. Second, others are biotransformed strictly through cometabolism. Third, some can be used as an electron acceptor, either for energy generation or through cometabolism. The potential for biotransformation of the individual CAHs by these different processes is summarized in Table 14.3. Several reviews on biological processes for CAH biotransformation provide more information (McCarty, 1997a; McCarty, 1999; Semprini, 1997a; Vogel et al., 1987; Rittmann et al., 1994).

Chloromethane and dichloromethane can be used by many microorganisms as primary substrates for energy and growth under aerobic or anaerobic conditions. Dichloroethane and vinyl chloride also can be used aerobically for energy and growth. When other conditions are suitable, these CAHs are readily biodegraded.

At the other extreme are two of the highly chlorinated compounds, carbon tetrachloride (CT) and tetrachloroethene (perchloroethylene or PCE). No aerobic biodegradation process is yet known for these two compounds. Carbon tetrachloride can be transformed anaerobically, but this appears to be through cometabolism. PCE can be used as an electron acceptor under anaerobic conditions, either through cometabolism or in energy generation.

Many of the other highly halogenated compounds—such as trichloroethene (TCE), dichloroethene (DCE), 1,1,1-trichloroethane (1,1,1-TCA), and chloroform (CF)—can be biodegraded aerobically through cometabolism or anaerobically through use as electron acceptors, which may either represent cometabolism or dehalorespiration. Thus, CAH biotransformation lends itself to no simple generalizations. The potential for biotransformation and the particular process involved are somewhat unique for each compound.

The kinetics of substrate utilization and organism growth for organic compounds used as primary electron donors was discussed in the earlier chapters of this book. On the other hand, the kinetics for cometabolism and/or for use of organic compounds as electron acceptors is a new topic that is included in this chapter.

CAHs as Electron Acceptors Chlorinated organic compounds can be used as electron acceptors, a process termed *reductive dehalogenation*. This is illustrated for PCE conversion to ethene:

$$CCl_2=CCl_2 \xrightarrow[H^+ + 2e^- \quad Cl^-]{} CHCl=CCl_2 \xrightarrow[H^+ + 2e^- \quad Cl^-]{} CHCl=CHCl$$
$$\text{PCE} \qquad\qquad \text{TCE} \qquad\qquad \text{cDCE}$$

$$\xrightarrow[H^+ + 2e^- \quad Cl^-]{} CH_2=CHCl \xrightarrow[H^+ + 2e^- \quad Cl^-]{} CH_2=CH_2$$
$$\text{VC} \qquad\qquad \text{Ethene}$$

[14.2]

Two electrons and a proton are accepted by PCE, which is converted to TCE, releasing a chloride ion to solution. TCE can then be an electron acceptor and is converted in a similar fashion to DCE. Of the three possible isomers of DCE, *cis*-1,2-dichloroethene (cDCE) is the product most commonly formed from biotransformation. DCE then can be reduced to VC, which in turn can be reduced to ethene. Even ethene can accept electrons and be converted to ethane, although this process is seldom observed. Reductive dehalogenation may be *cometabolic* (i.e., the organisms obtain no benefit from the process). Or, the CAH may be a terminal electron acceptor in energy metabolism through *dehalorespiration.*

The ability to dehalogenate CAHs is widespread among microorganisms, but the extent of the reaction and the particular CAHs acted upon vary widely. Generally, strictly anaerobic microorganisms bring about reductive dehalogenation, but facultative organisms can bring about reduction of PCE to TCE and cDCE. Some methanogens can convert PCE to TCE through cometabolism. Many other organisms can convert PCE to cDCE as part of energy metabolism. Most of the CAH reducers use hydrogen as an electron donor, but some can use organic electron donors, such as acetate and pyruvate.

The conversion of cDCE to VC and VC to ethene tends to be slower than the steps from PCE to TCE and TCE to cDCE. Although reductive dechlorination of cDCE and VC has been demonstrated frequently in mixed cultures, only one microorganism capable of these conversions has been isolated to date (Maymo-Gatell, Chien, Gossett, and Zinder, 1997). This organism also can reductively dehalogenate PCE and TCE. It uses only hydrogen as an electron donor. It can obtain energy while using PCE, TCE, and cDCE as electron acceptors, but not while using VC. However, a growing body of circumstantial evidence indicates that some H_2-oxidizing microorganisms can use cDCE and VC as electron acceptors in energy metabolism, but lack the ability to use PCE and TCE.

The hydrogen that serves as electron donor normally is produced through fermentation of organic compounds in the absence of other readily used electron acceptors, as illustrated in Figure 1.40. Nearly 30 percent of the electrons available in mixed organic material is converted to hydrogen. The percentage varies for different compounds, as is illustrated by a comparison between benzoate and propionate, two compounds that have been added to aquifers to stimulate reductive dehalogenation:

Benzoate $C_6H_5COOH + 6\ H_2O = 3\ CH_3COOH + 3\ H_2 + CO_2$ **[14.3]**

Propionate $CH_3CH_2COOH + 2\ H_2O = CH_3COOH + 3\ H_2 + CO_2$ **[14.4]**

Both compounds are fermented to 3 mol of H_2, but benzoate fermentation results in the production of two more moles of acetate (CH_3COOH), which ties up eight electrons per mole. Thus, the percentage of electrons converted to hydrogen with propionate (43 percent) is much higher than with benzoate (20 percent). On a weight basis, propionate would be the better source of hydrogen for reductive dehalogenation.

The H_2 released by organic fermentation can be used by many different organisms, including sulfate reducers, Fe(III) reducers, Mn(IV) reducers, and methanogens. As with methanogens, there is a thermodynamic threshold below which hydrogen cannot be used because of energy limitations. The threshold for hydrogen-using methanogens occurs at a hydrogen partial pressure of about 10^{-6} atm. Translated to hydrogen solution concentration, this is 8 to 10 nM. The threshold for dehalogenation is lower, about 1 to 2 nM (Yang and McCarty, 1998). The thresholds for sulfate, iron, and manganese reducers are similar to or a bit lower than that for dehalogenation; thus, these electron acceptors compete with and may limit reductive dehalogenation.

The H_2 thresholds suggest that, in the absence of other electron acceptors, dehalogenating organisms should win out over methanogens in a competition for hydrogen. This would be true if the hydrogen concentration could be maintained below 8 to 10 nM. This may be possible in a CSTR, but may not be realistic in a plug-flow reactor, where the concentration of the added electron donor (organic matter) is high near the inlet point. Rapid fermentation of the organic matter generally leads to hydrogen concentrations above methanogen threshold levels. Based upon Equation 14.2, e^- eq are required for the reduction of 1 mol PCE to ethene. This translates to 8 mmol H_2 or 64 mg COD to reduce 1 mmol or 161 mg PCE. However, for the above reasons, methanogens commonly are more efficient at using an available electron donor; hence, 20 to 100 times the theoretical amount of donor may be needed to achieve complete reductive dehalogenation of PCE. However, the situation can be improved greatly if slowly fermented organic donors are used for reductive dehalogenation, as the slow release of hydrogen that results helps to maintain hydrogen near the lower hydrogen threshold for dehalogenation (Yang and McCarty, 2000). Examples of slow-release compounds are fatty acids and complex materials, such as lignocellulosics and decaying bacterial biomass.

Kinetics of Reductive Dehalogenation   Models for the reductive dehalogenation process can be somewhat complicated. We consider that one group (or species) of microorganisms converts PCE and TCE to cDCE, and a second group converts cDCE and VC to ethene. The following equation is for the growth rate (μ, d^{-1}) of an organism using either PCE or TCE (or both) as its electron acceptor and hydrogen (S_H) as its electron donor (Haston, Yang, and McCarty, 2000):

$$\mu = \left[\frac{\hat{\mu}_i S_i}{S_i + K_i \left(1 + \dfrac{S_j}{K_j}\right)} + \frac{\hat{\mu}_j S_j}{S_j + K_j \left(1 + \dfrac{S_i}{K_i}\right)} \right] \left[\frac{(S_H - S_H^*)}{(S_H - S_H^*) + K_H} \right] - b$$

[14.5]

where subscript i indicates parameters for PCE and subscript j represents parameters for TCE. S_H^* is the hydrogen threshold for dehalogenation. This equation represents dual-substrate limitation on growth with either an electron acceptor or H_2 (or both) in limited supply. In addition, the denominators in the left bracketed portion of the equation indicate competitive inhibition between the two CAHs. A similar equation can be written for the conversion of cDCE to VC and then ethene, with parameters for cDCE represented by subscript i and for VC represented by subscript j. Reported values for the various factors are contained in Table 14.5. The utilization rate for the CAH and the rate of increase in cell mass are obtained as usual using the following

Table 14.5   Organism growth and substrate utilization kinetic coefficients for anaerobic dehalogenation of PCE, TCE, VC, and ethene

Coefficient	Description	Reported Values	Units
$\hat{q}_{PCE}$	Maximum specific dechlorination rate of PCE to TCE	12 2.0 2.7, 4.6	$\dfrac{\mu\text{mol PCE}}{\text{mg VSS}_a\text{-d}}$
$\hat{q}_{TCE}$	Maximum specific dechlorination rate of TCE to cDCE	1.6 7.7	$\dfrac{\mu\text{mol TCE}}{\text{mg VSS}_a\text{-d}}$
$\hat{q}_{cDCE}$	Maximum specific dechlorination rate of cDCE to VC	0.6 5.7	$\dfrac{\mu\text{mol cDCE}}{\text{mg VSS}_a\text{-d}}$
$\hat{q}_{VC}$	Maximum specific dechlorination rate of VC to ethene	0.6 4.7	$\dfrac{\mu\text{mol VC}}{\text{mg VSS}_a\text{-d}}$
K_{PCE}	Half-velocity coefficient for PCE dehalogenation	0.11 0.05 2.8 10 200	μM
K_{TCE}	Half-velocity coefficient for TCE dehalogenation	1.4 1.5 4 240	μM
K_{cDCE}	Half-velocity coefficient for cDCE dehalogenation	3.3 3.0	μM
K_{VC}	Half-velocity coefficient for VC dehalogenation	2.6 360	μM
K_{H}	Half-velocity coefficient for hydrogen consumption	9-21 100	nM
S_{H}^{*}	Hydrogen threshold for dehalogenation	2 2	nM
$\hat{\mu}_{PCE}$	Maximum growth rate on PCE dechlorination	0.9 0.9 6.7 2.0	d^{-1}
$\hat{\mu}_{cDCE}$	Maximum growth rate on cDCE dechlorination	0.23 0.30 0.45 0.44	d^{-1}
$\hat{\mu}_{VC}$	Maximum growth rate on VC dechlorination	0.21 0.18 0.45	d^{-1}

SOURCE: Haston, Yang, and McCarty, 2000.

equations:

$$-\frac{dS_i}{dt} = \left[\frac{\hat{q}_i S_i}{S_i + K_i\left(1 + \dfrac{S_j}{K_j}\right)} \right] \left[\frac{(S_H - S_H^*)}{(S_H - S_H^*) + K_H} \right] X_a \qquad \textbf{[14.6]}$$

and

$$\frac{dX_a}{dt} = \mu X_a = Y_i\left(-\frac{dS_i}{dt}\right) + Y_j\left(-\frac{dS_j}{dt}\right) - bX_a \qquad \textbf{[14.7]}$$

Figure 14.13 illustrates the modeled change in concentrations of PCE and daughter products for a batch reactor. The initial concentrations of PCE and cDCE dehalogenating organisms were assumed to be the same, 0.0001 mg/l, and the hydrogen concentration was maintained constant in solution at 0.008 nM. This figure illustrates the commonly observed phenomena that PCE is converted into TCE and then cDCE, here within about 100 d. However, the conversion of cDCE and VC to ethene requires much more time, a total of about 300 d. Even then, some VC remains, which may be of concern, since VC is a known human carcinogen.

The results illustrated in Figure 14.13 assume that hydrogen remains at 0.008 μM. If readily degradable organic materials were depleted, the hydrogen concentration would decrease, slowing the growth rate and dehalogenation rates further. To model the changes accurately, the amount of electron donor present and the resulting solution hydrogen concentration should be included. This complicates the kinetics considerably.

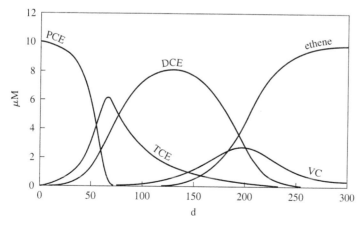

Figure 14.13 Transformation of PCE to ethene by reductive dehalogenation in a batch reactor and assuming two separate microbial populations, one that converts PCE into DCE, and the other, DCE to ethene. The starting concentration of each population is taken as 0.0001 mg/l.

Aerobic Cometabolism of Chlorinated Aliphatic Hydrocarbons

Several of the CAHs can be transformed aerobically through cometabolism. Wilson and Wilson (1985) first demonstrated this for transformation of TCE by methanotrophic bacteria. The two steps are shown here:

Methane Oxidation (Normal Metabolism):

$$CH_4 \xrightarrow[\text{2H, O}_2]{MMO} CH_3OH \longrightarrow H_2CO \longrightarrow HCOOH \longrightarrow CO_2$$

$$\qquad\qquad\qquad\qquad 2H \quad \text{Synthesis} \quad 2H \qquad\qquad 2H$$

$$[14.8]$$

TCE Epoxidation (Cometabolic Dechlorination):

$$CCl_2CHCl \xrightarrow[\text{2H, O}_2]{MMO} \underset{Cl_2C - CHCl}{\overset{O}{\triangle}} \longrightarrow \longrightarrow CO_2, Cl^-, H_2O$$

$$[14.9]$$

Here, methane monooxygenase (MMO) initiates the oxidation of methane by forming methanol. This also requires molecular oxygen and a supply of reducing power, noted as 2H. MMO also fortuitously oxidizes TCE, converting it into TCE epoxide, an unstable compound that degrades chemically to a variety of products that can be mineralized by other microorganisms. Thus, TCE biodegradation is effected through cometabolism.

TCE cometabolism does not benefit the methanotrophs; indeed, the degradation products of TCE epoxide are toxic to the methanotrophs. This realization led to the concept of *transformation capacity* and the related *transformation yield*. Transformation capacity (T_c) is the quantity of TCE that a given mass of microorganisms can degrade before they are killed by the transformation. Transformation yield (T_y) is the maximum mass of TCE degraded per unit mass of methane used to grow methanotrophs. The two are related $(T_y = Y T_c$, where Y is the yield constant for growth on methane). Another factor of importance is the reducing power available to the organism for cometabolism. Cometabolism requires NADH to provide the 2H in Equation 14.9. If the organism is not supplied with a donor substrate or does not have internal storage reserves to produce the required NADH, then cometabolism cannot occur. This results in a lower observed transformation capacity (Chang and Alvarez-Cohen, 1995; Criddle, 1993). Reported values of T_c for TCE are contained in Table 14.6 for situations when sufficient NADH is available and when it is not.

A broad group of primary substrates effect aerobic cometabolism of TCE (McCarty, 1999). In addition to methane, they include aliphatics such as ethane, ethene, propane, propene, butane, isoprene; aromatic compounds such as toluene, cresol, and phenol; and an inorganic compound, ammonium. An organism that oxidized vinyl chloride as a primary substrate for energy and growth also cometabolized several CAHs. The susceptibility to cometabolism is different for different CAHs, different cultures, and different primary substrates, as suggested by T_c values in Table 14.6. TCE, DCE, and VC are readily cometabolized by many different organisms, but TCA cometabolism is more restricted. Carbon tetrachloride and PCE are not aerobically cometabolized by any organisms so far identified. Often, if two CAHs are present, competitive inhibition between the two is involved. Also, since the primary substrate and CAH compete for the same oxygenase, competitive inhibition between the primary substrate and CAH is generally important.

The many interactions involved in cometabolism make the reaction kinetics complicated. Equation 14.10 contains many of the factors of importance for describing the specific growth rate (Anderson and McCarty, 1994; Chang and Alvarez-Cohen, 1995):

$$\mu = \left[\frac{1}{1 + \frac{S_d}{K_d} + \frac{S_c}{K_c}} \right] \left[\frac{Y\hat{q}_d S_d}{K_d} - \frac{\hat{q}_c S_c}{T_c K_c} \right] - b \qquad \textbf{[14.10]}$$

Here, the subscripts d and c represent the electron donor used for energy and growth and the CAH being biotransformed through cometabolism, respectively. The left bracketed term accounts for competitive inhibition. The right bracketed term contains the usual growth rate kinetic term (on the left), but also a term (on the right) that reduces growth rate because of toxicity of CAH degradation products. This equation illustrates that the organism's growth rate becomes negative when the CAH concentration (S_c) is too high. A high S_c also reduces the growth rate through competitive inhibition (left bracketed term).

The biotransformation of the CAH also is somewhat complicated, depending upon whether or not sufficient NADH is available for the epoxidation of the CAH to

Table 14.6 Mixed culture transformations capacities (mg CAH/mg VSS) for different CAHs and primary substrates (as determined with an external energy source supplied and without an external energy source supplied)

CAH	Methane		Propane		Toluene		Phenol		Ref.
	Energy	No Energy	Energy	No Energy	Energy	No Energy	Energy	No Energy	
TCE	0.11–0.53	0.05–0.25	0.014	0.0065	0.0073	0.0085	0.019–0.38	0.031–0.51	a
CF		0.023		0.0038					b
1,2-DCA		0.107		0.012					b
cDCE	0.50								c
tDCE	4.9								c
1,1-DCE	0.0019								c
VC	1.4								c

[a]Values as summarized from the literature by Semprini (1997b).
[b]Chang and Alvarez-Cohen (1995).
[c]Anderson and McCarty (1997).

occur. When it is, then the degradation rate for the CAH is a maximum and given simply by:

$$-\frac{dS_c}{dt} = \left(\frac{\hat{q}_c S_c}{S_c + K_c}\right) X_a \qquad \textbf{[14.11]}$$

In order to avoid competitive inhibition, a frequent proposal is to grow organisms in one reactor and then use them in a second reactor for CAH degradation. Since no growth substrate is present in the second reactor, the organisms are said to be "resting." Resting cells decrease in concentration due to normal decay (b), but also decrease while degrading TCE by transformation-product toxicity, the amount being indicated by the transformation capacity T_c. Here, the rate of decay is generally much less than the rate of death due to TCE transformation. An equation indicating the rate of TCE degradation in a batch reactor with resting cells is

$$-\frac{dS_c}{dt} = \left(\frac{\hat{q}_c S_c}{S_c + K_c}\right)\left(X_a^0 - \frac{S_c^0 - S_c}{T_c}\right) \qquad \textbf{[14.12]}$$

Figure 14.14 illustrates the course of TCE degradation in a batch reactor with resting cells for different transformation capacities. Greater TCE degradation is associated with a higher transformation capacity. For low transformation capacity, transformation products kill the microorganisms before much of the TCE can be degraded.

Chang and Alvarez-Cohen (1995) discuss the many complications that might result because of CAH transformation product toxicity and/or depletion of the or-

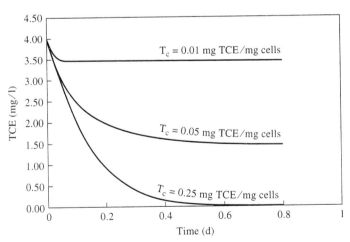

Figure 14.14   Effect of transformation capacity on the rate and extent of TCE biodegradation through aerobic cometabolism with resting cells. Data are representative of cometabolism by methanotrophs. Here, X_a = 50 mg/l, $\hat{q}$ is 1 mg TCE/mg cells-d, and K = 4 mg TCE/l.

ganisms' energy source. Successful cometabolism can be achieved by ensuring that primary substrate is well in excess of that needed to overcome toxicity from the CAH or its transformation products. An example is the *in situ* cometabolic biodegradation of TCE in an aquifer through the injection of toluene at Edwards Air Force Base (McCarty et al., 1998; National Research Council, 2000).

14.6.5 CHLORINATED AROMATIC HYDROCARBONS

The chlorinated aromatic compounds include single-ringed chlorinated benzenes, chlorinated phenols, and chlorinated benzoates; and two-ringed compounds such as the polychlorinated biphenyls. The single-ringed compounds generally are readily biodegraded and serve as primary substrates for organism energy and growth under aerobic conditions. Anaerobic biodegradation for these compounds is not as readily achieved. As indicated above for chlorobenzoic acid under the section on electron donors and acceptors, anaerobic degradation may involve reductive dehalogenation, in which the chlorine is removed from the ring. The chlorobenzoate is an electron acceptor. However, the benzoate becomes an electron donor.

Summarized here are three highly important members of this group: polychlorinated biphenyls, pentachlorophenol, and dioxins.

Polychlorinated Biphenyls (PCBs)   PCBs came into extensive use in the 1950s in electrical capacitors and transformers, hydraulic fluids and pump oils, and adhesives, dyes, and inks (Tiedje et al., 1993). Their highly desirable features were a great resistance to fire and explosive hazards and excellent electrical insulating properties (Boyle et al., 1992). Unfortunately, PCBs also are resistant to chemical and biological attack. Through inadvertent spillage and leakage, or uninformed disposal practices, PCBs were disseminated in the environment, affecting soils, rivers, lakes, estuaries, oceans, and sediments. Well after major contamination took place, widespread adverse effects were recognized in aquatic life, birds, and animals. As a result, the U.S. Congress banned the use of PCBs in 1976. Because of their great resistant to natural degradation processes, PCBs are still widely present in the environment. An estimated 10^8 kilograms reside in the biosphere (Boyle et al., 1992).

The backbone of the PCB molecule is biphenyl. The process for chlorination of biphenyl leads to the production of a mixture of chlorobiphenyls with different numbers of chlorine atoms per molecule (Boyle et al., 1992). Some examples are illustrated in Figure 14.15. By substitution of chlorine atoms on any of the 10 unlinked carbon atoms on biphenyl, it is possible to produce 209 different *congeners*. However, a typical synthetic PCB mixture contains between 60 and 80 different chlorinated biphenyl congeners. PCB mixtures commonly go by the trade name *Aroclor*, which is followed by a four-digit descriptor of the mixture composition. The first number (12) refers to the biphenyl parent structure, and the second two digits represent the average weight percent of chlorine in the mixture. Typical mixtures are Aroclor 1221, 1242, and 1254.

In spite of their highly resistant nature, PCBs can be biodegraded when the right microorganisms and environmental conditions occur. PCB degradation can occur

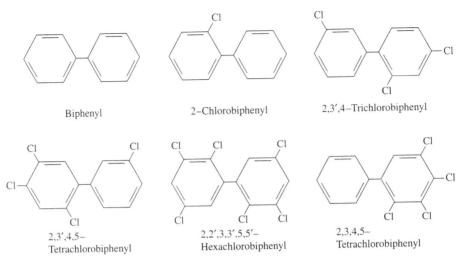

Biphenyl   2–Chlorobiphenyl   2,3′,4–Trichlorobiphenyl

2,3′,4,5–   2,2′,3,3′,5,5′–   2,3,4,5–
Tetrachlorobiphenyl   Hexachlorobiphenyl   Tetrachlorobiphenyl

Figure 14.15   Biphenyl and a few of the 209 possible chlorinated biphenyl congeners that comprise the PCBs.

under either aerobic or anaerobic conditions. As with chlorinated organics in general (Vogel et al., 1987), the less chlorinated PCBs are transformed most readily aerobically, while the more chlorinated ones are more readily transformed anaerobically.

Under aerobic conditions, the biphenyl molecule is degraded in a manner similar to other aromatic hydrocarbons: *oxygenase* addition of molecular oxygen to one of the aromatic rings, resulting in hydroxyl groups (Boyle et al., 1992). Further oxidation leads to the cleavage of one ring, and its further degradation eventually leads to benzoic acid. Benzoic acid then goes through ring oxidation, ring cleavage, and further oxidation through the usual pathways for aromatic hydrocarbons. PCBs with few chlorines on the molecule can easily enter the aerobic pathway for biphenyl oxidation, but greater chlorination makes enzymatic attack on the ring difficult. PCBs most subject to aerobic degradation are those containing one to three chlorine atoms on the molecule. The most difficult to degrade are PCBs with chlorine atoms in the *ortho* (or 2) position on the molecule.

Reductive dechlorination of PCBs was first noted in the Hudson River in the early 1980s (Brown et al., 1987) and has been studied is some detail since then. Figure 14.16 illustrates the common situation. Shown here in black is the usual distribution of PCBs in the original Aroclor 1260 mixture. After time in contact with actively dechlorinating sediments, the mixture composition is shifted towards less chlorinated congeners. As in other reductive dehalogenations, PCBs act as electron acceptors, and this requires the presence of electron donors (Tiedje et al., 1993). In stream sediments, organic detritus such as dead algae and organic soil particles likely serve this purpose. Dechlorination is faster for the more highly chlorinated congeners, and rates of dechlorination are usually *meta > para > ortho* (Tiedje et al., 1993).

Whether or not reductive dechlorination of PCBs is associated with energy production and growth has not been established. Undoubtedly, one of the major barriers

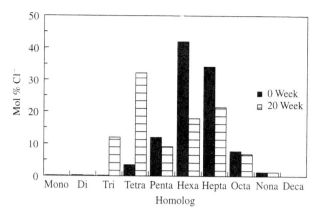

Figure 14.16   Dechlorination of 5,000 ppm of Arochlor 1260 in Hudson River sediment inoculated with a population enriched from an industrial sludge lagoon. SOURCE: Tiedje et al., 1993.

to more rapid dehalogenation is the insolubility of the higher chlorinated PCBs and their strong partitioning into sediment particles, rendering them less available for biodegradation. The realization that PCBs can be anaerobically transformed from more highly chlorinated congeners to less chlorinated ones having decreased toxicity and increased potential for aerobic biodegradation has lead to interest in using natural attenuation as a remediation alternative to dredging and other expensive approaches (National Research Council, 2000).

Pentachlorophenol   Pentachlorophenol (PCP) has been and is one of the most widely used wood preservatives (McAllister, Lee, and Trevors, 1996). As with most wood preservatives, it is toxic and listed as a priority pollutant by the U.S. Environmental Protection Agency. Because of its widespread use, PCP has contaminated many environments, such as soil and groundwater at wood preserving sites. Technical grade PCP contains 85–90 percent PCP, the other components being related compounds such as tri- and tetra-chlorophenols, predioxins, and iso-predioxins. The latter are formed from condensation of two molecules of PCP or other chlorinated phenol isomers.

Phenol is a weak acid ($pK_a = 9.82$) and thus not ionized at pH 7. However, chlorine is an electron-withdrawing group that makes the hydroxyl group on the chlorinated phenols a stronger acid. For example, 2,4,6-trichlorophenol has a pK_a of 6.13, and pentachlorophenol has a pK_a of 4.75 (Schwarzenbach et al., 1993). Thus, phenols with three or more chlorine atoms on the molecule are significantly ionized at neutral pH. Being ionizable, the phenols are quite soluble in water and are commonly found as groundwater contaminants. However, PCP also has a high octanol/water partition coefficient (over 10^5) and partitions quite strongly into high organic content soils and sediments. Partitioning is pH dependent and is greater at low pH. The sodium salt of PCP is the most common commercial form.

PCP and other chlorinated phenols are biodegradable aerobically and anaerobically. The aerobic biodegradation pathways were reviewed by McAllister et al. (1996), while Sanford and Tiedje (1997) and Cozza and Woods (1992) reported the potential for anaerobic degradation. Following the normal pattern, aerobic degradation occurs more readily with the less chlorinated phenols, and anaerobic dehalogenation occurs more readily with the more highly chlorinated phenols. Aerobic microbial degradation processes generally involve initial substitution of an $-OH$ group for $-Cl$. The anaerobic process is generally reductive dehalogenation.

PCP is toxic to microorganisms at concentrations as low as 1 mg/l, but tolerance to much higher concentrations (in the hundred milligram per liter range) can be obtained through acclimation (McAllister et al., 1996). The combined effect of strong partitioning to organic matter in soil and toxicity makes aerobic biodegradation of PCP somewhat difficult. However, when appropriate conditions are maintained, PCP and the less-chlorinated phenols can be readily degraded aerobically. Many pure cultures have been isolated, and most obtain cell carbon and energy from PCP oxidation and mineralization. With aerobic oxidation, a first step is *oxygenolytic dehalogenation,* or the substitution of an $-OH$ group for a $-Cl$. The first step could occur through hydrolysis or through a fortuitous action of an oxygenase enzyme, molecular oxygen, and two NAD(P)Hs. Different organisms appear to have different pathways for aerobic metabolism, including steps that involve hydrolysis and reductive dechlorination.

Anaerobic degradation of PCP also occurs. This is of critical importance, because groundwaters contaminated by PCP are generally anaerobic. Although pure cultures that can degrade PCP anaerobically are rare, anaerobic PCB degradation occurs readily in mixed cultures. PCP serves as an electron acceptor in reductive dehalogenation in the same manner as for other aliphatic and aromatic halogenated organic compounds. Thus, in order to obtain anaerobic dehalogenation of PCP, an electron donor, such as acetate or molecular hydrogen, needs to be present. Whether or not organisms obtain energy from anaerobic dehalogenation of PCP is still an open question, but the evidence suggests that it is likely (Sanford and Tiedje, 1997; Stuart and Woods, 1998). Reductive dehalogenation of PCP can take several pathways because of the many chlorine atoms on the molecule. Cozza and Woods (1992) reported a correlation between the largest negative value for the carbon-chlorine bond charge and the chlorine atoms removed by reductive dehalogenation. The correlation predicts the pathways shown in Figure 14.17. *Ortho* dechlorination occurs most readily, followed by *para* and then *meta* dechlorination. Complete anaerobic dehalogenation to phenol is possible, and phenol biodegrades anaerobically. Mineralization of PCP under anaerobic conditions can provide the electrons needed to drive reductive dechlorination.

Dioxins   Dioxins are chemicals of significant environmental concern, although they are not produced commercially as marketable products. They are formed as unwanted by-products (Halden and Dwyer, 1997) of pesticide manufacturing, combustion and incineration, chlorine bleaching and disinfection, and controlling dust and sediments. Major incidences of dioxin problems include contamination of 1.44 million tons of soil at Times Beach, Missouri, as a result of spreading dioxin-containing

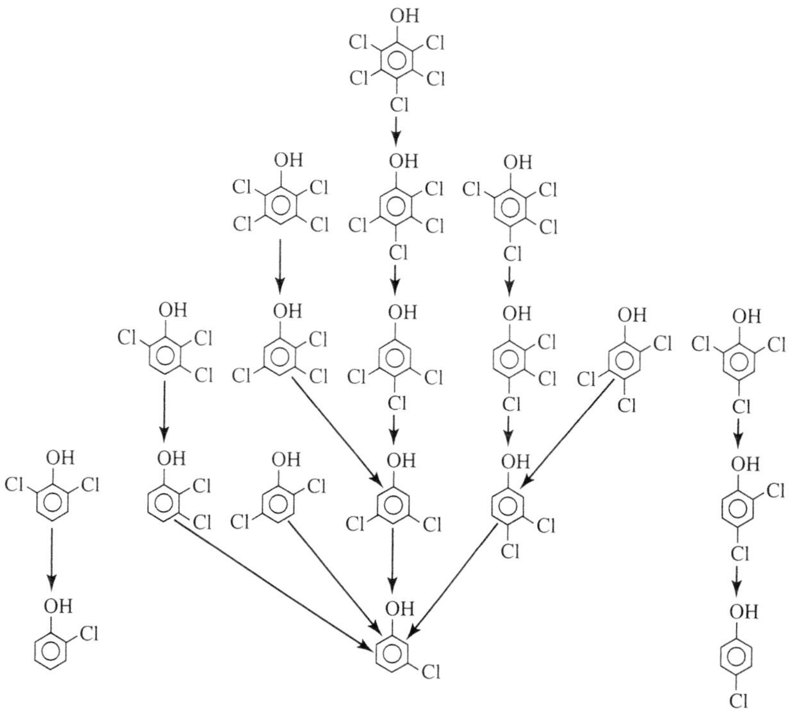

Figure 14.17 Likely pathways for anaerobic dehalogenation of chlorinated phenols based upon electronic properties of the molecule. SOURCE: Cozza and Woods, 1992.

oil on roads for dust control; widespread contamination of the countryside due to the explosion of a pesticide plant in Seveso, Italy, in 1976; and numerous human exposures due to the use of 2,4,5-T herbicide, which formerly contained high levels of dioxin as a by-product of manufacture.

Dioxins have many possible congeners, the most toxic of which is *2,3,7,8-tetrachlorodibenzo-p-dioxin (TCDD)* (Figure 14.18), which is thought to be the most poisonous synthetic chemical ever produced (Halden and Dwyer, 1997). Its biological effects, including teratogenicity and carcinogenicity, are species, sex, and exposure-route dependent. Human responses include chloroacne, weight loss, insomnia, liver dysfunction, spontaneous abortion, and possibly cancer. The toxicity of the different dioxin congeners varies widely from low to extreme. Thus, in order to evaluate possible toxic effects, congener-specific analysis is required.

A group of related compounds to dioxins are *dibenzofurans* and *diphenyl ethers* (Figure 14.18). Excavation and incineration of soils are cleanup procedures sometimes used for dioxins and related compounds, but they are expensive. Biodegradation is a desirable alternative, but dioxins and related compounds are highly resistant to biodegradation. Nevertheless, microorganisms can bring about their conversion under aerobic and anaerobic conditions (Halden and Dwyer, 1997). Whether or not

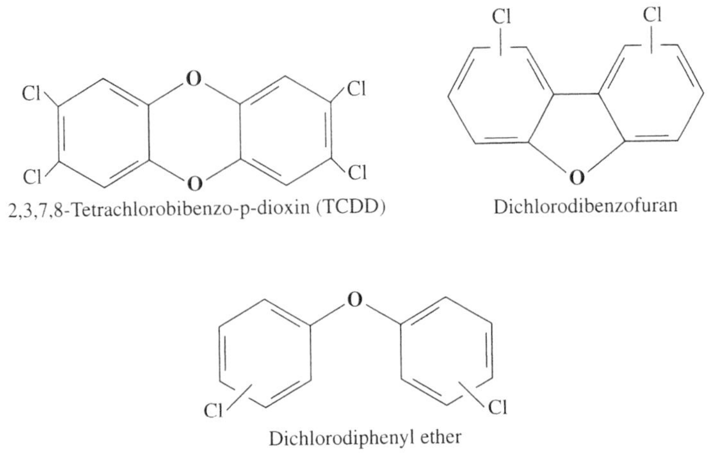

2,3,7,8-Tetrachlorobibenzo-p-dioxin (TCDD)

Dichlorodibenzofuran

Dichlorodiphenyl ether

Figure 14.18   Structure of chlorinated dioxin and related compounds.

the advantages of biodegradation for cleanup of dioxin-contaminated sites can be captured remains to be seen.

In many studies of aerobic biodegradation, only limited biotransformation of dioxins is observed. It generally involves the addition of oxygen with oxygenase enzymes, resulting in the unproductive formation of mono- and di-hydroxylated analogs, which tend to accumulate with no further degradation (Halden and Dwyer, 1997). However, bacteria that contain selective dioxygenases that simultaneously substitute an −OH group at an ether bond-carrying carbon and an adjacent unsubstituted carbon lead to the cleavage of ether bonds and compound mineralization.

Reductive dechlorination of dioxin has been reported under anaerobic conditions (Albrecht, Barkovskii, and Adriaens, 1999). Interestingly, the TCDD congener can be formed through reductive dehalogenation of other congeners, as well as being dechlorinated itself. As with PCBs, dehalogenation of dioxins is a slow process. Very low solution concentrations and the high tendency to partition to sediments reduces biological availability. Biodegradation studies suggest that natural attenuation of dioxins through biotransformation may be possible, but much needs to be learned about factors affecting their movement, fate, and effects in the environment.

14.6.6   EXPLOSIVES

The manufacture, use, and disposal of explosives from military operations have resulted in extensive contamination of soils and groundwater (Pennington, 1999; Spain, 1995). Figure 14.19 illustrates the most commonly used military explosives: TNT, RDX, and HMX. Treatment of wastewaters from the manufacture of explosives is a long-standing challenge. Interest in decommissioning military bases for other uses has led to an intensified need to remediate explosives-contaminated environments. While incineration is one of the options commonly considered for soils, it is im-

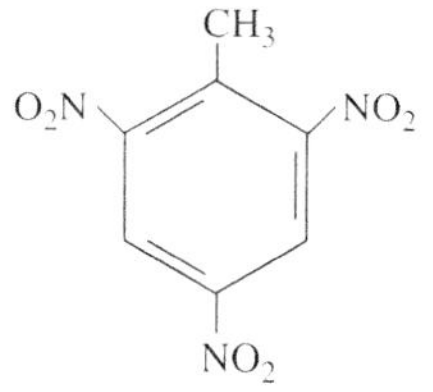

TNT
(2,4,6-Trinitrotoluene)

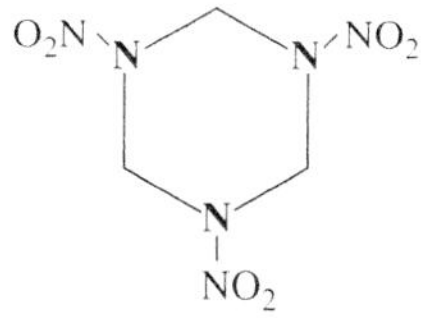

RDX
(Hexahydro-1,3,5-trinitro-
1,3,5-triazine)

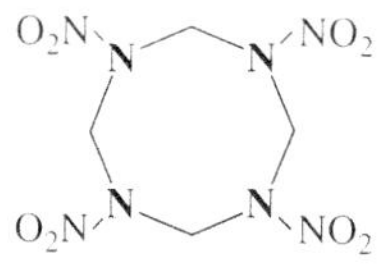

HMX
(Octahydro-1,3,5,7-tetra-
nitro-1,3,5,7-tetrazocine)

Figure 14.19 Common explosives of environmental concern.

possible for wastewaters and expensive in general. Research on the potential for bioremediation is an ongoing effort that is yielding some positive results.

A characteristic of explosives is the presence of nitro (-NO$_2$) groups on the molecule (Figure 14.19). Transformation of TNT generally results in the reduction of a nitro group to form an amino group. Thus, commonly found in TNT-contaminated soils are intermediate products such as 4-amino-2,6-dinitrotoluene (4ADNT) and 2-amino-4,6-dinitrotoluene (2ADNT). Less frequently found are the further reduction products, 2,4-diamino-6-nitrotoluene (2,4DANT), and 2,6-diamino-4-nitrotoluene (2,6DANT). The further reduction product 2,4,6-triaminotoluene (TAT) is found in the laboratory, but has not been reported in the environment (Pennington, 1999). The reductions can be enzymatically catalyzed or, under proper reducing conditions, abiotic.

The amino transformation products are subject to further interactions with each other and with components of soils, forming stable complexes through covalent bonding. Resulting products are large and insoluble. Thus, disappearance of TNT is often observed in biologically active environments, but mineralization does not commonly occur. Whether or not the products formed from transformation and interactions are harmful to humans or the environment is not known. Slow mineralization of TNT and its transformation products does appear to occur in soil in some instances.

While TNT biodegradation is problematic, toluene, benzene, and phenols with one or two nitro groups on the molecule are more subject to biodegradation and mineralization (Spain, 1995; Rittmann et al., 1994). Initial monooxygenation or dioxygenation reactions can lead to the release of $-NO_2$ groups and substitution of $-OH$. This can lead to mineralization. In some cases, the nitro group is reduced to an amino group, which accumulate as dead-end products.

RDX is more readily mineralized than TNT and is not subject to partial transformation and sequestration processes (Pennington, 1999). However, the rate at which RDX disappears usually is slower than for TNT. HMX biotransformation occurs primarily under anaerobic conditions, leading to the formation of mono- and di-nitroso intermediates. Its biotransformation is slower than that of RDX. The explosives are also transformed, at least partially, by plants, leading to some interest in the potential for phytoremediation (Pennington, 1999). Because of the complexity of the transformation processes involved, the unknown pathways for transformation, and the unknown environmental and health hazards of degradation products, much must be learned about the potential effectiveness of bioremediation for explosives.

14.6.7 GENERAL FATE MODELING FOR ORGANIC CHEMICALS

A recurring theme in this chapter is that many hazardous organic chemicals are hydrophobic, which means that they partition to solid or gas phases. Chemicals in these nonaqueous phases are not directly available for biodegradation. They also can come into contact with human or environmental receptors through contact with the contaminated gas or solids. We must account for the partitioning if we are to understand the fate of hydrophobic organic chemicals in treatment reactors or in the environment.

Concern about the release of volatile organic chemicals (VOCs) from wastewater treatment plants led to *general fate models* in the late 1980s and 1990s (e.g., Namkung and Rittmann, 1987b; Rittmann et al., 1988; Govind et al., 1991; Enviromega, 1994; Lee et al., 1998). These models are excellent examples of how phase-transfer processes are integrated with the microbial reactions presented in this book.

As an introduction to integration and general fate modeling, we consider an activated sludge treatment plant whose influent contains normal BOD and two types of hydrophobic organic chemicals: VOCs (like toluene and TCE) and strongly sorbable chemicals (like LAS). To keep the model as simple as possible, we assume that the hydrophobic compounds are present in the influent at low concentrations and are utilized by the microorganisms through secondary utilization. This assumption means that the biomass concentration is determined solely by the influent concentration of BOD, θ_x, and θ; the analysis techniques of Chapters 3, 5, and 6 are applicable here.

The steady-state mass balance on a hydrophobic chemical (indicated by concentration C) in the water is

$$0 = \text{Advection In} - \text{Advection Out} - \text{Volatilization} - \text{Sorption} - \text{Biodegradation}$$
[14.13]

Each of the terms is derived.

Advection in and out are simply stated by the product of the flow rate times concentration in that flow:

$$\text{Advection In} = Q \cdot C^0 \quad \textbf{[14.14]}$$
$$\text{Advection Out} = Q \cdot C \quad \textbf{[14.15]}$$

in which Q is the influent flow rate (L^3T^{-1}], C^0 is the influent concentration (M_cL^{-3}],
and C is the effluent soluble concentration [M_cL^{-3}].

Volatilization is comprised of two parts: volatilization at the surface and volatilization to bubbles in diffused aeration. Together, the volatilization rate is

$$\text{Volatilization} = k_La_cCV + Q_ACH_c \qquad \textbf{[14.16]}$$

in which k_La_c is the surface gas transfer rate coefficient [T^{-1}], V is the volume being aerated [L^3], Q_A is the aeration gas volumetric flow rate [L^3T^{-1}], and H_c is the Henry's constant in units L^3 water (L^3 gas)$^{-1}$.

The loss rate by sorption is equal to the density of the compound in the wasted sludge times the sludge-wasting rate, as long as adsorption is at equilibrium. In that case,

$$\text{Sorption} = X_vVK_pC/\theta_x \qquad \textbf{[14.17]}$$

in which X_v is the MLVSS concentration [M_xL^{-3}], V is the volume in which X_v is contained [L^3], θ_x is the solids retention time [T], and K_p is a linear partitioning coefficient [$L^3M_x^{-1}$].

Since the concentration of the target compound is low, we assume that the rate of biodegradation is first-order in C:

$$\text{Biodegradation} = k_1X_aCV \qquad \textbf{[14.18]}$$

in which k_1 is the mixed second-order rate coefficient [$L^3M^{-1}T^{-1}$], and X_a is the concentration of active biomass degrading the contaminant [M_xL^{-3}]. We shall assume that all of the active biomass utilizes the target compounds, which means that we can use the methods of Chapters 3, 5, and 6 to estimate X_a.

Substituting Equations 14.14 to 14.18 into Equation 14.13 gives

$$0 = QC^0 - QC - k_La_cCV - Q_ACH_c - X_vVK_pC/\theta_x - k_1X_aCV \quad \textbf{[14.19]}$$

Equation 14.19 can be rearranged to solve for C:

$$C = C^0/[1 + k_La_c\theta + Q_AH_c/Q + X_vK_p\theta/\theta_x + k_1X_a\theta] \qquad \textbf{[14.20]}$$

in which $\theta = V/Q$ is the hydraulic retention time [T]. Equation 14.20 is very simple to use, as long as we can measure or predict the parameters. Each of the five terms in the denominator is the relative rate compared to advection in the effluent. Thus, if we know k_1X_aV, we know immediately how important it is compared to advection (value of 1) or bubble volatilization (Q_AH_c/Q). The actual rates can be computed immediately by substituting the value of C into Equations 14.14 to 14.18.

The assumptions behind Equation 14.20 must not be overlooked, because they are not always valid. For instance, adsorption is not necessarily at equilibrium. If that is the case, then Equation 14.17 usually overestimates the removal in the waste-sludge flow. Likewise, a target compound may not be biodegraded by secondary-utilization alone or at all (Lee et al., 1998). Then, the fraction of the biomass active in degrading the target compound must be predicted separately from X_a. Furthermore, the biodegradation kinetics are not always first-order in C. Monod kinetics or some form of inhibition kinetics or cometabolic kinetics may be more appropriate. Even when the simple approach leading to Equation 14.20 is not completely correct, the general mass balance equation (Equation 14.13) remains the right starting point.

14.6.8 INORGANIC ELEMENTS

Microbial processes can affect the fate of inorganic environmental contaminants (National Research Council, 2000). Although inorganic elements cannot be destroyed, microorganisms can change their speciation in such a way that their mobility and toxicity are markedly altered. Precipitation, volatilization, sorption, and solubilization are the reactions that alter the mobility of the inorganic elements. Sometimes the microbially catalyzed redox reactions lead directly to the reactions that alter mobility. In other cases, the effects are indirect when the microorganisms alter the geochemistry of their environment. Efforts are now being made to take advantage of beneficial microbially mediated changes to inorganic contaminants. Reviews of inorganic interactions with microorganisms are available in Stumm and Morgan (1996), National Research Council (2000), Chappelle (1993), and Banaszak, Reed, and Rittmann (1999).

One method by which microorganisms can reduce concentrations of inorganic elements in water is through *immobilization in microbial biomass* and/or microbial exopolymers. Hydrophobic sorption, complexation, or incorporation as essential cofactors (Stumm and Morgan, 1996) are the mechanisms by which inorganic elements sometimes can be concentrated into biomass. The biomass must be removed and disposed of by some environmentally sound method in order for immobilization to work in the long term.

Behind the transformation of inorganic contaminants are oxidation-reduction reactions mediated by microorganisms. Figure 14.20 illustrates graphically the pH-dependent half-reaction potentials for various redox couples. Any one of the oxidized species (the left member of each couple) theoretically could be an electron acceptor in the oxidation of some organic or inorganic electron donor. Then, they would be converted to the reduced species (the right member of each couple). Note the location of the O_2/H_2O line. Oxygen is a more favored electron acceptor than any species with lines below the O_2/H_2O line. For example, the SO_4^{2-}/HS^- line lies well below the O_2/H_2O line, meaning that sulfate is much less energetically favorable than oxygen and can serve as an electron acceptor only under highly reducing conditions. Figure 14.20 illustrates important characteristics of many inorganic compounds of interest.

Perchlorate is used in some rocket fuels as an oxidant, and it contaminates waters around some manufacturing plants and rocket test facilities, causing widespread pollution of ground and surface waters in some cases. *Chlorate* (ClO_3^-) and *perchlorate* (ClO_4^-) lie well above oxygen; theoretically, they provide more energy for microorganisms if used as electron acceptors. Both are used by some microorganisms in energy metabolism of organic electron donors and H_2. Whether or not they would be selected as electron acceptors in the presence of oxygen is debatable, as organisms' regulatory mechanisms may prevent this, but many microorganisms can use both in the absence of O_2. One of the promising treatment processes for removal of perchlorate is biological reduction to chloride using readily available organic donors, such as acetate or H_2.

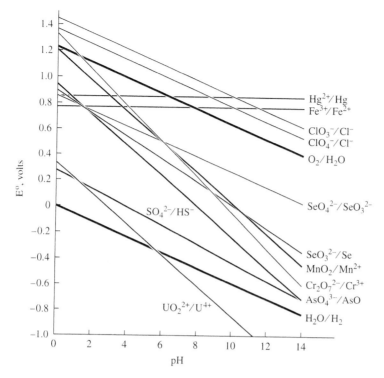

Figure 14.20   Effect of pH on reduction potentials for various inorganic species.

Mercury reduction may be a mechanism for detoxification of media containing mercury. The Hg^{2+}/Hg line intersects the oxygen line. Reduction of Hg^{2+} to elemental mercury occurs quite readily and is enhanced by bacterial enzymes. Energy is probably not obtained by use of Hg^{2+} as an electron acceptor.

Iron oxidation and reduction are very well known. The Fe^{3+}/Fe^{2+} line illustrates that, at high pH, Fe(III) could be a more favorable electron acceptor than oxygen. However, Fe(III) it is quite insoluble at high pH and not so available. However, at neutral or lower pH, where Fe(III) is fairly soluble, it does serve as an electron acceptor for energy metabolism under anaerobic conditions, resulting in the formation of Fe(II). Under aerobic conditions at neutral to lower pH, Fe(II) can serve as an electron donor and is biologically oxidized to Fe(III).

The potential for using Mn(IV) (e.g., $MnO_{2(s)}$) as an electron acceptor also is well known. Like for iron, the reduced manganese (e.g., Mn^{2+}) can be an electron donor. *Manganese* cycling between the (II) and (IV) states is important in sediments, lakes, and some groundwaters.

Particularly intriguing species illustrated in Figure 14.20 are the oxidized forms of *selenium, chromium, arsenic,* and *uranium.* Theoretically, all could serve as electron acceptors under anaerobic conditions. Indeed, they are acceptors under anaerobic conditions for some microorganisms. Whether or not the resulting reductions are cometabolic or energy yielding is not clear. In either case, an electron donor must be available for organism growth. In general, the reduced forms are much less soluble and mobile. Thus, potential exists for engineering a biological treatment system to reduce these elements and remove them from the water as solids.

Changes in the oxidation states of inorganic species also can be indirectly caused when the microorganisms create oxidizing or reducing conditions. For example, microbial reduction of sulfate produces sulfide, which acts as a chemical reductant for many species. Likewise, reduced iron and manganese, produced microbially, can be electron donors for abiotic reduction of other inorganic elements. These biotic/abiotic connections add complexity to the oxidation-reduction processes with inorganic elements, often making it difficult to determine just what are the specific fate mechanisms and their rates. On the other hand, the abiotic reactions expand the scope of ways in which inorganic elements can be transformed, often to our benefit.

Another factor of great importance for some inorganic species is the potential for microorganisms to methylate them, often yielding highly toxic and mobile compounds. Examples are the *methlyation* of mercury, arsenic, and selenium. Methylmercury and dimethylmercury are highly toxic compounds formed from Hg(II) under anaerobic conditions, generally by sulfate-reducing bacteria. The rate of the reaction is related to the rate of sulfate reduction, which depends on sulfate and electron-donor concentrations. Methylmercury is a detoxification method for the organisms, which want to move the mercury away from them. Hence, they form the more mobile methylated form, which can diffuse from the sediments into overlying waters. There, the methyl mercury can be concentrated by planktonic species and then concentrated further in the fatty tissue of fish. The mercury is further concentrated in the fatty tissues and brains of those consuming the fish, including humans, leading to Minamata disease. This disease of the central nervous systems claimed many victims around the Minamata River delta in Japan during the 1950s as a result of mercury discharges from an organic chemical plant.

The objective of most processes for treatment for inorganic hazardous materials is to concentrate and immobilize them in order to prevent their migration to human or environmental receptors. The biological reduction of Cr(VI) to Cr(III) results in the conversion of a highly toxic, soluble, and mobile species into a species having much lower toxicity and solubility. This desirable outcome of reduction reactions occurs for many metals, including Cu, Zn, Ni, Cd, As, U, Np, and Tc. The reduced forms of these metals form quite insoluble sulfide, hydroxide, or carbonate precipitates. A few species are more mobile in the reduced form. A classic case is biological reduction of Fe(III), which leads to the formation of the more soluble and mobile Fe(II). Plutonium is another species that is more mobile in the reduced state. Thus, creating reducing conditions makes most metals less mobile, but some become more mobile.

Microorganisms also cause other chemical changes that affect speciation and mobility of inorganic elements. Microbial reactions produce acids and bases, which alter the pH and affect speciation. For example, most cationic metals species form hydroxide or carbonate solids. A lower pH decreases concentrations of OH^- and CO_3^{2-}, thereby making the cationic metals more soluble and mobile. Microbial reactions also produce and consume ligands that complex metals. In general, formation of aqueous complexes competes with precipitation and enhances metal solubility and mobility. The complicated, but highly important interactions between microbial reactions and the water's geochemistry is reviewed in more detail by National Research Council (2000), Banaszak et al. (1999), VanBriesen and Rittmann (2000), and Rittmann and VanBriesen (1996).

In summary, the potential exists with many inorganic hazardous compounds for biological processes to bring about immobilization and hazard reduction. However, biological transformations can also do the reverse, cause them to become more mobile, more disperse, and more toxic. Conditions that lead to beneficial biologically mediated changes for one inorganic species may lead to harmful changes for another. Thus, one needs to consider all the hazardous chemicals at a site and to understand the interactions between microbial reactions and the geochemical reactions that affect the mobility of the inorganic elements.

14.7 SUMMARY

Detoxification of hazardous organic and inorganic compounds is often mediated naturally through the reactions catalyzed by microorganisms, and those reactions can be enhanced through engineered systems. Detoxification mechanisms vary widely depending upon the compound, the microbial species involved, and the environmental conditions present. Many organic and inorganic compounds are transformed as primary electron donors or electron acceptors; they are part of a cell's normal electron and energy metabolism. However, some compounds are transformed by cometabolism or by a change in environmental conditions mediated indirectly by microorganisms. These transformations are outside normal metabolism, but the cells still must carry out electron transfer and energy metabolism. At times, transformations lead to the formation of compounds that are more hazardous than the parent compounds. Thus, comprehensive knowledge of the range of contaminants present and their fate mechanisms in the system being considered is essential.

14.8 BIBLIOGRAPHY

Albrecht, I. D.; A. L. Barkovskii; and P. Adriaens (1999). "Production and Dechlorination of 2,3,7,8-Tetrachlorodibenzo-p-doxin in Historically-Contaminated Estuarine Sediments." *Environmental Sci. Technology*, 33:737–744.

Alexander, M. (1965). "Biodegradation: Problems of Molecular Recalcitrance and Microbial Fallibility." In *Advances in Applied Microbiology*. New York: Academic Press, pp. 35–80.

Alexander, M. (1994). *Biodegradation and Bioremediation*. San Diego: Academic Press.

Anderson, J. E. and P. L. McCarty (1994). "Model for Treatment of Trichloroethylene by Methanotrophic Biofilms." *Environmental Eng.* 120(2), pp. 379–400.

Anderson, J. E. and P. L. McCarty (1997). Transformation yields of chlorinated ethenes by a methanotrophic mixed culture expressing particulate methane monooxygenase. *Applied and Environmental Microbiology,* 63(2): pp. 687–93.

Ball, H. A.; M. Reinhard; and P. L. McCarty (1989). Biotransformation of Halogenated and Nonhalogenated Octylphenol Polyethoxylate Residues Under Aerobic and Anaerobic Conditions. *Environmental Science and Technology,* 23: pp. 951–961.

Banaszak, J. E.; D. T. Reed; and B. E. Rittmann (1999). "Subsurface Interactions of Actinide Species and Microorganisms: Implication on Bioremediation of Actinide-Organic Mixtures." *J. Radioanalytical and Nuclear Chemistry,* 241: pp. 385–435.

Beller, H. R. and A. M. Spormann (1997). "Benzylsuccinate Formation as a Means of Anaerobic Toluene Activation by Sulfate-Reducing Strain PRTOL1." *Applied and Environmental Microbiology* 68(9), pp. 3729–3731.

Bouwer, E. J. and P. L. McCarty (1984). "Modeling of Trace Organics Biotransformation in the Subsurface." *Ground Water,* 22, pp. 433–440.

Boyle, A. W., et al. (1992). "Bacterial PCB Biodegradation." *Biodegradation* 3(2/3), pp. 285–298.

Brown, J. F., et al. (1987). "Environmental Dechlorination of PCBs." *Environmental Toxicology and Chemistry* 6, pp. 579–593.

Carson, R. (1962). *Silent Spring.* Boston, Cambridge: Houghton Mifflin.

Cerniglia, C. E. (1992). "Biodegradation of Polycyclic Aromatic Hydrocarbons." *Biodegradation* 3(2/3), pp. 351–368.

Chang, H.-L. and L. Alvarez-Cohen (1995). "Model for the Cometabolic Biodegradation of Chlorinated Organics." *Environmental Sci. Technology* 29(9), pp. 2357–2367.

Chappelle, F. H. (1993). *Ground Water Microbiology and Geochemistry.* New York: John Wiley.

Cook, A. M. (1987). "Biodegradation of s-Triazine Xenobiotics." *FEMS Microbiol. Rev.* 46, pp. 93–116.

Cozza, C. L. and S. L. Woods (1992). "Reductive Dechlorination Pathways for Substituted Benzenes: A Correlation with Electronic Properties." *Biodegradation* 2(4), pp. 265–278.

Criddle, C. S. (1993). "The Kinetics of Cometabolism." *Biotechnology and Bioengineering* 41(11), pp. 1048–1056.

Dolfing, J. and J. M. Tiedje (1986). "Hydrogen Cycling in a Three-Tiered Food Web Growing on the Methanogenic Conversion of 3-chlorobenzoate." *FEMS Microbiology Ecology* 38, pp. 293–298.

Dolfing, J. and J. M. Tiedje (1987). "Growth Yield Increase Linked to Reductive Dechlorination in a Defined 3-chlorobenzoate Degrading Methanogenic Coculture." *Arch. Microbiology* 149, pp. 102–105.

Enviromega (1994). Toxchem (version 2) software developed and marketed by Enviromega, Inc., 7 Innovation Drive, Suite 245, Flanborough, Ontario, Canada L9H 7H9.

Govind, R.; L. Lai; and R. Dobbs (1991). "Integrated Model for Predicting the Fate of Organics in Wastewater Treatment Plants." *Environ. Prog.* 10(1), pp. 13–23.

Halden, R. U. and D. F. Dwyer (1997). "Biodegradation of Dioxin-Related Compounds: A Review." *Bioremediation J.* 1(1), pp. 11–25.

Haston, Z. C.; Y. Yang; and P. L. McCarty (2000). "Organism Growth and Substrate Utilization Kinetics for the Anaerobic Dehalogenation of cis-Dichloroethene and Vinyl Chloride." submitted.

Hill, D. W. and P. L. McCarty (1967). "Anaerobic Degradation of Selected Chlorinated Hydrocarbon Pesticides." *J. Water Pollution Control Federation* 39, pp. 1259–1277.

Key, B. D.; R. D. Howell; and C. S. Criddle (1997). "Fluorinated Organics in the Biosphere." *Environmental Sci. Technology* 31(9), pp. 2445–2454.

LaPat-Polasko, L. T.; P. L. McCarty; and A. J. B. Zehnder (1984). "Secondary Substrate Utilization of Methylene Chloride by an Isolated Strain of *Pseudomonas* Sp." *Appl. Environ. Microbiol.* 47, pp. 825–830.

Lee, K. C.; B. E. Rittmann; J. Shi; and D. McAvoy (1998). "Advanced steady-state model for the fate of hydrophobic and volatile compounds in activated sludge wastewater treatment." *Water Environment Res.* 70, pp. 1118–1131.

Lovley, D. R. and S. Goodwin (1988). "Hydrogen concentrations as an indicator of the predominant terminal electron-accepting reactions in aquatic sediments." *Geochimica et Cosmochimica Acta* 52, pp. 2993–3003.

Maymo-Gatell, X.; Y. Chien; J. M. Gossett; and S. H. Zinder (1997). "Isolation of a Novel Bacterium Capable of Reductively Dechlorinating Tetrachloroethene to Ethene." *Science* 276, pp. 1568–1571.

McAllister, K. A.; H. Lee; and J. T. Trevors (1996). "Microbial Degradation of Pentachlorophenol." *Biodegradation* 7(1), pp. 1–40.

McCarty, P. L. 1990. "Scientific Limits to Remediation of Contaminated Soils and Groundwater." In *Ground Water and Soil Contamination Remediation: Toward Compatible Science, Policy, and Public Perception.* Washington, D.C.: National Academy Press, pp. 38–52.

McCarty, P. L. (1997a). "Aerobic Cometabolism of Chlorinated Aliphatic Hydrocarbons." In C. H. Ward, J. A. Cherry, and M. R. Scalf Ed., *Subsurface Restoration.* Chelsea, MI: Ann Arbor Press, pp. 373–395.

McCarty, P. L. (1997b.) "Breathing with Chlorinated Solvents." *Science* 276, pp. 1521–1522.

McCarty, P. L. (1999). "Chlorinated Organics." In W. C. Anderson, R. C. Loehr, and B. P. Smith, Eds. *Environmental Availability of Chlorinated Organics, Explosives, and Heavy Metals on Soils and Groundwater,* American Academy of Environmental Engineers, Annapolis.

McCarty, P. L., et al. (1998). "Full-Scale Evaluation of *In Situ* Cometabolic Degradation of Trichloroethylene in Groundwater through Toluene Injection." *Environmental Sci. Technology* 32(1), pp. 88–100.

McCarty, P. L.; M. Reinhard; and B. E. Rittmann (1981). "Trace Organics in Groundwater." *Environmental Sci. & Technology* 15, pp. 40–51.

Mihelcic, J. R. and R. G. Luthy (1988). "Microbial-Degradation of Acenaphthene and Naphthalene under Denitrification Conditions in Soil-Water Systems." *Applied and Environmental Microbiology* 54(5), pp. 1188–1198.

Mikesell, M. D.; J. J. Kukor; and R. H. Olsen (1993). "Metabolic Diversity of Aromatic Hydrocarbon-Degrading Bacteria from a Petroleum-Contaminated Aquifer." *Biodegradation* 4(4), pp. 249–259.

Namkung, E. and B. E. Rittmann (1987a). "Evaluation of Bisubstrate Secondary Utilization Kinetics by Biofilms." *Biotechnol. Bioengr.* 29, pp. 335–342.

Namkung, E. and B. E. Rittmann (1987b). "Estimating Volatile Organic Compound (VOC) Emissions from Publicly Owned Treatment Works (POTWs)." *J. Water Pollution Control Federation* 59, pp. 670–678.

National Research Council (2000). *Natural Attenuation for Groundwater Remediation.* Washington, D.C.: National Academy Press.

Omenn, G. S., et al. (1988). *Environmental Biotechnology, Reducing Risks from Environmental Chemicals through Biotechnology.* New York: Plenum Press.

Pennington, J. C. (1999). "Explosives." In W. C. Anderson, R. C. Loehr, and B. P. Smith Ed. *Environmental Availability of Chlorinated Organics, Explosives, and Metals in Soils.* Annapolis: American Academy of Environmental Engineers, pp. 85–109.

Radosevich, M.; S. J. Traina; and O. H. Tuovinen (1996). "Biodegradation of Atrazine in Surface Soils and Subsurface Sediments Collected from an Agricultural Research Farm." *Biodegradation* 7(2), pp. 137–149.

Reinhard, M.; N. L. Goodman; and J. F. Barker (1984). Occurrence and Distribution of Organic Chemicals in Two Landfill Leachate Plumes. *Environmental Sci. & Technology* 18(12), pp. 953–961.

Rittmann, B. E. (1982). "Comparative Performance of Biofilm Reactor Types." *Biotechnol. Bioengr.* 24, pp. 1341–1370.

Rittmann, B. E.; D. Jackson; and S. L. Storck (1988). "Potential for Treatment of Hazardous Organic Chemicals with Biological Processes." In D. L. Wise, Ed., *Biotreatment Systems, Vol. III.* Boca Raton, FL: CRC Press, pp. 15–64.

Rittmann, B. E.; E. Seagren; B. A. Wrenn; A. J. Valocchi; C. Ray; and L. Raskin (1994). *In Situ Bioremediation,* 2nd ed. Park Ridge, NJ: Noyes Publishers.

Rittmann, B. E. and J. M. VanBriesen (1996). "Microbiological Processes in Reactive Modeling." *Rev. in Mineralogy* 34, pp. 311–334.

Rittmann, B. E. and R. Whiteman (1994). "Bioaugmentation Comes of Age." *Water Quality International* No. 1, pp. 22–26.

Rudel, R. A.; S. J. Melly; P. W. Geno; G. Sun; and J. G. Brody. (1998). "Identification of Alkylphenols and Other Estrogenic Phenonlic Compounds in Wastewater, Septage, and Groundwater on Cape Cod, Massachusetts." *Environmental Sci. Technology* 32(7), pp. 861–869.

Sanford, R. A. and J. M. Tiedje (1997). "Chlorophenol Dechlorination and Subsequent Degradation in Denitrifying Microcosms Fed Low Concentrations of Nitrate." *Biodegradation* 7(5), pp. 425–434.

Sawyer, C. N. (1958). "Effects of Synthetic Detergents on Sewage Treatment Processes." *Sewage and Industrial Wastes* 30(6), pp. 757–755.

Schwarzenbach, R. P.; P. M. Gschwend; and D. M. Imboden (1993). *Environmental Organic Chemistry.* New York: John Wiley.

Semprini, L. (1997a). "In Situ Transformation of Halogenated Aliphatic Compounds under Anaerobic Conditions." In C. H. Ward, J. A. Cherry, and M. R. Scalf Ed. *Subsurface Restoration.* Chelsea, MI: Ann Arbor Press, pp. 429–450.

Semprini, L. (1997b). "Strategies for the aerobic co-metabolism of chlorinated solvents." *Current Opinion in Biotechnology* 8(3), pp. 296–308.

Spain, J. C., Ed. (1995). *Biodegradation of Nitroaromatic Compounds.* New York: Plenum Press.

Spormann, A. and F. Widdel (2000). "Metabolism of Alkylbenzenes, Alkanes, and Other Hydrocarbons in Anaerobic Bacteria," *Biodegradation,* in press.

Strand, S. E. (1998). "Biodegradation of Bicyclic and Polycyclic Aromatic Hydrocarbons in Anaerobic Enrichments." *Environmental Sci. & Technology* 32(24), pp. 3962–3967.

Stuart, S. L. and S. L. Woods (1998). "Kinetic Evidence for Pentachlorophenol-Dependent Growth of a Dehalogenating Population in pentachlorophenol-Fed and Acetate-Fed Methanogenic Culture." *Biotechnology and Bioengineering* 57(4), pp. 420–429.

Stumm, W. and J. J. Morgan (1996). *Aquatic Chemistry,* 3rd ed. New York: John Wiley.

Tiedje, J. M.; J. F. Quensen III; J. Chee-Sanford; J. P. Schimel; and S. A. Boyd (1993). "Microbial Reductive Dechlorination of PCBs." *Biodegradation* 4(4), pp. 231–240.

VanBriesen, J. M. and B. E. Rittmann (2000). "Modeling Biodegradation in Mixed Waste System." *Biodegradation,* in press.

van Ginkel, C. G. (1996). "Complete Degradation of Xenobiotic Surfactants by Consortia of Aerobic Microorganisms." *Biodegradation* 7(2), pp. 151–164.

Vogel, T. M.; C. S. Criddle; and P. L. McCarty (1987). "Transformations of Halogenated Aliphatic Compounds." *Environmental Sci. Technology* 21, pp. 722–736.

Vogel, T. M. and D. Grbic-Galic (1986). "Incorporation of Oxygen from Water into Toluene and Benzene during Anaerobic Fermentative Transformation." *Applied and Environmental Microbiology* 5(1), pp. 200–202.

Wilson, J. T. and B. H. Wilson (1985). "Biotransformation of Trichloroethylene." *Applied and Environmental Microbiology* 49(1), pp. 242–243.

Wrenn, B. A. and B. E. Rittmann (1996). "Evaluation of a Mathematical Model for the Effects of Primary Substrates on Reductive Dehalogenation Linetics," *Biodegradation* 7, pp. 49–64.

Yang, Y. and P. L. McCarty (1998). "Competition for Hydrogen within a Chlorinated Solvent Dehalogenating Anaerobic Mixed Culture." *Environmental Sci. & Technology* 32(22), pp. 3591–3597.

Yang, Y. and P. L. McCarty (2000). "Biomass, Oleate, and Other Possible Substrates for Chloroethene Reductive Dehalogenation," *Bioremediation Journal,* 4(2), pp. 125–133.

14.9 PROBLEMS

14.1. Fluidized bed reactors can remove low concentrations of petroleum hydrocarbons from groundwater. They consist of a bed of sand or granular activated carbon with wastewater flowing in an upward direction with sufficient velocity to suspended the particles. Bacteria attach to the suspended particles as a biofilm, and in this way the wastewater is treated.

The object of this problem is to determine the detention time required for reducing the concentration of BTEX compounds (benzene, toluene, ethylbenzene, and xylene) from groundwater contaminated by a gasoline spill. Removal of the BTEX compounds down to the very low concentration of 0.1 mg/l is required.

Assume the following conditions apply:

$\hat{q} = 12$ mg/mg VS$_a$-d $X_f = 15,000$ mg VS$_a$/l

$K = 0.2$ mg/l $L = 0.005$ cm

$Y = 0.5$ mg VS$_a$/mg $D = 1.0$ cm^2/d

$b' = 0.14$/d $D_f = 0.8$ D

$b_{\mathrm{det}} = 0.05$/d

$Q = 100$ m^3/d $h = 0.75$ (liquid hold-up)

$S^0 = 4$ mg/l $a = 6.0$ cm^2/cm^3

$X_i^0 = 0$

Estimate the empty-bed detention time (V/Q) in minutes required to achieve the required treatment for a steady-state biofilm in a CMBR.

14.2. From biofilm kinetics, it was found that the flux into a biofilm at steady state is 0.15 mg/cm^2-d when the bulk liquid concentration of benzoate is 2 mg/l. What reactor volume is needed to treat 100 m^3/d of wastewater containing 50 mg/l benzoate if 96 percent removal is required, and if we assume completely mixed conditions and the specific surface area of the reactor is 3 cm^2/cm^3?

14.3. You have been asked to design a reactor to treat a waste stream of 10^4 m^3/d containing 150 mg/l of phenol. The following coefficients apply:

$Y = 0.6$ g VS$_a$/g phenol
$\hat{q} = 9$ g phenol/g VS$_a$-d
$b = 0.15$ d^{-1}
$K = 0.8$ mg/l

You have decided to evaluate use of a fixed-film reactor in which you apply sufficient recycle so that you can assume the system will act like a completely mixed reactor. Assume that deep biofilm kinetics apply $(S_w = 0)$, that the specific surface area of the reactor is 10^7 cm^2/m^3, the boundary layer thickness for the deep biofilm is 0.005 cm, X_f is 20 mg VS$_a$/cm^3, $D = 0.8$ cm^2/d, and $D_f = 0.8$D. Determine the volume V for this reactor in m^3.

14.4. In the biological oxidation of benzene by a fixed-film reactor, the flux into the deep biofilm $(S_w = 0)$ was found to be 5 mg per cm^2 of biofilm surface area per day when the benzene concentration at the biofilm surface was 15 mg/l. What do you estimate the flux will be if the benzene concentration at the biofilm surface is increased to 50 mg/l? Rate coefficients for benzene are as follows: $Y = 0.6$ mg VS$_a$ per mg benzene, $\hat{q} = 6$ mg benzene per mg VS$_a$ per day, $K = 2$ mg/l, and $b = 0.1$ per day.

14.5. A modification of the Monod reaction called the Haldane reaction is often used to describe the kinetics of substrate utilization when the substrate itself is toxic at high concentrations (phenol is an example of this). The Haldane reaction is as follows:

$$-\frac{dS}{dt} = \frac{\hat{q}S}{K + S + \dfrac{S^2}{K_i}} X_a$$

where K_i is a substrate inhibition constant. Calculate the effluent concentration for a substrate exhibiting Haldane kinetics from a CSTR and compare this with the value if the substrate were not inhibitory ($S^2/K_i = 0$) for the following conditions:

$$V = 25 \text{ m}^3 \qquad\qquad \hat{q} = 8 \text{ g/g VSS}_a\text{-d}$$
$$Q^0 = 10 \text{ m}^3/\text{d} \qquad\qquad K = 7 \text{ mg/l}$$
$$S^0 = 500 \text{ mg/l} \qquad\qquad b = 0.2/\text{d}$$
$$Y = 0.5 \text{ mg VSS}_a/\text{mg} \qquad K_i = 18 \text{ mg/l}$$

14.6. You have evaluated the rate of aerobic degradation of organic matter in an industrial wastewater and found that it does not follow normal Monod kinetics. After some testing, you have found that the substrate degradation rate follows the relationship:

$$-r_{ut} = k^{0.5} X_a S^{0.5}$$

where

$-r_{ut}$ = degradation rate in mg/L-day
$k = 4$ L/mg-d^2
S = substrate concentration in mg/L
X_a = active microorganism concentration in mg/L

You have also determined coefficients of bacterial growth on this substrate under aerobic conditions to be $Y = 0.25$ g VSS$_a$/g substrate and $b = 0.08/d$. Estimate the effluent substrate concentration when treating the above wastewater in a CSTR with recycle while operating at a θ_x of 4 d.

14.7. You are asked to design an activated sludge plant to treat 10^4 m^3/d of wastewater containing 150 mg/l of phenol. The regulatory authorities are requiring that the effluent phenol concentration from this plant not exceed 0.04 mg/l. For this problem assume the following biological coefficients apply, and that $X^0 = 0$:

$Y = 0.6$ g VSS$_a$/g phenol
$\hat{q} = 9$ g phenol/g VSS$_a$-d
$b = 0.15$ d^{-1}
$K = 0.8$ mg phenol/l

(*a*) What θ_x in days would you use for design of this plant? Be sure to list all assumptions used and show calculations that are appropriate for justifying your answer.

(*b*) Based upon your design θ_x, what would be the reactor volume in m^3? Specify additional assumptions you may have to make.

14.8. A biological treatment plant (CSTR with settling and recycle) is being operated with a θ_x of 8 d and is treating a waste consisting primarily of substrate A. The effluent concentration of A is found to be 0.3 mg/l, and its affinity constant (K) is known to be 1 mg/l. A relatively small concentration of compound B is then added to the treatment plant influent and θ_x is maintained constant. Compound B is partially degraded by the microorganisms present through cometabolism, but acts as a competitive inhibitor to substrate A degradation. If the effluent concentration of Compound B is then found to be 1.5 mg/l and its affinity constant (K_i) is 2 mg/l, what will now be the effluent concentration of Substrate A?

14.9. You have decided to design a CSTR with recycle to degrade TCE cometabolically by aerobic phenol-utilizing bacteria. To do this, you will feed phenol (C_6H_5OH) and oxygen, along with required nutrients, to the reactor in order to grow a population of bacteria that can fortuitously degrade TCE. The problem is to first determine what mass of bacteria is required in the reactor to effect the TCE destruction desired, and then to determine how much phenol and oxygen must be fed to the reactor to maintain that population.

The waste characteristics are as follows: $Q^0 = 5$ m^3/d, $C^0 = 100$ μg/l TCE (influent concentration), $X^0 = 0$ mg/l. The X_a selected for design is 1,500 mg VSS$_a$/l, and θ_x is 8 d. Kinetic coefficients for TCE decomposition are as follows: $\hat{q}_{TCE} = 0.015$ mg TCE/mg VSS$_a$, $K_{TCE} = 0.5$ mg/l. Kinetic coefficients for phenol degraders are $Y = 0.8$ g VSS$_a$/g phenol, $b = 0.1$/d, $\hat{q} = 1$ g phenol/g VSS$_a$-d, and $K_{(phenol)} = 1$ mg/l.

(a) Determine the reactor size required to effect 95 percent TCE removal.
(b) Determine the mass in kg/d of phenol that should be supplied to the reactor to maintain the desired bacterial population.
(c) Determine the masses (kg/d) of oxygen, ammonia-nitrogen, and phosphorus required.

14.10. An operator of an aerobic biological treatment system (CSTR with recycle) for an industrial wastewater is having difficulty in operating his system and has called upon you as the expert to help him out. The operating θ_x is 10 d. The plant is treating orthochlorophenol (OCP), among other things, and from measurements, you have found that the effluent concentration of orthochlorophenol from the reactor is 200 mg/l. From the literature you have found the following kinetic coefficients apply for this compound:

$Y = 0.67$ g VSS$_a$/g OCP

$\hat{q} = 12$ g OCP/g VSS$_a$-d

$K = 12$ mg OCP/l

$b = 0.1$/d

$K_i = 1.2$ mg OCP/l (Haldane)

(a) What do you expect the problem to be at this treatment plant? Give quantitative information to support your judgement.
(b) How do you suggest the operator proceed in order to overcome this problem?

14.11. An aerobic CSTR with recycle operating at θ_x of 8 d is treating a wastewater consisting primarily of soluble organic compound A. Under steady-state operation the effluent concentration is 1.4 mg/l and its K value is 250 mg/l. Compound B is then added to the wastewater and the reactor volume is increased appropriately to maintain the same θ_x. Different microorganisms use each of the two different substrates. Under the new steady-state conditions, the effluent concentration of B is found to equal 0.8 mg/l. What would you then expect the effluent concentration of compound A to be for the following conditions:

(a) Normal Monod kinetics apply for each substrate.
(b) Competitive inhibition between substrates occurs, and K_i (related to compound B concentration) is 0.8 mg/l.
(c) Noncompetitive inhibition between substrates occurs, and K_i (related to compound B concentration) is 0.8 mg/l.
(d) Compound B exhibits substrate toxicity (Haldane kinetics), and $K_i = 0.8$ mg/l.

14.12. Methane oxidizing bacteria (methanotrophs) can oxidize dichloroethene (DCE) and vinyl chloride (VC) by cometabolism. However, when both DCE and VC are present, the rate of oxidation of each is less than when they are present by themselves. What process is likely to be responsible for this phenomena?

14.13. Are aromatic hydrocarbons such as benzene biodegradable under anaerobic conditions?

14.14. Which of the chlorinated solvents has been found to be biodegradable under aerobic conditions, and what is the process involved?

14.15. (a) Is tetrachloroethene used as an electron donor or acceptor under anoxic conditions? (b) Write appropriate equations to illustrate your answer. (c) What is the ultimate end product of this process?

14.16. Three separate industries are potentially responsible for contamination of a municipal well supply with chlorinated aliphatic hydrocarbons (CAHs). Industry records indicate that Industry A was a large user of carbon tetrachloride (CT), Industry B was a large user of tetrachloroethene (PCE), and Industry C was a large user of 1,1,1-trichloroethane (TCA). As far as can be determined, each industry used only one chlorinated solvent. The only CAHs found present in the municipal well supply are vinyl chloride, 1,1-dichloroethene, and 1,1-dichloroethane. Since none of the industries used these compounds, they all claim they are innocent. You are called in by the regulatory authorities as an expert witness in this case to help shed some light on the possible source or sources of these contaminants.

(a) Does the evidence suggest the identified industries are innocent as claimed and that some unidentified industry (D) is responsible, or does it support the potential that one or all of the identified industries may be responsible?
(b) Who is the most likely culprit? Give technically sound reasons for your judgments.

14.17. A wastewater has a concentration of benzene equal to 30 mg/l, and no significant concentration of other biodegradable organic materials are present. The regulatory agencies require that the wastewater be treated to reduce the benzene concentration to 0.01 mg/l. Assume for benzene, that the following rate coefficients apply for aerobic treatment: $Y = 0.9$ g VSS_a/g benzene, $b = 0.2$/d, $\hat{q} = 8$ g benzene/g VSS_a-d, and $K = 5$ mg benzene/l.

(a) What is the minimum concentration of benzene that you could expect to achieve from biological treatment in a CSTR with recycle?

(b) Assuming the above does not meet regulatory compliance for benzene, describe another biological approach that is likely to have better potential for meeting the requirements. Describe the approach and indicate why it may be better than a CSTR. Calculations are not required to support your answer, a description of the concept is all that is needed.

14.18. For each of the following chemicals, indicate with an X, the process or processes by which biotransformation of the compounds might be accomplished based on currently available information.

Compound	Can be Used as a Primary Substrate		Can be Transformed by Cometabolism	
	Aerobic	**Anaerobic**	**Aerobic**	**Anaerobic**
Acetate				
Trichloroethene				
Tetrachloroethene				
PCB				
Glucose				
Protein				
Carbon Tetrachloride				
Benzene				
Vinyl Chloride				

14.19. Methanotrophic bacteria are being used to degrade a mixture of trichloroethene (TCE) and chloroform (CF) by cometabolism. Here, the enzyme MMO initiates the oxidation of these two compounds. Since they compete with one another for MMO, the rate of their degradation in the mixture is governed by competitive inhibition kinetics. Under the following conditions and with $X_a = 120$ mg/l, estimate the individual rates of transformation of TCE and CF in mg/l-d:

$$S_{TCE} = 6.4 \text{ mg TCE/l}$$
$$\hat{q}_{TCE} = 0.84 \text{ g TCE/g } VSS_a\text{-d}$$
$$K_{TCE} = 1.5 \text{ mg TCE/l}$$
$$S_{CF} = 5 \text{ mg CF/l}$$
$$\hat{q}_{CF} = 0.34 \text{ g CF/g } VSS_a\text{-d}$$
$$K_{CF} = 1.3 \text{ mg CF/l}$$

BIOREMEDIATION

One of the newest and fastest growing applications of environmental biotechnology is *bioremediation*. Unlike most of the applications described in earlier chapters, bioremediation most often addresses treatment of contaminated solids, such as groundwater aquifers, soils, and sediments. This focus on decontaminating solids occurs because flowing waters poorly mobilize most of the common contaminants that reach these environments. Because the contaminants adsorb to the amply present solid surfaces or form a separate liquid phase, they are "trapped" in the solid matrix. Cleanup strategies that rely on flushing the contaminants out with water and treating the water (i.e., so-called pump-and-treat strategies) often are unsuccessful, even after years or decades of remediation by flushing (Travis and Doty, 1990; National Research Council, 1994).

Bioremediation overcomes the main deficiency of conventional cleanup approaches that rely on water flushing: Dissolution or desorption of the contaminants into water is too slow, since the water's contaminant-carrying capacity is very limited. With bioremediation, high densities of active microorganisms locate themselves close to the nonaqueous source of contamination. Within a short distance from its point of dissolution into the aqueous phase, the contaminant molecule is biodegraded. If the biodegradation is fast and occurs close enough to the nonaqueous source, biodegradation replenishes the water's ability to accept more dissolving or desorbing contaminant, bringing about accelerated dissolution/desorption, which decontaminates the solid phase.

Although bacteria are present in all soils, sediments, and aquifers, their naturally occurring numbers may be too small to bring about the rapid reaction needed to have enhanced dissolution/desorption. In that case, the strategy is to add to the contaminated environment the materials needed to allow growth of the bacteria active in degrading the target contaminant. These materials provide the substrates and nutrients required for the growth and maintenance of a high-density microbial population. The most commonly added materials normally are selected from among an electron-acceptor substrate (like oxygen), an electron-donor substrate (like a sugar or natural gas), inorganic nutrients (like nitrogen and phosphorus), and materials to help dissolve/desorb immobile substrates (like surfactants).

Bioremediation can be divided into three main classes: engineered in situ, intrinsic in situ, and engineered ex situ. The distinction between in situ and ex situ indicates whether the contaminated solids remain in place (i.e., in situ) during the bioremediation or are excavated and transferred to an aboveground treatment system (i.e., ex situ). Engineered bioremediation refers to employing engineering tools to greatly increase the input rates for the stimulating materials. In contrast, intrinsic bioremediation relies on the intrinsically occurring rates of supply of substrates and nutrients, as well as the intrinsic population density of active microorganisms. The most important factor for any type of bioremediation is ensuring that the rate of biotransformation is fast enough to meet the cleanup objectives.

Because engineered bioremediation involves using engineered measures to add the materials necessary to *increase the biotransformation rates significantly,* the fast-enough criterion refers to increasing the biotransformation rates well above those that would occur without the engineered additions. In this way, the time required to clean up the contamination is substantially reduced. Engineered bioremediation is a cost-effective approach when one must meet a regulated deadline, minimize exposure risk, or facilitate a property sale.

Intrinsic bioremediation, on the other hand, relies on the intrinsic (or naturally developing) biological activity to *prevent the migration of contamination away from its source.* While intrinsic bioremediation does not accelerate the rate of cleanup, it prevents further spread of the contaminants. The natural supply of necessary materials, when coupled with the presence of appropriate microorganisms, must be fast enough to allow biodegradation of the contaminants before they can be transported a significant distance away from the source. A monitoring plan is essential to demonstrate that these intrinsic factors remain effective for preventing contaminant migration. It cannot be emphasized too much that intrinsic bioremediation is not a "do nothing" approach. Intrinsic bioremediation is sometimes called *natural attenuation,* a term that actually includes many naturally occurring mechanisms that can lead to a decrease in the concentration of a contaminant (National Research Council, 2000).

Prior to the design of a system for in situ bioremediation, several steps are required to ensure that the program is successful (Lee et al., 1988; National Research Council, 1993, 2000; Rittmann et al., 1994). These steps are needed to characterize the site in terms of hydrology, extent and type of contamination, intrinsic microbial activity, and intrinsic supply rates of key materials.

Although bioremediation is a new and uniquely challenging application of environmental biotechnology, it is founded on the same principles as are other applications. Because of the high specific surface area of aquifers, soils, and sediments, the microbial ecology is dominated by attached growth. Therefore, the principles and applications in Chapters 4 and 8 are particularly relevant here. Furthermore, the most common contaminants are the synthetic organic chemicals, whose biological detoxification is described in Chapter 14. The information in these closely related chapters is not repeated here, although short summaries are provided in order to provide the appropriate context for reviewing the other chapters.

This chapter focuses on the unique features of bioremediation. They are:

- The scope and characteristics of contamination
- Contaminant availability for biodegradation
- Treatability studies
- Engineering strategies for bioremediation
- Evaluating bioremediation

15.1   **SCOPE AND CHARACTERISTICS OF CONTAMINANTS**

The Office of Technology Assessment documented that over 200 substances have been detected in groundwater in the United States (OTA, 1984; Rittmann et al., 1994). These include organic chemicals, inorganic chemicals, biological organisms, and radionuclides. A summary is presented in Table 15.1.

Table 15.1 Summary of information presented on substances detected in groundwater

	Category of Compounds	Number of Entries	Examples[1]	Prevalent Uses[2]
1.	Aromatic Hydrocarbons	33	Benzene Ethylbenzene Toluene	• Dyestuffs • Solvents
2.	Oxygenated Hydrocarbons	24	Acetone 1,4 - Dioxane Phenols	• Solvents • Paints, varnishes
3.	Hydrocarbons with Specific Elements (e.g., with N, P, S, Cl, Br, I, F)	103	2,4-D Trichloroethene (TCE) Trichloroethanes (1,1,1 and 1,1,2 TCA) Trinitrotoluene (TNT)	• Pesticides • Solvents • Munitions
4.	Other Hydrocarbons	15	Kerosene Gasoline Lignin and Tannin	• Fuels
5.	Metals and Cations	27	Iron Chromium Arsenic	• Alloys • Electrical and electronics
6.	Nonmetals and Anions	10	Chloride Sulfate Ammonia	• Fertilizers • Foodadditives
7.	Microorganisms	2	Bacteria (e.g., coliforms) Viruses	—
8.	Radionuclides	19	Uranium 238 Tritium Radium 226	• Tracers • Medical applications

[1] The compounds taken for examples in each category are those with the highest concentrations.
[2] Uses listed are industrial applications that were commonly found under each category.
SOURCE: Rittmann et al., 1994.

Although the components listed in Table 15.1 are widely used by industry, agriculture, commerce, and households, poorly mobile contaminants trapped in the solid matrix may be underrepresented when only groundwater samples are taken. The fact that the groundwater is sampled tends to accentuate compounds that are more readily mobile. In addition, detection of substances in groundwater often is biased by sampling procedures, analytical detection limits, and the circumstances that prompted the detection and reporting (e.g., was it a planned activity, such as regulatory compliance, or in response to an apparent impact, such as a public complaint).

15.1.1 ORGANIC COMPOUNDS

This chapter focuses on organic compounds, which are most often amenable to bioremediation and are the most commonly detected contaminants in groundwater. The Council on Environmental Quality (1981) compiled a list of the 33 synthetic organic contaminants most frequently found in drinking-water wells (see Table 15.2).

Table 15.2   The 33 synthetic organic contaminants reported to be most frequently found in drinking-water wells

Trichloroethene (TCE)	Isopropyl benzene
Toluene	1,1-Dichloroethene
1,1,1-Trichloroethane	1,2-Dichloroethane
Acetone	Bis (2-ethylhexyl) Phthalate
Methylene chloride	DBCP (Dibromochloropropane)
Dioxane	Trifluorotrichloroethane
Ethyl benzene	Dibromochloromethane
Tetrachloroethene	Vinyl chloride
Cyclohexane	Chloromethane
Chloroform	Butyl benzyl-phthalate
Di-n-butyl-phthalate	gamma-BHC (Lindane)
Carbon tetrachloride	1,1,2-Trichloroethane
Benzene	Bromoform
1,2-Dichloroethene	1,1-Dichloroethane
Ethylene dibromide (EDB)	alpha-BHC
Xylene	Parathion
	delta-BHC

SOURCE: CEQ, 1981; Rittmann et al., 1994.

Organic chemicals are associated with a variety of sources, including: subsurface percolation; injection wells; land application of wastewater, wastewater by-products, and hazardous waste; landfills; open dumps; residential disposal; surface impoundments; underground and aboveground storage tanks; pipelines; materials transport and transfer operations; pesticide and fertilizer applications; urban runoff; and oil and gas production wells (OTA, 1984).

Information on the chemical and physical properties of these organic compounds was given in Table 14.2. A more extensive list of parameters is given in Rittmann et al. (1994). Many of the compounds are hydrophobic, which is indicated by a relatively large octanol/water partition coefficient (i.e., log $K_{ow} > 1$) and limited water solubility (i.e., $< 10,000$ mg/l). Some of the compounds are highly volatile (i.e., Henry's constant $> 10^{-3}$ atm-m^3/mol), which allows them to partition to a gas phase when it is present.

Included in Table 14.2 are highly hydrophobic contaminants whose water solubility is low enough that they seldom are high-risk factors in terms of water contamination. Chief among them is the polycyclic aromatic hydrocarbons (PAHs) derived from petroleum and the polychlorinated biphenyls (PCBs). For example, the log K_{ow} values and water solubilities (in mg/l) for the simpler PAHs are 3.3 and 31 for two-ringed naphthalene, 4.6 and 1.1 for three-ringed phenanthrene; 4.5 and 0.075 for four-ringed anthracene, and 6.1 and 0.0038 for five-ringed benzo(a)pyrene. For the PCBs, some representative solubility values are 0.24 mg/l for Arochlor 1242 and 0.057 mg/l for Arochlor 1254; the last two digits indicate the percent Cl in the PCB mixture. This brief listing makes it clear that PAHs and PCBs are much less mobile than are most of the common groundwater contaminants listed in Table 14.2. While their hydrophobicity greatly reduces their concentrations in the water, it also accentuates their longevity as trapped sources of contamination in the soil or sediment, as well as their ability to bioconcentrate in living organisms.

15.1.2 MIXTURES OF ORGANIC COMPOUNDS

In many situations, organic contaminants are found in mixtures. In many instances, the original contamination was a mixture of related components that co-exist normally in a commercial product. Key examples include the PCBs, various petroleum-distillation fractions used for a range of fuel and lubricating purposes, and PAHs found in tars, asphalts, and petroleum sludges. One of the most widely used PCB mixtures—Arochlor 1242—has 42 percent chlorine overall, but contains biphenyl congeners having 1 through 6 Cl substituents, with 80 percent having 3, 4, or 5 Cl substituents. The more lightly chlorinated congeners have higher water solubilities than do the more heavily chlorinated congeners, which means the "aged" PCB tends to gradually become enriched in more chlorinated components.

Table 15.3 lists the typical distillation fractions obtained from petroleum. Although the distillation fractions are much more homogenous than is crude petroleum, each fraction contains molecules having a substantial range of formula weights and the correlated large ranges of water solubilities and hydrophobicities.

Because of its widespread uses and improper disposal, gasoline is an excellent case to illustrate the complexity of these naturally occurring mixtures. Its chemical complexity depends on three variables: (1) the source of the crude oil from which the gasoline is refined, (2) how the gasoline is refined, and (3) the additives used. The total number of distinct hydrocarbons in a gasoline fraction might be of the order of 500; however, due to the predominance of certain hydrocarbons, a smaller number of components accounts for a relatively large portion of the gasoline fraction (Sachanen, 1954).

Table 15.3   Typical fractions obtained by distillation of petroleum

Boiling Range of Fraction	Number of Carbon Atoms per Molecule	Use
Below 20 °C	C_1–C_4	Natural gas, bottled gas, petrochemicals
20–60 °C	C_5–C_6	Petroleum ether, solvents
60–100 °C	C_6–C_7	Ligroin, solvents
40–200 °C	C_5–C_{10}	Gasoline (straight-run gasoline)
175–325 °C	C_{12}–C_{18}	Kerosene and jet fuel
250–400 °C	C_{12} and higher	Gas oil, fuel oil, and diesel oil
Nonvolatile liquids	C_{20} and higher	Refined mineral oil, lubricating oil, grease
Nonvolatile solids	C_{20} and higher	Paraffin wax, asphalt, and tar

SOURCE: Solomons, 1984; Rittmann et al., 1994.

Table 15.4 presents the composition of several typical straight-run gasolines (i.e., the gasoline fraction obtained by direct distillation of the crude petroleum) by class of hydrocarbon. Either alkanes or cycloalkanes dominate all of these straight-run gasolines, with those dominated by alkanes being more abundant; the aromatic hydrocarbons average below 20 percent by volume. In general, alkenes or unsaturated hydrocarbons are absent in straight-run gasolines. Despite these general trends, Table 15.4 shows that the source of the crude oil affects the chemical composition of the gasoline mixture.

The type of refining also plays a very important role in defining the gasoline mixture. Figure 15.1 shows that, in comparison with straight-run gasoline, thermal-cracked gasoline contains alkenes, increased aromatics, and decreased alkanes and cycloalkanes; catalytically cracked gasoline has alkenes, increased aromatics, and greatly decreased n-alkanes and cycloalkanes (Hill and Moxey, 1960). Further, un-leaded gasolines generally have a higher aromatic hydrocarbon fraction than do leaded gasolines (Cline, Delfino, and Rao, 1991).

The most troublesome of the gasoline hydrocarbons are the single-ring aromatics: benzene, toluene, ethylbenzene, and the xylenes (BTEX). Unleaded gasoline typically contains 2–5 percent (by volume) of benzene, 6–7 percent of toluene, 5 percent of ethylbenzene, and 6–7 percent of xylenes. Compared to almost all other gasoline hydrocarbons, BTEX have by far the highest water solubility. Although the total for

Table 15.4   Chemical composition of straight-run gasolines

Origin & Boiling Range, °C	Alkanes	% by Volume Cycloalkanes	Aromatics
Pennsylvania, 40–200	70	22	8
Oklahoma City, 40–180	62	29	9
Ponca (Oklahoma), 55–180	50	40	10
East Texas, 45–200	50	41	9
West Texas, 80–180	47	33	20
Conroe (Texas), 50–200	35	39	26
Hastings (Texas), 50–200	27	67	6
Michigan, 45–200	74	18	8
Rodessa (Louisiana), 45–160	72	20	8
Santa Fe Springs (Calif.), 45–150	41	50	9
Kettleman (Calif.), 45–150	48	45	7
Huntington Beach (Calif.), 50–210	35	54	11
Turner Valley (Canada), 45–200	51	35	14
Altamira (Mexico), 40–200	49	36	14
Portero (Mexico), 50–200	57	35	8
Bucsani (Romania), 50–150	56	32	12
Baku-Surachany, 60–200	27	64	9
Baku-Bibieibat, 60–200	29	63	8
Grozny, New Field, 45–200	64	29	7
Iran, 45–200	70	21	9
Kuwait, 40–200	72	20	8

SOURCE: Sachanen, 1954; Rittmann et al., 1994.

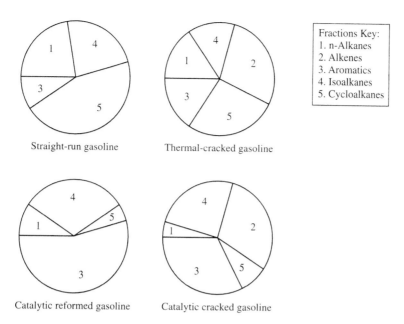

Fractions Key:
1. n-Alkanes
2. Alkenes
3. Aromatics
4. Isoalkanes
5. Cycloalkanes

Straight-run gasoline   Thermal-cracked gasoline

Catalytic reformed gasoline   Catalytic cracked gasoline

Figure 15.1   Hydrocarbon composition of gasoline components. SOURCE: Hill and Moxey, 1960.

BTEX is roughly 20 percent by volume, the water solubility of BTEX causes them to be the prime water pollutants.

The composition of gasoline is further complicated by additives, which are added for several purposes: (1) antiknock compounds, (2) antioxidants, (3) metal deactivators, (4) antirust agents, (5) antistall agents, (6) antipreignition agents, (7) upper-cylinder lubricants, (8) alcohols (e.g, ethanol), and (9) oxygenates (e.g., MTBE). Table 15.5 summarizes typical additives (Hill and Moxey, 1960; Lane, 1980; Cline et al., 1991; Rittmann et al., 1994). With the exception of alcohols used for gasohol,

Table 15.5   Additives typical in gasolines

Additive Type	Typical Chemicals Used	% by Volume
Antiknock agents	Tetraethyl lead, tetramethyl lead	0.08
	tert-butanol, methyl *tert*-butyl ether	< 3.5, < 15%
Antioxidants	Aminophenols, orthoalkylated phenols	< 0.006
Metal deactivators	N, N'-disalicylidene- 1,2-diaminopropane	0.0004
Antirust agents	Fatty acid amines, amine phosphates	0.0005
Antistall agents	Alcohols, glycols	1–2
Antipreignition agents	Alkyl phosphates	0.02
Upper-cylinder lubricants	Light mineral oils, Cyclo-paraffinic distillates	0.2–0.5
Alcohols	Ethanol	< 40
Oxygenates	MTBE (methyl-*tert*-butyl ether)	< 5

antistall, or antiknock purposes and methyl-*tert*-butyl ether (MTBE) used for antiknock and oxygenate purposes, the additives are present in very low percentages and seldom affect groundwater. On the other hand, the alcohols and MTBE can be major groundwater contaminants, along with BTEX. In fact, the massive introduction of MTBE in gasoline as an oxygenate to reduce smog in California has greatly worsened the problems of groundwater contamination in that state. MTBE is being removed from gasoline in the United States in order to prevent more groundwater contamination.

15.1.3   MIXTURES CREATED BY CODISPOSAL

In many instances, subsurface contamination is a mixture of disparate materials whose main (or only) connection is that they were codisposed of at a single site. These codisposal mixtures can include a wide range of organic compounds that do not necessarily occur together naturally, as well as inorganic chemicals.

A common situation of codisposal is the mixture of organic and inorganic materials in sanitary landfills and in their leachates. Table 15.6 presents a list of various inorganic constituents in leachate from sanitary landfills and representative ranges of concentrations. Large numbers of inorganic contaminants can be found in leachate,

Table 15.6   Representative ranges for various inorganic constituents in leachate from sanitary landfills

Parameter	Representative Range (mg/L)
K^+	200–1,000
Na^+	200–1,200
Ca^{2+}	100–3,000
Mg^{2+}	100–1,500
Cl^-	300–3,000
SO_4^{2-}	10–1,000
Alkalinity	500–10,000
Fe (total)	1–1,000
Mn	0.01–100
Cu	< 10
Ni	0.01–1
Zn	01–100
Pb	< 5
Hg	< 0.2
NO_3^-	0.1–10
NH_4^+	10–1,000
P as PO_4	1–100
Organic nitrogen	10–1,000
Total dissolved organic carbon	10–1,000
COD (chemical oxygen demand)	1,000–90,000
Total dissolved solids	5,000–40,000
pH	4–8

SOURCE: Freeze and Cherry, 1979; Rittmann et al., 1994.
The ranges for the constituents came from many leachate sources and are not necessarily consistent with each other.

some at high concentrations; many organic contaminants also can be found in leachate and cause the relatively high total dissolved organic carbon and COD values. The high concentration of inorganic ions can affect electron-acceptor availability (e.g., SO_4^{2-} and NO_3^-), precipitation potential (e.g., PO_4^{3-}, CO_3^{2-} in alkalinity), and toxicity (e.g., heavy metals, sulfides). A large fraction of the COD often is volatile acids, which tend to lower the pH, but also are good primary substrates. Biodegradation of the volatile acids removes their acidity and allows the pH to increase.

Robertson, Toussaint, and Jorque (1974) investigated the organic compounds leached into the groundwater from a landfill near Norman, Oklahoma. Refuse had been deposited below or near the water table of a sandy aquifer. They found low levels of more than 40 organic compounds in the groundwater and concluded that the source of many of these compounds was the leaching of manufactured products discarded within the landfill. In addition to constituents derived from solid wastes, many landfill leachates also contain toxic constituents emanating from liquid industrial wastes located within the landfill.

Many industrial, commercial, and military operations led to exotic mixtures of organic and inorganic contaminants. One example of groundwater pollution resulting from industrial/military operations is for the McClellan Air Force Base, near Sacramento, California. Table 15.7 shows a wide range of solvents, fuel components, and metals that resulted from wastes containing solvents, paint, fuel oil, and electroplating residues that were disposed of by landfilling and burning in pits.

Table 15.7 Volatile and nonvolatile organic compounds and trace metals found in groundwater at McClellan Air Force Base, Sacramento, CA

Compound	Concentration (mg/L)
1,1-Dichloroethene	60
Acetone	35
Methyl ethyl ketone	25
1,1,1-Trichloroethane	12
Trichloroethene	11
Methylene chloride	5
Methyl isobutyl ketone	3.7
Vinyl chloride	2.5
Dichlorobenzene	0.17
Benzene	0.68
1,1-Dichloroethane	0.25
Toluene	0.08
Phenols	0.5
Tetrachloroethane	0.07
Trans-1,2-dichloroethene	0.2
Chromium	0.12
Nickel	0.10
Zinc	0.073
Lead	0.093
Selenium	0.049
Cadmium	0.012

SOURCE: Pitra and McKenzie, 1990; Rittmann et al., 1994.

Major chemical-manufacturing facilities often have disposal areas contaminated over many years with waste solvents, sludges, unacceptable products, and other residues. The whole range of common contaminants often is present. In addition, very low solubility sludges and "bottoms" create a poorly characterized "goop" or "gumbo."

The Department of Energy (DOE) has responsibility for managing the sites where nuclear weapons were produced. DOE sites are unique in that the contamination of the subsurface often involves complex mixtures of organic and inorganic chemicals, including short- and long-lived radionuclides (USDOE, 1990). Important individual contaminants found at DOE sites are listed in Table 15.8. These inorganic and organic species have been found in a variety of combinations. An extremely important interaction is complexation between the heavy-metal radionuclides and the strong chelating agents, such as EDTA and NTA. The degree of complexation with

Table 15.8   DOE site contaminants

Inorganic Species	Organic Species
Radionuclides	Organic Contaminants
Plutonium-238,239	Chlorinated hydrocarbons
Americium-241	Methyl ethyl ketone
Thorium	Cyclohexanone
Uranium-232,234, 238	Tetraphenyl boron
Technetium-99	Polychlorinated biphenyls (PCBs)
Strontium-90	Select polycyclic aromatic
Cesium-134, 137	hydrocarbons (PAHs)
Cobalt-60	Tributyl phosphate
Europium-152, 154, 155	Toluene
Nickel-63	Benzene
Iodine-129	Kerosene
Neptunium-237	
Radium	
Metals	Facilitators[a]
Lead	Aliphatic Organic acids
Nickel	(citric, lactic, succinic, oxalic to octadecanoic)
Chromium	Chelating agents (EDTA, NTA, DTPA, HEDTA,
Copper	TTA, Di-2-ethyl HPA)
Mercury	Aromatic acids (humic acids and subunits)
Silver	Solvent, diluent, and chelate radiolysis
Bismuth	fragments
Palladium	
Aluminum	
Others	
Carbon-nitrogen compounds	
Nitrite	
Nitrate	

[a]Facilitators are organic compounds that interact with and modify metal/radionuclide geochemical behavior.
SOURCE: USDOE, 1990; Rittmann et al., 1994.

the chelators controls the mobility of the radionuclides, while the biodegradation of the chelators is affected by their complexation to the heavy metals.

15.2 BIODEGRADABILITY

Fortunately, essentially all of the common organic contaminants found in groundwater are biodegradable, provided the proper conditions exist. Chapter 14 reviewed the biochemical and physiological factors controlling the biodegradation of these kinds of hazardous organic pollutants.

A brief summation of items most relevant to bioremediation is:

• The aliphatic and aromatic hydrocarbons are readily biodegradable by a range of aerobic bacteria and fungi. The key is that molecular O_2 is needed to activate the molecules via initial oxygenation reactions.

• Evidence of anaerobic biodegradation of aromatic hydrocarbons is growing. Anaerobic biodegradation rates are slower than aerobic rates, but they can be important when fast kinetics are not essential.

• Most halogenated aliphatics can be reductively dehalogenated, although the rate appears to slow as the halogen substituents are removed.

• Highly chlorinated aromatics, including PCBs, can be reductively dehalogenated to less halogenated species.

• Lightly halogenated aromatics can be aerobically biodegraded via initial oxygenation reactions.

• Many of the common organic contaminants show inhibitory effects on microorganism growth and metabolism. Due to their strongly hydrophobic nature, many of the inhibitory responses are caused by interactions with the cell membrane. In some cases, intermediate products of metabolism can be more toxic than the original contaminant.

15.3 CONTAMINANT AVAILABILITY FOR BIODEGRADATION

A factor that frequently limits the rate of bioremediation is *substrate unavailability*, or when only a small fraction of all the substrate molecules is truly water soluble and, therefore, available to the microorganisms. Two phenomena that limit substrate availability are strong sorption to surfaces and formation of a nonaqueous phase. Both factors keep aqueous-phase concentrations low. Low concentration slows degradation kinetics in general and may also prevent biomass accumulation by being lower than S_{min} or by preventing enzyme derepression or induction. However, low aqueous-phase concentrations sometimes are beneficial for biodegradation, particularly when the compound is inhibitory.

15.3.1 SORPTION TO SURFACES

The effect of sorption on biodegradation in the subsurface can be complex. In some cases, sorption of substrates stimulates biodegradation, but in many cases it retards biodegradation. In order to come to some understanding of the interactions between biodegradation and sorption, it is helpful to review sorption and how solid surfaces affect microbial physiology.

What Controls Sorption   For most of the organic molecules of importance to bioremediation, their sorption to solid surfaces is controlled primarily by their hydrophobic nature. Being nonpolar and of low solubility, they are thermodynamically driven out of aqueous solution by entropy effects. In other words, having these nonpolar solutes dissolved in water requires a restructuring of the water molecules to a higher free-energy state. Ejecting the nonpolar solutes from water solution reduces the free energy of the system. This effect is magnified when the surface to which the nonpolar molecule adsorbs is itself hydrophobic.

A useful means for quantifying this adsorption effect for nonpolar, organic solutes is the linear partitioning model:

$$\psi = K_p C \qquad \textbf{[15.1]}$$

in which C is the aqueous-phase concentration of the solute $[ML^{-3}]$, ψ is the adsorption density of the sorbed solute $[M/M]$, and K_p is the linear partition coefficient $[L^3/M]$. The linear-partition model assumes equilibrium between C and ψ. The K_p value often is related directly to the hydrophobicity of the solute and the surface onto which it adsorbs. Based on the work of Karickhoff, Brown, and Scott (1979), K_p can be expressed as

$$K_p = 0.63\, K_{ow} f_{oc} \varepsilon / \rho_b \qquad \textbf{[15.2]}$$

in which K_{ow} is the solute octanol/water partition coefficient, which has units of volume water/volume octanol; f_{oc} is the fraction by mass of organic carbon in the solid phase; 0.63 is an empirical conversion factor that has units mass solid-volume octanol/(mass-volume water); ρ_b is the bulk density, or mass solids/total volume; and ε is the porosity, or water volume/total volume.

McCarty et al. (1985) derived a relationship for the retardation factor, R, which is the ratio of the advective water velocity to the velocity of the contaminant:

$$R = 1 + (\rho_b/\varepsilon)K_p = 1 + 0.63\, K_{ow} f_{oc} \qquad \textbf{[15.3]}$$

A large R value means that the solute is strongly immobilized by adsorption. For example, anthracene ($K_{ow} = 10^{4.54}$) would have an R of 1,100 in a highly organic soil or sediment ($f_{oc} = 0.05$). On the other hand, TCE ($K_{ow} = 10^{2.29}$) would have an R of only 1.12 in a sandy aquifer with a low organic content ($f_{oc} = 0.001$). Likewise, the adsorption densities (Q) would be quite different: A 10-μg/l aqueous-phase concentration would give Q values of 1,370 and 0.15 μg/g soil for anthracene and TCE, respectively ($\rho_b = 2$ kg/l and $\varepsilon = 0.25$).

Influences of Solid Surfaces on Microbial Physiology   Studies of the naturally occurring microbes in aquifers show that the numbers of microorganisms attached to aquifer solids usually are 10 to 1,000 times greater than those free-living in the groundwater. The ratio may be higher in soils and sediments. Although bacterial attachment to surfaces undoubtedly can affect the activity of the bacteria, what the effect will be cannot be predicted readily. Van Loosdrecht, Lyklema, Norde, and Zehnder (1990) summarized the apparently contradictory influences of solid surfaces on microbial behavior: increased growth rate, decreased growth rate, increased assimilation and decreased respiration rates, decreased assimilation, increased respiration, increased adhesion of active cells, higher activity of attached cells, decreased substrate utilization, lower substrate affinity, change in pH optimum, difference in fermentation pattern, increased productivity, decreased mortality, and no effect.

Some confusion can be attributed to a failure to distinguish between direct and indirect influences of solid surfaces. A direct influence results from the presence of a surface itself. An indirect influence occurs because the aggregation of microorganisms alters their environment, such as concentrations of substrate or pH. One reason for confusion comes about because bacteria attaching to living surfaces, such as with plants and animals, often attach at and respond to specific receptors. Whereas these highly specific attachments trigger profound alterations to the physiology of the attaching microbe, as well as its host, the nonliving surfaces of soil and sediments cannot evoke the same kind of reaction. Thus, any effects on microbial physiology in environmental settings are normally the result of indirect effects via changes to the cells' local environment. These changes in local environment can include concentration gradients for substrate and pH. They also can include signaling molecules that transfer between different microorganisms.

Biodegradation of Sorbed Molecules   Sorption can cause either decreased or increased biodegradation, and the observed effects depend on the characteristics of the compound, the solid phase, and the microorganisms. One obvious way in which sorption can slow biodegradation rates is simply by decreasing the aqueous-phase concentration. Since most bacteria utilize only dissolved solutes, adsorption becomes a competing sink. In such cases, compounds having large R values see a dramatic decrease in the biodegradation rate per unit of contaminant present, since most of the solute is adsorbed to the surface and inaccessible for direct biodegradation. If equilibrium partitioning holds, only $1/R$ of the total compound present is available for biodegradation. For example, if anthracene had a first-order degradation rate constant of 0.1/d for solute strictly in the dissolved state, its first-order rate for all anthracene would drop to only 0.000091/d if R were 1,100, as in the example above. Clearly, strong sorption can greatly slow degradation of highly hydrophobic chemicals.

When sorption is important, the rate of biodegradation can, in some cases, be controlled by the rate of desorption. As a first approximation, the desorption rate can be described by a single-resistance, mass-transport model,

$$J_d = k_{\mathrm{mt}}(\psi/K_p - C) \qquad\qquad \textbf{[15.4]}$$

in which J_d is the desorption flux $[ML^{-2} T^{-1}]$ and k_{mt} is a mass-transport coefficient $[LT^{-1}]$. In a general sense, biodegradation acts to increase J_d by keeping C low. When biodegradation (or another sink mechanism, such as advection or dispersion) forces C to approach zero, the rate of biodegradation then becomes mass-transport limited, and C and ψ are not in equilibrium (Seagren, Rittmann, and Valocchi, 1993, 1994).

When desorption is mass-transport limited, k_{mt} becomes the critical factor controlling the rate of bioremediation. Unfortunately, k_{mt} is a very poorly defined parameter for real aquifer, soil, or sediment systems. In many circumstances, k_{mt} is not controlled by a single mass-transfer step. Transport steps that may be acting include: movement from inside a porous solid (such as between the layers of clays or the interior of organic aggregates) to the water/solid interface, transfer across the interface, transfer across a nonturbulent boundary layer surrounding the solid particle, and transfer in the bulk liquid. This situation is complicated when the microorganisms are located at the solid/water interface and in the boundary layer; having biodegradation reactions in these locations violates the assumption of linear gradients that is inherent to Equation 15.4. This complication can be overcome by defining C to be adjacent to the interface. However, the interfacial C cannot be measured, and the complexity of the modeling increases greatly (Seagren et al., 1993, 1994, 2000).

Irreversible sorption is an extreme case of mass-transport limitation. Adsorption of solutes inside the crystalline lattice of clay minerals can lead to irreversible adsorption if the lattice structure collapses after adsorption takes place. Blocking of intra-aggregate pores by strongly adsorbed molecules also can lead to irreversible sorption. Polymerization of adsorbed organic molecules, particularly phenolics, also has been implicated as a means for making adsorption essentially irreversible. In this case, K_p effectively increases, and pore blockage also may result.

Bioremediation can be enhanced by adsorption when the compound is toxic. Sequestering the inhibitor and creating a nontoxic aqueous-phase concentration can lead to microbial growth and substrate metabolism. This beneficial sequestering is most effective when the solid has a strong intraparticle adsorption capacity. Sorption also can be beneficial when the sorbed substrate is stored and later released for biodegradation. This "warehousing" approach allows the microorganisms to accumulate to a high density and then modulate the desorption by creating a low value of C (van Loosdrecht et al., 1990; Chang and Rittmann 1987; Rittmann, Schwarz, and Sáez, 2000).

Although sorption can have various positive and negative influences on biodegradation, its relevance is accentuated for soils with high organic carbon or clay content and aggregates having internal porosity. However, in situ bioremediation generally requires fluid-permeable deposits that encompass materials ranging from silty sands to gravel. These materials may have little organic carbon content, cation exchange capacity, or internal pores.

15.3.2 FORMATION OF A NONAQUEOUS PHASE

A second way in which compounds show limited availability as substrates is formation into *nonaqueous phase liquids (NAPLs)*—fluids that are essentially immiscible with water and migrate through the subsurface as a separate phase. This category of fluids includes petroleum, its refined fluid derivatives, and aliphatic chlorinated hydrocarbons. These fluids become trapped as residual saturation and lenses in the unsaturated and saturated zone. In the case of NAPLs less dense than water (LNAPLs, e.g., petroleum products), the pools can form at the top of the water table. Pools or residuals migrate to the aquifer bottom in the case of NAPLs more dense than water (DNAPLs, e.g., aliphatic chlorinated hydrocarbons). This trapped NAPL serves as a long-term source of groundwater contamination via dissolution. The focus here is on petroleum-product NAPLs, although most of the subjects reviewed apply equally to other NAPLs or NAPL mixtures.

Solubility   Although NAPLs are defined as *immiscible* in water, they are, in general, slightly soluble in water. Table 15.9 lists selected gasoline hydrocarbons and their pure-compound water solubilities. The lowest solubilities (< 1 mg/l) are associated with the normal and branched alkanes. However, aromatics have solubilities greater than 100 mg/l.

Because a gasoline NAPL is a mixture, the actual equilibrium solubilities are substantially reduced from the pure-compound solubilities shown in Table 15.9. In most cases, the water solubility for mixtures of hydrocarbons follows ideal-solution theory, which says that the equilibrium solute concentration, C_i, of the ith component in an NAPL mixture is proportional to the product of its pure-compound solubility,

Table 15.9 Solubility of selected pure hydrocarbons in water at room temperature

Hydrocarbon	mg/L in Distilled Water
n-Alkanes	
n-propane	62
n-pentane	39
n-hexane	9.5
n-octane	0.7
Branched Alkanes	
2-methylbutane	48
2, 3-dimethylhexane	0.13
Cycloalkanes	
cyclopentane	156
cyclohexane	55
methylcyclopentane	42
methylcyclohexane	14
Aromatics	
benzene	1780
ethylbenzene	152
toluene	515
o-xylene	175
p-xylene	198

SOURCE: Verschueren, 1983; Rittmann et al., 1994.

S_i, and its mole fraction in the mixture,

$$C_i = X_i S_i \qquad\qquad \textbf{[15.5]}$$

in which X_i is the component's mole fraction in the NAPL.

Aromatic components in gasoline, such as BTEX, have the highest S_i values, and they also are present at relatively large mole fractions. Therefore, they dissolve most rapidly into the water. Although this causes BTEX to be the most prevalent water pollutant, it also makes BTEX the most accessible for biodegradation. When combined with the fact that BTEX compounds are readily biodegradable under aerobic conditions, this relative propensity to become water-soluble has made bioremediation of BTEX by far the most widely practiced type of bioremediation.

Because the C_i values for the non-BTEX hydrocarbons in NAPLs are very low, sequential dissolution and biodegradation of the truly dissolved component is a slow means for the hydrocarbon-degrading microorganisms to make their substrate accessible. To maximize their rate of access to hydrocarbon substrates, some bacteria and fungi able to degrade long-chain aliphatics adhere to the hydrocarbon-NAPL interface and participate in a form of "direct uptake" of the substrates that they can metabolize. [See Rittmann et al. (1994) for a review of the adherence phenomenon.] Adherent bacteria are predominant when the amount of surface area is small, a common situation when mechanical agitation cannot be provided to emulsify the NAPL. The exact mechanism by which adhesion promotes substrate accessibility is not clear. By being very close to the NAPL/water interface, adhering microorganisms minimize the diffusion distance from the NAPL to the cell, thereby reducing the mass-transport resistance. Some believe that the hydrocarbon substrate is transferred directly into the cell without dissolving into the liquid.

When direct contact is important for increasing hydrocarbon biodegradation rates, the amount of NAPL/water surface area is critical. Emulsification or pseudo-solubilization is a way in which the surface area can be increased. Agitation, use of synthetic surfactants, production of microbial metabolites having surface-active properties (bio surfactants), or a combination can bring them about. Ex situ technologies are most effective for this surface-enhancing approach. The lack of agitation and difficulties contacting the NAPL with the surfactant make in situ application of surface enhancement problematical.

For in situ bioremediation, surfactants can be used to mobilize residual NAPL. The hydrocarbons in the NAPL are extracted into surfactant micelles (i.e., colloids of nonaqueous-phase surfactant) formed when the surfactant's critical micelle concentration (CMC) is exceeded. Mobilization of hydrocarbons by micelles may increase or decrease biodegradation. Most field experiences indicate that the application of micellular surfactants has a negative effect on biodegradation (Staps, 1990). Reduced biodegradation can be attributed to toxicity from the surfactant or to a further reduction in the dissolved concentration.

The use of surfactants to enhance biodegradation has been confused by the unfortunate use of the term "solubilize" to refer to the partitioning of hydrocarbons into colloidal micelles of surfactant. The hydrocarbon molecules are mobilized with the micelles, which often move with the water phase. However, these components are not truly soluble, because they exist inside the surfactant phase. In fact, truly soluble components often have reduced concentrations due to their partitioning into the micelles. The term "solubilization" is false when biodegradation of soluble components is considered. The correct term is mobilization, and the components are mobilized in a colloidal, nonaqueous phase.

Mobilization in micelles increases biodegradation rates when the micellular hydrocarbon is more available to the microorganism than it would be if it remained in its original NAPL. Use of surfactants in ex situ bioremediation can greatly accelerate biodegradation when the nonaqueous-phase surface area is greatly increased. Most probably, the surface area is increased due to the combination of a decrease in surface tension and energy inputs, not from transfer of the hydrocarbons to surfactant micelles.

Surfactants, either biological or synthetic, could be used to mobilize hydrocarbons for the purpose of withdrawing them from the aquifer for treatment above ground (Lee et al., 1988). A danger concomitant with surfactant application to enhance contaminant removal in the subsurface is that the spreading of contaminants to previously uncontaminated parts of the aquifer could exacerbate the contamination problems.

15.4   TREATABILITY STUDIES

Before an in situ bioremediation project is attempted in the field, treatability studies may need to be performed (Lee et al., 1988; Thomas and Ward, 1989; Rittmann et al., 1994). The overall goal of treatability studies is to verify that the in situ approach has a high likelihood of succeeding. The number and type of treatability studies depend upon the characteristics of the site and the contaminants. In general, the level of effort that needs to be expended on treatability studies is proportional to the uncertainty about site and contaminant characteristics: When uncertainty is low, little effort may ensure success, but extensive efforts are needed when uncertainty is high.

Table 15.10 presents a hierarchical scheme for determining what kind of treatability study is needed. Each level is used to answer different kinds of questions. A treatability study should be carried out for each level for which answers are not known from other sources.

Level 1 is used to answer the fundamental question, "Is biodegradation a feasible option with the contaminants at the site?" Level 1 tests are not needed when the biodegradabilities of the contaminants are established from the literature or from previous bioremediation experience. Level 1 is needed most often when the contaminants are unusual, are not identified (e.g., they are only peaks on a chromatogram), or occur in mixtures that may be inhibitory (e.g., with toxic organic compounds, heavy metals, or high salts) or incompatible (e.g., some contaminants are known only to be degraded aerobically, while others are degraded only anaerobically).

The techniques for Level 1 studies ought to be as simple as possible. The tube microcosms of Wilson and Noonan (1984), serum bottles (Shelton and Tiedje, 1984), and hypovials (Bouwer, Rittmann, and McCarty, 1981) usually are satisfactory. The

Table 15.10   Hierarchical approach to determining what kind of treatability studies are needed to ensure success with in situ bioremediation

Level	Goals	Approach
1	Determine if biodegradation is an option	Simple laboratory microcosms, "yes" or "no" answer on biodegradability
2	Evaluate biodegradation nonidealities, such as concentration dependence of kinetics, inhibition, sorption, dissolution, mixed substrates, intermediates formation, and the need for other substrates or nutrients	Laboratory microcosms of all types, including hypovials, shake flasks, serum vials, columns, and slurries
3	Evaluate site-specific issues, such as hydrogeologic conditions and heterogeneity	Field pilot study

SOURCE: Rittmann et al., 1994.

keys are to avoid complications with volatilization and adsorption and to ensure that the reaction conditions in the microcosms are optimal for the reactions being evaluated. Usually, the disappearance of the target components, in comparison with abiotic controls, is sufficient to give the desired "yes" or "no" answer on the feasibility of biodegradation.

Once feasibility is established, Level 2 studies are needed to provide estimates of the rate of reaction and to identify complicating factors. Depending on the situation, Level 2 studies can address a wide range of questions. Some appropriate questions are:

- What is the functional relationship between the concentration of contaminant and its biodegradation rate?

- How do concentrations of other substrates, such as the primary electron donor and acceptor, affect the reaction rate?

- What is the stoichiometry needed for addition of substrates and nutrients? In particular, the stoichiometric needs should be established with knowledge of how the added material is going to be used; for example, as a growth-supporting substrate or nutrient, as a cosubstrate, or as an electron source or sink.

- Will environmental factors, such as low temperature, affect the rates?

- Do intermediates form, especially hazardous ones?

- Do the contaminants need to desorb or dissolve before they can be degraded? Can these rates be increased by addition of surface-active agents?

- Do toxicants, including intermediates and other substrates, interfere with biodegradation?

Level 2 experiments can be performed in many types of systems, including all those used for Level 1, columns, soil beds, and slurry reactors. The design of the system should depend on the specific question being asked and on sampling requirements. Even when the experimental system is the same as for Level 1, the intensity for sampling and analysis must be much greater for Level 2. Whenever possible, Level 2 studies should be performed with actual soil, aquifer, and/or groundwater samples.

Level 2 studies should yield reasonable estimates of the rate of reaction in situ; limitations from inhibition, solubility, or intermediates; and needs for substrates and nutrients. This information, when combined with knowledge of the site's hydrogeology and contamination, should allow design of a field program based on sound science and engineering.

Level 3, field pilot testing, is used to answer the question, "Can the in situ bioremediation succeed under the site-specific conditions actually existing?" Real-world factors of heterogeneity and poor permeability can prevent success in the field, even though microbiological factors are positive. Level 3 testing may not need to be performed when site characterization shows that the site is hydrogeologically simple and highly permeable.

A Level-3 test normally is performed on a small portion of the contaminated site. The pilot location should be representative of the overall site in terms of hydrogeology and contamination. In situ conditions for the pilot test should simulate the proposed bioremediation scheme. The pilot site must be large enough that realistic complications are encountered. Simulation modeling of the transport and microbiological aspects of the bioremediation is recommended as part of the design and evaluation of the pilot study.

15.5 ENGINEERING STRATEGIES FOR BIOREMEDIATION

15.5.1 SITE CHARACTERIZATION

The first step in any bioremediation is site characterization in terms of its geology, hydrology, geochemistry, type and distribution of contamination, and microbiology. Site assessments performed to meet regulatory requirements typically do not provide nearly enough or the right kind of information for evaluating the potential for bioremediation. Therefore, site characterization needs to be tailored for bioremediation.

Critical information includes:

- The nature of the geological deposits, including the presence of fractures, thickness and extent of aquifers, location of confining units, and particle-size distribution
- The porosity (ε), hydraulic conductivity (K_{hyd}), and the variability of these parameters for the water-transmitting units
- The direction and slope of the naturally occurring hydraulic gradient
- As much information as is possible to obtain on the location, type, and mass of the contamination source; and the extent and concentration distribution of a contaminant plume

- Intrinsic supply and consumption rates of electron acceptors, nutrients, alkalinity, and other geochemical species affecting biodegradation

- Information on microbiological activity at least at Level 1 under Treatability Studies

15.5.2   ENGINEERED IN SITU BIOREMEDIATION

Because engineered in situ bioremediation is used to accelerate biologically driven removal of contaminants trapped in the solid phase, its success depends upon being able to achieve substantially increased inputs of stimulating materials. Thus, the ideal site has these features:

1. The hydraulic conductivity is relatively homogeneous, isotropic, and large (i.e., greater than about 10^{-5} m/s).

2. Residual concentrations are not excessive; NAPL concentrations greater than about 10 g/kg reduce aquifer permeability and microorganism access to the biodegradable contaminants.

3. The contamination is relatively shallow, in order to minimize costs of drilling and sampling.

The permeability, which controls the hydraulic conductivity, probably is the most important factor. Most successful projects have been in sandy environments having K_{hyd} values of 10^{-5} to 10^{-3} m/s (Staps, 1990). Heterogeneity is the most prevalent deviation from the ideal situation. Even when a formation has a suitably large K_{hyd}, that K_{hyd} often is an "effective," or spatially averaged value. Heterogeneity in K_{hyd} can have a significant impact on engineered bioremediation. Contaminants trapped in regions of relatively low K_{hyd} tend to be by-passed by the fluid flows bringing stimulatory materials.

Engineered in situ bioremediation systems can be classified according to the means by which stimulatory materials are added. Although technical details depend strongly on site-specific conditions and are advancing rapidly, these broad classifications are valuable for identifying what approaches have merit. The following methods are being applied successfully for in situ bioremediation (mainly) of petroleum hydrocarbons.

Bioventing   When the contaminants are trapped in the unsaturated zone (also called the vadose zone) above the water table, the easiest way to supply O_2 is by pulling air through the unsaturated soil. This technique, called *bioventing,* is illustrated in Figure 15.2. Vacuum pumps create negative pressure that sweeps air through the soil and past the contaminated soil. The spacing of the vacuum points depends on the soil's permeability and the applied vacuum. Enough vacuum points are needed to ensure that O_2 is delivered to all contaminated units.

Passing air through the unsaturated zone evaporates moisture and can desiccate the soil enough that microbiological activity is slowed or even prevented. Therefore, Figure 15.2 illustrates how water is added to the *biologically active zone* (BAZ) by infiltrating water. Infiltration can be accomplished by spraying or flooding the surface

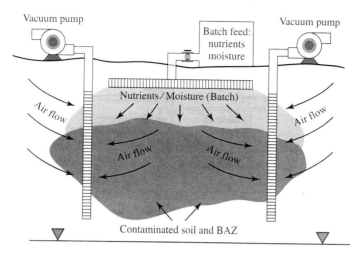

Figure 15.2 Bioventing is used to bioremediate contaminants trapped above the water table, which is indicated by the two triangles. The BAZ occurs where the water, nutrients, and air flow coincide with the contaminated soil. SOURCE: NRC, 1993.

for permeable soils or by injection through infiltration trenches, galleries, or dry wells for less permeable soils. The infiltration rate must be carefully controlled to preclude soil "flooding," which can decrease the soil's air permeability and leach contaminants to the water table.

When inorganic nutrients or other stimulants are required, they generally are added in soluble form (e.g., NH_4Cl, KNO_3, and NaH_2PO_4) in the infiltrating water. In some cases, nutrients can be added as gases, such as NH_3, N_2O, or triethylphosphate. Addition of methane can be employed to stimulate co-metabolic degradation of TCE. In general, gaseous additives are applied through trenches or wells located in line with the airflow routes.

Water-Circulation Systems When the contamination is partly or totally within the saturated zone, stimulatory materials can be applied by water circulation. Figure 15.3 illustrates the engineered bioremediation of an LNAPL that is above and below the water table. In this case, H_2O_2 is used as a dissolved source of oxygen substantially greater than can be achieved by air saturation. The figure shows a vertical well and a horizontal infiltration gallery; in most cases, one or the other approach is used to add the circulating water and stimulants.

A key to using water circulation is to recover all the circulating water. Recovery systems normally involve a series of extraction wells designed to extract more water than is injected. In this way, there is no export of contaminated groundwater from the site. The extracted water commonly is treated and recycled. Above-ground treatment can include air stripping (shown in Figure 15.3), biological treatment, activated carbon absorption, or a combination. If the extracted water is not

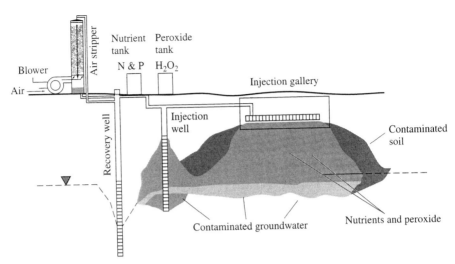

Figure 15.3   Schematic of a water-circulation system with H_2O_2 as an O_2 source. The BAZ is present where the water, nutrients, and oxygen intersect with contaminants dissolved from the contaminated soil. SOURCE: NRC, 1993.

recycled, drinking water or uncontaminated groundwater is amended and used for injection.

One problem that can occur as part of a water-circulation system is clogging near well screens and infiltration galleries (Staps, 1990; Wetzel et al., 1986). Localized growth of bacteria, sometimes coupled with chemical precipitation or gas evolution, reduces the soil's hydraulic conductivity (Rittmann et al., 1994). This can lead to decreased input rates and/or short-circuiting around clogged areas. Efforts to minimize the adverse impact of clogging include pulsing of various substrates and nutrients to "spread out" the BAZ, reduction of substrate loading, back-flushing to dislodge accumulated solids, periodic application of disinfectants (chlorine or additional H_2O_2) or acids, and adjustment of pH and inorganic ions to preclude precipitation.

Air Sparging   Originally developed in Europe, air sparging has become a popular means of engineered bioremediation in North America for strictly aerobic biodegradation. Injection of compressed air directly into the contaminated subsurface (Figure 15.4) is an efficient way to deliver oxygen to the BAZ. In addition, air sparging can strip volatile contaminants from the saturated zone into the unsaturated zone and a vapor-capture system. Air sparging is not effective when low-permeability geological zones trap or divert the gas flow.

Nutrients and other amendments can be added from injection wells or an infiltration gallery, as shown in Figure 15.4. Similar to bioventing, some nutrients can be added in a gaseous state. In most cases, a gas-recovery system is needed to capture volatile components and prevent off-site contaminant transport in the gas phase.

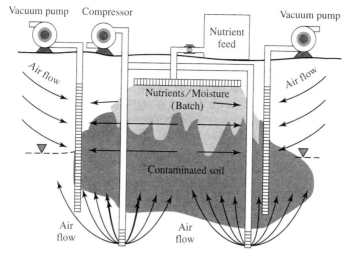

Figure 15.4 An air-sparging system for aerobic treatment of contamination below the water table. The BAZ occurs where the air flow contacts contaminated water. SOURCE: NRC, 1993.

15.5.3 INTRINSIC IN SITU BIOREMEDIATION AND NATURAL ATTENUATION

Intrinsic bioremediation relies on the intrinsic (i.e., naturally occurring) supplies of electron acceptors, nutrients, and other necessary materials to develop a BAZ and prevent the migration of contaminants away from the source. Although intrinsic bioremediation does not accelerate the rate of source cleanup, it can be an effective form of biological containment. Intrinsic bioremediation involves no engineered measures to increase the supply rates of oxygen, nutrients, or other stimulants. On the other hand, intrinsic bioremediation requires careful site characterization in terms of the extent and type of contaminant source and plume, the presence of microorganisms capable of degrading any mobile contaminants, and the intrinsic supply rates of electron acceptor and other required materials. A long-term monitoring program is essential to ensure that the intrinsic biological activity remains effective at preventing contaminant migration. The principles of evaluating bioremediation are given later in this chapter.

Intrinsic in situ bioremediation is best applied in situations in which the groundwater contains naturally high concentrations of electron acceptors and adequate concentrations of nutrients, consistent groundwater levels and flow velocity, and the presence of carbonate materials to buffer pH changes (NRC, 1993). Intrinsic bioremediation can be used alone or in concert with an engineered bioremediation or other technology. Because intrinsic bioremediation depends on naturally occurring supply rates, heavy source contamination can overwhelm the intrinsic supply rates. In such a case, an engineered technology can be used to reduce the size of the contaminant source until the rate of contaminant dissolution no longer exceeds the intrinsic supply rates.

Intrinsic bioremediation sometimes is confused with *natural attenuation,* which is defined as the reduction of contaminant concentrations due to naturally occurring processes of biodegradation, sorption, advection, dilution, dispersion, and chemical reaction. Intrinsic bioremediation is a more stringent subset of natural attenuation in which microbial transformation reactions are responsible for the decrease in concentration. The National Research Council (NRC, 2000) issued comprehensive guidance for natural attenuation. In principle, natural attenuation can minimize risks from a wide range of organic and inorganic contaminants (NRC, 2000). To date, natural attenuation has been proven to work reliably mainly for BTEX.

15.5.4   IN SITU BIOBARRIERS

A hybrid bioremediation technology is the in situ biobarrier, which is depicted schematically in Figure 15.5. Unlike the previous in situ methods, a biobarrier addresses mobile contaminants already being transported in a plume. The biobarrier is a containment method that prevents further transport of a plume. However, the biobarrier is an engineered system in which a BAZ is created in the path of the plume through injection of electron donor, electron acceptor, and/or nutrients. Hydraulic or physical controls on groundwater movement may be required to make certain that the plume passes through the BAZ. It seems reasonable that several different BAZs could be created sequentially by injection of different electron donors and acceptors. The Moffett Field experiment (Roberts, Hopkins, Mackay, and Semprini, 1990; Semprini, Roberts, Hopkins, and McCarty, 1990) is a small-scale example of creating a specialized BAZ by injecting methane (the carbon source and electron donor) and oxygen (the electron acceptor); cometabolic biodegradation of TCE was demonstrated.

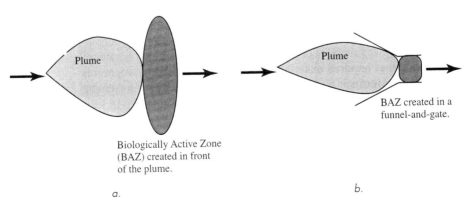

Biologically Active Zone
(BAZ) created in front
of the plume.

a.

BAZ created in a
funnel-and-gate.

b.

Figure 15.5   Schematic illustration of an in situ biobarrier that contains a mobile contaminant plume by intercepting it with a BAZ. SOURCE: Rittmann et al., 1994.

The materials needed to create a biobarrier can be injected in several ways. The most obvious way is to add them in liquid form through injection wells or trenches. An alternative to simple injection is the recirculating well, in which the injected materials and the plume water are mixed in and around wells that circulate the water between well screens placed at two different levels on the well casing. Full-scale implementation of this technology for cometabolism of TCE was conducted at Edwards Air Force Base, California (McCarty et al., 1998). Slow-release chemical sources of O_2 and H_2 can be placed in the flow path of the plume via wells or trenches sunk to the level of the plume. Iron filings also have been used to create an in situ barrier. The oxidation of the elemental iron can bring about reductive dechlorination of solvents, but it also may spur microbiologically catalyzed reductions.

15.5.5   EX SITU BIOREMEDIATION

Contaminated soils can be excavated and treated in aboveground, or ex situ, treatment systems. Aboveground bioremediation for contaminated solids include slurry reactors, composting, and land farming. Ex situ bioremediation is most applicable for small, heavily contaminated sources and when a rapid site cleanup is desired. Excavation incurs major costs and potentially increases exposure to workers and those who reside nearby.

Slurry reactors contain (typically) 5 percent solids and are vigorously agitated for mixing and aeration. They are advantageous for maximizing the rate of biodegradation. This rate increase is brought about by increased mass-transport rates due to agitation and the ease of adding oxygen, nutrients, and surface-active agents that emulsify NAPLs. The disadvantages of slurry systems are the added costs for capital, operation, and dewatering the decontaminated solids. Slurry reactors tend to be operated in a batch mode, but continuous feeding is possible.

Composting is an aerobic biodegradation scheme operated in a "solid state" format. Solid state means that the compost material behaves as a porous solid with a moisture content of 50 to 60 percent and through which air can be drawn to supply O_2 and to remove evaporated H_2O. Due to the high concentration of biodegradable organic material and the low moisture content of the compost mixture, the temperature rises to 60–70 °C during the time of peak biodegradation rates. In most cases, the air supply is determined by temperature considerations. If the temperature is too hot, more air circulation is needed to drive evaporative cooling. In parallel to control of air circulation, the moisture content must be controlled within the optimal range, 50–60 percent. Too much moisture prevents air circulation and slows microbial reactions. Too little moisture initially allows the temperature to rise above the optimal range; the excess temperature and concomitant desiccation of compost materials arrest biodegradation reactions.

Land farming is a simple, "low-tech" form of ex situ bioremediation. Contaminated solids are mixed into the surface layer of topsoil. Nutrients and moisture can be added initially and throughout the treatment period to optimize conditions for microbial growth. Land farming is most effective for solids contaminated by organic materials susceptible to aerobic biodegradation and where large tracts of land can be devoted to bioremediation.

15.5.6 PHYTOREMEDIATION

A new and rapidly developing form of bioremediation is *phytoremediation,* which uses green plants and their associated biota to destroy, remove, contain, or otherwise detoxify environmental contaminants (Cunningham, Anderson, Schwab, and Hsu, 1996). Four types of destruction and removal reactions can occur in phytoremediation. Definitions of the four pathways are:

- *Phytovolatilization* is an enhancement of the volatilization process from the soil or through the plant's roots or shoots. Enhanced volatilization can occur via plant transpiration of volatile compounds or transformation of contaminants to more volatile forms.
- *Phytodegradation* involves uptake by the plant and subsequent metabolism by plant enzymes to form benign products.
- *Phytoextraction* involves uptake by the plant and absorption of the contaminant into plant tissue, which subsequently is harvested. Hydrophobic contaminants are most susceptible to photoextraction.
- *Rhizo(sphere)degradation* to benign products is catalyzed by plant enzymes excreted by the roots or by the microorganisms found in the rhizosphere.

In addition, phytoremediation can sequester contaminants, a process called *phytostabilization* (Cunningham et al., 1996). The benefit of phytostabilization is that it makes the contaminants less bioavailable to humans and other receptors. Three forms of phytostabilization can be identified:

- *Humification* incorporates contaminants into the soil organic matter, or humus, through binding (polymerization) reactions of plant and microbial enzymes.
- *Lignification* incorporates the contaminant into plant cell-wall constituents.
- *Aging* slowly binds the compounds into the soil mineral fraction.

The distinctions among the various stabilization, destruction, and removal mechanisms often are not sharp. For example, rhizosphere peroxidases can catalyze oxidation reactions that lead to contaminant mineralization (degradation) or to polymerization (humificaction).

Phytoremediation is still in its early stages of research and development. Nonetheless, it has promise for the decontamination of petroleum hydrocarbons, a range of chlorinated aliphatics and aromatics (including PCBs), and pesticides (Cunningham et al., 1996). Phytoextraction of heavy metals is another promising application (Moffat, 1995).

Another aspect of phytoremediation is hydraulic control. During their growing season, plants transpire large volumes of water to the atmosphere. This transpiration can stop the downward flow of precipitation from the vadose zone to the saturated zone. It also can "pump" water from the top of the saturated zone to the atmosphere.

Transpiration is being exploited as a means to stop or slow the migration of contaminants that are in the vadose zone or the boundary between the vadose and saturated zones. Of course, this pumping action only occurs when the plants are active in photosynthesis. For example, deciduous trees are ineffective in the winter.

15.5.7 BIOREMEDIATION OF GAS-PHASE VOCs

Gases contaminated with volatile organic compounds can be generated in numerous ways:

- Bioventing and air sparging, as discussed above
- Vapor extraction of contaminated soils (NRC, 1994)
- Storage tanks or production facilities for volatile organic products, such as gasoline, petroleum, and solvents.
- Off gases from wastewater treatment processes
- Off gases from anaerobic treatment processes
- Off gases from sewers or preliminary treatment of sewage

Volatile organic compounds (VOCs) are organic chemicals that have a boiling point ≤ 100 °C and/or a vapor pressure ≥ 1 mm Hg at 25 °C. They include alkane gases (e.g., methane), alcohols (e.g., methanol), low-molecular-weight petroleum hydrocarbons (e.g., benzene and toluene), halogenated aromatics (e.g., chlorobenzenes), and halogenated aliphatic solvents (e.g., 1,1,1-TCA and TCE). These VOCs are of particular environmental concern because they are very mobile, may affect human senses through odor, may exert a narcotic effect, and may be toxic and carcinogenic. Some VOCs also react photochemically to form ground-level ozone, a key component of smog, while others are greenhouse gases. For these reasons, emissions of VOCs to the atmosphere are becoming a greater concern to regulatory agencies, industry, and the general public.

Biodegradation of VOCs is among a suite of technologies that can be used to treat gases containing VOCs (Moretti and Mukhopadhyay, 1993). Low capital and operating costs are the main advantages of biofiltration, the main type of biological treatment of gases. For example, Bohn (1992) found that the capital costs of a typical biofiltration system were only about 6 percent of those for incineration, 13 percent of those for ozone oxidation, and 40 percent of those for activated-carbon adsorption.

Ottengraf (1987) described the biological systems commonly being used or tested in Europe, Japan, and the USA for treatment of gas-phase VOCs, as well as NH_3, H_2S, and semi-volatile odor compounds. *Bioscrubbers* countercurrently contact the gas stream with circulating liquid in a spray column. The bacteria are mainly dispersed in the circulating liquid. In *trickling filters* and *biofilters,* the microorganisms are immobilized on a carrier or packing medium. In most cases, the water flow is downward by gravity, while the gas flow is countercurrent, or upward. Trickling filters are distinguished from biofilters by the amount of water applied. The trickling filter has a larger hydraulic loading so that the aqueous phase continuously flows downward. Because the hydraulic loading is less for biofilters, the water phase is essentially stationary.

Biological treatment of contaminated gas phases also can be accomplished by passing the gas through compost, peat, or soil beds (Ottengraf, Meeters, van den Oever, and Rozema, 1986; Wilson, Pogue, and Canter, 1988). These systems are very simple, although process control may be limited. More recent developments include using three-phase fluidized-bed biofilm reactors to create highly compact biofilm systems well suited to industrial settings.

Except for bioscrubbers, all of the gas-treatment systems are of the biofilm type. Therefore, the principles and design approaches provided in Chapters 4 and 8 are directly applicable here. The main difference is that the contaminants enter in a gas stream, which must be included in the mass balances. Because most of the contaminants are hazardous organic compounds, the information in Chapter 14 is of the utmost relevance.

15.6   EVALUATING BIOREMEDIATION

One of the factors that limit the widespread application of bioremediation is that its "success" has been difficult to "prove" in the field. Regulators, site owners, and the public are wary of a technology that requires a sophisticated appreciation for microbiology and how it connects to the hydrogeology and chemistry of contaminated soils. When measures for the technology's success are neither obvious nor agreed upon, the wary decision maker is likely to seek other solutions, including expensive alternatives, such as pump-and-treat or incineration.

The development and acceptance of evaluation protocols is a slow process for the following reasons:

*   The definition of success varies among the several parties involved. Regulators generally define success only in terms of meeting some compliance level for contaminant concentration in the water or soil. Buyers of bioremediation are acutely interested in cost-effectiveness. The public generally wants assurance that they are not being exposed to risk. Researchers and developers of bioremediation wish to demonstrate the cause-and-effect nature of the process. This means that agreement on evaluation protocols requires sustained communication among the many parties who are stakeholders and/or need to contribute key technical judgements: regulators, buyers, the public, engineers, microbiologists, hydrogeologists, chemists, and others.

*   Many different measures of success can be put forward, but no one measure is universally applicable for a type of contamination, soil conditions, or bioremediation approach. Expert and site-specific judgment always will play a role.

*   Contaminated soils are inherently heterogeneous. In situ technologies amplify the importance of heterogeneity, since sampling is difficult and expensive, while microorganisms often are highly localized. It is well known that finding residual NAPL contamination is extremely difficult. Because the microorganisms often accumulate on solids close to the trapped contaminants, they also are difficult to find.

Although evaluating bioremediation is challenging and not easily standardized, it can be accomplished if an evaluation program is properly designed (NRC, 1993, 2000). The key to an evaluation program is that the measures of "success" must directly link microbial activity to the observed loss of contaminant. To obtain this linkage, the NRC (1993) recommends an evaluation strategy based on three kinds of evidence:

1. documented loss of contaminants from the site

2. laboratory assays showing that microorganisms at the site have the *potential* to transform the contaminants under the expected site conditions

3. one or more pieces of evidence showing that the biodegradation potential is *actually realized* in the field

The third type of evidence is the most crucial and the most difficult to obtain. It links the loss of contaminants in the soil and water with the potential for them to be biodegraded. A wide range of measurement approaches is available, and the NRC (1993) discusses their scientific bases. Table 15.11 briefly summarizes some of the techniques and what they indicate. In all cases, the measurement techniques should be used to (1) show that characteristics of the site's chemistry or microbial population change in ways one would predict if bioremediation were occurring and (2) correlate these chemical and microbial changes with documented contaminant losses.

One key distinction is between techniques that can provide principal evidence versus those that provide only confirmatory evidence (Rittmann et al., 1994). Principal evidence should be equally capable of proving the success or failure of bioremediation. Among the various techniques in Table 15.11, good examples of principal evidence include stoichiometric consumption of electron acceptors, formation of inorganic carbon that originated in organic carbon, and increased degradation rates over time. Quantification of consumption or formation rates is of great importance. The mere demonstration that a reaction occurs is not always principal evidence; instead, the rate must commensurate with the loss of contamination.

Confirmatory evidence, on the other hand, usually only can support success, but its absence does not prove failure. Good examples of confirmatory evidence include increases in the populations of bacteria or protozoa (they are too hard to find to be principal), detection of intermediary metabolites (they can be degraded themselves), and an increase to the ratio of nondegradable to degradable components (other mechanisms also act differentially).

The complexities of the contaminated setting and the difficulties in obtaining good samples make it rare for one type of measurement to give unequivocal proof. Therefore, the three-part strategy relies upon building a consistent, logical case from convergent lines of independent evidence (NRC, 1993). A wide range of principal and confirmatory techniques is needed to create a strong case for proving (or disproving) the success of a bioremediation project.

Since the NRC's 1993 report, a number of public and private organizations have issued protocols for evaluating intrinsic bioremediation and natural attenuation. The protocols vary widely in terms of scope, detail, and adequacy. A critical evaluation of these protocols, carried out recently by a committee of the NRC (2000), provides

Table 15.11   Summary of techniques used to demonstrate biodegradation in the field

Technique	Indication of
Direct Measurements of Field Samples	
Number of bacteria	An increase in the population of contaminant-degrading bacteria over background conditions
Number of protozoa	An increase in the population of predators of the contaminant-degrading bacteria
Rate of bacterial activity in laboratory microcosms	Potential rates of biodegradation
Bacterial adaptation	An increase in the rates of biodegradation since bioremediation began
Inorganic carbon concentration	Formation of inorganic carbon by oxidation of the organic contaminant
Carbon isotope ratio	Inorganic carbon originating from organic contamination
Electron-acceptor concentration	A decrease in the electron acceptors used during contaminant oxidation
By-products of anaerobic activity	Formation of products generated when electron acceptors other than O_2 are used
Intermediary metabolites	Breakdown products of complex organic contaminants
Ratio of nondegradable to degradable components	Relative loss of biodegradable components
Experiments run in the field	
Stimulating bacteria within subsites	An increase in microbial activity with stimulation used in engineered bioremediation
Measuring electron-acceptor uptake rate	In situ rate of metabolism
Monitoring a conservative tracer	Contaminant losses due to abiotic mechanisms
Labeling contaminants	The fate of carbon contained in organic contaminants
Modeling Experiments	
Modeling abiotic mass loss	Potential losses through abiotic mechanisms
Direct modeling	In situ biodegradation rates

| SOURCE: Rittmann et al., 1994.

guidance on what should be included in an evaluation scheme. The foundation of the NRC's 2000 guidance is that the monitoring and evaluation program should document the presence of "footprints," which are the products of the biological reactions capable of destroying the contaminants. Footprints include:

- Loss of electron acceptors in proportion to loss of a contaminant that is oxidized
- Increases in inorganic carbon in proportion to the loss or organic contaminants that are mineralized

- Loss of an electron-donor substrate in parallel to the loss of a contaminant that is reduced, such as by reductive dechlorination

- Release of the chloride ion from organic chemicals that are dechlorinated

- Increases or decreases of alkalinity in proportion to reactions that release base or release acid, respectively

In some cases, confounding reactions or high background levels make it impossible to detect all the relevant footprints. Nevertheless, the detection of several footprints at levels commensurate to the loss of contaminant is necessary to ensure that natural attenuation is reliably protecting public health and the environment.

The level of effort to monitor a site depends on the likelihood that a natural-attenuation mechanism is able to destroy the contaminant in a given setting and the complexity of the setting. Reduced likelihood of destruction and/or a more complex site demands increases to the number, extent, and frequency of sampling, as well as more sophisticated evaluation of the data. These issues are discussed by NRC (2000).

15.7 BIBLIOGRAPHY

Bohn, N. (1992). "Consider Biofiltration for Decontaminating Gases." *Chemical Eng. Progress* 88 (4), pp. 34–40.

Bouwer, E. J.; B. E. Rittmann; and P. L. McCarty (1981). "Anaerobic Degradation of Halogenated 1- and 2-Carbon Organic Compounds." *Environmental Sci. & Technology* 15, pp. 40–49.

Chang, H. T. and B. E. Rittmann (1987). "Verification of the Model of Biofilm on Activated Carbon." *Environmental Sci. & Technology* 21, pp. 280–288.

Cline, P. V.; J. J. Delfino; and P. S. C. Rao (1991). "Partitioning of Aromatic Constituents into Water from Gasoline and Other Complex Solvent Mixtures." *Environmental Sci. & Technology* 25, pp. 914–920.

Council on Environmental Quality (CEQ) (1981). *Contamination of Groundwater by Toxic Organic Chemicals,* pp. 16–39, Washington, DC.

Cunningham, S. D.; T. A. Anderson; A. P. Schwab; and F. C. Hsu (1996). "Phytoremediation of Soils Contaminated with Organic Pollutants." *Advances in Agronomy.* New York: Academic Press.

Flathman, P. E., D. E. Jerger, and J. H. Exner, eds., (1994). *Bioremediation Field Practice.* Boca Raton, FL: Lewis Publ.

Freeze, R. A. and J. A. Cherry (1979). *Groundwater.* Englewood Cliffs, NJ: Prentice-Hall.

Hill, J. B. and J. G. Moxey, Jr. (1960). "Gasoline." *In* V. B. Guthrie, Ed. *Petroleum Products Handbook.* pp. 4-1–4-37. New York: McGraw-Hill.

Karickhoff, S. W.; D. S. Brown; and T. A. Scott (1979). "Sorption of Hydrophobic Pollutants on Natural Sediments." *Water Research* 13, pp. 241–248.

Lane, J. C. (1980). "Gasoline and Other Motor Fuels, *In* M. Grayson, Ed., *Kirk-Othmer Encyclopedia of Chemical Technology,* 3rd ed., Vol II., pp. 652–695. New York: John Wiley.

Lee, M. D.; J. M. Thomas; R. C. Borden; P. B. Bedient; C. H. Ward; and J. T. Wilson (1988). "Biorestoration of Aquifers Contaminated with Organic Compounds." *CRC Critical Reviews in Environmental Control* 18, pp. 29–89.

McCarty, P. L.; B. E. Rittmann; and M. Reinhard (1985). "Processes Affecting the Movement and Fate of Trace Organics in the Subsurface Environment. *In* T. Asano, Ed., *Artificial Recharge of Groundwater,* pp. 627–645. Boston: Butterworth Publishers.

McCarty, P. L.; M. N. Goltz; G. D. Hopkins; M. E. Dolan; J. P. Allan; B. T. Kawakami; and T. J. Carrothers (1998). "Full-Scale Evaluation of *In Situ* Cometabolic Degradation of Trichloroethylene in Groundwater through Toluene Injection." *Environ. Sci. Technol.* 32, pp. 88–100.

Moffat, A. S. (1995). "Plants Proving Their Worth in Toxic Metal Cleanup." *Science* 269, pp. 302–303.

Moretti, E. C. and N. Mukhopadhyay (1993). "VOC Control: Current Practices and Future Trends." *Chemical Engineering Progress* 89(7), pp. 20–26.

National Research Council (NRC) (1993). *In situ Bioremediation: When Does It Work?* B. E. Rittmann, Chairman, National Academy Press, Washington, DC.

National Research Council (NRC) (1994). *Alternatives for Ground Water Cleanup,* M. C. Kavanaugh, Chairman. Washington, DC: National Academy Press.

National Research Council (NRC) (2000). *Natural Attenuation for Groundwater Remediation.* B. E. Rittmann, Chairman. Washington, DC: National Academy Press.

Office of Technology Assessment (OTA) (1984). *Protecting the Nation's Groundwater from Contamination,* Vol. 1, pp. 19–60. OTA-0-233, Washington, DC.

Ottengraf., S. P. P. (1987). "Biological Systems for Waste Gas Elimination." *TIBTECH* 5(5), pp. 132–136.

Ottengraf, S. P. P.; J. J. P. Meeters; A. H. C. van den Oever; and H. R. Rozema (1986). "Biological Elimination of Volatile Xenobiotic Compounds in Biofilters." *Bioprocess Engineering* 1, pp. 61–69.

Pitra, R. and D. McKenzie (1990). "Groundwater Extraction/Treatment System. A Case History at McClellan Air Force Base, Sacramento, CA." Internal document, McClellan Air Force Base, Sacramento, CA.

Rittmann, B. E. (1993). "The Significance of Biofilm in Porous Media." *Water Resources Research* 29, pp. 2195–2202.

Rittmann, B. E.; E. Seagren; B. A. Wrenn; A. J. Valocchi; C. Ray; and L. Raskin (1994). *In Situ Bioremediation,* 2nd ed., Park Ridge, NJ: Noyes Publications.

Rittmann, B. E.; A. Schwarz; and P. B. Sáez (2000). "Biofilms Applied to Hazardous Waste Treatment." In *Biofilms,* 2nd edition, J. B. Bryers, Ed., New York: Wiley & Sons, pp. 207–234.

Roberts, P. V.; G. D. Hopkins; D. M. Mackay; and L. Semprini (1990). "A Field Evaluation of *in situ* Biodegradation of Chlorinated Ethenes: Part 1, Methodology and Field Site Characterization." *Ground Water* 28, pp. 591–604.

Robertson, J. M.; C. R. Toussaint; and M. A. Jorque (1974). "Organic Compounds Entering Groundwater from a Landfill." Environmental Protect. Technol. Ser., EPA 660/2-74-077.

Sachanen, A. N. (1954). "Hydrocarbons in Gasolines, Kerosenes, Gas Oils and Lubricating Oils." *In* B. T. Brooks, C. E. Boord, S. S. Kurtz, Jr., and L. Schmerling, Eds. *The Chemistry of Petroleum Hydrocarbons.* pp. 5–36. New York: Reinhold Publishing.

Schwille, F. (1984). "Migration of Organic Fluids Immiscible with Water in the Unsaturated Zone." *In* B. Yaron, G. Dagan, and J. Goldschmid, Eds. *Pollutants in Porous Media, the Unsaturated Zone between Soil Surface and Groundwater, Ecological Studies,* Vol. 47, pp. 27–48. Berlin: Springer-Verlag.

Seagren, E. A.; B. E. Rittmann; and A. J. Valocchi (1993). "Quantitative Evaluation of Flushing and Biodegradation for Enhancing *in situ* Dissolution of Nonaqueous-Phase Liquids." *J. Contaminant Hydrology* 12, pp. 103–132.

Seagren, E. A.; B. E. Rittmann; and A. J. Valocchi (1994). "Quantitative Evaluation of the Enhancement of NAPL-Pool Dissolution by Flushing and Biodegradation." *Environ. Sci. & Technol.* 28, pp. 833–839.

Seagren, E. A.; B. E. Rittmann; and A. J. Valocchi (2000). "Bioenhancement of NAPL-Pool Dissolution in Porous Medium," *J. Contam. Hydol.,* submitted.

Semprini, L.; P. V. Roberts; G. D. Hopkins; and P. L. McCarty (1990). "A Field Evaluation of *in situ* Biodegradation of Chlorinated Ethenes: Part 2, Results of Biostimulation and Biotransformation Experiments." *Ground Water* 28, pp. 715–727.

Shelton, D. R. and J. M. Tiedje (1984). "Isolation and Partial Characterization of Bacteria in an Anaerobic Consortium that Mineralizes 3-Chlorobenzoic Acid." *Applied and Environmental Microbiology* 48, pp. 840–848.

Solomons, T. W. G. (1984). *Organic Chemistry,* 3rd ed., New York: John Wiley & Sons.

Staps, J. J. M. (1990). "International Evaluation of *in situ* Biorestoration of Contaminated Soil and Groundwater." Report No. 738708006. National Institute of Public Health and Environmental Protection, Bilthoven, The Netherlands.

Thomas, J. M. and C. H. Ward (1989). "*In situ* Biorestoration of Organic Contaminants in the Subsurface." *Environmental Sci. & Technology* 23, pp. 760–766.

Travis, C. C. and C. B. Doty (1990). "Can Contaminated Aquifers at Superfund Sites be Remediated?" *Environmental Sci. & Technology* 23, pp. 1464–1466.

U.S. Coast Guard (1984). *Chemical Hazard Response Information System (CHRIS),* Vol. 2, U.S. Dept. of Transportation, Washington, DC.

U. S. Department of Energy (1990). *Subsurface Science Program, Program Overview and Research Abstracts,* p. 3-10, 47-55. Office of Health and Environmental Research, Office of Energy Research, DOE/ER-0432, Washington, DC, FY 1898–1990.

U.S. Environmental Protection Agency (1986). *Superfund Public Health Evaluation Manual.* EPA 540/1-86/060. Office of Emergency and Remedial Response, Office of Solid Waste and Emergency Response. Washington, DC.

Van Loosdrecht; M. C. M.; J. Lyklema; W. Norde; and A. J. B. Zehnder (1990). "Influences of interfaces on microbial activity." *Microbiology Review* 54, pp. 75–87.

Verschueren, K. (1983). *Handbook of Environmental Data on Organic Chemicals,* 2nd ed. New York, Van Nostrand Reinhold.

Wetzel, R. S.; C. M. Durst; P. A. Spooner; W. D. Ellis; D. J. Sarno; B. C. Vickers; J. R. Payne; M. S. Floyd; and Z. A. Saleem (1986). "*In Situ* Biological Degradation Test at Kelly Air Force Base, Volume I: Site Characterization, Laboratory Studies and Treatment System Design and Installation." Engineering & Services Laboratory, Air Force Engineering & Services Center, Tyndall Air Force Base, Florida, Report ESL-TR-85-52.

Wilson, J. and M. J. Noonan (1984). "Microbial Activity in Model Aquifer Systems." *In* G. Bitton, and C. P. Gerba, Eds. *Groundwater Pollution Microbiology.* New York: John Wiley.

Wilson, B. H.; D. W. Pogue; and L. W. Canter (1988). "Biological Treatment of Trichloroethylene and 1,1,1-Trichloroethane from Contaminated Streams." *Proc. Petroleum Hydrocarbons and Organic Chemicals in Ground Water,* Houston, TX.

15.8 PROBLEMS

15.1. What factors lead to recalcitrance of an organic molecule discharged to soil?

15.2. You are considering in situ biodegradation of the chlorinated solvent trichloroethene or TCE. Indicate whether aerobic or anaerobic treatment or both have good potential for degradation of this compound, and indicate the biological processes involved in the treatment or treatments with good potential and the intermediate compounds that are likely to be formed.

15.3. Three separate industries are potentially responsible for contamination of a municipal well supply with chlorinated aliphatic hydrocarbons (CAHs). Industry records indicate that Industry A was a large user of carbon tetrachloride (CT), Industry B was a large user of tetrachloroethene (PCE), and Industry C was a large user of 1,1,1-trichloroethane (TCA). Are far as can be determined, each industry used only one chlorinated solvent. The only CAHs found present in the municipal well supply are vinyl chloride, 1,1-dichloroethene, and 1,1-dichloroethane. Since none of the industries used these compounds, they all claim they are innocent. You are called in by the regulatory authorities as an expert witness in this case to help shed some light on the possible source or sources of these contaminants.

   (*a*) Does the evidence suggest the identified industries are innocent as claimed and that some unidentified industry (D) is responsible, or does it support the potential that one or all of the identified industries may be responsible?

   (*b*) Who is the most likely culprit?

   Give technically sound reasons for your judgments.

15.4. Describe biological transformations known to occur that might be used for in situ biodegradation of TCE? What chemicals would you need to add if not already present in order to bring about such TCE transformations?

15.5. You have been called in to consider *in situ* bioremediation of an earthen heavy equipment storage yard, the soil of which has been contaminated with oil leaking from heavy trucks, bulldozers, and other equipment that had been stored there for years. The average concentration of organic hydrocarbons (HC) in the upper 0.3 m of soil, where the contamination resides, has been measured to be 0.1 kg HC/kg soil. Describe the treatment you might consider giving to the contaminated soil, and quantify the amounts of materials, if any, that you might apply to help speed biodegradation.

15.6. You are called upon to investigate the sources of contamination of an aquifer and to recommend possible biological methods for degrading the contaminants found. For your investigation, you have found high concentrations (low mg/l range) of 1,1-dichloroethene (1,1-DCE); 1,2-*cis* dichloroethene (1,2-c-DCE);

1,2-*trans*-dichloroethene (1,2-t-DCE); and 1,1-dichloroethane (1,1-DCA); together with somewhat lower concentrations of 1,1,1-trichloroethane (TCA), trichloroethene (TCE), and vinyl chloride (VC).

(*a*) What might you expect to be the original contaminants that entered the subsurface system studied? Why?

(*b*) What different biological processes might bring about transformation of the contaminants? For each indicate respective electron donors and acceptors that would be required for the transformation.

(*c*) Considering the above one or more biological processes, indicate the possible advantages and disadvantages of each.

(*d*) Draw a pictorial flow sheet of a possible treatment scheme that you might consider for treating the above groundwater by one of the above biological processes. Show a profile of the surface and subsurface system, locations of points of injection or extraction, possible aboveground or subsurface reactors, locations of chemical injection, etc.

15.7. Compute the retardation factor for toluene (log $K_{ow} = 2.7$), vinyl chloride (log $K_{ow} = 0.6$), and phenanthrene (log $K_{ow} = 4.6$) for a sandy soil ($f_{oc} = 0.001$) and a peat soil ($f_{oc} = 0.1$) (Six answers).

15.8. For the six conditions of problem 15.7, determine what fraction of the solute's mass is present in the water phase when the porosity is 0.3 and the bulk soil density is 2 kg/l (Six answers).

15.9. Which is the better engineered bioremediation technology among air sparging, bioventing, and addition of glucose for these scenarios:

(*a*) Petroleum hydrocarbons strictly in the vadose zone

(*b*) Petroleum hydrocarbons with residual in the saturated zone

(*c*) TCE as a DNAPL.

15.10. Which of these scenarios is likely to be successful for intrinsic bioremediation?

(*a*) A leaking gasoline storage tank in a sandy soil

(*b*) A leaking storage tank of TCE in an otherwise clean aquifer

(*c*) A leaking storage tank of TCE next to a leaking storage tank of gasoline

(*d*) A deposit of polynuclear aromatic hydrocarbons in a sandy aquifer.

15.11. Compute the equilibrium concentration of toluene in groundwater if the water is in contact with gasoline having 6 percent toluene.

15.12. 18.4 mg/l of toluene is in a groundwater plume. Compute how much of the following would be necessary for full mineralization of the toluene if each reaction were the only reaction responsible for accepting the electrons from toluene. You may ignore biomass growth. Express the result as a concentration in the groundwater (mg/l).

(*a*) O_2 consumption in aerobic respiration, mgO_2/l

(*b*) SO_4^{2-} consumption in sulfate reduction to sulfides, mgS/l

(*c*) NO_3^- consumption by denitrification to N_2, mgN/l

(*d*) $Fe(OH)_3$ reduction to Fe^{2+}, mgFe/l

(*e*) CH_4 generation in methanogenesis, $mgCH_4/l$

A

FREE ENERGIES OF FORMATION
FOR VARIOUS CHEMICAL SPECIES, 25°C

Class	Substance	Form	kJ/mol	Reference
Hydrogen Ion	H^+	aq	0	Thauer et al. (1977)
Hydrogen Ion	$H^+(10^{-7})$	aq	−39.87	Thauer et al. (1977)
Hydrogen	H_2	g	0	Thauer et al. (1977)
Water	H_2O	l	−237.178	Thauer et al. (1977)
Hydrogen Peroxide	H_2O_2	aq	−134.097	Thauer et al. (1977)
Hydroxide	OH^-	aq	−157.293	Thauer et al. (1977)
Oxygen	O_2	g	0	Thauer et al. (1977)
Carbon Monoxide	CO	g	−137.15	Thauer et al. (1977)
Carbon Dioxide	CO_2	g	−394.359	Thauer et al. (1977)
Carbon Dioxide	CO_2	aq	−386.02	Thauer et al. (1977)
Carbon	C	c	0	Thauer et al. (1977)
Carbonic Acid	H_2CO_3	aq	−623.16	Thauer et al. (1977)
Bicarbonate	HCO_3^-	aq	−586.85	Thauer et al. (1977)
Carbonate	CO_3^{2-}	aq	−527.9	Thauer et al. (1977)
Chlorate	ClO_3^-	aq	−3.35	Thauer et al. (1977)
Chloride	Cl^-	aq	−133.26	Weast and Astle (1980)
Chlorite	ClO_2^-	aq	17.2	Thauer et al. (1977)
Ammonia	NH_3	aq	−26.57	Thauer et al. (1977)
Ammonium	NH_4^+	aq	−79.37	Thauer et al. (1977)
Ferrous Iron	Fe^{2+}	aq	−78.87	Thauer et al. (1977)
Ferric Iron	Fe^{3+}	aq	−4.6	Thauer et al. (1977)
Nitrite	NO_2^-	aq	−37.2	Thauer et al. (1977)
Nitrate	NO_3^-	aq	−111.34	Thauer et al. (1977)
Sulfur	S	c	0	Thauer et al. (1977)
Hydrogen Sulfide	H_2S	g	−33.56	Thauer et al. (1977)
Hydrogen Sulfide	H_2S	aq	−27.87	Thauer et al. (1977)
Bisulfide	HS^-	aq	12.05	Thauer et al. (1977)
Sulfite	SO_3^{2-}	aq	−486.6	Thauer et al. (1977)
Sulfate	SO_4^{2-}	aq	−744.63	Thauer et al. (1977)
Thiosulfite	$S_2O_3^{2-}$	aq	−513.4	Thauer et al. (1977)
Acid-dicarboxylic	Fumarate$(^{2-})$	aq	−604.21	Thauer et al. (1977)
Acid-dicarboxylic	Fumaric acid	aq	−647.14	Thauer et al. (1977)
Acid-dicarboxylic	α- Ketoglutarate	aq	−797.55	Thauer et al. (1977)

Class	Substance	Form	kJ/mol	Reference
Acid-dicarboxylic	L- Malate($^{2-}$)	aq	−845.08	Thauer et al. (1977)
Acid-dicarboxylic	Oxalacetate($^{2-}$)	aq	−797.18	Thauer et al. (1977)
Acid-dicarboxylic	Oxalate($^{1-}$)	aq	−698.44	Thauer et al. (1977)
Acid-dicarboxylic	Oxalate($^{2-}$)	aq	−674.04	Thauer et al. (1977)
Acid-dicarboxylic	Succinate ($^{2-}$)	aq	−690.23	Thauer et al. (1977)
Acid-dicarboxylic	Succinic acid	aq	−746.38	Thauer et al. (1977)
Acid-monocarboxylate	Glycollate	aq	−530.95	Thauer et al. (1977)
Acid-monocarboxylate	Glyoxylate	aq	−468.6	Thauer et al. (1977)
Acid-monocarboxylate	D- Gluconate	aq	[−1128.3]	Thauer et al. (1977)
Acid-monocarboxylic	Acetate	aq	−369.41	Thauer et al. (1977)
Acid-monocarboxylic	Acetic acid	lq	−392	Weast and Astle (1980)
Acid-monocarboxylic	Acrylate	aq	[−286.19]	Thauer et al. (1977)
Acid-monocarboxylic	Benzoic	aq	−245.6	Madigan et al. (1997)
Acid-monocarboxylic	Butyrate	aq	−352.63	Thauer et al. (1977)
Acid-monocarboxylic	Caproate	aq	[−335.96]	Thauer et al. (1977)
Acid-monocarboxylic	Crotonate	aq	−277.4	Thauer et al. (1977)
Acid-monocarboxylic	Formate	aq	−351.04	Thauer et al. (1977)
Acid-monocarboxylic	Formic acid	aq	−356.06	Weast and Astle (1980)
Acid-monocarboxylic	Gluconate	aq	[−1128.3]	Thauer et al. (1977)
Acid-monocarboxylic	Glycerate	aq	[−658.1]	Thauer et al. (1977)
Acid-monocarboxylic	Glycollate	aq	−530.95	Thauer et al. (1977)
Acid-monocarboxylic	β- Hydroxybutyrate	aq	[−506.3]	Thauer et al. (1977)
Acid-monocarboxylic	β- Hydroxypropionate	aq	−518.4	Thauer et al. (1977)
Acid-monocarboxylic	β- Ketobutyrate	aq	−493.7	Thauer et al. (1977)
Acid-monocarboxylic	Lactate	aq	−517.81	Thauer et al. (1977)
Acid-monocarboxylic	Palmitate	aq	[−309.5]	McCarty (1971)
Acid-monocarboxylic	Palmitic acid	c	−305	Thauer et al. (1977)
Acid-monocarboxylic	Propionate	aq	[−361.08]	Thauer et al. (1977)
Acid-monocarboxylic	Pyruvate	aq	−474.63	Thauer et al. (1977)
Acid-monocarboxylic	Valerate	aq	[−344.34]	Thauer et al. (1977)
Acid-tricarboxylic	cis- Aconitate($^{3-}$)	aq	−922.61	Thauer et al. (1977)
Acid-tricarboxylic	Citrate($^{3-}$)	aq	−1168.34	Thauer et al. (1977)
Acid-tricarboxylic	iso Citrate($^{3-}$)	aq	−1161.69	Thauer et al. (1977)
Alcohol	n- Butanol	aq	−171.84	Thauer et al. (1977)
Alcohol	Ethanol	aq	−181.75	Thauer et al. (1977)
Alcohol	Ethylene glycol	liq	−323.21	Thauer et al. (1977)
Alcohol	Ethylene glycol	aq	[−330.5]	Thauer et al. (1977)
Alcohol	Glycerol	liq	−477.06	Thauer et al. (1977)
Alcohol	Glycerol	aq	−488.52	Thauer et al. (1977)
Alcohol	Mannitol	aq	−942.61	Thauer et al. (1977)
Alcohol	Methanol	aq	−175.39	Thauer et al. (1977)
Alcohol	n- Propanol	aq	−175.81	Thauer et al. (1977)
Alcohol	iso Propanol	aq	−185.94	Thauer et al. (1977)
Alcohol	Sorbitol	aq	−942.7	Thauer et al. (1977)
Aldehyde	Acetaldehyde	aq	−139.9	Thauer et al. (1977)
Aldehyde	Acetaldehyde	g	−128.91	Thauer et al. (1977)
Aldehyde	Butyraldehyde	liq	−119.67	Thauer et al. (1977)
Aldehyde	Formaldehyde	aq	−130.54	Thauer et al. (1977)
Aldehyde	Formaldehyde	g	−112.97	Thauer et al. (1977)
Amino acid	L- Alanine	aq	−371.54	Thauer et al. (1977)
Amino acid	L- Arginine	c	−240.2	Thauer et al. (1977)
Amino acid	L- Asparagine × H_2O	aq	−763.998	Thauer et al. (1977)
Amino acid	L- Aspartate$^-$	aq	−700.4	Thauer et al. (1977)

Class	Substance	Form	kJ/mol	Reference
Amino acid	L- Aspartic acid	aq	−721.3	Thauer et al. (1977)
Amino acid	L- Cysteine	aq	−339.78	Thauer et al. (1977)
Amino acid	L- Cystine	aq	−666.93	Thauer et al. (1977)
Amino acid	L- Glutamate⁻	aq	−699.6	Thauer et al. (1977)
Amino acid	L- Glutamic acid	aq	−723.8	Thauer et al. (1977)
Amino acid	L- Glutamine	aq	−529.7	Thauer et al. (1977)
Amino acid	Glycine	aq	−370.788	Thauer et al. (1977)
Amino acid	L- Leucine	aq	−343.1	Thauer et al. (1977)
Amino acid	iso Leucine	aq	−343.9	Thauer et al. (1977)
Amino acid	L- Methionine	aq	−502.92	Thauer et al. (1977)
Amino acid	L- Phenylalanine	aq	−207.1	Thauer et al. (1977)
Amino acid	L- Serine	aq	−510.87	Thauer et al. (1977)
Amino acid	L- Threonine	aq	[−514.63]	Thauer et al. (1977)
Amino acid	L- Tryptophane	aq	−112.6	Thauer et al. (1977)
Amino acid	L- Tyrosine	aq	−370.7	Thauer et al. (1977)
Amino acid	L- Valine	aq	−356.9	Thauer et al. (1977)
Aromatic	Benzene	g	129.7	Perry et al.(1984)
Aromatic	Benzene	aq	[133.9]	*
Aromatic	Biphenyl	aq	275.2	Homes et al. (1993)
Aromatic	Ethylbenzene	g	130.6	Perry et al.(1984)
Aromatic	Ethylbenzene	aq	[136]	*
Aromatic	m-Xylene	g	118.8	Perry et al.(1984)
Aromatic	m-Xylene	aq	[122.9]	*
Aromatic	o-Xylene	g	122.1	Perry et al.(1984)
Aromatic	o-Xylene	aq	[124.6]	*
Aromatic	p-Xylene	g	121.1	Perry et al.(1984)
Aromatic	p-Xylene	aq	[125.2]	*
Aromatic	Toluene	g	122.3	Perry et al.(1984)
Aromatic	Toluene	aq	[127]	*
Carbohydrate	Dihydroxyacetone	aq	[−445.18]	Thauer et al. (1977)
Carbohydrate	D- Erythrose	aq	[−598.3]	Thauer et al. (1977)
Carbohydrate	D- Fructose	aq	−915.38	Thauer et al. (1977)
Carbohydrate	α-D- Galactose	aq	−923.53	Thauer et al. (1977)
Carbohydrate	α-D- Glucose	aq	−917.22	Thauer et al. (1977)
Carbohydrate	Glyceraldehyde	aq	[−437.65]	Thauer et al. (1977)
Carbohydrate	Glycogen(per glucose)	aq	−662.33	Thauer et al. (1977)
Carbohydrate	D- Heptose	aq	[−1,077]	Thauer et al. (1977)
Carbohydrate	α- Lactose	aq	−1515.24	Thauer et al. (1977)
Carbohydrate	β- Lactose	aq	−1570.09	Thauer et al. (1977)
Carbohydrate	β- Maltose	aq	−1497.04	Thauer et al. (1977)
Carbohydrate	D- Ribose	aq	[−757.3]	Thauer et al. (1977)
Carbohydrate	Sucrose	aq	−1551.85	Thauer et al. (1977)
Chlorite	ClO_2^-	aq	17.2	Thauer et al. (1977)
Chloroaliphatic	1,1,1,2-Tetrachloroethane	g	−80.26	Dolfing & Janssen (1994)
Chloroaliphatic	1,1,1,2-Tetrachloroethane	aq	−77.75	Dolfing & Janssen (1994)
Chloroaliphatic	1,1,1-Trichloroethane	g	−76.12	Dolfing & Janssen (1994)
Chloroaliphatic	1,1,1-Trichloroethane	aq	−69.04	Dolfing& Janssen (1994)
Chloroaliphatic	1,1,2,2-Tetrachloroethane	g	−85.48	Dolfing & Janssen (1994)
Chloroaliphatic	1,1,2,2-Tetrachloroethane	aq	−88.92	Dolfing & Janssen (1994)
Chloroaliphatic	1,1,2-Trichloroethane	g	−77.41	Dolfing & Janssen (1994)
Chloroaliphatic	1,1,2-Trichloroethane	aq	−77.64	Dolfing & Janssen (1994)
Chloroaliphatic	1,1-Dichloroethane	g	−73.23	Dolfing & Janssen (1994)

Class	Substance	Form	kJ/mol	Reference
Chloroaliphatic	1,1-Dichloroethane	aq	−68.69	Dolfing & Janssen (1994)
Chloroaliphatic	1,1-Dichloroethene	g	24.16	Dolfing & Janssen (1994)
Chloroaliphatic	1,1-Dichloroethene	aq	32.23	Dolfing & Janssen (1994)
Chloroaliphatic	1,2-Dichloroethane	g	−73.78	Dolfing & Janssen (1994)
Chloroaliphatic	1,2-Dichloroethane"	aq	−72.93	Dolfing & Janssen (1994)
Chloroaliphatic	1,2-Dichloropropane	g	−83.01	Dolfing & Janssen (1994)
Chloroaliphatic	1,2-Dichloropropane	aq	−79.88	Dolfing & Janssen (1994)
Chloroaliphatic	1-Chloropropane	g	−23.53	Dolfing & Janssen (1994)
Chloroaliphatic	1-Chloropropane	aq	−7.32	Dolfing & Janssen (1994)
Chloroaliphatic	Chloroethane	g	−43.01	Dolfing & Janssen (1994)
Chloroaliphatic	Chloroethane	aq	−36.83	Dolfing & Janssen (1994)
Chloroaliphatic	Chloroethene	g	51.46	Dolfing & Janssen (1994)
Chloroaliphatic	Chloroethene	aq	59.65	Dolfing & Janssen (1994)
Chloroaliphatic	Chloromethane	g	−58.4	Dolfing & Janssen (1994)
Chloroaliphatic	Chloromethane	aq	−52.71	Dolfing & Janssen (1994)
Chloroaliphatic	cis-1,2-Dichloroethene	g	24.33	Dolfing & Janssen (1994)
Chloroaliphatic	cis-1,2-Dichloroethene	aq	27.8	Dolfing & Janssen (1994)
Chloroaliphatic	Dichloromethane	g	−68.8	Dolfing & Janssen (1994)
Chloroaliphatic	Dichloromethane	aq	−66.11	Dolfing & Janssen (1994)
Chloroaliphatic	Hexachloroethane	g	−54.88	Dolfing & Janssen (1994)
Chloroaliphatic	Hexachloroethane	aq	−49.62	Dolfing & Janssen (1994)
Chloroaliphatic	Pentachloroethane	g	−70.18	Dolfing & Janssen (1994)
Chloroaliphatic	Pentachloroethane	aq	−68.26	Dolfing & Janssen (1994)
Chloroaliphatic	Tetrachloroethene	g	20.48	Dolfing & Janssen (1994)
Chloroaliphatic	Tetrachloroethene	aq	27.59	Dolfing & Janssen (1994)
Chloroaliphatic	Tetrachloromethane	g	−53.5	Dolfing & Janssen (1994)
Chloroaliphatic	Tetrachloromethane	aq	−45.1	Dolfing & Janssen (1994)
Chloroaliphatic	trans-1,2-Dichloroethene	g	26.54	Dolfing & Janssen (1994)
Chloroaliphatic	trans-1,2-Dichloroethene	aq	32.06	Dolfing & Janssen (1994)
Chloroaliphatic	Trichloroethene	g	19.86	Dolfing & Janssen (1994)
Chloroaliphatic	Trichloroethene	aq	25.41	Dolfing & Janssen (1994)
Chloroaliphatic	Trichloromethane	g	−70.06	Dolfing & Janssen (1994)
Chloroaliphatic	Trichloromethane	aq	−66.5	Dolfing & Janssen (1994)
Chloroaromatic	1,2-Dichlorobenzene	g	82.7	Dolfing & Harrison (1992)
Chloroaromatic	1,2-Dichlorobenzene	aq	84.3	Dolfing & Harrison (1992)
Chloroaromatic	1,3,5-Trichlorobenzene	g	52.6	Dolfing & Harrison (1992)
Chloroaromatic	1,3,5-Trichlorobenzene	aq	56.8	Dolfing & Harrison (1992)
Chloroaromatic	1,3-Dichlorobenzene	g	78.6	Dolfing & Harrison (1992)
Chloroaromatic	1,3-Dichlorobenzene	aq	81.8	Dolfing & Harrison (1992)
Chloroaromatic	1,4-Dichlorobenzene	g	77.2	Dolfing & Harrison (1992)
Chloroaromatic	1,4-Dichlorobenzene	aq	78.3	Dolfing & Harrison (1992)
Chloroaromatic	2,2',4,4',6-Pentachlorobiphenyl	aq	162.7	Homes et al. (1993)
Chloroaromatic	2,2',4,4'-Tetrachlorobiphenyl	aq	185.9	Homes et al. (1993)
Chloroaromatic	2,4'-Dichlorobiphenyl	aq	235	Homes et al. (1993)
Chloroaromatic	2,4,4'-Trichlorobiphenyl	aq	210.6	Homes et al. (1993)
Chloroaromatic	2,4-Dichlorobenzoate	aq	−276.4	Dolfing & Harrison (1992)
Chloroaromatic	2,4-Dichlorophenol	aq	−84.4	Dolfing & Harrison (1992)
Chloroaromatic	2,4-Dichlorophenate⁻	aq	−39.4	Dolfing & Harrison (1992)
Chloroaromatic	2-Chlorobenzoate	aq	−237.9	Dolfing & Harrison (1992)
Chloroaromatic	2-Chlorobiphenyl	aq	259.6	Homes et al. (1993)
Chloroaromatic	2-Chlorophenol	aq	−56.8	Dolfing & Harrison (1992)
Chloroaromatic	2-Chlorophenate⁻	aq	−8.1	Dolfing & Harrison (1992)

Class	Substance	Form	kJ/mol	Reference
Chloroaromatic	3-Chlorobenzoate	aq	−246	Dolfing & Harrison (1992)
Chloroaromatic	4-Chlorobenzoate	aq	−239.5	Dolfing & Harrison (1992)
Chloroaromatic	Chlorobenzene	g	99.2	Dolfing & Harrison (1992)
Chloroaromatic	Chlorobenzene	aq	102.3	Dolfing & Harrison (1992)
Chloroaromatic	Hexachlorobenzene	g	33.3	Dolfing & Harrison (1992)
Chloroaromatic	Hexachlorobenzene	aq	46	Dolfing & Harrison (1992)
Chloroaromatic	Pentachlorophenol	aq	−112.3	Dolfing & Harrison (1992)
Chloroaromatic	Pentachlorophenate$^-$	aq	−86.6	Dolfing & Harrison (1992)
Hydrocarbon	Acetylene	g	209.2	Thauer et al. (1977)
Hydrocarbon	Butane	aq	−0.21	Dolfing & Janssen (1994)
Hydrocarbon	Butane	g	−17.14	Dolfing & Janssen (1994)
Hydrocarbon	Ethane	aq	−17.43	Dolfing & Janssen (1994)
Hydrocarbon	Ethane	g	−32.77	Dolfing & Janssen (1994)
Hydrocarbon	Ethene	aq	81.43	Dolfing & Janssen (1994)
Hydrocarbon	Ethene	g	68.18	Dolfing & Janssen (1994)
Hydrocarbon	Methane	aq	−34.74	Dolfing & Janssen (1994)
Hydrocarbon	Methane	g	−50.79	Dolfing & Janssen (1994)
Hydrocarbon	Pentane	aq	9.24	Dolfing & Janssen (1994)
Hydrocarbon	Pentane	g	8.36	Dolfing & Janssen (1994)
Hydrocarbon	Propane	aq	−7.32	Dolfing & Janssen (1994)
Hydrocarbon	Propane	g	−23.53	Dolfing & Janssen (1994)
Ketone	Acetone	aq	−161.17	Thauer et al. (1977)
N-containing, other	Creatine	aq	−264.3	Thauer et al. (1977)
N-containing, other	Urea	c	−196.82	Thauer et al. (1977)
N-containing, other	Urea	aq	−203.76	Thauer et al. (1977)
Purine	Hypoxanthine	aq	89.5	Thauer et al. (1977)
Purine	Quanine	c	46.99	Thauer et al. (1977)
Purine	Urate-	aq	−325.9	Thauer et al. (1977)
Purine	Uric acid	aq	−356.9	Thauer et al. (1977)
Purine	Xanthine	c	−165.85	Thauer et al. (1977)

Dolfing, J. and B. K. Harrison. 1992. *Gibbs Free Energy of Formation of Halogenated Aromatic Compounds and Their Potential Role as Electron Acceptors in Anaerobic Environments.* Environmental Science and Technology, 26(11): 2213–2218

Dolfing, J. and D. B. Janssen. 1994. Estimates of Gibbs Free Energies of Formation of Chlorinated Aliphatic Compounds. Biodegradation, 5(1): 21–28.

Holmes, D. A., B. K. Harrison and J. Dolfing. 1993. Estimation of Gibbs Free Energies of Formation for Polychlorinated Biphenyls. Environmental Science & Technology, 27(4): 725–731.

Madigan, M. T., J. M. Martinko and J. Parker. 1997. Brock Biology of Microorganisms (8 ed.). Prentice Hall, Upper Saddle River, NJ.

McCarty, P. L. 1971. Energetics and Bacterial Growth. In S. D. Faust and J. V. Hunter ed., Organic Compounds in Aquatic Environments, Marcel Dekker, Inc, New York, pp. 495–531.

Perry, R. H., D. W. Green and J. O. Maloney. 1984. Perry's Chemical Engineering Handbook (6 ed.). McGraw-Hill Book Company, New York.

Thauer, R. K., K. Jungermann and K. Decker. 1977. Energy Conservation in Chemotrophic Anaerobic Bacteria. Bacteriological Reviews, 41(1): 100–180.

Weast, R. C. and M. J. Astle. 1980. CRC Handbook of Chemistry and Physics (61 ed.). CRC Press, Inc., Boca Raton, FL.

*Estimated from ΔG_f^0 for gas and Henry's law (K_H) values from W. J. Weber and F. A. DiGiano. 1996. Process Dynamics in Environmental Systems, John Wiley & Sons, Inc., New York, with ΔG_f^0 corrected by - RTln K_H.

Values in brackets were obtained by approximate calculation. The physical state of each substance is indicated in the "Form" column as crystalline (c), liquid (l), gas (g), or in a water solution (aq).

B

A Complete Set of Normalized Loading Curves for S^*_{MIN} and K^* Ranging from 0.01 to 100

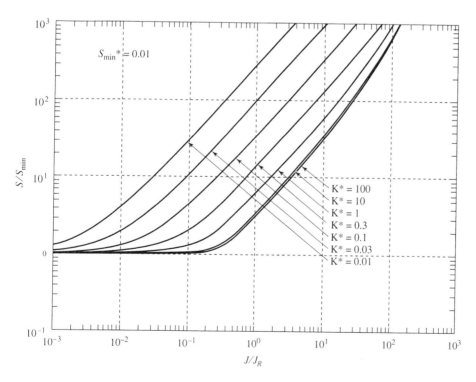

Figure B.1

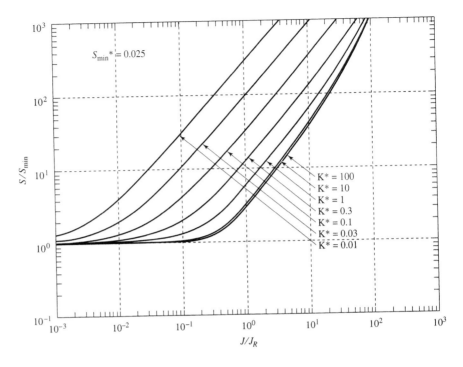

Figure B.2

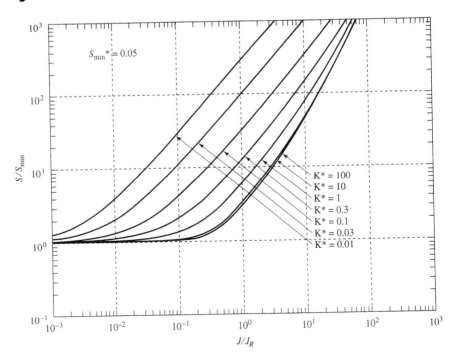

Figure B.3

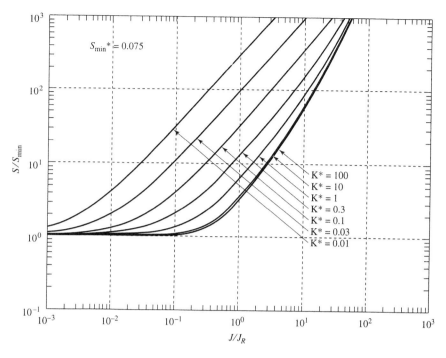

Figure B.4

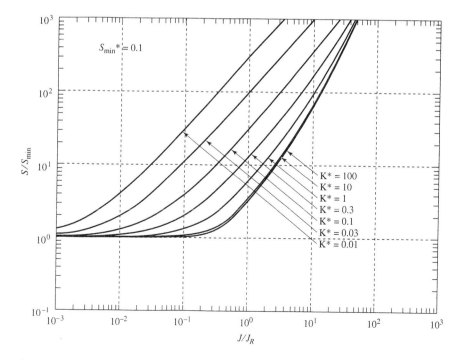

Figure B.5

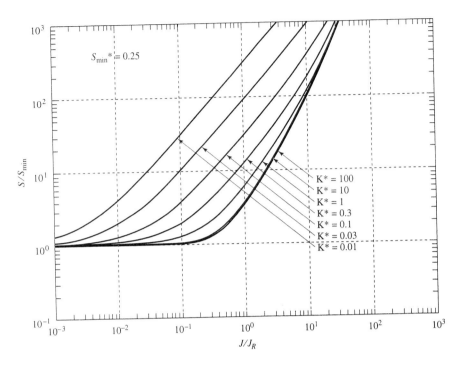

Figure B.6

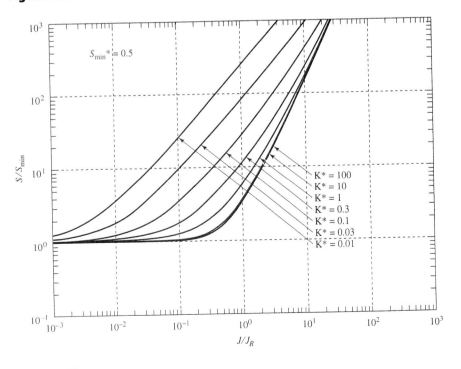

Figure B.7

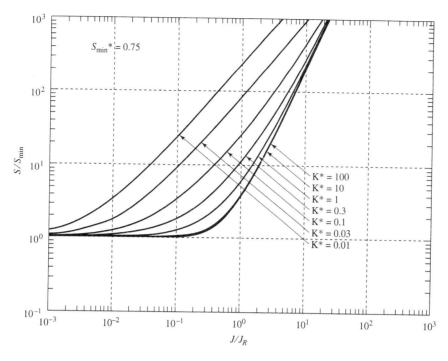

Figure B.8

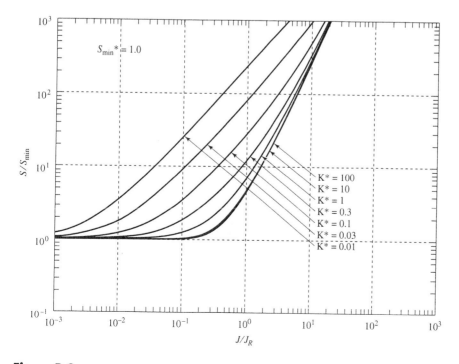

Figure B.9

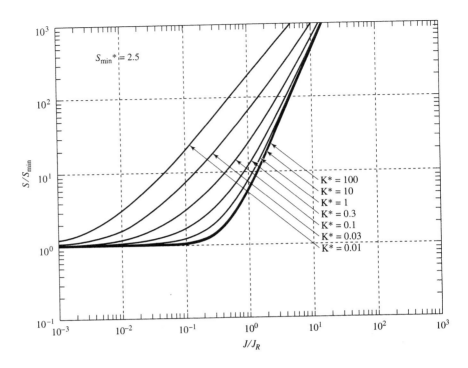

Figure B.10

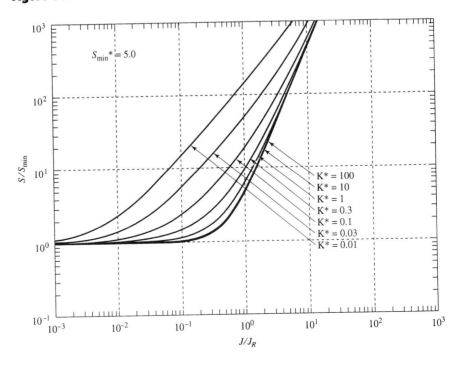

Figure B.11

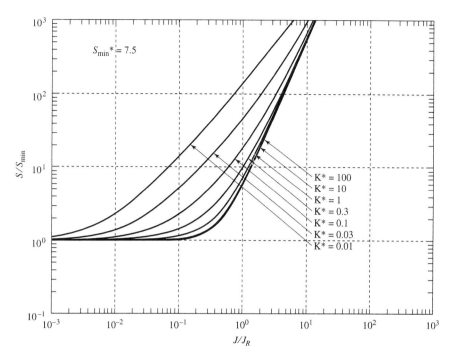

Figure B.12

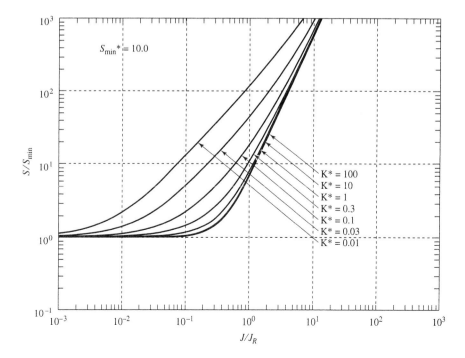

Figure B.13

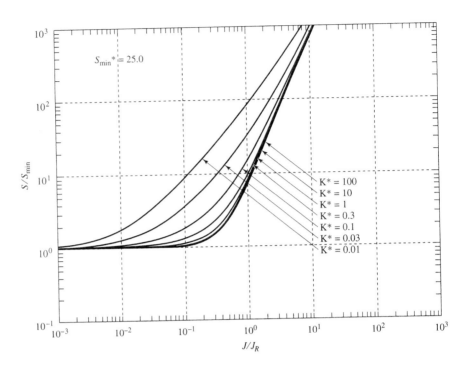

Figure B.14

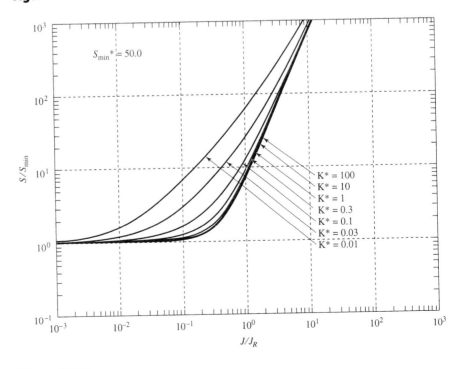

Figure B.15

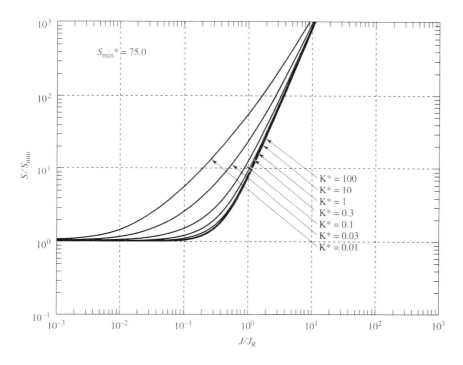

Figure B.16

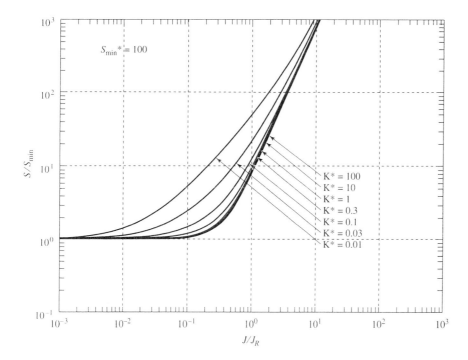

Figure B.17

INDEX

NAME & SUBJECT

A

ABF. *see* Activated Biofilter process
ABS. *see* Alkyl benzene sulfonate
Acclimation, adaptation and; 107
Acetyl CoA, synthesis; 65–67, 78–79
Achromobacter, 310
Acinetobacter, 346, 542
Activated Biofilter (ABF) process, 463–464
Activated sludge process
  aeration
    diffused aeration systems, 338–339
    field oxygen transfer efficiency, 337
    mechanical systems, 339–340
    oxygen-transfer and mixing rates, 335–338
    percent oxygen transfer, 337
    settler, 374
  aeration tank, 307, 374
  bacteria, 310–311
  centrifugal separations, 375
  characteristics
    microbial ecology, 308–311
    oxygen and nutrient requirements, 311
    solids retention time, 312–313, 326–329
  comparison
    aerated lagoon, 395
    biofilm process, 434
  denitrification, 503–506
  design
    analysis and, 346–353
    Eckenfelder and McKinney equations, 334–335
    food-to-microorganism ratio, 324–326
    in general, 323
    historical background, 324
    loading factors comparison, 329–330
    mixed-liquor suspended solids, 330–333
    solids retention time, 312–313, 326–329
  flocs, 307
  in general, 266–267, 294, 307–308
  loading factors
    critical loading, 364
    in general, 322–323, 329–330, 353, 360–362
    underloading, 365–366
    weir loading, 361

  membrane separations, 375–378
  physical configurations
    activated sludge with selector, 319
    completely mixed, 317–318
    contact stabilization, 318–319
    fill-and-draw, 317
    in general, 313–314
    plug-flow, 314–317
    step aeration, 317
  problems
    addition of polymers, 346
    biosolids separation problems, 341
    bulking sludge, 340–344
    dispersed growth, 345
    foaming and scum control, 344
    pinpoint floc, 345
    rising sludge, 345
    viscous bulking, 346
  process configurations
    in general, 313
    loading modifications, 322–323
    oxygen supply modifications, 319–322
  settler analysis and design
    flux theory, 362–368
    in general, 353
    loading criteria, 360–362
    properties, 353–355
    settler and aeration tank, 374
    state-point analysis, 368–373, 374
  settling tank, 307
    hindered settling, 354
    settler components, 355–359
  sludge age, 327
  sludge wasting line, 307
  solids retention time, 312–313, 326–329
Activator proteins, DNA, 94
Adaptation. *see also* Genetics
  in general, 107–110
    acclimation and, 107
    copiotrophy, 108
    diauxie, 109
    enzyme regulation, 109
    exchanged genetic material, 109
    inheritable genetic change, 109
    K-strategists, 108
    oligotrophs, 108
    r-strategists, 108
    selective enrichment, 108
    substrate inhibition, 109
Adaptation period, 107

Adenosine triphosphate (ATP), 52–53, 73, 74, 75
Aeration
  activated sludge process, 320, 322, 323
    diffused aeration systems, 338–339
    mechanical systems, 339–340
    oxygen-transfer and mixing rates, 335–338
  extended, 474
  lagoons, 394–397
Aeration tank, 307
Aerobacterium, chemical signals, 107
Aerobic biofilm process. *see also* Biofilm kinetics; Biofilm reactor
  circulating-bed, 457, 461–463
  fluidized-bed, 457–463
    bed stratification, 460
    three-phase system, 459–460
    two-phase system, 459
  in general, 434–437
  granular-media filters, 456–457
  hybrid biofilm/suspended-growth process, 463–464
    Activated Biofilter, 463–464
    moving-bed biofilm reactor, 463
    PACT process, 464
  rotating biological contactors, in general, 451–454
  trickling filters and biological towers, 438–451
    first-order model, 448
    fixed-nozzle distributors, 446
    flushing intensity, 444
    Galler-Gotaas formula, 447
    low-rate, 447
    NRC formula, 447
    rotating distributor arms, 446
    sloughing, 443–444
    solids contact, 446
    Spulkraft, 444
Aeromonas, 542
African sleeping sickness, 34
Aggregation. *see* Microbial aggregation
Air sparging, bioremediation, 715–716
Alanine, methanogenesis products, 140
Alcaligenes, 497
Alcohols, 62
  oxidation, 63
Aldehydes, 62–64
Algae, 6, 28, 29
  Chlorophyta, 28
  Chrysophyta, 28, 28
  Cyanophyta, 28, 30

Physical constants

Avogadro constant		6.022×10^{23} mole^{-1}
Gas constant	R	8.314×10^{7} ergs/K mol
		8.314 J/K mol
		1.99 cal/K mol
		82.054 cm^3 atm/K mol
Faraday constant	F	96,485 C/equiv
		96,485 J/V equiv
		23.06 kcal/V equiv
Speed of light in a vacuum	c	2.99792×10^{8} m/s
Gravitational constant	g	9.81 m/s^2
		32.2 ft/s^2

Conversion factors

1 in	= 2.54 cm
1 lb	= 453.59 g
1 gal (U.S.)	= 3.78 l
1 ft^3	= 28.32 l
1 mg/l	= 8.34 lb/million gal (U.S.) water
1 cal	= 4.184 J
	= 0.003968 Btu
	= 0.0413 l atm
1 J	= 1.000 VC
1 erg	= 10^{-7} J
1 atm	= 1.01325×10^{5} Pa
	= 14.7 lb/in^2

Fractions and multiples

atto	a	10^{-18}	milli	m	10^{-3}	kilo	k	10^{3}
femto	f	10^{-15}	centi	c	10^{-2}	mega	M	10^{6}
pico	p	10^{-12}	deci	d	10^{-1}	giga	G	10^{9}
nano	n	10^{-9}	deka	da	10	terra	T	10^{12}
micro	μ	10^{-6}	hecto	h	10^{2}			